Modern Applications of Automata Theory

IISc Research Monographs Series

World Scientific Publishing Company (WSPC), Singapore and Indian Institute of Science (IISc), Bangalore co-publish a series of state-of-the-art monographs written by experts in specific areas. They include, but are not limited to, the authors' own research work.

This pioneering collaboration aims to contribute significantly in disseminating current Indian scientific understanding worldwide. In addition, the collaboration also proposes to bring the best scientific thoughts and ideas across the world in areas of priority to India through specially designed India editions.

Books in the Series:

V. R. Voller, *Basic Control Volume Finite Element Methods for Fluids and Solids* (2009).

IISc Research Monographs Series

Modern Applications of Automata Theory

editors

Deepak D'Souza
Priti Shankar
Indian Institute of Science, India

NEW JERSEY • LONDON • SINGAPORE • BEIJING • SHANGHAI • HONG KONG • TAIPEI • CHENNAI

Published by

World Scientific Publishing Co. Pte. Ltd.
5 Toh Tuck Link, Singapore 596224
USA office: 27 Warren Street, Suite 401-402, Hackensack, NJ 07601
UK office: 57 Shelton Street, Covent Garden, London WC2H 9HE

British Library Cataloguing-in-Publication Data
A catalogue record for this book is available from the British Library.

IISc Research Monographs Series — Vol. 2
MODERN APPLICATIONS OF AUTOMATA THEORY

ISBN-13 978-981-4271-04-2
ISBN-10 981-4271-04-7

Printed in Singapore.

Priti Shankar, 1947–2011.

This book is dedicated to the memory of Professor Priti Shankar who passed away on the 17th of October 2011, while this book was in the final stages of production.

Foreword

Ask yourself what are indispensable parts of a core curriculum in computer science. A few highly hyped subjects will probably not make it into the core. Automata theory however has always been part of the core and, I believe, will always remain in the core. After all, our beloved machines on which we do our daily work are finite state machines albeit with quite a huge number of states.

Finite-state machines, originally invented by McCulloch and Pitts to describe the behavior of nervous systems, found their first formal treatment in Kleene's 1956 paper containing his famous theorem about the equivalence of finite-state machines and regular expressions. Rabin and Scott, in their joint paper, which earned them the Turing Award, introduced the notion of *nondeterministic* finite-state machines and presented the *subset construction* to constructively prove the equivalence of deterministic and nondeterministic finite-state machines.

Compiler writers have recognized quite early that different types of automata were ideally suited to understand and to implement several tasks in compilation: finite state machines for lexical analysis, pushdown automata for syntax analysis, and tree automata for code selection. The correspondence of these automata types to concepts on the formal language side was also recognized early. For the compiler writers this meant that automatic methods for the implementation of compiler tasks from given specifications was possible.

At some point in the development of computer hardware, engineers were overwhelmed by the task of verifying their hardware designs when feature size became smaller and the number of integrated transistors exceeded hundreds of thousands. Automatic or at least semi-automatic methods for the verification and the test of designs were needed. Automata theory profited from its connection to logic, in particular to temporal logic, imported into this field by Amir Pnueli in 1977. Model checking based on this connection was invented as a method of choice independently by Queille and Sifakis and by Clarke and Emerson at the beginning of the eighties. However, the problem of state-space explosion limited its applicability, despite the invention of ingenious data structures such as BDDs for the symbolic representation of the space. At the same time, automata with acceptance criteria defined by Büchi moved into the focus of research on automata. They were needed to reason about systems with infinite computations. In order to make model checking feasible, abstraction to systems with a sufficiently small number of states was necessary. The abstraction should preserve essential properties such that verifiable properties of the abstracted system would definitely hold on the original system. Much progress has been made here in recent times.

The connection to logic has always been a fruitful ground for research. Decidability, complexity, and expressiveness problems had to be solved. Characterizations of variants of automata in terms of logic or formal languages had to be found.

Modern applications of automata theory go far beyond compiler techniques or hardware verification. Automata are widely used for modelling and verification of software, distributed systems, real-time systems, or structured data. They have been equipped with features to model time and probabilities as well.

This volume tackles some of the challenges such as countering ever-growing state spaces by using finite-state abstractions, translations between several temporal logics and Büchi automata, equivalence between several types of automata and different types of regular and graph languages, decidability of the model-checking problem for infinite-transition systems, and the connection of variants of tree automata to monadic second order logic. As a practitioner, I would consider these to be *applications within theory.* However, this monograph also considers applications to *real-world problems.* These concern navigation in XML and type checking of XSLT, system verification, coordination in distributed systems, expression of concurrency, compression of tree-structured data and digital images, and access control in smartcard systems. I am happy to note that many of the contributors to this volume are leading experts in the areas mentioned above.

Dear Reader, have you noticed how many names of Turing Award winners were mentioned in this short preface? If not, here is the list: Rabin, Scott, Pnueli, Sifakis, Clarke, Emerson. This at least supports my claim that automata theory is considered a core part of computer science.

I congratulate the editors and the Indian Institute of Science for this fine collection of articles on modern applications of automata. I am happy to learn that the institute has recently celebrated its centenary, and that this volume is among the books published to mark this event.

Reinhard Wilhelm
Universität des Saarlandes.

Preface

Automata theory is perhaps one of the oldest and most researched areas in Computer Science. Over the past fifty years or so, numerous applications of automata have been developed in a wide spectrum of areas with a corresponding evolution of a variety of theoretical models. Early applications of automata theory included pattern matching, syntax analysis and software verification, where elegant theory was applied to real world problems, resulting in the generation of useful software tools significantly in the area of compilers. Deep connections between automata and logic continue to be discovered, and newer models of automata, for example, timed automata, hybrid automata, distributed automata and weighted automata have been proposed, all driven by specific applications. The main focus of this book is on verification which is, without a doubt, the crowning achievement of the area in recent years. Since a comprehensive coverage of applications in this area is not possible we have selected a few topics that are representative of recent trends in verification. In addition, a couple of chapters on the application of automata to the problems of image and tree compression have been included.

The mathematical prerequisites for understanding the contents of this book are relatively modest (mainly undergraduate courses on automata theory, formal languages and logic). Introductory chapters on advanced topics not typically covered in undergraduate courses have been added. We hope that the material in this book will be useful for an advanced course in automata theory and its applications.

The material in the book has been organised into four parts, based upon the kind of applications of automata discussed. We begin with the introductory chapters, followed by applications related to verification, then applications in logic, and finally applications in compression. We proceed below to describe the chapters in the light of this classification.

Part I comprises the introductory chapters and begins with the chapter *An Introduction to Finite Automata and their Connection to Logic* by Straubing and Weil. Apart from the basic results on finite-state automata like closure properties, the pumping lemma, and the Myhill-Nerode theorem, the chapter covers the important connections to logic, including Büchi's characterisation of regular languages via monadic second order logic (MSO), and the subsequent McNaughton–Schutzenberger characterisation of first-order definable languages. The chapter also introduces the algebraic view of regularity and the notion of the syntactic monoid of a language.

Chapter 2 on *Finite-State Automata on Infinite Inputs* by Mukund introduces automata that run on infinite words. Originally proposed by Büchi in 1960 to decide the truth of MSO-definable properties of discrete linear orders, these automata play an important role in automated verification, particularly with the development of temporal logic as a formalism for specifying and verifying properties of programs. The chapter contains a proof of Büchi's characterisation of the class of languages accepted by these automata in terms of MSO-definable properties, as well as a detailed account of Safra's construction to complement Büchi automata.

Chapter 3 titled the *Basics on Tree Automata* by Löding is a comprehensive introduction to the theory of automata over trees. The foundations of this theory were developed by Thatcher and Wright in the late sixties and by Doner in 1970. The chapter focuses mainly on ranked trees but also describes hedge automata for unranked trees and tree walking automata, the last two being of interest in the context of XML. The author demonstrates that much of the theory of automata over finite words can be adapted to the setting of ranked trees, in particular automata constructions for boolean operations on languages, minimization of automata, and the relation to monadic second-order logic.

Chapter 4 on *An Introduction to Timed Automata* by Pandya and Suman is about automata that run on "timed words," which are classical words along with a real-valued time-stamp for each letter. Timed automata were introduced by Alur and Dill in their seminal paper of 1994 to model and reason about the behaviour of real-time systems. The chapter gives a detailed account of the region construction for showing the regularity of the "untiming" of the language accepted by a timed automaton, closure and non-closure properties, and the undecidability of the universality problem for timed automata.

Moving on to Part II of this volume, we collect together chapters whose common theme is the verification problem. Simply stated, the problem of verification is about modelling and reasoning about the correctness of system models, in terms of satisfying various kinds of properties ranging from unreachability of certain configurations to patterns of behaviour specified by temporal logics. The chapters have been ordered according to the kind of system behaviour they consider, ranging from finite words to continuous dynamics.

This group of chapters begins with Chapter 5 titled *A Language-Theoretic View of Verification* by Lodaya, which views system behaviour as a finite sequence of actions or events, represented as finite words, and discusses the application of formal language theory to solve problems related to the verification of such systems. Chapter 6 by Chaturvedi et al. on *A Framework for Decentralized Access Control using Finite State Automata* continues this view of system behaviour, where system events comprise access requests and grants, and uses classical finite-state automata to implement MSO-based access control policies. In Chapter 7 by Chakraborty on *Reasoning about Heap Manipulating Programs using Automata Techniques*, a program's behaviour is viewed as a sequence of "states" that a program goes through during execution. Each state is modelled as a finite word representing the contents of the program's heap memory. In three different techniques the author explores the use of extensions of classical automata like transducers and counter automata to reason about the set of reachable states of a program. Chapter 8 titled *Chop Expressions and Discrete Duration Calculus* by Ajesh Babu and Pandya considers system behaviour as a sequence of states where each state is a valuation to a finite set of propositional variables. Properties are expressed in the Discrete Duration Calculus (DDC*). The chapter studies a regular-expression like formalism called chop expressions into which DDC* specifications can be compiled, enabling one to check equivalence and implication between such specifications.

The next couple of chapters address the verification of distributed systems, where it is convenient to view system behaviour as partially-ordered sets of events. Chapter 9 titled *Automata on Distributed Alphabets* by Mukund introduces asynchronous automata that accept such partially-ordered behaviours represented as "trace-closed" languages of finite words. The chapter contains a first-hand account of the important distributed time-stamping or "gossip" algorithm, which is then used to prove Zielonka's theorem characterising the class of languages accepted by asynchronous automata as trace-closed regular languages. In effect the chapter gives a way to synthesize a distributed finite-state implementation of any regular set of trace-closed behaviours. Chapter 10 on *The Theory of MSC Languages* by Narayan Kumar, is a comprehensive survey of formalisms to generate and reason about Message Sequence Charts which are a popular visual representation of patterns of distributed communication.

From partial-orders we move on to trees. Programs that transform XML documents can be viewed as tree walking transducers that take as input labelled trees of a certain type and output labelled trees of another type. Chapter 11 on *Type Checking of Tree Walking Transducers* by Maneth et al. considers the problem of verifying the type-correctness of such transducers. While the problem in its generality is computationally expensive to solve, the authors provide an efficient solution based on forward type inferencing for a subclass of tree walking transducers.

Chapter 12 titled *Three Case Studies on Verification of Infinite-State Systems* by Esparza and Kreiker once again views behaviour as a sequence of states, though now the number of states are potentially infinite, containing for example real-valued valuations for clock variables. Three example systems are considered which are modeled using timed, pushdown, and linear automata respectively. The properties considered are phrased in terms of reachability of certain states. Algorithms to solve the reachability problem for each of these models are described and then applied to the verification problem for the example systems.

We finally move to behaviours exhibited by continuous dynamical systems. Chapter 13 titled *Introduction to Hybrid Automata* by Gopinathan and Prabhakar, introduces hybrid automata for modelling dynamical systems with multiple modes of continuous evolution. The chapter describes algorithms for verifying reachability properties for the well-known subclasses of initialized rectangular hybrid automata and o-minimal hybrid automata. Chapter 14 by Agrawal et al. titled *The Discrete Time Behavior of Restricted Linear Hybrid Automata* studies the discrete time behaviour of a class of hybrid automata with a restricted form of linear dynamics. The authors show that if we consider the discrete time, finite precision semantics of this class of hybrid automata, the resulting language is regular and effectively computable. The results can be used to verify reachability properties of continuous systems modelled as such hybrid automata.

We now come to Part III which groups together material describing applications of automata in logic. Chapter 15 titled *Specification and Verification using Temporal Logics* by Demri and Gastin shows how to solve the satisfiability and model-checking problems for temporal logic. The authors describe in detail an elegant and simple translation from temporal logic to Büchi automata. Chapter 16 by Thomas on *Finite Automata and the Analysis of Infinite Transition Systems* explores an unusual application of automata. The author considers infinite-state transition systems that arise from automata-definable transition relations, and shows that the model-checking problem for these systems with respect to natural first-order and MSO logics is decidable. Chapter 17 titled *Automata over Infinite Alphabets* by Manuel and Ramanujam deals with automata and logics over "data words" which arise in modelling systems with unbounded data. The authors consider a range of automata models over data words with decidable emptiness problems, and use one of these classes of automata to show the decidability of a fragment of an MSO logic interpreted over data words. Chapter 18 by Chevalier et al. on *Automata and Logics over Signals* provides an application of automata to decide satisfiability and characterize the expressiveness of natural logics over finitely-varying discrete-valued functions or "signals."

Finally in Part IV we have two chapters on the application of automata for data compression. Chapter 19 by Shankar on *Syntax Directed Compression of Trees Using Pushdown Automata* describes the compression of ranked and unranked trees using pushdown automata to control the multiplexing of models for adaptive compressors. The first part of the chapter shows how ranked trees defined by a regular tree grammar can be compressed by using a modification of the LR parsing algorithm to achieve multiplexing. The second part describes a scheme for the compression of XML files (which are essentially unranked trees) using devices called recursive finite automata. Chapter 20 titled *Weighted Finite Automata and Digital Image Representation* by Krithivasan and Sivasubramanyam is on the application of weighted finite automata to the problem of image compression. Weighted finite automata were introduced in the nineties by Kulik and Kari for representing images. This chapter describes how such automata can be used in the representation, transformation, and compression of digital images.

We now address the issue of dependencies between chapters. While the chapters are largely self-contained, some chapters do make use of some concepts or results that are dealt with in more detail in other chapters. We have tried to capture some of these essential dependencies in the figure below.

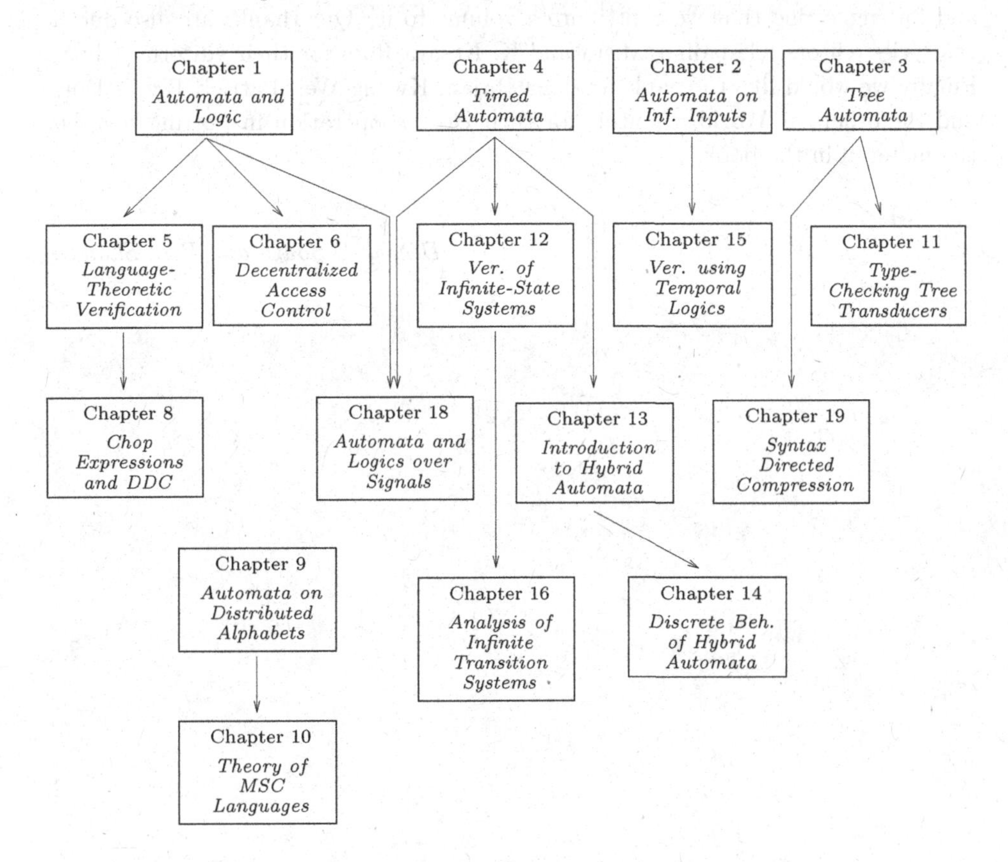

We will be happy to receive feedback from readers on typographical and other errors in this book. A list of errata will be maintained at http://www.worldscibooks.com/compsci/7237.html

Acknowledgments

We owe a huge debt of gratitude to all the authors, many of whom are well-known experts in their areas, for readily agreeing to contribute to this volume, and for having devoted a great deal of their time and energy to present their knowledge, experience, and insights in the chosen topics. We would also like to thank the authors for participating in a careful review of the chapters. We would like to specially thank Reinhard Wilhelm for readily agreeing to write the foreword for this volume. We are indebted to Eugene Asarin, Anca Muscholl, and V. Vinay for their painstaking and useful reviews that have helped to make the material in this book more complete. We would also like to thank Vijay Natarajan and Shantanu Choudhary for their help with generating the image for the book cover.

We would like to express our gratitude to our institute Director P. Balaram for initiating this series of books to commemorate the Centenary of the institute, and for suggesting that we contribute a volume to it. Our thanks are also due to the series editors, Gadadhar Misra and K. Kesava Rao, for their editorial advice. Finally we would like to thank Ranjana Rajan, Kwang Wei, Patrick Tay Zi Dong and the others at World Scientific for their close cooperation in putting together the material in the book.

Deepak D'Souza and Priti Shankar

Contents

Part I

Basic Chapters

Chapter 1

An Introduction to Finite Automata and their Connection to Logic

Howard Straubing*

Computer Science Department, Boston College,
Chestnut Hill, Massachusetts, USA
straubin@cs.bc.edu

Pascal Weil†

LaBRI, Université de Bordeaux and CNRS, Bordeaux, France
pascal.weil@labri.fr

This introductory chapter is a tutorial on finite automata. We present the standard material on determinization and minimization, as well as an account of the equivalence of finite automata and monadic second-order logic. We conclude with an introduction to the syntactic monoid, and as an application give a proof of the equivalence of first-order definability and aperiodicity.

1.1. Introduction

1.1.1. *Motivation*

The word *automaton* (plural: *automata*) was originally used to refer to devices like clocks and watches, as well as mechanical marvels built to resemble moving humans and animals, whose internal mechanisms are hidden and which thus appear to operate spontaneously. In theoretical computer science, the *finite automaton* is among the simplest models of computation: A device that can be in one of finitely many *states,* and that receives a discrete sequence of inputs from the outside world, changing its state accordingly. This is in marked contrast to more general and powerful models of computation, such as Turing machines, in which the set of global states of the device—the so-called *instantaneous descriptions*—is infinite. A finite automaton is more akin to the control unit of the Turing machine (or, for that matter, the control unit of a modern computer processor), in which the present state of the unit and the input symbol under the reading head determine the next state of the unit, as well as signals to move the reading head left or right

*Work partially supported by NSF Grant CCF-0915065
†Work partially supported by ANR 2010 BLAN 0202 01 FREC

and to write a symbol on the machine's tape. The crucial distinction is that while the Turing machine can record and consult its entire computation history, all the information that a finite automaton can use about the sequence of inputs it has seen is represented in its current state.

But as rudimentary as this computational model may appear, it has a rich theory, and many applications. In this introductory chapter, we will present the core theory: that of a finite automaton reading a finite *word,* that is, a finite string of inputs, and using the resulting state to decide whether to accept or reject the word. The central question motivating our presentation is to determine what properties of words can be decided by finite automata. Subsequent chapters will present both generalizations of the basic model (to devices that read infinite words, labeled trees, *etc.*) and to applications. An important theme in this chapter, as well as throughout the volume, is the close connection between automata and formal logic.

1.1.2. *Plan of the chapter*

In Section 1.2, we introduce finite automata as devices for recognizing formal languages, and show the equivalence of several variants of the basic model, most notably the equivalence of deterministic and nondeterministic automata. Section 1.3 describes Büchi's sequential calculus, the framework in predicate logic for describing properties of words that are recognizable by finite automata. In Section 1.4 we prove what might well be described as the two fundamental theorems of finite automata: that the languages recognized by finite automata are exactly those definable by sentences of the sequential calculus, and also exactly those definable by rational expressions (also called regular expressions). Section 1.5 presents methods that can be used to show certain languages cannot be recognized by finite automata. The last sections, 1.6 and 1.7, have a more algebraic flavor: we introduce both the minimal automaton and the syntactic monoid of a language, and prove the important McNaughton-Schützenberger theorem describing the languages definable in the first-order fragment of the sequential calculus.

1.1.3. *Notation*

Throughout this chapter, A denotes a finite *alphabet*, that is, a finite non-empty set. Elements of A are called *letters*, and a finite sequence of letters is called a *word*. We denote words simply by concatenating the letters, so, for example, if $A = \{a, b, c\}$, then $aabacba$ is a word over A. The *empty sequence* is considered a word, and we use ε to denote this sequence. The set of all words over A is denoted A^*, and the set of all nonempty words is denoted A^+. The *length* of the word w, that is, the number of letters in w, is denoted $|w|$.

If $u, v \in A^*$ then we can form a new word uv by concatenating the two sequences. Concatenation of words is obviously an associative and (unless A has a single element) noncommutative operation on A^*. We have

$$|uv| = |u| + |v|, \text{ and}$$
$$u\varepsilon = \varepsilon u = u.$$

(Other texts frequently use Λ or 1 to denote the empty word. The latter choice is justified by the second equation above.)

A subset of A^* is called a *language* over A.

1.1.4. *Historical note and references*

This chapter contains a modern presentation of material that goes back more than fifty years. The reader can find other accounts in classic papers and texts: The equivalence of finite automata and rational expressions given in Section 1.4 was first described by Kleene in [1]. The connection with monadic second-order logic was found independently by Trakhtenbrot [2] and Büchi [3].

Nondeterministic automata were introduced by Rabin and Scott [4], who showed their equivalence to deterministic automata. Minimization of finite-state devices (framed in the language of switching circuits built from relays) is due to Huffman [5]. The simple congruential account of minimization that we give originates with Myhill [6] and Nerode [7].

The equivalence of aperiodicity of the syntactic monoid with star-freeness is due to Schützenberger [8], and the connection with first-order logic is from McNaughton and Papert [9]. Our account of these results relies heavily on an argument given in Wilke [10].

Rational expressions, determinization and minimization have become part of the basic course of study in theoretical computer science, and as such are described in a number of undergraduate textbooks. Hopcroft and Ullman [11], Lewis and Papadimitriou [12] and the more recent Sipser [13] are notable examples. A more technical and algebraically-oriented account is given in the monograph by Eilenberg [14, 15]. An algebraic view of automata is developed by Sakarovitch [16]. Detailed accounts of the connection between automata, logic and algebra can be found in Straubing [17] and Thomas [18]. The state of the art, especially concerning the algebraic classification of automata, will appear in the forthcoming handbook [19].

1.2. Automata and rational expressions

1.2.1. *Operations on languages*

We describe here a collection of basic operations on languages, which will be building blocks in the characterization of the expressive power of automata.

Since languages over A are subsets of A^*, we may of course consider the boolean operations: union, intersection and complement. The product operation on words can be naturally extended to languages: if K and L are languages over A, we define their *concatenation product* KL to be the set of all products of a word in K followed by a word in L:

$$KL = \{uv \mid u \in K \text{ and } v \in L\}.$$

We also use the power notation for languages: if $n > 0$, L^n is the product $LL \cdots L$ of n copies of L. We let $L^0 = \{\varepsilon\}$. Note that if $n > 1$, L^n differs from the set of n-th powers of the elements of L. The *iteration* (or *Kleene star*) of a language L is the language $L^* = \bigcup_{n \geq 0} L^n$.

Finally, we introduce a simple rewriting operation, based on the use of morphisms. If A and B are alphabets, a *morphism* from A^* to B^* is a mapping $\varphi\colon A^* \to B^*$ such that

(1) $\varphi(\varepsilon) = \varepsilon$,
(2) for all $u, v \in A^*$, $\varphi(uv) = \varphi(u)\varphi(v)$.

To specify such a morphism, it suffices to give the images of the letters of A. Then the image of a word $u \in A^*$, say $u = a_1 \cdots a_n$, is obtained by taking the concatenation of the images of the letters, $\varphi(u) = \varphi(a_1) \cdots \varphi(a_n)$. That is, $\varphi(a_1 \cdots a_n)$ is obtained from $a_1 \cdots a_n$ by substituting for each letter a_i the word $\varphi(a_i)$. This operation naturally extends from words to languages: if $L \subseteq A^*$, then $\varphi(L) = \{\varphi(u) \mid u \in L\}$.

The consideration of these operations leads to the classical definition of *rational* languages (also called *regular* languages). The operations of union, concatenation and iteration are called the *rational operations*. A language over alphabet A is called rational if it can be obtained from the letters of A by applying (a finite number of) rational operations.

More formally, the class of rational languages over the alphabet A, denoted $\mathsf{Rat}A^*$, is the least class of languages such that

(1) the languages $\emptyset$ and $\{a\}$ are rational for each letter $a \in A$;
(2) if K and L are rational languages, then $K \cup L$, KL and L^* are also rational.

Example 1.1. The language $\left((a^*(ab)^*A^* \cap A^*(ba)^*)^2\right)^*$ is rational. (Note that in order to lighten the notation, we write a, b, etc., instead of $\{a\}$, $\{b\}$.)

The language $\{\varepsilon\}$, containing just the empty word, is rational. Indeed, it is equal to $\emptyset^*$.

Any finite language (that is, containing only finitely many words) is rational.

Let $a, b \in A$ be distinct letters. It is instructive to show that the following languages are rational: (a) the set of all words which do not contain two consecutive a; (b) the set of all words which contain the factor ab but not the factor ba.

We also consider the *extended rational operations*: these are the rational operations, and the operations of intersection, complement and morphic image. A language is said to be *extended rational* if it can be obtained from the letters of A

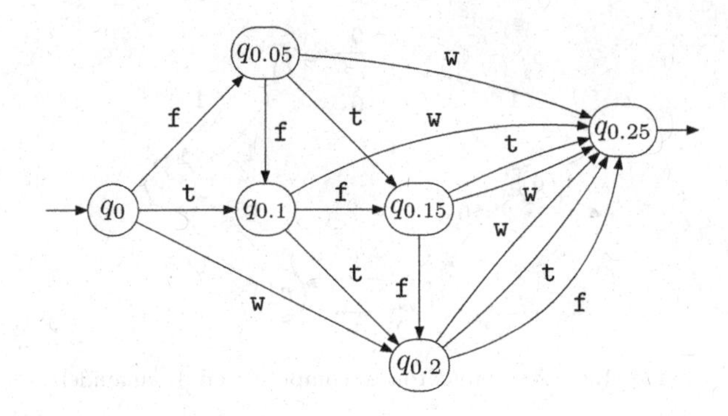

Fig. 1.1. The automaton of a (simplified) coffee machine.

by applying (a finite number of) extended rational operations. The class of extended rational languages over A is written X-RatA^*.

Of course, all rational languages are extended rational. The definition of extended rational languages offers more expressive possibilities but as we will see, they are not properly more expressive than rational languages.

1.2.2. *Automata*

Let us start with a couple of examples.

Example 1.2. A coffee machine delivers a cup of coffee for €.25. It accepts only coins of €.20, €.10 and €.05. While determining whether it has received a sufficient sum, the machine is in one of six states, q_0, $q_{0.05}$, $q_{0.1}$, $q_{0.15}$, $q_{0.2}$ and $q_{0.25}$. The names of the states correspond to the sum already received. The machine changes state after a new coin is inserted, and the new state it assumes is a function of the value of the new coin inserted and of the sum already received. The latter information is encoded in the current state of the machine.

Here, the input word is the sequence of coins inserted, and the alphabet consists of three letters, `w`, `t` and `f`, standing respectively for twenty cents, ten cents and five cents. The machine is represented in Figure 1.1.

The incoming arrow indicates the initial state of the machine (q_0), and the outgoing arrow indicates the only accepting state ($q_{0.25}$), that is, the state in which the machine will indeed prepare a cup of coffee for you. Notice that the machine does not return change, but that it will accept sums up to €.40.

Example 1.3. Our second example (Figure 1.2) reads an integer, given by its binary expansion and read from right to left, that is, starting with the bit of least weight. Upon reading this word on alphabet $\{0, 1\}$, the automaton decides whether the given integer is divisible by 3 or not.

For instance, consider the integer 19, in binary expansion `10011`: our input word is `11001`. It is read letter by letter, starting from the initial state (the state

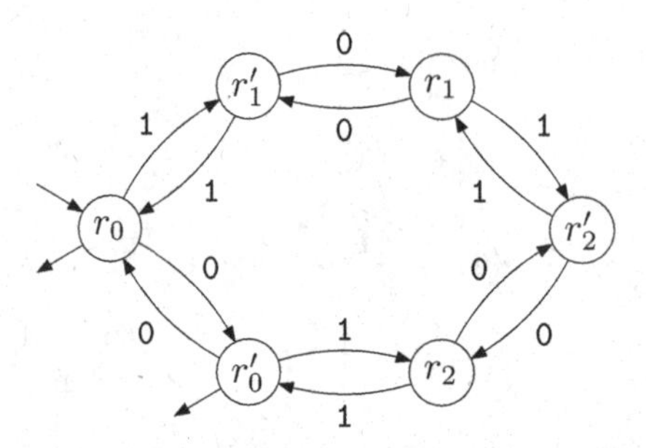

Fig. 1.2. An automaton to compute mod 3 remainders.

indicated by an incoming arrow, state r_0). After each new letter is read, we follow the corresponding edge starting at the current state. Thus, starting in state r_0, we visit successively the states r_1', r_0, r_0', r_0 again, and finally r_1'. This state is not accepting (it is not marked with an outgoing edge), so the word `11001` is not accepted by the automaton. And indeed, 19 is not divisible by 3.

In contrast, 93 is divisible by 3, which is confirmed by running its binary expansion, namely `1011101`, read from right to left, through the automaton: starting in state r_0, we end in state r_0'.

The reader will quickly see that this automaton is constructed in such a way that, if n is an integer and w_n is the binary expansion of n, then the state reached when reading w_n from right to left, starting in state r_0, is r_k (resp. r_k') if n is congruent to k (mod 3) and w_n has even (resp. odd) length.

We now turn to a formal definition. A (*finite state*) *automaton* on alphabet A is a 4-tuple $\mathcal{A} = (Q, T, I, F)$ where Q is a finite set, called the set of *states*, T is a subset of $Q \times A \times Q$, called the set of *transitions*, and I and F are subsets of Q, called respectively the sets of *initial states* and *final states*. Final states are also called *accepting states*.

For instance, the automaton of Example 1.2 uses a 3-letter alphabet, $A = \{\mathtt{f}, \mathtt{t}, \mathtt{w}\}$. Formally, it is the automaton $\mathcal{A} = (Q, T, I, F)$ given by $Q = \{q_0, q_{0.05}, q_{0.1}, q_{0.15}, q_{0.2}, q_{0.25}\}$, $I = \{q_0\}$, $F = \{q_{0.25}\}$ and T is a 15-element subset of $Q \times A \times Q$ containing such triples as $(q_0, \mathtt{f}, q_{0.05})$, $(q_{0.1}, \mathtt{t}, q_{0.2})$ or $(q_{0.2}, \mathtt{w}, q_{0.25})$.

As in our first examples, it is often convenient to represent an automaton $\mathcal{A} = (Q, T, I, F)$ by a labeled graph, whose vertices are the elements of Q (the states) and whoses edges are of the form $q \xrightarrow{a} q'$ if (q, a, q') is a transition, that is, if $(q, a, q') \in T$. The initial states are specified by an incoming arrow, and the final states are specified by an outgoing edge.

From now on, we will most often specify our automata by their graphical representations.

Example 1.4. Here, the alphabet is $A = \{a, b\}$. Figure 1.3 represents the automaton $\mathcal{A} = (Q, T, I, F)$ where $Q = \{1, 2, 3\}$, $I = \{1\}$, $F = \{3\}$ and

$$T = \{(1, a, 1), (1, b, 1), (1, a, 2), (2, b, 3), (3, a, 3), (3, b, 3)\}.$$

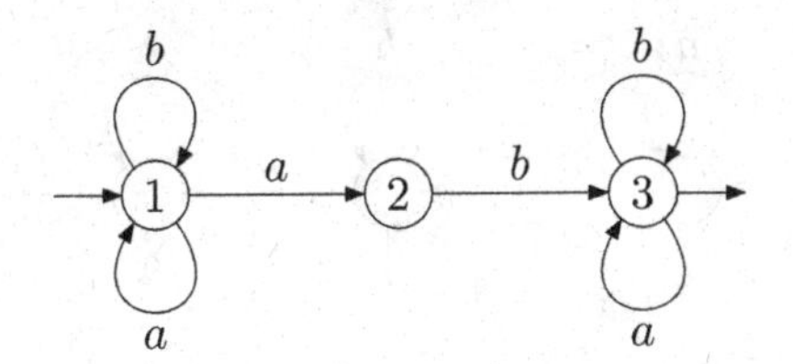

Fig. 1.3. An automaton accepting A^*abA^*.

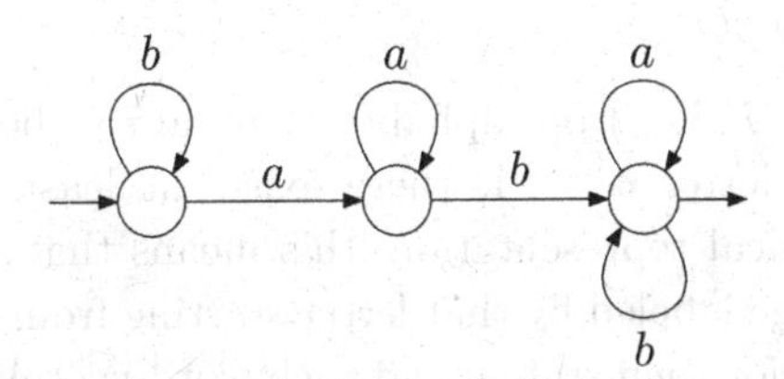

Fig. 1.4. Another automaton accepting A^*abA^*.

1.2.2.1. *The language accepted by an automaton*

A *path* in automaton $\mathcal{A}$ is a sequence of consecutive edges,

$$p = (q_0, a_1, q_1)(q_1, a_2, q_2) \quad \cdots \quad (q_{n-1}, a_n, q_n),$$

also drawn as

$$p = q_0 \xrightarrow{a_1} q_1 \xrightarrow{a_2} q_2 \quad \cdots \quad \xrightarrow{a_n} q_n.$$

Then we say that p is a path of *length* n from q_0 to q_n, *labeled* by the word $u = a_1 a_2 \cdots a_n$. By convention, for each state q, there exists an *empty path* from q to q labeled by the empty word.

For instance, in the automaton of Figure 1.3, the word a^3ba labels exactly four paths: from 1 to 1, from 1 to 2, from 1 to 3 and from 3 to 3.

A path p is *successful* if its initial state is in I and its final state is in F. A word w is *accepted* (or *recognized*) by $\mathcal{A}$ if there exists a successful path in the automaton with label w. And the *language accepted* (or *recognized*) by $\mathcal{A}$ is the set of labels of successful paths in $\mathcal{A}$. It is denoted by $L(\mathcal{A})$. We say that $\mathcal{A}$ *accepts* (or *recognizes*) $L(\mathcal{A})$.

For instance, the language of the automaton of Figure 1.1 is finite, with exactly 27 words. The automaton of Figure 1.3 accepts the set of words in which at least one occurrence of a is followed immediately by a b, namely A^*abA^*, where $A = \{a, b\}$.

Different automata may recognize the same language: if $\mathcal{A}$ and $\mathcal{B}$ are automata such that $L(\mathcal{A}) = L(\mathcal{B})$, we say that $\mathcal{A}$ and $\mathcal{B}$ are *equivalent*.

Example 1.5. The language A^*abA^*, accepted by the automaton in Figure 1.3, is also recognized by the automaton in Figure 1.4

A language L is said to be *recognizable* if it is recognized by an automaton.

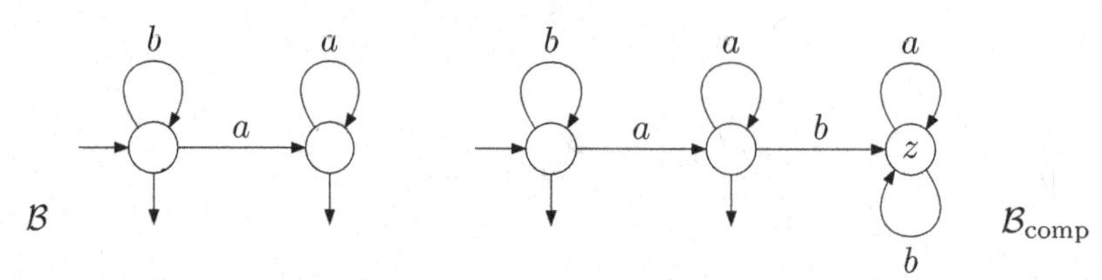

Fig. 1.5. Two automata accepting b^*a^*.

1.2.2.2. *Complete automata*

An automaton $\mathcal{A} = (Q, T, I, F)$ on alphabet A is said to be *complete* if, for each state $q \in Q$ and each letter $a \in A$, there exists at least one transition of the form (q, a, q'): in graphical representation, this means that, for each letter of the alphabet, there is an edge labeled by that letter starting from each state. Naturally, this easily implies that, for each state q and each word $w \in A^*$, there exists at least one path labeled w starting at q.

Every automaton can easily be turned into an equivalent complete automaton. If $\mathcal{A} = (Q, T, I, F)$ is not complete, the *completion* of $\mathcal{A}$ is the automaton $\mathcal{A}_{\text{comp}} = (Q', T', I, F)$ given by $Q' = Q \cup \{z\}$, where z is a new state not in Q, and T' is obtained by adding to T all triples (z, a, z) $(a \in A)$ and all triples (q, a, z) $(q \in Q, a \in A)$ such that there is no element of the form (q, a, q') in T.

If $\mathcal{A}$ is complete, we let $\mathcal{A}_{\text{comp}} = \mathcal{A}$. It is immediate that, in every case, $\mathcal{A}_{\text{comp}}$ is complete and $L(\mathcal{A}_{\text{comp}}) = L(\mathcal{A})$.

Example 1.6. Let $A = \{a, b\}$. The automaton $\mathcal{B}$ in Figure 1.5, which accepts the language b^*a^*, is evidently not complete. The automaton $\mathcal{B}_{\text{comp}}$ is represented next to it.

1.2.2.3. *Trim automata*

A complete automaton reads its entire input before deciding to accept or reject it: whatever input it receives, there is a transition that can be followed. However, we have seen that in the completion $\mathcal{A}_{\text{comp}}$ of a non-complete automaton $\mathcal{A}$, state z does not participate in any successful path: it is in a way a useless state. *Trimming* an automaton removes such useless states; it is, in a sense, the opposite of completing an automaton, and aims at producing a more concise device.

A state q of an automaton $\mathcal{A}$ is said to be *accessible* if there exists a path in $\mathcal{A}$ starting from some initial state and ending at q. State q is *co-accessible* if there exists a path in $\mathcal{A}$ starting from q and ending at some final state. Observe that a state is both accessible and co-accessible if and only if it is visited by at least one successful path.

The automaton $\mathcal{A}$ itself is *trim* if all its states are both accessible and co-accessible: in a trim automaton, each state is useful, in the sense that it is used in accepting some word of the language $L(\mathcal{A})$.

Of course, every automaton $\mathcal{A}$ is equivalent to a trim one, written $\mathcal{A}_{\text{trim}}$, obtained by restricting $\mathcal{A}$ to its accessible and co-accessible states and to the transitions between them.

Interestingly, $\mathcal{A}_{\text{trim}}$ can be constructed efficiently, using breadth-first search. One first computes the accessible states of $\mathcal{A}$, by letting $Q_0 = I$ (the initial states are certainly accessible) and by computing iteratively

$$Q_{n+1} = Q_n \cup \bigcup_{q \in Q_n, a \in A} \{q' \in Q \mid (q, a, q') \in T\}.$$

One verifies that the elements of Q_n are the states that can be reached from an initial state, reading a word of length at most n; and that if two consecutive sets Q_n and Q_{n+1} are equal, then $Q_n = Q_m$ for all $m \geq n$, and Q_n is the set of accessible states of $\mathcal{A}$. In particular, the set of accessible states is computed in at most $|Q|$ steps.

A similar procedure, starting from the final states instead of the initial states, and working in reverse, produces in at most $|Q|$ steps the set of co-accessible states of $\mathcal{A}$. The automaton $\mathcal{A}_{\text{trim}}$ is then immediately constructed.

Remark 1.1. The construction of $\mathcal{A}_{\text{trim}}$, or indeed, just of the set of accessible states of $\mathcal{A}$ provides an efficient solution of the *emptiness problem*: given an automaton $\mathcal{A}$, is the language $L(\mathcal{A})$ empty? that is, does $\mathcal{A}$ accept at least one word?

Indeed, $\mathcal{A}$ recognizes the empty set if and only if no final state is accessible: in order to decide the emptiness problem for automaton $\mathcal{A}$, it suffices to construct the set of accessible states of $\mathcal{A}$ and verify whether it contains a final state. This yields an $\mathcal{O}(|Q|^2|A|)$ algorithm.

1.2.2.4. *Epsilon-automata*

It is sometimes convenient to extend the notion of automata to the so-called ε-*automata*: the difference from ordinary automata is that we also allow ε-labeled transitions, of the form (p, ε, q) with $p, q \in Q$.

Proposition 1.1. *Every ε-automaton is equivalent to an ordinary automaton.*

Sketch of proof. Let $\mathcal{A} = (Q, T, I, F)$ be an ε-automaton, and let $\mathcal{R}$ be the relation on Q given by $p \mathrel{\mathcal{R}} q$ if there exists a path from p to q consisting only of ε-labeled transitions (that is: $\mathcal{R}$ is the reflexive transitive closure of the relation defined by the ε-labeled transitions of $\mathcal{A}$).

Let $\mathcal{A}'$ be the (ordinary) automaton given by the tuple (Q, T', I', F) with

$$T' = \big\{(p, a, q) \mid (p, a, q') \in T \text{ and } q' \mathrel{\mathcal{R}} q \text{ for some } q' \in Q\big\}$$
$$I' = \big\{q \mid p \mathrel{\mathcal{R}} q \text{ for some } p \in I\big\}.$$

Then $\mathcal{A}'$ is equivalent to $\mathcal{A}$. □

1.2.3. *Deterministic automata*

Example 1.7. Consider the automaton of Figure 1.3, say $\mathcal{A}$, and the automaton $\mathcal{B}$ of Figure 1.4. Both recognize the language, $L = A^*abA^*$, but there is an important, qualitative difference beween them.

We have defined automata as *nondeterministic* computing devices: given a state and an input letter, there may be several possible choices for the next state. Thus an input word might be associated with many different computation paths, and the word is accepted if one of these paths ends at an accepting state. In contrast, $\mathcal{B}$ has the convenient property that each input word labels at most one computation path.

These remarks are formalized in the following definition. An automaton $\mathcal{A} = (Q, T, I, F)$ is said to be *deterministic* if it has exactly one initial state, and if, for each letter a and for all states q, q', q'',

$$(q, a, q'),\ (q, a, q'') \in T \qquad \Longrightarrow \qquad q' = q''.$$

Thus, of the automata in Figures 1.3 and 1.4, the second one is deterministic, and the first is non-deterministic.

This definition imposes a certain condition of uniqueness on transitions, that is, on paths of length 1. This property is then extended to longer paths by a simple induction.

Proposition 1.2. *Let $\mathcal{A}$ be a deterministic automaton and let w be a word.*

(1) *For each state q of $\mathcal{A}$, there exists at most one path labeled w starting at q.*
(2) *If $w \in L(\mathcal{A})$, then w labels exactly one successful path.*

In particular, we can represent the set of transitions of a deterministic automaton $\mathcal{A} = (Q, T, I, F)$ by a *transition function*: the (possibly partial) function $\delta\colon Q \times A \to Q$ which maps each pair $(q, a) \in Q \times A$ to the state q' such that $(q, a, q') \in T$ (if it exists). This function is then naturally extended to the set $Q \times A^*$: if $q \in Q$ and $w \in A^*$, $\delta(q, w)$ is the state q' such that there exists a path from q to q' labeled by w in $\mathcal{A}$ (if such a state exists). In the sequel, deterministic automata will be specified as 4-tuples (Q, δ, i, F) instead of the corresponding $(Q, T, \{i\}, F)$. We note the following elementary characterization of δ.

Proposition 1.3. *Let $\mathcal{A} = (Q, \delta, i, F)$ be a deterministic automaton. Then we have*

$$\delta(q, \varepsilon) = q;$$

$$\delta(q, ua) = \begin{cases} \delta(\delta(q, u), a) & \textit{if both } \delta(q, u) \textit{ and } \delta(\delta(q, u), a) \textit{ exist,} \\ \textit{undefined} & \textit{otherwise;} \end{cases}$$

$$u \in L(\mathcal{A}) \textit{ if and only if } \delta(i, u) \in F.$$

for each state q, each word $u \in A^*$ and each letter $a \in A$.

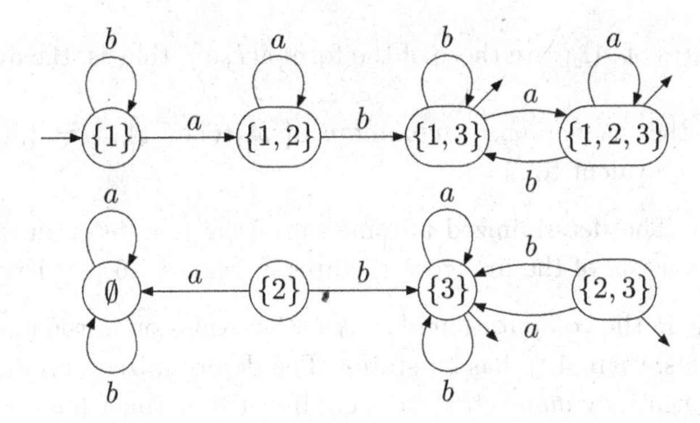

Fig. 1.6. The subset automaton of the automaton in Figure 1.3.

Again, it turns out that every automaton is equivalent to a deterministic automaton. This deterministic automaton can be effectively constructed, although the algorithm – the so-called *subset construction* – is more complicated than those used to construct complete or trim automata.

Let $\mathcal{A} = (Q, T, I, F)$ be an automaton. The *subset transition function* of $\mathcal{A}$ is the function $\delta\colon \mathcal{P}(Q) \times A \to \mathcal{P}(Q)$ defined, for each $P \subseteq Q$ and each $a \in A$ by

$$\delta(P, a) = \{q \in Q \mid \exists p \in P,\ (p, a, q) \in T\}.$$

Thus, $\delta(P, a)$ is the set of states of $\mathcal{A}$ which can be reached by an a-labeled transition, starting from an element of P. The *subset automaton* of $\mathcal{A}$ is $\mathcal{A}_{\text{sub}} = (\mathcal{P}(Q), \delta, I, F_{\text{sub}})$ where $F_{\text{sub}} = \{P \subseteq Q \mid P \cap F \neq \emptyset\}$.

The automaton $\mathcal{A}_{\text{sub}}$ is deterministic and complete by construction, and the subset transition function of $\mathcal{A}$ is the transition function of $\mathcal{A}_{\text{sub}}$. Moreover, if $\mathcal{A}$ has n states, then $\mathcal{A}_{\text{sub}}$ has 2^n states.

Example 1.8. The subset automaton of the non-deterministic automaton of Figure 1.3 is given in Figure 1.6. Notice that the states of the second row are not accessible.

Proposition 1.4. *The automata $\mathcal{A}$ and $\mathcal{A}_{\text{sub}}$ are equivalent.*

Sketch of proof. Let $\mathcal{A} = (Q, T, I, F)$. One shows by induction on $|w|$ that for all $P \subseteq Q$ and $w \in A^*$, $\delta(P, w)$ is the set of all states $q \in Q$ such that w labels a path in $\mathcal{A}$ starting at some state in P and ending at q.

Therefore, a word w is accepted by $\mathcal{A}$ if and only if at least one final state lies in the set $\delta(I, w)$, if and only if $\delta(I, w) \in F_{\text{sub}}$, if and only if w is accepted by $\mathcal{A}_{\text{sub}}$. This concludes the proof. □

In general, the subset automaton is not trim (see Example 1.8) and we can find a deterministic automaton smaller than $\mathcal{A}_{\text{sub}}$, which still recognizes the same language as $\mathcal{A}$, namely by trimming $\mathcal{A}_{\text{sub}}$. Observe that in the proof of Proposition 1.4, the only useful states of $\mathcal{A}_{\text{sub}}$ are those of the form $\delta(I, w)$, that is, the accessible states of $\mathcal{A}_{\text{sub}}$.

We define the *determinized automaton* of $\mathcal{A}$ to be $\mathcal{A}_{\text{det}} = (\mathcal{A}_{\text{sub}})_{\text{trim}}$. This automaton is equivalent to $\mathcal{A}$.

Example 1.9. The determinized automaton of the non-deterministic automaton of Figure 1.3 consists of the first row of states in Figure 1.6 (see Example 1.8).

An obstacle in the computation of $\mathcal{A}_{\text{det}}$ is the explosion in the number of states: if $\mathcal{A}$ has n states, then $\mathcal{A}_{\text{sub}}$ has 2^n states. The determinized automaton $\mathcal{A}_{\text{det}}$ may well have exponentially many states as well, but it sometimes has fewer. Therefore, it makes sense to try and compute $\mathcal{A}_{\text{det}}$ directly, in time proportional to its actual number of states, rather than first constructing the exponentially large automaton $\mathcal{A}_{\text{sub}}$ and then trimming it.

This can be done using the same ideas as in the construction of $\mathcal{A}_{\text{trim}}$ in Section 1.2.2.3. One first constructs $\mathcal{B}$, the accessible part of $\mathcal{A}_{\text{sub}}$, starting with the initial state of $\mathcal{A}_{\text{sub}}$, namely I. Then for each constructed state P and each letter a, we construct $\delta(P, a)$ and the transition $(P, a, \delta(P, a))$. And we stop when no new state arises this way.

The second step consists in finding the co-accessible part of $\mathcal{B}$, using the method in Section 1.2.2.3.

Example 1.10. Let $A = \{a, b\}$, let $n \geq 2$, and let $L = A^*aA^{n-2}$. Then L is accepted by a non-deterministic automaton $\mathcal{A}$ with n states. However, any deterministic automaton accepting L must have at least 2^{n-1} states. To see this, suppose that (Q, δ, i, F) is such a deterministic automaton. Let u, v be distinct words of length $n-1$. Then one of the words (let us say u) contains an a in a position in which v contains the letter b. Thus $u = u'ax$, $v = v'by$, where $|x| = |y|$. Let w be any word of length $n - 2 - |x|$. Then $uw \in L$, $vw \notin L$. It follows that $\delta(i, u) \neq \delta(i, v)$ and thus there are at least as many states as there are words of length $n-1$. This shows that the exponential blowup in the number of states in the subset construction cannot in general be reduced.

1.3. Logic: Büchi's sequential calculus

Let us start with an example.

Example 1.11. Recall that $\wedge$ is the logical conjunction, which reads "AND". And $\vee$ is the logical disjunction, which reads "OR". We will consider formulas such as

$$\exists x \exists y\ (x < y) \wedge R_a x \wedge R_b y.$$

This formula has the following interpretation on a word u: there exist two natural numbers $x < y$ such that, in u, the letter in position x is an a and the letter in position y is a b. Thus this formula specifies a language: the set of all words u in which this formula holds, namely $A^*aA^*bA^*$.

1.3.1. *First-order formulas*

Let us now formalize this point of view on languages.

1.3.1.1. *Syntax*

The formulas of Büchi's sequential calculus use the usual logical symbols ($\wedge$, $\vee$, $\neg$ for the negation), the equality symbol $=$, the constant symbol **true**, the quantifiers $\exists$ and $\forall$, variable symbols $(x, y, z, \ldots)$ and parentheses. They also use specific, non-logical symbols: binary relation symbols $<$ and S, and unary relation symbols R_a (one for each letter $a \in A$).

For convenience, we may assume that the variables are drawn from a fixed, countable, set of variables.

The *atomic formulas* are the formulas of the form **true**, $x = y$, $x < y$, $S(x, y)$, and $R_a x$, where x and y are variables and $a \in A$.

The *first-order formulas* are defined as follows:

- Atomic formulas are first-order formulas,
- If φ and ψ are first-order formulas, then $(\neg\varphi)$, $(\varphi\wedge\psi)$ and $(\varphi\vee\psi)$ are first-order formulas,
- If φ is a first-order formula and if x is a variable, then $(\exists x\ \varphi)$ and $(\forall x\ \varphi)$ are first-order formulas.

Remark 1.2. As is usual in logic, we will limit the usage of parentheses in our notation of formulas, to what is necessary for their proper parsing, writing for instance $\forall x\ R_a x$ instead of $(\forall x\ (R_a x))$.

Certain variables appear after a quantifier (existential or universal): occurrences of these variables within the scope of the quantifier are said to be *bound*. Other occurrences are said to be *free*. A precise, recursive, definition of the set $FV(\varphi)$ of the free variables of a formula φ is as follows:

- If φ is atomic, then $FV(\varphi)$ is the set of all variables occurring in φ,
- $FV(\neg\varphi) = FV(\varphi)$,
- $FV(\varphi \wedge \psi) = FV(\varphi \vee \psi) = FV(\varphi) \cup FV(\psi)$,
- $FV(\exists x\ \varphi) = FV(\forall x\ \varphi) = FV(\varphi) \setminus \{x\}$.

A formula without free variables is called a *sentence*.

1.3.1.2. *Interpretation of formulas*

In Büchi's sequential calculus, formulas are interpreted in words: each word u of length $n \geq 0$ determines a *structure* (which we abusively denote by u) with *domain* $\mathsf{Dom}(u) = \{0, \ldots, n-1\}$ ($\mathsf{Dom}(u) = \emptyset$ if $u = \varepsilon$). $\mathsf{Dom}(u)$ is viewed as the set of positions in the word u (numbered from 0).

The symbol $<$ is interpreted in $\mathsf{Dom}(u)$ as the usual order (as in $(2 < 4)$ and $\neg(3 < 2)$). The symbol S is interpreted as the *successor* symbol: if $x, y \in \mathsf{Dom}(u)$, then $S(x, y)$ if and only if $y = x + 1$. Finally, for each letter $a \in A$, the unary relation symbol R_a is interpreted as the set of positions in u that carry an a (a subset of $\mathsf{Dom}(u)$).

Example 1.12. If $u = abbaab$, then $\mathsf{Dom}(u) = \{0, 1, \ldots, 5\}$, $R_a = \{0, 3, 4\}$ and $R_b = \{1, 2, 5\}$.

A *valuation* on u is a mapping ν from a set of variables into the domain $\mathsf{Dom}(u)$. It will be useful to have a notation for small modifications of a valuation: if ν is a valuation and d is an element of $\mathsf{Dom}(u)$, we let $\nu[x \mapsto d]$ be the valuation ν' defined by extending the domain of ν to include the variable x and setting

$$\nu'(y) = \begin{cases} \nu(y) & \text{if } y \neq x, \\ d & \text{if } y = x. \end{cases}$$

If φ is a formula, $u \in A^*$ and ν is a valuation on u whose domain includes the free variables of φ, then we define $u, \nu \models \varphi$ (and say that the valuation ν *satisfies* φ in u, or equivalently u, ν satisfies φ) as follows:

- $u, \nu \models (x = y)$ (resp. $(x < y)$, $S(x, y)$, $R_a x$) if and only if $\nu(x) = \nu(y)$ (resp. $\nu(x) < \nu(y)$, $S(\nu(x), \nu(y))$, $R_a \nu(x)$) in $\mathsf{Dom}(u)$;
- $u, \nu \models \neg\varphi$ if and only if it is not true that $u, \nu \models \varphi$;
- $u, \nu \models (\varphi \vee \psi)$ (resp. $(\varphi \wedge \psi)$) if and only if at least one (resp. both) of $u, \nu \models \varphi$ and $u, \nu \models \psi$ holds (resp. hold);
- $u, \nu \models (\exists x\, \varphi)$ if and only if there exists $d \in \mathsf{Dom}(u)$ such that $u, \nu[x \mapsto d] \models \varphi$;
- $u, \nu \models (\forall x\, \varphi)$ if and only if, for each $d \in \mathsf{Dom}(u)$, $u, \nu[x \mapsto d] \models \varphi$.

Note that the truth value of $u, \nu \models \varphi$ depends only on the values assigned by ν to the free variables of φ. In particular, if φ is a sentence, then there is a valuation μ with an empty domain. We say that *φ is satisfied by u* (or *u satisfies φ*), and we write $u \models \varphi$ for $u, \mu \models \varphi$. Thus each sentence φ defines a language: the set $L(\varphi)$ of all words such that $u \models \varphi$. Note that this interpretation makes sense even if u is the empty word, for then the valuation μ is still defined: Every sentence beginning with a universal quantifier is satisfied by ε, and no sentence beginning with an existential quantifier is satisfied by ε. An early example was given in Example 1.11.

Remark 1.3. Two sentences φ and ψ are said to be *logically equivalent* if they are satisfied by the same structures. We will use freely the classical logical equivalence results, such as the logical equivalence of $\varphi \wedge \psi$ and $\neg(\neg\varphi \vee \neg\psi)$, or the logical equivalence of $\forall x\ \varphi$ and $\neg(\exists x\ \neg\varphi)$. We will also use the implication and bi-implication notation: $\varphi \rightarrow \psi$ stands for $\neg\varphi \vee \psi$ and $\varphi \leftrightarrow \psi$ stands for $(\varphi \rightarrow \psi) \wedge (\psi \rightarrow \varphi)$.

Example 1.13. Let φ and ψ be the following formulas.

$$\varphi = \exists x \left((\forall y \, \neg(y < x)) \wedge R_a x \right)$$

$$\psi = \forall x \left((\forall y \, \neg(y < x)) \to R_a x \right).$$

The sentence φ states that there exists a position with no strict predecessor, containing an a, while ψ states that every such position contains an a. The latter sentence, like all universally quantified first-order sentences, is vacuously satisfied by the empty string. Thus $L(\varphi) = aA^*$ and $L(\psi) = aA^* \cup \{\varepsilon\}$.

The *first-order logic* of the *linear order* (resp. of the *successor*), written $\mathsf{FO}(<)$ (resp. $\mathsf{FO}(S)$) is the fragment of the first-order logic described so far, where formulas do not use the symbol S (resp. $<$).

1.3.2. *Monadic second-order formulas*

In *monadic second-order logic*, we add a new type of variable to first-order logic, called *set variables* and usually denoted by upper case letters, e.g. $X, Y, \ldots$ The atomic formulas of monadic second-order are the atomic formulas of first-order logic, and the formulas of the form (Xy), where X is a set variable and y is an ordinary variable.

The recursive definition of *monadic second-order formulas*, starting from the atomic formulas, closely resembles that of first-order formulas: it uses the same rules given in Section 1.3.1, and the additional rule:

- If φ is a monadic second-order formula and X is a set variable, then $(\exists X \varphi)$ and $(\forall X \varphi)$ are monadic second-order formulas.

The notion of free variables is extended in the same fashion.

The interpretation of monadic second-order formulas also requires an extension of the definition of a valuation on a word u: a *monadic second-order valuation* is a mapping ν which associates with each first-order variable an element of the domain $\mathsf{Dom}(u)$, and with each set variable, a subset of $\mathsf{Dom}(u)$.

If ν is a valuation, X is a set variable, and R is a subset of $\mathsf{Dom}(u)$, we denote by $\nu[X \mapsto R]$ the valuation obtained from ν by mapping X to R (see Section 1.3.1.2).

With these definitions, we can recursively give a meaning to the notion that a valuation ν satisfies a formula φ in a word u ($u, \nu \models \varphi$): we use again the rules given in Section 1.3.1.2, to which we add the following:

- $u, \nu \models (Xy)$ if and only if $\nu(y) \in \nu(X)$;
- $u, \nu \models (\exists X \varphi)$ (resp. $(\forall X \varphi)$) if and only if there exists $R \subseteq \mathsf{Dom}(u)$ such that (resp. for each $R \subseteq \mathsf{Dom}(u)$) $u, \nu[X \mapsto R] \models \varphi$.

Note that the empty set is a valid assignment for a set variable: the empty word may satisfy monadic second order variables even if they start with an existential set quantifier.

Büchi's sequential calculus (see Section 1.3.1.2) is thus extended to include monadic second-order formulas. We denote by $\mathsf{MSO}(<)$ (resp. $\mathsf{MSO}(S)$) the fragment of monadic second-order logic, where formulas do not use the symbol S (resp. $<$). Of course, $\mathsf{FO}(<)$ and $\mathsf{FO}(S)$ are subsets of $\mathsf{MSO}(<)$ and $\mathsf{MSO}(S)$, respectively.

Example 1.14. Inspecting the following $\mathsf{MSO}(<)$ sentence,

$$\begin{aligned}\varphi = \exists X \quad & \big[\forall x\ (Xx \leftrightarrow ((\forall y\ \neg(x < y)) \vee (\forall y\ \neg(y < x)))) \\ & \wedge \forall x\ (Xx \to R_a x)\ \wedge\ \exists x\ Xx\big].\end{aligned}$$

one can see that the elements of X must be the first and last positions of the word in which we interpret φ, so $L(\varphi) = aA^* \cap A^*a$. This language can also be described by a first order sentence, see Example 1.13, that is: this formula is equivalent to a first-order formula.

Example 1.15. We now consider the more complex formula

$$\begin{aligned}\varphi = \exists X \quad & ((\forall x\ \forall y\ ((x < y) \wedge (\forall z\ \neg((x < z) \wedge (z < y)))) \to (Xx \leftrightarrow \neg Xy)) \\ & \wedge (\forall x\ (\forall y\ \neg(y < x)) \to Xx) \\ & \wedge (\forall x\ (\forall y\ \neg(x < y)) \to \neg Xx)\big).\end{aligned}$$

The formula φ states that there exists a set X of positions in the word, such that a position is in X if and only if the next position is not in X (so X has every other position), and the first position is in X, and the last position is not in X. Thus $L(\varphi)$ is the set of words of even length. It is an easy consequence of the results of Section 1.7 that this language cannot be described by a first-order formula.

The successor relation can be expressed in $\mathsf{FO}(<)$: $S(x, y)$ is logically equivalent to the following formula:

$$(x < y)\ \wedge\ \forall z\ ((x < z) \to ((y = z) \vee (y < z))).$$

In a weak converse, the order relation $<$ can be expressed in $\mathsf{MSO}(S)$: the formula $x < y$ is equivalent to:

$$\exists X\ (Xy \wedge \neg Xx\ \wedge [\forall z\ \forall t\ ((Xz \wedge S(z, t)) \to Xt)]).$$

It follows that $\mathsf{MSO}(<)$ and $\mathsf{MSO}(S)$ have the same expressive power.

Proposition 1.5. *A language can be defined by a sentence in $\mathsf{MSO}(S)$, if and only if it can be defined by a sentence in $\mathsf{MSO}(<)$.*

However, the order relation $<$ cannot be expressed in $\mathsf{FO}(S)$. This is a non-trivial result; for a proof, see [17].

Proposition 1.6. *If a language can be defined by a sentence in $\mathsf{FO}(S)$, then it can be defined by a sentence in $\mathsf{FO}(<)$. The converse does not hold.*

1.4. The Kleene-Büchi theorem

In this section, we prove the following theorem, a combination of the classical Kleene and Büchi theorems.

Theorem 1.1. *Let L be a language in A^*. The following conditions are equivalent:*

(1) L is defined by a sentence in $\mathsf{MSO}(<)$;
(2) L is accepted by an automaton;
(3) L is extended rational;
(4) L is rational.

1.4.1. *From automata to monadic second-order formulas*

Let $\mathcal{A} = (Q, i, \delta, F)$ be a deterministic automaton. The idea is to associate with each state $q \in Q$ a second order variable X_q, to encode the set of positions in which a given path visits state q. What we need to express about the sets X_q is the following:

- the sets X_q form a partition of the set of all positions (at each point in time, the automaton must be in one and exactly one state);
- if a path visits state q at time x, state q' at time $x + 1$ and if the letter in position $x + 1$ is an a, then $\delta(q, a) = q'$;

This analysis leads to the following formula. For convenience, let Q be the set $\{q_0, q_1, \ldots, q_n\}$, with initial state $i = q_0$. We also use the shorthand min and max to designate the first and last positions: this is acceptable as these positions can be expressed by $\mathsf{FO}(S)$-formulas. For instance, R_a min stands for $\forall x\, (\forall y\, \neg S(y, x) \rightarrow R_a x)$; and X max stands for $\forall x\, (\forall y\, \neg S(x, y) \rightarrow Xx)$.

$$\begin{aligned}
&\exists X_{q_0}\ \exists X_{q_1}\ \cdots\ \exists X_{q_n} \\
&\quad\Big(\bigwedge_{q \neq q'} \neg \exists x\, (X_q x \wedge X_{q'} x) \quad \wedge \quad \forall x \bigvee_q X_q x \\
&\quad \wedge\ \forall x\, \forall y\, \Big[S(x, y) \rightarrow \bigvee_{q \in Q,\ a \in A} (X_q x \wedge R_a y \wedge X_{\delta(q,a)} y) \Big] \\
&\quad \wedge\ \bigwedge_{a \in A} (R_a \min \rightarrow X_{\delta(q_0,a)} \min)\ \wedge\ \Big(\bigvee_{q \in F} X_q \max \Big) \Big).
\end{aligned}$$

This sentence is actually verified by the empty word, so the language it defines coincides with $L(\mathcal{A})$ on A^+. If $q_0 \in F$, it accurately defines $L(\mathcal{A})$. But if $q_0 \notin F$, we must consider the conjunction of this sentence with $\exists x$ **true**.

This is a sentence in $\mathsf{MSO}(S, <)$ but as we know, it is logically equivalent to one in $\mathsf{MSO}(<)$. Note that it is in fact an existential monadic second order sentence, that is, the second-order quantifications are all existential.

1.4.2. *From formulas to extended rational expressions*

The proof that an MSO(<)-definable language can be described by an extended rational expression, is more complex. The reasoning is by induction on the recursive definition of formulas. Instead of associating a language only with sentences (formulas without free variables), we will associate languages with all formulas but these languages will be over larger alphabets, which allow us to encode valuations.

1.4.2.1. *The auxiliary alphabets $B_{p,q}$*

Let $p, q \geq 0$ and let $B_{p,q} = A \times \{0,1\}^p \times \{0,1\}^q$. A word over the alphabet $B_{p,q}$ can be identified with a sequence $(u_0, u_1, \ldots, u_p, u_{p+1}, \ldots, u_{p+q})$ where $u_0 \in A^*$, $u_1, \ldots, u_p, u_{p+1}, \ldots, u_{p+q} \in \{0,1\}^*$ and all the u_i have the same length.

Let $K_{p,q}$ consist of the empty word and the words in $B^+_{p,q}$ such that each of the components $u_1, \ldots, u_p$ contains exactly one occurrence of 1. Thus each of these components really designates *one* position in the word u_0, and each of the components $u_{p+1}, \ldots, u_{p+q}$ designates a set of positions in u_0.

Example 1.16. If $A = \{a, b\}$, the following is a word in $K_{2,1}$:

$$
\begin{array}{ll}
u_0 & a\ b\ a\ a\ b\ a\ b \\
\\
u_1 & 0\,0\,0\,0\,1\,0\,0 \\
u_2 & 0\,0\,1\,0\,0\,0\,0 \\
\\
u_3 & 0\,1\,1\,0\,0\,1\,1
\end{array}
$$

Its components u_1 and u_2 designate positions 4 and 2, respectively, and its component u_3 designates the set $\{1, 2, 5, 6\}$.

The languages $K_{p,q}$ are extended rational. Indeed, for $1 \leq i \leq p$, let C_i be the set of elements $(b_0, b_1, \ldots, b_{p+q}) \in B_{p,q}$ such that $b_i = 1$. Then $K_{p,q}$ is the set of words in $B^*_{p,q}$ which contain at most one letter in each C_i:

$$K_{p,q} = \{\varepsilon\} \cup \bigcap_{1 \leq i \leq p} (B_{p,q} \setminus C_i)^* C_i (B_{p,q} \setminus C_i)^* = B^*_{p,q} \setminus \bigcup_{1 \leq i \leq p} B^*_{p,q} C_i B^*_{p,q} C_i B^*_{p,q}.$$

1.4.2.2. *The language associated with a formula*

Let now $\varphi(x_1, \ldots, x_r, X_1, \ldots, X_s)$ be a formula in which the free first order (resp. set) variables are $x_1, \ldots, x_r$ (resp. $X_1, \ldots, X_s$), with $r \leq p$ and $s \leq q$.

We interpret

- R_a as $R_a = \{i \in \mathsf{Dom}(u) \mid u_0(i) = a\}$;
- x_i as the unique position of 1 in u_i (if $u_i \neq \varepsilon$);
- X_j as the set of positions of 1 in u_{p+j}.

Note that if $p = q = 0$, then φ is a sentence and this is the usual notion of interpretation.

More formally, let $(u_0, u_1, \ldots, u_{p+q})$ be a non-empty word in $K_{p,q}$. Let n_i be the position of the unique 1 in the word u_i and let N_j be the set of the positions of the 1's in the word u_{p+j}. We say that $u = (u_0, u_1, \ldots, u_{p+q}) \in K_{p,q}$ satisfies φ if u_0, ν satisfy φ where ν is the valuation defined by

$$\nu(x_i) = n_i \text{ for } 1 \leq i \leq r \quad \text{and} \quad \nu(X_j) = N_j \text{ for } 1 \leq j \leq s.$$

We also say that the empty word (in $K_{p,q}$) satisfies φ if $\varepsilon \models \varphi$. We let $L_{p,q}(\varphi) = \{u \in K_{p,q} \mid u \text{ satisfies } \varphi\}$. Thus each *formula* φ defines a subset of $K_{p,q}$, and hence a language in $B_{p,q}^*$.

Example 1.17. Let $\varphi = \exists x\ (x < y \wedge R_a y)$. Then $FV(\varphi) = \{y\}$. And $L_{1,0}(\varphi)$ is the set of pairs of words (u_0, u_1) such that $u_0 \in A^*$, $u_1 \in \{0,1\}^*$, u_0 and u_1 have the same length, u_1 has a single 1, which is not the first position, and u_0 has an a in that position.

Let $\varphi = \forall x\ ((Xx \wedge x < y \wedge R_b y) \to R_a x)$. Then $L_{1,1}(\varphi)$ is the set of triples of words (u_0, u_1, u_2) with $u_0 \in A^*$, $u_1, u_2 \in \{0,1\}^*$, all three words have the same length, and either this length is zero, or u_1 has a single 1 such that:

> Let n be the position in u_1 which has a 1. If u_0 has a b in position n, then u_0 has an a in each position before n in which u_2 has a 1. If u_0 does not have a b in position n, then there is no constraint.

1.4.2.3. *The MSO(<)-definable languages are extended rational*

We first consider the languages associated with an atomic formula. Let $1 \leq i, j \leq p+q$ and let $a \in A$. Let

$$\begin{aligned} C_{j,a} &= \{b \in B_{p,q} \mid b_j = 1 \text{ and } b_0 = a\}, \\ C_{i,j} &= \{b \in B_{p,q} \mid b_i = b_j = 1\}, \\ \text{and } C_i &= \{b \in B_{p,q} \mid b_i = 1\}. \end{aligned}$$

Then we have

$$\begin{aligned} L_{p,q}(R_a x_i) &= K_{p,q} \cap B_{p,q}^* C_{i,a} B_{p,q}^* \\ L_{p,q}(x_i = x_j) &= K_{p,q} \cap B_{p,q}^* C_{i,j} B_{p,q}^* \\ L_{p,q}(x_i < x_j) &= K_{p,q} \cap B_{p,q}^* C_i B_{p,q}^* C_j B_{p,q}^* \\ L_{p,q}(X_i x_j) &= K_{p,q} \cap B_{p,q}^* C_{i+p,j} B_{p,q}^*. \end{aligned}$$

Thus, the languages defined by the atomic formulas, namely $L_{p,q}(R_a x)$, $L_{p,q}(x = y)$, $L_{p,q}(x < y)$ and $L_{p,q}(Xy)$, are extended rational.

Now let φ and ψ be formulas and let us assume that $L_{p,q}(\varphi)$ and $L_{p,q}(\psi)$ are extended rational. Then we have

$$L_{p,q}(\varphi \vee \psi) = L_{p,q}(\varphi) \cup L_{p,q}(\psi)$$
$$L_{p,q}(\varphi \wedge \psi) = L_{p,q}(\varphi) \cap L_{p,q}(\psi)$$
$$L_{p,q}(\neg\varphi) = K_{p,q} \setminus L_{p,q}(\varphi),$$

and hence these three languages are extended rational as well. We still need to handle existential quantification.

Let π_i be the morphism which deletes the i-th component in a word of $B_{p,q}^*$; that is: if $1 \leq i \leq p$, then $\pi_i \colon B_{p,q}^* \to B_{p-1,q}^*$, and if $p < i \leq p+q$, then $\pi_i \colon B_{p,q}^* \to B_{p,q-1}^*$. In either case, we have $\pi_i(b_0, b_1, \ldots, b_{p+q}) = (b_0, b_1, \ldots, b_{i-1}, b_{i+1}, \ldots, b_{p+q})$.

Now, observe that, for any formula $\varphi(x_1, \ldots, x_r, X_1, \ldots, X_s)$, and for $p \geq r$, $q \geq s$, $1 \leq i \leq p$ and $1 \leq j \leq q$ we have

$$L_{p-1,q}(\exists x_i \varphi) = \pi_i(L_{p,q}(\varphi)) \quad \text{and} \quad L_{p,q-1}(\exists X_j \varphi) = \pi_{p+j}(L_{p,q}(\varphi)).$$

This concludes the proof that $L_{p,q}(\varphi)$ is extended rational for any $p \geq r$, $q \geq s$.

In particular, if φ is a sentence in $\mathsf{MSO}(<)$ (that is, φ has no free variables), we may take $p = q = 0$. Then $L_{0,0}(\varphi)$ is extended rational – and we already noted that $L(\varphi) = L_{0,0}(\varphi)$.

1.4.3. *From extended rational expressions to automata*

It is immediately verified that the languages $\emptyset$, $\{\varepsilon\}$, $\{a\}$ ($a \in A$) are accepted by finite automata. We now need to show that if $K, L \subseteq A^*$ are recognizable and if $\pi \colon A^* \to B^*$ is a morphism, then $\overline{L}$, $K \cup L$, $K \cap L$, KL, K^* and $\pi(L)$ are recognizable.

Proposition 1.7. *If $L \subseteq A^*$ is recognizable, then the complement $\overline{L}$ of L is recognizable as well.*

Proof. Let $\mathcal{A} = (Q, \delta, i, F)$ be a deterministic complete automaton recognizing L. Then $\overline{\mathcal{A}} = (Q, \delta, i, \overline{F})$ recognizes $\overline{L}$ by Proposition 1.3. □

Example 1.18. The deterministic automata in Examples 1.5 and 1.6 confirm that, if $A = \{a, b\}$, then b^*a^* is the complement of A^*abA^*.

Note that the resulting procedure yields a deterministic automaton for $\overline{L}$. It is very efficient if L is given by a deterministic automaton, but may lead to an exponential growth in the number of states if L is given by a non-deterministic automaton.

Proposition 1.8. *If $K, L \subseteq A^*$ are recognizable, then $K \cup L$ and $K \cap L$ are recognizable as well.*

Proof. Let $\mathcal{A} = (Q, T, I, F)$ and $\mathcal{A}' = (Q', T', I', F')$ be automata recognizing L and L', respectively. We assume that the state sets Q and Q' are disjoint. Then it is readily verified that the automaton

$$\mathcal{A} \cup \mathcal{A}' = (Q \cup Q', T \cup T', I \cup I', F \cup F')$$

accepts $L \cup L'$. Thus $L \cup L'$ is recognizable, and hence so is $L \cap L' = \overline{\overline{L} \cup \overline{L'}}$, by Proposition 1.7. □

The construction in the above proof always yields a non-deterministic automaton for $L \cup L'$, even if we start from deterministic automata for L and L'. The product of automata provides an alternative construction which preserves determinism, avoids any exponentiation of the number of states, and works for both the union and the intersection.

Let $\mathcal{A} = (Q, T, I, F)$ and $\mathcal{A}' = (Q', T', I', F')$ be automata recognizing the languages L and L'. Their *cartesian product* is the automaton $\mathcal{A}'' = (Q \times Q', T'', I \times I', F \times F')$ where

$$T'' = \{((p, p'), a, (q, q')) \mid (p, a, q) \in T \text{ and } (p', a, q') \in T'\}.$$

Note that if $\mathcal{A}$ and $\mathcal{A}'$ are deterministic, then $\mathcal{A}''$ is deterministic as well. The main property of $\mathcal{A}''$ is the following: there exists a path $(p, p') \xrightarrow{u} (q, q')$ in $\mathcal{A}''$ if and only if there exist paths $p \xrightarrow{u} q$ and $p' \xrightarrow{u} q'$, in $\mathcal{A}$ and $\mathcal{A}'$ respectively. Therefore $\mathcal{A}''$ recognizes $L \cap L'$.

If we take $(F \times Q') \cup (Q \times F')$ as the set of final states, instead of $F \times F'$, and if the automata $\mathcal{A}$ and $\mathcal{A}'$ are complete, then the product automaton recognizes $L \cup L'$.

In practice, the cartesian product of $\mathcal{A}$ and $\mathcal{A}'$ may not be trim, and one may want to use the procedure in Section 1.2.2.3 to produce more concise automata for $L \cap L'$ and $L \cup L'$.

Remark 1.4. Let us record here an algorithmic consequence of Propositions 1.7 and 1.8: given two automata $\mathcal{A}$ and $\mathcal{B}$, it is decidable whether $L(\mathcal{A}) \subseteq L(\mathcal{B})$ and whether $L(\mathcal{A}) = L(\mathcal{B})$. Indeed, we can compute automata accepting $L(\mathcal{A}) \setminus L(\mathcal{B}) = L(\mathcal{A}) \cap \overline{L(\mathcal{B})}$ and $L(\mathcal{B}) \setminus L(\mathcal{A})$, and decide whether these languages are empty (see Remark 1.1).

Proposition 1.9. *If $L, L' \subseteq A^*$ are recognizable, then LL' and L^* are recognizable as well.*

Sketch of proof. Let $\mathcal{A} = (Q, T, I, F)$ and let $\mathcal{A}' = (Q', T', I', F')$ be automata accepting L and L', respectively, and let us assume that their state sets are disjoint.

It is easily verified that the ε-automaton

$$(Q \cup Q', T \cup T' \cup (F \times \{\varepsilon\} \times I'), I, F')$$

accepts LL' (see Section 1.2.2.4). Similarly, if j is a state not in Q, the ε-automaton

$$(Q \cup \{j\}, T \cup (F \times \{\varepsilon\} \times I), I \cup \{j\}, F \cup \{j\})$$

accepts L^*. □

Proposition 1.10. *If $L \subseteq A^*$ is recognizable and $\varphi\colon A^* \to B^*$ is a morphism, then $\varphi(L)$ is recognizable as well.*

Sketch of proof. Let $\mathcal{A} = (Q, T, I, F)$ be an automaton recognizing L. We let $\mathcal{A}'$ be the ε-automaton $\mathcal{A}' = (Q \cup Q', T', I, F)$, where the set T' consists of

- the transitions of the form (p, ε, q) such that $(p, a, q) \in T$ for some letter a with $\varphi(a) = \varepsilon$,
- the transitions occurring in the paths of the form

$$p \xrightarrow{b_1} q'_1 \xrightarrow{b_2} \cdots q'_{k-1} \xrightarrow{b_k} q$$

such that $(p, a, q) \in T$, $\varphi(a) = b_1 \cdots b_k \neq \varepsilon$ and $q'_1, \ldots, q'_{k-1}$ are new states that we adjoin for each such triple (p, a, q).

The set Q' contains all the new states that occur in the latter paths. It is elementary to verify that $\mathcal{A}'$ recognizes $\varphi(L)$. □

So far, we have shown that a language is recognizable, if and only if it is defined by a sentence in $\mathsf{MSO}(<)$, if and only if it is extended rational.

Remark 1.5. Note that the proofs of this logical equivalence are constructive, in the sense that given a sentence φ in $\mathsf{MSO}(<)$, we can construct an automaton $\mathcal{A}$ such that $L(\varphi) = L(\mathcal{A})$. It follows that $\mathsf{MSO}(<)$ is decidable: given an MSO sentence φ, we can decide whether φ always holds. Indeed, this is the case if and only if $L(\neg\varphi) = \emptyset$, which can be tested as discussed in Remark 1.1.

1.4.4. *From automata to rational expressions*

To complete the proof of the Kleene-Büchi theorem, it suffices to prove that every recognizable language is rational. For this, we use the *McNaughton-Yamada construction*.

Let $\mathcal{A} = (Q, T, I, F)$ be an automaton. For each pair of states $p, q \in Q$ and for each subset $P \subseteq Q$, let $L_{p,q}(P)$ be the set of all words $u \in A^*$ which label a path from state p to state q, such that the states visited internally by that path are all in P:

$$L_{p,q}(P) = \{a_1 a_2 \ldots a_n \in A^* \mid \text{there exists a path in } \mathcal{A} \quad p \xrightarrow{a_1} q_1 \xrightarrow{a_2} \ldots q_{n-1} \xrightarrow{a_n} q \text{ with } q_1, \ldots, q_{n-1} \in P\}.$$

Recall that, by convention, there always exists an empty path, labeled by the empty word, from any state q to itself. So $\varepsilon \in L_{p,q}(P)$ if and only if $p = q$.

We show by induction on the cardinality of P that each language $L_{p,q}(P)$ is rational. This will prove that $L(\mathcal{A})$ is rational, since $L(\mathcal{A}) = \bigcup_{i\in I,\ f\in F} L_{i,f}(Q)$.

If $P = \emptyset$, then $L_{p,q}(\emptyset) = \{a \in A \mid (p,a,q) \in T\}$ if $p \neq q$, and $L_{q,q}(\emptyset) = \{a \in A \mid (q,a,q) \in T\} \cup \{\varepsilon\}$. Thus $L_{p,q}(\emptyset)$ is always finite, and hence rational.

Now let $n > 0$ and let us assume that, for any $p, q \in Q$ and $P \subseteq Q$ containing at most $n-1$ states, the language $L_{p,q}(P)$ is rational. Let now $P \subseteq Q$ be a subset with n elements and let $r \in P$. Considering the first and the last visit to state r of a path from p to q, we find that

$$L_{p,q}(P) = L_{p,q}(P \setminus \{r\}) \ \cup \ L_{p,r}(P \setminus \{r\}) L_{r,r}(P \setminus \{r\})^* L_{r,q}(P \setminus \{r\}).$$

Since $P \setminus \{r\}$ has cardinality $n-1$, it follows from the induction hypothesis that $L_{p,q}(P)$ is rational.

This concludes the proof of the Kleene-Büchi theorem.

1.4.5. *Closure properties*

Rational languages enjoy many additional closure properties.

Proposition 1.11. *Let $\varphi\colon A^* \to B^*$ be a morphism and let $L \subseteq B^*$. If L is rational, then $\varphi^{-1}(L)$ is rational as well.*

Sketch of proof. Let $\mathcal{A} = (Q,T,I,F)$ be an automaton over B, recognizing L, and let $\mathcal{A}' = (Q,T',I,F)$ be the automaton over A where

$$T' = \{(p,a,q) \mid p \xrightarrow{\varphi(a)} q \text{ is a path in } \mathcal{A}\}.$$

It is readily verified that $\mathcal{A}'$ recognizes $\varphi^{-1}(L)$. □

Let $u \in A^*$ and $L \subseteq A^*$. The *left* and *right quotients* of L by u are defined as follows:

$$u^{-1}L = \{v \in A^* \mid uv \in L\};$$
$$Lu^{-1} = \{v \in A^* \mid vu \in L\}.$$

These notions are generalized to languages: if K and L are languages, the *left* and *right quotients* of L by K are defined as follows:

$$K^{-1}L = \{v \in A^* \mid \exists u \in K \text{ such that } uv \in L\} = \bigcup_{u\in K} u^{-1}L,$$

$$LK^{-1} = \{v \in A^* \mid \exists u \in K \text{ such that } vu \in L\} = \bigcup_{u\in K} Lu^{-1}.$$

Proposition 1.12. *If $L \subseteq A^*$ is rational and $K \subseteq A^*$ is any language (possibly not rational), then $K^{-1}L$ and LK^{-1} are rational as well.*

Sketch of proof. If $\mathcal{A} = (Q, T, I, F)$ is an automaton recognizing L. Let I' be the set of states of $\mathcal{A}$ which are accessible from an initial state of $\mathcal{A}$ following a path labeled by a word of K,

$$I' = \{q \in Q \mid \exists i \in I, \exists u \in K \text{ such that } i \xrightarrow{u} q\}.$$

Then one shows that $\mathcal{A}' = (Q, T, I', F)$ recognizes $K^{-1}L$. The proof for LK^{-1} is similar. □

Remark 1.6. The proof of Proposition 1.12 is not effective: we may not be able to construct the set of states I' associated with K. However, if K is rational too, then I' is effectively constructible.

Recall that a word u is a *prefix* of the word v if there exists a word $v' \in A^*$ such that $v = uv'$ (that is: v "starts" with u). Similarly, u is a *suffix* of v if there exists a word $v' \in A^*$ such that $v = v'u$. Finally u is a *factor* of v if there exist words $v', v'' \in A^*$ such that $v = v'uv''$.

If L is a language, we let $\text{Pref}(L)$ (resp. $\text{Suff}(L)$, $\text{Fact}(L)$) be the set of all prefixes (resp. suffixes, factors) of the words in L.

Proposition 1.13. *If $L \subseteq A^*$ is rational, then* $\text{Pref}(L)$, $\text{Suff}(L)$ *and* $\text{Fact}(L)$ *are rational as well.*

Proof. The result follows from Proposition 1.12, since $\text{Pref}(L) = L(A^*)^{-1}$, $\text{Suff}(L) = (A^*)^{-1}L$ and $\text{Fact}(L) = (A^*)^{-1}L(A^*)^{-1}$. □

We leave it to the reader to verify that the following operations also preserve rationality.

The *mirror image* of a word $u = a_1 \dots a_n \in A^*$ is the word $\tilde{u} = a_n \dots a_1$. The corresponding language operation is given by $\tilde{L} = \{\tilde{u} \mid u \in L\}$ for each $L \subseteq A^*$.

A word $u = a_1 \dots a_n \in A^*$ is a *subword* of a word $v \in A^*$ if there exist words $u_0, \dots, u_n \in A^*$ such that $v = u_0 a_1 u_1 \dots a_n u_n$. If $L \subseteq A^*$, we let $\text{SW}(L)$ be the set of all subwords of the words of L.

The *shuffle* of the words u and v is the set

$$u \sqcup\!\sqcup v = \{w \in A^* \mid \exists u_1, v_1, \dots, u_n, v_n \in A^* \text{ such that } u = u_1 \cdots u_n,\ v = v_1 \cdots v_n \text{ and } w = u_1 v_1 \cdots u_n v_n\}.$$

If K and L are languages, we let $K \sqcup\!\sqcup L = \bigcup_{u \in K,\ v \in L} u \sqcup\!\sqcup v$.

Proposition 1.14. *Let K, $L \subseteq A^*$ be rational languages. Then $\tilde{L}$, $SW(L)$ and $K \sqcup\!\sqcup L$ are rational as well.*

1.5. Pumping lemmas

The characterizations summarized in the Kleene-Büchi theorem are sufficient most of the time to show that a language is rational. Showing that a language is *not*

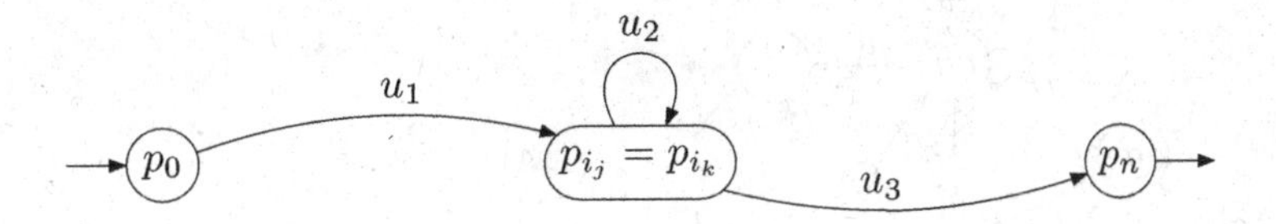

Fig. 1.7. Proof of the pumping lemma.

rational is a trickier problem. This short section presents the main tool for that purpose, namely the *pumping lemma*. We actually first present a rather abstract version of this statement, and then its more classical corollaries.

Theorem 1.2. *Let L be a rational language. There exists an integer $N > 0$ with the following property. For each word $w \in L$ and for each sequence of integers $0 \leq i_0 < i_1 < \ldots < i_N \leq |w|$, there exist $0 \leq j < k \leq N$ such that, if $w = u_1u_2u_3$ with $|u_1| = i_j$ and $|u_1u_2| = i_k$, then $u_1u_2^*u_3 \subseteq L$.*

Proof. Let $\mathcal{A}$ be an automaton recognizing L, and let N be the number of states of $\mathcal{A}$. Let $w = a_1a_2 \cdots a_n \in L$ and let

$$p_0 \xrightarrow{a_1} p_1 \xrightarrow{a_2} p_2 \cdots \xrightarrow{a_n} p_n$$

be a successful path in $\mathcal{A}$ labeled w. Let $0 \leq i_0 < i_1 < \cdots < i_N \leq n$ be a sequence of integers. Then two of the states $p_{i_0}, p_{i_1}, \ldots, p_{i_N}$ are equal, that is, there exist $0 \leq j < k \leq N$ such that $p_{i_j} = p_{i_k}$.

Let $u_1 = a_1 \cdots a_{i_j}$, $u_2 = a_{1+i_j} \cdots a_{i_k}$ and $u_3 = a_{1+i_k} \cdots a_n$. Of course, $w = u_1u_2u_3$, $|u_1| = i_j$, $|u_1u_2| = i_k$. The situation is summarized by Figure 1.7: we may iterate or skip the loop labeled u_2 and still retain a successful path, so $u_1u_2^*u_3 \subseteq L$. □

Corollary 1.1. *Let L be a rational language. There exists an integer $N > 0$ such that, for each word $w \in L$ with length $|w| \geq N$, we can factor w in three parts, $w = u_1u_2u_3$, with $u_2 \neq \varepsilon$ and $u_1u_2^*u_3 \subseteq L$.*

Corollary 1.2. *Let L be a rational language. There exists an integer $N > 0$ such that, for each word $w \in L$ with length $|w| \geq N$, we can factor w in three parts, $w = u_1u_2u_3$, with $u_2 \neq \varepsilon$, $|u_1u_2| \leq N$ (resp. $|u_2u_3| \leq N$) and $u_1u_2^*u_3 \subseteq L$.*

Sketch of proof. To prove Corollary 1.2, we apply Theorem 1.2 with $i_j = j$ (resp. $i_j = n - N + j$) for $0 \leq j \leq N$. And to prove Corollary 1.1, we take any sequence. □

Example 1.19. It is a classical application of Corollary 1.1 that $\{a^nb^n \mid n \geq 0\}$ is not rational: for each $N > 0$, the word a^Nb^N cannot be factored as $w = u_1u_2u_3$ with $u_2 \neq \varepsilon$ and $u_1u_2^*u_3 \subseteq \{a^nb^n \mid n \geq 0\}$.

Corollary 1.2 can be used to show that $\{u \in \{a,b\}^* \mid |u|_a = |u|_b\}$ is not rational (take again a^Nb^N); however, this language satisfies the necessary condition for rationality in Corollary 1.1, with $N = 2$.

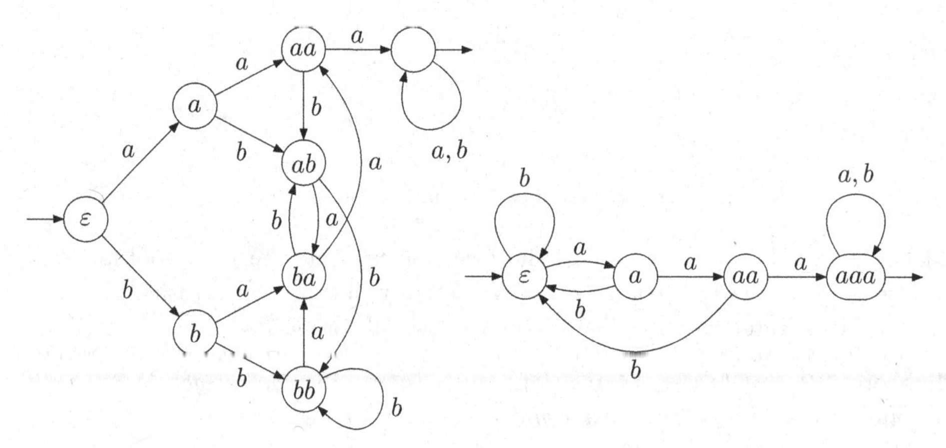

Fig. 1.8. Two different automata for A^*aaaA^*.

Consider now the following language over the alphabet $\{a, b, c, d\}$

$$\{(ab)^n(cd)^n \mid n \geq 0\} \cup A^*\{aa, bb, cc, dd, ac\}A^*$$

It satisfies the necessary condition for rationality in Corollary 1.2, but it is not rational, as can be proved using Theorem 1.2.

However, the pumping lemma as stated here may not be enough to prove that a given language is not rational. Let us say that a word *contains a square* if it can be written in the form $uvvw$ with $v \neq \varepsilon$. Then the language

$$\{udv \mid u, v \in \{a, b, c\}^* \text{ and either } u \neq v, \text{ or one of } u \text{ and } v \text{ contains a square}\}$$

satisfies the necessary condition for rationality in Theorem 1.2 (for $N = 4$). Yet it is not rational (the proof of that fact uses the existence of arbitrarily long words on the alphabet $\{a, b, c\}$ containing no square).

Ehrenfeucht, Parikh, Rozenberg gave a necessary and sufficient condition for rationality in the same style as the pumping lemma (see *e.g.* [16, Theorem I.3.3]).

1.6. Minimal automaton and syntactic monoid

Consider the two automata in Figure 1.8. Both are complete and deterministic, and both recognize the set of words over $A = \{a, b\}$ that contain some occurrence of the word aaa as a factor—that is, the language A^*aaaA^*. The two automata were designed using different intuitions about how to go about this task: In the first instance, the underlying algorithm is "keep track of the last two letters read from the input", as indicated by the state labels, while in the second automaton the algorithm is, "keep track of the length of the longest suffix of a's in the input". Thus the second automaton achieves the same result with a smaller number of states. It is easy to see that the second example is also optimal—no complete deterministic automaton recognizing this language can have a smaller number of states.

In this section we will see that for every rational language L there is a unique minimal complete deterministic automaton accepting L. We will also describe an efficient algorithm that takes as input an arbitrary complete deterministic automaton $\mathcal{A}$, and produces as output the minimal automaton for $L(\mathcal{A})$.

1.6.1. *Myhill-Nerode equivalence and the minimal automaton*

One way to see that there is something inefficient about the first automaton in the example above is to observe its behavior on the two input words $u = bab$ and $v = abb$. These words lead from the initial state to two different states. However, for purposes of recognizing words in L, there is no point in distinguishing between u and v, for no matter what the subsequent input w is, the result will be the same: either uw and vw are both in L or both outside of L.

To formalize this notion of inputs that are indistinguishable with respect to L, we make the following definitions: If $u, v \in A^*$ we define $u \equiv_L v$ if and only if $u^{-1}L = v^{-1}L$ (see Section 1.4.5). Obviously, $\equiv_L$ is an equivalence relation on A^*. We also note that if $u \equiv_L v$, and $w \in A^*$, then $uw \equiv_L vw$, since $(uw)^{-1}L = w^{-1}(u^{-1}L)$. An equivalence relation with this multiplicative property is said to be a *right congruence*. Further, L itself is a union of $\equiv_L$-classes, since $w \in L$ if and only if $\varepsilon \in w^{-1}L$.

We can accordingly define a complete deterministic automaton $\mathcal{A}_{\min}(L)$ by making the states these classes of equivalent words: We set $\mathcal{A}_{\min}(L) = (Q_L, \delta_L, i_L, F_L)$, where $Q_L = A^*/\equiv_L$, $i_L = [\varepsilon]_{\equiv_L}$, and F_L and $\delta_L \colon Q_L \times A \to Q_L$ are defined by

$$F_L = \{[v]_{\equiv_L} \mid v \in L\} \qquad \text{and} \qquad \delta([v]_{\equiv_L}, a) = [va]_{\equiv_L}.$$

We need to show that this is well-defined, since a state will in general have many different representations of the form $[v]_{\equiv_L}$. But well-definedness is an immediate consequence of our observation that $\equiv_L$ is a right congruence. We have the following result.

Theorem 1.3. *Let $L \subseteq A^*$.*

(1) $\mathcal{A}_{\min}(L)$ accepts L.
(2) L is rational if and only if $\equiv_L$ has finite index.

Proof. It follows at once by induction on $|w|$ that for all $w \in A^*$,

$$\delta_L([\varepsilon]_{\equiv_L}, w) = [w]_{\equiv_L}.$$

Since, as observed above, L itself is a union of $\equiv_L$-classes, it follows that w is accepted if and only if $w \in L$. This proves the first claim.

To prove the second claim in the theorem, note that if $\equiv_L$ has finite index, then $\mathcal{A}_{\min}$ is a finite automaton, and therefore by *(1)*, L is rational. Conversely, if L is rational, then it is accepted by some complete deterministic automaton (Q, δ, i, F)

with Q finite. Now suppose $u, v \in A^*$ and $\delta(i, u) = \delta(i, v)$. Then if $w \in A^*$ and $uw \in L$, we have

$$\delta(i, vw) = \delta(\delta(i, v), w) = \delta(\delta(i, u), w) = \delta(i, uw) \in F,$$

so $vw \in L$. Similarly, $vw \in L$ implies $uw \in L$, so $u \equiv_L v$. Thus the number of classes of $\equiv_L$ cannot be more than $|Q|$, so $\equiv_L$ has finite index. □

The proof of Theorem 1.3 shows that $\mathcal{A}_{\min}(L)$ has the least number of states among the complete deterministic automata accepting L. The automaton $\mathcal{A}_{\min}(L)$ is called the *minimal automaton* of L. We now give another, more algebraic justification for this terminology.

1.6.2. *Uniqueness and minimality of $\mathcal{A}_{\min}(L)$*

Let $\mathcal{A} = (Q, \delta, i, F)$ be a complete deterministic automaton over A, and let $L = L(\mathcal{A})$. We say that $p, q \in Q$ are *equivalent* states, and write $p \equiv q$, if

$$\{v \in A^* \mid \delta(p, v) \in F\} = \{v \in A^* \mid \delta(q, v) \in F\}.$$

Intuitively, this means that for purposes of recognizing words in L, p and q do the same job, and we might as well merge them into a single state.

We now repeat, in a somewhat different form, an observation made in the proof of Theorem 1.3: If $\delta(i, u) \equiv \delta(i, v)$, then

$$uw \in L \Longleftrightarrow \delta(\delta(i, u), w) \in L \Longleftrightarrow \delta(\delta(i, v), w) \in L \Longleftrightarrow vw \in L,$$

so that $u \equiv_L v$. In particular, if $\delta(i, u) = \delta(i, v)$, then $u \equiv_L v$, so we have a well-defined mapping $\delta(i, w) \mapsto [w]_{\equiv_L}$, from the set of *accessible* states of $\mathcal{A}$ onto the states of $\mathcal{A}_{\min}(L)$. Note that this mapping sends the initial state $i = \delta(i, \varepsilon)$ to $[\varepsilon]_{\equiv_L}$, final states of $\mathcal{A}$ to final states of $\mathcal{A}_{\min}(L)$, and respects the next-state function. We summarize these observations as follows.

Theorem 1.4. *Let $\mathcal{A} = (Q, \delta, i, F)$ be a complete deterministic automaton over A, and let $L = L(\mathcal{A})$. Then there is a map f from the set of accessible states in Q onto Q_L such that*

- *for all $a \in A$ and accessible $q \in Q$, $f(\delta(q, a)) = \delta_L(f(q), a)$,*
- *$f(i) = i_L$,*
- *$f(F) = F_L$.*

Moveover, $f(p) = f(q)$ if and only if $p \equiv q$.

In particular, if $\mathcal{A}$ has the same number of states as $\mathcal{A}_{\min}(L)$, then since f is onto, the two automata are isomorphic by Theorem 1.4.

1.6.3. *An algorithm for computing the minimal automaton*

Theorem 1.4 says that in principle we can compute the minimal automaton of a rational language L starting from any complete deterministic automaton (Q, δ, i, F) accepting L, first by removing the inaccessible states and then merging equivalent states. We have already seen how to compute the accessible states. How do we determine if two states are equivalent? If p, q are inequivalent states then there is a word $v \in A^*$ that distinguishes between these states in the sense that $\delta(p, v) \in F$ and $\delta(q, v) \notin F$, or vice-versa. It follows from a simple pumping argument that if such a distinguishing word exists, then it can be chosen to have length no more than $|Q|^2$. Thus we can effectively determine whether two states are equivalent by calculating $\delta(p, v)$ and $\delta(q, v)$ for all words up to this length.

Of course, this is a terrible algorithm, since there are $|A|^{|Q|^2}$ different words to check! In practice, we can proceed as follows: Let $m \geq 0$. We say $p \equiv_m q$ if for all $v \in A^*$ of length no more than m, $\delta(p, v) \in F$ if and only if $\delta(q, v) \in F$. This is clearly an equivalence relation on A^*, and $\equiv_{m+1}$ refines $\equiv_m$ for all m. The following lemma improves the $|Q|^2$ bound on the length of distinguishing words.

Lemma 1.1. *Let $p, q \in Q$. Then $p \equiv q$ if and only if $p \equiv_m q$ for $m = |Q| - 2$.*

Proof. First suppose that for some m, the equivalence relations $\equiv_m$ and $\equiv_{m+1}$ coincide. We claim that $\equiv_m$ and $\equiv$ coincide. To see this, suppose that p and q are inequivalent, and that w is a word of minimal length distinguishing them. If $|w| > m$, then we can write $w = uv$, where $|v| = m + 1$, so that $p' = \delta(p, u)$ and $q' = \delta(q, u)$ are inequivalent modulo $\equiv_{m+1}$. But this means that they are also inequivalent modulo $\equiv_m$, and thus distinguished by a word v' of length no more than m, and thus p and q are distinguished by the word uv' of length strictly less than that of w, a contradiction. Thus the minimal distinguishing word has length no more than m, so that $\equiv_m$ coincides with $\equiv$.

Now if $\equiv_{m+1}$ does not coincide with $\equiv_m$, then $\equiv_{m+1}$ has a larger number of classes. Since the number of classes can never exceed $|Q|$, and since $\equiv_0$ has two classes, the sequence $\{\equiv_m\}_{m \geq 0}$ will stabilize by the time m reaches $|Q| - 2$. □

Lemma 1.1 leads to the following practical algorithm for minimization. We begin with a list of all the pairs $\{p, q\}$ of distinct accessible states, and mark the pair if $p \in F$ and $q \notin F$, or vice-versa. In each phase of the algorithm, we visit each unmarked pair $\{p, q\}$ and each $a \in A$, we compute $\{p', q'\} = \{\delta(p, a), \delta(q, a)\}$, and we mark $\{p, q\}$ if $\{p', q'\}$ is marked. An easy induction shows that if a pair $\{p, q\}$ is distinguished by a word of length m, then it will be marked by the mth phase of the algorithm. Thus after no more than $|Q| - 2$ phases, the algorithm will not mark any new pairs, with the result that the algorithm terminates, and the unmarked pairs are exactly the pairs of equivalent states.

Example 1.20. Consider the first automaton in Figure 1.9. Initially we mark the pairs $\{i, j\}$, where $i \in \{1, 2, 3\}$ and $j \in \{4, 5, 6\}$. On the next pass, the pairs

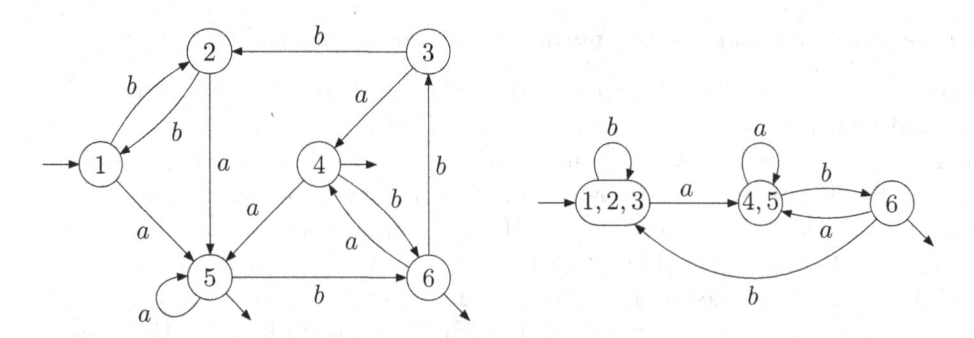

Fig. 1.9. The minimization algorithm.

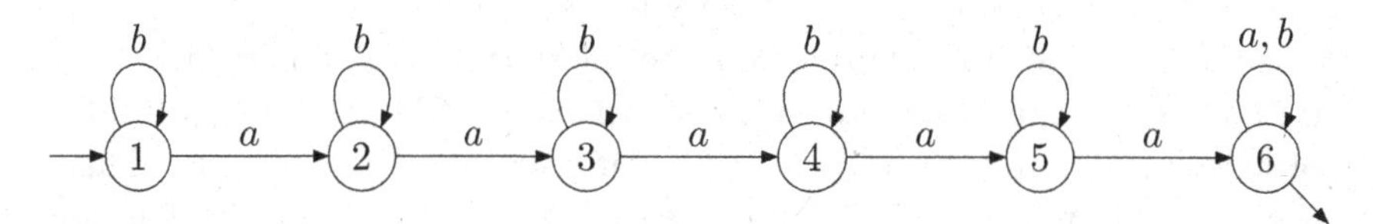

Fig. 1.10. A minimal automaton.

$\{4, 6\}$ and $\{5, 6\}$ are marked since applying b to these pairs gives the marked pair $\{3, 6\}$. No further pairs are marked on the next pass, so the algorithm terminates. Since the pairs $\{1, 2\}$ and $\{2, 3\}$ are unmarked, $\{1, 2, 3\}$ is an equivalence class, and since $\{4, 5\}$ is unmarked, it forms a second class. The remaining class is $\{6\}$. The resulting minimal automaton is pictured on the right-hand side of Figure 1.9.

Example 1.21. We now apply the algorithm to the automaton in Figure 1.10. Initially, the pairs $\{i, 6\}$ with $i < 6$ are marked. On the next pass the pairs $\{i, 5\}$ with $i < 5$ are marked, *etc.*, until on the fifth pass the pair $\{1, 2\}$ is marked. The result is that every pair of distinct states is marked: the automaton is already minimal.

The pair-marking implementation of the algorithm just illustrated is suitable for small examples worked by hand. In the worst case, shown in the last example, we check $\mathcal{O}(|Q|^2)$ unmarked pairs on each pass, and make $\mathcal{O}(|Q|)$ passes, with $|A|$ consultations of the state-transition table for each pair we inspect. Thus, the overall time complexity of the algorithm is $\mathcal{O}(|A| \cdot |Q|^3)$. More astute bookkeeping, in which we partition equivalence classes at each step, rather than marking pairs of inequivalent states, leads to a $\mathcal{O}(|A| \cdot |Q|^2)$ algorithm (Moore [20]). This can be further improved to $\mathcal{O}(|A| \cdot |Q| \cdot \log |Q|)$ (Hopcroft [21]).

1.6.4. *The transition monoid of an automaton*

Let $\mathcal{A} = (Q, \delta, i, F)$ be a complete deterministic automaton over an alphabet A. Let $w \in A^*$. We study the maps

$$f_w^{\mathcal{A}} : q \longmapsto \delta(q, w)$$

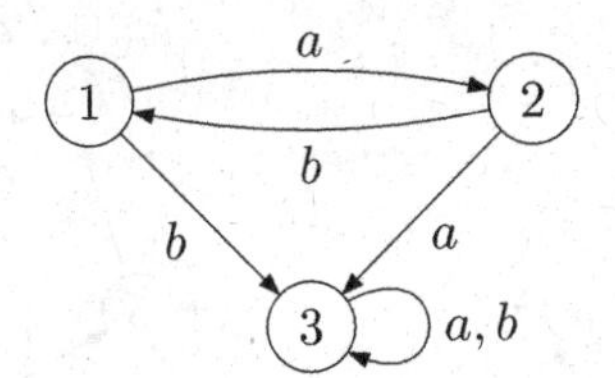

Fig. 1.11. The automaton $\mathcal{A}_1$, with no indication of initial or terminal states.

from Q into itself. We will write the image of a state q under $f_w^{\mathcal{A}}$ as $qf_w^{\mathcal{A}}$ rather than the more traditional $f_w^{\mathcal{A}}(q)$. We then have, for $v, w \in A^*$,

$$f_{vw}^{\mathcal{A}} = f_v^{\mathcal{A}} f_w^{\mathcal{A}},$$

where the product in the right-hand side of the equation is left-to-right composition of functions — that is, $q(f_v^{\mathcal{A}} f_w^{\mathcal{A}}) = (qf_v^{\mathcal{A}})f_w^{\mathcal{A}}$.

We will henceforth drop the superscript $\mathcal{A}$, except in situations where several different automata are involved. Observe that f_ε is the identity map on Q. Thus the set of maps

$$M(\mathcal{A}) = \{f_w \mid w \in A^*\}$$

forms an algebraic structure with an associative product and an identity element (usually denoted 1). Such a structure is called a *monoid*, and we call $M(\mathcal{A})$ the *transition monoid* of $\mathcal{A}$. Observe that if Q is finite, then $M(\mathcal{A})$ is finite, and that the structure of $M(\mathcal{A})$ depends only on the next-state function δ, and not at all on the initial or final states.

A^* is, of course, itself a monoid, with concatenation of words as the operation and the empty word ε as the identity. The map

$$\varphi\colon w \longmapsto f_w$$

is consequently a monoid *morphism* from A^* into $M(\mathcal{A})$; that is, it satisfies

$$\varphi(w_1w_2) = \varphi(w_1)\varphi(w_2)$$

for all w_1, w_2 in A^*, and it maps the identity element of A^* to the identity element of $M(\mathcal{A})$.

Example 1.22. In the diagrams in this example and in Examples 1.23 and 1.24, we indicate only the transitions between states, since, as we have observed, the initial and final states do not enter into the computation of the transition monoid of an automaton.

First, consider the automaton $\mathcal{A}_1$ in Figure 1.11. We will write an element f_w of $M(\mathcal{A}_1)$ as a vector $f_w = (1f_w\ 2f_w\ 3f_w)$. We can then begin enumerating the

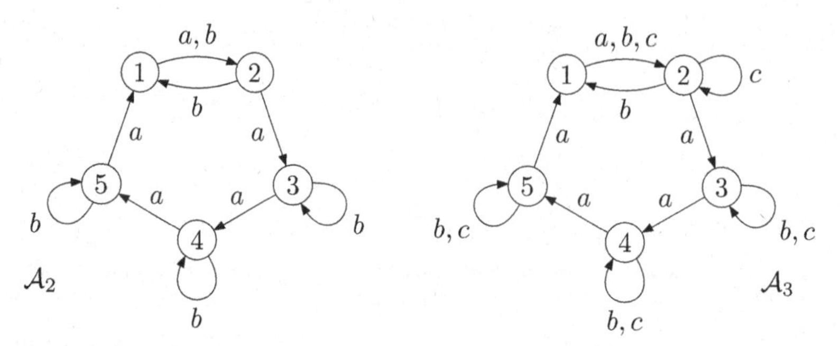

Fig. 1.12. The automata $\mathcal{A}_2$ and $\mathcal{A}_3$.

elements of $M(\mathcal{A}_1)$:

$$f_\varepsilon = (1\ 2\ 3)$$

$$f_a = (2\ 3\ 3) \qquad f_b = (3\ 1\ 3)$$

$$f_{aa} = (3\ 3\ 3) \qquad f_{ab} = (1\ 3\ 3) \qquad f_{ba} = (3\ 2\ 3) \qquad f_{bb} = (3\ 3\ 3)$$

We could continue enumerating like this, but instead we note that $f_{aba} = f_a$, $f_{bab} = f_b$, and for all other $w \in A^*$ of length 3, $f_w = (3\ 3\ 3)$. Thus the inventory above is the entire transition monoid, since any transition induced by a word of length greater than 2 is equal to one induced by a shorter word. Thus $M(\mathcal{A})$ has 6 elements $1, \alpha = f_a, \beta = f_b, \alpha\beta, \beta\alpha$, and 0. The multiplication is then determined by the laws $\alpha\alpha = \beta\beta = 0$, $\alpha = \alpha\beta\alpha$, and $\beta = \beta\alpha\beta$. The complete multiplication table is shown below:

$\cdot$	1	α	β	$\alpha\beta$	$\beta\alpha$	0
1	1	α	β	$\alpha\beta$	$\beta\alpha$	0
α	α	0	$\alpha\beta$	0	α	0
β	β	$\beta\alpha$	0	β	0	0
$\alpha\beta$	$\alpha\beta$	α	0	$\alpha\beta$	0	0
$\beta\alpha$	$\beta\alpha$	0	β	0	$\beta\alpha$	0
0	0	0	0	0	0	0

This example illustrates an important general point: There is an effective procedure for computing the multiplication table of the transition monoid of a complete deterministic finite automaton. We enumerate the maps f_w until we find that all words of some length induce the same maps as shorter words.

Example 1.23. Consider the automaton $\mathcal{A}_2$ in Figure 1.12. The transition monoid is generated by the two permutations f_a and f_b, both of which are permutations of the set of states: f_a cycles the five states and f_b transposes a pair of adjacent states. It is well known from elementary group theory that we can obtain all transpositions t of adjacent elements by repeated conjugation with the cycle (the map $t \mapsto f_a^4\, t\, f_a$), and that all permutations of the states can be obtained by composing transpositions of pairs of adjacent elements. So $M(\mathcal{A})$ consists of all the permutations of

$\{1, 2, 3, 4, 5\}$, and is consequently the symmetric group of degree 5, with $5! = 120$ elements. Of course, we can do likewise with any finite set of states.

Example 1.24. Now consider the effect of adding a third input letter to the preceding example, obtaining the automaton $\mathcal{A}_3$ in Figure 1.12. It is not hard to show that *every* map from $\{1, 2, 3, 4, 5\}$ into itself can be obtained by repeatedly composing f_c with permutations. Thus $M(\mathcal{A}_3)$ is the *full transformation monoid* on 5 states, which has $5^5 = 3125$ elements. We can similarly generate a transition monoid with n^n elements using an n-state automaton.

1.6.5. *The syntactic monoid*

Now let $L \subseteq A^*$, and consider the transition monoid of the minimal automaton $\mathcal{A}_{\min}(L) = (Q_L, \delta_L, i_L, F_L)$. Let $u, v \in A^*$. When are the two elements f_u, f_v of this monoid the same? If they are different, then there is some state q such that $qf_u \neq qf_v$. Since the automaton is minimal, there is a word $y \in A^*$ distinguishing these two states, so that $qf_uf_y \in F_L$ and $qf_vf_y \notin F_L$, or vice-versa. Since every state is accessible, there is also a word x such that $q = if_x$, so that either $xuy \in L$ and $xvy \notin L$, or vice-versa. Conversely, if such a pair of words x, y exists, then f_u and f_v cannot be equal. We thus have:

Theorem 1.5. *Let $L \subseteq A^*$, and let $u, v \in A^*$. Let $\mathcal{A} = \mathcal{A}_{\min}(L)$. Then $f_u^{\mathcal{A}} = f_v^{\mathcal{A}}$ if and only if for all $x, y \in A^*$*

$$xuy \in L \iff xvy \in L.$$

If the conditions in this theorem are satisfied, then we write $u \cong_L v$. The equivalence relation $\cong_L$ is called the *syntactic congruence* of L, and the transition monoid of $\mathcal{A}_{\min}(L)$ is called the *syntactic monoid* of L. We denote the syntactic monoid of L by $M(L)$. In algebraic terms, $M(L) = M(\mathcal{A}_{\min}(L)) = A^*/{\cong_L}$, that is $M(L)$ is the quotient monoid of A^* by the syntactic congruence. The morphism mapping each $w \in A^*$ to its $\cong_L$-class is called the *syntactic morphism* of L, and is denoted μ_L.

The syntactic congruence is a two-sided congruence on A^*; that is, if $u \cong_L v$ and $u' \cong_L v'$, then $uu' \cong_L vv'$. Compare this to the Myhill-Nerode congruence $\equiv_L$, which, as we noted, is a right congruence. The equivalence $\cong_L$ refines $\equiv_L$.

Transition monoids, and, in particular, the syntactic monoid, allow us to place many questions about the behavior of automata in a purely algebraic setting. For instance, we have the following algebraic characterization of rationality: Let M be a monoid and $\varphi\colon A^* \to M$ a morphism. We say that φ *recognizes* $L \subseteq A^*$ if and only if there is a subset X of A^* such that $L = \varphi^{-1}(X)$. We also say in this situation that M recognizes L.

Theorem 1.6. *Let $L \subseteq A^*$. The following are equivalent:*

(1) L is rational.

(2) $M(L)$ is finite.
(3) L is recognized by a finite monoid.

Proof. To show *(1)* implies *(2)*, note that if L is rational, then $\mathcal{A}_{\min}(L)$ has a finite set of states, and thus its transition monoid, $M(L)$, is finite. For *(2)* implies *(3)*, if $u \in L$ and $u \cong_L v$, then $v = \varepsilon\, v\, \varepsilon$ is also in L. Thus L is a union of equivalence classes of $\cong_L$, so that $L = \mu_L^{-1}(X)$, where $X = \{f_w \in M(\mathcal{A}_{\min}(L)) \mid w \in L\}$. Finally, to show *(3)* implies *(1)*, suppose $\varphi\colon A^* \to M$, where M is finite, and that $L = \varphi^{-1}(X)$. Then L is accepted by the complete deterministic automaton $\mathcal{A}(M) = (M, \delta, 1, X)$, where for $m \in M$ and $a \in A$,

$$\delta(m, a) = m\,\varphi(a).$$

Since M is finite, L is rational. □

Remark 1.7. Observe that if M is a finite monoid and $\mathcal{A}(M)$ is the automaton defined in the proof of Theorem 1.6, then the transition monoid of $\mathcal{A}(M)$ is M itself.

The syntactic monoid plays the same role in this algebraic view of rational languages that the minimal automaton plays in the automaton-theoretic view. Here we make this precise: We say that a monoid N *divides* a monoid M, and write $N \prec M$, if there is a submonoid M' of M and a surjective morphism $\varphi\colon M' \to N$. It is easy to see that $\prec$ is a transitive relation on monoids.

Theorem 1.7. *Let $L \subseteq A^*$. Then M recognizes L if and only if $M(L) \prec M$.*

Proof. First suppose M recognizes L, so that there is a morphism $\varphi\colon A^* \to M$ such that $L = \varphi^{-1}(X)$ for some $X \subseteq M$. We claim that if $\varphi(u) = \varphi(v)$, then $u \cong_L v$. To see this, suppose that $xuy \in L$ for some $x, y \in A^*$. Then $\varphi(xuy) \in X$, and since $\varphi(u) = \varphi(v)$, we have $\varphi(xvy) \in X$, so that $xvy \in L$. By the same argument, if $xvy \in L$ then $xuy \in L$, so that $u \cong_L v$.

Now let $M' = \varphi(A^*)$. We define a map $\psi\colon M' \to M(L)$ by $\psi(\varphi(u)) = \mu_L(u)$. By the remark just made, ψ is well-defined, since the value of ψ only depends on $\varphi(u)$ and not on u. Moreover ψ is clearly a morphism, and it is surjective because μ_L is, so $M(L) \prec M$.

Conversely, suppose $M(L) \prec M$, so that there is a morphism ψ from a submonoid M' of M onto $M(L)$. For $a \in A$, we set $\varphi(a)$ to be any $m \in M'$ for which $\psi(m) = \mu_L(a)$. This can be extended to a unique morphism $\varphi\colon A^* \to M$ such that $\mu_L = \psi \circ \varphi$. Let $X = \varphi(L)$. If $\varphi(u) \in X$ then $\varphi(u) = \varphi(v)$ for some $v \in L$, and thus $\mu_L(u) = \mu_L(v)$, so $u \cong_L v$. Since, as noted in the proof of Theorem 1.6, L is a union of $\cong_L$-classes, this implies $u \in L$, so that $L = \varphi^{-1}(X)$, and thus M recognizes L.□

Example 1.25. Consider the transition monoid of the automaton $\mathcal{A}$ in Figure 1.13. We can fairly easily determine the elements of this monoid without doing an exhaustive tabulation: First, if a word w has even length, then it maps $\{1, 3\}$ into $\{1, 3\}$, and $\{2, 4\}$ into $\{2, 4\}$, while if w has odd length, then it interchanges these

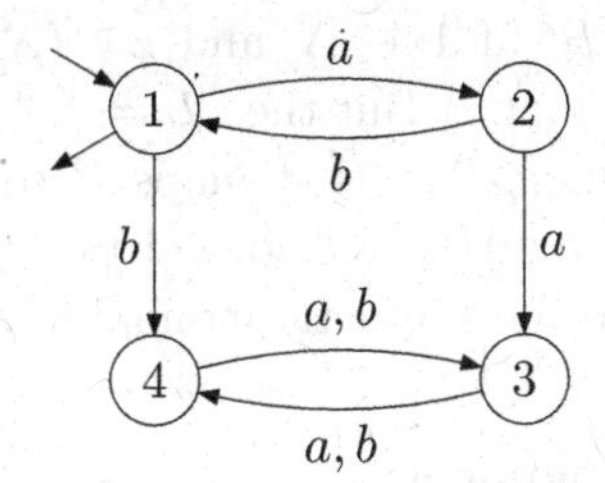

Fig. 1.13. An automaton, whose transition monoid contains a non-trivial group.

two sets. Second, if w contains aa or bb as a factor, then the image of f_w is contained in $\{3, 4\}$. Finally, if the letters of w alternate, then f_w maps either 1 or 2 to $\{1, 2\}$, but not both, depending on whether the first letter of w is a or b. We thus get these elements:

$$1 = f_\varepsilon = (1\ 2\ 3\ 4) \qquad \gamma = f_a = (2\ 3\ 4\ 3) \qquad \delta = f_b = (4\ 1\ 4\ 3)$$
$$\gamma\delta = f_{ab} = (1\ 4\ 3\ 4) \qquad \delta\gamma = f_{ba} = (3\ 2\ 3\ 4)$$
$$\gamma^2 = f_{aa} = (3\ 4\ 3\ 4) \qquad \gamma^3 = f_{aaa} = (4\ 3\ 4\ 3)$$

Observe that $\{\gamma^2, \gamma^3\}$ forms a group, permuting the states 3 and 4. This automaton accepts the language $(ab)^*$. In algebraic terms, the morphism $\varphi\colon w \mapsto f_w$ from A^* into $M(\mathcal{A})$ recognizes this language with $(ab)^* = \varphi^{-1}(X)$, where

$$X = \{f \in M(\mathcal{A}) \mid f \text{ maps state 1 to itself }\}.$$

The states 3 and 4 are equivalent, and the minimal automaton of L is obtained by merging these states: it is the automaton examined in Example 1.22 (with 1 as initial and final state), where we computed its transition monoid, namely $M(L)$. According to Theorem 1.7, $M(L) \prec M(\mathcal{A})$, and, indeed, the map sending 1 to 1, $\gamma, \delta, \gamma\delta, \delta\gamma$ to $\alpha, \beta, \alpha\beta, \beta\alpha$, respectively, and γ^2 and γ^3 both to 0, is a morphism from $M(\mathcal{A})$ onto $M(L)$.

Example 1.26. Let us take the automaton of Example 1.24 and specify 1 as both the initial state and the unique accepting state. With these choices, the automaton is the minimal automaton of the language it accepts, since every state is accessible and no two distinct states are equivalent. This shows that the syntactic monoid of a language accepted by an n-state automaton can have as many as n^n elements.

Example 1.27. Not every finite monoid is the syntactic monoid of a rational language. Consider, for instance, the monoid $M = \{1, \alpha, \beta, \gamma\}$ with multiplication $m_1 m_2 = m_2$ for $m_2 \neq 1$. Suppose A is a finite alphabet and $\varphi\colon A^* \to M$ is a morphism. Let $X \subseteq M$. We partition A into three subsets, B, C, and D,

$$B = \{a \in A \mid \varphi(a) = 1\}$$
$$C = \{a \in A \mid \varphi(a) \in X \setminus \{1\}\}$$
$$D = A \setminus (B \cup C)$$

Then $\varphi^{-1}(X) = B^* \cup A^*CB^*$ if $1 \in X$ and $\varphi^{-1}(X) = A^*CB^*$ otherwise. (Observe that B or C might be empty.) But then $L = \varphi^{-1}(X)$ is recognized by the submonoid $\{1, \alpha, \beta\}$, using the morphism that maps B to 1, C to α and D to β. Thus every language recognized by M is recognized by a strictly smaller monoid, so by Theorem 1.7, M cannot be the syntactic monoid of any language.

1.7. First-order definable languages

This section is devoted to proving one of the earliest and most important applications of the syntactic monoid: the characterization of the languages definable in FO(<).

A finite monoid M can contain a nontrivial group, as for example the group $\{\gamma^2, \gamma^3\}$ in the monoid $M(\mathcal{A})$ of Example 1.25. If there is no nontrivial group in M, we say that M is *aperiodic*.

Lemma 1.2. *Let M be a finite monoid. Then the following are equivalent:*

(1) M is aperiodic.
(2) There is an integer $n > 0$ such that $m^n = m^{n+1}$ for all $m \in M$.

Proof. Suppose M is aperiodic. Let $m \in M$, and consider the sequence $1, m, m^2, \ldots$ Since M is finite, if we take $n = |M|$, we have $m^r = m^n$ for some $r < n$. Take the largest such r, and consider the set $G = \{m^k \mid r \leq k < n\}$. Observe that for all $g \in G$, $gG = Gg = G$, since

$$m^{r+t}m^s = m^{r+[(t+s) \bmod (n-r)]}$$

for all $s, t \geq 0$. This implies that G is a group, so that $|G| = 1$, and thus $r = n - 1$ and $m^r = m^{r+1}$. Conversely, if M is not aperiodic, then M contains a nontrivial group G, and an element $g \in G$ different from the identity element e of G. Then $g^k = e$ for some $k > 1$, so that $g^n \neq g^{n+1}$ for all $n \geq 0$. □

Note that the proof shows that we can choose n in condition *(2)* of Lemma 1.2 to be $|M| - 1$.

We say that a language $L \subseteq A^*$ is *star-free* if it can be defined by an extended rational expression without the $*$ operation or morphic images. The Schützenberger-McNaughton-Papert Theorem offers the following characterization.

Theorem 1.8. *Let $L \subseteq A^*$ be a rational language. Then the following are equivalent.*

(1) L is star-free.
(2) L is definable by a sentence of FO(<).
(3) L is recognized by an aperiodic finite monoid.
(4) $M(L)$ is aperiodic.

Before we turn to the proof of this theorem, we give an important corollary, and an example.

Corollary 1.3. *It is decidable whether a rational language (given by a rational expression or an accepting automaton) is definable by a sentence of first-order logic.*

Proof. As we have seen, we can compute $\mathcal{A}_{\min}(L)$ from any automaton or expression for L, and thence compute the multiplication table of $M = M(L)$. We can then test for all $m \in M$ whether $m^{|M|-1} = m^{|M|}$, and thus, by Lemma 1.2 determine whether $M(L)$ is aperiodic. By Theorem 1.8, this decides whether L is first-order definable. □

In fact, the proof of Theorem 1.8 will show that if $M(L)$ is aperiodic, then we can effectively construct both a star-free expression and a first-order sentence for L from an automaton that recognizes L.

Example 1.28. Let $L = (ab)^*$. We computed $M(L)$ in Example 1.22. We have $\alpha^2 = \beta^2 = 0 = \alpha^3 = \beta^3$, and $(\alpha\beta)^2 = \alpha\beta$, $(\beta\alpha)^2 = \beta\alpha$, so by Lemma 1.2, $M(L)$ is aperiodic. Theorem 1.8 says that L is definable by a star-free extended rational expression, and also by a sentence of $\mathsf{FO}(<)$. Let us exhibit such expressions.

First, note that membership of a word w in L is equivalent to saying that w contains no occurrence of either aa or bb as a factor, and that the first letter of w (if there is one) is a, and the last letter is b. We thus have

$$L = \{\varepsilon\} \cup (aA^* \cap A^*b \cap \overline{A^*(aa \cup bb)A^*}).$$

witnessing the fact that L is star-free (note that A^* is star-free, since $A^* = \overline{\emptyset}$).

To obtain a first-order sentence defining L, we use the same characterization of words in L. We say there is no occurrence of aa as a factor using the following sentence:

$$\neg\exists x \exists y (R_a x \wedge R_a y \wedge S(x, y)).$$

This uses the successor predicate S, but as we noted earlier, S can be expressed in $\mathsf{FO}(<)$. We can likewise write a sentence saying that there is no occurrence of bb. An FO-sentence stating that the first letter of a word is a was given in Example 1.13. A similar sentence can be formed to say that the last letter is b. Note that all these sentences are satisfied by the empty word as well, so that the conjunction of the four sentences defines the language $(ab)^*$.

This language is also recognized by the first monoid that we exhibited in Example 1.25, which is not aperiodic. This in no way contradicts Theorem 1.8, which only says that *some* aperiodic monoid recognizes L.

Remark 1.8. The decision procedure outlined in the proof of Corollary 1.3 may take exponential time in the size of an automaton accepting L, since it involves computing the syntactic monoid of L (see Example 1.24). While this procedure may be improved, this decision problem is intrinsically difficult. In fact, it is known to be PSPACE-complete (Cho and Huynh [22]).

We now turn to the proof of Theorem 1.8. We will show (4) $\Leftrightarrow$ (3) $\Rightarrow$ (1) $\Rightarrow$ (2) $\Rightarrow$ (4). By Theorem 1.7, every language is recognized by its syntactic monoid. Also every divisor of an aperiodic monoid is aperiodic, since the property $m^n = m^{n+1}$ for all elements m in a monoid is inherited by morphic images and submonoids. Thus (3) and (4) are equivalent.

The most difficult part of the proof is (3) $\Rightarrow$ (1). To prove this, we suppose $L \subseteq A^*$ is recognized by a finite aperiodic monoid. This is equivalent to L being accepted by a complete deterministic automaton $\mathcal{A} = (Q, \delta, i, F)$ whose transition monoid is aperiodic (see the proof of Theorem 1.6 and Remark 1.7). We will show that for all $q, q' \in Q$, the set $L^{\mathcal{A}}_{q,q'} = \{w \mid qf^{\mathcal{A}}_w = q'\}$ is a star-free language. Since L is a finite union of such languages, L is star-free.

The proof is by induction on the pair $(|Q|, |A|)$: the induction hypothesis is that the claim holds for all automata with a strictly smaller state set, or with the same size state set and a strictly smaller input alphabet. In the case $|Q| = 1$, L is either A^* or $\emptyset$, which are star-free. In the case $|A| = 1$, so that $A = \{a\}$, aperiodicity implies that L is a finite union of singleton sets $\{a^k\}$, possibly together with the language $a^r a^*$, where $r = |Q| - 1$, which is also star-free, since $a^* = \overline{\emptyset}$.

We thus assume both $|Q| > 1$ and $|A| > 1$. First suppose that for every $a \in A$, $Qf^{\mathcal{A}}_a = Q$, so that $f^{\mathcal{A}}_a$ is a permutation of Q. Aperiodicity implies $(f^{\mathcal{A}}_a)^r = (f^{\mathcal{A}}_a)^{r+1}$ for some r, and thus $f^{\mathcal{A}}_a$ is the identity map on Q. Consequently $f^{\mathcal{A}}_w$ is the identity map for all $w \in A^*$, and thus the claim holds trivially. We can therefore assume that there is some $a \in A$ such that

$$Qf^{\mathcal{A}}_a = Q' \subsetneq Q.$$

We now define two new automata $\mathcal{B}$ and $\mathcal{C}$. Automaton $\mathcal{B}$ has state set Q and next-state function $\delta|_{Q\times B}$, where $B = A \setminus \{a\}$. We need not define initial and final states for $\mathcal{B}$, because we are only interested in the state transitions $f^{\mathcal{B}}_w$. Automaton $\mathcal{C}$ has state set Q', input alphabet

$$C = \{(f^{\mathcal{B}}_w, a) \mid w \in B^*, a \in A\},$$

and next-state function

$$\delta' \colon (q, (f^{\mathcal{B}}_w, a)) \longmapsto q \cdot f^{\mathcal{A}}_{wa}.$$

This makes sense, because $Qf^{\mathcal{A}}_a = Q'$ and because $f^{\mathcal{B}}_w = f^{\mathcal{B}}_{w'}$ implies that $f^{\mathcal{A}}_{wa} = f^{\mathcal{A}}_{w'a}$. The inductive hypothesis applies to both $\mathcal{B}$ and $\mathcal{C}$. (The transition monoids of these automata inherit the aperiodicity of $\mathcal{A}$, because every transition in them is the restriction of a transition in $\mathcal{A}$.)

A word in $L^{\mathcal{A}}_{q,q'}$ can contain either no occurrences of a, a single occurrence of a, or two or more occurrences of a. We can accordingly write $L^{\mathcal{A}}_{q,q'}$ as a finite union of sets of the form

$$L^{\mathcal{B}}_{q,q'}, L^{\mathcal{B}}_{q,p}aL^{\mathcal{B}}_{p',q'}, L^{\mathcal{B}}_{q,p}aT_{p',q''}L^{\mathcal{B}}_{q'',q'},$$

where $p \in Q$, $p' = p \cdot f^{\mathcal{A}}_a \in Q'$, and $T_{p,q''} = L^{\mathcal{A}}_{p,q''} \cap A^*a$.

By the inductive hypothesis all the sets of the form $L^{\mathcal{B}}_{s,t}$ are star-free, so it remains to show that $T_{p',q''}$ is a star-free language. We can factor any $w \in A^*a$ uniquely as

$$w = v_1 a \cdots v_k a,$$

where $v_1, \ldots, v_k \in B^*$. Let us associate to w the word

$$w_C = c_1 \cdots c_k \in C^*,$$

where $c_j = (f^{\mathcal{B}}_{v_j}, a) \in C$. By the inductive hypothesis, the language $L^{\mathcal{C}}_{p',q''}$ is star-free. So we need to show that if $R \subseteq C^*$ is star-free, then $\Psi(R) = \{w \in A^*a \mid w_C \in R\}$ is also star-free, since $T_{p'q''} = \Psi(L^{\mathcal{C}}_{p',q''})$. It is thus enough to show

(i) If $c \in C$, then $\Psi(\{c\})$ is star-free.
(ii) If $\Psi(R)$ is star-free, then $\Psi(C^* \setminus R)$ is star-free.
(iii) If $\Psi(R_1), \Psi(R_2)$ are star-free, then $\Psi(R_1 \cup R_2)$ is star-free.
(iv) If $\Psi(R_1), \Psi(R_2)$ are star-free, then $\Psi(R_1R_2)$ is star-free.

For *(i)*, note that $\Psi(\{c\}) = Sa$, where $S = \{v \in B^* \mid c = (f^{\mathcal{B}}_v, a)\}$. Since S is a boolean combination of languages of the form $L^{\mathcal{B}}_{p,p'}$, Sa is star-free. For the other assertions, we clearly have $\Psi(C^* \setminus R) = A^*a \cap (A^* \setminus \Psi(R))$, $\Psi(R_1 \cup R_2) = \Psi(R_1) \cup \Psi(R_2)$, and $\Psi(R_1R_2) = \Psi(R_1)\Psi(R_2)$. This completes the proof that *(3) ⇒ (1)*.

To prove (1) ⇒ (2), we need to show that every star-free language is first-order definable. Since the singleton sets $\{a\}$ for $a \in A$ are clearly first-order definable, and since the boolean operations are part of first-order logic, this reduces to showing that if $L_1, L_2 \subseteq A^*$ are first-order definable, then so is L_1L_2. To do this, we introduce the notion of *relativizing* a first-order sentence. Let φ be a sentence of $\mathsf{FO}(<)$ and x a variable symbol that does not occur in φ. We define a formula $\varphi_{<x}$ with one free variable with the following property: Let ν be an interpretation mapping x to $i \in \mathsf{Dom}(u)$, and let v be the prefix v of u with domain $\{0, \ldots, i-1\}$. Then $u, \nu \models \varphi_{<x}$ if and only if $v \models \varphi$. To construct $\varphi_{<x}$, we simply work from the outermost quantifier of φ inward, replacing each quantified subformula $\exists y\, \alpha$ by $\exists y\, ((y < x) \wedge \alpha)$. We define $\varphi_{>x}$ and $\varphi_{\leq x}$ analogously.

Now suppose φ, ψ are first-order sentences defining L_1 and L_2, respectively. Let x be a variable symbol that does not occur in φ or ψ. We have L_1L_2 defined by the sentence

$$\begin{array}{ll} \exists x\ (\varphi_{\leq x} \wedge \psi_{>x}) & \text{if } \varepsilon \notin L_1, \\ \exists x\ (\varphi_{\leq x} \wedge \psi_{>x}) \vee \psi & \text{if } \varepsilon \in L_1. \end{array}$$

To prove (2) $\Rightarrow$ (4), we need to show that the syntactic monoid of every first-order definable language in A^* is aperiodic. We will proceed as in Section 1.4.2, and treat a first-order formula with free variables contained in $\{x_1, \ldots, x_p\}$ as defining a language over the extended alphabet $B_p = A \times \{0,1\}^p$. We will show by induction on the quantifier depth that every first-order definable language $L \subseteq B_p^*$ in this extended sense has an aperiodic syntactic monoid. More precisely, we will show that for each such L there exists an integer $q > 0$ such that for all $v \in B_p^*$, $v^q \cong_L v^{q+1}$. By Lemma 1.2, this implies aperiodicity.

First suppose L is defined by one of the atomic formulas $x_1 < x_2$ or $R_a x_1$. Let $u, v, w \in B_p^*$. If v has a letter with a 1 in one of its last p components, then neither uv^2w nor uv^3w can be in L, since only one letter of a word in L can have a 1 in a given component. If v has no such letter, then membership of uvw in L is witnessed by the relative positions and values of letters in u and w, so that $uvw \in L$ if and only if $uv^2w \in L$. Thus in all cases, we have $uv^2w \in L$ if and only if $uv^3w \in L$, so that $v^2 \cong_L v^3$.

Now suppose the claim is true for $L_1, L_2 \subseteq B_p^*$ defined by formulas φ_1, φ_2, and suppose L is defined by $\varphi_1 \vee \varphi_2$. We have, by assumption, $v^q \cong_{L_1} v^{q+1}$, and $v^q \cong_{L_2} v^{q+1}$, for some $q > 0$. (The exponents for these two languages are, *a priori*, different, but we can then choose q to be the maximum of the two exponents.) Now $\varphi_1 \vee \varphi_2$ defines $L_1 \cup L_2$, and we have directly $uv^qw \in L_1 \cup L_2$ if and only if $uv^{q+1}v \in L_1 \cup L_2$.

Care must be taken with the negation operator, since it does not exactly correspond to the boolean complement. We can assume that the exponent q for L_1 is at least 2. Let L_1' be the language defined by $\neg\varphi_1$. Suppose $uv^qw \in L_1'$. Then $uv^qw \notin L_1$, and thus $uv^{q+1}w \notin L_1$. Further v cannot contain a 1 in the last p components of any of its positions, so $uv^{q+1}w$ has exactly one occurrence of 1 in each of the last p positions, and thus is in L_1'. The same argument shows that if $uv^{q+1}w \in L_1'$, then so is uv^qw. Thus $v^q \cong_{L_1'} v^{q+1}$.

So now let $K \subseteq B_{p-1}^*$ be the language defined by $\exists x_p \varphi_1$. Let $v \in B_{p-1}^*$. We will show $v^{2q+1} \cong_K v^{2q+2}$. Suppose $uv^{2q+1}w \in K$. Let us extend each letter in this word by adding a pth component with 0. We will still denote the resulting word as $uv^{2q+1}w$. Since K is defined by $\exists x_p \varphi_1$, we can switch the pth component of some letter to obtain a word $z \in B_p^*$ such that $z \in L_1$. Now, wherever the position in which we switched the pth component is located, at least q consecutive occurrences of v will be left intact. We thus find that z can be written in the form xv^qy, for some $x, y \in B_p^*$. (The extreme case is when the position is within the middle occurrence of v, in which case we get two factors of the form v^q.) Thus $xv^{q+1}y \in L_1$. If we now switch the changed 1 back to 0, we find $uv^{2q+2}w \in K$. The identical argument shows $uv^{2q+2}w \in K$ implies $uv^{2q+1}w \in K$. Thus $v^{2q+1} \cong_K v^{2q+2}$, as claimed.

Remark 1.9. Interesting presentations of proofs of all or part of Theorem 1.8 can be found, for instance, in the work of Perrin [23], Straubing [17] and Diekert and Gastin [24].

References

[1] S. C. Kleene, Representation of events in nerve nets and finite automata. In C. E. Shannon and J. McCarthy (Eds.), *Automata Studies*, in *Annals of Mathematics Studies*, vol. 40, 3–40. Princeton University Press, (1956).

[2] B A. Trakhtenbrot, The synthesis of logical nets whose operators are described in terms of monadic predicates, *Doklady AN SSR.* **118**, 646–649, (1958).

[3] J. R. Büchi, Weak second-order arithmetic and finite automata, *Z. Math. Logik Grundlagen Math..* **6**, 66–92, (1960).

[4] M. O. Rabin and D. Scott, Finite automata and their decision problems, *IBM Journal of Research and Development.* **3**, 114–125, (1959). Reprinted in E. F. Moore (Ed.), *Sequential Machines: Selected Papers.* (Addison-Wesley, 1964).

[5] D. A. Huffman, The synthesis of sequential switching circuits, *J. Franklin Institute.* **257**, 161–190, 275–303, (1954).

[6] J. R. Myhill. Finite automata and the representation of events. Technical Report 57-624, Wright Airport Development Command, (1957).

[7] A. Nerode, Linear automaton transformations, *Proc. AMS.* **9**, 541–544, (1958).

[8] M. P. Schützenberger, On finite monoids having only trivial subgroups, *Inform. and Comput.* **8**, 190–194, (1965).

[9] R. McNaughton and S. Papert, *Counter-Free Automata.* (MIT Press, 1971).

[10] T. Wilke. Classifying discrete temporal properties. In *Proc. 16th STACS*, vol. 1443 *LNCS*, pp 32–46, Springer, (1999).

[11] J. E. Hopcroft and J. D. Ullman, *Introduction to Automata Theory, Languages, and Computation.* (Addison-Wesley, 1979).

[12] H. R. Lewis and C. H. Papadimitriou, *Elements of the Theory of Computation.* (Prentice-Hall, 1981).

[13] M. Sipser, *Introduction to the Theory of Computation.* (Course Technology, 2006), 2nd edition.

[14] S. Eilenberg, *Automata, Languages, and Machines*, volume A. (Academic Press, 1974).

[15] S. Eilenberg, *Automata, Languages, and Machines*, volume B. (Academic Press, 1976).

[16] J. Sakarovitch, *Elements of Automata Theory.* (Cambridge University Press, 2009). Translated from the 2003 French original by Reuben Thomas.

[17] H. Straubing. *Finite Automata, Formal Logic, and Circuit Complexity.* (Birkhäuser, 1994).

[18] W. Thomas. Languages, automata and logic. In *Handbook of Formal Languages*, volume 3, Beyond Words. Springer, (1997).

[19] J. -E. Pin, (Ed.), *Automata: from Mathematics to Applications.* (European Mathematical Society, 2011).

[20] E. F. Moore. Gedanken experiments on sequential machines. In *Automata Studies*, pp. 129–153. (Princeton University Press, 1956).

[21] J E. Hopcroft. An $n \log n$ algorithm for minimizing the states in a finite automaton. In *The Theory of Machines and Computations*, pp. 189–196. Academic Press, (1971).

[22] S. Cho and D. T. Huynh, Finite automaton aperiodicity is PSPACE-complete, *Theoretical Computer Science.* **88**, 99–116, (1991).

[23] D. Perrin. Finite automata. In *Handbook of Theoretical Computer Science, Vol. B*, pp. 1–57. Elsevier, (1990).

[24] V. Diekert and P. Gastin. First-order definable languages. In *Logic and Automata: History and Perspectives*, Texts in Logic and Games, pp. 261–306. Amsterdam University Press, (2008).

Chapter 2

Finite-State Automata on Infinite Inputs

Madhavan Mukund

Chennai Mathematical Institute
H1 SIPCOT IT Park, Padur PO
Siruseri 603103, India
madhavan@cmi.ac.in

This article is a self-contained introduction to the theory of finite-state automata on infinite words. The study of automata on infinite inputs was initiated by Büchi in order to settle certain decision problems arising in logic. Subsequently, there has been a lot of fundamental work in this area, resulting in a rich and elegant mathematical theory. In recent years, there has been renewed interest in these automata because of the fundamental role they play in the automatic verification of finite-state systems.

Introduction

Büchi initiated the study of finite-state automata working on infinite inputs in [1]. He was interested in showing that the monadic second order logic of infinite sequences (S1S) was decidable. Büchi discovered a deep and elegant connection between sets of models of formulas in this logic and ω-regular languages, the class of languages over infinite words accepted by finite-state automata.

A few years later, Muller independently proposed an alternative definition of finite-state automata on infinite inputs [2]. His work was motivated by questions in switching theory.

The theory of ω-regular languages and automata on infinite words is substantially more complex than the corresponding theory for finite words. This was evident from Büchi's initial work, where he showed that non-deterministic automata over infinite inputs are strictly more powerful than deterministic automata. This means that basic constructions like complementation are correspondingly more intricate for this class of automata.

During the 1960's, fundamental contributions were made to this area. McNaughton proved that with Muller's definition, deterministic automata suffice for recognizing all ω-regular languages [3]. Later, Rabin extended Büchi's decidability result for S1S to the monadic second order of the infinite binary tree (S2S) [4]. The logical theory S2S is extremely expressive and Rabin's theorem can be used to settle a number of decision problems in logic.

Despite this strong connection between automata on infinite inputs and the decidability of logical theories, there was a lull in the area during the 1970's. One reason for this was Meyer's negative result about the complexity of the automata-theoretic decision procedure for S1S and S2S [5]—he showed that, in the worst case, the automata that one constructs would be hopelessly large and impossible to use in practice.

Since the 1980s, however, there has been renewed interest in applying automata on infinite words to solve problems in logic. To a large extent, this is a consequence of the development of temporal logic as a formalism for specifying and verifying properties of programs [6, 7]. It turns out that automata on infinite words (and trees) can be *directly* applied to settle important questions in temporal logic, without invoking S1S and S2S. In conjunction with these new applications, there has been greater emphasis on evaluating the complexity of different constructions on these automata [8–11].

In this article, we present a self-contained introduction to the theory of finite-state automata on infinite words. We begin with some preliminaries on the notation we will use in the paper. In Section 2.1, we introduce Büchi automata and ω-regular languages and prove some basic results. The next section describes in detail the connection between ω-regular languages and formulas of S1S. In Section 2.3, we look at stronger definitions of automata, proposed by Muller, Rabin and Streett. The last technical section, Section 2.4, describes in detail a determinization construction for Büchi automata. We conclude with a quick summary of various aspects of the theory that could not be discussed in this article.

For a more detailed introduction to the area, the reader is encouraged to consult the excellent survey by Thomas [12]. This chapter—especially Sections 2.1 and 2.2—draws heavily on the material presented in [12].

Notation

Throughout this article, Σ denotes a finite set of symbols called an *alphabet.* A *word* is a sequence of symbols from Σ. The set Σ^* denotes the set of finite words over Σ while the set Σ^ω is the set of infinite words over Σ. A *language* is a set of words. Every language we consider either consists exclusively of finite words or exclusively of infinite words.

Typically, elements of Σ will be denoted $a, b, c, \ldots$, finite words will be denoted $u, v, w, \ldots$, and infinite words will be denoted $\alpha, \beta, \ldots$. We use $U, V, \ldots$ to denote languages of finite words—that is, subsets of Σ^*. L will be reserved for languages consisting exclusively of infinite words.

An infinite word $\alpha \in \Sigma^\omega$ is an infinite sequence of symbols from Σ. We shall represent α as a function $\alpha : \mathbb{N}_0 \to \Sigma$, where $\mathbb{N}_0$ is the set $\{0, 1, 2, \ldots\}$ of natural numbers. Thus, $\alpha(i)$ denotes the letter occurring at the ith position. For natural numbers m and n, $m \leq n$, $[m..n]$ will denote the set $\{m, m+1, \ldots, n\}$ and $[m..]$ the infinite set $\{m, m+1, \ldots\}$. We let $\alpha[m..n]$ denote the finite word $\alpha(m)\alpha(m+1)\cdots\alpha(n-1)\alpha(n)$ occurring between positions m and n and $\alpha[m..]$ denote the infinite $\alpha(m)\alpha(m+1)\ldots$ starting at position m.

In general, if S is a set and σ an infinite sequence of symbols over S—in other words, $\sigma : \mathbb{N}_0 \to S$—then $\mathsf{inf}(\sigma)$ denotes the set of symbols from S that occur infinitely often in σ. Formally, $\mathsf{inf}(\sigma) = \{s \in S \mid \exists^\omega n \in \mathbb{N}_0 : \sigma(n) = s\}$, where $\exists^\omega$ denotes the quantifier "there exist infinitely many".

2.1. Büchi automata

A Büchi automaton is a non-deterministic finite-state automaton that takes infinite words as input. A word is accepted if the automaton goes through some designated "good" states infinitely often while reading it.

Automata An *automaton* is a triple $\mathcal{A} = (S, \to, S_{in})$ where S is a finite set of *states*, $S_{in} \subseteq S$ is a set of *initial states* and $\to \subseteq S \times \Sigma \times S$ is a *transition relation.* Normally, we write $s \xrightarrow{a} s'$ to denote that $(s, a, s') \in \to$.

The automaton is said to be *deterministic* if S_{in} is a singleton and $\to$ is a function from $S \times \Sigma$ to S.

We could have weakened the condition for deterministic automata by permitting $\to$ to be a *partial* function from $S \times \Sigma$ to S. A "weak" deterministic automaton can always be converted to a "strong" deterministic automaton by adding a "dump" or "reject" state to take care of all missing transitions. Since it is often convenient to assume that deterministic automata are "complete" and never get stuck when reading their input, we shall stick to the stronger definition in this paper.

If u is a finite *non-empty* word, we write $s \xrightarrow{u}^+ s'$ to denote the fact that there is a sequence of transitions labelled by u leading from s to s'. In other words, if $u = a_1 a_2 \cdots a_m$, then $s \xrightarrow{u}^+ s'$ if there exist states $s_0, s_1, \ldots, s_m$ such that $s = s_0 \xrightarrow{a_1} s_1 \xrightarrow{a_2} \cdots \xrightarrow{a_m} s_m = s'$.

Runs Let $\mathcal{A} = (S, \to, S_{in})$ be an automaton and $\alpha : \mathbb{N}_0 \to \Sigma$ an input word. A *run* of $\mathcal{A}$ on α is a infinite sequence $\rho : \mathbb{N}_0 \to S$ such that $\rho(0) \in S_{in}$ and for all $i \in \mathbb{N}_0$, $\rho(i) \xrightarrow{\alpha(i)} \rho(i+1)$.

A run of $\mathcal{A}$ on the finite word $w = a_0 a_1 \ldots a_m$ is a sequence of states $s_0 s_1 \ldots s_{m+1}$ such that $s_0 \in S_{in}$ and for all $i \in [0..m]$, $s_1 \xrightarrow{\alpha(i)} s_{i+1}$.

So, a run is just a "legal" sequence of states that an automaton can pass through while reading the input. In general, an input may admit many runs because of non-determinism. Since a non-deterministic automaton may have states where there are no outgoing transitions corresponding to certain input letters, it is also possible that an input admits *no* runs—in this case, every potential run leads to a state from where there is no outgoing transition enabled for the next input letter. If the automaton is deterministic, each input admits precisely one run.

Automata on finite words A finite-state automaton on finite words is a structure $(\mathcal{A}, F)$ with $\mathcal{A} = (S, \rightarrow, S_{in})$ and $F \subseteq S$. The automaton $\mathcal{A}$ accepts an input $w = a_0 a_1 \dots a_m$ if w admits a run $s_0 s_1 \dots s_{m+1}$ such that $s_{m+1} \in F$. The language *recognized by* $(\mathcal{A}, F)$, $L(\mathcal{A}, F)$, is the set of all finite words accepted by $(\mathcal{A}, F)$.

Throughout this article, we shall refer to languages recognized by finite-state automata on finite words as *regular languages*. In other words, a set $U \subseteq \Sigma^*$ is regular iff there is an automaton $(\mathcal{A}, F)$ such that $U = L(\mathcal{A}, F)$.

Our goal is to study automata that recognize languages of infinite words. The first definition of such automata was proposed by Büchi [1].

Büchi automata A *Büchi automaton* is a pair $(\mathcal{A}, G)$ where $\mathcal{A} = (S, \rightarrow, S_{in})$ and $G \subseteq S$. G denotes a set of *good states.* The automaton $(\mathcal{A}, G)$ *accepts* an input $\alpha : \mathbb{N}_0 \rightarrow \Sigma$ if there is a run ρ of $\mathcal{A}$ on α such that $\mathsf{inf}(\rho) \cap G \neq \emptyset$. The language *recognized by* $(\mathcal{A}, G)$, $L(\mathcal{A}, G)$, is the set of all infinite words accepted by $(\mathcal{A}, G)$. A set $L \subseteq \Sigma^\omega$ is said to be *Büchi-recognizable* if there is a Büchi automaton $(\mathcal{A}, G)$ such that $L = L(\mathcal{A}, G)$.

According to the definition, a Büchi automaton accepts an input if there is a run along which some subset of G occurs infinitely often. Since G is a finite set, it is easy to see that there must actually be a state $g \in G$ that occurs infinitely often along σ. In other words, if we regard the state space of a Büchi automaton as a graph, an accepting run traces an infinite path that starts at some state s in S_{in}, reaches a good state $g \in G$ and, thereafter, keeps looping back to g infinitely often (see Figure 2.1).

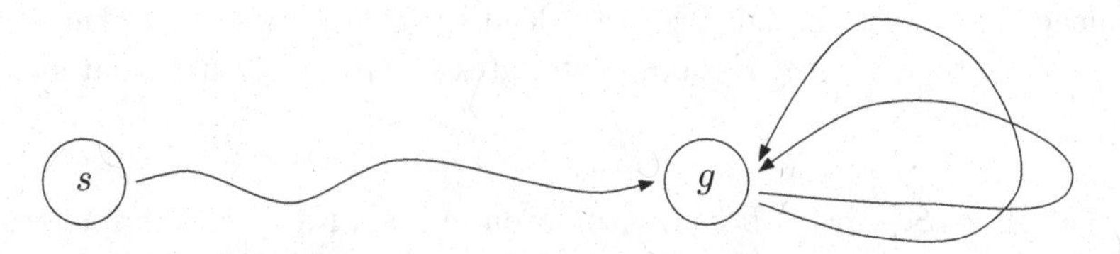

Fig. 2.1. A typical accepting run of a Büchi automaton, with $s \in S_{in}$ and $g \in G$.

Example 2.1. Consider the alphabet $\Sigma = \{a, b\}$. Let $L \subseteq \Sigma^\omega$ consist of all infinite words α such that there are infinitely many occurrences of a in α. Figure 2.2 shows a Büchi automaton recognizing L. The initial state is marked by an incoming arrow. There is only one good state, which is indicated with a double circle. In this automaton, all transitions labelled a lead into the good state and, conversely, all transitions coming into the good state are labelled a. From this, it follows that the automaton accepts an infinite word iff it has infinitely many occurrences of a.

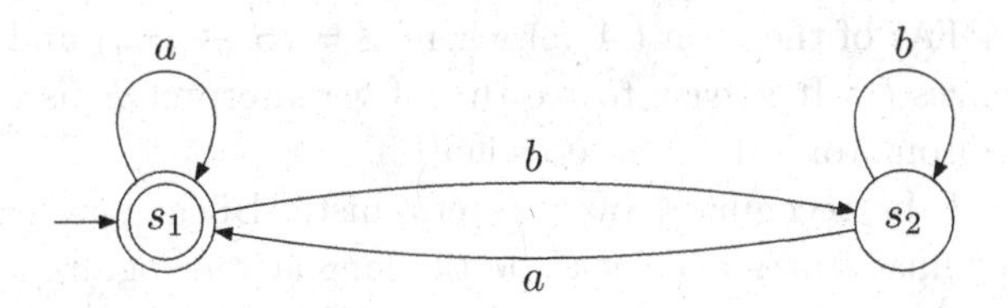

Fig. 2.2. A Büchi automaton for L (Example 2.1).

The complement of L, which we denote $\overline{L}$, is the set of all infinite words α such that α has only finitely many occurrences of a. An automaton recognizing $\overline{L}$ is shown in Figure 2.3. The automaton guesses a point in the input beyond which it will see no more a's—such a point must exist in any input with only a finite number of a's. Once it has made this guess, it can process only b's—there is no transition labelled a from the second state—so if it reads any more a's it gets stuck.

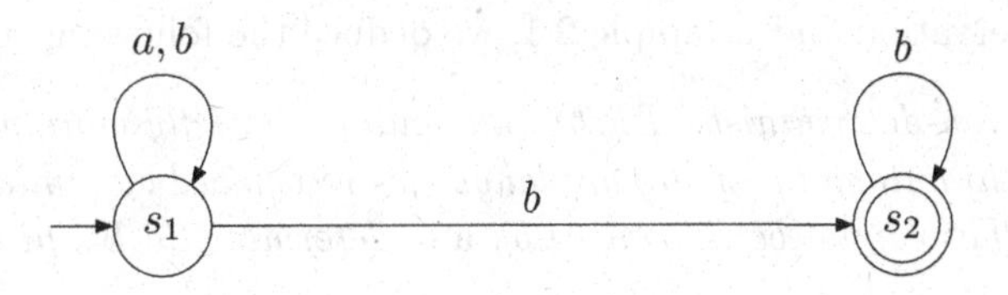

Fig. 2.3. A Büchi automaton for $\overline{L}$ (Example 2.1).

In the example, notice that the automaton recognizing L is deterministic while the automaton for $\overline{L}$ is non-deterministic. We now show that the non-determinism in the second case is unavoidable—that is, there is *no* deterministic automaton recognizing $\overline{L}$. This means that Büchi automata are fundamentally different from their counterparts on finite inputs: we know that over finite words, deterministic automata are as powerful as non-deterministic automata.

Limit languages Let $U \subseteq \Sigma^*$ be a language of finite strings. The limit of U, $\mathsf{lim}(U)$ is the set

$$\{\alpha \in \Sigma^\omega \mid \exists^\omega n \in \mathbb{N}_0 : \alpha[0..n] \in U\}.$$

So, a word belongs to $\mathsf{lim}(U)$ iff it has infinitely many prefixes in U. We then have the following characterization of languages recognized by deterministic Büchi automata.

Theorem 2.2. *A language $L \subseteq \Sigma^\omega$ is recognizable by a deterministic Büchi automaton iff L is of the form* $\mathsf{lim}(U)$ *for some regular language $U \subseteq \Sigma^*$.*

Proof. Let U be a regular language. Then, there exists a deterministic finite-state automaton (DFA) of the form $(\mathcal{A}, F)$ where $\mathcal{A} = (S, \rightarrow, S_{in})$ and $F \subseteq S$ such that $(\mathcal{A}, F)$ recognizes U. It is easy to see that if we interpret F as a set of good states, the Büchi automaton $(\mathcal{A}, F)$ accepts $\mathsf{lim}(U)$.

Conversely, let L be recognized by a deterministic Büchi automaton $(\mathcal{A}, G)$. Treat G as a set of final states and let U be the language recognized by the DFA $(\mathcal{A}, G)$. Once again, it is easy to see that $L = \mathsf{lim}(U)$. □

We now show that the language $\overline{L}$ of Example 2.1 is not of the form $\mathsf{lim}(U)$ for *any* language U. Recall that $\overline{L}$ is the set of all infinite words α over the alphabet $\Sigma = \{a, b\}$ such that α contains only finitely many occurrences of a.

Suppose that $\overline{L} = \mathsf{lim}(U)$ for some $U \subseteq \Sigma^*$. Since $b^\omega \in \overline{L}$, there must be some finite prefix $b^{n_1} \in U$. Since, $b^{n_1}ab^\omega \in \overline{L}$, we can then find a prefix $b^{n_1}ab^{n_2} \in U$. From the fact that $b^{n_1}ab^{n_2}ab^\omega \in \overline{L}$, we obtain a prefix $b^{n_1}ab^{n_2}ab^{n_3} \in U$. Proceeding in this way, we get an infinite sequence of words $\{b^{n_1}, b^{n_1}ab^{n_2}, b^{n_1}ab^{n_2}ab^{n_3}, \ldots\} \subseteq U$. From this it follows that the infinite word $\beta = b^{n_1}ab^{n_2}ab^{n_3}a \cdots ab^{n_i}a \cdots$ belongs to $\mathsf{lim}(U)$. But β has infinitely many occurrences of a, so it certainly does not belong to $\overline{L}$, thus contradicting the assumption that $\overline{L} = \mathsf{lim}(U)$.

From this observation and Example 2.1, we deduce the following corollary.

Corollary 2.3. *Non-deterministic Büchi automata are strictly more powerful than deterministic Büchi automata—there are languages recognized by non-deterministic Büchi automata that cannot be recognized by* any *deterministic Büchi automaton.*

2.1.1. *Characterizing Büchi-recognizable languages*

For finite words, we can characterize the class of languages recognized by non-deterministic finite-state automata in a number of ways—for instance, in terms of regular expressions, or in terms of syntactic congruences. In the same spirit, we now describe a characterization of Büchi-recognizable languages of infinite words. We first need to define the ω-iteration of a set of finite words. Let $U \subseteq \Sigma^*$. Then

$$U^\omega = \{\alpha \in \Sigma^\omega \mid \alpha = u_0u_1u_2\cdots \text{ where } u_i \in U \text{ for all } i \in \mathbb{N}_0\}.$$

Also, we observe that if U is a language of finite words and L is a language of infinite words, we can define the language UL of infinite words obtained by concatenating each finite word in U with an infinite word from L. Formally, $UL = \{\alpha \mid \exists u \in U : \exists \beta \in L : \alpha = u\beta\}$.

ω-regular languages A language $L \subseteq \Sigma^\omega$ is said to be *ω-regular* if it is of the form $\bigcup_{i\in[1..n]} U_iV_i^\omega$, where each U_i and V_i is a regular language of finite words.

Theorem 2.4. *A language is Büchi-recognizable iff it is ω-regular.*

Proof.
($\Rightarrow$): Let L be recognized by a Büchi automaton $(\mathcal{A}, G)$, where $\mathcal{A} = (S, \rightarrow, S_{in})$. We have observed earlier that each infinite word $\alpha \in L$ admits an accepting run ρ that begins in an initial state, reaches a good state g, and then loops back through g infinitely often. For $s, s' \in S$, let $V_{ss'} = \{w \in \Sigma^* \mid s \xrightarrow{w}{}^{+} s'\}$ denote the set of finite words that can lead from s to s'. It is easy to see that $V_{ss'}$ is regular—to recognize this set, use the non-deterministic automaton $(S, \rightarrow, \{s\})$ with $\{s'\}$ as the set of final states. From our observation about accepting runs, it follows that we can write $L(\mathcal{A}, G)$ as $\bigcup_{s \in S_{in},\ g \in G} V_{sg} V_{gg}^{\omega}$.

($\Leftarrow$): It is not difficult to show that the set of Büchi-recognizable languages satisfies the following closure properties:

(i) If U is regular, then U^{ω} is Büchi-recognizable.
(ii) If U is regular and L is Büchi-recognizable then UL is Büchi recognizable.
(iii) If $L_1, L_2, \ldots, L_n$ are Büchi-recognizable, so is $\bigcup_{i \in [1..n]} L_i$.

From this, it follows that every language of the form $\bigcup_{i \in [1..n]} U_i V_i^{\omega}$, where each U_i and V_i is regular, is Büchi-recognizable. □

2.1.2. *Constructions on Büchi automata*

It turns out that the class of Büchi-recognizable languages is closed under boolean operations and projection. These operations will be crucial when applying Büchi automata to settle decision problems in logic.

Union To show closure under finite union (which we have already assumed when proving the previous theorem!), let $(\mathcal{A}_1, G_1)$ and $(\mathcal{A}_2, G_2)$ be two Büchi automata. To construct an automaton $(\mathcal{A}, G)$ such that $L(\mathcal{A}, G) = L(\mathcal{A}_1, G_1) \cup L(\mathcal{A}_2, G_2)$, we take $\mathcal{A}$ to be the *disjoint* union of $\mathcal{A}_1$ and $\mathcal{A}_2$. Since we are permitted to have a *set* of initial states in $\mathcal{A}$, we retain the initial states from both copies. If a run of $\mathcal{A}$ starts in an initial state contributed by $\mathcal{A}_1$, it will never cross over into the state space contributed by $\mathcal{A}_2$ and vice versa. Thus, we can set the good states of $\mathcal{A}$ to be the union of the good states contributed by both components.

Complementation Showing that Büchi-recognizable languages are closed under complementation is highly non-trivial. One problem is that we cannot determinize Büchi automata, as we have observed in Corollary 2.3. Even if we could work with deterministic automata, the formulation of Büchi acceptance is not symmetric with respect to complementation in the following sense. Suppose $(\mathcal{A}, G)$ is a deterministic Büchi automaton and α is an infinite word that does not belong to $L(\mathcal{A}, G)$. Then, the (unique) run ρ_α of $\mathcal{A}$ on α is such that $\inf(\rho_\alpha) \cap G = \emptyset$. Let $\overline{G}$ denote the complement of G. It follows that $\inf(\rho_\alpha) \cap \overline{G} \neq \emptyset$, since *some* state must occur infinitely often in ρ_α. It would be tempting to believe that the automaton $(\mathcal{A}, \overline{G})$ recognizes $\Sigma^\omega - L(\mathcal{A}, G)$. However, there may be words that admit runs which visit both G and $\overline{G}$ infinitely often. These words will be including both in $L(\mathcal{A}, \overline{G})$ as well as in $L(\mathcal{A}, G)$. So, there is no convenient way to express the complement of a Büchi condition again as a Büchi condition.

We shall postpone describing a complementation construction for Büchi automata until Section 2.4. Till then we shall, however, assume that we can complement these automata.

Intersection Turning to intersection, the natural way to intersect automata $\mathcal{A}_1$ and $\mathcal{A}_2$ is to construct an automaton whose state space is the cross product of the state spaces of $\mathcal{A}_1$ and $\mathcal{A}_2$ and let both copies process the input simultaneously. For finite words, the input is accepted if each copy can generate a run that reaches a final state at the end of the word.

For infinite inputs, we have to use a more sophisticated product construction. An infinite input α should be accepted by the product system provided both copies generate runs that visit good states infinitely often. Unfortunately, there is no guarantee that these runs will ever visit good states simultaneously—for instance, it could be that the first run goes through a good state after $\alpha(0)$, $\alpha(2)$, ... while the second run enters good states after $\alpha(1)$, $\alpha(3)$, ... So, the main question is one of identifying the good states of the product system.

The key observation is that to detect that both components of the product visit good states infinitely often, one need not record *every* point where the copies visit good states; in each copy, it suffices to observe an infinite subsequence of the overall sequence of good states. So, we begin by focusing on the first copy and waiting for its run to enter a good state. When this happens, we switch attention to the other copy and wait for a good state there. Once the second copy reaches a good state, we switch back to the first copy and so on. Clearly, we will switch back and forth infinitely often iff both copies visit their respective good states infinitely often. Thus, we can characterize the good states of the product in terms of the states where one switches back and forth.

Formally, the construction is as follows. Let $(\mathcal{A}_1, G_1)$ and $(\mathcal{A}_2, G_2)$ be two Büchi automata such that $\mathcal{A}_i = (S_i, \rightarrow_i, S^i_{in})$ for $i = 1, 2$. Define $(\mathcal{A}, G)$, where $\mathcal{A} = (S, \rightarrow, S_{in})$, as follows:

- $S = S_1 \times S_2 \times \{1, 2\}$
- The transition relation $\rightarrow$ is defined as follows:

$$
\begin{array}{ll}
(s_1, s_2, 1) \xrightarrow{a} (s_1', s_2', 1) & \text{if } s_1 \xrightarrow{a}_1 s_1',\ s_2 \xrightarrow{a}_2 s_2' \text{ and } s_1 \notin G_1. \\
(s_1, s_2, 1) \xrightarrow{a} (s_1', s_2', 2) & \text{if } s_1 \xrightarrow{a}_1 s_1',\ s_2 \xrightarrow{a}_2 s_2' \text{ and } s_1 \in G_1. \\
(s_1, s_2, 2) \xrightarrow{a} (s_1', s_2', 2) & \text{if } s_1 \xrightarrow{a}_1 s_1',\ s_2 \xrightarrow{a}_2 s_2' \text{ and } s_2 \notin G_2. \\
(s_1, s_2, 2) \xrightarrow{a} (s_1', s_2', 1) & \text{if } s_1 \xrightarrow{a}_1 s_1',\ s_2 \xrightarrow{a}_2 s_2' \text{ and } s_2 \in G_2.
\end{array}
$$

- $S_{in} = \{(s_1, s_2, 1) \mid s_1 \in S^1_{in} \text{ and } s_2 \in S^2_{in}\}$
- $G = S_1 \times G_2 \times \{2\}$.

In the automaton $\mathcal{A}$, each product state carries an extra tag indicating whether the automaton is checking for a good state on the first or the second component. The automaton accepts if it switches focus from the second component to the first infinitely often. (Notice that we could equivalently have defined G to be the set $G_1 \times S_2 \times \{1\}$.) It is not difficult to verify that $L(\mathcal{A}, G) = L(\mathcal{A}_1, G_1) \cap L(\mathcal{A}_2, G_2)$.

Projection Let Σ_1 and Σ_2 be alphabets such that $|\Sigma_2| \leq |\Sigma_1|$. A *projection function* from Σ_1 to Σ_2 is a surjective map $\pi : \Sigma_1 \to \Sigma_2$. We can extend π from individual letters to words as usual: if $\alpha \in \Sigma_1^\omega$, $\pi(\alpha)$ denotes the word β where $\beta(i) = \pi(\alpha(i))$ for all i in $\mathbb{N}_0$.

Let $L \subseteq \Sigma_1^\omega$. Then $\pi(L)$, the projection of L via π, is the language $\{\beta \in \Sigma_2^\omega \mid \exists \alpha \in L : \beta = \pi(\alpha)\}$. It is easy to verify that if L is Büchi-recognizable, then so is $\pi(L)$. Let $(\mathcal{A}_1, G_1)$ be an automaton recognizing L, where $\mathcal{A}_1 = (S_1, \to_1, S^1_{in})$. We construct an automaton $\mathcal{A}_2 = (S_2, \to_2, S^2_{in})$ over Σ_2 as follows: set $S_2 = S_1$, $S^2_{in} = S^1_{in}$ and $s \overset{b}{\to}_2 s'$ iff $s \overset{a}{\to}_1 s'$ for some $a \in \Sigma_1$ such that $\pi(a) = b$. It is easy to verify that $(\mathcal{A}_2, G_1)$ recognizes $\pi(L)$.

Emptiness In applications, we will need to be able to check whether the language accepted by a Büchi automaton is empty. To do this, we recall our observation that any accepting run of a Büchi automaton must begin in an initial state, reach a final state g and then cycle back to g infinitely often.

If we ignore the labels on the transitions, we can regard the state space of a Büchi automaton $(\mathcal{A}, G)$ as a directed graph $G_\mathcal{A} = (V_\mathcal{A}, E_\mathcal{A})$ where $V_\mathcal{A} = S$ and $(s, s') \in E_\mathcal{A}$ iff for some $a \in \Sigma$, $s \overset{a}{\to} s'$. Recall that a set of vertices X in a directed graph is a *strongly connected component* iff for every pair of vertices $v, v' \in X$, there is a path from v to v'. Clearly, $L(\mathcal{A}, G)$ is non-empty iff there is a strongly connected component X in $G_\mathcal{A}$ such that X contains a vertex g from G and X is reachable from one of the initial states. We thus have the following theorem.

Theorem 2.5. *The emptiness problem for Büchi automata is decidable.*

Notice that it is sufficient to analyze *maximal* strongly connected components in $G_\mathcal{A}$ in order to check that $L(\mathcal{A}, G) \neq \emptyset$. Computing the maximal strongly connected components of a directed graph can be done in time linear in the size of the graph [13], where the size of a graph $G = (V, E)$ is, as usual, given by $|V| + |E|$. Checking reachability can also be done in linear time. So, if $\mathcal{A}$ has n states, checking that $L(\mathcal{A}, G) \neq \emptyset$ can be done in time $O(n^2)$.

2.2. The logic of sequences

Büchi's original motivation for studying automata on infinite inputs was to solve a decision problem from logic. He discovered a deep and beautiful connection between ω-regular languages and sets of models of formulas in certain logics.

S1S

The logic that Büchi considered was the *monadic second-order theory of one successor*, abbreviated as S1S. This logic is interpreted over the set $\mathbb{N}_0$ of natural numbers. In general, second-order logic permits quantification over relations and functions, unlike first-order logic, which permits quantification over just individual elements. However, the fact that we are dealing with a "monadic" second-order logic restricts this extra power to quantification over one-place relations. Since a one-place relation is just a subset, this effectively means that we can quantify over individual elements of $\mathbb{N}_0$ and subsets of $\mathbb{N}_0$. The fact that we are dealing with "one successor" means we are talking about $\mathbb{N}_0$ with the usual ordering where each element has a unique successor. Permitting two successors, for instance, would produce the infinite binary tree which has countably many nodes but has two successors for each node.

Formally, the logical language S1S is defined as follows.

Terms A *term* in S1S is built up from the constant 0 and *individual variables* $x, y, \ldots$ by application of the *successor function succ*. Thus, the following are terms: 0, $succ(x)$, $succ(succ(succ(0)))$, $succ(succ(y))$, ...

Atomic formulas Let $t, t', \ldots$ be terms. An *atomic formula* is of the form $t = t'$ or $t \in X$, where X is a *set variable*.

Formulas A *formula* is built up from atomic formulas using the boolean connectives $\neg$ (not) and $\vee$ (or), together with the existential quantifier $\exists$. The quantifier $\exists$ can be applied to *both* individual and set variables—one can write $\exists x$ and $\exists X$. In other words, if φ and ψ are inductively assumed to be formulas, so are $\neg\varphi$, $\varphi \vee \psi$, $(\exists x)\ \varphi$ and $(\exists X)\ \varphi$.

In addition, we can define the remaining boolean connectives like $\wedge$ (and), $\Rightarrow$ (if-then), $\Leftrightarrow$ (iff) as usual, in terms of $\neg$ and $\vee$: for instance, $\varphi \Rightarrow \psi$ is defined as $(\neg\varphi \vee \psi)$. We also have the universal quantifier $\forall$ which is the dual of $\exists$: $(\forall x)\ \varphi \stackrel{def}{=} \neg((\exists x)\ \neg\varphi)$ and $(\forall X)\ \varphi \stackrel{def}{=} \neg((\exists X)\ \neg\varphi)$.

Assigning truth values to formulas Formulas are interpreted over $\mathbb{N}_0$. The constant 0 denotes the number 0. Individual variables $x, y, \ldots$ are interpreted as natural numbers—that is, elements of $\mathbb{N}_0$. The function *succ* corresponds to adding one: $succ(x)$ denotes the number that is one greater than the interpretation of x. Thus, the term $succ(succ(succ(0)))$ represents the number 3. And, if the current interpretation of x is the number 47 then $succ(x)$ denotes 48.

The connective $=$ used in defining atomic formulas denotes equality, as usual. Thus $t = t'$ is true provided t and t' denote the same natural number.

Set variables like $X, Y, \ldots$ are interpreted as subsets of $\mathbb{N}_0$. The atomic formula $t \in X$ is true iff the number denoted by t belongs to the set denoted by X.

Once the interpretation of atomic formulas has been fixed, the meaning of compound formulas involving $\neg$, $\vee$ and $\exists$ is the "natural" one.

Let φ be a formula. A variable is said to occur free in φ if it is not within the scope of a quantifier. For instance, in the formula $(\exists x)(\forall Y)\ (0 \in Y) \vee (x = y) \vee (x \in X)$, the variables y and X occur free. Variables that do not occur free are said to be bound. In the preceding formula, x and Y are bound. We write $\varphi(x_1, x_2, \ldots, x_k, X_1, X_2, \ldots, X_\ell)$ to indicate that all the variables that occur free in φ come from the set $\{x_1, x_2, \ldots, x_k, X_1, X_2, \ldots, X_\ell\}$. Let $\overrightarrow{X} = (x_1, x_2, \ldots, x_k, X_1, X_2, \ldots, X_\ell)$. To assign a truth value to the formula $\varphi(\overrightarrow{X})$, we have to first fix an interpretation of the variables in $\overrightarrow{X}$. In other words, we must map each individual variable x_i to a natural number $m_i \in \mathbb{N}_0$ and each set variable X_j to a subset $M_j \subseteq \mathbb{N}_0$. Let $\overrightarrow{M} = (m_1, m_2, \ldots, m_k, M_1, M_2, \ldots, M_\ell)$. We write $\overrightarrow{M} \models \varphi(\overrightarrow{X})$ to denote that φ is true under the interpretation $\{x_i \mapsto m_i\}_{i \in [1..k]}$ and $\{X_i \mapsto M_i\}_{i \in [1..\ell]}$. Rather than go into formal details, we look at some illustrative examples.

Example 2.6.

(i) Let $\mathsf{Sub}(X, Y) = (\forall x)\ x \in X \Rightarrow x \in Y$.
Then $(M, N) \models \mathsf{Sub}(X, Y)$ iff $M \subseteq N$.

(ii) Let $\mathsf{Zero}(X) = (\exists x)\ [x \in X \wedge \neg(\exists y)(y < x)]$.
This formula asserts that X contains an element that has no predecessors in $\mathbb{N}_0$. Thus, $M \models \mathsf{Zero}(X)$ iff $0 \in M$.

(iii) Let $\mathsf{Lt}(x, y) = (\forall Z)[succ(x) \in Z \wedge (\forall z)(z \in Z \Rightarrow succ(z) \in Z)] \Rightarrow (y \in Z)$.
Then $(m, n) \models \mathsf{Lt}(x, y)$ iff $m < n$. What the formula asserts is that any set Z that contains $x{+}1$ and is closed with respect to the successor function must also contain y.

(iv) Let $\mathsf{Sing}(X) = (\exists Y)\ [\mathsf{Sub}(Y, X) \wedge (Y \neq X) \wedge$
$\neg(\exists Z)\ (\mathsf{Sub}(Z, Y) \wedge (Z \neq X) \wedge (Z \neq Y))]$.
In this formula, $X \neq Y$ abbreviates $\neg(X = Y)$, where $X = Y$ is itself an abbreviation for $\mathsf{Sub}(X, Y) \wedge \mathsf{Sub}(Y, X)$. The formula asserts that X has only one proper subset, which is Y. This is true only for singletons, where Y is the empty set. So, $M \models \mathsf{Sing}(X)$ iff M is a singleton $\{m\}$.

A *sentence* is a formula in which no variables occur free. A sentence φ is either true or false—we do not have to interpret any variables to assign meaning to φ. For instance consider the sentence

$$(\forall X)\ [0 \in X \wedge (\forall x)\ (x \in X \Rightarrow succ(x) \in X)] \Rightarrow (\forall x)\ x \in X.$$

This sentence is true: it expresses the familiar property of mathematical induction for subsets of $\mathbb{N}_0$—if a set of natural numbers contains 0 and is closed with respect to the successor function, then the set in fact includes all of $\mathbb{N}_0$.

Satisfiability An S1S formula $\varphi(x_1, \ldots, x_k, X_1, \ldots, X_\ell)$ is said to be *satisfiable* if we can choose $\overrightarrow{M} = (m_1, \ldots, m_k, M_1, \ldots, M_\ell)$ such that $\overrightarrow{M} \models \varphi(\overrightarrow{X})$.

Büchi showed how to associate an ω-regular language L_φ with each S1S formula φ, such that every word in L_φ represents an interpretation for the free variables in φ under which the formula φ evaluates to true. Moreover, every interpretation that makes φ true is represented by some word in L_φ. Thus, φ is satisfiable iff there is some interpretation that makes it true iff L_φ is non-empty. The language L_φ is defined over the alphabet $\{0,1\}^m$, where m is the number of free variables in φ.

In fact, Büchi showed that the converse is also true. Let us say that a language $L \subseteq (\{0,1\}^m)^\omega$ is S1S-definable if $L = L_\varphi$ for some S1S formula φ. We can always embed an arbitrary alphabet Σ as a subset of $\{0,1\}^m$ for some suitable choice of m. In this way, any language $L \subseteq \Sigma^\omega$ can be converted into an equivalent language $L_{\{0,1\}}$ over $\{0,1\}^m$. Büchi showed that if L is ω-regular, then $L_{\{0,1\}}$ is S1S-definable.

Thus, the notions of S1S-definability and ω-regularity are equivalent. The rest of this section will be devoted to formally stating and proving this result.

We begin by defining L_φ for an S1S formula $\varphi(x_1, x_2, \ldots, x_k, X_1, X_2, \ldots, X_\ell)$. Let $\overrightarrow{M} = (m_1, m_2, \ldots, m_k, M_1, M_2, \ldots, M_\ell)$ such that $\overrightarrow{M} \models \varphi(\overrightarrow{X})$. We can associate with $\overrightarrow{M}$ an infinite word α_M over $\{0,1\}^{k+\ell}$ that represents the characteristic function of $\overrightarrow{M}$. For $i \in \mathbb{N}_0$, and $j \in [1..k+\ell]$, let $\alpha_M(i)(j)$ denote the jth component of $\alpha_M(i)$. Then for $i \in \mathbb{N}_0$ and $j \in [1..k]$, $\alpha_M(i)(j) = 1$ iff $i = m_j$ and $\alpha_M(i)(j) = 0$ iff $i \neq m_j$. Similarly, for $i \in \mathbb{N}_0$, and $j \in [k+1..k+\ell]$, $\alpha_M(i)(j) = 1$ iff $i \in M_j$ and $\alpha_M(i)(j) = 0$ iff $i \notin M_j$. Then

$$L_\varphi = \{\alpha_M \mid \overrightarrow{M} \models \varphi(\overrightarrow{X})\}.$$

Next we define the $\{0,1\}$-image $L_{\{0,1\}}$ corresponding to a language L over an arbitrary alphabet Σ. Let $\Sigma = \{a_1, a_2, \ldots, a_m\}$. Then, each word $\alpha \in \Sigma^\omega$ can be represented by a word $\alpha_{\{0,1\}}$ over $\{0,1\}^m$, where for all $i \in \mathbb{N}_0$ and $j \in [1..m]$, $\alpha_{\{0,1\}}(i)(j) = 1$ if $\alpha(i) = a_j$ and $\alpha_{\{0,1\}}(i)(j) = 0$ if $\alpha(i) \neq a_j$. Then

$$L_{\{0,1\}} = \{\alpha_{\{0,1\}} \mid \alpha \in L\}.$$

We can now state Büchi's result more precisely.

Theorem 2.7.

(i) Let φ be an S1S formula. Then L_φ is an ω-regular language.

(ii) Let L be an ω-regular language. Then $L_{\{0,1\}}$ is S1S-definable.[*]

Proof.

(i) To show that L_φ is ω-regular, we proceed by induction on the structure of φ. To do this, it will be convenient to cut down the language S1S to an equivalent language $\mathrm{S1S}_0$ that has a simpler syntax.

Formally, in $\mathrm{S1S}_0$ we do not have individual variables x_i—there are *only* set variables X_j. The atomic formulas are of the form $X \subseteq Y$ and $succ(X, Y)$. The first formula is true if X is a subset of Y while the second is true if X and Y are singletons $\{x\}$ and $\{y\}$ respectively and $y = x+1$.

We now argue that every S1S formula φ can be converted to an $\mathrm{S1S}_0$ formula φ_0 such that $L_\varphi = L_{\varphi_0}$.

We begin by eliminating nested applications of the successor function. For instance, if the S1S formula contains the atomic formula $succ(succ(x)) \in X$, we write instead

$$(\exists y)(\exists z)\ y = succ(x) \wedge z = succ(y) \wedge z \in X.$$

We then eliminate formulas of the form $0 \in X$ using the formula $\mathsf{Zero}(X)$ defined in Example 2.6.

Finally, we eliminate singleton variables using the formula Sing from Example 2.6. For instance, we rewrite $(\forall x)(\exists y)\ succ(x) = y \wedge y \in Z$ as

$$(\forall X)\ (\mathsf{Sing}(X) \Rightarrow [(\exists Y)\ \mathsf{Sing}(Y) \wedge succ(X, Y) \wedge Y \subseteq Z]).$$

Notice that we can uniformly replace $\mathsf{Sub}(X, Y)$ by $X \subseteq Y$ in Sing since $X \subseteq Y$ is an atomic formula in SIS_0.

We now construct for each $\mathrm{S1S}_0$ formula φ, a Büchi automaton $(\mathcal{A}_\varphi, G_\varphi)$ recognizing L_φ.

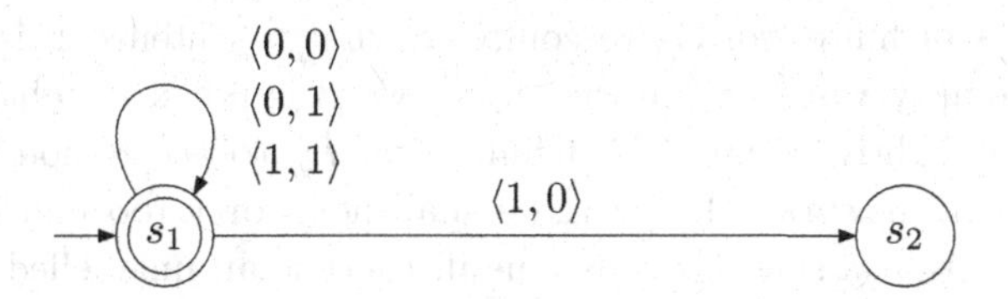

Fig. 2.4. Büchi automaton for the atomic formula $X \subseteq Y$.

For the atomic formula $X \subseteq Y$, the corresponding automaton over $\{0,1\}\times\{0,1\}$ is shown in Figure 2.4. This automaton accepts any input word that does not contain $\langle 1, 0\rangle$—if $\alpha(i) = \langle 1, 0\rangle$, in the corresponding interpretation, $i \in X$ but $i \notin Y$, thus violating the requirement that $X \subseteq Y$.

The other atomic formula is $succ(X, Y)$. The corresponding automa-

[*]We have fixed a apecific embedding of Σ into $\{0,1\}^m$ that is relatively easy to describe in S1S. In general, we can choose *any* embedding for defining $L_{\{0,1\}}$ and the result will still go through.

ton is shown in Figure 2.5. This automaton accepts inputs of the form $\langle 0,0\rangle^i\langle 1,0\rangle\langle 0,1\rangle\langle 0,0\rangle^\omega$, $i \in \mathbb{N}_0$, corresponding to the interpretation where $X = \{i\}$ and $Y = \{i+1\}$.

For the induction step, we need to consider the connectives $\neg$, $\vee$ and $\exists X$.

Let $\varphi = \neg\psi$. Then, L_φ is the complement of L_ψ. By the induction hypothesis, there exists a Büchi automaton $(\mathcal{A}_\psi, G_\psi)$ that recognizes L_ψ. As we mentioned earlier, there is an effective way to construct an automaton $(\mathcal{A}_\varphi, G_\varphi)$ recognizing the complement of L_ψ. The details will be described in Section 2.4.

If $\varphi = \varphi_1 \vee \varphi_2$, then $L_\varphi = L_{\varphi_1} \cup L_{\varphi_2}$. By the induction hypothesis, there exist automata $(\mathcal{A}_{\varphi_1}, G_{\varphi_1})$ and $(\mathcal{A}_{\varphi_2}, G_{\varphi_2})$ such that $L(\mathcal{A}_{\varphi_i}, G_{\varphi_i}) = L_{\varphi_i}$ for $i = 1, 2$. We have seen in Section 2.1.2 how to construct an automaton $(\mathcal{A}_\varphi, G_\varphi)$ such that $L(\mathcal{A}_\varphi, G_\varphi) = L_{\varphi_1} \cup L_{\varphi_2}$.

Finally, if $\varphi = (\exists X_1)\, \psi(X_1, X_2, \ldots X_m)$, the language L_φ corresponds to the projection of L_ψ via the function $\pi : \{0,1\}^m \to \{0,1\}^{m-1}$ that erases the first component of each m-tuple in $\{0,1\}^m$. A word of $(m-1)$-tuples belongs to L_φ if it can be padded out with an extra component so that the resulting word over m-tuples is in L_ψ. This padding operation corresponds to guessing a witness for the set X_1. The automaton $(\mathcal{A}_\varphi, G_\varphi)$ recognizing L_φ can be obtained from $(\mathcal{A}_\psi, G_\psi)$, the automaton recognizing L_ψ, as described in Section 2.1.2.

In this way, we inductively associate with each S1S$_0$ formula φ, a Büchi automaton $(\mathcal{A}_\varphi, G_\varphi)$ such that $L_\varphi = L(\mathcal{A}_\varphi, G_\varphi)$.

There is a slight technicality involved when we deal with sentences. Notice that every time we encounter an existential quantifier, we eliminate one component from the input alphabet of $\mathcal{A}_\varphi$. If φ is a sentence—that is, all variables in φ are bound—we would have erased *all* components of the input by the time we construct $\mathcal{A}_\varphi$. In other words, $\mathcal{A}_\varphi$ will be an input-free automaton whose states and transitions define an unlabelled directed graph. If we were dealing with languages of finite words, we could say that a sentence φ is true iff L_φ contains the empty word. Since the empty word is not a member of Σ^ω, we would have to slightly modify our definition of L_φ to accommodate this case cleanly in our framework. However, we shall not worry too much about this since it is clear that a sentence φ is true iff there is an unlabelled path in the graph corresponding to $\mathcal{A}_\varphi$ that begins at some initial state and visits a good state infinitely often.

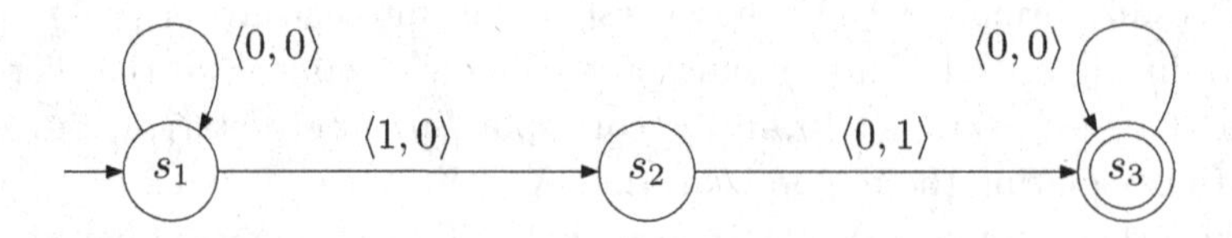

Fig. 2.5. Büchi automaton for the atom formula *succ*(X, Y).

(ii) Let $(\mathcal{A}, G)$ be a Büchi automaton recognizing $L \subseteq \Sigma^\omega$, where $\Sigma = \{a_1, a_2, \ldots, a_m\}$ and $\mathcal{A} = (S, \rightarrow, S_{in})$, with $S = \{s_1, s_2, \ldots, s_k\}$. We use free variables $A_1, A_2, \ldots, A_m$ to describe each infinite word over Σ—the variable A_i describes the positions in the input where letter a_i occurs. We then use existentially quantified variables $S_1, S_2, \ldots, S_k$ to describe runs of the automaton over the input—the variable S_j describes the positions in the run where the automaton is in state s_j.

The formula φ_L can then be written as follows:

$$
\begin{array}{llll}
(\exists S_1)(\exists S_2)\cdots(\exists S_k) & & & \\
(\forall x) & \bigvee\limits_{i\in[1..m]} & (x \in A_i) \wedge & \bigwedge\limits_{i\in[1..m]} \left(x \in A_i \Rightarrow \bigvee\limits_{j\neq i} x \notin A_j \right) \\
\wedge\ (\forall x) & \bigvee\limits_{i\in[1..k]} & (x \in S_i) \wedge & \bigwedge\limits_{i\in[1..k]} \left(x \in S_i \Rightarrow \bigvee\limits_{j\neq i} x \notin S_j \right) \\
\wedge & \bigvee\limits_{s_i\in S_{in}} & (0 \in S_i) & \\
\wedge\ (\forall x) & \bigvee\limits_{(s_i,a_j,s_k)\in\rightarrow} & (x \in S_i) \wedge (x \in A_j) \wedge (succ(x) \in S_k) & \\
\wedge & \bigvee\limits_{s_i\in G} & (\forall x)(\exists y)\ (x < y) \wedge (y \in S_i) &
\end{array}
$$

The first two lines capture the fact that with each position $x \in \mathbb{N}_0$ we associate precisely one input letter and one state from the run. The third line asserts that the position 0 corresponds to an initial state. The fourth line guarantees that the sequence of states represented by $S_1, S_2, \ldots, S_k$ is a run. The last line then says that the run is good—that is, some final state appears infinitely often. It is not difficult to verify that $L_{\varphi_L} = L_{\{0,1\}}$. □

2.3. Stronger acceptance conditions

As we saw earlier, deterministic Büchi automata cannot recognize all ω-regular languages (Corollary 2.3). It turns out that we can define classes of deterministic automata that recognize ω-regular languages by strengthening the acceptance criterion. We begin with the definition proposed by Muller [2].

Muller automata A *Muller automaton* is a pair $(\mathcal{A}, \mathcal{T})$ where $\mathcal{A} = (S, \rightarrow, S_{in})$ is an automaton, as before, and $\mathcal{T} = \langle F_1, F_2, \ldots, F_k \rangle$ is an *acceptance table* with $F_i \subseteq S$ for $i \in [1..k]$.

The automaton $(\mathcal{A}, \mathcal{T})$ accepts an input $\alpha : \mathbb{N}_0 \rightarrow \Sigma$ if there is a run ρ of $\mathcal{A}$ on α such that $\mathsf{inf}(\rho) \in \mathcal{T}$—that is, $\mathsf{inf}(\rho) = F_i$ for some $i \in [1..k]$.

The acceptance table of a Muller automaton places a much more stringent requirement on runs than a Büchi condition does. The table entry F_i makes a positive demand on the states in F_i, as well as a negative demand on the states in $S - F_i$—states in F_i must all be visited infinitely often while states outside F_i must be visited only finitely often. In other words, for a run ρ to satisfy the table entry F_i, after some point it must "settle down" in the set F_i and visit all the states in this set infinitely often without making transitions to any state outside F_i.

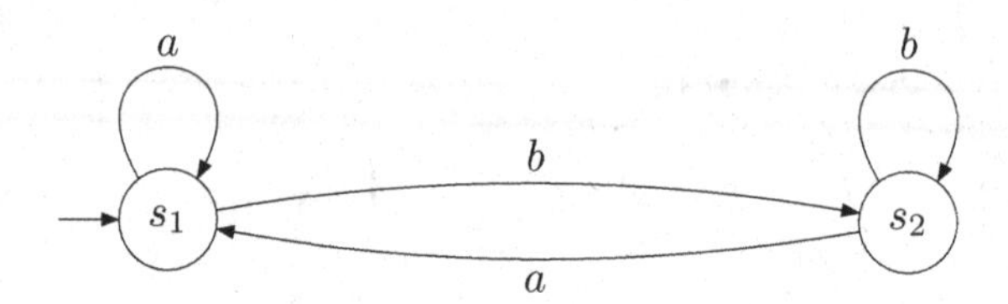

Fig. 2.6. Automaton for L and $\overline{L}$ (Example 2.8).

Example 2.8. Recall the language L over $\{a, b\}$ defined in Example 2.1—L contains all words that contain infinitely many occurrences of a. We saw that the deterministic automaton with two states shown in Figure 2.6 recognizes L with a Büchi condition $\{s_1\}$.

To accept L using a Muller condition, we retain the same automaton and set the acceptance table to $\langle\{s_1\}, \{s_1, s_2\}\rangle$.

We also saw that $\overline{L}$, the complement of L could not be recognized by *any* deterministic Büchi automaton. However, it is easy to verify that $\overline{L}$ can be recognized with a Muller condition by using the *same* automaton as for L, but with the acceptance table given by $\langle\{s_2\}\rangle$.

Simulations The example shows that deterministic Muller automata are strictly more powerful than deterministic Büchi automata. It is quite straightforward to simulate a Büchi automaton by a Muller automaton—we construct an entry in the Muller table for each subset of states that contains a good state. Formally, let $(\mathcal{A}, G)$ be a Büchi automaton, where $\mathcal{A} = (S, \rightarrow, S_{in})$. The corresponding Muller automaton is given by $(\mathcal{A}, \mathcal{T}_G)$ where $\mathcal{T}_G = \{F \subseteq S \mid F \cap G \neq \emptyset\}$. It is easy to see that $L(\mathcal{A}, G) = L(\mathcal{A}, \mathcal{T}_G)$—any successful run of the Büchi automaton will satisfy one of the entries in the Muller table. Conversely, any run that satisfies an entry in $\mathcal{T}_G$ must visit a good state infinitely often. Notice that the Muller automaton $(\mathcal{A}, \mathcal{T}_G)$ is deterministic iff the original automaton $(\mathcal{A}, G)$ was deterministic: this simulation neither introduces nor removes any non-determinism.

Conversely, any Muller automaton can be simulated by a *non-deterministic* Büchi automaton. Let $(\mathcal{A}, \mathcal{T})$ be a Muller automaton, where $\mathcal{T} = \langle F_1, F_2, \ldots, F_k\rangle$. For each $i \in [1..k]$, we construct a Büchi automaton $(\mathcal{A}_i, G_i)$ such that $(\mathcal{A}_i, G_i)$

accepts an input α iff there is a run ρ of $(\mathcal{A}, \mathcal{T})$ on α with $\mathsf{inf}(\rho) = F_i$. It is easy to see that $L(\mathcal{A}, \mathcal{T}) = \bigcup_{i\in[1..k]} L(\mathcal{A}_i, G_i)$. As described in Section 2.1.2, we can then construct a Büchi automaton $(\mathcal{A}_\mathcal{T}, G_\mathcal{T})$ that recognizes $L(\mathcal{A}, \mathcal{T})$.

To construct $(\mathcal{A}_i, G_i)$ we proceed as follows. When reading an input α, $\mathcal{A}_i$ simulates a run of $\mathcal{A}$. At some point, $\mathcal{A}$ non-deterministically decides that no more states from $S - F_i$ will occur along the run being simulated. After this guess is made, $\mathcal{A}_i$ will only simulate moves that stay within F_i. At the same time, $\mathcal{A}_i$ repeatedly cycles through F_i, checking that all states from F_i are seen infinitely often.

Let $\mathcal{A} = (S, \rightarrow, S_{in})$ and $F_i = \{s_{i_1}, s_{i_2}, \ldots, s_{i_m}\}$. Then $(\mathcal{A}_i, G_i)$, with $\mathcal{A}_i = (S_i, \rightarrow_i, S^i_{in})$, is defined as follows:

- $S_i = \{(s, \mathsf{finite}) \mid s \in S\} \cup \{(s, \mathsf{infinite}, j) \mid s \in F_i, j \in [0..m{-}1]\}$.
- The transition relation $\rightarrow_i$ is given as follows:

  $(s, \mathsf{finite}) \xrightarrow{a}_i (s', \mathsf{finite})$ if $s \xrightarrow{a} s'$.
  $(s, \mathsf{finite}) \xrightarrow{a}_i (s', \mathsf{infinite}, 0)$ if $s \xrightarrow{a} s'$ and $s' \in F_i$.
  $(s, \mathsf{infinite}, k) \xrightarrow{a}_i (s', \mathsf{infinite}, k)$ if $s \xrightarrow{a} s'$, $s' \in F_i$ and $s \neq s_{i_{k+1}}$.
  $(s, \mathsf{infinite}, k) \xrightarrow{a}_i (s', \mathsf{infinite}, (k{+}1) \bmod m)$ if $s \xrightarrow{a} s'$, $s' \in F_i$ and $s = s_{i_{(k+1) \bmod m}}$.

- $S^i_{in} = \{(s, \mathsf{finite}) \mid s \in S_{in}\}$.
- $G_i = \{(s_{i_m}, \mathsf{infinite}, m{-}1)\}$.

These two simulation constructions show that the class of Muller-recognizable languages coincides with the class of Büchi-recognizable languages. In other words, Muller automata also recognize ω-regular languages.

However, Example 2.8 suggests that *deterministic* Muller automata may suffice for recognizing all ω-regular languages. In fact, this is the case—this non-trivial result was proved by McNaughton [3].

Theorem 2.9. *Every ω-regular language is recognized by a* deterministic *Muller automaton.*

We shall prove McNaughton's result indirectly in Section 2.4. Notice that McNaughton's theorem, combined with the simulation constructions described above, yields a complementation construction for Büchi automata. This is because complementing deterministic Muller automata is easy. Let $(\mathcal{A}, \mathcal{T})$ be a deterministic Muller automaton, where $\mathcal{A} = (S, \rightarrow, S_{in})$. Let $\overline{\mathcal{T}} = \{F \subseteq S \mid F \notin \mathcal{T}\}$. It is straightforward to verify that $L(\mathcal{A}, \overline{\mathcal{T}}) = \Sigma^\omega - L(\mathcal{A}, \mathcal{T})$. So, to complement a Büchi automaton $(\mathcal{A}, G)$, we first convert it into an equivalent deterministic Muller automaton $(\mathcal{A}, \mathcal{T})$ using McNaughton's theorem. We then simulate $(\mathcal{A}, \overline{\mathcal{T}})$ using the construction described earlier to get a Büchi automaton $(\mathcal{A}_{\overline{\mathcal{T}}}, G_{\overline{\mathcal{T}}})$ that accepts the complement of $L(\mathcal{A}, G)$.

Rather than follow this route, we shall describe an alternative determinization construction due to Safra [8]. Safra's construction converts a Büchi automaton to a deterministic automaton with a pairs table. Acceptance in terms of a pairs table was first described by Rabin [4].

Rabin automata A *Rabin automaton* is a structure $(\mathcal{A}, \mathcal{PT})$ where $\mathcal{A} = (S, \rightarrow, S_{in})$ is an automaton, as before, and $\mathcal{PT} = \langle (G_1, R_1), (G_2, R_2), \ldots, (G_k, R_k) \rangle$ is a *pairs table* with $G_i, R_i \subseteq S$ for $i \in [1..k]$.

The automaton $(\mathcal{A}, \mathcal{PT})$ accepts an input $\alpha : \mathbb{N}_0 \rightarrow \Sigma$ if there is a run ρ of $\mathcal{A}$ on α such that for some $i \in [1..k]$, $\mathsf{inf}(\rho) \cap G_i \neq \emptyset$ and $\mathsf{inf}(\rho) \cap R_i = \emptyset$.

Thus each pair (G_i, R_i) in the pairs table of a Rabin automaton specifies a positive and a negative requirement on the run, as in the acceptance table of a Muller automaton. The positive entry G_i is just a Büchi condition while the negative entry R_i is like the one specified for $S - F_i$ by an entry F_i in a Muller acceptance table. If we think of G_i and R_i as "green lights" and "red lights", a run ρ satisfies (G_i, R_i) if some green light from G_i flashes infinitely often and no red light from R_i flashes infinitely often.

Returning to Example 2.8, the language L is accepted by the automaton of Figure 2.6 with the pairs table $\langle (\{s_1\}, \emptyset) \rangle$, while $\overline{L}$ is accepted by the same automaton with the pairs table $\langle (\{s_2\}, \{s_1\}) \rangle$.

Büchi automata can be simulated trivially by Rabin automata—if $(\mathcal{A}, G)$ is a Büchi automaton, the corresponding Rabin automaton is $(\mathcal{A}, \mathcal{PT}_G)$, where $\mathcal{PT}_G = \langle \{G, \emptyset\} \rangle$.

Conversely, we can simulate Rabin automata by Büchi automata using a construction similar to the one for simulating Muller automata by Büchi automata. As before, it suffices to construct a separate Büchi automaton $(\mathcal{A}_i, G'_i)$ for each entry (G_i, R_i) in the pairs table of a Rabin automaton $(\mathcal{A}, \mathcal{PT})$. The automaton $(\mathcal{A}_i, G'_i)$ simulates a run of $\mathcal{A}$ and guesses when no more states from R_i will be seen. It then checks that states from G_i occur infinitely often.

Let $\mathcal{A} = (S, \rightarrow, S_{in})$ and $\mathcal{PT} = \langle (G_1, R_1), (G_2, R_2), \ldots, (G_k, R_k) \rangle$. Then $(\mathcal{A}_i, G'_i)$, with $\mathcal{A}_i = (S_i, \rightarrow_i, S^i_{in})$, is defined as follows:

- $S_i = \{(s, \mathsf{finite}) \mid s \in S\} \cup \{(s, \mathsf{infinite}, j) \mid s \in (S - R_i), j \in \{0, 1\}\}$.
- The transition relation $\rightarrow_i$ is given as follows:

  $(s, \mathsf{finite}) \xrightarrow{a}_i (s', \mathsf{finite})$ if $s \xrightarrow{a} s'$.
  $(s, \mathsf{finite}) \xrightarrow{a}_i (s', \mathsf{infinite}, 0)$ if $s \xrightarrow{a} s'$ and $s' \notin R_i$.
  $(s, \mathsf{infinite}, 0) \xrightarrow{a}_i (s', \mathsf{infinite}, 0)$ if $s \xrightarrow{a} s'$, $s' \notin R_i$ and $s \notin G_i$.
  $(s, \mathsf{infinite}, 0) \xrightarrow{a}_i (s', \mathsf{infinite}, 1)$ if $s \xrightarrow{a} s'$, $s' \notin R_i$ and $s \in G_i$.
  $(s, \mathsf{infinite}, 1) \xrightarrow{a}_i (s', \mathsf{infinite}, 0)$ if $s \xrightarrow{a} s'$ and $s' \notin R_i$.

- $S^i_{in} = \{(s, \mathsf{finite}) \mid s \in S_{in}\}$.
- $G'_i = \{(s, \mathsf{infinite}, 1) \mid s \in (S - R_i)\}$.

Notice that a Rabin automaton can also simulated by a Muller automaton in quite a straightforward manner. Let $(\mathcal{A}, \mathcal{PT})$ be a Rabin automaton, where $\mathcal{A} = (S, \rightarrow, S_{in})$

and $\mathcal{PT} = \langle (G_1, R_1), (G_2, R_2), \ldots, (G_k, R_k) \rangle$. Each pair (G_i, R_i) generates a Muller table $\mathcal{T}_i = \{F \subseteq (S - R_i) \mid F \cap G_i \neq \emptyset\}$. Let $\mathcal{T} = \bigcup_{i \in [1..k]} \mathcal{T}_i$. It is easy to see that $(\mathcal{A}, \mathcal{T})$ recognizes $L(\mathcal{A}, \mathcal{PT})$. Once again, since we have not modified $\mathcal{A}$, the simulating automaton is deterministic iff the original automaton was.

To simulate Muller automata using Rabin automata one has to use a construction that is pretty much the same as the one for simulating Muller automata by Büchi automata. Such a simulation introduces non-determinism: there is no straightforward way to directly simulate a deterministic Muller automaton by a deterministic Rabin automaton even though deterministic Rabin automata *do* recognize all ω-regular languages, as we shall see in the next section.

The last acceptance condition we look at is obtained by interpreting the pairs table of a Rabin automaton in a complementary fashion.

Streett automata A *Streett automaton* is a structure $(\mathcal{A}, \mathcal{PT})$ where $\mathcal{A} = (S, \rightarrow, S_{in})$ and $\mathcal{PT} = \langle (G_1, R_1), (G_2, R_2), \ldots, (G_k, R_k) \rangle$ are defined in the same way as for Rabin automata.

The Streett automaton $(\mathcal{A}, \mathcal{PT})$ accepts an input $\alpha : \mathbb{N}_0 \rightarrow \Sigma$ if there is a run ρ of $\mathcal{A}$ on α such that for every $i \in [1..k]$, if $\mathsf{inf}(\rho) \cap G_i \neq \emptyset$ it is also the case that $\mathsf{inf}(\rho) \cap R_i \neq \emptyset$.

These automata were defined by Streett in [14]. They are useful for describing *fairness conditions* in infinite computations—for instance, conditions of the form "if a request for a resource is made infinitely often, then the system grants access to the resource infinitely often". The following observation is immediate from the close connection between Rabin and Streett automata.

Proposition 2.10. *Let $(\mathcal{A}, \mathcal{PT})$ be a deterministic automaton with a pairs table. Let L_R be the language accepted by $(\mathcal{A}, \mathcal{PT})$ when $\mathcal{PT}$ is interpreted as a Rabin condition and L_S be the language accepted by $(\mathcal{A}, \mathcal{PT})$ when $\mathcal{PT}$ is interpreted as a Streett condition. Then L_S is the complement of L_R.*

As usual, simulating a Büchi automaton $(\mathcal{A}, G)$ by a Streett automaton is easy. Let $\mathcal{A} = (S, \rightarrow, S_{in})$. Construct an automaton $(\mathcal{A}, \mathcal{PT}_G)$ where $\mathcal{PT}_G = \langle (S, G) \rangle$. Since $\mathsf{inf}(\rho) \cap S$ must be non-empty for any run ρ of $\mathcal{A}$, it follows that a run ρ satisfies the pair (S, G) iff $\mathsf{inf}(\rho) \cap G \neq \emptyset$, which is precisely what the Büchi condition demands.

In the converse direction, Safra describes a construction due to Vardi that shows that Streett automata can be efficiently simulated by Büchi automata [8].

Lemma 2.11. *Let $(\mathcal{A}, \mathcal{PT})$ be a Street automaton where $\mathcal{A} = (S, \rightarrow, S_{in})$. Let $n = |S|$ and let k be the number of pairs in $\mathcal{PT}$: that is, $\mathcal{PT} = \langle (G_1, R_1), (G_2, R_2), \ldots, (G_k, R_k) \rangle$. Then, we can construct a Büchi automaton $(\mathcal{A}', G')$ with $\mathcal{A}' = (S', \rightarrow', S'_{in})$ such that $L(\mathcal{A}', G') = L(\mathcal{A}, \mathcal{PT})$ and $|S'| = n \cdot 2^{O(k)}$.*

Proof. The automaton $\mathcal{A}'$ simulates $\mathcal{A}$. As usual, $\mathcal{A}'$ guesses an initial prefix of the run after which every state that is visited by the run will in fact be visited infinitely often. After making this guess, $\mathcal{A}$ checks that the acceptance criterion is met for each pair $(G_i, R_i) \in \mathcal{PT}$. In other words, for every i such that some state from G_i appears in the infinite portion of the run, $\mathcal{A}'$ ensures that some state from R_i also appears infinitely often. To do this, $\mathcal{A}'$ maintains two sets as part of its state. The first set accumulates the list of indices corresponding to pairs (G_i, R_i) where some element of G_i occurs infinitely often. The second set repeatedly accumulates indices of pairs (G_i, R_i) for which some element of R_i has been visited. Each time the second set becomes as large as the first, it is reset to empty. It is not difficult to see that the acceptance criterion specified by $\mathcal{PT}$ is met iff the second set is reset to empty infinitely often during the simulation.

Formally, we construct $(\mathcal{A}', G')$ as follows:

- $S' = \{(s, \mathsf{finite}) \mid s \in S\} \cup \{(s, X_1, X_2) \mid s \in S \text{ and } X_1, X_2 \subseteq [1..k]\}$.
- The transition relation $\rightarrow'$ is defined as follows:

$$(s, \mathsf{finite}) \xrightarrow{a}' (s', \mathsf{finite}) \text{ if } s \xrightarrow{a} s'.$$
$$(s, \mathsf{finite}) \xrightarrow{a}' (s', \emptyset, \emptyset) \text{ if } s \xrightarrow{a} s'.$$
$$(s, X, Y) \xrightarrow{a}' (s', X \cup G_{s'}, Y \cup R_{s'}) \text{ if } s \xrightarrow{a} s' \text{ and } X \cup G_{s'} \not\subseteq Y \cup R_{s'},$$
$$\text{where } G_{s'} = \{i \in [1..k] \mid s' \in G_i\} \text{ and}$$
$$R_{s'} = \{i \in [1..k] \mid s' \in R_i\}.$$
$$(s, X, Y) \xrightarrow{a}' (s', X \cup G_{s'}, \emptyset) \text{ if } s \xrightarrow{a} s' \text{ and } X \cup G_{s'} \subseteq Y \cup R_{s'}.$$

- $S^i_{in} = \{(s, \mathsf{finite}) \mid s \in S_{in}\}$.
- $G_i = \{(s, X, \emptyset) \mid s \in S, X \subseteq [1..k]\}$.

□

2.4. Determinizing Büchi automata

We now describe an elegant construction due to Safra for determinizing Büchi automata [8]. Safra's construction converts a non-deterministic Büchi automaton $(\mathcal{A}, G)$ into a deterministic Rabin automaton $(\mathcal{A}_G, \mathcal{PT}_G)$ such that $L(\mathcal{A}_G, \mathcal{PT}_G) = L(\mathcal{A}, G)$. If we regard $(\mathcal{A}_G, \mathcal{PT}_G)$ as a Streett automaton, we get a deterministic automaton recognizing the complement of $L(\mathcal{A}, G)$. By Lemma 2.11, we can simulate the Streett automaton $(\mathcal{A}_G, \mathcal{PT}_G)$ by a Büchi automaton. Thus Safra's construction also solves the complementation problem for Büchi automata.

Also, recall that it is easy to convert a deterministic Rabin automaton into a deterministic Muller automaton. As a consequence, Safra's construction gives an indirect proof of McNaughton's Theorem (Theorem 2.9).

Safra's determinization construction for Büchi automata is a clever extension to the infinite word case of the classical subset construction for determinizing automata on finite words. In order to motivate the construction, we begin with the subset construction for finite words and enhance it in a graded manner to achieve the final result.

Subset construction For automata on finite words, the subset construction is the standard way to eliminate non-determinism. If the original automaton is $(\mathcal{A}, F)$, with $\mathcal{A} = (S, \rightarrow, S_{in})$, each state of the *subset automaton* $(S_{sub}, \rightarrow_{sub}, S_{in}^{sub})$ is a subset of S. The (single) initial state of the subset automaton is the set S_{in} of initial states of $\mathcal{A}$. The subset automaton's transition relation $\rightarrow_{sub}$ is defined as follows:

$$X \xrightarrow{a}_{sub} Y \text{ iff } Y = \{y \in S \mid \exists x \in X : x \xrightarrow{a} y\}.$$

Henceforth, we use $\delta_{sub}(X, a)$ to denote the set Y such that $X \xrightarrow{a}_{sub} Y$.

The subset automaton satisfies the following property: If $X \xrightarrow{w}^{+}_{sub} Y$ then for each state y in Y, there is an state $x \in X$ such that $x \xrightarrow{w}^{+} y$ in the original automaton.

From this, it follows that if we set the set of final states of the subset automaton to be $F_{sub} = \{X \subseteq S \mid S \cap F \neq \emptyset\}$, then $(\mathcal{A}_{sub}, F_{sub})$ recognizes the same set of words as the original automaton.

Let $(\mathcal{A}, G)$ be a non-deterministic Büchi automaton, with $\mathcal{A} = (S, \rightarrow, S_{in})$. The natural extension of the subset construction to Büchi automata would set the good states of the subset automaton to $G_{sub} = \{X \subseteq S \mid X \cap G \neq \emptyset\}$.

It is easy to see that if $(\mathcal{A}, G)$ accepts an input α, so will $(\mathcal{A}_{sub}, G_{sub})$. Unfortunately, the converse is not true—the subset automaton will accept words that are not part of the original language.

Example 2.12. Consider the automaton of Example 2.1 (Figure 2.3) over $\{a, b\}$ that recognizes $\overline{L} = \{\alpha \mid \alpha$ has only a finite number of occurrences of $a\}$.

In this example, G_{sub}, the set of good states of the extended subset automaton, is given by $\{\{s_1, s_2\}, \{s_2\}\}$. On the input $(ab)^\omega = ababab\cdots$, the (unique) run of the subset automaton is $\{s_1\}(\{s_1\}\{s_1, s_2\})^\omega = \{s_1\}\{s_1\}\{s_1, s_2\}\{s_1\}\{s_1, s_2\}\{s_1\}\{s_1, s_2\}\cdots$. Since this run visits G_{sub} infinitely often, the automaton $(\mathcal{A}_{sub}, G_{sub})$ accepts the word $(ab)^\omega$, even though a occurs infinitely often in this word.

The problem with the subset construction is best brought out by drawing all the "threads" between the individual subsets in this run of the subset automaton—see Figure 2.7.

As we can see, every finite run of the original automaton that reaches a good state actually dies out at that point. In general, all that this subset construction guarantees is that the original automaton has arbitrarily long *finite* runs that visit good states.

Marked subset construction We next try to strengthen the subset construction so that it explicitly keeps track of the threads between subsets. In the marked

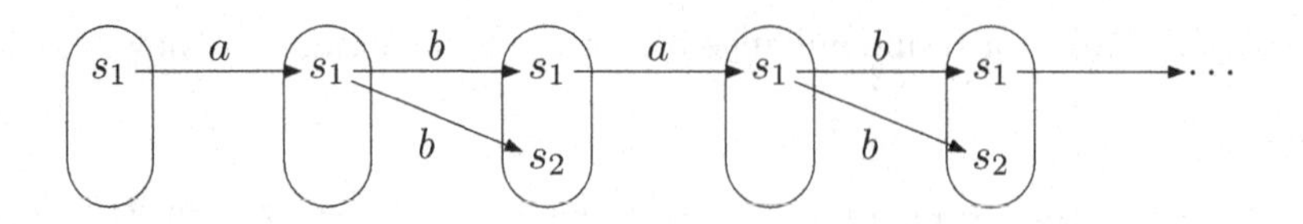

Fig. 2.7. A run of the extended subset automaton for $\overline{L}$ on input $(ab)^\omega$ (Example 2.12).

subset construction, in addition to keeping a subset of states, the subset automaton also has the ability to "mark" each state in the subset. A state in the current subset is marked if it satisfies one of two conditions: either it is a good state, or it has a marked predecessor in the previous subset. However, if all the states in the previous subset are marked, then only good states are marked in the current subset—no marks are inherited from a fully marked state. The good states in the marked subset automaton are those where the entire subset is marked.

Concretely, let $(\mathcal{A}, G)$ be the original non-deterministic Büchi automaton with $\mathcal{A} = (S, \rightarrow, S_{in})$. Then, the marked subset automaton $(\mathcal{A}_M, G_M)$, with $\mathcal{A}_M = (S_M, \rightarrow_M, S_{in}^M)$, is given as follows.

- $S_M = \{(X, f) \mid X \subseteq S, f : X \rightarrow \{\mathsf{marked}, \mathsf{unmarked}\}\}$.
- The transition function $\rightarrow_M$ is as follows:
  $(X, f) \xrightarrow{a}_M (Y, g)$ iff
  - $Y = \delta_{sub}(X, a)$.
    (Recall that $Y = \delta_{sub}(X, a)$ iff in the normal subset automaton, $X \xrightarrow{a}_{sub} Y$.)
  - If $f(x) = \mathsf{marked}$ for all $x \in X$
    then
    $$\forall y \in Y : g(y) = \begin{cases} \mathsf{marked} & \text{if } y \in G \\ \mathsf{unmarked} & \text{otherwise} \end{cases}$$
    else
    $$\forall y \in Y : g(y) = \begin{cases} \mathsf{marked} & \text{if } y \in G \text{ or} \\ & (\exists x \in X : f(x) = \mathsf{marked} \text{ and } x \xrightarrow{a} y) \\ \mathsf{unmarked} & \text{otherwise} \end{cases}$$
- $S_{in}^M = \{(S_{in}, f) \mid \forall s \in S_{in} : f(s) = \mathsf{marked}\}$.
- $G_M = \{(X, f) \mid \forall x \in X : f(x) = \mathsf{marked}\}$.

The main property satisfied by this automaton is the following.

> Let ρ be a run of $(\mathcal{A}_M, G_M)$ on an input α such that $\rho(i) = (X_i, f_i)$ for all $i \in Nat$. Suppose that $(X_k, f_k) \in G$ for some $k > 0$. Let j be the largest natural number less than k such that $(X_j, f_j) \in G$—such a number j must exist because the initial state of $\mathcal{A}_M$ belongs to G.

Then, for each state $y \in Y_k$, there is a state $x \in X_j$ such that $x \xrightarrow{\alpha[j..k-1]}{}^{+} y$ in the original automaton and, moreover, when going from x to y on reading $\alpha[j..k-1]$, the original automaton goes through some good state.

From this observation, we can deduce that the marked subset construction is sound.

Proposition 2.13. *Let $(\mathcal{A}_M, G_M)$ be the marked subset automaton that corresponds to the Büchi automaton $(\mathcal{A}, G)$. Then, if $(\mathcal{A}_M, G_M)$ accepts an input α, so does $(\mathcal{A}, G)$.*

Proof. Let ρ_α be the (unique) run of $\mathcal{A}_M$ on an input α with $\rho_\alpha(i) = (X_i, f_i)$ for $i \in \mathbb{N}_0$. If $(\mathcal{A}_M, G_M)$ accepts α, there must be an infinite sequence of positions $\{i_0, i_1, \ldots\} \subseteq \mathbb{N}_0$ such that $0 = i_0 < i_1 < \cdots$ and $(X_j, f_j) \in G$ for all $j \in \{i_0, i_1, \ldots\}$.

From our previous observation about the marked subset construction, we know that for each index i_{k+1} in the set $\{i_0, i_1, \ldots\}$ and for each state $x \in X_{i_{k+1}}$, there is a state $y \in X_{i_k}$ such that in the original automaton, there is a sequence of transitions leading from y to x on the input $\alpha[i_k..i_{k+1}-1]$ that passes through some good state. Let us call such a state y a *good predecessor* of x.

We construct an infinite tree T_α as follows. The root of the tree is the set S_{in} of initial states. At each level k of the tree, $k \geq 1$, we have a node $n_{(x,i_k)}$ corresponding to each state $x \in X_{i_k}$. The parent of a node $n_{(x,j)}$ at level j, $j > 1$, is a node $n_{(y,j-1)}$ at level $j-1$ such y is a good predecessor of x. (Of course, x may have more than one good predecessor. If this is the case, we arbitrarily select one of them and make the corresponding node the parent of $n_{(x,j)}$ in the tree.)

The tree T_α is finitely branching and has an infinite number of nodes. By König's lemma, it must have an infinite path. Each infinite path in T_α corresponds to a run of the original automaton $\mathcal{A}$ on α. By construction, such a run must pass through a good state between each level in the tree. Thus, $\mathcal{A}$ has at least one run on α that meets G infinitely often. □

Unfortunately, though the marked subset construction is sound, it is not complete—there may be inputs accepted by $(\mathcal{A}, G)$ that are not accepted by $(\mathcal{A}_M, G_M)$. Consider the following example.

Example 2.14.

In the Büchi automaton shown in Figure 2.8, the input a^ω generates the sequence of subsets $\{s_0\}(\{s_1, s_2\})^\omega$. Since s_2 is not a final state, the subset $\{s_1, s_2\}$ never becomes fully marked. Thus, though the original automaton has an accepting run $s_0 s_1^\omega$ on this input, the marked subset construction fails to detect this.

The problem is that the marked subset construction demands too much from the underlying runs. As the example shows, it should be sufficient to identify a portion of the subset that is marked and can infinitely often regenerate its marks.

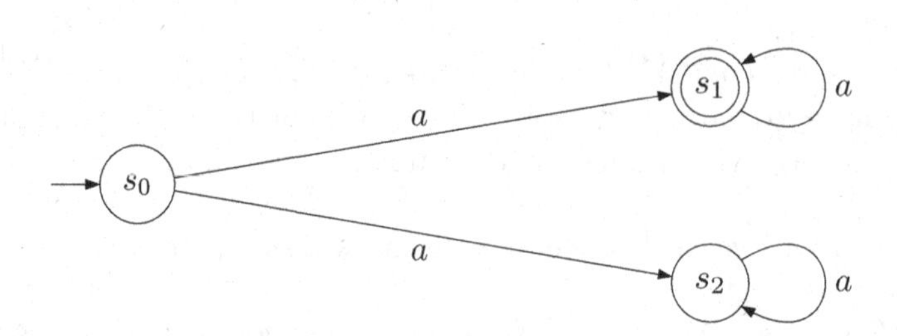

Fig. 2.8. The marked subset construction is not complete (Example 2.14).

Hierarchical Marked Subset Construction A first attempt to weaken the marked subset construction would be to have a hierarchy of marks. At the base level, the subset automaton runs the marked subset construction and marks states using a level 1 mark. The states that have level 1 marks then start off a nested copy of the marked subset construction with level 2 marks. Similarly, the states that have level 2 marks start off a marked subset construction with level 3 marks. What we would like to detect is whether some level i can get completely marked. This corresponds to checking if the set of nodes marked at level i is equal to the set of nodes marked at level $i+1$. If so, we reset all marks at levels greater than i and continue.

Since the number of nodes marked at level i is always strictly greater than the set of nodes marked at level $i+1$, there can be at most as many levels as there are states in the original automaton.

To specify the acceptance condition, we need to verify that some level $i+1$ gets set to empty infinitely often *and* that level i does not get set to empty in between. In other words, level i denotes a permanent thread through the subset construction that gets marked infinitely often. To do this, we have to pass from a Büchi condition to a Rabin condition—for each i, a positive condition for level $i+1$ has to be qualified by a negative condition for level i. (Note that this transition to a stronger acceptance condition was inevitable, since we have already seen that deterministic Büchi automata cannot recognize all ω-regular languages.)

Here is a formal description of a hierarchical marked subset construction that attempts to achieve this goal. Let $(\mathcal{A}, G)$ be a Büchi automaton, with $\mathcal{A} = (S, \rightarrow, S_{in})$. Define $(\mathcal{A}_H, \mathcal{PT}_H)$, with $\mathcal{A}_H = (S_H, \rightarrow_H, S_{in}^H)$, as follows:

- Let $|S| = n$. S_H consists of pairs of the form (σ, χ) where:
  - $\sigma : [1..n] \rightarrow 2^S$ is a *subset list* satisfying the condition that $\sigma(i+1)$ is a *proper* subset of $\sigma(i)$ whenever $\sigma(i)$ is non-empty.
  - $\chi : [1..n] \rightarrow \{\mathsf{white}, \mathsf{green}\}$ is a *colour list.*
- $S_{in}^H = (\sigma_0, \chi_0)$, where $\sigma_0(1) = S_{in}$, $\sigma_0(i) = \emptyset$ for all $i \in [2..n]$ and $\chi(i) = \mathsf{white}$ for all $i \in [1..n]$.

- The transition function $\rightarrow_H$ performs the following sequence of actions. Initially, each level runs the subset construction locally. Next, any final states appearing in the new subset at level i are added to the subset at level $i+1$—this corresponds to generating fresh marks at level i. We now look for the smallest level i whose subset is the same as that at level $i+1$. If such an i exists, we "clear out" the subset list from level $i+1$ onwards and set the colour of level i to green.
  More formally, on reading an input a, the state (σ, χ) generates a new state (σ', χ') as follows:

  (i) Let $\sigma_1 : [1..n+1] \rightarrow 2^S$ be defined as follows:

  - $\sigma_1(1) = \delta_{sub}(\sigma(1), a)$.
  - For $i \in [2..n]$, $\sigma_1(i) = \delta_{sub}(\sigma(i), a) \cup (\delta_{sub}(\sigma(i-1), a) \cap G)$.
  - $\sigma_1(n+1) = \delta_{sub}(\sigma(n), a) \cap G$.

  (ii) If there is no index $i \in [1..n]$ such that $\sigma_1(i) = \sigma_1(i+1)$, then

  - $\sigma'(i) = \sigma_1(i)$ for all $i \in [1..n]$.
  - $\chi'(i) = \mathsf{white}$ for all $i \in [1..n]$.

  else, let m be the smallest index such that $\sigma_1(m) = \sigma_1(m+1)$. Then,

  - $\sigma'(i) = \sigma_1(i)$ for all $i \in [1..m]$ and $\sigma'(i) = \emptyset$ for all $i > m$.
  - $\chi'(m) = \mathsf{green}$ and $\chi'(i) = \mathsf{white}$ for all $i \neq m$.

- The acceptance table $\mathcal{PT}_H$ consists of n pairs $\langle (G_1, R_1), (G_2, R_2), \ldots, (G_n, R_n) \rangle$ where:

  - $R_i = \{(\sigma, \chi) \mid \sigma(i) = \emptyset\}$
  - $G_i = \{(\sigma, \chi) \mid \chi(i) = \mathsf{green}\}$

In this construction, the list σ implicitly records the levels of marks associated with the states in the current subset—a state s belongs to $\sigma(i+1)$ iff s has a level i mark in the current subset. It is not difficult to show that this construction is complete.

Proposition 2.15. *Let $(\mathcal{A}_H, \mathcal{PT}_H)$ be the hierarchical marked subset automaton that corresponds to the Büchi automaton $(\mathcal{A}, G)$. If $(\mathcal{A}, G)$ accepts an input α, so does $(\mathcal{A}_H, \mathcal{PT}_H)$.*

Proof. Suppose $(\mathcal{A}, G)$ accepts α. Then, α admits a run ρ that visits G infinitely often. We must show that the unique run ρ_α of $\mathcal{A}_H$ on α satisfies some entry in $\mathcal{PT}_H$.

For $j \in \mathbb{N}_0$, let $\rho_\alpha(j) = (\sigma_j, \chi_j)$. We know that for all $j \in \mathbb{N}_0$, $\sigma_j(1)$ is the set of states maintained by the subset construction. Since ρ is a valid run of $\mathcal{A}$ on α, $\sigma_j(1)$ is always non-empty. So, ρ_α satisfies condition R_1. If it also satisfies G_1 then $(\mathcal{A}_H, \mathcal{PT}_H)$ accepts α and we are done.

If ρ_α does not satisfy G_1, let k_0 be the last position where the colour of the first level is green. We wait for the first position $i_0 > k_0$ where ρ, the accepting run of $\mathcal{A}$ on α, next visits a good state. We know that $\sigma_j(2)$ is non-empty for all $j \geq i_0$—once the good state seen at position i_0 gets pushed to level 2, the accepting run ρ will be part of the subset construction maintained at level 2, thus guaranteeing that at least one valid state is generated at each point. So, ρ_α satisfies R_2. If it also satisfies G_2 we are done.

Otherwise, we repeat the argument above and deduce that ρ_α satisfies R_3, with the accepting run ρ a part of $\sigma_j(3)$ for all j greater than a finite index i_1. We can repeat this argument only a finite number of times, till we reach level n. The subset maintained at level n can never be more than a singleton. If the accepting run ρ is part of the subset construction at level n, it must generate the signal green infinitely often in which case ρ_α satisfies the pair (G_n, R_n). □

Unfortunately, the hierarchical subset construction is *not* sound. Consider this example.

Example 2.16.

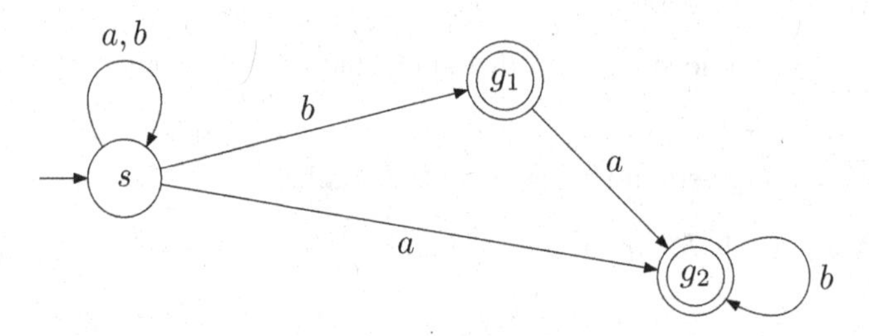

Fig. 2.9. The hierarchical marked subset construction is not sound (Example 2.16).

The automaton shown in Figure 2.9 does not accept the input $(ba)^\omega$. However, the run of the hierarchical marked subset automaton on this input is the following:

1	$(\{s\}, \text{white})$
2	$(\emptyset, \text{white})$
3	$(\emptyset, \text{white})$

$\xrightarrow{b}_H$

1	$(\{s, g_1\}, \text{white})$
2	$(\{g_1\}, \text{white})$
3	$(\emptyset, \text{white})$

$\xrightarrow{a}_H$

1	$(\{s, g_2\}, \text{white})$
2	$(\{g_2\}, \text{green})$
3	$(\emptyset, \text{white})$

$\xrightarrow{b}_H$

1	$(\{s, g_1, g_2\}, \text{white})$
2	$(\{g_1, g_2\}, \text{white})$
3	$(\{g_2\}, \text{white})$

$\xrightarrow{a}_H$

1	$(\{s, g_2\}, \text{white})$
2	$(\{g_2\}, \text{green})$
3	$(\emptyset, \text{white})$

$\xrightarrow{b}_H \cdots$

Since level 2 remains populated forever and turns green infinitely often, the hierarchical construction incorrectly accepts this input.

In the preceding example, the problem is that the good state g_2 that appears to be permanently part of level 2 is actually a transient state. Each time an a is read, the g_2 state at level 2 disappears, only to be replaced by a fresh copy of g_2 that is pushed from level 1.

To rectify this defect, we distinguish new copies and old copies of a state at each level by attaching a label to each fresh subset of states that is generated. If an older copy of a state continues to exist, we remove the new copy. The labels partition the states at each level into disjoint subsets. A label that persists corresponds to an infinite run whereas when a run dies out, as in the example above, its label disappears.

Safra's construction implements such a labelling scheme. The hierarchy of subsets is represented as a tree. Each node in the tree is a collection of states with the same label, corresponding to a set of runs that were initiated at the same time. The root of the tree contains the subset at the first level of the hiearchical construction. The parent-child relation in the tree accurately records how subsets at each level arise from subsets at the previous level. Only the oldest copy of each active state is retained, bounding the size of the tree. This allows a fixed set of labels to be recycled, making the overall construction finite-state.

Safra's Construction Before presenting Safra's construction, we review some terminology regarding trees. A tree is a structure $T = (V, v_r, \pi)$ where V is a set of nodes, $v_r \in T$ is a special node known as the *root* and for all $v \in V - \{v_r\}$, $\pi(v) \in T$ fixes the *parent* of the node. If $v = \pi^i(v')$ for some $i > 0$, we say that v is an *ancestor* of v'. The root v_r is an ancestor of every other node. If $v' = \pi(v)$ then v is said to be a *child* of v'. We assume that for any node v, all the children of v are ordered so that we can talk of one child being to the *left* of another. This generates a total order on nodes—if v and v' are nodes, we say that $v < v'$ if v is an ancestor of v' or if there is a common ancestor u of v and v' such that v is in the subtree rooted at a child u_1 of v, v' is in the subtree rooted at a child u_2 of v and u_1 is to the left of u_2.

Given a Büchi automaton $(\mathcal{A}, G)$, with $\mathcal{A} = (S, \rightarrow, S_{in})$, Safra's construction produces a Rabin automaton $(\mathcal{A}_G, \mathcal{PT}_G)$, with $\mathcal{A}_G = (S_G, \rightarrow_G, S^G_{in})$. The automaton $(\mathcal{A}_G, \mathcal{PT}_G)$ is as follows:

- Each state in S_G is a structure $(T, \sigma, \chi, \lambda)$ where
  - $T = (V, v_r, \pi)$ is a tree.
  - $\sigma : V \rightarrow 2^S$ associates a set of states of $\mathcal{A}$ with each node in V in such a way that:
    - The union of the sets associated with the children of a node v is a *proper* subset of $\sigma(v)$.
    - If v and v' are two nodes such that v is not an ancestor of v' and v' is not an ancestor of v then $\sigma(v)$ is disjoint from $\sigma(v')$.

  * If $\sigma(v) = \emptyset$, then v is the root v_r.

  It is not difficult to verify that the conditions imposed on the function σ ensure that $|V|$ can be no larger than n, where n is the number of states in S.

- $\chi : V \rightarrow \{\mathsf{white}, \mathsf{green}\}$ fixes a colour for each node.
- $\lambda : V \rightarrow \mathcal{L}$ is an injective function that attaches a *label* from the set $\mathcal{L} = \{\ell_1, \ell_2, \ldots, \ell_{2n}\}$ to each node. Notice that $\mathcal{L}$ has $2n$ elements.

As we mentioned earlier, each layer of the tree corresponds to one level of the hierarchical marked subset construction, partitioned into disjoint subsets. The tree structure records how the partitions at each level are connected to the partitions at the previous level.

- On reading an input a, the state $(T, \sigma, \chi, \lambda)$ is transformed to the state $(T', \sigma', \chi', \lambda')$ as follows:

(i) Let $T = (V, v_r, \pi)$. Expand the T to a tree $T_1 = (V_1, v_r, \pi_1)$ as follows: For each $v \in V$, if $\sigma(v) \cap G \neq \emptyset$, add a node v' such that $\pi_1(v') = v$ and v' is the right-most child of v.

(ii) Extend σ and λ to functions σ_1 and λ_1 over T_1 as follows:
For all nodes v in $V_1 \cap V$, let $\sigma_1(v) = \sigma(v)$. For a new node $v \in V_1 - V$, $\sigma_1(v) = \sigma(\pi_1(v)) \cap G$.
All nodes v in $V_1 \cap V$ inherit the label $\lambda(v)$. For each node in $V_1 - V$, choose a new label from $\mathcal{L}$ that is not assigned to any other node. Since there $2n$ labels to choose from, this is always possible—each node in V generates at most one new child in V_1 and there were not more than n nodes in V.

(iii) For every node v, apply the subset construction locally. In other words, define a new function $\sigma'_1 : V_1 \rightarrow 2^S$ such that $\sigma'_1(v) = \delta_{sub}(\sigma_1(v), a)$ for all $v \in V_1$.

At this stage, we have to "clean up" T_1 and σ'_1 so that the structure once again satisfies the conditions specified for states of $\mathcal{A}_G$.

(iv) For every node $v \in V_1$, if $s \in \sigma'_1(v)$ and s also belongs to $\sigma'_1(v')$ for some node v' such that v' is not an ancestor of v but $v' < v$, (recall the total order on all nodes in a tree) remove s from $\sigma'_1(v)$. This corresponds to retaining only the "oldest" copy of each active state in the simulation.

(v) Remove all nodes v such that $\sigma'_1(v) = \emptyset$ and v is not the root v_r.

(vi) For each node v such that $\sigma'_1(v)$ is equal to $\bigcup\{\sigma'_1(v') \mid v = \pi_1(v')\}$, remove all the children of v and set $\chi_1(v) = \mathsf{green}$. For all other nodes, set $\chi_1(v) = \mathsf{white}$.

(vii) Let the set of nodes remaining be V'. For $v \in V'$, $\sigma'(v)$ is that part of $\sigma'_1(v)$ that remains after discarding states which already appear to the left, as specified in step (iv) above. The label $\lambda'(v)$ of a node v is retained from T_1. Finally, set $\chi' = \chi_1$.

- The initial state of $\mathcal{A}_G$ is the tree $(\{v_r\}, v_r, \emptyset)$ where $\sigma(v_r) = S_{in}$, $\chi(v_r) = \mathsf{white}$ and $\lambda(v_r) = \ell_1$.
- The pairs table $\mathcal{PT}_G = \langle (G_1, R_1), (G_2, R_2), \ldots, (G_{2n}, R_{2n}) \rangle$ is defined as follows:

  - $R_i = \{(T = (V, v_r, \pi), \sigma, \chi, \lambda) \mid \forall v \in V : \lambda(v) \neq \ell_i\}$.
  - $G_i = \{(T = (V, v_r, \pi), \sigma, \chi, \lambda) \mid \exists v \in V : \lambda(v) = \ell_i \text{ and } \chi(v) = \mathsf{green}\}$.

The labelling procedure guarantees that the labels of new nodes added at each stage are disjoint from the labels of the existing nodes. In other words, if a node labelled ℓ_i is deleted from the tree during a transition, the label ℓ_i temporarily disappears from the tree.

Thus, an entry (G_i, R_i) in the pairs table specifies the following condition. The condition R_i is satisfied if at some stage a node labelled ℓ_i is added to the tree and it is never deleted henceforth. The condition G_i then says that this node turns green infinitely often.

Figure 2.10 describes the run generated by Safra's construction on the input $(ba)^\omega$ for the automaton shown in Figure 2.9. In the figure, each node of a tree is denoted by a circle, with the label indicated inside the circle and the associated subset written by its side. Nodes coloured green are drawn as double circles, while white nodes are drawn as single circles. When selecting labels for new nodes added to the tree at each stage, we have followed the policy of using the first "free" label in $\mathcal{L}$.

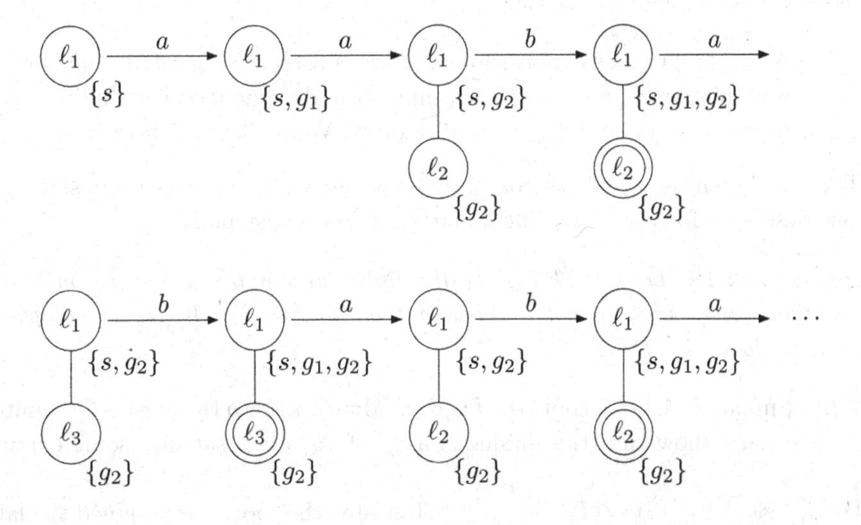

Fig. 2.10. A run generated by Safra's construction.

At the second level of the tree, nodes labelled ℓ_2 and ℓ_3 turn green infinitely often, so the run satisfies G_2 and G_3. However, since both these labels also disappear from the tree infinitely often, the run does not satisfy R_2 or R_3, thus ensuring that the automaton rejects this input.

It is not difficult to see that Safra's construction satisfies a property similar to the one described for the marked subset construction:

Let ρ be a run of $(\mathcal{A}_G, \mathcal{PT}_G)$ on an input α such that $\rho(i) = (T_i, \sigma_i, \chi_i, \lambda_i)$ for all $i \in \mathbb{N}_0$. Let $j, k \in \mathbb{N}_0$ with $j < k$ and ℓ be a label from $\mathcal{L}$ such that:

- For all positions $i \in [j..k]$, there is a node v in the tree T_i such that $\lambda_i(v) = \ell$.
- $\chi_j(v) = \chi_k(v) =$ green and for all i such that $j < i < k$, $\chi_i(v) =$ white.

Then, for each state $y \in \sigma_k(v)$, there is a state $x \in \sigma_j(v)$ such that $x \xrightarrow{\alpha[j..k-1]}{}^{+} y$ in the original automaton and, moreover, when going from x to y on reading $\alpha[j..k-1]$, the original automaton goes through some good state.

Once we have this property, the soundness of Safra's construction follows from an argument very similar to the one described for the marked subset construction in Proposition 2.13. In other words, we can show the following.

Proposition 2.17. *Let $(\mathcal{A}_G, \mathcal{PT}_G)$ be the Rabin automaton generated by Safra's construction, corresponding to the Büchi automaton $(\mathcal{A}, G)$. Then, if $(\mathcal{A}_G, \mathcal{PT}_G)$ accepts an input α, so does $(\mathcal{A}, G)$.*

Proof. As in the proof of Proposition 2.13, we construct a finitely branching tree T_α with an infinite set of nodes for each input α and argue that each infinite path in T_α corresponds to an accepting run of $\mathcal{A}$ on α. We omit the details. □

The completeness of Safra's construction is shown by an argument similar to the one described for the hierarchical marked subset construction.

Proposition 2.18. *Let $(\mathcal{A}_G, \mathcal{PT}_G)$ be the Rabin automaton generated by Safra's construction corresponding to the Büchi automaton $(\mathcal{A}, G)$. If $(\mathcal{A}, G)$ accepts an input α, so does $(\mathcal{A}_G, \mathcal{PT}_G)$.*

Proof. Suppose $(\mathcal{A}, G)$ accepts α. Then, α admits a run ρ that visits G infinitely often. We must show that the unique run ρ_α of $\mathcal{A}_G$ on α satisfies some entry in $\mathcal{PT}_G$.

For $j \in \mathbb{N}_0$, let $\rho_\alpha(j) = (T_j, \sigma_j, \chi_j, \lambda_j)$. Initially, the root v_r is assigned the label ℓ_1. Since the root is never removed, the run ρ_α satisfies R_1. If it also satisfies G_1 then $(\mathcal{A}_G, \mathcal{PT}_G)$ accepts α and we are done.

If ρ_α does not satisfy G_1, let k_1 be the last position at which the root is coloured green. Let i_1 be the first position after k where ρ, the accepting run of $\mathcal{A}$ on α, visits G. At this point, a child v_1 of the root comes into existence.

We know that the accepting run ρ is part of the overall subset construction maintained by the root node. Once v_1 is created, we know that ρ is also being maintained at the first level. It would appear that the run is maintained by v_1 itself, but there is a subtle complication to be taken into account. Since we only retain the left-most copy of each state, the run ρ may be passed on by v_1 to some sibling on the left. In any case, it can only move left a finite number of times. Let us suppose it eventually settles down at some node v_1'.

It is not difficult to verify that the node v_1' must already have been in the tree when v_1 was added. Let $\ell_{i_1} = \lambda_{i_1}(v_1')$ be the label of v_1'. Since α never dies out, v_1' will never be deleted from the tree. In other words, ρ_α satisfies the condition R_{i_1}. If it satisfies the corresponding condition G_{i_1} we are done.

Otherwise, let k_2 be the last time where v_1' turns green. As before, we wait for i_2, the next time ρ visits a good state, and look at the child v_2 of v_1' that is created at this point. The run ρ is copied into the subset maintained by v_2 and passed on left a finite number of times till it settles down at a node v_2'. If ℓ_{i_2} is the label of v_2', ρ_α must satisfy R_{i_2}. If ρ_α does not also satisfy G_{i_2} we push ρ down one more level.

Since there are only n levels in the tree, we cannot do this indefinitely. Thus, we must eventually find a node v_m' labelled ℓ_{i_m} such that ρ_α satisfies the pair (G_{i_m}, R_{i_m}). □

The complexity of Safra's construction The automaton $\mathcal{A}_G$ has $2^{O(n \log n)}$ states, where n is the number of states in $\mathcal{A}$. To see this, we estimate the number of bits required to write down a typical state of $\mathcal{A}_G$. We have to specify the structure $(T, \sigma, \lambda, \chi)$.

Since T has at most n nodes, we can "name" the nodes $[1..n]$, with $v_r = 1$. The structure (V, v_r, π) can then be written down as a list of the form $\{\pi(i)\}_{i \in [1..n]}$. Since $\pi(i) \in [1..n]$ requires $\log n$ bits to write down, T can be described using $n \log n$ bits. Similarly, λ and χ can be written as lists of length n with each entry made up of $\log n$ bits and 1 bit, respectively.

The only catch is with σ—if we naïvely represent the function $\sigma : V \to 2^S$ as a list of subsets, we will need n bits to represent each entry, resulting in n^2 bits overall. However, notice that if a state s belongs to $\sigma(v)$ and $\sigma(v')$ for two different nodes v and v' it must be the case that v is an ancestor of v' or that v' is an ancestor of v. Also, if $s \in \sigma(v)$, s must belong to $\sigma(v')$ for *every* ancestor v' of v, all the way upto the root. Thus, we can characterize the set of nodes where s appears in terms of the lowest node v_s such that $s \in \sigma(v_s)$: if $s \in \sigma(v_s)$ then $s \in \sigma(v')$ for any other node v' iff v' is an ancestor of v_s. In this way, σ can also be represented as a list of length n by matching each state s in S to its corresponding node v_s in V. Each entry in this list can be written down using $\log n$ bits.

Since we can characterize a state of $\mathcal{A}_G$ using $O(n \log n)$ bits, it follows that the number of distinct states is bounded by $2^{O(n \log n)}$.

The number of pairs in $\mathcal{PT}_G$ is $O(n)$—by construction, there is a pair (G_i, R_i) for each label $\ell_i \in \mathcal{L}$ and $\mathcal{L}$ contains exactly $2n$ elements.

By Lemma 2.11, we can simulate the Streett automaton $(\mathcal{A}_G, \mathcal{PT}_G)$ by a Büchi automaton with $2^{O(n \log n)}$ states. Thus, complementing Büchi automata using Safra's construction results in the state space blowing up from n to $2^{O(n \log n)}$. Recall that for automata on finite words, the number of states in the complement (via the subset construction) is $2^{O(n)}$. It has been shown that the bound achieved by Safra's construction is optimal [15].

Why complement Büchi automata? We have seen that if we work with Muller, Rabin or Streett conditions, we can in fact accept all ω-regular languages using deterministic automata. So, why do we bother about complementing non-deterministic Büchi automata?

The reason is that the natural translation of logical questions into automata necessarily introduces non-determinism. For instance, when we constructed the automaton $(\mathcal{A}_\varphi, G_\varphi)$ corresponding to an S1S formula φ in Section 2.2, non-determinism was unavoidable in the inductive step for handling existential quantification. This non-determinism arises regardless of what type of acceptance condition we choose to work with. Surprisingly, determinizing Muller or Rabin automata directly is no easier than first converting them to Büchi automata and then applying Safra's construction [8].

Arguably, for our purposes it should suffice to complement Büchi automata— determinization is a stronger construction that yields complementation as a corollary. In fact, Klarlund [16] has shown that it *is* possible to directly complement non-deterministic Büchi automata without determinizing them and without sacrificing the optimal $2^{O(n \log n)}$ bound achieved by Safra's construction. However, there are applications where determinization is crucial—for instance, in the game-theoretic analysis of automata on infinite trees [17].

2.5. Discussion

Our main focus in this survey has been in describing how Büchi automata can be used to settle decision problems in logic. On the way, we have proved some simple results about ω-regular languages. There has also been a lot of work on the algebraic and topological aspects of ω-regular languages that we have not even touched on. A detailed introduction can be found in the survey [12].

As mentioned in the Introduction, Meyer showed in [5] that the decision procedure for S1S has a *non-elementary* complexity. A formula of length n may generate an automaton with $2^{2^{\cdot^{\cdot^{\cdot^{2}}}}}$ states, where the tower of exponentials is of height n. In other words, the size of the automaton cannot be bounded by a function of constant exponential height. This result appears to make it impossible to use this elegant theory in a practical setting for verifying properties of programs.

However, for temporal logics, it turns out that there are direct ways to construct a Büchi automaton $(\mathcal{A}_\varphi, G_\varphi)$ recognizing L_φ for a temporal logic formula φ, such that the size of $\mathcal{A}_\varphi$ is exponential in the length of the formula [18]. As shown in [18], Büchi automata also provide a clean solution to the *model-checking problem* for finite-state systems. The model checking problem is the following—given a finite-state program P and a temporal logic formula φ, do all the computations of P satisfy φ?

Rabin showed that Büchi's decidability result for S1S could be extended to the logic S2S, the monadic second order theory of the infinite binary tree [4]. The logic S2S is very powerful—for instance, it is powerful enough to embed the logic SωS, the monadic second order theory of the infinite *countably branching* tree.

Rabin's results are proved by extending the techniques developed for automata on infinite words to automata operating on infinite trees. These extensions are highly non-trivial—especially the result that automata on infinite trees are closed under complementation. A number of attempts have been made to simplify Rabin's difficult proof. A very readable account can be found in [19].

The theory of Büchi automata and ω-regular languages has also been lifted to the setting of concurrent programs [20–22]. When dealing with concurrent programs, it is often advantageous to regard the runs of the system as partial orders rather than as sequences. Two actions in such a run are unordered if they occur independently. A sequential description of a concurrent program will generate a number of equivalent interleavings for each partially ordered run. To verify properties of a concurrent program in terms of such a sequential description, we have to check all these equivalent interleavings when, in principle, it should suffice to check one representative interleaving for each partially ordered computation. By directly working with infinite labelled partial orders rather than infinite sequences, we can avoid some of this duplication of effort. A challenging open problem is to extend the work of [20–22] to branching structures with concurrency, corresponding to the case of infinite trees for sequential systems.

Acknowledgments

Some of this material was originally presented at the *National Seminar on Theoretical Computer Science* held at Banasthali Vidyapeeth, Rajasthan in August, 1996. This write-up is based on material from a graduate course at Chennai Mathematical Institute attended by Deepak D'Souza, P. Madhusudan, Samik Sengupta and Barbara Sprick, who gave valuable feedback. In particular, Example 2.16 is due to Madhu and Barbara. An earlier version of this chapter has been available as a technical report. I thank Mohsin Ahmed, Swarup Mohalik and Milind Sohoni for their constructive criticism based on a careful reading of this technical report that has led to some improvements in the presentation.

References

[1] J.R. Büchi, On a decision method in restricted second order arithmetic, *Z. Math. Logik Grundlag. Math.* **6**, 66–92, (1960).

[2] D.E. Muller. Infinite sequences and finite machines. In *Proc. 4th Symp. on Switching Circuit Theory and Logical Design*, pp. 3–16. IEEE, (1963).

[3] R. McNaughton, Testing and generating infinite sequences by a finite automaton, *Inform. Control.* **9**, 521–530, (1966).

[4] M.O. Rabin, Decidability of second order theories and automata on infinite trees, *Trans. AMS.* **141**, 1–37, (1969).

[5] A. Meyer. Weak monadic second order theory of successor is not elementary recursive. In *Proc. Logic Colloquium*, vol. 453, *LNM*, pp. 132–154. Springer, (1975).

[6] E.A. Emerson. Temporal and modal logic. In *Handbook of Theoretical Computer Science: Volume B*, pp. 995–1072, North-Holland, (1990)

[7] A. Pnueli. The temporal logic of programs, In *Proc. 18th FOCS*, pp. 46–57. IEEE, (1977).

[8] S. Safra. On the complexity of ω-automata. In *Proc. 29th FOCS*, pp. 319–327. IEEE, (1988).

[9] S. Safra and M. Vardi. On ω-automata and temporal logic. In *Proc. 21st STOC*, pp. 127–137. ACM, (1989).

[10] A.P. Sistla, M. Vardi and P. Wolper, The complementation problem for ω-automata with applications to temporal logic, *Theoret. Comput. Sci.* **49**(2-3), 217–237, (1987).

[11] M. Vardi and P. Wolper, Reasoning about infinite computations, *Inform. Comput.* **115**(1), pp. 1–37, (1994).

[12] W. Thomas. Automata on infinite objects. In *Handbook of Theoretical Computer Science, Volume B*, pp. 133–191, North-Holland, (1990).

[13] A.V. Aho, J.E. Hopcroft and J.D. Ullman, *The Design and Analysis of Algorithms.* (Addison-Wesley, 1974).

[14] R. Streett, Propositional dynamic logic of looping and converse is elementarily decidable, *Inform. Control.* **54**(1-2), 121–141, (1981).

[15] M. Michel, Complementation is more difficult with automata on infinite words, Manuscript (1988).

[16] N. Klarlund. Progress measures for complementation of ω-automata with applications to temporal logic. In *Proc. 32nd FOCS*, pp 358–367. IEEE, (1991).

[17] Y. Gurevich and L. Harrington. Trees, automata and games. In *Proc. 14th STOC*, pp. 60–65. ACM, (1982).

[18] M. Vardi and P. Wolper. An automata theoretic approach to automatic program verification. In *Proc. 1st LICS*, pp 332–345. IEEE, (1986).

[19] W. Thomas. Languages, automata, and logic. In *Handbook of Formal Languages, Volume III*, pp. 389–455. Springer, (1997).

[20] P. Gastin and A. Petit. Asynchronous Cellular Automata for Infinite Traces. In *Proc. 19th ICALP*, vol. 623, *LNCS*, pp. 583–594. Springer, (1992).

[21] W. Ebinger and A. Muscholl. Logical Definability on Infinite Traces. In *Proc. 20th ICALP*, vol. 700, *LNCS*, pp. 335–346. Springer, (1993).

[22] N. Klarlund, M. Mukund and M Sohoni. Determinizing Büchi Asynchronous Automata. In *Proc. 15th FSTTCS*, vol. 1026, *LNCS*, pp. 456–470. Springer, (1995).

Chapter 3

Basics on Tree Automata

Christof Löding
Informatik 7, RWTH Aachen, Germany
loeding@informatik.rwth-aachen.de

This article is an introduction to the theory of finite tree automata for readers who are familiar with standard automata theory on finite words. It covers basic constructions for ranked tree automata as well as minimization, algorithmic questions, and the connection to monadic second-order logic. Further, we present hedge automata for unranked trees and the model of tree-walking automaton.

3.1. Introduction

The theory of tree automata was established in the late sixties by Thatcher and Wright [1] and Doner [2]. They showed that the basic logical and algorithmic properties of standard automata theory can be transferred from the domain of finite words to the domain of finite trees (or terms), leading to a theory of regular tree languages. Quoting from the introduction of [1], where the theory of tree automata is called generalized finite automata theory: "*...the results presented here are easily summarized by saying that conventional finite automata theory goes through for the generalization – and it goes through quite neatly!*"

In parallel, Rabin [3] developed a theory of automata on infinite trees, which serves as a basis for the analysis of specification logics in the context of the verification of state based systems with non-terminating behavior. The theory of automata on infinite trees is quite different from the one on finite trees and in this chapter we exclusively consider finite trees.

Besides their application for showing the decidability of certain logical theories [1, 2], tree automata found their first applications in the area of term rewriting, e.g., for the automated termination analysis of certain rewriting systems, which are documented in the electronic book [4]. Nowadays, tree automata are used in various fields, e.g., in verification to model the state space of parametric systems [5] or as the underlying formalism for a logic that allows to specify properties of heap manipulating programs [6].

All these algorithmic applications of tree automata are based on the following two facts: The class of regular tree languages has strong closure properties (e.g., it is closed under boolean operations and projection), and the main algorithmic problems like emptiness and inclusion are decidable. This article is an introduction to the theory of tree automata, written for readers familiar with conventional automata theory on finite words, where we present basic facts as the ones mentioned above. Of course, we can only cover some aspects of the whole theory in this chapter. The focus is on standard automata for ranked trees (terms) because these are the starting point and the central model of the whole area, but we also mention some other models.

As opposed to ranked trees, where the label of a node determines the number of child nodes, there is no bound on the number of children of a node in an unranked tree. This makes unranked trees suitable to model the structure of XML documents. There are various languages for specifying properties of XML documents, e.g., DTD (document type definition), XML schema, and RELAX NG (see [7]). The core of all these formalisms can be seen as a mechanism for defining languages of unranked trees. Motivated by this new area of application, a theory of automata for unranked trees has been developed (often referred to as hedge automata). Although this model was already mentioned in early papers on tree automata (e.g., [8] and [9]), a systematic study of hedge automata has only been started in the nineties, and is documented in [10]. We present the model of hedge automaton and give a brief overview of their basic properties and their relation to ranked tree automata.

We further present tree-walking automata, which have already been introduced in the seventies in [11] but also have gained new interest in recent years in the context of XML. Tree-walking automata process their input in a sequential way by navigating through the tree, and it turns out that they have rather different properties from standard tree automata.

The structure of the chapter is as follows. In Section 3.2 we present the basic terminology and introduce ranked and unranked trees. Section 3.3 presents the model of ranked tree automata, followed by hedge automata in Section 3.4. In Section 3.5 we briefly mention some results on tree-walking-automata, and we conclude in Section 3.6.

3.2. Trees

We start with some basic notations. By $\mathbb{N}$ we denote the set of natural numbers, i.e., the set of non-negative integers. For a set X we denote by X^* the set of all finite words (or sequences) over X and by X^+ all finite nonempty words over X. The empty word is denoted by ε. A word u is a prefix of a word w, written as $u \sqsubseteq w$, if there is some word v such that $w = uv$. By $u \sqsubset w$ we denote the strict prefix relation.

There are two types of trees that we consider in this article: ranked and unranked trees. Ranked trees correspond to terms built up from function symbols with fixed arities. Therefore, in a ranked tree, the number of children of a node is determined by the label of the node. Unranked trees do not have this restriction. They can be used to model ordered hierarchical structures as, e.g., XML documents or recursive computations.

3.2.1. *Ranked trees*

A *ranked alphabet* Σ is a finite set of symbols together with an arity $|a| \in \mathbb{N}$ for each $a \in \Sigma$. The set of symbols of arity i is denoted by Σ_i.

The set $\mathcal{T}_\Sigma$ of finite *trees* over Σ is the least set containing for each $a \in \Sigma$ and each $t_1, \ldots, t_{|a|} \in \mathcal{T}_\Sigma$ also $a(t_1, \ldots, t_{|a|})$. In case of $|a| = 0$ we simply write a instead of $a()$. If we want to make the alphabet of labels explicit we also refer to a tree as a Σ-tree. Note that in this way we represent trees as words (with a special structure) over the alphabet containing all letters from Σ and additionally the parentheses and the comma. An alternative view of trees makes their hierarchical structure more explicit, as explained in the following.

The *domain* $dom(t) \subseteq \mathbb{N}^*$ of a tree $t = a(t_1, \ldots, t_{|a|})$ is defined inductively as $dom(t) = \{\varepsilon\} \cup \bigcup_{i=1}^{|a|} i \cdot dom(t_i)$, where $\cdot$ denotes the concatenation of words, and we view t as a mapping from its domain to the alphabet Σ (see Example 3.1). The elements of the domain are called the *nodes* of the tree. Nodes that are labeled with symbols of arity 0 are called *leafs*, the other ones are called *inner nodes*, and ε is the *root* of the tree. For $u \in \mathbb{N}^*$ and $i \in \mathbb{N}$ we call $u \cdot i$ the ith *child* or the ith *successor* of u.

Example 3.1. The left-hand side of Figure 3.1 shows a graphical representation of the tree $t = a(b(a(c, d)), c)$ over the ranked alphabet $\Sigma = \{a, b, c, d\}$ with $\Sigma_2 = \{a\}$, $\Sigma_1 = \{b\}$, and $\Sigma_0 = \{c, d\}$, i.e., $|a| = 2$, $|b| = 1$, $|c| = |d| = 0$. The children of a node are ordered from left to right, i.e., the ith edge going downwards from a node leads to the ith child. This is indicated by the labels on the edges. In future drawings we omit these labels because they can be derived from the left-to-right ordering of the nodes.

On the right-hand side the domain of the tree is shown. One can see that the name of a node corresponds to the concatenation of the edge labels leading to this node. Viewing t as a mapping from its domain to Σ, we have $t(\varepsilon) = a$, $t(1) = b$, $t(2) = c$, $t(11) = a$, $t(111) = c$ and $t(112) = d$. ◁

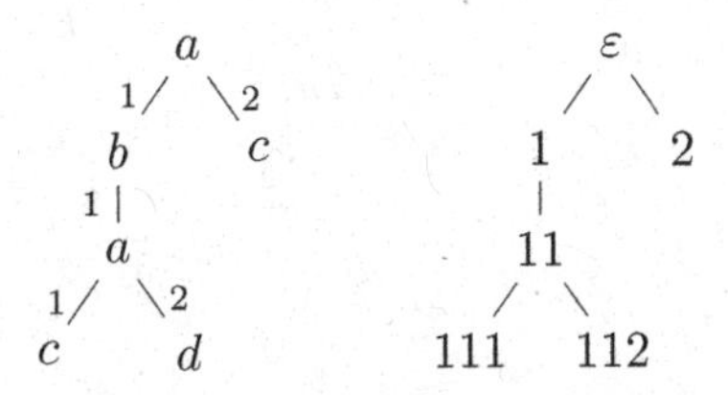

Fig. 3.1. A graphical representation of a tree and its domain.

Given a tree t and a node $u \in dom(t)$, the *subtree* t_u of t at node u is the tree naturally obtained when removing all nodes of which u is not a prefix. Formally t_u is the tree with domain $dom(t_u) = \{v \in \mathbb{N}^* \mid uv \in dom(t)\}$ such that $t_u(v) = t(uv)$. In the above example (the tree t in Figure 3.1) for $u = 11$ we get the subtree $t_u = a(c,d)$.

The *height* of a tree corresponds to the length of a longest path from the root to a leaf. Defined inductively, the height of a for $a \in \Sigma$ is 0, and for a tree $a(t_1, \ldots, t_{|a|})$ the height is the maximal height of one of the trees t_i plus 1.

3.2.2. *Hedges and unranked trees*

Unranked trees are defined in a similar way as ranked ones but over a simple alphabet where the symbols do not have arities. Let Σ be an (unranked) alphabet, i.e., just a finite set of symbols. The set $\mathcal{T}_\Sigma^{\mathrm{unr}}$ of unranked trees over Σ is the least set containing for each $a \in \Sigma$ and each $t_1, \ldots, t_n \in \mathcal{T}_\Sigma^{\mathrm{unr}}$ with $n \geq 0$ also $a(t_1, \ldots, t_n)$.

Note that unranked trees are unbounded in two dimensions, vertically and horizontally. While there is only a finite number of ranked trees of a fixed height (for a fixed alphabet), there are infinitely many unranked trees of height one.

All the terminology and definitions from ranked trees, such as domain, nodes, leafs, etc. can be directly transferred to the unranked setting.

Finite sequences of unranked trees are called *hedges*. We denote the set of all hedges over the alphabet Σ by $\mathcal{H}_\Sigma$.

Because unranked trees are unbounded vertically and horizontally, it is often useful to encode them by ranked trees. In particular, when dealing with automata, a lot of results for ranked tree automata can be transferred to the unranked setting using such encodings. We present here two such encodings.

The *first-child-next-sibling encoding*, abbreviated as FCNS encoding, is very natural when unranked trees have to be represented as data structures using pointers (see the textbook [12]). For each node we have one pointer to its first (left-most) child, and one pointer to its next (right) sibling. This is illustrated in Figure 3.2. The symbols from the unranked alphabet all become binary symbols in the ranked

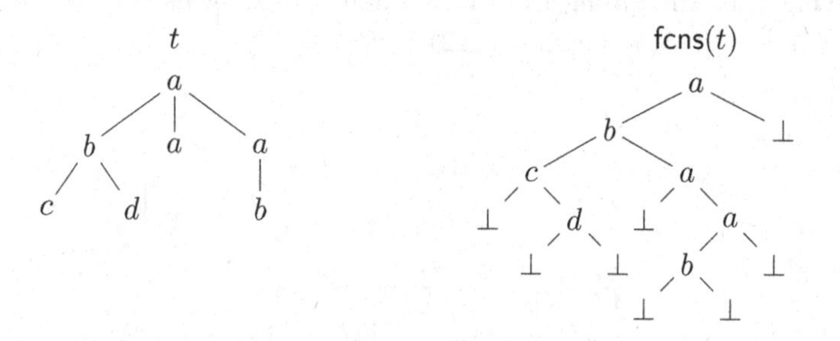

Fig. 3.2. An unranked tree and its FCNS representation.

alphabet, and a new symbol $\perp$ of arity 0 is added. The first child of a node in the FCNS encoding corresponds to its first child in the unranked tree, and the second child of a node in the FCNS encoding corresponds to its right sibling in the unranked tree. If the corresponding nodes (first child or right sibling) do not exist in the unranked tree, then we put the symbol $\perp$ in the FCNS encoding. The encoding should be clear from the example and we do not give a formal definition here. For a tree t we denote its FCNS encoding by $\mathsf{fcns}(t)$.

The *extension encoding* is a natural way of representing unranked trees by building them up from the basic trees of height zero, using a binary operation. This operation is called the extension operation. Applied to two trees t and t', the operations attach t' as the right-most subtree of the root of t. It appears explicitly in [13] as an encoding, but already in [9] it is used in a proof (where it is denoted "1" and oriented the other way, the left operand is added as left-most subtree of the root of the right operand).

Consider, for example, the tree on the left-hand side of Figure 3.2. On the left-hand side of Figure 3.3 it is shown how it is built up from two other trees by one application of the extension operator @. These two trees can further be constructed from smaller trees by this operation. This yields a coding as a binary tree where the leafs are trees of height 0, and the inner nodes indicate how to combine them by the extension operation. The right-hand side of Figure 3.3 shows the resulting coding for our example tree.

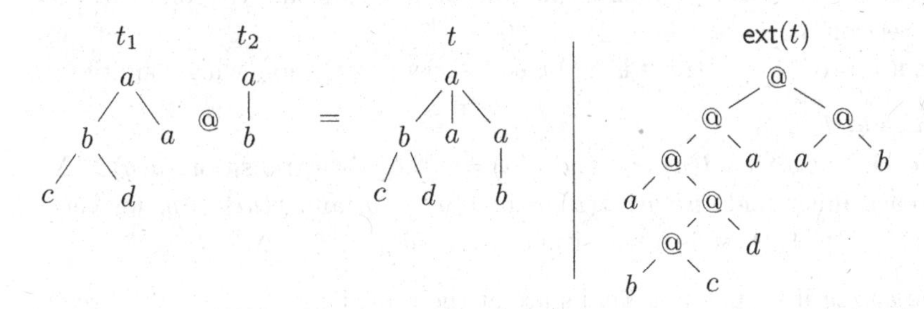

Fig. 3.3. The extension operation @ and the corresponding encoding of the tree from the left-hand side of Figure 3.2.

Looking at the picture it seems to be difficult to extract the unranked tree from the encoding. But there is a rather simple way of doing it: The left most leaf corresponds to the root of the unranked tree t. Walking up from this leaf to the root of $\mathsf{ext}(t)$ we meet three times @. The right subtrees of these @ correspond to the three subtrees of the root in t. Now we can proceed in the same way for decoding these subtrees.

In Section 3.4 we show how these encodings can be used to transfer results for automata on ranked trees to automata on unranked trees.

3.3. Ranked Tree Automata

The model of finite automaton on finite words is one of the many ways to characterize the class of regular word languages. In the following we develop a notion of tree automaton that has many of the good properties making finite automata so attractive in various fields of computer science.

In the whole section we only consider ranked trees and fix a ranked alphabet Σ. A *nondeterministic finite tree automaton* (NFTA) over Σ is a tuple $\mathcal{A} = (Q, \Sigma, \Delta, F)$, where Q is a finite set of states, $F \subseteq Q$ is the set of final states, and $\Delta \subseteq \bigcup_{i=0}^{k} Q^k \times \Sigma_i \times Q$ is the transition relation. Transitions are of the form $(q_1, \dots, q_{|a|}, a, q)$ with $a \in \Sigma$, and $q_1, \dots, q_{|a|}, q \in Q$. If a is of arity 0, then the transitions are written as (a, q) and are called *leaf transitions*.

Intuitively, such an automaton works as follows on an input tree. It starts at the leafs and labels them by states according to the leaf transitions. For an inner node u whose children are already labeled by some state, the transition relation determines which states q can be put at u, depending on its label a, and the states $q_1, \dots, q_{|a|}$ at its children. For this reason, these automata are called bottom-up or frontier-to-root automata because they start at the leafs (which are at the bottom in the graphical representations), and work their way upwards to the root. One should note that we can also view the automaton as working top-down. Then F is interpreted as the set of initial states, and the transitions (a, q) for a of arity zero are interpreted as allowed or accepting pairs at the leafs. However, the bottom-up and top-down views lead to two different notions of determinism. We come back to this issue in Section 3.3.1.

Formally, a *run* of $\mathcal{A}$ on a tree t is a Q-tree ρ satisfying the following conditions:

- $dom(\rho) = dom(t)$.
- For each leaf u with $t(u) = a$ and $\rho(u) = q$ there is a transition $(a, q) \in \Delta$.
- For each inner node u with $t(u) = a$, $\rho(u) = q$, and $\rho(ui) = q_i$ for each $i \in \{1, \dots, |a|\}$, there is a transition $(q_1, \dots, q_{|a|}, a, q) \in \Delta$.

A run ρ is *accepting* if it ends in a final state at the root, i.e., if $\rho(\varepsilon) \in F$. A tree t is accepted by $\mathcal{A}$ if there is an accepting run of $\mathcal{A}$ on t. The set of all trees accepted by $\mathcal{A}$ is called the *language* of $\mathcal{A}$ and is denoted by $T(\mathcal{A})$. In general, we call a set of trees a *tree language* or simply language. A language is called *regular* if it is the language of some NFTA. We call two NFTAs *equivalent* if they accept the same language.

Example 3.2. A typical example is an NFTA that evaluates boolean terms built from the constants 0, 1, the binary operations $\wedge$, $\vee$, and the unary operation $\neg$. The automaton uses two states q_0 and q_1 with leaf transitions $(0, q_0)$, $(1, q_1)$, and for each inner symbol of the term applies its semantics: $(q_i, q_j, \wedge, q_{i \wedge j})$ for $i, j \in \{0, 1\}$, and similarly for $\vee$. For the negation it simply flips the state: $(q_i, \neg, q_{\neg i})$. If we are interested in all terms evaluating to 1, then we let q_1 be the only final state. $\triangleleft$

Example 3.3. As second example we consider a language over the ranked alphabet from Example 3.1. We want to construct an NFTA recognizing all trees with the following property: There is a leaf such that the sequence of labels from this leaf to the root is of the from da^*ba^*, and there is a leaf such that the label sequence to the root is of the form ca^*. The tree depicted in Figure 3.1 satisfies the property because the path from the middle leaf to the root is labeled *daba*, and the path from the right leaf to the root is labeled *ca*.

The NFTA that we construct, nondeterministically guesses the two leafs and then verifies the required property on the corresponding paths. We use states q_{db}, q_{da}, q_c, where q_{db} is used on the path from the d-leaf before the b and q_{da} after the b, and q_c is used on the path from the c-leaf. An additional state q is used on the parts of the tree that do not belong to one of the two paths, and a state q_a is intended for the positions where the two paths have merged. The transition relation contains the following transitions

$$\begin{aligned}
&(c, q_c),\ (d, q_{db}),\ (c, q),\ (d, q),\\
&(q_{db}, q, a, q_{db}),\ (q, q_{db}, a, q_{db}),\ (q_{db}, b, q_{da}),\\
&(q_c, q, a, q_c),\ (q, q_c, a, q_c),\ (q_{da}, q, a, q_{da}),\ (q, q_{da}, a, q_{da}),\\
&(q_{da}, q_c, a, q_a),\ (q_c, q_{da}, a, q_a),\ (q_a, q, a, q_a),\ (q, q_a, a, q_a).
\end{aligned}$$

The final state is q_a. Figure 3.4 shows an accepting run of this NFTA on the tree from Figure 3.1. ◁

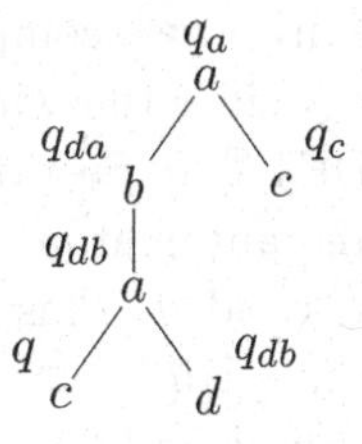

Fig. 3.4. A run of the automaton from Example 3.3.

An NFTA is called *complete* if for each $a \in \Sigma$ and all $q_1, \ldots, q_{|a|} \in Q$ there is at least one $q \in Q$ such that $(q_1, \ldots, q_{|a|}, a, q) \in \Delta$. One can easily turn each NFTA into an equivalent one that is complete. But one should note that for alphabets of high arity this might require to add many transitions.

3.3.1. *Determinization and closure properties*

On finite words it is well known that nondeterministic automata can be transformed into equivalent deterministic automata using the subset construction. We show here that the same holds for tree automata.

An NFTA is called *deterministic* (DFTA) if for each $a \in \Sigma$ and all $q_1, \ldots, q_{|a|} \in Q$ there is at most one $q \in Q$ such that $(q_1, \ldots, q_{|a|}, a, q) \in \Delta$. For this definition it is essential that we take the bottom-up view of the automata. Given the states at the children of the node and the label of the node there is at most one transition that can be executed. For DFTAs we can view the transition relation as a (partial) function, and we denote it by δ instead of Δ, i.e., $\delta(q_1, \dots, q_{|a|}, a) = q$ instead of $(q_1, \dots, q_{|a|}, a, q) \in \Delta$.

Theorem 3.1 ([1, 2]). *For each NFTA one can construct an equivalent complete DFTA. In the worst case, the number of states of the DFTA is exponential in the number of states of the given NFTA.*

Proof. The proof is done by an adaption of the classical subset construction for word automata from [14] (see also [15]). Let $\mathcal{A} = (Q, \Sigma, \Delta, F)$ be an NFTA. We construct the subset automaton $\mathcal{P}(\mathcal{A}) = (2^Q, \Sigma, \delta, F')$, where 2^Q denotes the power set of Q, and δ and F' are defined as follows:

- $\delta(R_1, \dots, R_{|a|}, a) = R$ for every $R_1, \dots, R_{|a|} \in 2^Q$ and
$$R = \{q \in Q \mid \exists q_1 \in R_1, \dots, q_{|a|} \in R_{|a|} \,:\, (q_1, \dots, q_{|a|}, a, q) \in \Delta\}.$$
- $F' = \{R \subseteq Q \mid R \cap F \neq \emptyset\}$.

It is straightforward to prove that $\mathcal{P}(\mathcal{A})$ is equivalent to $\mathcal{A}$. □

For this generic construction presented in the proof of Theorem 3.1 the resulting DFTA always has exponentially many states compared to the given NFTA. But in many cases it is not necessary to consider the whole power set 2^Q because not all of these sets are reachable in the DFTA. In Section 3.3.2 we show how to compute the set of reachable states of a tree automaton. However, there are cases where the exponential blow-up cannot be avoided. This follows from the lower bound for determinization of word automata (see [15]).

It is also possible to consider top-down determinism. We call an NFTA *top-down deterministic* if $|F| = 1$ (F is interpreted as set of initial states and there should be only one such initial state), and for each $a \in \Sigma$ and each $q \in Q$ there is at most one transition of the form $(q_1, \dots, q_{|a|}, a, q) \in \Delta$. It is rather easy to see that top-down determinism is a strong restriction. Not even all finite tree languages can be accepted by top-down deterministic automata. For example, each top-down deterministic automaton accepting the trees $a(b, c)$ and $a(c, b)$ will also accept the trees $a(b, b)$ and $a(c, c)$. One can show that top-down deterministic automata can only define so-called path-closed languages. Intuitively, this means that if we take several trees accepted by a top-down deterministic automaton, decompose these trees into their paths, and build a new tree out of (some of) these paths, then the resulting tree is also accepted by the automaton. In the above example we can use the left-most path of $a(b, c)$ and the right-most path of $a(c, b)$ to obtain $a(b, b)$. For more results and references on top-down deterministic tree automata we refer the reader to [16].

For the closure properties of regular tree languages we observe that complete deterministic automata can be complemented by exchanging final and non-final states. For the intersection we can apply a standard product construction simulating both automata, and accepting if both of them reach a final state. For the union we can simply take the disjoint union of the components of two NFTAs. In case we want to take the union of two DFTAs and want to obtain a DFTA again we also use a product construction.

Theorem 3.2. *The class of regular tree languages is closed under union, intersection, and complement.*

3.3.2. *Decision problems and algorithms*

We now turn to algorithmic questions for tree automata. For estimating the complexity of algorithms we have to define the size of a tree automaton. A natural and common way of doing this is to take the number of states and for each transition the number of its entries. More precisely, let $\mathcal{A} = (Q, \Sigma, \Delta, F)$ be an NFTA. The size of a transition $\tau = (q_1, \ldots, q_n, a, q) \in \Delta$ is $|\tau| = n + 2$, and the size of $\mathcal{A}$ is

$$|\mathcal{A}| = |Q| + \sum_{\tau \in \Delta} |\tau| .$$

We consider the following decision problems.

- The *membership problem*[a]

  Given: An NFTA $\mathcal{A}$ and a tree t
  Question: Is $t \in T(\mathcal{A})$?

- The *emptiness problem*

  Given: An NFTA $\mathcal{A}$
  Question: Is $T(\mathcal{A}) = \emptyset$?

- The *universality problem*

  Given: An NFTA $\mathcal{A}$
  Question: Is $T(\mathcal{A}) = \mathcal{T}_\Sigma$?

- The *inclusion problem*

  Given: NFTAs $\mathcal{A}_1$ and $\mathcal{A}_2$
  Question: Is $T(\mathcal{A}_1) \subseteq T(\mathcal{A}_2)$?

---

[a]Sometimes, the membership problem is defined for a fixed automaton and the problem where the tree and the automaton are part of the input is called the uniform membership problem.

- The *equivalence problem*

  Given: NFTAs $\mathcal{A}_1$ and $\mathcal{A}_2$
  Question: Is $T(\mathcal{A}_1) = T(\mathcal{A}_2)$?

A detailed complexity analysis of the membership problem is given in [17]. Here we just show how to solve the problem in polynomial time.

Theorem 3.3. *The membership problem can be solved in polynomial time.*

Proof. Let $\mathcal{A} = (Q, \Sigma, \Delta, F)$ be an NFTA and $t \in \mathcal{T}_\Sigma$. The algorithm computes for each node u of t the set Q_u of states that the automaton can reach in this node. For a leaf u this set is given by $Q_u = \{q \mid (t(u), q) \in \Delta\}$. For an inner node u with k children the set Q_u can easily be computed from the transitions and the sets $Q_{u1}, \ldots, Q_{uk}$:

$$Q_u = \{q \mid \exists q_1 \in Q_{u1}, \ldots, q_k \in Q_{uk} : (q_1, \ldots, q_k, t(u), q) \in \Delta)\}.$$

Each set Q_u can be computed in polynomial time and hence the whole algorithm is polynomial in the size of $\mathcal{A}$ and t. □

The decidability of the emptiness problem is already shown in [1] and [2]. In both papers the argument is the following: If an NFTA accepts a tree, then there is also an accepted tree whose height can be bounded by the number of states of the NFTA. Hence, one can do the membership test for all the trees up to this height. Of course this algorithm is not very efficient.

To obtain efficient algorithms, the emptiness problem for tree automata can be reduced to many other problems that are efficiently solvable. For example, it can easily be transformed into the satisfiability problem for Horn-formulas as described in [4]. The latter problem can be solved in linear time[b] by so-called unit propagation. Another closely related problem is alternating graph reachability, a problem known to be P-complete [18]. To be self contained we present an algorithm directly working on tree automata. The algorithm is explained in the proof of Theorem 3.4.

Theorem 3.4. *The emptiness problem for NFTA can be solved in linear time.*

Proof. We solve the problem by computing the set of reachable states, i.e., the set of states that can be reached at the root in a run on some tree. If one the final states is reachable, then the language of the automaton is not empty, otherwise it is empty.

Figure 3.5 shows an algorithm to compute the set R of reachable states. The underlying idea is a simple fixpoint computation: All states that occur in a leaf transition are reachable, and if there is a transition $(q_1, \ldots, q_{|a|}, a, q)$ such that $q_1, \ldots, q_{|a|}$ are already in R, then q can be added to R. This is exactly what is done in the algorithm in Figure 3.5 but in such a way that it can be implemented to run

[b]For the complexity estimations we refer to the RAM model of computation.

INPUT: NFTA $\mathcal{A} = (Q, \Sigma, \Delta, F)$
1: $R = \{q \in Q \mid \exists a \in \Sigma : (a, q) \in \Delta\}$
2: **for all** $\tau = (q_1, \ldots, q_{|a|}, a, q) \in \Delta$ **do**
3: $\quad pre(\tau) = \{q_1, \ldots, q_{|a|}\}$
4: $\quad dest(\tau) = q$
5: **end for**
6: $M := R$
7: **while** $M \neq \emptyset$ **do**
8: $\quad$ Choose and remove p from M
9: $\quad$ **for all** τ with $p \in pre(\tau)$ **do**
10: $\quad\quad$ Remove p from $pre(\tau)$
11: $\quad\quad$ **if** $pre(\tau) = \emptyset$ **then**
12: $\quad\quad\quad$ Add $dest(\tau)$ to M and to R
13: $\quad\quad$ **end if**
14: $\quad$ **end for**
15: **end while**
OUTPUT: R

Fig. 3.5. Algorithm for computing the reachable states of a tree automaton.

in linear time by choosing the appropriate data structures. As described above, we have to check for a transition $\tau = (q_1, \ldots, q_{|a|}, a, q)$ whether the states $q_1, \ldots, q_{|a|}$ are already in R. We do this by first defining the set $pre(\tau) = \{q_1, \ldots, q_{|a|}\}$, and whenever we add a state q to R, we remove it from all *pre*-sets it occurs in. If $pre(\tau)$ becomes empty, we add $q = dest(\tau)$ to R.

The set M is used to keep track of which states still have to be removed from the *pre*-sets. To obtain a linear time algorithm, the set M can be implemented as a FIFO-queue, and to quickly access the transitions in the loop of line 9 one has to store for each state q a list of the transitions τ with $q \in pre(\tau)$. □

Similar to the case of word automata, where emptiness is easy to decide and universality is PSPACE-hard, we obtain a much higher complexity for the universality problem. It is difficult to attribute the following result to a specific paper. But [19] shows a similar result for the equivalence problem, and those techniques can also be used for the universality problem.

Theorem 3.5. *The universality problem for tree automata is* EXPTIME*-complete.*

Proof. Checking universality can be done in EXPTIME by determinizing the automaton and then checking the complement for emptiness.

A standard way to show EXPTIME-hardness is to use a reduction from the word problem for alternating polynomial space bounded Turing machines. Alternating machines are natural in connection with tree automata because computations of these machines are trees. Proofs based on alternating machines can, e.g., be found in [19] and [20] (not for the universality problem but related decision problems).

Besides alternating machines there is another formalism closely related to tree automata: Strategies in games can be represented by trees (the branching coming from the different options of the opponent). Hence, it is natural to express problems concerning the existence of winning strategies by means of tree automata. We use here the *two-person corridor tiling problem*, known to be EXPTIME-hard [21].

A *corridor tiling system* is of the form $\mathcal{D} = (D, H, V, \bar{b}, \bar{f}, n)$, where D is a finite set of tiles (or dominos), $H, V \subseteq D \times D$ are horizontal and vertical constraints, $\bar{b}, \bar{f} \in D^n$ are n-tuples of tiles corresponding to the initial row and the final row of the corridor tiling, and $n \in \mathbb{N}$ is the width of the corridor. Intuitively, the goal is to cover a corridor of width n (and arbitrary length m) with the tiles such that the horizontal and vertical constraints are respected, and the corridor starts with $\bar{b}$ and ends with $\bar{f}$.

Formally, a corridor tiling is a mapping $C : \{1, \ldots, m\} \times \{1, \ldots, n\} \to D$ for some $m \in \mathbb{N}$ such that

(1) $(C(1,1), \ldots, C(1,n)) = \bar{b}$ (it begins with the initial row),
(2) $(C(i,j), C(i,j+1)) \in H$ for all $i \in \{1, \ldots, m\}$ and $j \in \{1, \ldots, n-1\}$ (it respects the horizontal constraints),
(3) $(C(i,j), C(i+1,j)) \in V$ for all $i \in \{1, \ldots, m-1\}$ and $j \in \{1, \ldots, n\}$ (it respects the vertical constraints),
(4) $(C(m,1), \ldots, C(m,n)) = \bar{f}$ (it ends with the final row).

The problem of deciding for a given corridor tiling system whether there exists a corridor tiling for it is PSPACE-hard. We consider the game variant of the problem. The game board is the infinite corridor $\mathbb{N}_1 \times \{1, \ldots, n\}$, where $\mathbb{N}_1$ denotes the natural numbers without 0. The starting configuration is the one where the first row of the corridor is covered with the initial row $\bar{b}$. Now the two players Eva and Adam play tiles in turn on successive positions, Eva starting in position $(2, 1)$. Both players have to respect the horizontal and vertical constraints.

Thus, a configuration in the game corresponds to a partial function $C : \mathbb{N}_1 \times \{1, \ldots, n\} \to D$ whose domain is an initial segment of $\mathbb{N}_1 \times \{1, \ldots, n\}$, i.e., if $C(i,j)$ is defined, then also $C(i',j')$ is defined if $i' < i$, or $i' = i$ and $j' < j$.

Eva wants to construct a corridor tiling, i.e., she wins if at some point the mapping C is a corridor tiling. Note that the rules of the game ensure that conditions 1–3 are always satisfied. Hence, her goal is to construct the final row. Adam wins otherwise, i.e., either if the play gets stuck because there is no next move respecting the constraints, or if the game goes on forever and no corridor tiling is ever reached.

The problem of two-player corridor tiling is to decide for a given tiling system $\mathcal{D}$ whether Eva has a winning strategy in the game described above. This problem is EXPTIME-hard [21].

The basic idea for the reduction to the universality of NFTAs is simple: We code strategies of Eva as trees and then construct an NFTA of polynomial size that accepts all trees that do *not* code a winning strategy for Eva.

The alphabet that we use is $\Sigma = D \times \{\mathsf{E}, \mathsf{A}, \perp, !\}$, where the elements of $D \times \{\mathsf{E}\}$ are of arity 1, those of $D \times \{\mathsf{A}\}$ are of arity $|D|$, and those of $D \times \{\perp, !\}$ are of arity 0. A path through such a tree corresponds in a natural way to a mapping $C : \mathbb{N}_1 \times \{1, \ldots, n\} \to D$ whose domain is an initial segment of $\mathbb{N}_1 \times \{1, \ldots, n\}$ (just place the tiles in the order they appear, starting from the position $(2, 1)$ as in a play). The idea is that the first component corresponds to the tile that has been played, and the second component indicates whose turn it is (E for Eva, A for Adam, $\perp$ in case the play is over because a constraint was not respected, and ! in the case a corridor tiling has been completed).

A tree of the above shape represents a winning strategy for Eva if the following conditions are satisfied

(1) The root is labeled by (d, A) for some $d \in D$ (d is the first tile that is played by Eva, and A indicates that it is now Adam's turn).
(2) The nodes from $D \times \{\mathsf{A}\}$ and $D \times \{\mathsf{E}\}$ alternate on each path (Eva and Adam play their tiles in turn).
(3) There is no node that has only $\perp$ successors (the play never gets stuck).
(4) At each node not labeled $\perp$ the tile respects the constraints associated to its position.
(5) At each node labeled $\perp$ the tile does not respect one of the constraints associated to its position.
(6) A node is labeled ! if the mapping C corresponding to the path to the root ends with the row $\bar{f}$ (and hence corresponds to a corridor tiling in combination with the previous conditions).

It is not difficult to see that for each of the above conditions one can construct a small NFTA checking whether the condition does *not* hold. For the last three items the automata need to guess the nodes where the condition is not satisfied. For verifying that a constraint is not satisfied it is sufficient that the automaton can remember a tile and can count n steps. The details of these constructions are left to the reader.

Taking the union of these NFTAs results in an NFTA of polynomial size that accepts all trees iff Eva does not have a winning strategy. □

Since universality is a special case of the inclusion and equivalence problems, we can easily transfer the lower bound.

Theorem 3.6 ([19]). *The inclusion and the equivalence problem for tree automata are* EXPTIME*-complete.*

Proof. Since $T(\mathcal{A}_1) \subseteq T(\mathcal{A}_2)$ iff $L(\mathcal{A}_1) \cap (\mathcal{T}_\Sigma \setminus L(\mathcal{A}_2)) = \emptyset$, we can reduce the inclusion problem to the emptiness problem with an exponential cost. Because emptiness is decidable in polynomial time (Theorem 3.4), we obtain membership in EXPTIME for the inclusion problem. Equivalence can be decided by checking both inclusions.

For the EXPTIME-hardness we observe that $L(\mathcal{A})$ is universal iff $\mathcal{T}_\Sigma \subseteq T(\mathcal{A})$ and apply Theorem 3.5. The same argument works for equivalence with $=$ instead of $\subseteq$. □

It is easy to see that the problems of inclusion and equivalence are decidable in polynomial time for DFTAs (the exponential step for NFTAs is the complementation). In [19] it is shown that equivalence can also be decided in polynomial time if the automata are m-ambiguous for some constant m, where m-ambiguous means that there are at most m accepting runs of the automaton for each input.

3.3.3. *Congruences and minimization*

On finite words deterministic automata can be efficiently minimized. The minimization algorithm identifies equivalent states of the automaton and merges them. In Section 3.3.1 we have already seen that the determinization technique for automata on finite words can be generalized to (bottom-up) tree automata. In this section we show that this also works for minimization.

The background for minimization of automata on finite words is the right-congruence that defines two words w_1, w_2 to be equivalent w.r.t. some language L of finite words if every word w either leads both w_1 and w_2 into the language when appended as a suffix, or leads both outside of the language: $w_1 w \in L$ iff $w_2 w \in L$. This defines an equivalence relation on the set of finite words that is compatible with concatenation to the right. The index of the equivalence, i.e., the number of its classes is finite iff the language L is regular, and it naturally induces an automaton for L whose states are the equivalence classes. This relation is often referred to as Myhill/Nerode-congruence referring to the first papers where it has been used [22, 23].

The goal of this section is to lift these results to ranked trees. The first step is to define a concatenation operation for trees. For this we need the notion of context. In our setting a context is a tree in which exactly one of the leafs is labeled by a variable.[c] This leaf is then used for the concatenation operation.

Let $X \notin \Sigma$ be a variable. A *context* C is a tree in $\mathcal{T}_{\Sigma \cup \{X\}}$ (where the arity of X is 0), in which exactly one of the leafs is labeled by a variable X. We denote the set of contexts by $\mathcal{C}_\Sigma$.

For a context $C \in \mathcal{C}_\Sigma$ and a tree $t \in \mathcal{T}_\Sigma$ we write $C[t]$ for the tree that is obtained by replacing the X-labeled leaf in C by t. This is illustrated in Figure 3.6.

Based on this concatenation operation we can now define an equivalence relation on trees. Let $T \subseteq \mathcal{T}_\Sigma$ be a tree language. We say that two trees $t_1, t_2 \in \mathcal{T}_\Sigma$ are

[c] In general, a context can have several different variables at the leafs. A context with n variables is called an n-context. We are using 1-contexts here.

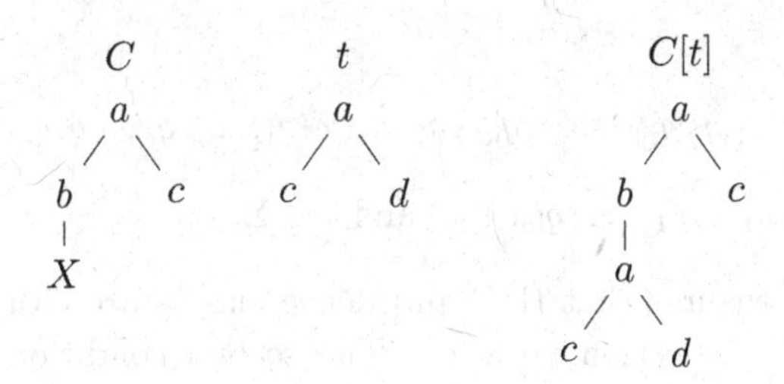

Fig. 3.6. Concatenation of a context and a tree.

equivalent w.r.t. T if

$$\forall C \in \mathcal{C}_\Sigma : C[t_1] \in T \Leftrightarrow C[t_2] \in T .$$

We denote this equivalence by $\sim_T$. As for finite words there is a close correspondence between this relation and deterministic automata. In [24] it is mentioned that it is difficult to attribute the following result to a specific paper. We refer the reader to [24] for references.

Theorem 3.7. *The index of $\sim_T$ is finite iff T is regular, and the number of equivalence classes of $\sim_T$ corresponds to the number of states of a minimal DFTA accepting T.*

Proof. Assume that T is regular and let $\mathcal{A} = (Q, \Sigma, \delta, F)$ be a complete DFTA for T. Denote by δ^* the function $\delta^* : \mathcal{T}_\Sigma \to Q$ that assigns to each tree the state that is at the root of the unique run of $\mathcal{A}$ on t. If $\delta^*(t_1) = \delta^*(t_2)$ for two trees t_1 and t_2, then it is easy to see that $t_1 \sim_T t_2$. Since Q is finite the index of $\sim_T$ is finite.

Now assume that $\sim_T$ has only finitely many equivalence classes. By $[t]_T$ we denote the $\sim_T$ equivalence class of t. We define the complete DFTA $\mathcal{A}_T = (Q_T, \Sigma, \delta_T, F_T)$ as follows:

- $Q_T = \{[t]_T \mid t \in \mathcal{T}_\Sigma\}$,
- $\delta_T([t_1]_T, \ldots, [t_{|a|}]_T, a) = [a(t_1, \ldots, t_{|a|})]_T$,
- $F_T = \{[t]_T \mid t \in T\}$.

Note that this definition of the transition function is possible because the equivalence class of $a(t_1, \ldots, t_{|a|})$ does not depend on the choice of the representatives $t_1, \ldots, t_{|a|}$. A straightforward induction shows that $\delta_T^*(t) = [t]_T$, and hence $T(\mathcal{A}_T) = T$. □

To compute from a given DFTA an equivalent one that is minimal we introduce an equivalence relation on states. Let $\mathcal{A} = (Q, \Sigma, \delta, F)$ be a DFTA. The relation $\sim_\mathcal{A} \subseteq Q \times Q$ is the smallest relation satisfying the following:

(1) If $p \sim_\mathcal{A} q$, then $p \in F \Leftrightarrow q \in F$.

(2) If $p \sim_{\mathcal{A}} q$, then

$$\delta(q_1, \ldots, q_{i-1}, p, q_{i+1}, \ldots, q_{|a|}, a) \sim_{\mathcal{A}} \delta(q_1, \ldots, q_{i-1}, q, q_{i+1}, \ldots, q_{|a|}, a)$$

for all $q_1, \ldots, q_{i-1}, q_{i+1}, \ldots, q_{|a|} \in Q$ and $a \in \Sigma$.

The first condition ensures that the equivalence classes are either contained in F or have an empty intersection with F. The second condition ensures that the equivalence relation is compatible with the transition function. Hence, one can built the quotient DFTA $\mathcal{A}_{/\sim}$ that uses the equivalence classes as states. It is left as an exercise for the reader to show that $\mathcal{A}_{/\sim}$ is isomorphic to $\mathcal{A}_T$ from the proof of Theorem 3.7.

Computing the relation $\sim_{\mathcal{A}}$ can be done by successively marking those pairs of states that do not satisfy one of the two conditions. Concerning the complexity of minimization, [25] presents an implementation of the algorithm that runs in quadratic time in the size of the given tree automaton.

3.3.4. *Logic*

Instead of taking the view of trees as mappings from a domain to a finite set of labels we can also view them as relational structures and use predicate logic to express properties of such structures. Using first-order logic extended with quantification over monadic predicates (sets of elements) one obtains a formalism that is expressively equivalent to tree automata. Such a relation has first been established for finite words [26–28], and was lifted to trees in [1, 2].

For a ranked alphabet Σ with symbols of maximal arity k we consider the signature consisting of the relational symbols $S_1, \ldots S_k$ for the binary successor (child) relations, $\sqsubseteq$ for the prefix relation on nodes, and $(P_a)_{a \in \Sigma}$ for the unary relations "the node is labeled a". A tree $t \in \mathcal{T}_\Sigma$ corresponds to a relational structure

$$\underline{t} = \langle dom(t), S_1^t, \ldots, S_k^t, \sqsubseteq^t, (P_a^t)_{a \in \Sigma} \rangle$$

over the universe $dom(t)$ and the relations defined in the natural way as

- $S_i^t(u, v)$ if v is the ith child of u, i.e., $v = ui$,
- $u \sqsubseteq^t v$ if $u \sqsubseteq v$, and
- $P_a^t(u)$ if $t(u) = a$.

In *monadic second-order logic* (MSO) we use two types of variables: first-order variables interpreted by elements and denoted by small letters (e.g., x, y), and monadic second-order variables interpreted by sets of elements and denoted by capital letters (e.g., X, Y).

MSO-formulas are built up from atomic formulas $S_i(x, y)$, $x \sqsubseteq y$, $P_a(x)$, $x \in X$, and $x = y$ by boolean combinations and quantifications over the two types of variables. Let us consider two simple examples.

- $leaf(x) = \neg\exists y : S_1(x, y)$ expresses that x does not have a first child, and hence is a leaf (if it does not have a first child it has no children).
- The following formula expresses that the set X is a cut through the tree, i.e., a maximal set of nodes that are pairwise incomparable w.r.t. $\sqsubseteq$:

$$cut(X) = \forall y \exists x : x \in X \wedge (y \sqsubseteq x \vee x \sqsubseteq y) \wedge \\ \forall x, y : x \sqsubset y \rightarrow (x \notin X \vee y \notin X)$$

- Using the above formula one can easily express that the nodes labeled a form a cut: $\exists X : cut(X) \wedge \forall x : (P_a(x) \leftrightarrow x \in X)$

For a *sentence* φ, i.e., a formula without free variables we write $\underline{t} \models \varphi$ if φ holds in $\underline{t}$. Given such a sentence φ it defines the tree language

$$T(\varphi) = \{t \in \mathcal{T}_\Sigma \mid \underline{t} \models \varphi\}.$$

We call a tree language $T \subseteq \mathcal{T}_\Sigma$ *MSO-definable* if it is the language of some MSO-sentence φ. Now we can state the equivalence theorem.

Theorem 3.8 ([1, 2]). *A tree language $T \subseteq \mathcal{T}_\Sigma$ is MSO-definable iff it is regular.*

Proof. We only sketch the idea: For the direction from logic to automata one proceeds by induction on the structure of the formula. To deal with free variables, one annotates the nodes of the tree with tuples of 0 and 1 to encode the interpretations of the free variables. For example, an automaton for the atomic formula $S_1(x, y)$ accepts trees over the alphabet $\Sigma \times \{0, 1\} \times \{0, 1\}$ where exactly one node u has label 1 in the second component (the interpretation of x), and exactly one node v has label 1 in the third component (the interpretation of y), and v is the first child of u. It is easy to see that one can build such automata for all atomic formulas. Then one uses the closure properties of tree automata to deal with boolean combinations (union, intersection, and complement), and quantifications (projection).

For the direction from automata to logic one constructs a formula describing an accepting run of the given automaton. For each state q one introduces a set variable X_q that is interpreted by the positions in the run in which the automaton is in state q. If the state set is $\{q_1, \ldots, q_n\}$ then the formula has the shape

$$\exists X_{q_1} \cdots \exists X_{q_n} : AccRun(X_{q_1}, \ldots, X_{q_n})$$

where the formula $AccRun(X_{q_1}, \ldots, X_{q_n})$ expresses that the sets indeed code a run (each node is in exactly one set, the root is in a set X_q for a final state q, and the local constraints of the transition relation are respected). □

The translation between MSO-formulas and automata is effective. As a consequence, decision problems for MSO logic on finite trees can be reduced to decision problems for tree automata. For example, the satisfiability problem "Given an MSO-sentence, does it have a model?" reduces to the emptiness problem for tree automata, which is decidable according to Theorem 3.4.

Corollary 3.1. *The satisfiability problem for MSO logic over finite trees is decidable.*

One should note however that the translation from formulas to automata is non-elementary because each negation requires complementation of the automaton, which is exponential. In the worst case this cannot be avoided (see [29] for a proof in the case of words). Nevertheless, the construction has been implemented in the tool MONA [30, 31] and finds applications, e.g., in program verification [6].

On finite words there is also a nice characterization of languages that can be described in first-order logic (FO), i.e., MSO without set variables. These are exactly the star-free languages (see [32] for an overview). Over trees such a characterization is still missing (see [33] for a recent paper on the subject).

3.4. Hedge automata

In this section we adapt the automaton model such that we can handle unranked trees (and hedges). We keep the principle that the state at a node should only depend on the node label and the states at the children. For this purpose we have to define the transitions in such a way that they can deal with child sequences of arbitrary length. Since we have to deal with state sequences, we can rely on formalisms for defining word languages. To obtain an automaton model that has good closure and algorithmic properties we use regular languages of words to specify the transitions.

A *nondeterministic finite hedge automaton* (NFHA) over Σ is a tuple $\mathcal{A} = (Q, \Sigma, \Delta, F)$, where Q is a finite set of states, $F \subseteq Q$ is the set of final states, and Δ is a finite set of transitions of the form (L, a, q) with $a \in \Sigma$, $q \in Q$, and $L \subseteq Q^*$ a regular language over the alphabet Q.

NFHAs work on unranked trees in a similar way as NFTAs work on ranked trees: Each node is labeled by a state, where the possible states at a node are determined by its label from Σ, and by the states at the children. Formally, a *run* of $\mathcal{A}$ on a tree t is a Q-tree ρ satisfying the following conditions:

- $dom(\rho) = dom(t)$.
- For each node u with $t(u) = a$, $\rho(u) = q$, and $\rho(ui) = q_i$ for each $i \in \{1, \ldots, k\}$, where k is the number of children of u, there is a transition $(L, a, q) \in \Delta$ with $q_1 \cdots q_k \in L$. In particular, if $k = 0$ then the empty words has to be in L.

We use the same terminology as for ranked trees: A run ρ is *accepting* if it ends in a final state at the root, i.e., if $\rho(\varepsilon) \in F$. A tree t is accepted by $\mathcal{A}$ if there is an accepting run of $\mathcal{A}$ on t. The set of all trees accepted by $\mathcal{A}$ is called the *language* of $\mathcal{A}$ and is denoted by $T(\mathcal{A})$.

Example 3.4. We construct an NFHA over $\Sigma = \{a, b, c, d\}$ that accepts the trees that contain the following pattern: there is an a labeled node and two levels below this node there is a d labeled node. The NFHA guesses the d labeled node and verifies that the node two levels above is labeled a. For this purpose we use the states q_d for guessing the d labeled node, q'_d at the node above q_d, $q_\top$ if the pattern has been identified, and q for the other parts of the tree. The transitions are given below, where e denotes an arbitrary symbol from Σ:

$$(q^*, e, q),\ (q^*, d, q_d),\ (q^* q_d q^*, e, q'_d),\ (q^* q'_d q^*, a, q_\top),\ (q^* q_\top q^*, e, q_\top).$$

The only final state is $q_\top$. ◁

3.4.1. *Relation to ranked tree automata*

In Section 3.2 we have shown two possibilities how to code unranked trees by ranked ones. It is of course desirable that such encodings preserve recognizability of languages by the automaton models that we have presented. This allows to transfer results for ranked tree automata to hedge automata.

The following equivalence theorem can be shown by simple manipulations of automata. Therefore we do not attribute the results to specific papers but consider them to be folklore.

Theorem 3.9. *Given a language $T \subseteq \mathcal{T}_\Sigma^{\mathrm{unr}}$, the following conditions are equivalent:*

- *T is recognizable by a hedge automaton.*
- *$\mathsf{fcns}(T)$ is regular.*
- *$\mathsf{ext}(T)$ is regular.*

Proof. The equivalence can be shown by direct automaton constructions. We illustrate this by showing how to transform a hedge automaton into an automaton for $\mathsf{ext}(T)$. The other constructions follow a similar principle and are left to the reader.

Let $\mathcal{A} = (Q, \Sigma, \Delta, F)$ be an NFHA. We assume that for each a and q there is exactly one rule $(L_{a,q}, a, q)$. This is no restriction because several rules can be merged by taking the union of the horizontal languages. Furthermore, assume that each language $L_{a,q}$ is given by a deterministic finite (word) automaton $\mathcal{B}_{a,q} = (P_{a,q}, Q, p_{a,q}, \delta_{a,q}, F_{a,q})$ over the input alphabet Q. We assume that all state sets are pairwise disjoint.

For simulating a run of $\mathcal{A}$ on an unranked tree t by a run on $\mathsf{ext}(t)$ we only use the states of the automata $\mathcal{B}_{a,q}$. A subtree of $\mathsf{ext}(t)$ that is rooted at a node u that is the right child of some other node corresponds to a subtree in t, say at node u'. The simulation is implemented such that on $\mathsf{ext}(t)$ at u there is a state from $F_{a,q}$ iff in the corresponding run on t the state q is at node u', and u' is labeled a.

The NFTA $\mathcal{A}' = (Q', \Sigma', \Delta', F')$ for $\mathsf{ext}(T)$ is defined as follows:

- $Q' = \bigcup_{a,q} P_{a,q}$.
- Δ' contains the following transitions:

  - $(a, p_{a,q})$ for each $a \in \Sigma$ and $q \in Q$. These transitions are used to guess at each leaf the state q that is used in a run on the unranked tree at the node corresponding to this leaf.
  - $(p, p', @, p'')$ if p, p'' are both in some $P_{b,q}$, $p' \in F_{a,q'}$ for some q', and $\delta_{b,q}(p, q') = p''$.
- $F' = \bigcup_{a \in \Sigma, q \in F} F_{a,q}$.

The construction is illustrated in Figure 3.7. On the left hand side a run on the unranked tree $b(c, d)$ is shown. The states of the form p_* are the states of the automata $\mathcal{B}_{a,q}$ for the horizontal languages. In the picture it is shown how they are used to process the states of the form q_*. Note that the initial states p_{c,q_1} and p_{d,q_2} of $\mathcal{B}_{c,q_1}$ and $\mathcal{B}_{d,q_2}$ must also be final states because q_1 and q_2 are assigned to the two leafs and hence $\varepsilon \in L_{c,q_1}$ and $\varepsilon \in L_{d,q_2}$. Furthermore p_2 is a final state of $\mathcal{B}_{b,q_3}$. If $q_3 \in F$, then both runs are accepting because p_2 is a final state of $\mathcal{A}'$. □

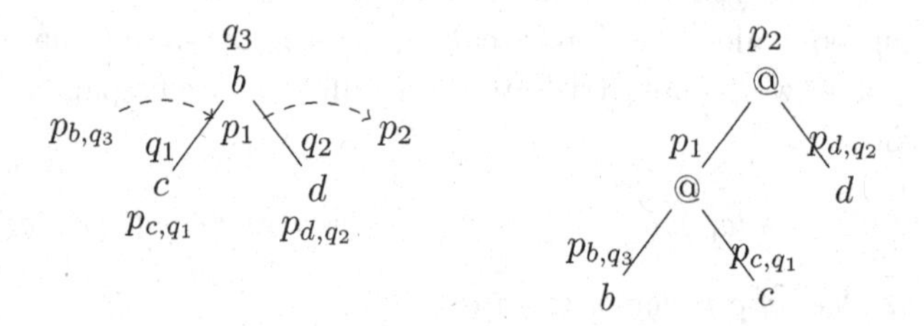

Fig. 3.7. Transferring runs from an unranked tree t to $\mathsf{ext}(t)$.

Based on this theorem we call a language of unranked trees regular if it is the language of some NFHA.

Theorem 3.10. *The class of regular unranked tree languages is closed under union, intersection, and complement.*

Proof. Let $T_1, T_2 \subseteq \mathcal{T}_\Sigma^{\mathrm{unr}}$ be two regular languages of unranked trees. According to Theorem 3.9 we obtain that $\mathsf{ext}(T_1)$ and $\mathsf{ext}(T_2)$ are regular. Since ext is a bijection between unranked trees and ranked trees we obtain $\mathsf{ext}(T_1) \cap \mathsf{ext}(T_2) = \mathsf{ext}(T_1 \cap T_2)$. By Theorem 3.2 $\mathsf{ext}(T_1) \cap \mathsf{ext}(T_2)$ is regular and therefore also $\mathsf{ext}(T_1 \cap T_2)$. Another application of Theorem 3.9 yields that $T_1 \cap T_2$ is regular.

Similar arguments work for union and complement. □

It is also possible to show Theorem 3.10 directly by giving automaton constructions that are similar to the ones for the ranked case. But because of the horizontal languages these constructions are more technical to write.

Finally, we mention that we can also use the encodings to solve decision problems for hedge automata using the results from Section 3.3.2. The translations forth and back from hedge automata to binary tree automata are polynomial, and thus we obtain the same complexity bounds as in Section 3.3.2. Note however, that there is a variety of formalisms for representing the horizontal languages in hedge automata which depend on the application at hand. Of course, this has an influence on the complexity. The above statement on the transfer of complexity bounds from NFTAs assumes that the horizontal languages are represented by regular expressions or nondeterministic finite automata. A more detailed analysis of algorithms for hedge automata can be found in [4].

3.4.2. *Grammar based formalisms*

Document type definitions for XML documents can be seen on an abstract level as definitions of regular tree languages. However, the common formalisms for such specifications are not based on automata but rather on grammars. Intuitively, a grammar generates trees instead of processing and then rejecting or accepting them. The mechanism is the same as for grammars generating word languages: A set of production rules specifies how nonterminal symbols can be replaced, where the new object that is substituted for a nonterminal can contain further nonterminals. This process is repeated until (after finitely many steps) an object without nonterminal symbols is reached.

Depending on the kind of production rules that are allowed, one obtains classes of grammars with different expressive power. On words this leads to the well-known Chomsky hierarchy. The lowest level consists of the regular languages generated, e.g., by right linear grammars, where the production rules are of the form $X \to wY$ for nonterminals X, Y, and a terminal word w, i.e., the nonterminals are only allowed to appear at the end of the word. In the tree setting this corresponds to the nonterminals only occurring at the leafs. This leads to rules of the form $X \to t$, where X is a nonterminal, and t is a tree in which nonterminal symbols only occur at leafs. Because we are interested in unranked trees, we also need a mechanism for horizontal recursion allowing to generate trees of unbounded width.

One possibility is to allow hedges on the right-hand sides of the rules. For example, the rules $X_1 \to a(X_2)$ and $X_2 \to b(c)\,X_2 \mid \varepsilon$ would allow to generated trees of the form $a(b(c) \cdots b(c))$. The hedge $b(c)\,X_2$ on the right-hand side of the second rule allows to produce an unbounded number of $b(c)$ below the root. To ensure that we only obtain regular languages one has to restrict this usage of nonterminals in hedges to right linear rules (as indicated above for words). We do not give formal definitions here because we use another popular variant of regular tree grammars that directly allow regular expressions in the rules and thus avoid the recursive usage of nonterminals on the horizontal level.

A *regular tree grammar* is of the form $G = (N, \Sigma, S, P)$, where Σ is the unranked alphabet (the letter are also called terminal symbols), N is a finite set of nonterminals, $S \in N$ is the start symbol, and P is a finite set of rules of the form $X \to a(r)$, where $a \in \Sigma$ and r is a regular expression over N.

The right-hand side of each rule defines a set of trees of height 1 or height 0 in case ε is in the language defined by the regular expression. A derivation of such a grammar starts with the nonterminal S and in each step replaces a nonterminal X with a tree $a(w)$ if there is a rule $X \to a(r)$, and $w \in N^*$ is in the language defined by r. The language $T(G)$ defined by G is the set of all trees in $\mathcal{T}_\Sigma^{\text{unr}}$ that can be generated in this way (in finitely many steps).

Example 3.5. The language from Example 3.4 is generated by the grammar with start symbol $S_\top$ and the following rules:

$$\begin{array}{lll} S_\top \to e(X^* S_\top X^*) & Y_d \to e(X^* X_d X^*) & X \to e(X^*) \\ S_\top \to a(X^* Y_d X^*) & X_d \to d(X^*) & \end{array}$$

for each $e \in \Sigma$. ◁

The translation between regular tree grammars and hedge automata is rather simple. In particular, the definition for regular tree grammars that we have given here is very close to the definition of hedge automata: the nonterminals are in correspondence to the states of the automaton, and the production rules correspond to the transitions (compare Examples 3.4 and 3.5).

Remark 3.1. The languages that can be generated by regular tree grammars are exactly the regular tree languages.

By imposing restrictions on the grammars one can obtain common languages that are used for defining types of XML documents. Such an analysis is given in [7] (see also [4]). *Document type definitions* (DTD) correspond to local grammars, in which nonterminals can be identified with the symbols from Σ, and the rules are of the form $a \to a(r)$, i.e., each nonterminal generates the terminal symbol it corresponds to.

In *XML Schema* the nonterminals are typed versions of the terminals, and the rules are of the form $a^{(i)} \to a(r)$, where the superscript on the left-hand side of the rule indicates the type of the nonterminal. This itself is not yet a true restriction for regular tree grammars, only a naming convention for nonterminals. Regular grammars following this convention are also referred to as extended DTDs (EDTDs). XML Schema corresponds to so-called single type EDTDs: the regular expressions on the right-hand side of the rules are restricted such that they do not produce words that contain two different types of the same symbol. For example, the rule $a^{(1)} \to a(b^{(1)}b^{(2)})$ is not allowed because the word $b^{(1)}b^{(2)}$ contains two different types of b. The reason for imposing these restrictions is to allow efficient membership tests. A detailed analysis of XML Schema and single type EDTDs is given in [34] (where extended DTDs are called specialized DTDs).

3.5. Tree-walking automata

In the previous sections we have seen automata models that work in parallel spread over the whole input tree. The transitions define how to merge the information computed on different subtrees. This section deals with a sequential automaton model introduced in [11]. A run of such an automaton (on a ranked tree) starts at the root. Based on the current state, the label of the current node, and the information on the position of the current node among its siblings, the automaton can move to one of its children or to its parent node (and change its state at the same time). In general, a run can cross a node of a tree several times. The automaton accepts by switching to a final state.

Let Σ be a ranked alphabet and let k be the maximal arity of a symbol from Σ. A *tree-walking automaton* (TWA) is of the form $\mathcal{A} = (Q, \Sigma, q^{\mathrm{i}}, \Delta, F)$, where Q is a finite set of states, $q^{\mathrm{i}} \in Q$ is the initial state, $F \subseteq Q$ is the set of final states, and Δ is the transition relation. A transition depends on the current state as well as the label and type of the current node. The type of the current node is either *root* or a number in $\{1, \ldots, k\}$, where number i means that the node is the ith child of its parent. We let $\mathit{Types} = \{\mathit{root}, 1, \ldots, k\}$. Based on this information, a transition determines a new state and a direction in which to move. In general, the possible directions are $\uparrow$ (move up to the parent node), ε (stay at the current node), and $1, \ldots, k$ (move to the the corresponding child). The set of directions is denoted by $\mathit{Dir} = \{\uparrow, \varepsilon, 1, \ldots, k\}$.

The transition relation is of the form $\Delta \subseteq Q \times \mathit{Types} \times \Sigma \times Q \times \mathit{Dir}$, where of course the direction specified in a transition has to be compatible with the other information: If $(q, \mathit{type}, a, q', d) \in \Delta$, then $d \in \{\uparrow, \varepsilon, 1, \ldots, |a|\}$, and if $\mathit{type} = \mathit{root}$ then $d \neq \uparrow$.

A configuration of a TWA is a pair (u, q) of a node and a state. A *run* of a TWA $\mathcal{A}$ on a tree t is a sequence $(u_0, q_0)(u_1, q_1)\cdots$ of configurations that starts in the initial state at the root, i.e., $u_0 = \varepsilon$ and $q_0 = q^{\mathrm{i}}$, and each two successive configurations are related by a transition in Δ. A formal definition of the latter statement requires a lot of case distinctions. We only give two typical examples:

- If u_{i+1} is the jth child of u_i, i.e., $u_{i+1} = u_i j$, and u_i is the root, then there must be a transition $(q_i, \mathit{root}, t(u_i), q_{i+1}, j) \in \Delta$. This is shown on the left-hand side of Figure 3.8.
  If u_i is not the root but the ℓth child of some other node, then the type *root* in the transition must be replaced by ℓ.
- If u_i is the jth child of u_{i+1}, i.e., $u_i = u_{i+1} j$, then there must be a transition $(q_i, j, t(u_i), q_{i+1}, \uparrow) \in \Delta$. This is shown on the right-hand side of Figure 3.8.

A run is accepting if it ends in a final state at the root. Note that runs of a TWA might be infinite because they start looping at some point. These runs are of course non-accepting. The language $T(\mathcal{A})$ of a TWA is defined as usual as the set of trees for which there exists an accepting run of $\mathcal{A}$ on t.

The possibility of circular runs that do not terminate are one of the reasons that make TWAs difficult to analyze. For example, this prevents easy complementation of deterministic TWAs by exchanging final and non-final states. However,

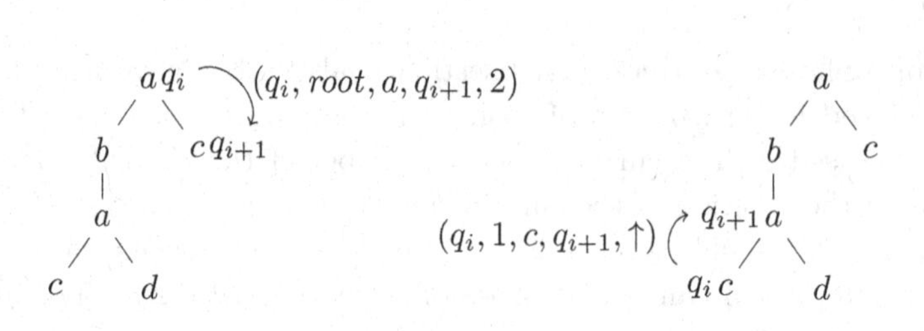

Fig. 3.8. Illustration of transitions of a TWA.

in [35] it is shown that every deterministic TWA can be turned into a deterministic TWA that does not admit circular runs (it follows that deterministic TWAs can be complemented).

Example 3.6. The most basic example is a TWA performing a depth-first left-to-right search in the tree. For simplicity we assume that we are working on an alphabet with one binary symbol a, and one constant c. We use states q_l, q_r, q_u (for left, right, and up), and one final state q_f. The initial state is q_l, and the transitions are

$$
\begin{array}{l}
(q_l, type, a, q_l, 1) \text{ for all } type \in Types \\
(q_l, 1, c, q_r, \uparrow) \\
(q_l, 2, c, q_u, \uparrow) \\
(q_r, type, a, q_l, 2) \text{ for all } type \in Types \\
(q_u, 1, a, q_r, \uparrow) \\
(q_u, 2, a, q_u, \uparrow) \\
(q_u, root, a, q_f, \varepsilon)
\end{array}
$$

An run of this automaton is depicted in Figure 3.9. Note that the automaton from this example simply traverses the whole tree and then accepts at the root, i.e., its language is the set of all trees. But now one can combine this generic automaton with other automata that test certain properties while traversing the tree. ◁

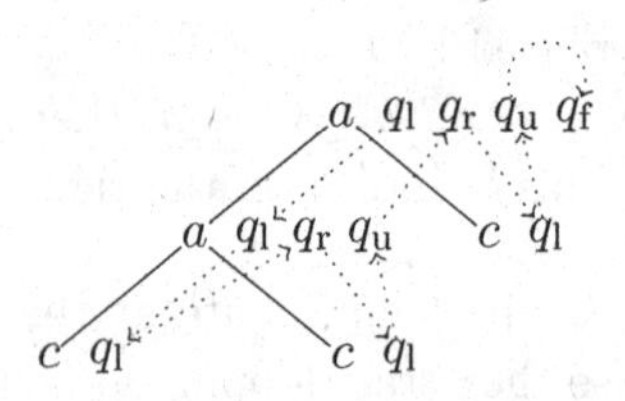

Fig. 3.9. Run of a TWA doing a depth-first search (Example 3.6).

The way TWAs process an input tree is completely different from the way it is done by NFTAs. A first question that comes up is whether the ability of visiting parts of the tree several times gives more power to TWAs. But similar to the case of two-way automata on finite words this is not the case.

Theorem 3.11. *TWAs only recognize regular languages.*

Proof. It is not difficult to show that $\sim_T$ has finite index for a TWA language T. Fix a TWA $\mathcal{A}$ with the usual components for the language T and consider for each tree t and each type *type* the possible behaviors of $\mathcal{A}$ on t if t is a subtree of type *type*. For the type *root* the behavior is simply the information whether t is accepted by $\mathcal{A}$ or not. If t is of type i (i.e., it is the ith subtree of some node), then the behavior is a relation containing all pairs (p, q) of states such that $\mathcal{A}$ started at the root of t in state p has a computation that remains on t and reaches q when exiting t.

If the behavior of $\mathcal{A}$ on two trees t_1 and t_2 is the same, then $t_1 \sim_T t_2$ because

- for the trivial context $C = X$ it follows that $C[t_1] \in T \Leftrightarrow C[t_2] \in T$ because the behavior of $\mathcal{A}$ includes the information whether the trees are accepted or not, and
- for a nontrivial context C, an accepting run of $\mathcal{A}$ on $C[t_1]$ can be turned into an accepting run of $\mathcal{A}$ on $C[t_2]$ (and vice versa): whenever the run enters the subtree t_1 with state p and exists with state q at the parent node, then we know that this is also possible on t_2.

Since there are only finitely many possible behaviors, T is regular according to Theorem 3.7. □

It has been open for a long time whether TWAs can recognize all regular languages. Recently, it has been shown that this is not the case. Before that it has been shown for some restricted models of TWA that they are weaker than NFTAs (e.g. [36]).

Theorem 3.12 ([37]). *There exists a regular tree language that cannot be recognized by a TWA.*

As opposed to parallel tree automata, where the nondeterministic and the deterministic model have the same expressive power, determinism is a true restriction for TWAs.

Theorem 3.13 ([38]). *There exists a tree language that can be accepted by a TWA but not by a deterministic TWA.*

This already illustrates the difference between NFTAs and TWAs. The two models also differ w.r.t. algorithmic questions, as witnessed, e.g., by the following theorem.

Theorem 3.14. *The emptiness problem for TWAs is* EXPTIME*-complete.*

The reason for this cost is the ability of TWAs to move up and down in the tree. This allows a TWA to sequentially check certain properties that have to be checked in one pass by an NFTA. The EXPTIME lower bound can be shown by a reduction from the two-person corridor tiling problem used in the proof of Theorem 3.5: One can construct a TWA of polynomial size (even a deterministic one) that verifies that a given tree represents a winning strategy for Eva in a given two-person corridor tiling system. The upper bound can be derived from an exponential translation of TWAs into standard tree automata and the polynomial time emptiness test for the latter (Theorem 3.4). Such an exponential translation can, e.g., be derived from a corresponding result for the more general model of two-way alternating tree automata [39]. That this exponential blow-up cannot be avoided, in general, easily follows from results on the succinctness of two-way automata on words (see [40]).

For more references on TWAs (in particular on extensions of TWAs by pebbles) we refer the reader to the abstract[d] [41].

3.5.1. *Streams*

Because of their sequential behavior, tree-walking automata can be seen as automata working on a linearized version of the tree. However, for navigating inside the linear representation they still use the tree structure. In the context of XML document processing there is a natural setting in which the automaton cannot work on the document as a tree but has to work on a linearized version: when documents are exchanged over a network they arrive as a stream. If large documents are exchanged, then it is desirable to process them online such that it is not necessary to store them completely before analyzing them.

Given a tree, there are various ways of presenting it as a linear structure. In XML documents opening and closing tags are used. We use the symbol itself as opening tag and an over-lined version of the symbol as closing tag. The coding is obtained by a depth-first traversal of the tree, putting an opening tag when entering a subtree, and a closing tag when leaving the subtree.

For example, the tree from the left-hand side of Figure 3.1 and the tree from Figure 3.2 are, respectively, coded as

$$abac\bar{c}d\bar{d}\bar{a}\bar{b}c\bar{c}\bar{a} \quad \text{and} \quad abc\bar{c}d\bar{d}\bar{b}a\bar{a}ab\bar{b}\bar{a}\bar{a}.$$

For a tree t (ranked or unranked) we refer to the above linearization as $lin_{\mathrm{XML}}(t)$, and for a tree language T we denote the set of all linearizations obtained from trees in T by $lin_{\mathrm{XML}}(T)$.

Languages of this kind have already been analyzed in [9] under the name of nest sets, and languages of the form $lin_{\mathrm{XML}}(T)$ for regular tree languages T are characterized by a specific type of context-free grammars. These grammars are studied in detail in [42].

[d] An extended version of this abstract can be found on the web page of the author.

It is not difficult to show that $lin_{\mathrm{XML}}(T)$ is not a regular word language in general for regular T. The reason is that a finite automaton cannot check whether all opening tags are closed correctly. In the setting where we are interested in verifying streaming objects, it is often reasonable to assume that the object is well-formed (i.e., all opening tags are closed correctly). The question whether a finite automaton can test for the membership in $lin_{\mathrm{XML}}(T)$ when the inputs are assumed to be well-formed is much more interesting. Consider, for example, the (unranked) tree language T that contains all trees whose right-most branch is completely labeled by a. Obviously, $lin_{\mathrm{XML}}(T)$ is not regular but it contains exactly those words from $\Sigma^* a\bar{a}^+$ that correspond to well-formed inputs.

The problem of deciding for a regular tree language whether $lin_{\mathrm{XML}}(T)$ can be verified by a finite automaton in the above sense is still open in its full generality. Partial results have been obtained in [43, 44].

The problem becomes different when considering another coding with only one type of parenthesis: We again traverse the tree by a depth first search. When entering a subtree we put the symbol and an opening parenthesis, when leaving the subtree we put a closing parenthesis. For the same example trees used above we obtain the following linearizations

$$a(b(a(c()d()))c()) \text{ and } a(b(c()d())a()a(b())).$$

Because this notation corresponds to the standard way how trees are written as terms, we refer to this coding of t by $lin_{\mathrm{term}}(t)$ and $lin_{\mathrm{term}}(T)$ for languages T. One can easily see that $lin_{\mathrm{term}}(T)$ can be obtained from $lin_{\mathrm{XML}}(T)$ by the simple morphism that maps each a to $a($ and each $\bar{a}$ to $)$. Therefore, an automaton accepting $lin_{\mathrm{XML}}(T)$ when restricted to well-formed inputs, can easily be turned into an automaton for $lin_{\mathrm{term}}(T)$ on well-formed inputs.

In the other direction such a translation does not work. For the above example language T containing those trees whose right-most branch is completely labeled by a, there is no regular language of words that yields $lin_{\mathrm{term}}(T)$ when restricted to well-formed inputs (this can be shown, e.g., using the techniques developed in [45]).

It turns out that this difference makes the corresponding decision problem simpler: In [45] it is shown that it is decidable for a regular tree language T whether $lin_{\mathrm{term}}(T)$ can be recognized by a finite automaton when the inputs are restricted to be well-formed. But for the application to processing streaming XML documents this result is less interesting because it does not use the standard coding.

Coming back to the first encoding, an automaton model that captures the languages $lin_{\mathrm{XML}}(T)$ for regular T can be derived from extended versions of tree-walking automata that are powerful enough to capture the regular tree languages. In [46] and [47] such extensions are mentioned. In the first paper it is shown that TWAs with colored marbles that can be droped when entering a subtree and have to be picked up when leaving the subtree recognize all regular languages. In the second paper marbles are called invisible pebbles. Basically, for these extended TWAs it is enough to make one depth-first traversal of the tree. The marbles are used to give the TWA access to the parts of the run it has already computed. These automata are very closely related to specific kinds of pushdown automata as they are used, e.g., in [48] and [49].

3.6. Conclusion

In this chapter we have presented basic definitions and results in the field of tree automata. The material presented in Section 3.3 shows that a lot of concepts for the theory of automata over finite words can be adapted to the setting of ranked trees, in particular automata constructions for boolean operations on languages, minimization of automata, and the relation to monadic second-order logic. Also from an algorithmic point of view there are a lot of similarities to word automata, where in most cases the cost for algorithms on tree automata is one exponential higher than for word automata. Section 3.4 illustrates that it is possible to develop a theory of automata for unranked trees along the same lines as for ranked trees. Using encodings of unranked trees by ranked ones it is possible to lift many of the results from ranked tree automata. Tree-walking automata as presented in Section 3.5 process the input tree in a sequential way by navigating inside the tree. We have seen that this leads to a weaker expressive power and different algorithmic properties. Nevertheless, extensions of this model are useful, e.g., when dealing with logical formalisms like (Core)XPath [50] or first-order logic with transitive closure (see [47] and [51]).

There are various active research areas on different new aspects of tree automata, some of which are treated in other chapters of this volume. We have already mentioned the problem of processing streaming XML documents in Section 3.5.1. Another current topic that is also motivated from the XML world is the extension of automata to deal with data values from infinite domains (see Chapter 17 of this volume). Related to this are automaton models that allow the comparison of subtrees for equality and disequality (see [4] for an overview and [52, 53] for more recent developments).

Various applications do not just require to classify trees into accepted and rejected but are concerned with transformations of trees. There is a wide range of literature on tree transducers and we cannot survey all the suggested models here. We refer the reader to Chapter 11 of this volume, which deals with a specific kind of transducers with tree-walking automata as underlying mechanism.

Finally, for further reading we mention some recent surveys on different aspects of tree automata: The electronic book [4] covers a lot of material related to ranked tree automata and hedge automata. Automata related formalisms that are interesting in the context of XML are presented in [54] and logics for unranked trees are surveyed in [55].

Acknowledgements

I thank the editors for the invitation to contribute to this volume, and the anonymous referee for the helpful remarks and suggestions.

References

[1] J. W. Thatcher and J. B. Wright, Generalized finite automata theory with an application to a decision problem of second-order logic, *Mathematical Systems Theory.* **2** (1), 57–81, (1968).

[2] J. Doner, Tree acceptors and some of their applications, *J. Computer and System Sciences.* **4**, 406–451, (1970).

[3] M. O. Rabin, Decidability of second-order theories and automata on infinite trees, *Transactions of the American Mathematical Society.* **141**, 1–35 (July, 1969).

[4] H. Comon, M. Dauchet, R. Gilleron, F. Jacquemard, C. Löding, D. Lugiez, S. Tison, and M. Tommasi. *Tree Automata Techniques and Applications.* (`http://tata.gforge.inria.fr/`). Last Release: October 12, 2007.

[5] P. A. Abdulla, A. Legay, J. d'Orso, and A. Rezine, Tree regular model checking: A simulation-based approach, *J. Log. Algebr. Program.* **69**(1-2), 93–121, (2006).

[6] A. Møller and M. I. Schwartzbach. The pointer assertion logic engine. In *Proc. PLDI 2001*, pp. 221–231. ACM, (2001).

[7] M. Murata, D. Lee, M. Mani, and K. Kawaguchi, Taxonomy of XML schema languages using formal language theory., *ACM Trans. Internet Technology.* **5**(4), 660–704, (2005).

[8] J. W. Thatcher, Characterizing derivation trees of context-free grammars through a generalization of finite automata theory, *J. Computer and System Sciences.* **1**, 317–322, (1967).

[9] M. Takahashi, Generalizations of regular sets and their application to a study of context-free languages, *Information and Control.* **27**, 1–36, (1975).

[10] A. Brüggemann-Klein, M. Murata, and D. Wood. Regular tree and regular hedge languages over unranked alphabets. Technical Report HKTUST-TCSC-2001-05, HKUST Theoretical Computer Science Center Research, (2001).

[11] A. V. Aho and J. D. Ullman, Translations on a context-free grammar, *Information and Control.* **19**(5), 439–475, (1971).

[12] D. E. Knuth, *The Art of Computer Programming. Volume 1: Fundamental Algorithms.* (Addison-Wesley, 1997), 3rd edition.

[13] J. Carme, J. Niehren, and M. Tommasi. Querying unranked trees with stepwise tree automata. In *Proc. 15th RTA 2004*, vol. 3091, *LNCS*, pp. 105–118. Springer, (2004).

[14] M. O. Rabin and D. Scott, Finite automata and their decision problems, *IBM J. Research and Development.* **3**, 114–125 (April, 1959).

[15] J. E. Hopcroft, R. Motwani, and J. D. Ullman, *Introduction to Automata Theory, Languages, and Computation.* (Addison-Wesley, 2001), 2nd edition.

[16] W. Martens, F. Neven, and T. Schwentick. Deterministic top-down tree automata: past, present, and future. In *Logic and Automata - History and Perspectives*, vol. 2, *Texts in Logic and Games*, pp. 505–530. Amsterdam University Press, (2007).

[17] M. Lohrey. On the parallel complexity of tree automata. In *Proc. 12th RTA*, vol. 2051, *LNCS*, pp. 201–215. Springer, (2001).

[18] R. Greenlaw, H. J. Hoover, and W. L. Ruzzo, *Limits to Parallel Computation: P-Completeness Theory.* (Oxford University Press, 1995).

[19] H. Seidl, Deciding equivalence of finite tree automata, *SIAM J. Comput.* **19**(3), 424–437, (1990).
[20] M. Veanes. On computational complexity of basic decision problems of finite tree automata. Technical Report UPMAIL Technical Report 133, Uppsala University, Computing Science Department (Jan., 1997).
[21] B. S. Chlebus, Domino-tiling games, *J. Computer and System Sciences.* **32**(3), 374–392, (1986).
[22] J. Myhill. Finite automata and the representation of events. Technical Report WADC TR-57-624, Wright Patterson Air Force Base, Ohio, USA, (1957).
[23] A. Nerode, Linear automata transformations, *Proc. Amer. Math. Soc.* **9**, 541–544, (1958).
[24] D. Kozen, On the Myhill-Nerode theorem theorem for trees, *Bulletin of the EATCS.* **47**, 170–173, (1992).
[25] R. C. Carrasco, J. Daciuk, and M. L. Forcada. An implementation of deterministic tree automata minimization. In *Proc. 12th CIAA*, vol. 4783, *LNCS*, pp. 122–129. Springer, (2007).
[26] J. R. Büchi, Weak second order logic and finite automata, *Zeitschrift für mathematische Logik und Grundlagen der Mathematik.* **6**, 66–92, (1960).
[27] C. C. Elgot, Decision problems of finite automata design and related arithmetics, *Trans. American Mathematical Society.* **98**, 21–51, (1961).
[28] B. A. Trakhtenbrot, Finite automata and the logic of one-place predicates, *Sibirian Mathematical J.* **3**, 103–131, (1962). English translation in: AMS Transl. 59 (1966) 23-55.
[29] K. Reinhardt. The complexity of translating logic to finite automata. In *Automata, Logics, and Infinite Games*, vol. 2500, *LNCS*, pp. 231–238. Springer, (2002).
[30] N. Klarlund. Mona & Fido: The logic-automaton connection in practice. In *Proc. 11th CSL*, vol. 1414, *LNCS*, pp. 311–326. Springer, (1997).
[31] N. Klarlund and A. Møller. *MONA Version 1.4 User Manual.* BRICS, Department of Computer Science, University of Aarhus (January, 2001). Notes Series NS-01-1. Available from `http://www.brics.dk/mona/`. Revision of BRICS NS-98-3.
[32] V. Diekert and P. Gastin. First-order definable languages. In *Logic and Automata - History and Perspectives*, vol. 2, *Texts in Logic and Games*, pp. 261–306. Amsterdam University Press, (2007).
[33] M. Benedikt and L. Segoufin, Regular tree languages definable in FO and in FO_{mod}, *ACM Trans. Comput. Log.* **11**(1), (2009).
[34] W. Martens, F. Neven, T. Schwentick, and G. J. Bex, Expressiveness and complexity of XML schema., *ACM Trans. Database Systems.* **31**(3), 770–813, (2006).
[35] A. Muscholl, M. Samuelides, and L. Segoufin, Complementing deterministic tree-walking automata, *Inf. Process. Lett.* **99**(1), 33–39, (2006).
[36] F. Neven and T. Schwentick, On the power of tree-walking automata, *Inf. Comput.* **183**(1), 86–103, (2003). Conference version in ICALP'00.
[37] M. Bojanczyk and T. Colcombet, Tree-walking automata do not recognize all regular languages, *SIAM J. Comput.* **38**(1), 658–701, (2008). Conference version in STOC'05.
[38] M. Bojanczyk and T. Colcombet, Tree-walking automata cannot be determinized, *Theor. Comput. Sci.* **350**(2-3), 164–173, (2006). Conference version in ICALP'04.
[39] S. S. Cosmadakis, H. Gaifman, P. C. Kanellakis, and M. Y. Vardi. Decidable optimization problems for database logic programs (preliminary report). In *Proc. 20th STOC*, pp. 477–490. ACM, (1988).
[40] C. A. Kapoutsis. Removing bidirectionality from nondeterministic finite automata. In *Proc. 30th MFCS*, vol. 3618, *LNCS*, pp. 544–555. Springer, (2005).

[41] M. Bojańczyk. Tree-walking automata. In *Proc. 2nd LATA*, vol. 5196, *LNCS*, pp. 1–2. Springer, (2008).
[42] J. Berstel and L. Boasson, Formal properties of XML grammars and languages, *Acta Inf.* **38**(9), 649–671, (2002).
[43] L. Segoufin and V. Vianu. Validating streaming XML documents. In *Proc. 21st PODS*, pp. 53–64. ACM, (2002).
[44] L. Segoufin and C. Sirangelo. Constant-memory validation of streaming xml documents against dtds. In *Proc. 11th ICDT*, vol. 4353, *LNCS*, pp. 299–313. Springer, (2007).
[45] V. Bárány, C. Löding, and O. Serre. Regularity problems for visibly pushdown languages. In *Proc 23rd STACS*, vol. 3884, *LNCS*, pp. 420–431. Springer, (2006).
[46] J. Engelfriet, H. J. Hoogeboom, and J.-P. V. Best, Trips on trees, *Acta Cybern.* **14** (1), 51–64, (1999).
[47] J. Engelfriet, H. J. Hoogeboom, and B. Samwel. XML transformation by tree-walking transducers with invisible pebbles. In *Proc. 26th PODS'07*, pp. 63–72. ACM, (2007).
[48] C. Koch and S. Scherzinger. Attribute grammars for scalable query processing on XML streams. In *Proc. 9th DBPL*, vol. 2921, *LNCS*, pp. 233–256. Springer, (2003).
[49] R. Alur and P. Madhusudan. Visibly pushdown languages. In *Proc. 36th STOC*. ACM, (2004).
[50] G. Gottlob, C. Koch, and R. Pichler, Efficient algorithms for processing XPath queries, *ACM Trans. Database Systems.* **30**(2), 444–491, (2005).
[51] B. ten Cate and L. Segoufin. XPath, transitive closure logic, and nested tree walking automata. In *Proc. 27th PODS*, pp. 251–260. ACM, (2008).
[52] W. Karianto and C. Löding. Unranked tree automata with sibling equalities and disequalities. In *Proc. 34th ICALP*, vol. 4596, *LNCS*, pp. 875–887. Springer, (2007).
[53] E. Filiot, J.-M. Talbot, and S. Tison. Tree automata with global constraints. In *Proc. 12th DLT*, vol. 5257, *LNCS*, pp. 314–326. Springer, (2008).
[54] T. Schwentick, Automata for XML - a survey, *J. Computer System Sciences.* **73**(3), 289–315, (2007).
[55] L. Libkin, Logics for unranked trees: An overview, *Logical Methods in Computer Science.* **2**(3), (2006).

Chapter 4

An Introduction to Timed Automata

Paritosh K. Pandya and P. Vijay Suman*

Tata Institute of Fundamental Research,
Homi Bhabha Road, Mumbai 400005, India
pandya@tifr.res.in
vsuman@tifr.res.in

Timed automata have emerged as a prominent model for representation and analysis of real-time systems. The study of closure properties and decision problems for timed automata provides key insights into their computational power. The resulting algorithms have influenced the development of automatic analysis tools for model checking real-time systems. In this chapter we provide an introduction to the theory of timed automata.

4.1. Introduction

Real-time systems must operate under stringent timing constraints. Such programs can change their behaviour in response to passing of time by making use of internal timers. In order to formalize the behaviours of real-time systems a notion of timed word is used: a *timed word* is a finite sequence of letters where each letter also carries a real-valued time-stamp giving the time of its occurrence. Moreover, finite state automata can be enriched with clocks which all grow at a fixed rate. The clocks can be reset at events (letters) and the automaton transitions can carry constraints (guards) on the current values of clocks to restrict the time of occurrence of transitions. Using these ideas, an extension of finite state automata to *Timed Automata* has been defined by Alur-Dill and a rich study of *timed regular languages* has been carried out [1, 2]. Moreover, the decision procedures for analyzing the properties of these timed automata have resulted into model checking tools for real-time systems [3–5].

In this chapter, we provide an introduction to the theory of timed automata. Such automata recognize timed words and the set of timed words accepted by a timed automaton constitutes a *timed regular language* which is the central object of study in this article. We also study the closure and decidability properties such as

*This work was partially supported by the project 11P-202 at TIFR titled *Laboratory for Construction, Analysis and Verification of Embedded Systems.*

determinizable, language emptiness and language inclusion. The language emptiness essentially captures reachability of locations in timed automaton, and hence it captures safety properties of timed systems. Language inclusion relates to the model checking question. These questions have shed light upon the trade-offs between the expressive power of various classes of timed automata and the checkability of their desirable properties using algorithms.

This chapter deals with timed automata recognizing finite timed words. These automata have guards with limited expressive power: a clock value can only be constrained to be in a fixed interval with specified rational end-points. Moreover, the clocks are allowed to be reset only to the value 0 in transitions. With these restrictions, the emptiness of timed language of a timed automata can be shown to be decidable using the well-known *region automaton* construction of Alur and Dill [1]. This construction is at the heart of several model checking tools for timed systems. Unfortunately, the technique does not extend to language inclusion checking and the language inclusion between timed automata turns out to be undecidable. It can also be shown that non-deterministic timed automata are considerably more powerful than their deterministic counter-part. In the article, we give rigorous proofs of these and related results. We also explore some determinizable classes. We also explore how the expressive power of the automata changes with the number of clocks.

Timed automata are rather limited in their modellng abilities. They can be enhanced in many ways. For example, Alur and Dill originally formulated timed automata as recognizers for infinite timed words with Buchi and Muller acceptance conditions. They also allowed as guards interval constraints on difference between two clocks (these are called diagonal constraints). It can be shown that most of the properties of the simpler timed automata studied in this chapter carry over to these extensions. However, more general guards consisting of arbitrary linear constraints on clock values, or more general notions of clock updates which decrement clocks seem to render the emptiness checking problem undecidable. We give an overview of these results.

One of the fundamental issues when modelling timed behaviour is the underlying nature of time. The timed automaton model deals with instantaneous events (transitions) unfolding in real-valued time. This is often called point wise or sampled notion of time. Alternate studies range from automata with integer valued time, to automata with continuous time where events (conditions) last for time intervals. The expressive power of the automata and their decidability properties change quite significantly with the change in the nature of time. In the chapter, we describe properties of integer timed automata and their improved decidability. Moreover, we discuss techniques for approximating dense timed automata with discrete timed automata using a notion of threshold based sampling called digitization [6]. This provides us with a partial but practically usable technique for solving many instances of the language inclusion question.

A salient feature of timed automata is that the time is considered global with all clocks perfectly synchronized and growing at the same fixed rate. This can be conveniently be denoted by the invariant $\dot{x} = 1$ stating that the rate of change of a clock x in any state is 1. Embedded systems typically consist of entities which continuously evolve with time according to specified dynamics, say, a differential equation. They also include digital control which cause mode changes in this dynamics. To model such situations, timed automata have been generalized to hybrid automata where the rate of change of any variable is state dependent and it is specified by associating to each state a differential equation for every variable. Some later chapters in this volume deal with such hybrid automata.

The rest of this chapter is organized as follows. Section 4.2 gives the basic model of timed automata and explores its closure properties. Section 4.3 gives the region automaton construction showing that the language emptiness of timed automata is decidable. The section following it shows undecidability of checking universality and language inclusion for timed automata. Section 4.5 deals with the expressive power of deterministic timed automata and the vexed issue of determinizability in timed automata. The expressive power of timed automata with fixed number of clocks is explored in Section 4.6. Section 4.7 presents the digitization technique for approximating the behaviour of dense timed automata with simpler integer timed automata. The last section explores extensions to the basic model of timed automata.

The material covered in this introductory chapter draws heavily upon published papers and some excellent surveys on the theory of timed and hybrid automata. The bibliography and the citations within the text provide pointers to these sources. We very gratefully acknowledge the influence of these original works.

4.2. Timed Automata

In any system, there are some observable events whose occurrences in time constitutes the observed behaviour of the system. We define timed words to model such behaviours. A timed word is a sequence of pairs wherein each pair consists of an event and its real-valued time of occurrence. The set of observable events constitutes the alphabet.

Definition 4.1 (Timed Word). *A finite timed word over a finite alphabet Σ is defined as $\rho = (\sigma, \tau)$, where $\sigma = \sigma_1 \dots \sigma_n$ with $0 \leq n$ is a finite sequence of symbols in Σ and $\tau = \tau_1 \dots \tau_n$ is a weakly monotone sequence of real numbers which represent the time stamps of the corresponding events in σ. The weak monotonicity of time is captured by requiring that for $1 \leq i < n$, $\tau_i \leq \tau_{i+1}$. We may alternatively write ρ also as $(\sigma_1, \tau_1) \dots (\sigma_n, \tau_n)$. A timed language is any set of timed words. The set of all timed words over Σ is denoted by $T\Sigma^*$.* □

The *timed automaton* (TA) model was introduced in the seminal paper by Alur and Dill [1] where its formal properties were also studied. A timed automaton is basically a finite state automaton adjuncted with real valued clocks (or variables) which increment at the uniform rate of the global time. Such an automaton acts as a device which accepts or rejects time words. Its formal definition is given below.

Let C be the set of clocks of a timed automaton. A state of a timed automaton is composed of the current location of the automaton and the current value of each of its clocks. A clock valuation (denoted by symbols $\nu, \nu', \ldots$) gives the value of each clock. We use $\mathbb{Z}$, $\mathbb{Q}$ and $\mathbb{R}$ to represent the set of natural numbers, the set of rational numbers and the set of real numbers, respectively, and $\mathbb{Z}_0$, $\mathbb{Q}_0$ and $\mathbb{R}_0$ to denote their subsets with non-negative values.

Definition 4.2 (Clock Valuation). *An interpretation/valuation ν is a map from C to $\mathbb{R}_0$. The set of all such maps is denoted by $\mathbb{R}_0^C$. Given a clock valuation ν, a non-negative real value t and $\lambda \subseteq C$ we define $(\nu + t)(x) = \nu(x) + t$, and*

$$\nu[\lambda := 0] = \begin{cases} 0 & \forall x \in \lambda, \\ \nu(x) & \forall x \in (C \setminus \lambda). \end{cases}$$

Moreover, let ν_{init} be a valuation such that $\nu_{init}(x) = 0$ for any clock x. □

A transition of a timed automaton is guarded by a constraint on the values of the clocks. The constraint specifies the condition under which the transition is allowed to occur. Let $\Phi(C)$ be the set of constraints φ over the set of clocks C whose syntax given below. Let x range over C and c range over non-negative rational constants in $\mathbb{Q}_0$. Let φ range over $\Phi(C)$. Then,

$$\varphi \quad := \quad x \leq c \mid x \geq c \mid x < c \mid x > c \mid \varphi_1 \wedge \varphi_2$$

Notation $\nu \models \varphi$ denotes that the constraint φ evaluates to *true* in the valuation ν.

Definition 4.3 (Timed Automaton). *A timed automaton $\mathcal{A}$ is a tuple $(L, L_0, \Sigma, C, E, F)$ where,*

- *L is a finite set of locations,*
- *$L_0 \subseteq L$ is the set of initial locations,*
- *Σ is a finite set of symbols (alphabet),*
- *C is a finite set of real valued (variables called) clocks,*
- *$E \subseteq L \times L \times \Sigma \times \Phi(C) \times 2^C$ gives the set of transitions. An edge $e = (l, l', a, \varphi, \lambda)$ represents a transition from the source location l to the target location l' on input symbol a. Here, φ is a guard over C and $\lambda \subseteq C$ gives the set of clocks to be reset to value 0 when the transition is taken, and*
- *$F \subseteq L$ is the set of final locations.* □

Example 4.1. The timed automaton given in Figure 4.1 models a system which acknowledges a request within 2 time units of receiving it. Further requests are not accepted till the acknowledgment is sent. Here, $L = \{l_0, l_1\}$, $L_0 = \{l_0\}$, $F = \{l_0\}$, $\Sigma = \{req, ack\}$; $C = \{x\}$, and $E = \{(l_0, l_1, req, True, \{x\}), (l_1, l_0, ack, x \leq 2, \emptyset)\}$. □

Note that when the constraint is *True* or when the set of clocks to be reset over an edge is $\emptyset$, these are omitted in the pictorial depiction.

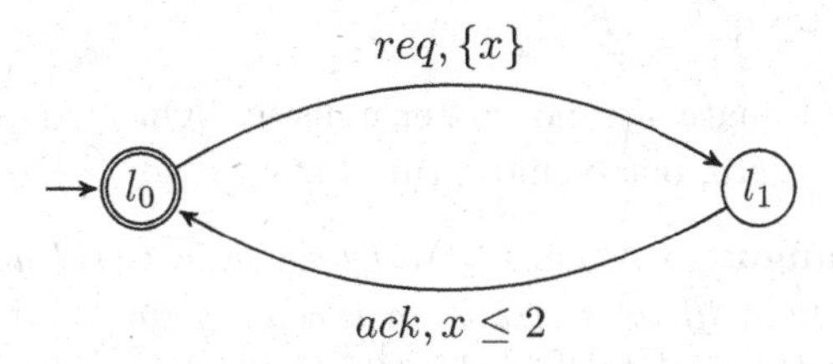

Fig. 4.1. Timed Automaton $\mathcal{A}_1$.

Remark 4.1. The reader may observe that we do not allow disjunctions in our definition of clock constraints. However, an edge $e = (l, l', a, \varphi_1 \vee \varphi_2, \lambda)$ with disjunction can be represented by having two edges $e_1 = (l, l', a, \varphi_1, \lambda)$ and $e_2 = (l, l', a, \varphi_2, \lambda)$ with the same effect. We also disallow negations in clock constraints. However, using de Morgan's laws, every constraint with negation can be rewritten as a constraint with disjunction and conjunctions where negation occurs only before an atomic constraint. The negation of an atomic constraint can be rewritten as a disjunction of some atomic constraints. For example, $\neg(x > c) = (x \leq c)$. Thus, our timed automata have sufficient expressive power to represent more general timing constraints with boolean operations.

4.2.1. *Semantics*

A *state* of a timed automaton $\mathcal{A}$ is a pair (l, ν) such that l is a location of $\mathcal{A}$ and ν is a clock interpretation over the set C of the clocks of $\mathcal{A}$. Hence the state space of $\mathcal{A}$ is $L \times \mathbb{R}_0^C$. The state of a timed automaton can change in two ways:

(1) Due to elapse of time: for a state (l, ν) and a real-number $t \geq 0$, $(l, \nu) \xrightarrow{t} (l, \nu + t)$. This kind of transition is called a *time elapse transition.*
(2) Due to a location-switch: for a state (l, ν) and an edge $(l, l', a, \varphi, \lambda)$ such that $\nu \models \varphi$, $(l, \nu) \xrightarrow{a} (l', \nu[\lambda := 0])$. This kind of transition is known as a *discrete transition* or a *Σ-transition* or a *switch.*

Note that we consider weakly monotonic semantics of time. That is, the time distance between two events is allowed to be 0 or more. This is very useful in modelling concurrent functioning of timed components by interleaving their actions without passing of time in between. We will point out explicitly wherever strong monotonicity is required.

Definition 4.4 (Timed Run). *Let $\mathcal{A}$ be a timed automaton as given in Definition 4.3 and let $\rho = (\sigma, \tau)$ be a timed word as given in Definition 4.1. A run r of a timed automaton on the timed word ρ is a sequence of alternating time elapse and discrete transitions* $(l_0, \nu_0) \xrightarrow{\tau_1} (l_0, \nu_1) \xrightarrow{\sigma_1} (l_1, \nu_1') \xrightarrow{\tau_2-\tau_1} (l_1, \nu_2) \xrightarrow{\sigma_2} \cdots (l_{n-1}, \nu_{n-1}') \xrightarrow{\tau_n-\tau_{n-1}} (l_{n-1}, \nu_n) \xrightarrow{\sigma_n} (l_n, \nu_n')$ *with $l_0 \in L_0$ and $\nu_0 = \nu_{init}$, i.e. $\nu_0(x) = 0$ for each $x \in C$. The run r is accepting iff $l_n \in F$.* □

Note that timed automata are non-deterministic. On a given timed word, the automaton may have zero, one or more runs.

Definition 4.5 (Language Accepted). *Let $\mathcal{A}$ be a timed automaton. A timed word ρ is accepted by $\mathcal{A}$ iff there exists an accepting run of $\mathcal{A}$ over ρ. The timed language $L(\mathcal{A})$ accepted by $\mathcal{A}$ is defined as the set of all timed words accepted by $\mathcal{A}$, i.e. $L(\mathcal{A}) \triangleq \{\rho \in T\Sigma^* \mid \mathcal{A}$ has an accepting run on $\rho\}$.* □

A timed language L is called *timed regular* provided $L = L(\mathcal{A})$ for some timed automaton $\mathcal{A}$.

Definition 4.6 (Untimed Language). *Let L be a timed language. Define Untime $(L) = \{\sigma \in \Sigma^* \mid (\sigma, \tau) \in L$ for some $\tau\}$.* □

Example 4.2. The state space corresponding to the timed automaton $\mathcal{A}_1$ in Figure 4.1 is $\{l_0, l_1\} \times \mathbb{R}_0$. The following is a run of $\mathcal{A}_1$. We abbreviate a valuation $(x \mapsto c)$ by c.

$$(l_0, 0) \xrightarrow{6.4} (l_0, 6.4) \xrightarrow{req} (l_1, 0) \xrightarrow{1.9} (l_1, 1.9) \xrightarrow{ack} (l_0, 1.9)$$

The timed language accepted by this automaton is given by $L(\mathcal{A}_1) = \{(\sigma, \tau) \mid \sigma = (req\ ack)^k$ for some integer $k \geq 0$, and $\tau_{2i} - \tau_{2i-1} \leq 2$ for all $1 \leq i \leq k\}$. Moreover, *Untime* $(L(\mathcal{A}_1)) = \{(req\ ack)^k$ for some integer $k \geq 0\}$. □

Timed automata have the following linearity property.

Lemma 4.1. *Let $\mathcal{A}$ be a timed automaton with rational time constants occurring in guards and let ρ be a finite (nonempty) sequence of configurations (l_i, ν_i). Let d be a rational constant.*

- *Let $d * \mathcal{A}$ denote the timed automaton obtained by replacing each time constant c in (the guards of) $\mathcal{A}$ by $d * c$.*
- *If $\tau = \tau_1 \ldots \tau_n$ then let $d*\tau$ be the sequence of time stamps $d*\tau_1, d*\tau_2, \ldots, d*\tau_n$. Also let $d * (\sigma, \tau) = (\sigma, d * \tau)$ and $d * L = \{d * \rho \mid \rho \in L\}$.*
- *Given a valuation ν let $d*\nu$ denote the valuation such that $(d*\nu)(x) = d*(\nu(x))$ for each clock x. Given a run r let $d * \rho$ denote the sequence of configurations obtained by replacing each configuration (l_i, ν_i) by $(l_i, (d * \nu_i))$.*

Then, r is a run of $\mathcal{A}$ on timed word (σ, τ) iff $d * r$ is a run of $d * \mathcal{A}$ on the timed word $(\sigma, d * \tau)$. Hence, $L(d * \mathcal{A}) = d * L(\mathcal{A})$. □

Several properties of timed automata such as emptiness, universality and language inclusion are preserved under such time scaling.

So far, in the semantics of timed automata, the clocks took real values; i.e. the time domain was taken to be the set of non-negative reals $\mathbb{R}_0$. However, the time domain $\mathbb{D}$ need not always be the set of reals. Typically, $\mathbb{D}$ is a linearly ordered set which is closed with respect to addition. (Care must be taken to ensure that the rational constants used in the automaton constraints are also from $\mathbb{D}$.) We indicate the time domain under which a timed automaton functions by adding a suffix $\mathbb{D}$. For example, the language accepted by an automaton $\mathcal{A}$ under rational numbered timed would be represented by $L(\mathcal{A})_{\mathbb{Q}_0}$. Also, $L(\mathcal{A})_{\mathbb{R}_0}$ will be abbreviated by $L(\mathcal{A})$ under the convention that the default time domain is $\mathbb{R}_0$ unless and otherwise specified. The decidability of important questions about timed automata such as emptiness, universality and language inclusion as well as the algorithmic techniques applicable for the analysis of timed automata depend quite significantly on the underlying time domain. Section 4.7 of this chapter will deal with timed automata with integer time.

4.2.2. *Closure Properties*

A timed system has several components working under a uniform global time. Such a system is conveniently modelled as a network of timed automata where each automaton models a component of the system. Time elapses simultaneously for each component. Moreover, the component automata are synchronized by simultaneously executing an edge in each component with a common label. An important result is the expansion theorem stating that such a network of synchronized automata can be represented by a single equivalent automaton which is constructed using the well-known product construction. Such a product automaton recognizes the intersection of the languages of its components. A construction can also be given for recognizing the union of the languages of the components.

In verifying properties of a timed system (automaton), the desired property can often be specified as *reachability* of some good/bad location(s). In a more general setting, we can often use a timed automaton to specify the set of acceptable behaviours (timed words) of the system. A formal verification of such a property reduces to showing that the language of the system automaton is contained within the language of the property automaton. Closure under the boolean operations of union, intersection and complementation and the ability to check for the reachability of locations gives a useful handle for solving these verification problems. In this section, we investigate the closure properties of timed automata.

4.2.2.1. *Union and Intersection*

Theorem 4.1. *Timed automata (over the same alphabet) are closed under union and intersection.* □

Proof. Let $\mathcal{A}_i = (L_i, L_{i_0}, \Sigma, C_i, E_i, F_i)$, for $1 \leq i \leq n$ be a given set of timed automata with the same alphabet Σ. Without loss of generality we assume that the sets of clocks C_i, C_j and respectively the sets of locations L_i, L_j are pairwise disjoint. Below, we construct two automata whose languages are, respectively, the union and the intersection of $L(\mathcal{A}_i)$.

Union Let $\mathcal{A} = (L, L_0, \Sigma, C, E, F)$ be the timed automaton defined as follows.

- $L = \bigcup_{1\leq i\leq n} L_i$,
- $L_0 = \bigcup_{1\leq i\leq n} L_{i_0}$,
- $C = \bigcup_{1\leq i\leq n} C_i$,
- $E = \bigcup_{1\leq i\leq n} E_i$, and
- $F = \bigcup_{1\leq i\leq n} F_i$.

It is easy to see that $L(\mathcal{A}) = \bigcup_{1\leq i\leq n} L(\mathcal{A}_i)$. We denote $\mathcal{A}$ as $\mathcal{A}_1 \cup \ldots \cup \mathcal{A}_n$.

Intersection Intersection is implemented by a simple modification of the product construction for finite automata. Let $\mathcal{A} = (L, L_0, \Sigma, C, E, F)$ be the timed automaton defined as the following.

- $L = \prod_{1\leq i\leq n} L_i$ is the cross product of sets L_i. Thus, each location in L is a tuple of the form $(l_1, \ldots, l_n)$ where each $l_i \in L_i$,
- $L_0 = \prod_{1\leq i\leq n} L_{i_0}$ and $F = \prod_{1\leq i\leq n} F_i$ and $C = \bigcup_{1\leq i\leq n} C_i$,
- $((l_1, \ldots, l_n), (l'_1, \ldots, l'_n), a, \varphi, \lambda) \in E$ iff
  - $\forall 1 \leq i \leq n$, there is an edge $(l_i, l'_i, a, \varphi_i, \lambda_i) \in E_i$, and
  - $\varphi = \bigwedge_{1\leq i\leq n} \varphi_i$ and $\lambda = \bigcup_{1\leq i\leq n} \lambda_i$.

It is easy to see that $L(\mathcal{A}) = \bigcap_{1\leq i\leq n} L(\mathcal{A}_i)$. We denote $\mathcal{A}$ as $\mathcal{A}_1 \times \ldots \times \mathcal{A}_n$. □

The reader will observe that the union and intersection constructions are the extensions of disjoint union and product constructions on finite state automata. Some more general products such as synchronized product and shuffle product of finite automata where the component alphabets are not required to be the same, are discussed in Chapter 9. These can also be adapted to timed automata and they are quite useful in modeling of timed systems.

4.2.2.2. *Complementation*

For a given timed language L let $\overline{L}$ denote its complement $T\Sigma^* - L$. Moreover, for a given timed automaton $\mathcal{A}$ let $\overline{\mathcal{A}}$ represent a timed automaton such that

$L(\overline{\mathcal{A}}) = \overline{L(\mathcal{A})}$, provided such an automaton exists. Note that the language inclusion problem $L(\mathcal{A}) \subseteq L(\mathcal{B})$ can be reduced to language emptiness with the following equivalences.

$$L(\mathcal{A}) \subseteq L(\mathcal{B}) \Leftrightarrow L(\mathcal{A}) \cap \overline{L(\mathcal{B})} = \emptyset \Leftrightarrow L(\mathcal{A} \times \overline{\mathcal{B}}) = \emptyset$$

The product of timed automata is easy to calculate, and we shall show in Section 4.3 that the language emptiness problem is decidable for timed automata. We shall also show that for every timed regular language L its untimed language *Untime* (L) is regular. Using these facts, we now prove that the timed automata are not closed under complementation.

Theorem 4.2 ([1, 2]). *Timed automata are not closed under complementation.* □

Proof. We give a one clock timed automaton for which there exists no timed automaton accepting its complement language. Consider the timed automaton $\mathcal{A}_2$ over the alphabet $\{a, b\}$ given in Figure 4.2. The language $L(\mathcal{A}_2)$ accepted by this automaton consists of timed words $(\sigma, \vec{\tau})$ containing an a event at some time t such that no event occurs at time $t+2$. We claim that $\overline{L}$ is not timed regular. Assume to the contrary that $\overline{L}$ is timed regular. Consider the timed language L' consisting of timed words w such that *Untime*(L') is a^*b^*. Also, all the a events happen before time 2 and no two a events happen at the same time in a word in L'. It is easy to see (by constructing a timed automaton for it) that L' is timed regular. Observe that a word of the form $a^n b^m$ belongs to *Untime*$(\overline{L} \cap L')$ iff $m \geq n$. Since timed regular languages are closed under intersection, the language $\overline{L} \cap L'$ is timed regular. As we shall show in the next section, untime of any timed regular language is regular. This implies that the untimed language *Untime*$(\overline{L} \cap L') = \{a^n b^m : m \geq n\}$ is regular, which is a contradiction. Hence $\overline{L}$ is not timed regular. □

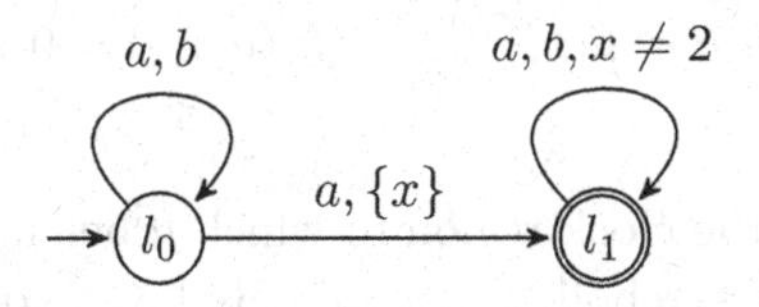

Fig. 4.2. Noncomplementable One Clock Timed Automaton $\mathcal{A}_2$ (adapted from [2]).

4.3. Language Emptiness and Location Reachability

The location reachability problem for timed automata is to decide whether there exists a run of a given timed automaton which ends in one of the given set of locations. The language non-emptiness of a timed automaton is equivalent to the location reachability of the set of its final locations. Since the set of configurations of a timed automaton is infinite (even uncountable), naive explicit state search is not very effective for location reachability. A symbolic search technique based on partitioning the state space into regions was used by Alur & Dill to solve the location reachability problem [1]. We now give an account of the result and the technique used.

As shown in Lemma 4.1, the language of every timed automaton $\mathcal{A}$ can be isomorphically represented by the language the timed scaled automaton $d * \mathcal{A}$ where d is the least common multiple of the denominators of all the rational constants occurring in the guards of $\mathcal{A}$. This automaton $d * \mathcal{A}$ has only integer constants in its guards. Moreover, the location reachability, language emptiness and language inclusion properties are preserved by this time scaled representation. Hence, in the rest of this chapter we shall assume (unless stated otherwise) that a timed automaton has only integer constants in its guards. This causes no loss in the generality of the results.

4.3.1. *Region Equivalence*

Let $\mathcal{A}$ be a timed automaton with the set of clocks $C = \{x_1, \ldots, x_k\}$. Moreover, for a clock x let c_x represent the maximum constant with which the clock x is compared within the guards of the given timed automaton. For a real number t, let $int(t)$ and $fr(t)$ represent the integral and the fractional parts of t. Given a valuation ν, a clock x is called saturated in ν if $\nu(x) > c_x$.

Definition 4.7 (Region equivalence). *Given a set of clocks C and maximum constants c_x for each $x \in C$, let two clock interpretations ν and ν' be defined to be region equivalent, denoted $\nu \simeq \nu'$, iff all the following conditions hold:*

- *For all $1 \leq i \leq k$, either $int(\nu(x_i)) = int(\nu'(x_i))$ or both exceed c_{x_i}.*
- *For all $1 \leq i, j \leq k$, such that $\nu(x_i) \leq c_{x_i}$ and $\nu(x_j) \leq c_{x_j}$,*
  $$fr(\nu(x_i)) \leq fr(\nu(x_j)) \textit{ iff } fr(\nu'(x_i)) \leq fr(\nu'(x_j)).$$
- *For all $1 \leq i \leq k$ such that $\nu(x_i) \leq c_{x_i}$, $fr(\nu(x_i)) = 0$ iff $fr(\nu'(x)) = 0$.*

□

We use $reg(\nu)$ to denote the clock region to which ν belongs. It is easy to see that each region $reg(\nu)$ can be symbolically represented as a conjunction of constraints fixing (a) the set of saturated clocks, (b) the integer part of each unsaturated clock, (c) the relative ordering (and equality) of the fractional parts of unsaturated clocks, and (d) whether an unsaturated clock has fractional part zero.

The region equivalence partitions the set of clock interpretations into a finite number of equivalence classes which are called *regions*.The set of regions of a timed automaton $\mathcal{A}$ is denoted by $Reg_{\mathcal{A}}$. The region equivalence can be extended to the states of the automaton $\mathcal{A}$ using the definition $(l, \nu) \simeq (l', \nu')$ iff $(l = l' \wedge \nu \simeq \nu')$. Figure 4.3 illustrates the region partitioning with two clocks x, y assuming that $c_x = 3$ and $c_y = 2$. The constraints defining some sample regions are also given.

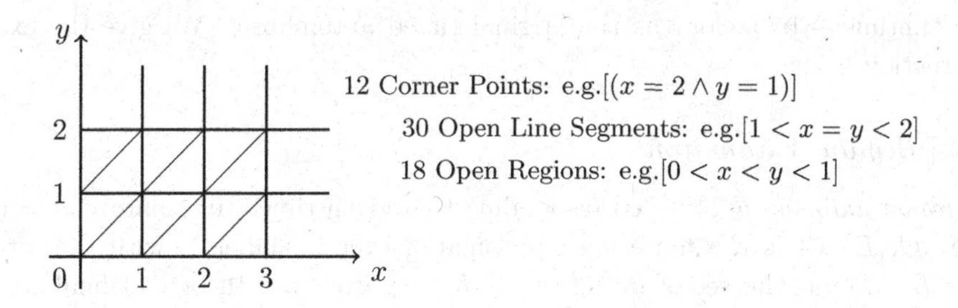

Fig. 4.3. Region partitioning for $c_x = 3$ and $c_y = 2$ (adapted from [1]).

It is informative to visualize the geometric locus of how a valuation changes with time elapse and discrete transitions. Each point represents a clock valuation. Passing of time causes the valuation to move in NE direction along the diagonal as all clocks increment by the same amount. Resetting of a clock x causes a valuation to be projected onto x-axis. Note that each point within a fixed region passes through the same sequence of regions with elapse of time. Moreover clock reset causes each valuation of a region to go to the same target region. These properties show that the region partitioning is stable with respect to time elapse and clock resets. The key properties of region equivalence are stated below.

Lemma 4.2. *Let $\mathcal{A}$ be a timed automaton. Let φ be any clock constraint where any clock x is only compared with integer constants which are at most c_x.*

(1) If $\nu_1 \simeq \nu_2$ then $\nu_1 \models \varphi \Leftrightarrow \nu_2 \models \varphi$.
(2) If $\nu_1 \simeq \nu_2$ and $(l, \nu_1) \xrightarrow{t_1} (l, \nu_1')$ then there exists a time t_2 such that $(l, \nu_2) \xrightarrow{t_2} (l, \nu_2')$ and $\nu_1' \simeq \nu_2'$.
(3) If $\nu_1 \simeq \nu_2$ then $(\nu_1[\lambda := 0]) \simeq (\nu_2[\lambda := 0])$.
(4) If $\nu_1 \simeq \nu_2$ and $(l_1, \nu_1) \xrightarrow{a} (l_2, \nu_1')$ then there exists ν_2' such that $(l_1, \nu_2) \xrightarrow{a} (l_2, \nu_2')$ and $\nu_2 \simeq \nu_2'$.

□

The property (1) follows from the definition of the regions and the restrictions on φ that it compares a clock only with integer constants which are at most c_x. Properties (2) and (3) are easy to see from the geometric interpretation of how valuation changes with the passing of time and the clock resets. Property (4) follows from properties (1) and (3) which together state that exactly the same transitions are enabled from region equivalent states (l, ν_1) and (l, ν_2) and that the clock resets preserve region equivalence.

Formally, the properties (2) and (4) of the above lemma show that the region equivalence is a time abstract bisimulationover the state-space of the timed automaton (where the value of time elapse before a discrete move are ignored). By quotienting the state space of $\mathcal{A}$ under this bisimulation, we can construct the region automaton. This untimed automaton with finite number of states has the same "untimed" behaviour as the original timed automaton. We give the exact construction below.

4.3.2. *Region Automaton*

The *region automaton*, $\mathcal{A}^{Reg}$ corresponding to a given timed automaton, $\mathcal{A} = (L, L_0, \Sigma, C, E, F)$ is the finite state automaton over alphabet Σ with the set of states $L \times Reg_{\mathcal{A}}$, the set of initial states $L_0 \times \{r_{init}\}$ and the set of final states $F \times Reg_{\mathcal{A}}$. Here, r_{init} is the region corresponding to the singleton set of valuation $\{\nu_{init}\}$ mapping each clock to zero. Let $r \preceq r'$ iff there exist ν and $t \geq 0$ such that $\nu \in r \wedge \nu + t \in r'$. Using the symbolic constraints defining r and r' it is easy to check whether $r \preceq r'$. Let $r[\lambda := 0]$ be the region such that $\nu \in r$ implies $\nu[\lambda := 0] \in r[\lambda := 0]$. Again, it is easy to compute this region given the symbolic constraints for r. The time elapse and the discrete transitions in $\mathcal{A}$ respectively give rise to the ϵ-transitions and Σ labelled transitions in $\mathcal{A}^{Reg}$. Thus, in $\mathcal{A}^{Reg}$, we have $(l, r) \xrightarrow{\epsilon} (l, r')$ iff $r \preceq r'$. Also, $(l, r) \xrightarrow{a} (l', r')$ iff there exists an edge $(l, l', a, \varphi, \lambda) \in E$ such that $r \models \varphi$ and $r' = r[\lambda := 0]$. From Lemma 4.2 it follows that $\mathcal{A}^{reg}$ is *time abstract bisimilar* to $\mathcal{A}$. For every run of $\mathcal{A}$ on a timed word $\rho = (\sigma, \tau)$ there exists a run of A^{Reg} on σ going through exactly the corresponding edges, and vice verse. This gives us the following theorem.

Theorem 4.3 ([1]). *$L(\mathcal{A}^{Reg}) = Untime\ (L(\mathcal{A}))$. Hence, $Untime\ (L(\mathcal{A}))$ is regular.* □

Note that $\mathcal{A}^{reg}$ can be effectively constructed and that $L(\mathcal{A})$ is empty iff *Untime* $(L(\mathcal{A}))$ is empty. Also, a location l is reachable in $\mathcal{A}$ iff it is reachable in $\mathcal{A}^{Reg}$. Now, location reachability is easy to check in the finite state automaton $\mathcal{A}^{Reg}$. This gives us the following theorem.

Corollary 4.1. *The location reachability and language emptiness problems are decidable for timed automata.* □

4.3.3. *Complexity*

Let $\mathcal{A}$ be a TA with k clocks and n locations where c_x is the largest integer constant with which a clock x is compared in any guard. Let $c_m = max\{c_{x_i}\}$. Then, it is easy to see that the number of clock regions is upper bounded by $k! \cdot 2^k \cdot \Pi_{x \in C}(2c_x + 2)$. This can be upper bounded by $k! \cdot 2^k \cdot (2c_m + 2)^k$. Then, the number of states in the region automaton $\mathcal{A}^{Reg}$ is upper bounded by $n \cdot k! \cdot 2^k \cdot (2c_m + 2)^k$. Non-emptiness can be decided by solving the location reachability problem for the final states in $\mathcal{A}^{reg}$. Location reachability can be checked using the depth-first search algorithm over the region automaton graph with time complexity quadratic in the number of states of $\mathcal{A}^{Reg}$. Theoretically, it can been shown that the location reachability problem for timed automata is PSPACE-complete (see [1]).

In practice, extended form of the region automata called the *zone automata* are used in model checking tools [3, 4]. The reachability search in a zone automaton typically terminates much faster than in the corresponding region automaton. Moreover, the zone automaton is constructed on-the-fly during the search. Zones are symbolic representations for convex sets of clock valuations given as conjunctions of interval constraints over clock-differences [7]. Several data structures such as the difference-bound matrices and Bellman-Ford inequalities are used to efficiently represent and manipulate zones during the symbolic search.

4.4. Universality and Language Inclusion

A timed automaton $\mathcal{A}$ over an alphabet Σ is called universal if $L(A) = T\Sigma^*$. In this section we give a proof that the problem of checking universality for TA is undecidable. We do so by reducing the problem of checking "does the computation of a deterministic 2-counter machine fail to halt" to the checking of "universality of a corresponding timed automaton". The result was originally proved in [1] for timed Buchi automata and it is presented here after adapting it to timed automata over finite words.

A deterministic 2-counter machine M consists of two counters C and D, and a sequence of n instructions numbered $0, 1, \ldots, (n-1)$. Implicitly, we assume that index n contains the "halt" instruction. Instructions are normally executed in sequence. Each instruction may increment or decrement one of the counters (instructions $incr(r)$, $decr(r)$), or it may jump conditionally upon one of the counters being zero to the instruction at index m (instruction $JumpZ(r, m)$). A configuration of M, represented by $\langle i, c, d\rangle$ with $c \geq 0$ and $d \geq 0$, gives the index i of the next instruction to be executed and, c and d, the current values of the counters C and D respectively. Decrementing a counter which is zero leaves it at zero. The initial configuration of M is $\langle 0, 0, 0\rangle$. The machine halts by reaching (or jumping to) the index n.

For a 2-counter machine M, a *halting computation* of M is a finite sequence of configurations $\langle i_0, c_0, d_0\rangle\langle i_1, c_1, d_1\rangle \ldots \langle i_k, c_k, d_k\rangle$ which begins with the initial configuration $\langle 0, 0, 0\rangle$ and ends in a halting configuration, i.e. $i_k = n$. For each $j : 0 \leq j < k$, the configuration $\langle i_{j+1}, c_{j+1}, d_{j+1}\rangle$ results from the configuration $\langle i_j, c_j, d_j\rangle$ by the execution of the instruction indexed i_j. A non-halting computation of M is an infinite sequence of such configurations where a halting configuration is never encountered. Each 2-counter machine, because of determinism and fixed starting configuration, has a unique computation which is either halting or non-halting.

Given (the instruction sequence of) a deterministic 2-counter machine M, we come up with a corresponding timed language $L_{undec}(M)$ whose words encode the halting computations of M. Then, we construct a timed automaton $\mathcal{A}_{undec}(M)$ which accepts the complement of $L_{undec}(M)$. Clearly, the language of $\mathcal{A}_{undec}(M)$ is universal iff the (unique) computation of M does not halt.

The alphabet of $L_{undec}(M)$ is $\{b_0,\ldots,b_n, a_1, a_2\}$. A computation of a 2-counter machine is represented as concatenation of sequences of the form $b_i a_1^c a_2^d$ each of which represent a configuration $\langle i, c, d\rangle$. For example, the first sequence is just b_0 (corresponding to the initial configuration $\langle 0, 0, 0\rangle$). We require that the subsequence occurring in the time interval $[j, j+1)$ encodes the jth configuration of the computation with symbol b_{i_j} occurring at time j. We also require that two adjacent configurations should have corresponding a_1 symbols and a_2 symbols exactly one time unit apart. Formally,

$(\sigma, \tau) \in L_{undec}(M)$ iff

- $\sigma = b_{i_0} a_1^{c_0} a_2^{d_0} \ldots b_{i_k} a_1^{c_k} a_2^{d_k}$ such that $\langle i_0, c_0, d_0\rangle \ldots \langle i_k, c_k, d_k\rangle$ is a halting computation of M. In particular, $i_k = n$ and $i_0 = 0, c_0 = 0, d_0 = 0$.
- $\forall j \geq 1$, the time stamp of b_{i_j} is j.
- Between any two letters there is non-zero time distance.
- $\forall j \geq 1$,
  - if $c_{j+1} = c_j$ then for every a_1 at time t in the interval $(j, j+1)$ there is an a_1 at time $t+1$, and for every a_1 at time t in the interval $(j+1, j+2)$ there is an a_1 at time $t-1$.
  - if $c_{j+1} = c_j + 1$ then for every a_1 at time t in the interval $(j, j+1)$ there is an a_1 at time $t+1$, and for every a_1 at time t in the interval $(j+1, j+2)$ except the last a_1, there is an a_1 at time $t-1$.
  - if $c_{j+1} = c_j - 1$ then for every a_1 at time t in the interval $(j, j+1)$ except the last one, there is an a_1 at time $t+1$, and for every a_1 at time t in the interval $(j+1, j+2)$ there is an a_1 at time $t-1$.
- Similar conditions apply to the times of occurrences of the letter a_2.

The desired automaton $\mathcal{A}_{undec}$, recognizing the complement of L_{undec}, is a disjunction of several timed automata, each of which encodes *violation* of one of the conditions for L_{undec} given above.

Let $\mathcal{A}_s$ be a timed automaton which accepts (σ, τ) iff for some integer $j \geq 0$, either there is no b symbol at time j, or the subsequence of σ having time stamps in the time interval $(j, j+1)$ is not of the form $a_1^* a_2^*$. It is easy to construct such a deterministic automaton. Let $\mathcal{A}_t$ be the (deterministic) timed automaton which accepts iff there are two consecutive letters with the same time stamp.

For any timed word (σ, τ) in L_{undec}, its fragment in time interval $[0, 1)$ must encode the initial configuration. Let $\mathcal{A}_{init}$ be the (deterministic) timed automaton which accept the timed word whose fragments corresponding to the interval $[0, 1)$ are from $(b_0, 0)$.

For each instruction indexed $0 \leq i < n$,we construct a timed automaton $\mathcal{A}_i$. This $\mathcal{A}_i$ accepts (σ, τ) iff there exists an occurrence of letter b_i in σ at, say, time t and the configuration corresponding to the subsequence of σ occurring in interval $[t+1, t+2)$ does not follow from the configuration corresponding to the subsequence of σ occurring in the interval $[t, t+1)$ by executing the instruction indexed i. We illustrate this by constructing $\mathcal{A}_i$ for the instruction $incr(D)$. The automaton $\mathcal{A}_i$ is the disjunction of the following six timed automata $\mathcal{A}_i^1, \ldots, \mathcal{A}_i^6$.

Let $\mathcal{A}_i^1$ be the nondeterministic timed automaton which accepts (σ, τ) iff for some $j \geq 1$, $\sigma_j = b_i$ and there is no occurrence of b_{i+1} at the time $\tau_j + 1$. It is easy to construct such an automaton. (The automaton in Figure 4.2 illustrates the required technique.)

Let $\mathcal{A}_i^2$ be the non-deterministic timed automaton in Figure 4.4. In this figure, if the label is omitted from a transition then it can be taken on any symbol in the alphabet of the automaton. The automaton $\mathcal{A}_i^2$ accepts (σ, τ) iff there is a b_i at some time t followed by an a_1 at a time $t' < (t+1)$ such that there is no matching a_1 at the time $(t'+1)$. The complement of the language of $\mathcal{A}_i^2$ contains words where for every occurrence of b_i say at time t, each a_1 in the interval $(t, t+1)$ has a matching a_1 in the interval $(t+1, t+2)$ exactly one time unit later.

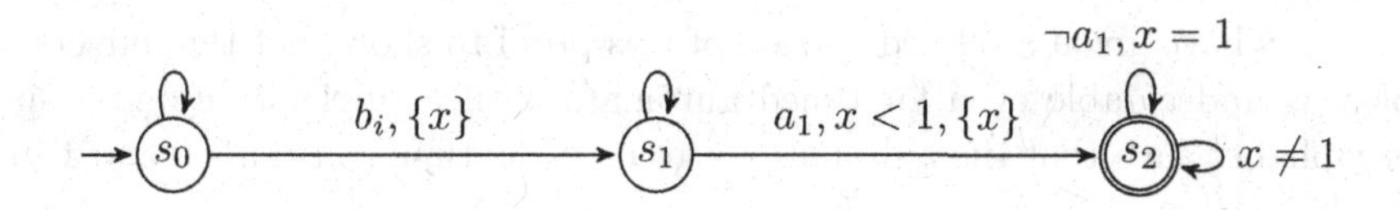

Fig. 4.4. Timed Automaton $\mathcal{A}_i^2$ (adapted from [1]).

Similarly, we can construct $\mathcal{A}_i^3$ which accepts (σ, τ) iff there is b_i at a time t, and for some t' in interval $(t, t+1)$ there is no a_1 at t' but there is an a_1 at time $t'+1$. This is checked by non-deterministically matching b_i at some time t, nondeterministically matching two *successive* symbols at times say t_1 and t_2 in the interval $[t, t+1]$ and then accepting iff there an occurrence of a_1 in the interval (t_1+1, t_2+1). This witnesses an unmatched a_1 in the interval $(t+1, t+2)$. The complement of the language of $(\mathcal{A}_i^2 \cup \mathcal{A}_i^3)$ contains timed words where for every occurrence of b_i say at a time t, an a_1 occurs at a time t' in the interval $(t, t+1)$ iff a_1 occurs at the time $t'+1$ in the interval $(t+1, t+2)$.

In a similar fashion, we ensure a proper matching of a_2 symbols after an increment in their numbers. Let $\mathcal{A}_i^4$ be the automaton which requires that for some b_i at a time t, there is an a_2 at a time $t' \in (t, t+1)$ with no matching a_2 at the time $(t'+1)$. The construction of this automaton is similar to the automaton in Figure 4.4. Let $\mathcal{A}_i^5$ be the automaton which checks that for some b_i at a time t, there are two a_2's in the interval $(t+1, t+2)$ without matching a_2's in the interval $(t, t+1)$. Let $\mathcal{A}_i^6$ be the automaton which accepts if for some b_i at time t the last a_2 in the interval $(t+1, t+2)$ has a matching a_2 in the interval $(t, t+1)$ exactly one time unit earlier.

The reader should convince himself that $\mathcal{A}_i^1 \cup \ldots \cup \mathcal{A}_i^6$ exactly accepts those words where for some occurrence of b_i at time t, the encodings in intervals $[t, t+1)$

and $[t+1, t+2)$ do not match as required by the execution of the instruction $incr(D)$.

The requirement that the computation does not end in a halting configuration translates to the requirement that b_n is not the last state. This can be checked by a deterministic automaton $A_{nonhalt}$. The union of all these automata $A_s, A_t, A_{init}, (A_i)_{0 \leq i < n}, A_{nonhalt}$ together constitutes the automaton $\mathcal{A}_{undec}(M)$ which accepts the complement of $L_{undec}(M)$.

From the construction, it is clear that $L(\mathcal{A}_{undec}(M))$ is universal iff $L_{undec}(M)$ is empty iff the computation of M is non-halting. Note that $\mathcal{A}_{undec}(M)$ can be effectively constructed for any given deterministic 2-counter automaton M. It is well known from the classical work of Turing and Minsky that checking whether the computation of an arbitrary deterministic 2-counter automaton is non-halting is undecidable. This immediately gives us the following result.

Theorem 4.4. *The universality problem for TA is undecidable.* □

Adams *et al* have given a refined version of this proof to show that the universality problem is undecidable even for timed automata with a single state and a single letter alphabet such that the automaton uses only the time constants 0 and 1 in its guards [8].

The language inclusion problem for TA means, given any two timed automata $\mathcal{A}$ and $\mathcal{B}$, checking whether $L(\mathcal{A}) \subseteq L(\mathcal{B})$. Note that a timed automaton $\mathcal{U}$ (without clocks) can be constructed over any given alphabet Σ to accept the universal timed language $T\Sigma^*$. Hence the universality problem for a timed automaton $\mathcal{A}$ is a special case of the language inclusion problem $L(\mathcal{U}) \subseteq L(\mathcal{A})$. Hence,

Corollary 4.2. *The language inclusion problem for TA is undecidable.* □

Ouaknine and Worrell [9] as well as Abdulla *et al* have studied the special case of 1-clock timed automata.

Theorem 4.5 ([9]). *The language inclusion problem $L(\mathcal{A}) \subseteq L(\mathcal{B})$ is decidable if $\mathcal{A}$ is a timed automaton and $\mathcal{B}$ is a timed automaton with one clock. However, the complexity of this algorithm is nonprimitive recursive [10].* □

Note that this makes the checking of universality of 1-clock timed automata decidable. The results holds in spite of the fact that there exist one clock timed automata which are not complementable. Figure 4.2 gives one such automaton.

4.5. Deterministic Timed Automata (DTA)

A deterministic timed automaton is a timed automaton such that it has a unique start state, and from any state of the automaton, given the next input symbol and its time of occurrence, the next automaton state is uniquely determined. Formally,

Definition 4.8. *A timed automaton $\mathcal{A} = (L, L_0, \Sigma, C, E, F)$ is called deterministic iff*

- *The set of initial locations is a singleton. i.e., $|L_0| = 1$.*
- *$\forall l \in L$ and $\forall a \in \Sigma$, if there are two distinct edges $(l, l_1, a, \varphi_1, \lambda_1)$ and $(l, l_2, a, \varphi_2, \lambda_2)$ in E, then the guards φ_1 and φ_2 are mutually exclusive, i.e., $\varphi_1 \wedge \varphi_2$ is not satisfiable.*

We sometimes represent a deterministic timed automaton by $(L, l_0, \Sigma, C, E, F)$, where l_0 is the unique initial location. □

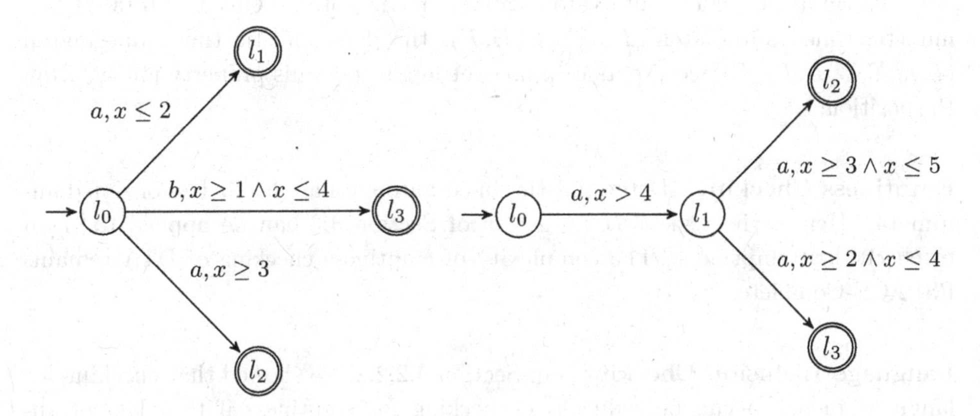

Fig. 4.5. A Deterministic TA. Fig. 4.6. A Nondeterministic TA.

Figure 4.5 gives a deterministic timed automaton. Note that the Definition 4.8 is only a syntactic characterization and not rich enough to capture the set of all TA which have at most one run over any given timed word. For instance, in the automaton shown in Figure 4.6 the edges from l_1 to l_2 and l_1 to l_3 have constraints which are not disjoint, and hence the automaton is not deterministic. However, the edge from l_1 to l_3 is never taken and the automaton has at most one run on any timed word.

4.5.1. *Closure Properties, Decision Problems and Expressiveness*

A timed automaton is called *total* if for any location l and any letter a, the disjunction $\vee_i \phi_i$, of the guards ϕ_i of the transitions from l labelled a, is equivalent to *true*. Notice that this is a syntactic requirement. Given a timed automaton $\mathcal{A}$, we can easily construct a total timed automaton $\mathcal{A}'$ such that $L(\mathcal{A}) = L(\mathcal{A}')$. The construction involves adding a new (non-final) sink state s and adding a transition from l to s labelled a and having guard $\neg \vee \phi_i$. Also for each letter there is a transition from s to itself with the guard *true*. While negation and disjunction ($\neg$, $\vee$) operators are not included in our definition of clock constraints, by the remark 4.1, we can always construct a timed automaton which represents this more general constraint.

Also note that the above construction gives a total DTA when applied to a DTA. It is easy to check that a total deterministic automaton has the following property.

Proposition 4.1. *A total deterministic timed automaton has exactly one run over any given timed word.* □

Corollary 4.3. *DTA are closed under intersection and complementation.* □

Proof. It is easy to check that the intersection construction for timed automata (see Theorem 4.1), when applied to DTA results in a DTA. As we have shown, any deterministic timed automaton can be made total. Given a total deterministic timed automaton $(L, l_0, \Sigma, C, E, F)$, the deterministic timed automaton $(L, l_0, \Sigma, C, E, L \setminus F)$ accepts its complement language. This property follows from Proposition 4.1. □

Emptiness Checking Deterministic timed automata are a subclass of timed automata. Hence, the regions construction of Section 4.3 can be applied to them to check their emptiness. The complexity of emptiness checking of DTA remains PSPACE-Complete.

Language Inclusion Checking In Section 4.2.2.2, we showed that checking for language inclusion can be reduced to checking for emptiness if the class of automata considered are closed under intersection and complementation. Since DTA are closed under intersection, and they can be complemented and their language emptiness is also decidable, it follows that the language inclusion of DTA is decidable. In fact, we get the following more general result.

Theorem 4.6 **([1]).** *For a timed automaton $\mathcal{A}$ and a deterministic timed automaton $\mathcal{B}$, checking whether $L(\mathcal{A}) \subseteq L(\mathcal{B})$ is PSPACE-complete.* □

We now show that deterministic timed automata are strictly less expressive than nondeterministic timed automata..

Theorem 4.7. *Timed automata are not closed under determinization.* □

Proof. Closure under determinization implies closure under complementation which does not hold for timed automata. Concretely, there is no deterministic timed automaton which accepts the language accepted by the timed automaton $\mathcal{A}_2$ in Figure 4.2. Assume to contrarty that there exists a language equivalent deterministic timed automaton $\mathcal{A}'$ then we can easily complement $\mathcal{A}'$ as in Corollary 4.3. This would mean there exists a (deterministic) timed automaton $\overline{\mathcal{A}'}$ recognizing the complement of the language of $\mathcal{A}_2$. But in the proof of Theorem 4.2 we have already shown that this is impossible. Hence, $\mathcal{A}_2$ is not complementable. □

4.5.2. *Determinizability*

A timed automaton $\mathcal{A}$ is called determinizable if there exists a deterministic timed automaton recognizing the language $L(\mathcal{A})$. The timed automaton $\mathcal{A}$ is called complementable if there exists a timed automaton $\mathcal{B}$ recognizing the complement of the language of $\mathcal{A}$; that is $L(\mathcal{B}) = \overline{L(\mathcal{A})}$. Tripakis [11] showed that that the problem of checking whether a timed automaton is determinizable with construction is undecidable. By construction we mean that the checking algorithm is required to give a suitable deterministic automaton in positive instances. Similarly, the problem of checking whether a timed automaton is complementable with construction is also undecidable. This result was further generalized by Finkel [12] to show that checking determinizability and complementability of a timed automaton (without construction) are both undecidable. We give an account of this proof.

Definition 4.9 (Concatenation). *Let $\rho = (\sigma_1, \tau_1), \ldots, (\sigma_n, \tau_n)$ and $\rho' = (\sigma'_1, \tau'_1), \ldots, (\sigma'_m, \tau'_m)$ be two timed words over the alphabets Σ_1 and Σ_2 respectively. The concatenation $\rho \cdot \rho'$ of ρ and ρ' is defined as the timed word $(\sigma_1, \tau_1), \ldots, (\sigma_n, \tau_n)(\sigma'_1, \tau''_1), \ldots, (\sigma'_m, \tau''_m)$, where for each $1 \leq i \leq m$, $\tau''_i = \tau'_i + \tau_n$. The alphabet of $\rho \cdot \rho'$ is $\Sigma_1 \cup \Sigma_2$. Intuitively, the time at which the events in ρ' occur are adjusted relative to the time of occurrence of the last event in ρ. For two timed languages L and L', their concatenation $L \cdot L'$ is defined as the language $\{\rho \cdot \rho' : \rho \in L$ and $\rho' \in L'\}$.* □

We state some useful properties of the catenation of timed languages.

Lemma 4.3.

- *If L_1 and L_2 are timed regular languages then $L_1 \cdot L_2$ is also timed regular.*
- *If L_1 is a finite collection of timed words and $L_1 \cdot L_2$ is timed regular then L_2 is timed regular. Moreover, L_2 can be recognized by a timed automaton with no more clocks than those in $L_1 \cdot L_2$.*

□

Proof. For the proof of the first part, note that given automata for L_1 and L_2, it is we can construct a nondeterministic timed automaton recognizing $L_1 \cdot L_2$. This automaton consists of the disjoint union of the states and transitions of the two automata. Moreover, for every transition going to a final state of the first automaton and for each of the initial states of the second automaton, we add a transition (with same label and guard) which goes to the initial state of the second automaton. It is easy to see that the language of the resulting automaton is $L_1 \cdot L_2$.

The second part of the lemma follows by noting that for any given nondeterministic timed automaton $\mathcal{A}$ and a timed word ρ, the set of runs of of $\mathcal{A}$ on ρ is finite and hence the set of last states after ρ is also finite. Consider a timed word $\rho \in L_1$. Then, the behaviour of an automaton recognizing $L_1 \cdot L_2$ starting with any of a the finite number of states (location and valuation pairs) can be modelled by a timed automaton (with no additional clocks) which will precisely recognizes L_2. □

Theorem 4.8 ([12]). *It is undecidable to determine, for a given timed automaton $\mathcal{A}$, whether*

- $L(\mathcal{A})$ *is accepted by a deterministic timed automaton.*
- $\overline{L(\mathcal{A})}$ *is accepted by a timed automaton.*

□

Proof. We reduce the problem of checking universality of a given TA to the problem of checking whether a TA is determinizable/complementable. Since checking universality of TA is undecidable we conclude that determining whether a TA is determinizable/complementable are also both undecidable.

Given a timed regular language L over an alphabet Σ (as a timed automaton), we effectively construct another timed regular language L' below. We show that we can check whether L is universal exactly by asking whether L' is determinizable/ complementable.

We assume without loss of generality that $a, b \in \Sigma$ and d is a fresh letter, i.e. $d \notin \Sigma$. Let $\Sigma' = \Sigma \cup \{d\}$. Let C be the language accepted by the automaton given in Figure 4.2. We have shown in Theorem 4.2 that the complement of C is not timed regular. Recall that $T\Sigma^*$ denotes the set of all timed words over Σ.

Given a timed regular language L over Σ, the constructed language L' over Σ' is the union of the following three languages.

- $L_1 = L \cdot \{(d,t) : t \in \mathbb{R}^+\} \cdot (T\Sigma^*)$,
- L_2 is the set of timed words over Σ' having no d's or having at least two d's, and
- $L_3 = (T\Sigma^*) \cdot \{(d,t) : t \in \mathbb{R}^+\} \cdot C$.

It is easy to see that L' is timed regular as it is obtained from timed regular languages using catenation and union (which preserve timed regularity). We consider two cases.

(1) $L = T\Sigma^*$. Then it is easy to see that $L' = T\Sigma'^*$, the set of all timed words over Σ'. Therefore L' is accepted by a deterministic timed automaton (without any clock). Moreover its complement $\overline{L'}$ is empty which is trivially timed regular. Hence L' is complementable.

(2) L is strictly included in L_Σ. Then, there is a timed word $u = (\sigma_1, \tau_1), \ldots, (\sigma_n, \tau_n) \in \overline{L}$ which does not belong to L. We show that the complement language $\overline{L'}$ is not timed regular. Assume to the contrary that $\overline{L'}$ is timed regular, and hence, $L'' = \overline{L'} \cap \{u \cdot (d,1) \cdot (T\Sigma^*)\}$ is also timed regular. Observe that $L'' = \{u \cdot (d,1)\} \cdot \overline{C}$. By the lemma 4.3, this implies that $\overline{C}$ is timed regular. But this is a contradiction as we have already shown in Theorem 4.2 that $\overline{C}$ is not timed regular. Hence, we conclude that L' is not complementable. But

this also establishes that L' is not determinizable as all determinizable timed regular languages are closed under complementation. □

4.5.3. *Event-Recording Automata (ERA)*

In general, finding determinizable subclasses of TA is an interesting but difficult quest. Alur *et al* [13] have shown that a subclass of TA called event-recording automata (ERA) can indeed be determinized with one exponential blowup in the automaton size. ERA are closed under all boolean operations and hence the language inclusion problem is decidable for such automata.

ERA are a special subclass of timed automata wherein for every event a there is a clock (x_a) that records the time since the last occurrence of a. An event-recording automaton is a tuple $(L, L_0, \Sigma, C, E, F)$ where each component is defined as in the case of timed automata. However, the set of clocks C is fixed as $C = \{x_a : a \in \Sigma\}$. Moreover, any transition labelled with letter a exactly resets the (singleton) set of clocks $\{x_a\}$.

The event recording automaton in Figure 4.7 accepts the words $((abcd)^n, \tau)$ where n is a non-negative integer and $\forall j \geq 0$, $\tau_{4j+3} < \tau_{4j+1}+2$ and $\tau_{4j+4} > \tau_{4j+2}+4$.

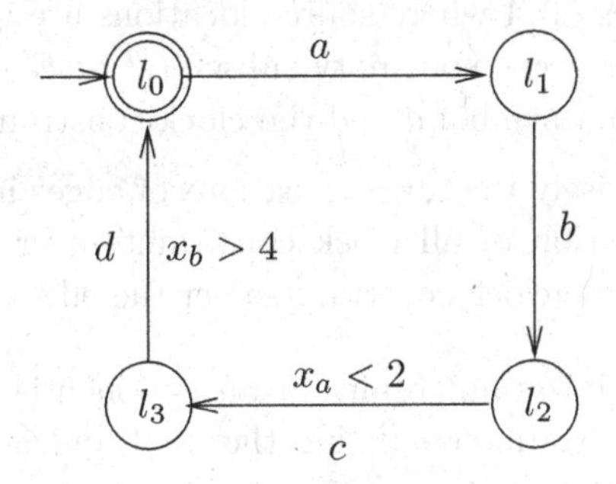

Fig. 4.7. An Event Recording Automaton (adapted from [13]).

Note that ERA can be non-deterministic. However, in ERA, the clock values are driven by the timed word rather than the (non-deterministic) choice transitions as in the case of TA. In this sense ERA are "input determined" [14]. The following lemma captures the essential nature of these automata.

Lemma 4.4. *Let r_1 and r_2 be two (nondeterministically occurring) runs of an ERA $\mathcal{A}$ on a given timed word (σ, τ). Let the runs r_1 and r_2 respectively end in configurations (l_1, ν_1) and (l_2, ν_2). Then, $\nu_1 = \nu_2$.* □

Proof. In both the configurations (l_i, ν_i) with $1 \leq i \leq 2$, the value $\nu_i(x_b)$ is given by $\tau_n - \tau_j$ where j is the last index in the timed word with $a_j = b$. In case there is no occurrence of letter b in the time word, then $\nu_i(x_b)$ has the value τ_n. This depends only on the input timed word (σ, τ) and not on the non-deterministic choice of the transitions that led to different runs. □

From this lemma, we can see that the non-determinism in ERA affects only the choice of location (and not the valuation).

4.5.3.1. *Determinization of ERA*

Since, ERA are a subclass of timed automata, their language emptiness is decidable using the regions automaton construction for TA. It is easy to check that ERA are closed under union and intersection using the constructions for these operations for TA given earlier. We now consider the question of determinzation of ERA.

The determinization follows the standard subset construction [13]. Let $\mathcal{A} = (L, L_0, \Sigma, C, E, F)$ be an event-recording automaton. For simplicity assume that $\mathcal{A}$ is total. The language equivalent deterministic event-recording automaton $Det(\mathcal{A})$ is defined as the following.

- The locations of $Det(\mathcal{A})$ are nonempty subsets of L.
- The only start location is L_0.
- A location $L' \subseteq L$ is an accepting location iff $L_f \cap L'$ is nonempty.
- Consider a location $L' \subseteq L$ of $Det(\mathcal{A})$ and an input symbol $a \in \Sigma$. Let $E' \subseteq E$ be the set of all edges of $\mathcal{A}$ where source locations are in L' and whose input symbol is a. Then for every nonempty subset E'' of E', there is an edge from L' to L'' with the input symbol a and the clock constraints φ such that
  - L'' contains precisely the target locations of edges in E'', and
  - φ is the conjunction of all clock constraints over the edges in E'' and negations of all the clock constraints over the edges in $(E' \setminus E'')$.

Note that the determinization results in an exponential blowup in the number of locations. Recall that interchanging the final and non-final locations of a deterministic timed (event-recording) automaton gives a timed (event-recording) automaton which accepts the complement language. Thus, ERA are closed under complement. Moreover, since an ERA can be determinized, by Theorem 4.6, the language inclusion between a TA and an ERA is decidable. We have the following result.

Theorem 4.9 ([13]).

(1) *ERA are closed under determinization and complementation.*
(2) *For a timed automaton $\mathcal{A}$ and an event-recording automaton $\mathcal{B}$, the language inclusion problem $L(\mathcal{A}) \subseteq L(\mathcal{B})$ is decidable and PSPACE-Complete.* □

Note that this determinization technique cannot be applied to timed automata in general. In TA, two transitions originating from a state and having the same label can have different sets of clocks to be reset leading to separate valuations after these transitions. In case of ERA, all such transitions have the same reset set, leading to the same valuation.

4.5.3.2. *ERA and DTA*

From Theorem 4.9 it follows that, from the point of view of expressiveness, ERA are a subclass of DTA. We now prove that they form a proper subclass of DTA.

Theorem 4.10. *ERA* $\subsetneq$ *DTA*. □

Proof. Consider the language $L = \{(aaa, \tau) : \tau_3 - \tau_1 = 1\}$ over the alphabet $\{a\}$. L is accepted by the deterministic timed automaton given in Figure 4.8. We prove that L is not accepted by any event-recording automaton. Assume to the contrary that $\mathcal{A}$ is an event-recording automaton accepting L. For generality assume that $\mathcal{A}$ uses rational constants in its guards. Let k be the least common multiple of the denominators occurring in $\mathcal{A}$. Consider the two timed words $\rho_1 = \langle(a,0),(a,1-\frac{1}{2k}),(a,1)\rangle$ and $\rho_2 = \langle(a,0),(a,1-\frac{1}{2k}),(a,1-\frac{1}{4k})\rangle$. Then, $\rho_1 \in L$ and $\rho_2 \notin L$. In any run over $\mathcal{A}$, $\nu_3^{\rho_1}(x_a) = \frac{1}{2k}$ and $\nu_3^{\rho_2}(x_a) = \frac{1}{4k}$. But we know that $\frac{1}{2k}$ and $\frac{1}{4k}$ cannot be differentiated by any clock constraint in $\mathcal{A}$ as they compare clock values with rational constants at least as large as $\frac{1}{k}$. Hence, either both ρ_1 and ρ_2 belong to $L(\mathcal{A})$ or both of them do not, which is a contradiction. □

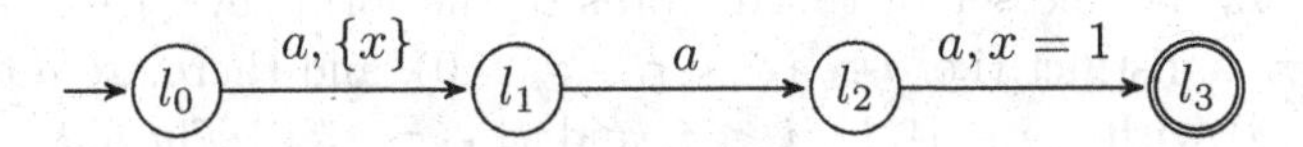

Fig. 4.8. A Deterministic Timed Automaton.

Other Determinizable Classes of Timed Auotmata Alur and Henzinger have considered a further extension of ERA called event clock automata (ECA) where apart from having event recording clocks of the form x_a which measures the time since last occurrence of letter a, the automata also have event predicting clocks of the form y_a which essentially measures the time till the next future occurrence of the letter a. It has been shown that all ECA are determinizable (to equivalent deterministic ECA). Moreover, the language accepted by an ECA is timed regular (i.e. nondeterministic TA recognizable) but it is not necessarily recognized by a deterministic timed automaton. Thus, ECA are expressively a subclass of TA. Event recording automata are input determined in the sense that the valuation of the clocks is uniquely determined by the input word. D'Souza et al [14] have showed that any input determined automaton can be determinized using the subset construction. The subset construction can also be applied to determinize event-recording stopwatch automata as well as event recording linear-hybrid automata [15, 16]. Other determinizable classes of timed automata are discussed in some recent papers (see [17–19]).

4.6. Clock Reduction

The number of clocks used in a timed automaton is an important measure of its complexity. For example, the number of states in the region automaton of a timed automaton grows exponentially with the number of clocks. Most tools for model checking of timed automata are unable to deal with models with say 50 clocks. Hence, techniques for reducing the number of clocks used in a timed automaton have also been explored. In this section, we study the effect of number of clocks on expressiveness and decidability of timed automata.

It has been shown that the the expressiveness of timed automata strictly increases when the number of clocks allowed is increased [20]. At one end of this spectrum are the timed automata with 1-Clock (1C-TA from hereon) which are known to be well behaved compared to the general class of timed automata [9, 21]. We shall now give an acoount of some of these results.

Theorem 4.11 **([20]).** *For every $n \in \mathbb{N}$, the class of timed languages accepted by timed automata with n clocks is properly contained in the class of languages accepted by timed automata with $n+1$ clocks.* □

Proof. Let L_n be the set of timed words of the form $(a, \tau_1), \cdots (a, \tau_k)$, where, $\tau_1 > 0$, each τ_i is distinct (i.e. $i > j \Rightarrow \tau_i - \tau_j > 0$), and there are n distinct pairs of integers (i, j) with $i, j \in [1, k]$, $i < j$, and $\tau_j - \tau_i = 1$. Thus, L_n is the set of words over $\{a\}$ where there are at least n distinct pairs of a's separated exactly by the time distance 1. One can easily construct a non-deterministic timed automaton with n clocks which accepts L_n. The key to this proof is that though L_{n+1} is timed regular it can not be accepted by any timed automaton with only n clocks. To see this, consider a timed word ρ in L_{n+1} with a occurring at time $1/i$ as well as $1 + 1/i$ for $1 \leq i \leq (n+1)$. Clearly this timed word is in L_{n+1}. To accept this word $\rho = (\sigma, \tau)$, but no other word $\rho' = (\sigma, \tau')$ where τ and τ' differ somewhere during the time interval $(1, 2]$, a timed automaton must make transitions on a's exactly at times $1 + 1/i$. This can only be enforced if some clock reaches an *integer* value exactly at that time point. Since, a clock can reach an integer value only once in interval $(1, 2]$ the automaton requires at least $n+1$ clocks. □

In other words, timed regular languages can be arranged in a strict hierarchy based on the minimum number of clocks needed to model the language. Unfortunately, given a timed automaton, it is not possible to effectively determine whether there exists a timed automaton with less number of clocks accepting the same timed language.

Theorem 4.12 **([12]).** *Let $n \geq 2$ be a positive integer. It is undecidable to determine, for a given TA $\mathcal{A}$ with n clocks, whether there exists a TA $\mathcal{B}$ with $n-1$ clocks, such that $L(\mathcal{B}) = L(\mathcal{A})$.* □

Proof. We reduce the problem of checking universality of a given TA with n clocks to the problem of checking whether the language of a TA is accepted by some TA with n clocks. Since checking universality of TA with n clocks is undecidable for $n \geq 2$, we conclude that determining whether the language of a TA is accepted by a TA with n clocks is also undecidable. The proof has the same structure as the proof of the Theorem 4.8 given earlier.

Given a timed automaton with n clocks recognizing a timed regular language L over an alphabet Σ, we effectively construct another timed regular language L'' below. We show that we can check whether L is universal by asking whether L'' is acceptable by a timed automaton with n clocks.

We assume without loss of generality that $a \in \Sigma$ and d is a fresh letter, i.e. $d \notin \Sigma$. Let $\Sigma' = \Sigma \cup \{d\}$. Let L_n be be the timed language defined in the proof of the Theorem 4.11. We have shown that L_n cannot be accepted by any TA with less than n clocks. Recall that $T\Sigma^*$ denotes the set of all timed words over Σ.

Given the timed regular language L over Σ as a timed automaton with n clocks, we first construct a language L' over Σ' as the union of the following three languages.

- $L_1 = L \cdot \{(d,t) : t \in \mathbb{R}^+\} \cdot (T\Sigma^*)$,
- L_2 is the set of timed words over Σ' having no d's or having at least two d's, and
- $L_3 = (T\Sigma^*) \cdot \{(d,t) : t \in \mathbb{R}^+\} \cdot L_n$.

It is easy to see that L' is timed regular as it is obtained from timed regular languages using catenation and union (which preserve timed regularity). Moreover, L' can be accepted by a TA with n clocks. We consider two cases.

(1) $L = T\Sigma^*$. Then it is easy to see that $L' = T\Sigma'^*$, the set of all timed words over Σ'. Therefore L' is accepted by a deterministic timed automaton without any clock. Choose $L'' = L'$ in this case.

(2) L is strictly included in L_Σ. Then, there is a timed word $u = (\sigma_1, \tau_1) \ldots (\sigma_n, \tau_n) \in \overline{L}$ which does not belong to L. Without loss of generality, assume that all the time stamps are rational. Let $L'' = \{u \cdot (d,1)\} \cdot L_n$. Then, $\{u \cdot (d.1)\}$ can be recognized by a single clock automaton, and, by reusing this clock, we can easily construct a timed automaton accepting L'' using n-clocks. We show that L'' cannot be accepted by a timed automaton with at most $n-1$ clocks. Assume to the contrary that L'' is accepted by a timed automaton $\mathcal{A}''$ with $n-1$ clocks. By the lemma 4.3, this implies that L_n is accepted by a timed automaton with at most $n-1$ clocks. But this is a contradiction as we have already shown in Theorem 4.2. Hence, we conclude that L'' is not accepted by any TA with at most $n-1$ clocks.

□

The reader may wonder why the condition $n \geq 2$ is needed in the above theorem. Below, we consider the special cases of 1-clock and 2-clock timed automata.

1C-TA and 2C-TA We have seen that location reachability (non-emptiness) of a timed automaton is PSPACE complete in general. We can consider the specific case of timed automata with 1-clock and 2-clocks, respectively called 1C-TA and 2C-TA. Larrousinni *et al* have shown that location reachability is NLOGSPACE-complete for 1C-TA and it is NP-hard for 2C-TA where as for 3C-TA this complexity becomes PSPACE-complete [21]. Ouaknine, Worrell *et al* have shown that the language inclusion problem $L(\mathcal{A}) \subseteq L(\mathcal{B})$ is decidable with non-primitive-recursive complexity when $\mathcal{B}$ is a one clock timed automaton [9, 10, 22]. This also shows that the universality of 1C-TA is decidable and the proof of the Theorem 4.12 is not valid for $n = 1$. Indeed, the question whether the language of a 1C-TA is accepted by a clock-less timed automaton is decidable. Suman *et al* have identified a subclass of timed automata which can be effectively reduced to 1-clock deterministic timed automata [17, 18].

4.7. Integer Timed Automata and Digitization

The undecidability of language inclusion checking for timed automata led to a search for subclasses where the problem becomes solvable. ERA as well as 1C-timed automata presented earlier are such subclasses. Additionally, some approximation techniques have also been considered. Henzinger, Manna and Pnueli came up with one such widely applicable technique called *Digitization* [6]. The digitization technique can conservatively under-approximate the dense time language inclusion problem to the discrete time language inclusion problem such that a yes answer in discrete time guarantees the yes answer in dense time. The digitization technique can also over-approximate the language inclusion problem such that a no answer in discrete time guarantees a no answer in dense time. Together, these approximations give a partial but often a practical method for solving instances of language inclusion problem over timed automata. Moreover, under appropriate conditions it can be shown that this digitization approximation is precise. We present an overview of the digitization approach following its formalization by Ouaknine and Worrell [23]. For simplicity, we shall assume that (in place of rational) only integer valued constants are used in the guards of timed automata.

4.7.1. *Integer Timed Automata*

A timed word $\rho = (\sigma, \tau)$ with $\tau = \tau_1\tau_2\ldots\tau_n$ is called an integer timed word (or integral timed word) if each τ_i is integer valued, i.e. $\tau_i \in \mathbb{Z}_0$ for all i. Let $T\Sigma^*_{\mathbb{Z}_0}$ denote the set of integer timed words, and let $\mathbb{Z}L \triangleq L \cap T\Sigma^*_{\mathbb{Z}_0}$ denote the set of integer timed words in L. A timed (regular) language consisting only of integer timed words is called an integer timed (regular) language.

A timed automaton over the time domain $\mathbb{Z}_0$ only accepts integer timed words. Such an automaton is called an *Integer Timed Automaton* (ITA). We denote its language by $L(\mathcal{A})_{\mathbb{Z}_0}$. It is easy to check that $L(\mathcal{A})_{\mathbb{Z}_0} = (L(\mathcal{A})_{\mathbb{R}_0} \cap T\Sigma^*_{\mathbb{Z}_0}) = \mathbb{Z}(L(\mathcal{A})_{\mathbb{R}_0})$.

We shall abbreviate this by $\mathbb{Z}L(\mathcal{A})$.

Note that, technically, an integer timed word is a finite sequence over the alphabet $\Sigma \times \mathbb{Z}_0$ which is an infinite alphabet. Hence, in a strict sense, an integer timed regular language is not a regular language accepted by a finite automaton with a finite alphabet. However, we can represent a timed word symbolically as a word over the finite alphabet $\Sigma \cup \{\checkmark\}$, where every event $\checkmark$ represents the passage of one time unit. For instance the integral timed word $(a, 3)(b, 5)$ is represented by the untimed $\checkmark$-word $\checkmark\checkmark\checkmark a\checkmark\checkmark b$. We shall denote the symbolic representation of a timed word ρ by $\rho^{\checkmark}$. For an integer timed language L, define $L^{\checkmark} \triangleq \{\rho^{\checkmark} \mid \rho \in L\}$. Ouaknine and Worrell have shown that the $\checkmark$-word representation of an integer timed regular language is regular.

Proposition 4.2 ([23]). *For a given timed automaton $\mathcal{A}$, we can effectively construct an untimed finite state automaton $\mathcal{A}^{\checkmark}$ such that $L(\mathcal{A}^{\checkmark}) = (\mathbb{Z}L(\mathcal{A}))^{\checkmark}$.* □

Proof. We modify the region automaton $\mathcal{A}^{reg}$ defined in Section 4.3 to obtain $\mathcal{A}^{\checkmark}$. The set of states consists only of "integral" regions. An integral clock region is one whose projection on any axis is either a (integer valued) singleton or it is unbounded, denoting saturation. For example, assuming that $c_x = 2, c_y = 3$, regions $x = 1 \wedge y = 0$ as well as $x = 1 \wedge y > 3$ are integral but $0 < x = y < 1$ is not an integral region. The Σ-transitions of $\mathcal{A}^{\checkmark}$ are exactly the Σ transitions $\mathcal{A}^{Reg}$ between integral regions. The ϵ transitions are discarded. Instead, we have transitions of $\mathcal{A}^{\checkmark}$ labelled $\checkmark$ given by $(l, r) \xrightarrow{\checkmark} (l, r')$ iff $\nu \in r$ implies $\nu + 1 \in r'$. It is easy to check that the resulting automaton accepts $(\mathbb{Z}L(\mathcal{A}))^{\checkmark}$. □

Corollary 4.4. *The language emptiness and the language inclusion problems for integer timed automata are decidable.* □

4.7.2. *Digitization*

Definition 4.10. *Let $t \in \mathbb{R}_0$ and let $0 \le \epsilon < 1$ be a real number. Let $\lfloor t \rfloor$ denote the largest integer less than or equal to t, and let $\lceil t \rceil$ denote the smallest integer greater than or equal to t. Then, the ϵ-digitization of t is defined as,*

$$[\![t]\!]_\epsilon \triangleq \begin{cases} \lfloor t \rfloor & \textit{if } fr(t) < \epsilon, \\ \lceil t \rceil & \textit{otherwise.} \end{cases}$$

□

We can extend the definition of $[\![\cdot]\!]_\epsilon$ to timed words and timed languages by point-wise application. Specifically, $[\![\tau_1 \cdots \tau_n]\!]_\epsilon \triangleq [\![\tau_1]\!]_\epsilon \cdots [\![\tau_n]\!]_\epsilon$, and $[\![(\sigma, \tau)]\!]_\epsilon = (\sigma, [\![\tau]\!]_\epsilon)$. For example, if $\rho = (a_1, 1.5) \longrightarrow (a_2, 4.35) \longrightarrow (a_3, 5.0)$, then $[\![\theta]\!]_{0.0} = (a_1, 2) \longrightarrow (a_2, 5) \longrightarrow (a_3, 5)$, and $[\![\theta]\!]_{0.4} = (a_1, 2) \longrightarrow (a_2, 4) \longrightarrow (a_3, 5)$ where as $[\![\theta]\!]_{0.6} = (a_1, 1) \longrightarrow (a_2, 4) \longrightarrow (a_3, 5)$. Note that any $0 \le \epsilon < 1$ gives rise to only one of these three digitizations. Digitization can result into a weakly monotonic timed word even when the original word is strictly monotonic. Each digitization is an integer timed word. It can be considered as an integer timed approximation of the original timed word.

Definition 4.11. *For a timed word ρ, the set of its digitizations is defined to be $[\![\rho]\!] \triangleq \{[\![\rho]\!]_\epsilon \mid 0 \leq \epsilon < 1\}$. For a timed language L, the set of its digitizations is defined to be $[\![L]\!] \triangleq \{[\![\rho]\!] \mid 0 \leq \epsilon < 1 \wedge \rho \in L\}$.* □

Definition 4.12 (Closure under Digitization). *A timed language L is called* closed under digitization *if $\rho \in L$ implies that $[\![\rho]\!] \subseteq L$. The timed language L is called* closed under inverse digitization *if for any $\rho \in T\Sigma^*$, if $[\![\rho]\!] \subseteq L$ then $\rho \in L$.* □

Theorem 4.13 ([6]). *Let L be a set of timed words closed under digitization and let L' be a set of timed words closed under inverse digitization. Then $L \subseteq L'$ iff $\mathbb{Z}L \subseteq \mathbb{Z}L'$.* □

Proof. Proving that $L \subseteq L' \Rightarrow \mathbb{Z}L \subseteq \mathbb{Z}L'$ is trivial. To prove the other way, assume that $\rho \in L$. Since L is closed under digitization, by its definition, $[\![\rho]\!] \subseteq L$. Note that $[\![\rho]\!] \subseteq T\Sigma^*_{\mathbb{Z}_0}$. Hence, $[\![\rho]\!] \subseteq \mathbb{Z}L$. And since $\mathbb{Z}L \subseteq \mathbb{Z}L'$, we have $[\![\rho]\!] \subseteq \mathbb{Z}L'$ and hence $[\![\rho]\!] \subseteq L'$. Since L' is closed under inverse digitization, by its definition, we get that $\rho \in L'$. □

Corollary 4.5. *Let $\mathcal{A}$ and $\mathcal{B}$ be two TA with $L(\mathcal{A})$ closed under digitization and $L(\mathcal{B})$ closed under inverse digitization. Then the timed language inclusion problem $L(\mathcal{A}) \subseteq L(\mathcal{B})$ is decidable.*

Proof. By Theorem Theorem 4.13, checking $L(\mathcal{A}) \subseteq L(\mathcal{B})$ reduces to checking whether $L(\mathcal{A})_{\mathbb{Z}_0} \subseteq L(\mathcal{B})_{\mathbb{Z}_0}$. This is decidable according to Corollary 4.4 by checking whether $L(\mathcal{A}^{\checkmark}) \subseteq L(\mathcal{B}^{\checkmark})$. □

4.7.3. *Trace-Regions*

Given a timed automaton $\mathcal{A}$ and timed words $\rho_1 = (\sigma_1, \tau_1)$ and $\rho_2 = (\sigma_2, \tau_2)$, let r_1 and r_2 be runs of $\mathcal{A}$ on ρ_1 and ρ_2. The runs r_1 and r_2 are called *path-equivalent* if they go through exactly the same sequence of discrete edges. For this to happen, it is necessary that $\sigma_1 = \sigma_2$. We now study some conditions under which different timed words give rise to path equivalent runs.

Consider a timed word (σ, τ) of length n. Here, $\tau = \tau_1\tau_2 \ldots \tau_n$ is a weakly monotonic sequence of time stamps. We can geometrically represent τ by a point $(\tau_1, \ldots, \tau_n)$ in the n-dimensional hyperplane $T_n = \{(x_1, \ldots, x_n) \in (\mathbb{R}_0)^n \mid x_1 \leq \cdots \leq x_n\}$. Indeed, the points of this hyperplane and the n-length weakly monotonic time stamp sequences are in one-to-one correspondence. This hyperplane T_n can be divided into trace-regions using the following equivalence.

Definition 4.13. *Two weakly monotonic sequences of time stamps τ and τ' are called trace-equivalent (written $\tau \simeq_{Tr} \tau'$) iff*

- *Both τ and τ' are of same length, say n,*
- $\forall 1 \le i \le n$, $int(\tau_i) = int(\tau'_i)$,
- $\forall 1 \le i, j \le n$, $fr(\tau_i) \le fr(\tau_j)$ *iff* $fr(\tau'_i) \le fr(\tau'_j)$, *and*
- $\forall 1 \le i \le n$, $fr(\tau_i) = 0$ *iff* $fr(\tau'_i) = 0$. □

Definition 4.14. *Two timed words $\rho = (\sigma, \tau)$ and $\rho' = (\sigma', \tau')$ are called trace-equivalent (written $\rho \cong_{Tr} \rho'$) iff $\sigma = \sigma'$ and $\tau \simeq_{Tr} \tau'$.* □

The trace-equivalence partitions the set of timed words $T\Sigma^*$ into regions called *trace-regions.* Note that even for the set of timed words of length n, the trace-equivalence does not result into a finite number of partitions. Given an integral timed word $\rho \in T\Sigma^*_{\mathbb{Z}_0}$, its trace-region contains only the timed word ρ. Such a trace-region will be called an *integral trace-region.*

Lemma 4.5. *Let ρ, ρ' be a pair of timed words such that $\rho \cong_{Tr} \rho'$. Then, for any timed automaton $\mathcal{A}$ we have that $\rho \in L(\mathcal{A})$ iff $\rho' \in L(\mathcal{A})$.* □

Proof. Assume that $\rho = (\sigma_1, \tau_1) \cdots (\sigma_n, \tau_n) \in L(\mathcal{A})$. Let r be a run of $\mathcal{A}$ on the timed word ρ. Let

$$r = (l_0, \nu_0) \xrightarrow{t_1} (l_0, \nu_1) \xrightarrow{e_1} (l_1, \nu'_1) \xrightarrow{t_2} \cdots \xrightarrow{e_n} (l_n, \nu'_n),$$

with $e_j = (l_{j-1}, l_j, \varphi_j, \sigma_j, \lambda_j)$. For any clock x and index j, $\nu_j(x) = \tau_j - \tau_i$, where $i \le j$ is the index of the last transition which resets the clock x, or $\nu_j(x) = \tau_0$ if x was never reset in the run r. Consider $\rho' = (\sigma_1, \tau'_1) \cdots (\sigma_n, \tau'_n)$. Since $\rho \cong_{Tr} \rho'$, one readily verifies that, for any $1 \le i, j \le n$ and $k \in \mathbb{N}$, whenever $\tau_j - \tau_i \sim k$ we must have $\tau'_j - \tau'_i \sim k$, for any $\sim \in \{<, \le, >, \ge, =\}$. It follows immediately that $\mathcal{A}$ accepts ρ' with a run r' which is path equivalent to r (i.e. r' has the same sequence of discrete edges as r). □

4.7.4. *Open And Closed Timed Automata*

We now define some syntactic subclasses of timed automata by restricting the form of clock constraints allowed in the guards. Recall the definition of clock constraints given in Section 4.2.

Definition 4.15. *Open clock constraints are constraints that are generated by the grammar $\varphi := x < k | x > k | \varphi \wedge \varphi$. Thus, only strict comparisons with integer constants are allowed in open clock constraints. Similarly, closed clock constraints are constraints that are generated by the grammar $\varphi := x \le k | \ge k | x = k | \varphi \wedge \varphi$. Only non-strict comparisons with integer constants are allowed in closed clock constraints. An open timed automaton is a timed automaton in which all the clock constraints occurring in its transitions are open. A closed timed automaton is a timed automaton in which all the clock constraints are closed.* □

4.7.5. *Topology And Closure*

We study how openness and closedness of the clock constraints affects the class of timed regular languages. A metric topology over the set of timed words provides a useful insight in this investigation [23, 24]. The presentation here follows the treatment of Ouaknine and Worrell [23].

Definition 4.16 ([24]). *Let $\rho = (\sigma, \tau)$ and $\rho' = (\sigma', \tau')$ be two timed words. Let the distance d between these words be defined as follows:*

$$d(\rho, \rho') \triangleq \begin{cases} \infty & \textit{if } \sigma \neq \sigma', \\ max\{|\tau_i - \tau'_i| : 1 \leq i \leq n\} & \textit{otherwise.} \end{cases}$$

□

The set of timed words $T\Sigma^*$ forms a complete metric space under this (extended) metric d which satisfies the conditions *(a)* $d(\rho, \rho') \geq 0$, *(b)* $d(\rho, \rho) = 0$, *(c)* $d(\rho, \rho') = d(\rho', \rho)$, and *(c)* $d(\rho, \rho') \leq d(\rho, \rho'') + d(\rho'', \rho')$. We briefly remind the reader of some standard concepts from metric topology. A sequence of timed words $\langle \rho_i \rangle_{i \geq 1}$ is said to converge to a limit ρ if for any $\epsilon > 0$ there exists j such that for all $k > j$, $d(\rho_k, \rho) < \epsilon$. Also, given a timed word ρ and $\epsilon > 0$, let the open ball of radius ϵ around ρ be defined as $B(\rho, \epsilon) = \{\rho' \mid d(\rho, \rho') < \epsilon\}$. A set of timed words L is called d-open if for each $\rho \in L$ there exists $\epsilon > 0$ such that $B(\rho, \epsilon) \subseteq L$. Intuitively, around each member of an open L we can have an open ball of non-zero radius such that all words within the ball are in L. Complements of d-open sets are called d-closed sets. It can be shown that a set L is d-closed iff for every converging sequence $\langle \rho_i \rangle_{i \geq 1} \subseteq L$ its limit is included in L. Closure of a set S is denoted by S^{Cl}; this is the set obtained by adding the limit points of the converging sequences in S to S.

Lemma 4.6. *If $\mathcal{A}$ is an open timed automaton, then $L(\mathcal{A})$ is d-open.* □

Proof. Given an open timed automaton $\mathcal{A}$, consider the timed run $r = (l_0, \nu_0) \xrightarrow{t_1} (l_0, \nu_1) \xrightarrow{e_1} (l_1, \nu'_1) \cdots \xrightarrow{e_n} (l_n, \nu'_n)$ that accepts the timed word ρ. Note that for each $1 \leq i \leq n$, the constraint φ_i corresponding to discrete transition e_i is open and $\nu_i \models \varphi_i$. Hence, for each i, there exists an $\epsilon_i > 0$ such that all the interpretations in the open interval $(\nu_i - \epsilon_i,\ \nu_i + \epsilon_i)$ will satisfy φ_i. All these valuations may not be reachable but the clock constraints are sensitive only to the differences between the time stamps of the timed word. Define $\epsilon = min\{\frac{\epsilon_i}{n} : 0 \leq i \leq n\}$. Then, it can be shown that any timed word in the open ball $B(\rho, \epsilon)$ around ρ has a run which is path equivalent to r (i.e. it has the same sequence of discrete edges at r). □

Lemma 4.7. *If $\mathcal{A}$ is a closed timed automaton, then $L(\mathcal{A})$ is d-closed.* □

Proof. Let $\mathcal{A}$ be a closed timed automaton. Let $\langle \rho_i \rangle_{i \geq 1}$ be a sequence of time words in $L(\mathcal{A})$ which converge to a timed word ρ. Let $\langle r_i \rangle_{i \geq 1}$ be a sequence of accepting runs of $\mathcal{A}$ on the timed words $\langle \rho_i \rangle_{i \geq 1}$. There exists an infinite subsequence of these runs denoted $\langle r_{j_i} \rangle_{i \geq 1}$ which are which are all path equivalent, since there are only finitely many transitions. Clearly, the corresponding subsequence of timed words $\langle \rho_{j_i} \rangle_{i \geq 1}$ also converges to ρ. Each r_{j_i} is a n-length sequence of valuations each of which assigns a real value to the k clocks. We identify each r_{j_i} with an element of $(\mathbb{R}_0)^l$ with $l = n \times k$. The set of runs of $\mathcal{A}$ which are path equivalent to (any one of the) $\langle r_{j_i} \rangle_{i \geq 1}$ form a compact subset of $\mathbb{R}^l$, say W, because the constraints satisfied by each guard are all closed. Clearly, the converging sequence $\langle r_{j_i} \rangle_{i \geq 1}$ is itself within W and it must have its limit point r in W. This run r, with timed word ρ, (being a member of W) is a valid run of $\mathcal{A}$ implying that $\rho \in L(\mathcal{A})$. □

Topology of Trace-Regions

It is informative to visualize the topological properties of trace-regions of timed words of length n. The n-dimensional hyperplane T_n given in Section 4.7.3 has grid points in $(\mathbb{Z}_0)^n$ with integral coordinates. The hyperplane T_n is partitioned into infinitely many trace-regions each of which is either an n-dimensional open triangle between between adjacent grid points, or one of its (lower-dimensional) faces/vertices. These are similar to the clock regions shown in Figure 4.3 except that the hyperplane is divided into infinitely many regions without saturating at any maximum constant. The open triangles constitute d-open sets of timed words and are called d-open trace-regions. The only trace-regions constituting d-closed sets are the grid points each of which contains a single integer timed word. (Earlier, we term these grid-points as integral trace-regions.) Thus, the integral timed words $T\Sigma^*_{\mathbb{Z}_0}$ are in one-to-one correspondence with the d-closed trace-regions. All other trace-regions are neither d-closed nor d-open. The closure r^{Cl} of a trace-region r extends r with lower-dimensional trace-regions which form the boundaries of r. In particular, r^{Cl} contains all the integral timed traces (grid points) bordering the region r. Note that r^{Cl} is not necessarily a trace-region (it is a union of trace-regions).

Lemma 4.8. *Let ρ be a timed word of length n, and let Reg_ρ be its corresponding trace-region. The reader can easily visualize that the digitizations $[\![\rho]\!]$ of ρ are exactly the set of integral timed words (or integral trace-regions) bordering Reg_ρ. Formally, $[\![\rho]\!] = Reg_\rho^{Cl} \cap T\Sigma^*_{\mathbb{Z}_0}$.* □

Thus, digitization takes every timed word in a trace-region r to every integral trace (grid point) on the boundary of r.

Theorem 4.14. *Let L be a timed language. If L is d-open then L is closed under inverse digitization.* □

Proof. Let $\rho \in T\Sigma^*$ be an arbitrary timed word. Assume that $[\![\rho]\!] \subseteq L$. Pick a $\rho' \in [\![\rho]\!]$. (In fact, it is sufficient to assume that there exists $\rho' \in ([\![\rho]\!] \cap L)$). Since L is d-open, there exists $\epsilon > 0$ such that the open ball $B(\rho', \epsilon) \subseteq L$. By Proposition 4.8, we have that $[\![\rho]\!] \subseteq Reg_\rho^{Cl}$, and ρ' is one of the vertices boardering Reg_ρ. Hence, $B(\rho', \epsilon) \cap Reg_\rho \neq \emptyset$. Lemma 4.5 then gives us that $Reg_\rho \subseteq L$. Hence, $\rho \in L$. □

Theorem 4.15. *Let L be a timed language. If L is d-closed, then L is closed under digitization.* □

Proof. Let $\rho \in L$. Proposition 4.5 implies that $Reg_\rho \subseteq L$. Moreover, since L is d-closed, we have $(Reg_\rho)^{Cl} \subseteq L(\mathcal{A})$. Then, using Lemma 4.8, we get that $[\![\rho]\!] \subseteq L(\mathcal{A})$. □

Combining Theorems 4.14 and 4.15 with Lemmas 4.6 and 4.7, we obtained the following corollary.

Corollary 4.6. *If $\mathcal{A}$ is an open timed automaton then $L(\mathcal{A})$ is closed under inverse digitization. If $\mathcal{A}$ is a closed automaton then $L(\mathcal{A})$ is closed under digitization.* □

Combining the above Corollary with Theorem 4.13, we obtain the following useful result.

Corollary 4.7. *If $\mathcal{A}$ is a closed timed automaton and $\mathcal{B}$ is an open timed automaton then the language inclusion question $L(\mathcal{A}) \subseteq L(\mathcal{B})$ is decidable.* □

Every timed automaton $\mathcal{A}$ can be syntactically approximated to the automaton $\mathcal{A}^{Op}_{\leq}$ (or $\mathcal{A}^{Cl}_{\leq}$) by replacing each guard in $\mathcal{A}$ by a stronger guard which is open (or closed). Clearly, this gives an automaton with a smaller language which is d-open (or d-closed). Analogously, we can approximate the automaton $\mathcal{A}$ by weakening its guards to obtain $\mathcal{A}^{Op}_{\geq}$ (or $\mathcal{A}^{Cl}_{\geq}$) having a larger d-open (or d-closed) language. These approximations can be used together with the corollary 4.7 for approximately solving the language inclusion problem.

Ouaknine and Worrell have also investigated the decidability of digitization closures [23]. We state some of their results without giving proofs.

- The universality of closed timed automata is undecidable.
- Given a timed automaton $\mathcal{A}$ we can effectively construct an automaton $\mathcal{A}^{Cl}$ such that $L(\mathcal{A}^{Cl}) = (L(\mathcal{A}))^{Cl}$.
- Given a timed automaton $\mathcal{A}$ it is decidable whether $L(\mathcal{A})$ is closed under digitization.
- Given a closed timed automaton $\mathcal{A}$, checking whether $L(\mathcal{A})$ is closed under inverse digitization is undecidable.

4.8. Extensions of TA

Timed automata can be seen to be highly restrictive. In guards, the clocks can only be compared with rational constants. Moreover, in a transition a clock can be reset (updated) only to the value zero. These conditions are delicately chosen to allow the emptiness of timed automaton to be decidable using the region-automaton construction. In this section, we consider some more general forms of timed automata and we also comment on their properties. While we have omitted detailed definitions and proofs, we do provide references to the relevant papers where these extensions are studied.

4.8.1. *Diagonal Constraints*

The constraints allowed over a timed automaton can be generalized to a class called the diagonal constraints. These constraints are defined by

$$\varphi^d := x \sim c \mid x - y \sim c \mid \varphi_1 \wedge \varphi_2$$

where $\sim$ is a comparison operator in $\{\leq, <, =, >, \geq\}$.

Every automaton with diagonal constraints can be modelled by a timed automaton with simple constraints by introducing additional clocks [1, 25]. Hence, this class of timed automata has the same properties as the standard timed automata and the diagonal constraints do not add to the expressive power of basic timed automata.

4.8.2. *Location Invariants and Urgency*

Timed automata can be extended to allow constraints on clocks to be associated with locations (see [1]). These constraints essentially prescribe the condition under which the automaton can remain in a location. These constraints are popularly known as *location invariants.* Nevertheless, a timed automaton with location invariants can be effectively transformed to an equivalent timed automaton without location invariants [7].

An *urgent* location is a location from which a discrete transition should be taken instantaneously. Thus, only non-zero time is allowed to pass when a run of the automaton steps into an urgent location [5].

Location invariants and urgency can enforce deadlines on how long an automaton can stay in a state. They are a useful modelling feature. However, they do not add to the expressive power of the basic timed automata. Every timed automaton with location invariants and urgent transitions can be reduced to a language equivalent timed automaton without location invariants or urgent transitions by introducing an additional clock [26].

4.8.3. *Updatable Timed Automata*

Basic timed automata only allow clock updates (resets) of the form $x := 0$ on any transition; i.e. clocks can only be reset to the value zero. Timed automata with more general forms of update have been studied by Bouyer *et al* [27]. We state some of the main results from this study.

- Extending timed automata with clock updates to constants, i.e. $x := c$, does not change the expressive power of basic timed automata whether diagonal constraints are allowed or not. Hence, emptiness remains decidable with such updates.

- Timed automata with updates of the form $x := x - 1$ (i.e. decrement the clock by 1 (saturating at 0 if value becomes negative) are strictly more expressive than basic timed automata and their language emptiness is undecidable.
- Timed automata with updates of the form $x := x + 1$ (i.e. increment the clock by 1) are strictly more expressive than basic timed automata. Moreover, their language emptiness is decidable. However, in presence of incrementing updates and diagonal constraints, the language emptiness is undecidable.

4.8.4. *Infinite words*

An ω-word is an infinite sequence of letters. Timed ω-words were introduced by Alur and Dill [1]. Infinite runs of timed automata on such runs were considered with either the Büchi acceptance condition or the Muller acceptance condition. Note that an infinite timed word needs to satisfy the progress condition stating that for every $t \in \mathbb{R}_0$, there exists i such that $\tau_i > t$. It can be shown that almost all of the constructions, properties and the decidability/undecidability results for finite word timed automata presented in this chapter extend to these timed ω-word automata (see [1]).

4.8.5. *Periodic guards and silent actions*

Many activities in timed systems are periodic. To model such activities the transition guards can be extended to constraints of the form $\exists k \in \mathbb{N} : x \in \langle a + kp, b + kp \rangle$ where the notation $\langle a, b \rangle$ denotes an open, closed or half-open interval depending on whether $\langle$ is [or (and $\rangle$ is] or). Such automata are called timed automata with periodic guards [25, 28]. An alternative technique is to add an unobservable action ϵ to the set Σ of transition labels giving TA with silent transitions (or ϵ-TA) [29]. The occurrences of such unobservable transitions in a run are not recorded in the timed word. Most of the closure and decidability properties of timed automata carry over to periodic and ϵ-TA. It has been shown that that periodic TA are more expressive than TA, and ϵ-TA are more expressive than periodic TA.

4.8.6. *Additive and linear constraints*

Let x_i range over clocks and a_i, c range over rational constants. The atomic constraints in the guards of timed automata can be extended to linear inequalities of the form $\Sigma_i\, a_i x_i \sim c$ with $\sim \in \{\leq, <, =, >, \geq\}$. It can be shown that even with simple additive constraints of the form $x_i + x_j \sim c$ the problem of language emptiness is undecidable for timed automata with four or more clocks [30]. The problem becomes decidable for automata with two clocks and its status is still open for automata with three clocks.

Acknowledgment

We would like to thank Deepak D'Souza, Priti Shankar and the anonymous reviewers for their careful reviewing and constructive comments on this article.

References

[1] R. Alur and D. L. Dill, A theory of timed automata, *Theoretical Computer Science*. **126**(2), 183–235, (1994).
[2] R. Alur and P. Madhusudan. Decision problems for timed automata: A survey. In *Proc. SFM-RT*, vol. 3185, *LNCS*, pp. 1–24. Springer, (2004).
[3] J. Bengtsson, K. G. Larsen, F. Larsson, P. Pettersson, and W. Yi. UPPAAL - a Tool Suite for Automatic Verification of Real-Time Systems. In *Proc. Hybrid Systems III: Verification and Control*, vol. 1066, *LNCS*. Springer, (1996).
[4] C. Daws, A. Olivero, S. Tripakis, and S. Yovine. The tool KRONOS. In *Proc. Hybrid Systems III: Verification and Control*, vol. 1066, *LNCS*. Springer, (1996).
[5] G. Behrmann, A. David, and K. G. Larsen. A tutorial on uppaal. In *Proc. SFM-RT*, vol. 3185, *LNCS*, pp. 200–236. Springer, (2004).
[6] T. A. Henzinger, Z. Manna, and A. Pnueli. What good are digital clocks? In *Proc. 19th ICALP*, vol. 623, *LNCS*, pp. 545–558. Springer, (1992).
[7] J. Bengtsson and W. Yi. Timed automata: Semantics, algorithms and tools. In *Lectures on Concurrency and Petri Nets*, pp. 87–124. Springer, (2004).
[8] S. Adams, J. Ouaknine, and J. Worrell. Undecidability of universality for timed automata with minimal resources. In *Proc. 5th FORMATS*, vol. 4763, *LNCS*, pp. 25–37. Springer, (2007).
[9] J. Ouaknine and J. Worrell. On the language inclusion problem for timed automata: Closing a decidability gap. In *Proc. 19th LICS*, pp. 54–63. IEEE Computer Society, (2004).
[10] P. A. Abdulla, J. Deneux, J. Ouaknine, and J. Worrell. Decidability and complexity results for timed automata via channel machines. In *Proc. 32nd ICALP*, vol. 3580, *LNCS*, pp. 1089–1101. Springer, (2005).
[11] S. Tripakis, Folk theorems on the determinization and minimization of timed automata, *Information Processing Letters*. **99**(6), 222–226, (2006).
[12] O. Finkel. Undecidable problems about timed automata. In *Proc. 4th FORMATS*, vol. 4202, *LNCS*, pp. 187–199. Springer, (2006).
[13] R. Alur, L. Fix, and T. A. Henzinger, Event-clock automata: a determinizable class of timed automata, *Theoretical Computer Science*. **211**(1-2), 253–273, (1999).
[14] F. Chevalier, D. D'Souza, and P. Prabhakar. On continuous timed automata with input-determined guards. In *Proc. 26th FSTTCS*, vol. 4337, pp. 369–380. Springer, (2006).
[15] P. K. Pandya, Interval duration logic: Expressiveness and decidability, *Electr. Notes Theor. Comput. Sci.* **65**(6), 254–272, (2002).
[16] P. V. Suman. *Determinization, Clock reduction and Experimental Analysis of Timed Systems.* PhD thesis, Tata Institute of Fundamental Research, (2009).
[17] P. V. Suman, P. K. Pandya, S. N. Krishna, and L. Manasa. Timed automata with integer resets: Language inclusion and expressiveness. In *Proc. 6th FORMATS*, vol. 5215, *LNCS*, pp. 78–92. Springer, (2008).
[18] P. V. Suman and P. K. Pandya. Determinization and expressiveness of integer reset

timed automata with silent transitions. In *Proc. 3rd LATA*, vol. 5457, *LNCS*. Springer, (2009).

[19] C. Baier, N. Bertrand, P. Bouyer, and T. Brihaye. When are timed automata determinizable? In *Proc. 36th ICALP (Part 2)*, vol. 5556, *LNCS*, pp. 43–54. Springer, (2009).

[20] T. A. Henzinger, P. W. Kopke, and H. Wong-Toi. The expressive power of clocks. In *Proc. 22nd ICALP*, vol. 944, *LNCS*, pp. 417–428. Springer, (1995).

[21] F. Laroussinie, N. Markey, and Ph. Schnoebelen. Model checking timed automata with one or two clocks. In *Proc. 15th CONCUR*, vol. 3170, *LNCS*, pp. 387–401. Springer, (2004).

[22] P. A. Abdulla, J. Deneux, J. Ouaknine, K. Quaas, and J. Worrell, Universality analysis for one clock timed automata, *Fundamenta Informaticae.* **89**(4), 419–450, (2009).

[23] J. Ouaknine and J. Worrell. Revisiting digitization, robustness, and decidability for timed automata. In *Proc. 18th LICS.* IEEE Computer Society, (2003).

[24] V. Gupta, T. A. Henzinger, and R. Jagadeesan. Robust timed automata. In *Proc. Workshop on Hybrid and Real-Time Systems*, vol. 1201, *LNCS*, pp. 331–345. Springer, (1997).

[25] B. Berard, A. Petit, P. Gastin, and V. Diekert, Characterization of the expressive power of silent transitions in timed automata, *Fundamenta Informaticae.* **36**(2-3), 145–182, (1998).

[26] M. Adélaïde and C. Pagetti. On the urgency expressiveness. In *Proc. 24th FSTTCS*, vol. 3328, *LNCS*, pp. 71–83. Springer, (2004).

[27] P. Bouyer, C. Dufourd, E. Fleury, and A. Petit, Updatable timed automata, *Theoretical Computer Science.* **321**(2-3), 291–345, (2004).

[28] C. Choffrut and M. Goldwurm, Timed automata with periodic clock constraints, *Journal of Automata, Languages and Combinatorics.* **5**(4), 371–404, (2000).

[29] P. Bouyer, S. Haddad, and P.-A. Reynier. Undecidability results for timed automata with silent transitions. Research Report LSV-07-12, Laboratoire Spécification et Vérification, ENS Cachan, France, (2007).

[30] B. Bérard and C. Dufourd, Timed automata and additive clock constraints, *Information Processing Letters.* **75**(1-2), 1–7, (2000).

Part II

Verification

Chapter 5

A Language-Theoretic View of Verification

Kamal Lodaya

The Institute of Mathematical Sciences, Chennai 600 113, India
kamal@imsc.res.in

In this chapter, we motivate the verification of properties of systems in terms of membership, nonemptiness and inclusion problems of regular languages. These techniques were pioneered by Edmund Clarke and Allen Emerson, and independently by Jean-Pierre Queille and Joseph Sifakis. Clarke, Emerson and Sifakis won the Turing award in 2007 for their work. A key construction was outlined in a paper by Moshe Vardi and Pierre Wolper, which won the Gödel prize in 2000. The system properties are described in different kinds of "temporal logics", which were first introduced in this setting by Amir Pnueli and Zohar Manna. Pnueli won the Turing award in 1996 for his work. We discuss computation tree logic (CTL), linear temporal logic (LTL), interval temporal logic (ITL) and a few variants.

5.1. Setting the stage

The use of automata in verification goes back a long way, to Büchi [1], Elgot [2] and Trakhtenbrot [3] who, in the early 1960s, used the theory of automata on finite words to give an algorithm to check whether a sentence of monadic second order logic on such structures is valid, true in all models. They showed that this validity problem can be reduced to checking whether the language accepted by such a finite automaton is empty. Büchi went on to develop [4] a theory of finite automata on infinite words and proved the same result (which was considerably harder). This would come into use in verification nearly 25 years later!

These days, the verification question is posed as follows. We are given a model of a system (which can be hardware, software, or even mixed), typically as some kind of transition system. More precisely, we might be provided with a run of the system, modelled as a word over a suitable alphabet. We are given a system property, specified by a logical formula, typically in a propositional framework such as temporal logic (rather than monadic second order logic). Does the run satisfy this property? Or, do all runs of the system satisfy this property? These **truth checking** and **model checking** problems can be posed as membership or inclusion problems in automata theory, since the set of models of a temporal logic formula is a set of words—a language.

One of the highlights of the modern approach to automata theory is a careful consideration of algorithmic questions and the determination of their precise complexity. We refer the reader to the many textbooks available on complexity theory, but we give here a primer on complexity classes pertaining to this chapter. The reader familiar with basic complexity notions is advised to skip to the next section.

5.1.1. *A primer on complexity classes*

The complexity of a problem can be measured in terms of the number of computation steps (this is abbreviated as time) when the input is of a particular size, say n, but also in terms of the memory used (abbreviated space) as a function of n.

Nondeterministic algorithms. The algorithms we consider can be either deterministic, which is usual, or nondeterministic, which is not so usual. A standard paradigm of how a nondeterministic algorithm works is exemplified by:

Problem 1 (Directed graph accessibility, DGAP).
Instance: A directed graph G with n vertices, a source vertex s and a target vertex t
Question: Is there a path from s to t?
Complexity: NLOGSPACE

Here is how a nondeterministic algorithm might work: it writes out a string of vertices $v_1 \dots v_m$ of length $m \leq n$. This is called a guess. Now the algorithm checks whether the first vertex v_1 is s, each pair of vertices (v_i, v_{i+1}) is an edge in the graph and if the last vertex v_m is t. If this is indeed the case, the algorithm says "Yes" (and returns the guess if required), otherwise it says "No" (there is no path from s to t).

Surprisingly, this is indeed a nondeterministic *algorithm* because if there is a path from s to t, there is *one* run of the algorithm which is correct, and if there is no path from s to t, *all* the possible runs put together have examined all the possible paths of length $\leq n$ from s and not reached t, which is sufficient to conclude the "No" answer.

The time complexity of the algorithm depends upon the way the graph is represented, but certainly it can be done in time polynomial in n. We can say the algorithm is nondeterministic polynomial time or in NPTIME.

If we analyze the memory usage, we find that we can optimize the algorithm a little: instead of guessing the whole path, the algorithm first guesses the pair (v_1, v_2) and verifies that this is an edge with $v_1 = s$, then it erases v_1 and guesses v_3 to form the pair (v_2, v_3), verifies that this is an edge, and so on. (v_3 might be the same as v_1, but the existence of a walk of length $< n$ guarantees a path of length $< n$.) Since writing out a vertex of a graph with n nodes and managing a counter takes $O(\log n)$ bits, this is a nondeterministic logarithmic space or NLOGSPACE algorithm. Now we have a complexity upper bound of NLOGSPACE for DGAP, which is better than NPTIME.

Upper bounds. Complexity theorists have defined classes of algorithms like we did above, and shown that the following inclusions hold between them. None of them is known to be strict, but we do know of exponential separation between classes (for example, NLOGSPACE $\neq$ NPSPACE, PTIME $\neq$ EXPTIME).

ALOGTIME $\subseteq$ LOGSPACE $\subseteq$ NLOGSPACE $\subseteq$ PTIME $\subseteq$ NPTIME $\subseteq$ PSPACE = NPSPACE
PSPACE = NPSPACE $\subseteq$ EXPTIME $\subseteq$ NEXPTIME $\subseteq \cdots \subseteq$ ELEMENTARY $\subset$ COMPUTABLE

More precisely, complexity theory classifies *problems* rather than *algorithms*, since it is possible to write very inefficient algorithms! So the class LOGSPACE consists of all problems which have deterministic algorithms using space bounded by a polynomial in $\log n$, where n is the size of the input of the problem instance. The even more restrictive class ALOGTIME is sometimes also called NC1, but there are some technicalities involved and it is best to consult a textbook for the details. We will use this class only for the reason that it exactly matches the complexity of evaluation of propositional logic (or Boolean sentences), a result shown by Buss [5].

Problem 2 (Boolean sentence value, BSVP).
Instance: A tree with vertices labelled from $\{\wedge, \vee, 0, 1\}$*, where the vertices labelled* 0 *or* 1 *are sources.*
Question: Does this "formula", with the vertex labels being given their usual boolean interpretation, evaluate to 1 *at the root?*
Complexity: ALOGTIME

The class NLOGSPACE has all problems which have *nondeterministic* algorithms with this space requirement. Since a deterministic algorithm is a special case of a nondeterministic one, the inclusion of a deterministic class in a nondeterministic class with the same requirements follows. What about the other way? Savitch [6] gave a clever divide-and-conquer technique which shows that any NLOGSPACE algorithm can be converted to one which is deterministic, but the space usage is squared.

The classes PTIME and NPTIME contain deterministic and nondeterministic algorithms whose *time* requirement is a polynomial in n. If an algorithm takes space $s(n)$, after time exponential in $s(n)$ it will return to a configuration which it had already seen earlier. A more formal proof based on this argument, which can be found in a textbook, shows the inclusion of a class taking space $O(s(n))$ in a deterministic class taking time $2^{O(s(n))}$.

The class PSPACE contains problems which have algorithms taking space polynomial in n. Applying Savitch's theorem, we see that NPSPACE turns out to be the same as PSPACE. An algorithm taking time order $t(n)$ can only use space order $t(n)$, since it does not have the time to reach more space! This gives inclusion of the time classes in space classes with the same requirements.

The classes EXPTIME and NEXPTIME have algorithms whose time requirement is $2^{p(n)}$, for some function $p(n)$ polynomial in n. Higher classes can also be defined similarly, using towers of exponentials. The class ELEMENTARY includes algorithms whose space is a tower of exponentials of any constant height. Finally, the class COMPUTABLE includes all problems which have any kind of algorithm at all.

Lower bounds. We are not yet done with DGAP, we want to find a lower bound for it. Complexity theory really took off in the 1970s when Steve Cook discovered how to give lower bounds to a problem.

A problem P is said to reduce to a problem Q (we write $P \leq Q$) if, suppose we are given an algorithm for Q, we can find a "preprocessing" function f which will take any instance I of the problem P and translate it into an instance $f(I)$ of the problem Q, such that $f(I)$ has a solution if, and only if, I has a solution. Since it does not make sense for f to take more time or space than the algorithm for Q itself, we will assume that f is itself computed by an algorithm which takes logarithmic space. A textbook on complexity would call this a "many-one logspace reduction".

Here is Cook's insight. Suppose we find a problem Q in NLOGSPACE such that *every* problem in NLOGSPACE reduces to Q (we write NLOGSPACE $\leq Q$). Indeed, DGAP is such a problem, which is said to be NLOGSPACE-hard. To prove that NLOGSPACE is a lower bound for some problem R, it is sufficient to show that DGAP reduces to R. In symbols, NLOGSPACE $\leq$ DGAP and DGAP $\leq R$ shows NLOGSPACE $\leq R$.

DGAP is also NLOGSPACE-complete: in NLOGSPACE as well as NLOGSPACE-hard, giving both a lower and an upper bound. Cook's theorem showed that NPTIME $\leq$ SAT, the problem of checking whether a propositional logic formula is satisfiable. (Since SAT is also in NPTIME, SAT is NPTIME-complete.) We can say that the complexity of DGAP *is* NLOGSPACE, since NLOGSPACE is both an upper and a lower bound. Similarly, the monotone circuit value problem MCVP defined below is a problem whose complexity *is* PTIME.

It is not known whether there is a *deterministic* LOGSPACE algorithm for DGAP. By a simple analysis, you can work out that showing that there is a LOGSPACE algorithm for the DGAP problem amounts to showing that LOGSPACE=NLOGSPACE. This question, like the more famous PTIME = NPTIME question, is yet unsolved.

Problem 3 (Monotone circuit value, MCVP).
Instance: A directed acyclic graph G with vertices labelled from $\{\wedge, \vee, 0, 1\}$ and a sink vertex v, where the vertices labelled 0 or 1 are sources.
Question: Does this "circuit", with the vertex labels being given their usual boolean interpretation, evaluate to 1 at the output v?
Complexity: PTIME

The circuit may be assumed to be layered, with alternate layers being labelled $\wedge$ and $\vee$, until we reach the input layer which has 0 or 1 nodes.

Lower bounds using simulations. To give a PSPACE lower bound, we again need a "hard" problem as earlier. We show next how this is done using a "simulation" technique.

A computation of a nondeterministic algorithm can be thought of as a sequence of configurations $\#c_0\#c_1\#\ldots$. We consider the case of a *linear space* computation. That is, given an input w of size n to the algorithm, over an alphabet A, each configuration c_i is a word of length $n+1$ over a finite alphabet (the extra letter encodes some "state" information).

Now comes the simulation: *the whole computation itself* is a word over a larger alphabet $B \supset A$ (B includes symbols like $\#$ and state information). This word can be of arbitrary length. The initial configuration consists of an *initial state* (of the computation) and the input w provided to the computation. The final configuration is decided by a *final state* of the computation. Moving from the configuration c_i of the computation to c_{i+1}, the next configuration in the sequence, is represented by changing just three consecutive letters in the word c_i in accordance with the algorithm. This surprising representation was discovered by Alan Turing and formalized in his now-famous Turing machine. A textbook would call our description that of a "linear bounded machine". The problem which we now consider is whether such a simulation can be found.

Problem 4 (Valid computation for a linear space algorithm).
Instance: A word w over a finite alphabet $A \subset B$
Question: Is there a valid linear space computation, as a word over B, of the algorithm on input w?
Complexity: PSPACE

Notice that the algorithm itself is implicit in the problem. It might be desirable to have a fixed alphabet. This can be easily done using a two-letter alphabet since the encoding only adds a constant factor.

5.2. Membership, emptiness and inclusion

This section is an introduction to the complexity of three fundamental problems of language theory. We sketch a couple of lower bound proofs which are published but might not be easily accessible. We also cast our results in a logical setting by introducing an interval logic.

5.2.1. *The membership problem*

In formal language theory, when we specify a membership problem between a word and a language, we have to present the input language, which might be infinite, in a finite way. For example, one might provide a machine description.

Problem 5 (NFA membership).
Instance: A finite word w of length n and a nondeterministic finite automaton M of size m
Size of instance: $n + m$
Question: Is w accepted by M?
Complexity: NLOGSPACE

In this case, since a nondeterministic finite automaton is a labelled directed graph, it is an exercise to work out that the complexity of this problem matches that of the graph accessibility problem (DGAP) and the membership problem NFA is NLOGSPACE-complete.

5.2.2. *Different kinds of regular expressions*

Specifying the input language by a syntactic entity like a regular expression, rather than a graphical one such as an automaton, must have been a natural afterthought of the development of program verification in the late 1960s and early 1970s.

First attempts at formulating the problem of verifying a program were made by Naur [7] and Floyd [8], and Hoare developed the first programming logic [9] where the program was presented as a syntactic expression. Soon after, Stockmeyer and Meyer observed that the membership problem for a language given by a regular expression (including complement operations) is in PTIME [10]. This is a textbook exercise (for example, in Hopcroft and Ullman) using a dynamic programming algorithm.

Here are the definitions for these extended regular expressions (ERE), along with regular expressions (RE) and starfree expressions (SF)—the last-named are regular expressions with complement, but without the Kleene star operator.

$e ::= a \mid e_1 \cdot e_2 \mid e_1 + e_2 \mid e_1^* \mid \overline{e_1}$
$r ::= \emptyset \mid a \mid r_1 \cdot r_2 \mid r_1 + r_2 \mid r_1^*$
$s ::= \varepsilon \mid a \mid s_1 \cdot s_2 \mid s_1 + s_2 \mid \overline{s_1}$

As usual, $e_1 \cdot e_2$ is written $e_1 e_2$ where no confusion arises. $L(e)$ is the language defined by the expression e: thus all words are in $L(\overline{\emptyset})$. We will use the notation $e_1 \cap e_2$ for the derived operator $\overline{\overline{e_1} + \overline{e_2}}$.

Next we enumerate, following Schützenberger [11], all ways in which words in the language of a starfree expression, of length at least two, can be broken up into a prefix and a suffix, again described using a pair of starfree expressions.

$Br(\varepsilon) = Br(a) = \emptyset$
$Br(s_1 s_2) = \{(s_1, s_2)\} \cup \{(s_{11}, s_{12}s_2) \mid (s_{11}, s_{12}) \in Br(s_1)\}$
$\qquad\qquad\quad \cup \{(s_1 s_{21}, s_{22}) \mid (s_{21}, s_{22}) \in Br(s_2)\}$
$Br(s_1 + s_2) = Br(s_1) \cup Br(s_2)$
Suppose $Br(s) = \{(e_1, f_1), \ldots, (e_k, f_k)\}$. Then:
$Br(\overline{s}) = \bigcup_{I \subseteq \{1,\ldots,k\}} \{(\bigcap_{i \in I} \overline{e_i}, \bigcap_{j \notin I} \overline{f_j})\}$

By induction on expressions s, for any word w of length at least two, it can be shown that $w \in L(s)$ if and only if there is an (e, f) in $Br(s)$ such that $w \in L(e \cdot f)$.

The tricky case is complementation. Suppose $Br(s) = \{(e_1, f_1), \ldots, (e_k, f_k)\}$. Let $w = uv \in L(\overline{s})$, that is, $uv \notin L(s)$. Using the induction hypothesis on s, for $1 \leq i \leq k$, either $u \notin L(e_i)$ or $v \notin L(f_i)$. Let $I = \{i \mid u \notin L(e_i)\}$. Then $u \in L(\bigcap_{i \in I} \overline{e_i})$ and $v \in L(\bigcap_{j \notin I} \overline{f_j})$.

Conversely suppose some I, $u \in L(\bigcap_{i \in I} \overline{e_i})$ and $v \in L(\bigcap_{j \notin I} \overline{f_j})$. For contradiction suppose $uv \notin L(\overline{s})$, that is, $uv \in L(s)$. By the induction hypothesis there is some $1 \leq l \leq k$ such that $u \in L(e_l)$ and $v \in L(f_l)$ which contradicts our supposition.

5.2.3. *The membership problem for regular expressions*

Having introduced the different kinds of regular expressions, we now proceed to consider their membership problems.

Problem 6 (ERE, RE, SF membership).
Instance: A finite word w and a regular expression e in the class ERE/RE/SF
Question: Is w in the language defined by e?
Complexity: NLOGSPACE *for RE,* PTIME *for ERE and SF*

The membership problem RE is NLOGSPACE-complete. This was shown by Jiang and Ravikumar [12]. It is in NLOGSPACE since the usual inductive Kleene construction of a finite automaton from a regular expression can be thought of as a logspace reduction from RE to FA. That NLOGSPACE is a lower bound (even without using Kleene star and complement) is established by a reduction from DGAP.

Consider the graph accessibility problem (DGAP) where the input graph G has self-loops for all nodes. This subcase is also NLOGSPACE-complete. Encode the vertices of the graph in unary as $a, a^2, \ldots, a^n$. Let the complement $\overline{a^i}$ of the node a^i be defined as a^{n-i}.

Define words $w_1 \ldots w_n$ by taking edges (u, v) as substrings $u \cdot b \cdot \overline{v}$. A path from s to t is described by the expression $e = \mathit{start} \cdot (\mathit{middle})^{n-2} \cdot \mathit{end}$, with $\mathit{start} = s \cdot b \cdot (\Sigma_{(s,u) \in G} \overline{u})$, $\mathit{middle} = \Sigma_{(u,v) \in G}(u \cdot b \cdot \overline{v})$ and $\mathit{end} = (\Sigma_{(v,t) \in G} v) \cdot b \cdot \overline{t}$.

Now a word $w = s \cdot (ba^n)^{n-2} \cdot \overline{t}$ is in the language defined by e iff G has a path from s to t. The expression e and word w can be constructed from the description of DGAP in LOGSPACE, so we have that NLOGSPACE is a lower bound for RE membership.

Petersen showed that the complexity of ERE and SF membership is PTIME [13], by a reduction from MCVP. Assume that the gates in the circuit are enumerated so that the inputs to each gate occur earlier in the enumeration. The reduction inductively constructs starfree expressions s_k whose language L_k is such that for every $j \leq k$, the word a^{2j} is in L_k if and only if the output of gate j is 1. Let All_k be an expression whose language contains all the a^{2j} for $j \leq k$.

For the base case, $s_0 = \emptyset$. The induction step for gate k being labelled $\vee$ and having inputs from the gates i, j is

$$s_k = All_k \cap (((s_{k-1}(a^{2(k-i)-1} + a^{2(k-j)-1} + \varepsilon)) \cap (All_k + a^{2k-1}))(a + \varepsilon)).$$

The idea is that all words in L_{k-1} have even length, and so a^{2k-1} can be formed only by concatenating $a^{2(k-i)-1}$ or $a^{2(k-j)-1}$ with some word in L_{k-1}. Solving the linear equations thus formed will be possible only if a^{2i} or a^{2j} are in L_{k-1}. The remaining part of s_k adds a^{2k} to L_k if and only if the previous part generated a^{2k-1}.

A similar expression (using negations) takes care of the $\wedge$ case. The reduction can be carried out in LOGSPACE by reading the circuit description from right to left, generating the parentheses in the expression, and then filling in the rest of the expression from left to right. Storing the numbers and expanding them out as a's in unary notation can be done in logarithmic space.

5.2.4. *The nonemptiness and inclusion problems*

The other problems typically considered in formal language theory are emptiness and inclusion.

Problem 7 (RE nonemptiness).
Instance: An expression $e \in RE$
Question: Is the language defined by e nonempty?
Complexity: ALOGTIME

Problem 8 (NFA nonemptiness).
Instance: A finite automaton M
Question: Is the language defined by M nonempty?
Complexity: NLOGSPACE

Nonemptiness checking of finite automata is another textbook construction, and the problem again matches DGAP. With reference to emptiness, regular expressions can be mapped rather directly to boolean formulae by translating a letter (or the empty word) to 1 and the empty set to 0, with concatenation serving as $\wedge$, union as $\vee$ and star as a constant function returning 1. Now we have that a regular expression is nonempty if and only if the translated formula has value 1.

Problem 9 (RE/NFA inclusion).
Instance: An expression $e_1 \in RE$ or a finite automaton M_1, another expression $e_2 \in RE$ or a finite automaton M_2
Question: Is L_1, the language defined by e_1 or M_1, included in L_2, the language defined by e_2 or M_2?
Complexity: PSPACE

We use a reduction to emptiness. Since $L_1 \subseteq L_2$ is equivalent to $L_1 \cap \overline{L_2} = \emptyset$ we use the Rabin-Scott subset construction to complement an automaton, causing an exponential blow-up in size, and solve the problem in PSPACE, and because of the Kleene reduction it does not matter whether the input is a regular expression in RE or an automaton.

Meyer and Stockmeyer observed [14] that those words which are *not* valid computations of a linear space nondeterministic algorithm can be described by a regular expression which says that either the initial configuration is wrong, or the final configuration is wrong, or a move from some c_i to c_{i+1} is wrong. This last bit can be done since the distance in the word between the three letters which change from c_i to c_{i+1} is exactly n. Hence an $O(n)$ regular expression r can be written to describe these three cases. The expression constructed satisfies $Lang(r) = B^*$ (which is the same as $B^* \subseteq Lang(r)$) if and only if the computation is not valid. This is a reduction from the validity problem of linear space computations to the inequivalence and non-inclusion problems of regular expressions, making them PSPACE-complete.

Things change dramatically when the language L_2 is presented as a regular expression with complement (ERE or SF).

Problem 10 (ERE/SF nonemptiness).
Instance: *An expression $e \in ERE$*
Question: *Is the language defined by e nonempty?*
Complexity: *above* ELEMENTARY, *even if $e \in SF$*

Problem 11 (ERE/SF inclusion).
Instance: *An expression $e_1 \in ERE$ or a finite automaton M_1, another expression $e_2 \in ERE$*
Question: *Is L_1, the language defined by e_1 or M_1, included in L_2, the language defined by e_2?*
Complexity: *above* ELEMENTARY, *even if $e_2 \in SF$*

By the Kleene construction we know the problem can be solved by constructing a finite automaton for L_2 whose size is a tower of exponentials of height n determined by the nesting depth of negations in the expression for L_2, but the ELEMENTARY class is a *lower bound* for this problem, and no algorithm with space bounded by a tower of exponentials of constant height suffices! A sketch of the proof ideas is given in the book of Aho, Hopcroft and Ullman [15]. The details are in the PhD thesis of Stockmeyer [16], and a proof for the related satisfiability problem of monadic second order logic is in [17].

5.2.5. *Interval temporal logic*

Since a syntax such as the starfree expressions has union as well as complementation, it can be reformulated as a logic. We describe below a variant of the propositional interval temporal logic (ITL) defined by Moszkowski and Manna [18], where the propositions are related to the letters of the alphabet. B stands for a set of letters from the alphabet A, and the special proposition pt denotes a point, a single letter. The ";" operator is called chop (sometimes also called fusion) and is defined slightly differently from concatenation in starfree expressions. A closely related syntax is called starfree chop expressions (SFCE) in Chapter 8 by Ajesh Babu and Pandya.

$$\phi ::= \lceil\lceil B \rceil\rceil, B \subseteq A \mid pt \mid \neg \phi \mid \phi \vee \psi \mid \phi; \psi$$

Temporal logics are interpreted on state sequences $\pi = s_1 s_2 \ldots$, where each state s_i is a set of atomic propositions. In the present case, since we want to mimic the starfree expressions, we let $\pi : N \to A$ be a nonempty word over the alphabet A.

For an interval logic, when a word satisfies a formula is inductively defined using intervals $[b, e]$ where $b \leq e$ are natural numbers. Alternatively they could be thought of as substrings $\pi(b) \ldots \pi(e)$ of a word.

$$\pi, [b, e] \models \lceil\lceil B \rceil\rceil \text{ iff for all } m : b \leq m \leq e : \pi(m) \in B$$
$$\pi, [b, e] \models pt \text{ iff } b = e$$
$$\pi, [b, e] \models \phi; \psi \text{ iff } \exists m : b \leq m \leq e : \pi, [b, m] \models \phi \text{ and } \pi, [m, e] \models \psi$$

We now restrict ourselves to finite word models and say that $\pi \models \phi$ if $\pi, [1, |\pi|] \models \phi$ holds. $Lang(\phi)$, the language defined by ϕ, is the set of words $\{\pi \mid \pi \models \phi\}$. Now we consider the truth checking or membership problem for this logic.

Problem 12 (ITL truth checking).
Instance: A finite word u over alphabet A and an ITL formula ϕ
Question: Does $u \models \phi$? Alternately, is u in $Lang(\phi)$?
Complexity: PTIME

Because of their similarity, the dynamic programming algorithm for the membership problem of starfree expressions can be used for the membership problem of a word against an ITL formula. Now we use the results of Petersen [13] to show that the complexity of this problem *is* PTIME. Similarly ITL satisfiability corresponds to the nonemptiness problem of starfree expressions and we can use the results of [16].

Problem 13 (SAT(ITL), ITL satisfiability).
Instance: An ITL formula ϕ
Question: Does ϕ have a word model? Alternately, is $Lang(\phi)$ nonempty?
Complexity: above ELEMENTARY

5.3. The membership problem for linear temporal logic

The propositional temporal logic of linear time LTL was introduced in the verification setting by Pnueli [19]. Let us backtrack a bit to establish how this came about.

We saw that Hoare introduced a programming logic [9]. Hoare's notation $\alpha\{Program\}\beta$ specified the property that all runs of *Program*, when started in a state satisfying the pre-assertion α, would end in a state satisfying the post-assertion β, provided the run terminated.

These "invariant" assertions were complemented by Burstall's "intermittent" assertions for proving termination itself [20], and Manna and Pnueli realized that the program could be kept implicit, its run described as a sequence of states, and an assertion α could be made in a logic of sequences, namely, linear temporal logic (LTL) in which both kinds of assertions could be described.

Over the same period, several people working on extending Hoare logic to concurrent programs (Owicki and Gries [21]; Apt, Francez and de Roever [22]; Misra and Chandy [23]; Soundararajan [24]) realized that the completeness proofs for their logics demanded an encoding of "histories" or sequences inside the assertion language. Meanwhile, Pnueli demonstrated [19] that interesting properties of concurrent protocols (such as mutual exclusion) could be stated and proved in temporal logic. His book with Manna [25] is a repository of such properties. Hence model checking came to be based on temporal logic.

5.3.1. *LTL*

We consider a basic temporal logic LTL. The operator X below is read as "next", and the operator U as "until". We refer to the book of Huth and Ryan [26] for motivating examples.

$$\alpha ::= p \in Prop \mid \neg\alpha \mid \alpha \vee \beta \mid \mathsf{X}\alpha \mid \alpha \mathsf{U} \beta$$

The formula $\mathsf{F}\alpha$ is defined as $true\mathsf{U}\alpha$ and $\mathsf{G}\alpha$ as $\neg\mathsf{F}\neg\alpha$.

LTL is interpreted on an infinite state sequence $\pi = s_1 s_2 \ldots$, where each state s_i is labelled by a function $\pi : N \to \wp(Prop)$ with a set of atomic propositions. Alternatively π could be read as an infinite word over the alphabet $A = \wp(Prop)$. One can also consider nonempty finite words as models of LTL, but we will not need them in this chapter.

When a word satisfies a formula is defined traditionally. Instead of using the interval $[k, \infty)$ we just use the index k.

$\pi, k \models p$ iff $p \in \pi(k)$
$\pi, k \models \mathsf{X}\alpha$ iff $\pi, k+1 \models \alpha$
$\pi, k \models \mathsf{F}\alpha$ iff for some $m \geq k : \pi, m \models \alpha$
$\pi, k \models \alpha\ \mathsf{U}\ \beta$ iff for some $m \geq k : \pi, m \models \beta$
and for all $l : k \leq l < m : \pi, l \models \alpha$

This time we say $\pi \models \alpha$ holds when $\pi, 1 \models \alpha$ holds, that is, the whole infinite word is a model of the formula. This is the same thing as whether an infinite word is a member of a language of infinite words, the latter given by a logical formula.

We next indicate how LTL could be translated into starfree expressions. To be more precise, we should use starfree expressions which describe languages of infinite words, but here we only sketch the ideas using the usual starfree expressions over finite words.

$$sf(p) = (\Sigma_{p \in a} a) \cdot \overline{\emptyset}$$
$$sf(\mathsf{X}\alpha) = (\Sigma_{a \in A} a) \cdot sf(\alpha)$$

The translation of until requires more work. From the semantics of the operator, we know that a word $w[1, \infty)$ satisfies the formula $\alpha \mathsf{U} \beta$ if it has a suffix $w[m, \infty)$ which satisfies β and the "intermediate" suffixes $w[l, \infty)$ for $1 \leq l < m$ satisfy α.

So consider that $w[1, \infty)$ satisfies α. Inductively, the starfree expression $sf(\alpha)$ will describe such words. But we have to restrict this set of words to those where α holds for some suffixes after which β holds. To describe this, Pnueli and Zuck [27] found a technique of forcing prefixes which satisfy α until a suffix which satisfies β is reached. The translation of an until formula uses Schützenberger's breakups which we saw earlier. The expression $e \cap \overline{(\overline{\emptyset} \cdot \overline{e})}$ denote all words described by e all whose suffixes are again in $L(e)$.

$$sf(\alpha \mathsf{U} \beta) = \Sigma_{B \subseteq Br(sf(\alpha))} \Sigma_{(e,f) \in B} (e \cap \overline{(\overline{\emptyset} \cdot \overline{e})})(f \cap sf(\beta))$$

This translation is not terribly efficient, hopefully a better one is yet to be found.

5.3.2. *Truth checking a path as membership*

We now consider the truth checking question for this logic, which has also been given the name of *path checking* [28]. Since we are dealing with infinite words, we restrict ourselves to ultimately periodic words uv^ω which can be described by giving the pair of finite words u and v as input.

Problem 14 (LTL truth checking).
Instance: Two finite words u and v over the alphabet A and an LTL formula ϕ
Question: Does $uv^\omega \models \phi$ hold?
Complexity: lower bound ALOGTIME, *upper bound* AC1(LOGDCFL)

Since ALOGTIME lower bounds the evaluation of propositional logic formulas, it also serves as a lower bound for LTL. A polynomial time dynamic programming algorithm taking time $O(|w||\phi|)$ for the problem is easy to see and is sketched below. An exciting recent theoretical development was a more efficient parallel algorithm (for LTL truth checking over finite words) found by Kuhtz and Finkbeiner [29]. The complexity class mentioned above for the upper bound is defined using circuit families which use oracle gates: for an input of length n they construct a circuit of depth order $\log n$ which uses LOGDCFL oracle gates. The class LOGDCFL consists of problems which reduce in logarithmic space to membership in a deterministic context-free language. It is known that

LOGSPACE $\subseteq$ LOGDCFL $\subseteq$ AC1(LOGDCFL) $\subseteq$ PTIME. Hopefully this can be extended to general LTL truth checking over infinite words as well.

The PTIME algorithm proceeds by successively, at all states, extending the labelling $\pi : N \to A$ to a new labelling $\widehat{\pi} : N \to B$, where $B = \wp(Sub(\phi))$ and $Sub(\phi)$ is the set of subformulas of ϕ. The number of subformulas of ϕ is linear in the size of ϕ. The algorithm starts with the smallest subformulas and works outwards towards ϕ.

The inductive construction can be described as follows. If the formula being tackled is

p: nothing needs to be done since the state s is already labelled if $p \in \pi(s)$.
$\alpha \wedge \beta$: label s with $\alpha \wedge \beta$ if s is already labelled with α and β.
$\neg\alpha$: label s with $\neg\alpha$ if s is not already labelled with α.
$\mathsf{X}\alpha$: label s with $\mathsf{X}\alpha$ if its successor is already labelled with α.
$\alpha\mathsf{U}\beta$: do a backward pass through the word, labelling states already labelled with β with $\alpha\mathsf{U}\beta$, and propagating this formula backwards so long as a state is labelled with α and its successor is labelled with $\alpha\mathsf{U}\beta$.

5.4. The model checking problem for computation tree logic

The motivation for the logic CTL defined below came from the notion of "safety" and "liveness" properties, due to Lamport [30]. Two groups working independently, Emerson and Clarke in the US [31], and Queille and Sifakis in France [32], came up with algorithms for checking temporal logic properties of a system, described as a tree. (We used their algorithm in the previous section, restricted to a word.) CTL is due to Emerson and Clarke and, like Queille and Sifakis's logic, it was based on UB, another of Manna and Pnueli's temporal logics, developed with Ben-Ari [33]. The advantage of these historically first approaches is that the model checking algorithm continues to be linear time $O(|M||\phi|)$.

$\alpha ::= p \in Prop \mid \neg\alpha \mid \alpha \vee \beta \mid \mathsf{EX}\alpha \mid \mathsf{EG}\alpha \mid \mathsf{E}[\alpha\mathsf{U}\beta]$

Similar to before, we define $\mathsf{EF}\alpha = \mathsf{E}[true\mathsf{U}\alpha]$ (these are called weak liveness properties) and its dual $\mathsf{AG}\neg\alpha = \neg\mathsf{EF}\alpha$ (safety properties, which say that no "bad" α happens in any run). $\mathsf{EG}\neg\alpha$ is called a weak safety property and the dual liveness property $\mathsf{AF}\alpha = \neg\mathsf{EG}\neg\alpha$ says that some "good" α happens in every run. We also define $\mathsf{A}[\alpha\mathsf{U}\beta] = \neg(\mathsf{EG}\neg\beta \vee \mathsf{E}[\neg\beta\mathsf{U}\neg\alpha \wedge \neg\beta])$.

Again we refer to Huth and Ryan's book [26] for motivating examples. The idea is that a system is modelled as a tree $\tau = (T, t_0, \to, L)$ of all its runs, together with a state labelling function. $(T, \to)$ is the tree, t_0 the root and $L : T \to A$ is the labelling function which specifies which atomic propositions hold at the tree nodes. This semantics is what gives the name computation tree logic. The definition of satisfaction formalizes the intuition.

$\tau, t \models p$ iff $p \in L(t)$
$\tau, t \models \mathsf{EX}\alpha$ iff for some t' such that $t \to t'$, we have $\tau, t' \models \alpha$

$\tau, t \models \mathsf{EG}\alpha$ iff for some path $\pi = t_1 \to t_2 \to t_3 \to \ldots$ where $t = t_1$, $\pi, 1 \models \mathsf{G}\alpha$
$\tau, t \models \mathsf{E}[\alpha \mathsf{U} \beta]$ iff for some path $\pi = t_1 \to t_2 \to t_3 \to \ldots$ where $t = t_1$, $\pi, 1 \models \alpha \mathsf{U} \beta$

From its definition, you can work out that $\mathsf{A}[\alpha \mathsf{U} \beta]$ checks the until property along *all* paths from the root. It should be clear that Lamport's criteria of describing safety and liveness properties of a system are met by this logic.

But there is a question: how is a tree with infinite paths to be presented as input to an algorithm? The solution is to again restrict ourselves to regular infinite trees, which have finitely many subtrees upto isomorphism. Such trees can be described by a **transition system** $M = (S, s_1, \to, L)$ (also called a Kripke structure), where S is a finite set of states with a labelling function $L : S \to A$ and s_1 the root state from which infinite paths ("runs") of the system start off. The transition system M with start state s_1 *unfolds* into a tree model $Unf(M)$ with $root(M)$ as its root.

Problem 15 (MC(CTL), CTL model checking).
Instance: A transition system M over the alphabet A and a CTL formula ϕ
Question: Does $Unf(M), root(M) \models \phi$ hold?
Complexity: PTIME

Observe that this is an alternate way of framing the question of checking whether a given infinite tree (represented as a finite structure) is a member of a language of infinite trees (described again by a finite logical formula). So this could be called the truth checking, or membership, problem for CTL. But by now the name *model checking* is firmly established for this problem when the model is given as a transition system, and we use the popular terminology.

To solve the problem we look at extending the algorithm for truth checking a word, and this can be done with minor changes. As before, the states of M are labelled by the alphabet B of sets of subformulas of a formula ϕ. Tarjan's algorithm is used to find strongly connected components (SCCs) for EG subformulas, thus maintaining linear time.

If the formula being tackled is

$\mathsf{EX}\alpha$: label s with $\mathsf{EX}\alpha$ if it has a successor which is already labelled with α

$\mathsf{E}[\alpha \mathsf{U} \beta]$: label states already labelled with β with $\mathsf{E}[\alpha \mathsf{U} \beta]$, and propagate this formula using a backwards breadth-first search, that is, so long as a state is labelled with α and has a successor state labelled with $\mathsf{E}[\alpha \mathsf{U} \beta]$

$\mathsf{EG}\alpha$: to deal with this case efficiently, restrict the graph to states satisfying α (consider other states and their transitions to be deleted), find the maximal SCCs, and use backwards breadth-first search to find states which can reach an SCC, all of these states are labelled with $\mathsf{EG}\alpha$

While CTL model checking is in $O(|S||\phi|)$ time, one can also verify that it has a PTIME lower bound, by a reduction from the monotone (layered) circuit value problem MCVP. For example, if the layers begin with "and"s and end with "or"s, the CTL formula (which does not even use until) $\mathsf{AXEX}\ldots\mathsf{AXEX}\mathit{true}$ holds iff the circuit evaluates to true. Thus membership in a tree language, or model checking a transition system against a formula, is a PTIME-complete problem.

Inclusion of tree languages is a much harder problem, and we refer to Chapter 3 by Christof Löding for details. The satisfiability question of CTL (which can also be seen as nonemptiness of the tree language) is correspondingly hard, as was shown by Fischer and Ladner [34].

Problem 16 (SAT(CTL), CTL satisfiability).
Instance: A CTL formula ϕ
Question: Is there a tree model for ϕ?
Complexity: EXPTIME

The equivalences below can be used to handle boolean equivalences over paths, but we cannot reduce nested path formulas.

$\mathsf{E}[\neg\mathsf{X}\alpha] \equiv \mathsf{EX}\neg\alpha,\ \mathsf{E}[\neg(\alpha\mathsf{U}\beta)] \equiv \mathsf{EG}\neg\beta \vee \mathsf{E}[\neg\beta\mathsf{U}\neg\alpha \wedge \neg\beta]$

$\mathsf{E}[(\alpha_1\mathsf{U}\beta_1) \wedge (\alpha_2\mathsf{U}\beta_2)] \equiv \mathsf{E}[(\alpha_1 \wedge \alpha_2)\mathsf{U}(\beta_1 \wedge \mathsf{E}[\alpha_2\mathsf{U}\beta_2])] \vee \mathsf{E}[(\alpha_1 \wedge \alpha_2)\mathsf{U}(\beta_2 \wedge \mathsf{E}[\alpha_1\mathsf{U}\beta_1])]$

The "formula" $\mathsf{E}[\mathsf{GF}p]$, which intuitively says that there is a path along which p is infinitely often true, does not have any CTL equivalent at all.

5.5. Model checking temporal logics

Once algorithms for CTL model checking were developed, those for checking LTL properties were not far behind. Instead of interpreting a transition system $M = (S, s_1, \rightarrow, L)$ by its tree unfolding, one could consider $Lang(M)$, the set of words which are runs from its initial state, and ask whether all these are models of an LTL formula. We return to our LTL notation, and extend it momentarily:

$\alpha ::= p \in Prop \mid \neg\alpha \mid \alpha \vee \beta \mid \mathsf{X}\alpha \mid \alpha\mathsf{U}\beta$

$M, s \models \mathsf{A}[\alpha]$ iff for all paths $\pi = s_1 \rightarrow s_2 \rightarrow \ldots$ where $s = s_1$, we have $\pi, 1 \models \alpha$

Manna and Pnueli took the decision to interpret *every* LTL formula using such a $\forall$path-semantics, hence we do not explicitly write the $\mathsf{A}[\ ]$-operator and continue to use the old LTL syntax.

Problem 17 ($\mathbf{MC^{\forall}}$(LTL), LTL model checking).
Instance: A transition system M and an LTL formula ϕ
Question: Does $Unf(M), root(M) \models \mathsf{A}[\phi]$ hold?
Question: Alternately, does $\pi, 1 \models \phi$ hold for every $\pi \in Lang(M)$?
Complexity: PSPACE

The basic strategy for the LTL model checking problem is different from that for CTL: it changes from checking membership of a tree (presented finitely) in a language of trees (described by a formula) to checking inclusion of a set of words

$Lang(M)$ (implicitly presented by the transition system M, assuming all states are accepting) in another set of words (described by a formula).

Historically, a key rôle here was played by dynamic logic (PDL), invented by Pratt [35] as a generalization of traditional modal logic with program-indexed modalities. PDL was proved decidable by a filtration argument of Fischer and Ladner [34]. A *tour de force* completeness-cum-decidability proof of PDL by Kozen and Parikh [36] and the tight "tableau" construction of Pratt [37] laid the seeds of the "formula automaton" construction of Vardi and Wolper [38], which is today the standard way of implementing LTL model checking algorithms. We only sketch this theory, more details can be found in Chapter 15 by Stéphane Demri and Paul Gastin.

We first describe how to construct a formula automaton, also called a tableau, for a formula ϕ, which is exponential in the size of ϕ, with the property that the language accepted by this automaton is precisely the set of models of ϕ.

Since a formula automaton is a finite automaton (over infinite words) and the given transition system is also a finite automaton, we can perform a "product" construction to recognize the intersection of the accepted paths. Since we are concerned with infinite paths, the product is of automata over infinite words (here we witness the return of the automata of Büchi [4]) rather than the usual one of automata over finite words. Chapter 2 by Madhavan Mukund describes this construction.

The model checking algorithm consists of taking the product of the automaton for the *complement* formula $\neg\phi$ with M, and checking the product for *nonemptiness.* If there is a path in the product of this kind, we exhibit this as a counterexample to M satisfying ϕ; if there isn't, we declare that M is a model for ϕ. In effect, we have checked whether the language of M is included in the "language" of ϕ by checking the emptiness of $L(M) \cap L(\neg\phi) = L(M) \cap \overline{L(\phi)}$.

The states of the formula automaton are all subsets a of $Sub(\phi)$ which could be potential states of the transition system, in the following sense:

Consistent: For every subformula ψ, both ψ and $\neg\psi$ are not in a.
Downward saturated: If $\alpha \vee \beta$ is in a, one of α or β is in a.
Maximal: For every subformula ψ, either ψ or $\neg\psi$ is in a.

Let $\widehat{a}$ be the conjunction of formulas in a. The transitions are defined so that $\widehat{a} \wedge \mathsf{X}\widehat{b}$ is consistent when there is a transition from a to b. Since there are exponentially many subsets of $Sub(\phi)$, which is itself linear in the size of ϕ, it is possible to give an exponential $|M|.2^{O(|\phi|)}$ time model checking algorithm. Using the fact that nonemptiness of the language of a finite state Büchi automaton can be checked in nondeterministic logarithmic space, we get an NPSPACE algorithm, which we know is the same as PSPACE, so that is an upper bound for the complexity of model checking LTL.

Before considering the lower bound, we introduce another problem:

Problem 18 (SAT(LTL), LTL satisfiability).
Instance: An LTL formula ϕ
Question: Is there a word model for ϕ?
Complexity: PSPACE

PSPACE is a lower bound for both model checking and satisfiability, as was shown by Sistla and Clarke [39]. We refer to their paper, or the survey by Schnoebelen [40], for proofs. Here is an idea of how a reduction from the valid linear space computation problem to either model checking or satisfiability can be done.

A letter of the word representing a configuration c_i can take as many possible values as the size of the alphabet B, which is represented by transitions branching to that many states where a proposition coding that letter is true, and then coming together again, forming a diamond-shaped structure. The whole word c_i of length n can be coded as that many "diamonds" in sequence. After the last "diamond" we have a transition cycling back to the beginning. The LTL formula now describes the initial configuration, and specifies using $2n$ nested X operators, how a configuration c_i can change to the next one c_{i+1}, and asserts using an F operator the existence of a final configuration. A model of this formula (notice that it does not even use the until operaor) describes a valid computation and conversely, every valid computation is described by a model of this formula.

Sistla and Clarke showed that if an LTL formula is satisfiable, it is satisfiable in a word uv^ω where u and v are at most exponential in the size of the formula [39]. Guessing this model and model checking ϕ along it would give an NEXPTIME algorithm. But as in the case of our algorithm for DGAP, instead of guessing the whole model, it can be guessed and verified on the fly. This gives an NPSPACE algorithm, and by Savitch's theorem the problem is in PSPACE.

5.5.1. *CTL**

Recall that the formula $\mathsf{E}[\mathsf{GF}p]$ says there is a path along which p is infinitely often true. This does not have an equivalent in the LTL $\forall$-semantics either. This motivated the development of the logic CTL*.

$$\alpha ::= p \in \mathit{Prop} \mid \neg\alpha \mid \alpha \vee \beta \mid \mathsf{X}\alpha \mid \alpha \mathsf{U} \beta$$
$$\phi ::= \mathsf{A}[\alpha] \mid \neg\phi \mid \phi \vee \psi$$

Now we can define $\mathsf{E}[\alpha] = \neg\mathsf{A}[\neg\alpha]$, yielding the semantics:

$M, s \models \mathsf{E}[\alpha]$ iff for some path $\pi = s_1 \to s_2 \to s_3 \to \ldots$ where $s = s_1$, we have $\pi, 1 \models \alpha$

The PSPACE model checking complexity of LTL can be lifted to CTL* as well. Chapter 15 by Demri and Gastin has the details.

5.5.2. *Model checking ITL*

It is possible to describe paths much more succinctly. We return to the interval logic defined in Section 5.2.5 and recall its syntax and semantics.

$\phi ::= \lceil\lceil B \rceil\rceil, B \subseteq A \mid pt \mid \neg\phi \mid \phi \vee \psi \mid \phi;\psi$

$\pi, [b,e] \models \lceil\lceil B \rceil\rceil$ iff for all $m : b \leq m \leq e : \pi(m) \in B$

$\pi, [b,e] \models pt$ iff $b = e$

$\pi, [b,e] \models \phi;\psi$ iff $\exists m : b \leq m \leq e : \pi, [b,m] \models \phi$ and $\pi, [m,e] \models \psi$

Problem 19 ($\mathbf{MC^{\forall}}$(ITL), ITL model checking).
Instance. *A transition system M and an ITL formula* ψ
Question: *Does* $\pi, 1 \models \phi$ *hold for every* $\pi \in Lang(M)$*?*
Complexity: *above* ELEMENTARY

ITL formulas are succinct compared to LTL formulas, although the class of languages covered is the same. Theoretically, this power derives from the negation $\neg(\phi;\psi)$ of a chop formula. As can be seen from the semantics, this ranges over all possible ways of breaking up an interval into subintervals. In fact, this combination means that we can obtain lower bounds from the inclusion problem for starfree expressions, and we have that $MC^{\forall}$(ITL), model checking an ITL formula against (all runs of) a transition system, has an ELEMENTARY lower bound. It is difficult to find natural properties which require several nestings of such operators and model checkers for ITL [41] work reasonably well in practice.

5.5.3. *Unambiguous interval logic*

Recently, motivated by the good performance of an ITL model checker [42], with Pandya and Shah we considered a logic UITL [43] where the chops are "marked" in a deterministic manner. In the syntax below, a stands for a letter and B for a set of letters from the alphabet.

$\phi ::= \lceil\lceil B \rceil\rceil \mid pt \mid \neg\phi \mid \phi \vee \psi \mid \phi F_a \psi \mid \phi L_a \psi \mid \oplus\phi \mid \ominus\phi$

The operators F_a and L_a can be read "first a" and "last a". They chop an interval into two subintervals at a point determined by either the first or the last occurrence of the specified letter a in the interval (provided it exists). The last two operators provide successor and predecessor modalities at the interval level.

$\pi, [b,e] \models \phi F_a \psi$ iff for some $m : b \leq m \leq e.\ \ \pi(m) = a$ and
(for all $l : b \leq l < m.\ \pi(l) \neq a$) and
$\pi, [b,m] \models \phi$ and $\pi, [m,e] \models \psi$

$\pi, [b,e] \models \phi L_a \psi$ iff for some $m : b \leq m \leq e.\ \ \pi(m) = a$ and
(for all $l : m < l \leq j.\ \pi(l) \neq a$) and
$\pi, [b,m] \models \phi$ and $\pi, [m,e] \models \psi$

$\pi, [b,e] \models \oplus\phi$ iff $b < e$ and $w, [b+1,e] \models \phi$

$\pi, [b,e] \models \ominus\phi$ iff $b < e$ and $w, [b,e-1] \models \phi$

This logic is much less expressive than full ITL, and matches a fragment of LTL. Interestingly, the complexity of model checking comes down: checking a UITL formula ϕ against a word w is in $O(|w||\phi|^2)$ and also in the complexity class LOGDCFL (which we saw earlier, recall that LOGDCFL $\subseteq$ PTIME), while checking whether all runs of a transition system satisfy the formula is in NPTIME. We refer to the paper [43] for some of the details.

Problem 20 (UITL truth checking).
Instance: *A word w and a UITL formula ϕ*
Question: *Does $w \models \phi$ hold?*
Complexity: *lower bound* ALOGTIME, *upper bound* LOGDCFL

Problem 21 ($MC^{\forall}$(UITL), UITL model checking).
Instance: *A transition system M and a UITL formula ϕ*
Question: *Does $\pi, 1 \models \phi$ hold for every $\pi \in Lang(M)$?*
Complexity: NPTIME

Problem 22 (SAT(UITL), UITL satisfiability).
Instance: *A UITL formula ϕ*
Question: *Does ϕ have a word model?*
Complexity: NPTIME

5.6. Reading ahead

In this chapter we considered the simple case where the system model is finite. The more difficult case when the model is infinite is considered in the articles by Javier Esparza and Jörg Kreiker (Chapter 12) and by Wolfgang Thomas (Chapter 16) in this volume.

The alphabet we consider for labelling the transitions is just a finite set. A more distributed structure on the alphabet is considered by Madhavan Mukund in Chapter 9, whereas the representation of abstract data values is in Chapter 17 by Amaldev Manuel and R. Ramanujam.

As should be clear by now, verification theory using automata is developing in many directions. Our article is conceptually closer to the early theory of the 1980s and we have used the chapter in the *Handbook of Theoretical Computer Science* by Emerson [44]. The material on model checking is covered in detail in the books by Huth and Ryan [26] and by Clarke, Grumberg and Peled [45]. The book by Stirling [46] uses an alternative treatment using a rich formalism called the modal μ-calculus, and reduces model checking to problems of finding the winner in a suitable game. The complexity results for most of these logics and more detailed fragments are surveyed in the long article by Schnoebelen [40], and the journal article by Demri and Schnoebelen [47].

Acknowledgements

Thanks to Simoni Shah and Paul Gastin for their detailed comments, and to Nutan Limaye and A.V. Sreejith for their help.

References

[1] J. R. Büchi, Weak second-order arithmetic and finite automata, *Z. Math. Logik Grundl. Math.* **6**, 66–92, (1960).
[2] Calvin C. Elgot, Decision problems of finite automata design and related arithmetics, *Trans. AMS.* **98**, 21–52, (1961).
[3] B. A. Trakhtenbrot, Finite automata and the logic of monadic predicates, *Dokl. Akad. Nauk SSSR.* **140**, 326–329, (1961).
[4] J. R. Büchi. On a decision method in restricted second-order arithmetic. In *Proc. 1960 Congr. Logic, Methodology, Philosophy and Science*, pp. 1–11, Stanford Univ Press, (1962).
[5] S. R. Buss. The Boolean formula value problem is in ALOGTIME, In *Proc. 19th STOC*, pp. 123–131, ACM, (1987).
[6] W. J. Savitch, Relationship between nondeterministic and deterministic tape classes, *J. Comp. Syst. Sci.*. **4**, 177–192, (1970).
[7] P. Naur, Proof of algorithms by general snapshots, *BIT.* **6**(4), 310–316, (1966).
[8] R. W. Floyd. Assigning meanings to programs. In *Mathematical aspects of computer science, Proc. Symp. Appl. Math.*, vol. 19, pp. 19–32, AMS, (1967).
[9] C. A. R. Hoare, An axiomatic basis for computer programming, *Commun. ACM.* **12**(10), 576–580, (1969).
[10] L. J. Stockmeyer and A. R. Meyer, Word problems requiring exponential time, In *Proc. 5th STOC*, pp. 1–9, ACM, (1973).
[11] M.-P. Schützenberger, On finite monoids having only trivial subgroups, *Inf. Control.* **8**(2), 190–194, (1965).
[12] T. Jiang and B. Ravikumar, A note on the space complexity of some decision problems for finite automata, *Inf. Proc. Lett.* **40**(1), 25–31, (1991).
[13] H. Petersen. Decision problems for generalized regular expressions. In *Proc. Descr. Compl. Aut. Gram. Related Str.*, pp. 22–29, (2000).
[14] A. R. Meyer and L. J. Stockmeyer. The equivalence problem for regular expressions with squaring requires exponential space. In *Proc. Switch. Autom. Theory*, pp. 125–129, IEEE, (1972).
[15] A. V. Aho, J. E. Hopcroft and J. D. Ullman, *The design and analysis of computer algorithms.* (Addison-Wesley, 1974).
[16] L. J. Stockmeyer. *The complexity of decision problems in automata theory and logic.* PhD thesis, MIT, (1974).
[17] A. R. Meyer. Weak monadic second order theory of successor is not elementary recursive. In *Proc. Logic Colloq.*, vol. 453, *LNM*, pp. 132–154, Springer, (1975).
[18] B. C. Moszkowski and Z. Manna. Reasoning in interval temporal logic. In *Proc. Logics of programs*, vol. 164, *LNCS*, pp. 371–382, Springer, (1983).
[19] A. Pnueli. The temporal logic of programs, In *Proc. 18th FOCS*, pp. 46–57, IEEE, (1977).
[20] R. M. Burstall. Program proving as hand simulation with a little induction. In *Proc. 6th IFIP Congress*, pp. 308–312, North-Holland, (1974).
[21] S. S. Owicki and D. Gries, An axiomatic proof technique for parallel programs I, *Acta Inf.* **6**, 319–340, (1976).

[22] K. R. Apt, N. Francez and W.-P. de Roever, A proof system for communicating sequential processes, *Trans. Prog. Lang. Syst.* **2**(3), 359–385, (1980).
[23] J. Misra and K. M. Chandy, Proofs of networks of processes, *Trans. Softw. Engg.* **7**(4), 417–426, (1981).
[24] N. Soundararajan, Correctness proofs of CSP programs, *Theoret. Comp. Sci.* **24**, 131–141, (1983).
[25] Z. Manna and A. Pnueli, *The temporal logic of reactive and concurrent systems: specification.* (Springer, 1992).
[26] M. R. A. Huth and M. D. Ryan, *Logic in computer science: modelling and reasoning about computer systems.* (Cambridge, 2000).
[27] A. Pnueli and L. D. Zuck. In and out of temporal logic. In *Proc. 8th LICS*, pp. 124–135, IEEE, (1993).
[28] N. Markey and P. Schnoebelen. Model checking a path. In *Proc. 14th CONCUR*, vol. 2761, *LNCS*, pp. 251–265, Springer, (2003).
[29] L. Kuhtz and B. Finkbeiner. LTL path checking is efficiently parallelizable. In *Proc. 36th ICALP, Part II, LNCS* 5556, Springer, 235–246, 2009.
[30] L. Lamport. 'Sometime' is sometimes 'not never'. In *Proc. 7th POPL*, pp. 174–185, ACM, (1980).
[31] E. A. Emerson and E. M. Clarke Jr, Using branching time temporal logic to synthesize synchronization skeletons, *Sci. Comp. Program.* **2**, 241–266, (1982).
[32] J.-P. Queille and J. Sifakis. Specification and verification of computer systems in CESAR. In *Proc. Symp. Program.*, vol. 137, *LNCS*, pp. 337–351, Springer, (1982).
[33] M. Ben-Ari, Z. Manna and A. Pnueli, The temporal logic of branching time, *Acta Inf.* **20**, 207–226, (1983).
[34] M. J. Fischer and R. E. Ladner, Propositional dynamic logic of regular programs, *J. Comput. Syst. Sci.* **18**(2), 194–211, (1979).
[35] V. R. Pratt. Semantical considerations on Floyd-Hoare logic. In *Proc. 17th FOCS*, pp. 109–121, IEEE, (1976).
[36] D. C. Kozen and R. J. Parikh, An elementary proof of the completeness of PDL, *Theoret. Comp. Sci.* **14**, 113–118, (1981).
[37] V. R. Pratt, A near-optimal method for reasoning about action, *J. Comput. Syst. Sci.* **20**(2), 231–254, (1980).
[38] M. Y. Vardi and P. Wolper, Reasoning about infinite computations, *Inf. Comput.* **115**(1), 1–37, (1994).
[39] A. P. Sistla and E. M. Clarke Jr., The complexity of propositional linear temporal logics, *J. ACM.* **32**(3), 733–749, (1985).
[40] Philippe Schnoebelen. The complexity of temporal logic model checking. In *Proc. Adv. Modal Log.*, pp. 393–436, King's College, (2003).
[41] P. K. Pandya. Specifying and deciding quantified discrete-time duration calculus formulae using DCVALID: an automata theoretic approach. In *Proc. RTTOOLS*, (2001).
[42] S. N. Krishna and P. K. Pandya. Modal strength reduction in quantified discrete duration calculus. In *Proc. FSTTCS*, vol. 3821, *LNCS*, pp. 444–456, Springer, (2005).
[43] K. Lodaya, P. K. Pandya and S. S. Shah. Marking the chops: an unambiguous temporal logic. In *Proc. IFIP TCS, IFIP Series*, vol. 273, pp. 461–476, Springer, (2008).
[44] E. Allen Emerson, Modal and temporal logics, in *Handbook of theoretical computer science, volume B*, 995–1072, Elsevier, (1990).
[45] E. M. Clarke Jr., O. Grumberg and D. Peled. *Model checking*, MIT Press, 1999.
[46] C. P. Stirling, *Modal and temporal properties of processes.* (Springer, 2001).
[47] S. Demri and P. Schnoebelen, The complexity of propositional linear temporal logic in simple cases, *Inf. Comput.* **174**(1), 84–103, (2002).

Chapter 6

A Framework for Decentralized Physical Access Control using Finite State Automata

Namit Chaturvedi, Atish Datta Chowdhury and B. Meenakshi

Honeywell Technology Solutions – Research
151/1, Doraisanipalya, Bannerghatta Road
Bangalore-560226, India
*meenakshi@iiitb.ac.in**

We present a decentralized authorization framework for physical access control, using finite state automata on a system of networked controllers and smart cards. Access to individual rooms is guarded by context-dependent policies that are dynamically evaluated. Policies are specified using a logical language parameterized by events which are then converted into equivalent executable automata. Storage and evaluation of policy automata is done in a distributed manner. We include illustrative examples of context sensitive physical access control policies that are derived from actual applications.

6.1. Introduction

The domain of access control involves solutions to the problems of authorization, validation, and authentication. The goal of authorization is to specify and evaluate a set of policies that control the access of users to resources, resulting in grant or denial of access. Validation usually refers to securely verifying the authorized privileges, and the goal of authentication is to prove the identity that a user claims. In this paper, we present a framework of decentralized authorization for physical access control, using finite state automata, where access to resources is guarded by context-dependent policies that are dynamically evaluated. For this purpose, resources are rooms, which can be defined as enclosures guarded by entrances or doors.

In traditional implementations[a] of physical access control, the doors are equipped with card readers, which are either connected to a central controller, or a locally cached image of the central controller. When a user presents the access card to the reader, the latter communicates the card information to the controller and waits for a reply instructing whether or not to allow access. In these systems card readers are passive devices without any processing power of their own. The

*Contact author
[a] As exemplified in several current commercial physical access control solutions.

central controller, on the other hand, is a well designed sophisticated device with fail-over capabilities, and advanced hardware and algorithms to enable fast decision making. Policies dictating access are predominantly specified using Access Control Lists (ACLs) that describe static policies, e.g. "user X is not allowed in room R". The decision making process of the central controller includes look-up of these lists to determine whether or not a user has privileges to enter a room.

These static policies are in contrast with context dependent policies where the conditions that determine access change dynamically. They must be evaluated at the time of request. For example, a policy may prohibit a user from entering into the lobby if all the rooms accessible from the lobby have reached their maximum limits. In such a situation, a central controller will have to collate information from each of the rooms before making a decision about access to the lobby. It is envisioned that physical access control in buildings, facilities, and townships of the future will have a significant reliance on context dependent policies. Therefore, the central controller will be burdened with information from all over the facility, and each piece of information may possibly contribute toward constructing many kinds of contexts for different users. In the centralized systems of today, particular solutions enabling dynamic context handling are built as instances of specific programs or triggers on the central controller, and such solutions will not scale with increasing number of users in the facility.

State of the art physical access control solutions are moving toward using general purpose building networks for communications between various card readers and the central controller. Reasons of flexibility and ease of installation will keep driving this trend. However, a general purpose broadcast network (e.g. Ethernet) is not very suitable for a centralized access control solution. This unsuitability is due to the inherent dependency on the central controller for every decision.

In this paper we present a decentralized framework to support dynamic authorizations in physical access control, which addresses the issues mentioned above. We replace the central controller with a network of smaller controllers that maintain system information (context) and introduce per user writable memory devices including active devices with processing capabilities. Memory cards, smart cards, Java enabled smart buttons, SIM cards fit precisely in our definition of small, portable, and intelligent user devices. Policies are specified using a logical language parameterized by events which are then converted into equivalent executable automata. Storage and evaluation of automata is done in a distributed manner. Controller and the user card come together to execute the automaton corresponding to the appropriate policy, and decide whether to grant or deny access based on the context information.

We provide an overview of our framework for decentralized access control in the next section. Section 6.3 provides a study of related works. Section 6.4 gives details on the formal logical language used to specify the policies. Compilation of specification into executable models is described in Section 6.5. Section 6.6

summarizes our approach toward realizing this framework. We then look at the benefits of our approach, followed by our conclusion and identification of areas for future efforts.

6.2. Decentralized access control framework — Overview

We present a framework for access control applications where decisions of nearly all policies depend upon context. The framework is scalable, flexible and provides quick responses to access requests. Figure 6.1 presents the proposed framework. The components of this framework are:

User-carried devices These are writable memory devices, possibly with built in computational capabilities. Every user carries one such device and uses it to make access requests. In this paper we use smart cards [1] as representatives of these devices.

Room A room is an enclosed space in a facility, access to which is obtained through doors. Rooms constitute *resources* to which users of the system request and obtain access to. Resources also include collection of rooms (for example, a zone including a lobby and some rooms) and specific devices (e.g. those issuing critical assets).

Door A door refers to an element from the set of physical entry/exit points to rooms, which are access control enabled, and hence act as access agents for rooms. Each door has a reader on either side, and an actuator responsible for opening the door.

Reader This is a device installed on a door, which can read from and write to the memory cards. Users request access by either bringing their cards in close vicinity of the readers, or by swiping or inserting them in appropriate reader slots.

Controller Readers of each door are connected to microcontroller based devices, called controllers. Readers from many doors may be connected to the same controller. A controller is expected to be connected over a network with other controllers, and have reasonable processing capabilities and memory. Controllers instruct the actuators to release the door locks to grant access to rooms.

Interconnect Controllers are connected to each other through a network installation that is referred to as the interconnect, e.g. the IP network of a facility. It may include wired or wireless communication components. Devices like special purpose servers, required for log keeping etc, are also connected with the interconnect.

Users are grouped into *roles* that dictate their access privileges to resources in a facility. *Access control policies* precisely define the access privileges of each role, for each resource in the facility. As illustrated in Figure 6.1, policies (per

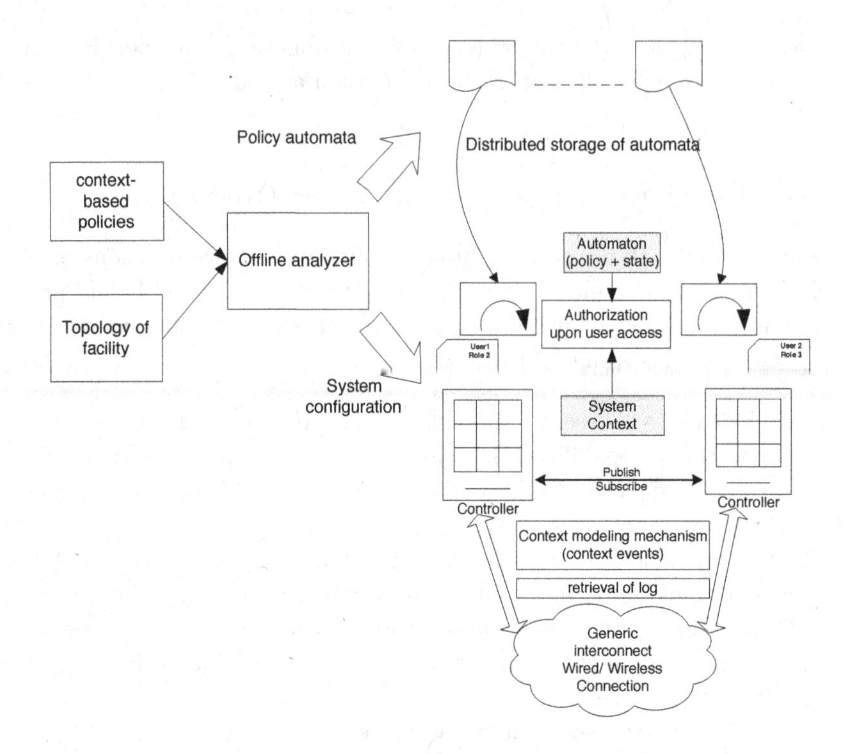

Fig. 6.1. Decentralized access control framework: Overview.

role, per resource) are executed as automata amongst the user carried cards and the controllers in the framework. The controllers, in collaboration or isolation, maintain information about the system context. Upon an access request, a decision is taken locally by the virtue of interaction of smart card and a controller. The interconnect is used to transfer system-level context information among pertinent set of controllers.

Contexts are typically constructed from information about user actions (from the same role or different roles) and user movements, and other events occurring in the system (e.g. raised security levels, alerts etc.). In order to take access decisions this information must be available with the authorizing entities. For example, an access control policy for the role that authorizes visitors of a facility might mandate the presence of an employee in order for a visitor to enter the lab area. The context in this case is the presence of an employee and needs to be checked while access is being granted to a visitor.

We adopt a model theoretic approach in the framework, and introduce mechanisms therein to address the requirements mentioned above. *Behavior* of the application is modeled as a sequence of events in the order of their occurrence. The set of events primarily includes *user actions* and *application actions*. User actions include access requests by users and allowing/denial of access. Application actions (also referred to as context events) include actions that constitute contexts for policies like the number of users in a room etc. We use the terms event and action interchangeably. Access policies are *specified* as Monadic Second Order (MSO) logic [2] formulae over these events. That is, with logic formulae, we specify what conditions or contexts must be met before the model can accept an access grant event in the behavior.

Access control policies specified using MSO logic are *executed* using finite state automata [3]. An offline analyzer module takes in context based policies along with the information regarding topology of the facility and generates executable automata. With events and actions supplied to it as literals we check whether or not access events take it to accepting state and respectively decide whether or not to allow. As we describe in later sections, finite state automata provide for a constant time decision mechanism.

Policies are written in our framework as:

access is acceptable now *only if* required context is satisfied already.

This manner of writing policies facilitates policy execution by ensuring that the corresponding automata enter rejecting states only when access events occur in the absence of correct contexts. Access events are discarded in such cases, consumed otherwise, hence the decision. It suffices to express policies with access as the antecedent and contexts as consequent rather than in the form of a strict biconditional (*if and only if*), where presence of contexts would mandate an access, thereby making policies impractical. On the other hand, it enables us to conclude a denial in the absence of correct context.

We would like to emphasize that the policies use context information that is dynamically varying to take a decision on whether to allow access or not. The policies themselves are not dynamic. Changes in policies applicable to a role are made by re-programming the cards or the readers that the policies are stored in. Context definitions can be changed independent of policies, again by re-programming appropriate components that store/compute the context information.

A complete access control framework also needs to support collecting log of events that occur in the system. The controllers buffer and forward their local audit trail of request and responses at a central server for log keeping. This communication may be given a lower priority as compared to that for event exchange, hence it will not put any undue constraint on maintaining the context.

6.3. Logics for Access Control

There has been plenty of work done on languages and frameworks for access control. When it comes to access control, there is the problem of specifying access control policies and the problem of devising suitable formalisms that model the environment or context that influence the access control policies. Logics have been used for modeling both policies and the contexts. We provide an overview of the use of logics in the domain of access control and also explain how the bridge between logic and automata theory can be exploited in this application area. An early survey of logic based policy specification and evaluation in access control is presented in [4]. Later, Samarati and di Vimercati [5] provided a comprehensive survey of various access control policies, models and mechanisms. Almost all the works cited in these surveys follow proof-theoretic mechanisms. We begin by providing a perspective of work that is closer to ours presented in this paper.

Halpern and Weissman [6] also use first order logic to reason about access control policies for digital content. They consider two access control problems: (i) given a set of policies and an environment that provides all necessary facts, the problem of taking an allow or deny decision upon request for access and (ii) the problem of checking if a given set of policies is consistent. They show how general first order logic, when used as a specification language for policies and for the environment, yields undecidability for these access control problems. The authors then define a restricted subset of first order logic to get tractable algorithms for these problems using proof-theoretic approaches. It is also worth noting that the restricted subset of first order logic is expressive enough to specify many interesting access control policies.

Bauer et al [7] also use a formal logical language to specify access control policies and present a distributed algorithm that assembles a proof from various pieces of proofs provided by entities involved in their physical access control system. Each user carried "smart phone" is equipped with a theorem prover, which assists in distributed proof generation for authorization decisions. The distributed proof generation algorithm is guaranteed to terminate whenever a centralized prover system also terminates.

There are also works focusing on collaborative environments and on the problem of dynamically granting authorizations. Early work in this regard is by Bertino et al [8] wherein they use workflow management system as an example to underline the need for a formal model to specify constraints on policies, including constraints that arise dynamically at runtime. Jajodia et al [9] extend this approach in a more generic and flexible way. They introduce a formal mechanism for expressing authorization specifications which is rich enough to incorporate any application-specific requirements within the fold of the authorization framework. Bertino et al [10] specifically focus on the problem of modeling dynamism in terms of time within access control and define a logic-based model to reason about access control policies involving time.

First order logic has also been used to specify context in generic context-aware frameworks that cater to various applications. Ranganathan and Campbell's first order logic based context modeling [11] deploys ontology based specification of context predicates such that "it allows different components in the system to have a common understanding of the semantics of different contexts." Their generic and expressive formalism based on first order logic makes the framework amenable to various applications.

All of these approaches concentrate more on access control and/or contexts as modeled on computer systems in general and not on physical access control in buildings in particular. Consequently, their focus is on languages that provide flexibility in specifying role based policies, and on guaranteeing unambiguous evaluation (decision) with feasible bounds on the run time. While the specific problem of physical access control also demands all these features, there is an additional challenge of devising a framework that can be implemented in a distributed way using a network of readers and/or controllers. This is mainly due to the changing trend in the access control hardware, from big centralized controllers to a network of intelligent controllers. The functional architecture of many of the above-mentioned approaches assumes a centralized authorization resolver that would decide on a grant/denial, given an access request, a history of the system and the authorization specification policies.

We develop a new policy language based on MSO logic, which gives us many advantages over other frameworks. First is that MSO logic provides a formal groundwork to support powerful policies, and can uniformly accommodate the changing nature of policy requirements. We illustrate this with an example involving a physical access control policy that exploits the power of MSO logic (beyond first order quantifiers), especially for specifying certain contexts in physical access control. Secondly, MSO logic representation make it easy to translate policies into finite state automata for deployment on small devices. Finite state automata that are used for policy execution turn out to be beneficial in terms of distributed storage and execution for access control decision. Finite state automata can be stored in a distributed manner exploiting the available access control infrastructure as we will explain in later sections. Taking an access control decision using finite state automata executing policies involves checking for one state change in the appropriate automaton, a constant time operation. This is a big advantage when compared to proof-theoretic approaches that are not guaranteed to terminate in constant time.

6.4. Policy language and specification

The Decentralized Access Control (DAC) framework includes a language to define complex policies with features to handle various parameters of dynamism, like user's history of movement, context induced by events in various rooms of the facility etc. Typical examples of context sensitive policies are:

- Supervisor required: A user may enter the room only if a supervisor or a security guard is already present.
- Escorted access: A user may enter a room (through a door) only if a designated escort has entered the room (through the same door) no longer than, say 7 seconds ago.

- Interlocking of doors: Certain rooms can only be accessed if a surrounding protective space is first closed.
- Anti-passback: Users who don't have a record of making a legal exit from a room will be denied next time they try to access the same and/or any other room in the facility.
- Guard tour: Guards are required to tour a facility at specified intervals. He/she must visit the rooms in a specific order, and must not spend any longer than specified amounts of time in each.
- Room count: Certain users cannot enter a room if the number of users reaches a certain value.
- Connected count: If the number of users in a set of rooms have collectively reached a stipulated number, access to certain other rooms or specific areas (e.g. lobby) should be denied, thereby ensuring regulated movement.

We use *Monadic Second Order (MSO) Logic* [2] parameterized by events of the system as the formal language for policies.

6.4.1. *Syntax*

Syntax of the logic is built over a set of first and second order variables, which are used to quantify over events of the system. First order variables quantify over individual events, while second order variables are used to quantify over a set of events. We represent a first order variable in lower case, e.g. x, and a second order variable in upper case, e.g. X. Policies are formulae of the logic, and are built on top of certain elementary relations over these first and second order variables using Boolean operations and quantification. Relations talk about how a first order variable represents a particular event and about the order of occurrence of events.

Events in our framework are typically composed of many *parameters*. Parameters describe the components of the access control framework (like room, user-id etc.) and actions of individual users (like request). For example, an event representing a user's request to access a particular room, say, *roomA*, will have the user-id and the room-id as its parameters among others. Such an event will be represented as *request-roomA-user1* with parameters *request*, *roomA*, and *user1* respectively representing the action of the user, the room for which access is being requested and the id of the requesting user. We would like to note that parameters may be common to many events.

Let Σ be the set of all events of the framework as defined above. Events are used to construct *atomic formulae*. The atomic formulae are given by:

- For each parameter p of some event in Σ, we have a predicate $p(x)$ which represents the fact that the event represented by the variable x has a parameter p. Given that, in general, an event can have more than one parameter, we would like to observe that there will be more than one such predicate $p(x)$ to completely represent the event occurring at x.

- For first order variables x, y the predicate $x \leq y$ represents the fact that the event corresponding to y occurs after the event corresponding to x, or x refers to the same event as y in a behavior of the system. We define $x < y$ analogously.
- For a first order variable x and a second order variable X, the atomic formula $x \in X$ represents the fact that the event corresponding to the variable x belongs to the set of events corresponding to X.

Formulae depicting policies are built from the atomic formulae using the following connectives:

- Boolean operators representing *negation* and *disjunction* are $\neg$ and $\vee$ respectively. The operators for *conjunction* ($\wedge$), *implication* ($\Rightarrow$), and *equivalence* ($\equiv$) can be derived from negation and disjunction.
- The operators *for all* ($\forall$) and *there exists* ($\exists$) will be used to quantify over first and second order variables.

To summarize, the syntax of the policy language is MSO logic tuned for physical access control by parameterizing over an appropriate set of events. Each access control policy is defined as a formula using the above syntax.

6.4.2. *Semantics*

Semantics of policies will be defined using *words* over the alphabet Σ. Words are finite sequences of events from Σ that represent behaviors. Words can also be thought of as finite sequences of positions (numbers) labeled by events. With this understanding, we use the term variable to quantify over events and positions interchangeably.

Consider a formula φ. φ is interpreted over a word w as follows: An *interpretation* of first and second order variables is a function I that assigns an event of Σ to each first order variable and a finite set of events of Σ to each second order variable. These letters occur as labels of positions in a word when a formula (policy) is interpreted over it.

The notion of when a word w *satisfies* a formula φ, under an interpretation I is given by $w \models_I \varphi$ and is defined inductively as follows:

- $w \models_I p(x)$ iff p is one of the parameters of the event represented by x.
- $w \models_I x \leq y$ iff either $I(x) = I(y)$ or $I(x)$ occurs before $I(y)$ in the word w.
- $w \models_I x \in X$ iff $I(x) \in I(X)$.
- $w \models_I \neg\varphi$ iff it is not the case that $w \models_I \varphi$.
- $w \models_I \varphi_1 \vee \varphi_2$ iff $w \models_I \varphi_1$ or $w \models_I \varphi_2$.
- $w \models_I \exists x\ \varphi$ iff there exists an interpretation function I' that extends I by assigning an event to the variable x such that $w \models_{I'} \varphi$.

- $w \models_I \exists X\ \varphi$ iff there exists an interpretation function I' that extends I by assigning a set of events to the variable X such that $w \models_{I'} \varphi$.

All our policies are *sentences* in MSO logic.

6.4.3. *Policy specification*

Formulae that constitute the policy that regulates access of users of *role 2* to *room A* are shown in Figure 6.2. The policy requires that "a user of *role 2* can access *room A* *only if* a member of *role 1* accessed it within the specified time interval and both *room A* and *room B* are completely closed." Note that we specify the two independent contexts, timed presence of a member of *role 1* and interlocked rooms, with the help of separate predicates. We then make use of these predicates in the final policy.

predicate *timed-r1*(p, q) :	predicate *inter-locked*(p, q) :
$\exists x\ (p \leq x < q \ \wedge\ timer(x)$	$\exists x, y\ (p \leq x < q \ \wedge\ p \leq y < q$
$\wedge$	$\wedge$
$access(x) \ \wedge\ role1(x) \ \wedge\ roomA(x)$	$roomA(x) \ \wedge\ close(x)$
$\wedge$	$\wedge$
$\neg\exists y\ (x \leq y < q \wedge same\text{-}user(x, y)$	$roomB(y) \ \wedge\ close(y)$
$\wedge$	$\wedge$
$(exit\text{-}roomA(y) \ \vee\ overflow(y))$	$\neg\exists z(x < z < q \ \wedge\ roomA(z) \ \wedge\ open(z))$
$)$	$\wedge$
$)$	$\neg\exists z(y < z < q \ \wedge\ roomB(z) \ \wedge\ open(z))$
	$)$

policy *role2-roomA* :

$$
\begin{aligned}
&\forall x\ (access(x) \ \wedge\ role2(x) \ \wedge\ roomA(x) \\
&\quad \Rightarrow \\
&\quad \exists w\ (w \leq x \wedge timed\text{-}r1(w, x)) \\
&\quad \wedge \\
&\quad \exists w\ (w \leq x \wedge inter\text{-}locked(w, x)) \\
&)
\end{aligned}
$$

Fig. 6.2. MSO policy formula.

The access policy says that every position x where a user of *role 2* accesses *room A* must be preceded by:

(1) a position w such that predicate *timed-r1* is satisfied between w and x[b], and
(2) a different position w such that predicate *inter-locked* is satisfied between w and x.

[b] $p \leq x \leq y < q$ is short hand for $(p \leq x) \wedge (x \leq y) \wedge (y < q)$

Both the predicates used in this policy take two arguments, which again represent positions. Predicate *timed-r1*, for example, makes sure that given positions p and q the following is true:

(1) There exists a position x, where a member of *role1* accesses *roomA* and the timer is set. That is, *timer* event also happens at x.
(2) There does not exist a position y between x and q where, corresponding to the same user as above, neither the timer overflows nor the user makes an exit. This ensures that within the scope of the predicate the timer is still running and the user is still in the room. Note that we maintain a timer for every user.
(3) Lastly, we don't want the same person who entered *roomA* at position y to exit the room (within the scope of this predicate). This part will be satisfied as long as there is one member of *role1* who satisfies this condition, which is ensured by the *same-user* predicate; while there may be many who entered and left the room before q.

In this predicate we ensure that there is at least one user who recently entered the room in role *role*1 and is still present there. The *same-user* predicate ensures that exit and timer overflow is associated with the correct user who initially entered. While making the predicate call in the policy formula, we pass arguments w and x to stand for p and q respectively. This is a way to make sure that the predicate, and hence the required context, is satisfied at the moment when access is made. The other predicate can also be understood along the same lines.

The context part of this policy is defined by the predicates *timed-r1* and *interlocked* (at positions labeled by w before x) and the access part is the event of requesting access to *roomA* by a member of *role2*.

6.4.4. *Expressibility of contexts*

The contexts presented in Section 6.4.3 are written as formulae in MSO logic. However, it may be noted that arbitrary counts or comparisons cannot be expressed with MSO logic. Consider a context that allows a role to access a room if there are not more than 10 persons occupying it. Note that this context does not restrict an arbitrary number of members of *other* roles from entering the room, hence it is not sufficient for the occupant count to stop at 10. However arbitrary counts are not expressible in MSO logic. Therefore, for the sake of our policies we assume a high enough upper limit, e.g. 100, on the number of people in a room. Similarly, such upper limits can also be assumed for the number of roles, rooms, doors etc.

Counting policies are possible with the help of second order variables. One such policy that allows no more than two people inside a room is shown in Figure 6.3. It makes use of a predicate *count-property* that works over second order variables, X_0 to X_3 which store the sets of positions in the word where the count of user accesses 0 to 3 respectively. Owing to lack of space, we skip the details of this predicate, but it is important to note that the first position of any word is always in X_0.

```
policy role2-roomB :
∃X0, X1, X2, X3 count-property(X0, X1, X2, X3)
∧
pos_0 ∈ X0
∧
∀x (access(x) ∧ role2(x) ∧ roomB(x)
    ⇒
    x ∉ X3
   )
```

Fig. 6.3. Restricting the number of *role*2 users in *roomB*.

So far, context expressions exclude situations when the context is not expressible in MSO logic at all. For example, consider the count example provided above and extend the system context predicate with an additional requirement of a sliding time window of a fixed duration, e.g. "no more than 10 persons having accessed the room *in the last one hour.*" This takes the context beyond the scope of MSO logic. In general, one may consider a number of interesting contexts that fall outside their expressive power. Should such contexts prove critical to the application, they can be supported with the help of general-purpose context evaluators (i.e. not necessarily limited to automata) that provide evaluations of contexts to policy automata in the form of system events, thereby hiding the implementation of those events. However, we observe that MSO logic based formulae suffice in practice.

6.5. Compilation

In order for the policies specified in MSO to be operational in terms of enforcing access control rules, they have to be converted into computational/executable models. These models can then be stored at appropriate locations for execution. We work with conventional finite state automata [3] as machine models for executing policies. Formal techniques are available for converting MSO formulae into automata [2] such that they accept precisely those behaviors that *satisfy* the policies.

As described in Section 6.4.3, policies describe when members of different roles can access different resources. Given policies for all roles belonging to the set, $\mathcal{R}$, of roles, and all the resources from set $\mathcal{O}$, the policy automaton is constructed as a conjunction: $\mathfrak{A} = \bigwedge_{r \in \mathcal{R},\ o \in \mathcal{O}} \texttt{policy}\ r\text{-}o$.

The language accepted by this automaton defines the valid behavior of the application. The automaton makes a state transition upon every event of the system, and only those events that take it to an accepting state are allowed. Given the nature of policies, it is easy to see that only access-events that violate one or more policies take the automaton to a rejecting state. These events are therefore forbidden, thereby resulting in denial of accesses.

We use MONA [12] as the policy analyzer to translate the MSO logic formulae into minimal Deterministic Finite State Automata (DFA). Several steps are involved in converting a given MSO formula into minimal DFA. Firstly, MONA rewrites the formula to obtain a compact representation. This is followed by an inductive translation of the rewritten formula into an automaton. Finally, the automaton obtained from the above step is minimized so that an effective decision procedure can be obtained [13]. Shared multi-terminal Binary Decision Diagrams (referred to as BDDs in this paper) are generated by MONA to store the transition tables of the automata. This data structure is efficient in terms of its size and amenability to fast execution [12].

The policy is translated by MONA to a corresponding finite state automaton that accepts only those sequences of events as expressed in the formula. MONA works with binary encodings of events in the form of *literals* in the corresponding automata. The corresponding BDD encapsulates a state transition by considering each bit of the literal and guiding a traversal over its nodes. The encoding of events into literals is explained in the following section.

6.5.1. *Encoding events into literals*

The translation of events into literals may be done in a straightforward manner by encoding them as bit sequences $\{0,1\}^k$ where k is smallest integer such that $|\Sigma| \leq 2^k$. However, it is judicious to enforce a structure on the encoded bit sequences in order to optimize the sizes of generated BDDs. This is because a structured encoding of literals can result in compact node traversals in the BDD, as compared to traversals over unstructured literal encodings. Overall, this results in lesser number of internal nodes, and hence smaller BDDs [12].

We consider an encoded literal as a concatenation of several predefined sub-sets of bits of fixed lengths, each representing the value of a specific parameter of the corresponding event. For example, the literal corresponding to the *access-role1-roomA* event can be encoded as $|01|101|1100|000111|$, where the two most significant bits represent the *action* parameter (say with possible values of *request*(0), *access*(1), *deny*(2) etc.), the next 3 bits represent the *role* parameter, the next 4 bits represent the *room* parameter and the final 6 bits represent the user id of the user who caused that specific event. In order to obtain a uniform encoding for all the events, we construct literals in a manner that they can accommodate the union of all the parameters that can possibly occur in different kinds of events.

6.5.2. *Optimizing automaton state-space*

In order to make this framework viable we must make it tenable on the state-of-art in access control hardware. We observe that the policy automaton, $\mathfrak{A}$, swells in state-space very quickly with the number of roles and resources, even with simple contexts like the ones shown in Section 6.4.3.

We make a further observation: the antecedents of all our policies are independent of each other. That is, the words rejected by a given policy are solely those words that violate it, which are precisely the words that end with an access event while the prefix does not satisfy the context. Formally, the set of words rejected by $\mathfrak{A}$ is a disjoint union of the sets of words rejected by individual policies. That is,

$$L^c(\mathfrak{A}) = \bigsqcup_{r \in \mathcal{R}, o \in \mathcal{O}} (\Sigma^* \setminus L(\texttt{policy}\ r\text{-}o)).$$

We therefore work with individual policy automata, $\mathfrak{A}_{r,o}$ =policy r-o : $r \in \mathcal{R}, o \subset \mathcal{O}$. It suffices to check each policy individually for access events, and if policy r-o rejects the word w then $w \notin L(\mathfrak{A}_{r,o})$, and hence[c] $w \notin L(\mathfrak{A})$. By breaking up policies in such a manner we produce an automaton for every role-resource pair in the application. These automata are capable of arriving at access decisions for respective role-resource pairs independently.

We identify an important class of policies that enable us to achieve further optimization over role-resource automata. These are policies whose contexts involve actions of only the users who seek accesses. For example, consider the policy of Figure 6.4 that specifies anti-passback prevention [cf. Section 6.4] for users of *role*3 in *roomA*.

```
policy role3-roomA :
∀x (access(x) ∧ role3(x) ∧ roomA(x)
    ⇒
    ¬(∃y (y ≤ x ∧ access(y) ∧ roomA(x) ∧ same-user(x, y)
          ∧
          ∄z (y ≤ z ≤ x ∧ exit-roomA(z) ∧ same-user(z, y))
          )
      )
   )
```

Fig. 6.4. Anti-passback protection in *roomA*.

Such policies depend only on user specific contexts or user specific histories. That is, in order to take an access decision for an individual, automata corresponding to such policies will require events corresponding to the same individual only. Other examples of such policies include restricted pathways, where a user must have accessed previous objects in a specified order before he/she can be granted access; mandatory action sequences where, say, users must first issue anti-corrosive suits from the issuance room before they can be granted access to laboratory, etc. In such

[c]Though it is possible that there exist $r' \in \mathcal{R}, o' \in \mathcal{O}$, where either $r' \neq r$ or $o' \neq o$ or both, such that $w \in L(\mathfrak{A}_{r',o'})$.

```
policy role3-roomA :
 ∀x (access(x) ∧ role3(x) ∧ roomA(x) ∧ user1(x)                  ⎫
     ⇒                                                            ⎪
     ¬(∃y (y ≤ x ∧ access(y) ∧ roomA(y) ∧ user1(y)                ⎪
              ∧                                                   ⎪
              ∄z (y < z < x ∧ exit-roomA(z) ∧ user1(z))           ⎬ policy role3-roomA-user1
              )                                                   ⎪
          )                                                       ⎪
    )                                                             ⎭
∧
 ∀x (access(x) ∧ role3(x) ∧ roomA(x) ∧ user2(x)                  ⎫
     ⇒                                                            ⎪
     ¬(∃y (y ≤ x ∧ access(y) ∧ roomA(y) ∧ user2(y)                ⎪
              ∧                                                   ⎪
              ∄z (y < z < x ∧ exit-roomA(z) ∧ user2(z))           ⎬ policy role3-roomA-user2
              )                                                   ⎪
          )                                                       ⎪
    )                                                             ⎭
∧
⋮
```

Fig. 6.5. Identifying user-resource policies.

policies, *same-user* predicates in the consequent tie the context with the user in the antecedent. Taking *role3-roomA* policy as an example, we rewrite a role-resource policy in the form of user-resource policies by instantiating the user ID parameters and considering per-user conjunctions as shown in Figure 6.5.

Then, we obtain $\mathfrak{A}_{r,o,u}$ automata for all users of the role for this kind of policies. The argument for evaluating these automata individually is the same as one discussed above. Note that each of the conjuncts is still a sentence in MSO logic.

In general, reducing the overall length of the formula translates in reduction of state-space in the corresponding automaton. Our empirical observations confirm that there is an order of magnitude reduction in number of states when we distribute the overall policy automaton, $\mathfrak{A}$, to role-resource automata, $\mathfrak{A}_{r,o}$. However, in the case of creation of user-resource policies [cf. Figure 6.5], the reduction in number of states is in fact quantifiable. Let m be the length of user-specific history, that is the number of events that occur in a sequence while describing the context. For example, in Figure 6.4, $m = 1$ since an anti-passback violation is implied by the presence of a single previous access. If history is the only context present then the role resource automaton will contain $(m+1)^n$ states for n users that belong to the role. However, once this policy is distributed over individual users of the role, each resulting automaton contains $m+1$ states. Since we create n automata as a result of distribution, the number total of states across all automata is $n(m+1)$.

6.6. Decentralized execution

This section explains how the principle of decomposition mentioned in Section 6.5.2, naturally extends to a set of design guidelines towards a framework for *decentralized* physical access control. The key observation is that the automaton for each of the (`policy` r-o) depends only upon the events that are part of the corresponding policy specification. Accordingly, the distributed architecture is based upon (i) a mechanism of mapping the execution of the (`policy` r-o) automata to the set of available physical compute nodes (ii) a mechanism of composing the events and propagating the events of interest to the appropriate subset of automata and (iii) a mechanism of arriving at access decisions based on the execution of the individual automata, so that $L^c(\mathfrak{A}) = \bigsqcup_{r \in \mathcal{R}, o \in \mathcal{O}} (\Sigma^* \setminus L(\texttt{policy}\ r\text{-}o))$.

6.6.1. *Distributed execution of automata*

The set of distributed compute nodes consists of embedded door controllers and user carried smart cards, as detailed in Section 6.1 and Section 6.2. We map a role-resource automaton to a controller associated with the resource. In the context of our application, an association would typically imply a physical control over the access to the resource. E.g. a microcontroller attached to a door of a *roomA* may be selected to run the automaton *role1-roomA*. This controller stores and executes the BDD (Section 6.5) corresponding to the automaton.

If a single resource o contains multiple controllers (e.g, a room with multiple doors), then, following this principle, all the controllers need to agree upon a leader for each role r, which will run the r-o automaton. Clearly, a single controller may be the leader for multiple roles for a given resource or even across resources (e.g. for topologically adjacent resources).

For the above scenario, an automaton may even be replicated across multiple controllers associated with a resource. Since an access request will be typically made to any one of these controllers, hence having the rules duplicated across the controllers will potentially enable quick parallel local decision making, without the controllers having to talk to and fro with the designated leader. This however, introduces the need for maintaining synchronized states across the replicated automata, thereby negating this advantage.

For the class of policies whose contexts only involve actions of the users who seek access (Section 6.5.2), an elegant workaround follows from the definition of $\mathfrak{A}_{r,o,u}$ if there exists hardware support in the form of memory cards or smart cards. Since the cards are typically issued per-user, hence, the states of the $\mathfrak{A}_{r,o,u}$ automaton can be stored in the user carried cards and the transition rules be replicated across controllers.

Smart cards offer an additional flexibility. Along with storing the per-user states, even the transition rules and execution mechanism of the automata can be stored and executed from within the smart cards. I.e., instead of storing the $\mathfrak{A}_{r,o,u}$ automata on the controller associated with object o, as described above, the automata can be stored and executed on the smart cards themselves. Since the only access requests to these automata will come from the user requests (by definition), hence they may very well be stored on the user carried smart cards and be *virtually* made available when the smart card *comes to* the associated controller. In this sense, it is an extension of the same principle, with the controller *virtually* storing the policy automata.

Both the card based solutions offer tremendous benefits in the face of disconnects in the underlying network of controllers. The smart card based approach, in addition, alleviates the memory/lookup/ and processing requirements at the controllers and helps avoid re-programming of controllers for policy changes.

6.6.2. *Event subscription*

Each of the $\mathfrak{A}_{r,o}$ or $\mathfrak{A}_{r,o,u}$ automaton requires events from a subset of Σ. As and when these events occur in the system, they must be provided to all the automata that require them. The *user action* events are generated by components which interface with the hardware (typically, controllers) and generate an event corresponding to the physical action of the user (e.g. request for access). *Access decision* events refer to access grants (or denials) generated by decision evaluation components (i.e. automata running on controllers or smart cards). *System notifications* are events generated outside of our framework, corresponding to physical happenings like door ajar or alert level high or intrusion detected etc.

Note that for its execution, a role-resource automaton may require both user action events or system notification (access decision) events from any number of controllers (or automata executing on controllers/smart cards). A formalism of this dependency can be conveniently abstracted as a publish-subscribe requirement. The following observation greatly simplifies the construction of a publish-subscribe map. Every unique decision event is always generated by a single automata (and by extension, a single computing node), and every unique user event is always observed and generated by a single controller. This compute node, or the *owner*, is responsible for communicating this event to an *event management middleware.*

The publish-subscribe mechanism relies on event subscription maps, comprising of (i) a pair wise mapping of automata such that output literal(s) of one is (are) the input literal(s) of the other, (ii) a pair wise mapping of automata with system notification events. The event subscription maps exploit the fact that each event has a unique owner (producer/publisher) but may have multiple users (consumers/subscribers).

Event subscription maps may be statically compiled, or dynamically generated after the system is powered-on. For the latter, each controller is initially configured with the enumeration of the events it generates, and the events it requires for the

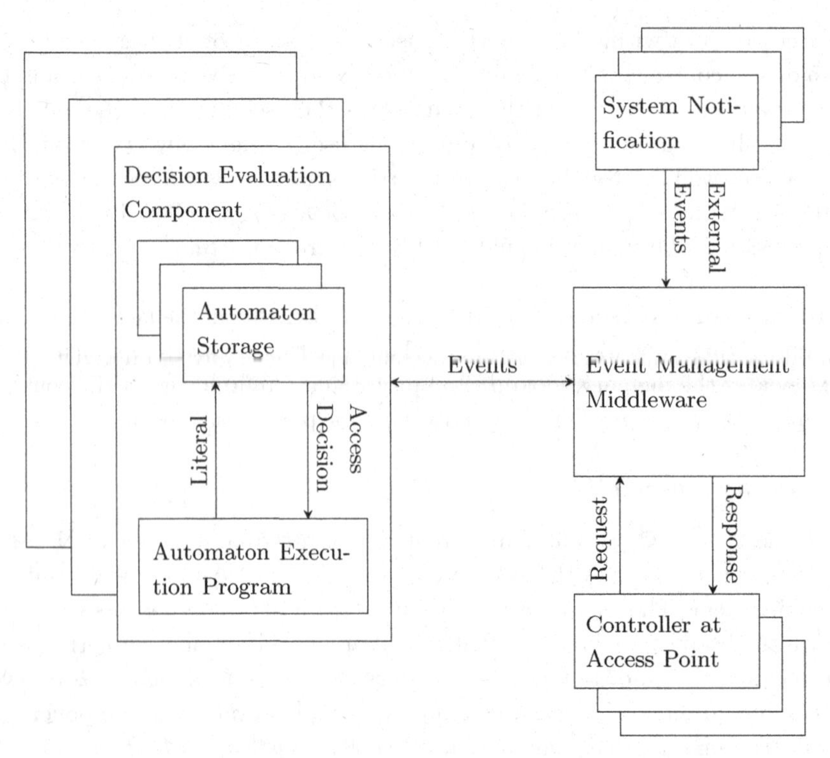

Fig. 6.6. Execution architecture.

sake of transitions of its automata. Subscription maps are then built as a result of controllers exchanging this information.

Events are provided to the automata through an event management middleware as shown in Figure 6.6.

The event management middleware gets events from all event owners and maintains a mapping of the automata to the compute nodes. It also encodes events as literals before forwarding them to appropriate nodes. Depending on the implementation, the event literals may be pulled by the consumer components or be pushed by the event management middleware to the components.

E.g. whenever a user presents an access card at a door, an access-request event is generated by the controller connected to the card reader. It is then propagated to all the controllers that need it through the event management middleware. At the very least, it is provided to the compute node that stores the policy automaton for that user (role) for that room. If the automaton indeed grants an access then the grant event is generated and subsequently propagated to other controllers (whose automata may have subscribed for this event), otherwise a deny event may be similarly generated and propagated.

In addition to supporting the publish-subscribe mechanism, the event management middleware must synchronize the propagation of events across controllers such that all of them view the same order of occurrence of events, that is *all automata must view the same 'word'*. Choosing a single controller for a given resource ensures that all requests by the users of same resources are queued at the same controller. However, access decisions and other events occurring in the rest of the system system still must be uniformly propagated through the entire system. The basic problem is that of two or more policy automata taking mutually contradicting decisions as a result of observing the same events in different orders, owing to the underlying communication infrastructure. Treating each access decision as a transaction, with cascading effects on the execution of other dependent automata - may be an useful approach in addressing this problem.

6.6.3. *Access decision*

When an access request is made, the user request literal is first applied to the appropriate policy automaton (i.e. it undergoes a state transition). A literal corresponding to the access-grant is then composed and *provisionally* applied to the automaton. If the automaton makes a transition to an accepting state then it is said to have allowed the access. In this case the transition is finalized (or committed). If however, this takes the automaton to a non-accepting state then it is said to have denied the access. The access-grant literal is then rolled back.

The combined application of the above decision procedure by each role-resource automaton ensures that none of them ever accepts a *word* which is rejected by the individual role-resource policies. This is equivalent to the original policy automaton $\mathfrak{A}$ never making a transition to a non-accepting state (Section 6.5.2). This further means that only words that are accepted by the policies are ever allowed in the run of events, i.e. an erroneous grant is never generated.

6.7. Benefits

We introduce an authorization framework for physical access control using finite state automata that supports context based access control policies. Our mechanism attempts to enforce a valid behavior by ensuring that access events only follow a certain sequence, as permitted by the policies. Granting an access upon request only involves checking for an accepting transition in the automaton, and hence decision to grant/deny access is taken in constant time, unlike that in most proof theoretic approaches. This is owing to the fact that our interpretation is fixed over a set of events, whereas other proof theoretic approaches use first order logic without a fixed interpretation. This is especially important for an access control application where low response latency is crucial. This, coupled with the fact that in the context of the application it is possible to optimize the size of the automaton, makes the framework appealing for enterprise physical access control solutions.

In the context of our application, a centralized architecture does not scale with the number of users. Local decision making capacity at points of access effectively decentralizes the process of dynamic authorization. This not only reduces the per-user per-request round-trip communication to a central server, but also cuts down on the processing involved to evaluate contexts with respect to users' policies. Further, effective decentralization and localization of policy decision enables meaningful enforcement of some access control policies even if there is a partial disconnection in the network of the controllers. For example, policies depending only on a user's past behavior, and not on other system context, can be enforced even if a controller is totally disconnected from the system. Even otherwise, as long as dependent events (usually originating in spatial proximity) are available to the automata, the access control solution can continue to function in a partitioned network, given some local pockets of connectivity. This reduced dependency on a dedicated reliable underlying network makes it attractive to use the general purpose enterprise data network for access control as well.

Since access control rules are specified using MSO logic, we can exploit standard formal verification algorithms to check for various interesting properties on the behavior that the policies are intended to enforce. It may be required to ensure, e.g. that certain undesirable situations can never arise or certain necessary checks are never averted. Mutually conflicting roles or conflicting contexts can be considered as examples of such properties. The proposed solution opens up the possibility of applying formal verification algorithms in this context - which can bring a lot of richness in the domain of physical access control policy specification, by virtue of both detecting such properties and illustrating specific scenarios of conflicts.

6.8. Conclusion and future work

We focus on a decentralized policy evaluation framework for dynamic authorizations. As an initial approach, we simplify our assumptions about the relationship between users and roles (no role hierarchies), and derive the requirements from a specific application, namely physical access control. We observe that our specification model is useful for capturing fairly involved context-sensitive physical access control scenarios, including connecting physical access to enterprise logical access control solutions.

However, we note that a complete solution needs to address many other related issues. Validation of user credentials [14] and security of data transfer among various system entities, for example, are important considerations for commercial access control solutions today. With distributed policy execution, the issue of compromise of nodes becomes important. With respect to smart cards, tamper proofing is one way to ensure that stored data are safe from manipulation. Using cryptographic techniques to establish trust on smart cards poses another significant challenge. A full-fledged solution will also require a verification framework, possibly exploiting model checking approaches to ensure that policies are internally consistent and *well-defined* and result in a meaningful behavior in the system.

Acknowledgment

The authors are grateful to their colleagues Abhishek Bhatnagar[d] and V. Senthilnathan for their valuable suggestions and contribution toward prototyping this framework.

References

[1] Z. Chen, *Technology for Smart Cards: Architecture and Programmer's Guide.* (Addison Wesley, 2000).
[2] W. Thomas, *Languages, automata, and logic*, In *Handbook of formal languages*, vol. 3, pp. 389–455. Springer, (1997).
[3] D. Kozen, *Automata and Computability.* (Springer, 1997).
[4] M. Abadi. Logic in access control. In *Proc. 18th LICS*, pp. 228–233. IEEE, (2003).
[5] P. Samarati and S. De Capitani di Vimercati. Access control: Policies, models, and mechanisms. In *Proc. IFIP WG 1.7 International School on Foundations of Security Analysis and Design*, pp. 137–196. Springer, (2001).
[6] J. Y. Halpern and V. Weissman, Using first-order logic to reason about policies, *Trans. Information and System Security.* **11**(4), 1–41, (2008).
[7] L. Bauer, S. Garriss, and M. K. Reiter. Distributed proving in access-control systems. In *Proc. Symposium on Security and Privacy*, pp. 81–95. IEEE, (2005).
[8] E. Bertino, E. Ferrari, and V. Atluri. A flexible model supporting the specification and enforcement of role-based authorization in workflow management systems. In *Proc. Workshop on Role-Based Access Control*, pp. 1–12. ACM, (1997).
[9] S. Jajodia, P. Samarati, M. L. Sapino, and V. S. Subrahmanian, Flexible support for multiple access control policies, *Trans. Database Systems.* **26**(2), 214–260, (2001).
[10] E. Bertino, C. Bettini, E. Ferrari, and P. Samarati, An access control model supporting periodicity constraints and temporal reasoning, *Trans. Database Systems.* **23**(3), 231–285, (1998).
[11] A. Ranganathan and R. H. Campbell, An infrastructure for context-awareness based on first order logic, *Personal Ubiquitous Comput.* **7**(6), 353–364, (2003).
[12] N. Klarlund and A. Møller. *MONA Version 1.4 User Manual.* BRICS Notes Series NS-01-1, Department of Computer Science, University of Aarhus, (2001).
[13] J. G. Henriksen, J. Jensen, M. Jørgensen, N. Klarlund, B. Paige, T. Rauhe, and A. Sandholm. MONA: Monadic Second-Order Logic in Practice. In *Proc. 1st TACAS*, vol. 1019, *LNCS*. Springer, (1995).
[14] S. Micali. NOVOMODO: Scalable certificate validation and simplified PKI management. In *Proc. PKI Research Workshop*, (2002).

[d]Currently at Qualcomm Inc.

Chapter 7

Reasoning about Heap Manipulating Programs using Automata Techniques

Supratik Chakraborty

Dept. of Computer Science and Engineering,
I.I.T. Bombay, Powai,
Mumbai 400076, India
supratik@cse.iitb.ac.in

Automatically reasoning about programs is of significant interest to the program verification, compiler development and software testing communities. While property checking for programs is undecidable in general, techniques for reasoning about specific classes of properties have been developed and successfully applied in practice. In this article, we discuss three automata based techniques for reasoning about programs that dynamically allocate and free memory from the heap. Specifically, we discuss a regular model checking based approach, an approach based on storeless semantics of programs and Hoare-style reasoning, and a counter automaton based approach.

7.1. Introduction

Automata theory has been a key area of study in computer science, both for the theoretical significance of its results as well as for the remarkable success of automata based techniques in diverse application areas. Interesting examples of such applications include pattern matching in text files, converting input strings to tokens in lexical analyzers (used in compilers), formally verifying properties of programs, solving Presburger arithmetic constraints, machine learning and pattern recognition, among others. In this article, we focus on one such class of applications, and discuss how automata techniques can be used to formally reason about computer programs that dynamically allocate and free memory from the heap.

Formally reasoning about programs has interested computer scientists since the days of Alan Turing. Among the more difficult problems in this area is analysis of programs that manipulate dynamic linked data structures. This article is an overview of three important automata based techniques to address this problem. It is not meant to be an exhaustive overview of all automata-based techniques for reasoning about programs. Instead, we look at three different and interesting approaches, and explain them in some detail. To lend concreteness to the problem, we wish to answer the following question: *Given a sequential program that manipulates dynamic linked data structures by means of creation and deletion of memory cells and updation of links between them, how do we prove assertions about the resulting structures in heap memory (e.g. linked lists, trees, etc.)?* This problem, also commonly called *shape analysis*, has been the subject of extensive research over the last few decades. Simple as it may seem, answering the above question in its complete generality is computationally impossible or undecidable. Nevertheless, its practical significance in optimization and verification of programs has motivated researchers to invest significant effort in studying special classes of programs and properties. The resulting advances in program analysis techniques have borrowed tools and techniques from different areas of computer science and mathematics. In this article, we restrict our discussion to a subset of these techniques that are based on automata theory. Specifically, we discuss the following three techniques for shape analysis: (i) regular model checking, (ii) Hoare-style reasoning using a storeless semantics, and (iii) a counter automaton based abstraction technique. These techniques also provide insights into how more general cases of the problem might be solved in future.

The remainder of this article is organized as follows. Section 7.2 presents notation, definitions and some key automata-theoretic results that are used subsequently. Section 7.3 discusses a simple imperative programming language equipped with constructs to dynamically allocate and free memory, and to update selectors (fields) of dynamically allocated memory locations. The example programs considered in this article are written in this language. Section 7.4 provides an overview of the challenges involved in reasoning about heap manipulating programs. Sections 7.5, 7.6 and 7.7 describe three automata based techniques for reasoning about programs manipulating heaps. Specifically, we discuss finite word based regular model checking in Section 7.5, and show how this can be used for shape analysis. Section 7.6 presents a regular language (automaton) based storeless semantics for our programming language, and a logic for reasoning about programs using this semantics. We show in this section how this logic can be used in Hoare-style reasoning about programs. A counter automaton based abstraction of programs manipulating singly linked lists is discussed in Section 7.7. Finally, section 7.8 concludes the article.

7.2. Automata notation and preliminaries

Let Σ be a finite alphabet and let Σ^* denote the set of all finite words on Σ. Note that Σ^* contains ε — the empty word of length 0. A language over Σ is a (possibly empty) subset of Σ^*. A finite-state transition system over Σ is a 4-tuple $\mathcal{B} = (Q, \Sigma, Q_0, \delta)$ where Q is a finite set of states (also called control locations), $Q_0 \subseteq Q$ is the set of initial states, and $\delta \subseteq Q \times \Sigma \times Q$ is the transition relation. If $|Q_0| = 1$ and δ is a function from $Q \times \Sigma$ to Q, we say that $\mathcal{B}$ is a *deterministic* finite-state transition system. Otherwise, we say that $\mathcal{B}$ is non-deterministic. A finite-state automaton $\mathcal{A}$ over Σ is a finite-state transition system equipped with a set of designated final states. Thus, $\mathcal{A} = (Q, \Sigma, Q_0, \delta, F)$, where $\mathcal{B} = (Q, \Sigma, Q_0, \delta)$ is a finite-state transition system and $F \subseteq Q$ is a set of final states. The notation $|\mathcal{A}|$ is often used to refer to the number of states (i.e., $|Q|$) of automaton $\mathcal{A}$.

The transition relation δ induces a relation $\widehat{\delta} \subseteq Q \times \Sigma^* \times Q$, defined inductively as follows: (i) for every $q \in Q$, $(q, \varepsilon, q) \in \widehat{\delta}$, and (ii) for every $q_1, q_2, q_3 \in Q$, $w \in \Sigma^*$ and $a \in \Sigma$, if $(q_1, w, q_2) \in \widehat{\delta}$ and $(q_2, a, q_3) \in \delta$, then $(q_1, w.a, q_3) \in \widehat{\delta}$, where "." denotes string concatenation. A word $w \in \Sigma^*$ is said to be *accepted* by the automaton $\mathcal{A}$ iff $(q, w, q') \in \widehat{\delta}$ for some $q \in Q_0$ and $q' \in F$. The set of all words accepted by $\mathcal{A}$ is called the language of $\mathcal{A}$, and is denoted $L(\mathcal{A})$. A language that is accepted by a finite-state automaton is said to be *regular*.

Given languages L_1 and L_2, we define the language concatenation operator as $L_1 \cdot L_2 = \{w \mid \exists x \in L_1, \exists y \in L_2, w = x.y\}$. For a language L, the language L^i is defined as follows: $L^0 = \{\varepsilon\}$, and $L^i = L^{i-1} \cdot L$, for all $i \geq 1$. The Kleene-closure operator on languages is defined as $L^* = \{w \mid \exists i \geq 0, w \in L^i\}$. We define the left quotient of L_2 with respect to L_1 as $L_1^{-1} L_2 = \{w \mid w \in \Sigma^* \text{and } \exists v \in L_1, v.w \in L_2\}$. If $L_1 = \emptyset$, we define $L_1^{-1} L_2 = \emptyset$ in all cases, including when $L_2 = \emptyset$.

The following results from automata theory are well-known and their proofs can be found in Hopcroft and Ullman's book [1].

(1) If $\mathcal{A}_1$ and $\mathcal{A}_2$ are finite-state automata on an alphabet Σ, there exist effective constructions of finite-state automata accepting each of $L(\mathcal{A}_1) \cup L(\mathcal{A}_2)$, $L(\mathcal{A}_1) \cap L(\mathcal{A}_2)$, $\Sigma^* \setminus L(\mathcal{A}_1)$, $L(\mathcal{A}_1) \cdot L(\mathcal{A}_2)$, $L^*(\mathcal{A}_1)$ and $L(\mathcal{A}_1)^{-1} L(\mathcal{A}_2)$.
(2) For every non-deterministic finite-state automaton $\mathcal{A}_1$, there exists a deterministic finite-state automaton $\mathcal{A}_2$ such that $L(\mathcal{A}_1) = L(\mathcal{A}_2)$ and $|\mathcal{A}_2| \leq 2^{|\mathcal{A}_1|}$.
(3) For every deterministic finite-state automaton $\mathcal{A}$, there exists a minimal deterministic finite-state automaton $\mathcal{A}_{min}$ that is unique up to isomorphism and has $L(\mathcal{A}) = L(\mathcal{A}_{min})$. Thus, any deterministic finite-state automaton accepting $L(\mathcal{A})$ must have at least as many states as $|\mathcal{A}_{min}|$.

Let L be a language, and let R_L be a binary relation on $\Sigma^* \times \Sigma^*$ defined as follows: $\forall x, y \in \Sigma^*$, $(x, y) \in R_L$ iff $\forall z \in \Sigma^*$, $x.z \in L \Leftrightarrow y.z \in L$. The relation R_L is easily seen to be an equivalence relation. Therefore, R_L partitions Σ^* into a set of equivalence classes. A famous theorem due to Myhill and Nerode (see [1] for a nice exposition) states that a language L is regular iff the index of R_L is finite. Furthermore, there is no deterministic finite-state automaton that recognizes L and has fewer states than the index of R_L.

A *finite state transducer* over Σ is a 5-tuple $\tau = (Q, \Sigma_\varepsilon \times \Sigma_\varepsilon, Q_0, \delta_\tau, F)$, where $\Sigma_\varepsilon = \Sigma \cup \{\varepsilon\}$ and $\delta_\tau \subseteq Q \times \Sigma_\varepsilon \times \Sigma_\varepsilon \times Q$. Similar to the case of finite state automata, we define $\widehat{\delta_\tau} \subseteq Q \times \Sigma^* \times \Sigma^* \times Q$ as follows: (i) for every $q \in Q$, $(q, \varepsilon, \varepsilon, q) \in \widehat{\delta_\tau}$, and (ii) for every $q_1, q_2, q_3 \in Q$, $u, v \in \Sigma^*$ and $a, b \in \Sigma_\varepsilon$, if $(q_1, u, v, q_2) \in \widehat{\delta_\tau}$ and $(q_2, a, b, q_3) \in \delta_\tau$, then $(q_1, u.a, v.b, q_3) \in \widehat{\delta_\tau}$. The transducer τ defines a regular binary relation $R_\tau = \{(u, v) \mid u, v \in \Sigma^* \text{ and } \exists q \in Q_0, \exists q' \in F, (q, u, v, q') \in \widehat{\delta_\tau}\}$. For notational convenience, we will use τ for R_τ when there is no confusion. Given a language $L \subseteq \Sigma^*$ and a binary relation $R \subseteq \Sigma^* \times \Sigma^*$, we define $R(L) = \{v \mid \exists u \in L,\ (u, v) \in R\}$. Given binary relations R_1 and R_2 on Σ^*, we use $R_1 \circ R_2$ to denote the composed relation $\{(u, v) \mid u, v \in \Sigma^* \text{ and } \exists x \in \Sigma^*,\ ((u, x) \in R_1 \text{ and } (x, v) \in R_2)\}$. Let $id \subseteq \Sigma^* \times \Sigma^*$ denote the identity relation on Σ^*. For every relation $R \subseteq \Sigma^* \times \Sigma^*$, we define $R^0 = id$, and $R^{i+1} = R \circ R^i$ for all $i \geq 0$.

With this background, we now turn our attention to reasoning about heap manipulating programs using automata based techniques.

7.3. A language for heap manipulating programs

Memory locations accessed by a program can be either statically allocated or dynamically allocated. Storage represented by statically declared program variables are allocated on the *stack* when the program starts executing. If the program also dynamically allocates memory, the corresponding storage comes from a logical pool of free memory locations, called the *heap*. In order for a program to allocate, de-allocate or access memory locations from the heap, special constructs are required in the underlying programming language. We present below a simple imperative programming language equipped with these constructs. Besides supporting allocation and de-allocation of memory from the heap, our language also supports updating and reading from selectors (or fields) of allocated memory locations. This makes it possible to write interesting heap manipulating programs using our language. In order to keep the discussion focused on heaps, we will henceforth be concerned only with link structures between allocated memory locations. Therefore we restrict our language to have a single abstract data type, namely pointer to a memory location. All other data-valued selectors (or fields) of memory locations are assumed to be abstracted away, leaving only pointer-valued selectors.

Dynamically allocated memory locations are also sometimes referred to as *heap objects* in the literature. Similarly, selectors of such memory locations are sometimes referred to as *fields* of objects. In this article, we will consistently use the terms *memory locations* and *selectors* to avoid confusion with objects and fields in the sense of object-oriented programs. The syntax of our language is given

Table 7.1. Syntax of our programming language.

`PVar`	::=	`u` | `v` | ... (pointer-valued variables)
`FName`	::=	`n` | `f` | ... (pointer-valued selectors)
`PExp`	::=	`PVar` | `PVar->FName`
`BExp`	::=	`PVar = PVar` | `Pvar =` **nil** | **not** `BExp` | `BExp` **or** `BExp` | `BExp` **and** `BExp`
`Stmt`	::=	`AsgnStmt` | `CondStmt` | `LoopStmt` | `SeqCompStmt` | `AllocStmt` | `FreeStmt`
`AsgnStmt`	::=	`PExp := PVar` | `PVar := PExp` | `PExp :=` **nil**
`AllocStmt`	::=	`PVar :=` **new**
`FreeStmt`	::=	**free**`(PVar)`
`CondStmt`	::=	**if** `(BoolExp)` **then** `Stmt` **else** `Stmt`
`LoopStmt`	::=	**while** `(BoolExp)` **do** `Stmt`
`SeqCompStmt`	::=	`Stmt ; Stmt`

in Table 7.1. Here, `PExp` represents a pointer expression obtained by concatenating at most one selector to a pointer-valued variable. `BExp` represents Boolean expressions on pointer variables, and are constructed using two basic predicates: the "=" predicate for checking equality of two pointer variables, and the "= **nil**" predicate for checking if a pointer variable has the **nil** value. `AllocStmt` represents a statement for allocating a fresh memory location in the heap. A pointer to the freshly allocated location is returned and assigned to a pointer variable. `FreeStmt` represents a statement for de-allocating a previously allocated heap memory location pointed to by a pointer variable. The remaining constructs are standard and we skip describing their meanings.

We restrict the use of long sequences of selectors in our language. This does not sacrifice generality since reference to a memory location through a sequence of k selectors can be effected by introducing $k - 1$ fresh temporary variables, and using a sequence of assignment statements, where each statement uses at most one selector. Our syntax for assignment statements also disallows statements of the form `u->f := v->n`. The effect of every such assignment can be achieved by introducing a fresh temporary variable `z` and using a sequence of two assignments: `z := v->n; u->f := z` instead. For simplicity of analysis, we will further assume that assignment statements of the form `u := u` are not allowed. This does not restrict the expressiveness of the language, since `u := u` may be skipped without affecting the program semantics. Assignment statements of the form `u := u->n` frequently arise in programs that iterate over dynamically created linked lists. We allow such assignments in our language for convenience of programming. However, we will see later that for purposes of analysis, it is simpler to replace every occurrence of `u := u->n` by `z := u->n; u := z` where `z` is a fresh temporary variable.

Example 7.1. The following program written in the above language searches a linked list pointed to by `hd` for the element pointed to by `x`. On finding this element, the program allocates a new memory location and inserts it as a new element in the list immediately after the one pointed to by `x`. The relative order of all other elements in the list is left unchanged.

Table 7.2. A program manipulating a linked list.

L1:	`t1 := hd;`	L6:	`t2->n := t3;`
L2:	**while** (**not** (`t1 = nil`)) **do**	L7:	`x->n := t2;`
L3:	**if** (`t1 = x`) **then**	L8:	`t1 := t1->n;`
L4:	`t2 :=` **new**;	L9:	**else** `t1 := t1->n`
L5:	`t3 := x->n;`	L10:	`// end if-then-else, end while-do`

7.4. Challenges in reasoning about heap manipulating programs

Given a heap manipulating program such as the one in Example 7.1, there are several interesting questions that one might ask. For example, can the program de-reference a null pointer, leading to memory access error? Or, if `hd` points to an (a)cyclic linked list prior to execution of the program, does it still point to an (a)cyclic linked list after the program terminates? Alternatively, can executing the program lead to memory locations allocated in the heap, but without any means of accessing them by following selectors starting from program variables? The generation of such "orphaned" memory locations, also called *garbage*, is commonly referred to as *memory leak* . Yet other important problems concern finding pairs of pointer expressions that refer to the same memory location at a given program point during some or all executions of the program. This is also traditionally called *may-* or *must-alias analysis*, respectively.

Unfortunately, reasoning about heap manipulating programs is difficult. A key result due to Landi [2] and Ramalingam [3] shows that even a basic problem like may-alias analysis admits undecidability for languages with if statements, while loops, dynamic memory allocation and recursive data structures (like linked lists and trees). Therefore any reasoning technique that can be used to identify may-aliases in programs written in our language must admit undecidability. This effectively rules out the existence of exact algorithms for most shape analysis problems. Research in shape analysis has therefore focused on sound techniques that work well in practice for useful classes of programs and properties, but are conservative in general.

A common problem that all shape analysis techniques must address is that of representing the heap in a succinct yet sufficiently accurate way for answering questions of interest. Since our language permits only pointer-valued selectors, the heap may be viewed as a set of memory locations with a link structure arising from values of selectors. A natural representation of this view of the heap is a labeled directed graph. Given a program P, let Σ_p and Σ_f denote the set of variables and set of selectors respectively in P. We define the *heap graph* as a labeled directed graph $G_H = (V, E, v_{\mathrm{nil}}, \lambda, \mu)$, where V denotes the set of memory locations allocated by the program and always includes a special vertex v_{nil} to denote the **nil** value of pointers, $E \subseteq (V \setminus \{v_{\mathrm{nil}}\}) \times V$ denotes the link structure between memory locations, $\lambda : E \to 2^{\Sigma_f} \setminus \{\emptyset\}$ gives the labels of edges, and $\mu : \Sigma_p \hookrightarrow V$ defines the (possibly partial) mapping from pointer variables to memory locations in the heap. Specifically, there exists an edge (u, v) with label $\lambda((u, v))$ in graph G_H iff for every $f \in \lambda((u, v))$, selector f of memory location u points to memory location v, or to **nil** (if $v = v_{nil}$). Similarly, for every variable $x \in \Sigma_p$, $\mu(x) = v$ iff x points to memory location v, or to **nil** (if $v = v_{\mathrm{nil}}$).

Since a program may allocate an unbounded number of memory locations, the size of the heap graph may become unbounded in general. This makes it difficult to use an explicit representation of the graph, and alternative finite representations must be used. Unfortunately, representing unbounded heap graphs finitely comes at the cost of losing some information about the heap. The choice of representation formalism is therefore important: the information represented must be sufficient for reasoning about properties we wish to study, and yet unnecessary details must be abstracted away. In order to model the effect of program statements, it is further necessary to define the operational semantics of individual statements in terms of the chosen representation formalism. Ideally, the representation formalism should be such that the operational semantics is definable in terms of efficiently implementable operations on the representation. A reasoning engine must then use this operational semantics to answer questions pertaining to the state of the heap resulting from execution of an entire program. Since the choice of formalism for representing the heap affects the complexity of analysis, a careful balance must be struck between expressiveness of the formalism and decidability or complexity of reasoning with it. In the following three sections, we look at three important automata based techniques for addressing the above issues.

7.5. Shape analysis using regular model checking

Model checking refers to a class of techniques for determining if a finite or infinite state model of a system satisfies a property specified in a suitable logic [4]. The state transition model is usually obtained by defining a notion of system state, and by defining a transition relation between states to represent the small-step operational semantics of the system. In symbolic model checking [4], sets of states are represented symbolically, rather than explicitly. Regular model checking, henceforth called *RMC*, is a special kind of symbolic model checking, in which words or trees over a suitable alphabet are used to represent states. Symbolic model checking using word based representation of states was first introduced by Kesten et al. [5] and Fribourg [6]. Subsequently, significant advances have been made in this area (see [7] for an excellent survey). While RMC is today used to refer to a spectrum of techniques that use finite/infinite words, trees or graphs to represent states, we will focus on finite word based representation of states in the present discussion. Specifically, the works of Jonsson, Nilsson, Abdulla, Bouajjani, Moro, Touilli, Habermehl, Vojnar and others [7–14] form the basis of our discussion on RMC.

If individual states are represented as finite words, a set of states can be represented as a language of finite words. Moreover, if the set is regular, it can be represented by a finite-state automaton. In the remainder of this section, we will refer to a state and its word-representation interchangeably. Similarly, we will refer to a set of states and its corresponding language representation interchangeably. The small-step operational semantics of the system is a binary relation that relates pairs of words representing the states before and after executing a statement. The state transition relation can therefore be viewed as a word transducer. For several classes of systems, including programs manipulating singly linked lists, the state transition relation can indeed be modeled as a finite state transducer. Given a regular set I of words representing the initial states, and a finite state transducer τ, automata theoretic constructions can be used to obtain finite state representations of the sets (languages) $R_\tau^i(I)$, $i \geq 1$, where R_τ is the binary relation defined by τ and $R_\tau^i = \underbrace{R_\tau \circ (R_\tau \circ (\cdots (R_\tau \circ R_\tau) \cdots))}_{i}$. For notational convenience, we will use $\tau^i(I)$ to denote $R_\tau^i(I)$, when there is no confusion. The limit language $\tau^*(I)$, defined as $\bigcup_{i \geq 0} \tau^i(I)$, represents the set of all states reachable from some state in I in finitely many steps. Given a regular set Bad of undesired states, the problem of determining if some state in Bad can be reached from I therefore reduces to checking if the languages Bad and $\tau^*(I)$ have a non-empty intersection. Unfortunately, computing $\tau^*(I)$ is difficult in general, and $\tau^*(I)$ may not be regular even when both I and τ are regular. A common approach to circumvent this problem is to use an upper approximation of $\tau^*(I)$ that is both regular and efficiently computable. We briefly survey techniques for computing such upper approximations later in this section.

7.5.1. *Program states as words*

To keep things simple, let us consider the class of programs that manipulate dynamically linked data structures, but where each memory location has a single pointer-valued selector. The program in Example 7.1 belongs to this class. We will treat creation of garbage as an error, and will flag the possibility of garbage creation during our analysis. Hence, for the remainder of this discussion, we will assume that no garbage is created. Under this assumption, the heap graph at any snapshot of execution of a program (in our chosen class) consists of singly linked lists, with possible sharing of elements and circularly linked structures. Figure 7.1 shows three examples of such heap graphs.

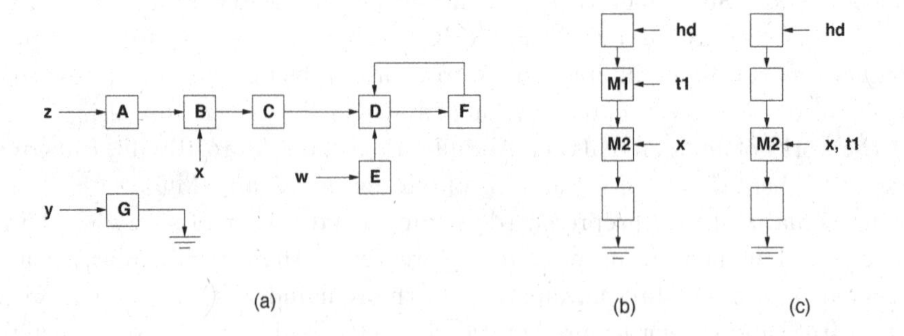

Fig. 7.1. Shared lists.

Adapting the terminology of Manevich et al. [15], we say that a node v in the heap graph is *heap-shared* if either (i) there are two or more distinct nodes with edges to v, or (ii) v is pointed to by a program variable and there is a node with an edge to v. Furthermore, a node v is called an *interruption* if it is either heap-shared or pointed to by a program variable. As an example, the heap graph depicted in Figure 7.1a has two heap-shared nodes (B and D) and five interruptions (A, B, D, E and G). It can be shown that for a program (in our class) with n variables, the number of heap-shared nodes and interruptions in the heap graph is bounded above by n and $2n$, respectively [15]. The heap graph can therefore be represented as a set of at most $2n$ *uninterrupted list segments*, where each uninterrupted list segment has the following properties: (i) the first node is an interruption, (ii) either the last node is a heap-shared node, or the selector of the last node is uninitialized or points to **nil**, and (iii) no other node in the uninterrupted list segment is an interruption. As an example, the heap graph in Figure 7.1a has five uninterrupted list segments: $A \to B$, $B \to C \to D$, $D \to F \to D$, $E \to D$ and $G \to$ **nil**.

The above observation motivates us to represent a heap graph as a set of uninterrupted list segments. To represent a list segment, we first assign a unique name to every heap-shared node, and rank the set of all names (heap-shared node names and program variable names). Note that names are assigned only to heap-shared nodes and not to all nodes in the heap graph. Since the number of heap-shared nodes is bounded by the number of program variables, a finite number of names suffices for our purpose. Ranking names allows us to represent a set of names uniquely as a rank-ordered sequence of names. An uninterrupted list segment with r nodes, each having a single selector named n, can then be represented by listing the set of names (program variable names and/or heap-shared node name) corresponding to the first node in the list segment, followed by r copies of $.n$ (selector name), followed by M, $\top$, or $\bot$, depending on whether the last node is heap-shared with name M, or the selector of the last element is uninitialized or points to **nil**, respectively. A heap graph can then be represented as a word obtained by concatenating the representation of each uninterrupted list segment, separated by a special symbol, say $|$. For example, if the selector in the heap graph shown in Figure 7.1a is named n, then this graph can be represented by the word $z.nB \,|\, xB.n.nD \,|\, D.n.nD \,|\, w.nD \,|\, y.n\bot$, where we have assumed that names in lower case (e.g., x) are ranked before those in upper case (e.g., B). Note that the order in which the list segments are enumerated is arbitrary. Hence a heap graph may have multiple word representations. Since the number of heap-shared nodes is bounded above by the number of program variables, it is useful to have a statically determined pool of ranked names for heap-shared nodes of a given program. Whenever a new heap-shared node is created (by the action of a program statement), we can assign a name to it from the set of unused names in this pool. Similarly, when a heap-shared node ceases to be heap-shared (by the action of a program statement), we can add its name back to the pool of unused names. While this allows us to work with a bounded number of names for heap-shared nodes, it also points to the need for reclaiming names for reuse. We will soon see details of how special modes of computation are used to reclaim unused names for heap-shared nodes.

In order to represent the state of a heap manipulating program, we need to keep track of some additional information beyond representing the heap graph. Given a program with k variables, let $\Sigma_M = \{M_0, M_1, M_2, \ldots M_k\}$ be a set of $k+1$ rank-ordered names for heap-shared nodes of the program, and let Σ_L be the set of program locations. We follow the approach of Bouajjani et al. [12], and represent the program state as a word $w = |w_1|w_2|w_3|w_4|w_5|$, where w_5 is a word representation of the heap graph as described above, and $|$ does not appear in any of w_1, w_2, w_3 or w_4. Sub-word w_1 contains the current program location ($\in \Sigma_L$) and a flag indicating the current *mode of computation*. This flag takes values from the set $\Sigma_C = \{C_N, C_0, C_1, C_2, \ldots C_k\}$, where k is the number of program variables. A value of C_N for the flag denotes normal mode of computation. A value of C_i $(0 \leq i \leq k)$ for the flag denotes a special mode of computation used to reclaim M_i as an unused name for heap-shared nodes. Sub-word w_2 contains a (possibly empty) rank-ordered sequence of unused names for heap-shared nodes. Sub-words w_3 and w_4 contain (possibly empty) rank-ordered sequences of variable names that are uninitialized and set to **nil**, respectively. Using this convention, every program state can be represented as a finite word over the alphabet $\Sigma = \Sigma_C \cup \Sigma_L \cup \Sigma_M \cup \Sigma_p \cup \{\top, \bot, |, .n\}$, where Σ_p is the set of all program variables, and all selectors have the name n. We also restrict every heap-shared node in a program state to have exactly one name from the set Σ_M.

As an example, suppose Figure 7.1b represents the heap graph when the program in Example 7.1 is at location `L9` during the second iteration of the while loop, and suppose variables `t2` and `t3` are uninitialized. Since there are 5 program variables, $\Sigma_M = \{M_0, M_1, M_2, M_3, M_4, M_5\}$. The state of the program at this point of execution can be represented by the word $\alpha = |\, C_N\, L9 \,|\, M_0\, M_3\, M_4\, M_5 \,|\, t2\, t3 \,|\,|\, hd.nM_1 \,|\, t1\, M_1.nM_2 \,|\, x\, M_2.n.n\bot \,|$. The sub-word $t2\, t3$ of α encodes the fact that both `t2` and `t3` are uninitialized. Moreover, since there are no variables with the **nil** value, we have an empty list between a pair of consecutive separators (i.e., $|\,|$) after $t2\, t3$. Similarly, if Figure 7.1c represents the heap graph when the program is at location `L10` in the second iteration of the while loop, the corresponding program state can be represented by the word $\alpha' = |\, C_N\, L10 \,|\, M_0\, M_1\, M_3\, M_4\, M_5 \,|\, t2\, t3 \,|\,|\, hd.n.nM_2 \,|\, x\, t1\, M_2.n.n\bot|$. Note that M_1 has been reclaimed as an unused name for heap-shared nodes in α', since the node named M_1 in Figure 7.1b is no longer heap-shared in Figure 7.1c.

7.5.2. *Operational semantics as word transducers*

Having seen how program states can be represented as finite words, we now discuss how operational semantics of program statements can be represented as non-deterministic finite state word transducers. Given a program in our language, we assume without loss of generality that each program location is associated with at most one primitive statement, i.e., `AsgnStmt`, `AllocStmt` or `FreeStmt` as described in Table 7.1. Each compound statement, i.e. `CondStmt`, `LoopStmt` or `SeqCompStmt` as described in Table 7.1, is assumed to be split across multiple program locations to ensure that no program location is associated with more than one primitive statement. Specifically, for conditional statements, we assume that the first program location is associated with the partial statement "**if** (`BoolExp`) **then**", while for loop statements, the first program location is assumed to be associated with the partial statement "**while** (`BoolExp`) **do**". Example 7.1 illustrates how a program with multiple compound statements can be written in this manner. Given a program such that no program location is associated with more than one primitive statement, we construct a separate transducer for the statement (or part of it) at each program location. We also construct transducers for reclaiming names of heap-shared nodes without changing the actual heap graph, program location or values of variables. Finally, these individual transducers are non-deterministically combined to give an overall word transducer for the entire program. For notational convenience, we will henceforth refer to both the statement and part of a statement at a given program location as the statement at that location.

Given a word $w = |w_1|w_2|w_3|w_4|w_5|$ representing the current program state, it follows from the discussion in Section 7.5.1 that (i) the sub-word $|w_1|w_2|w_3|w_4|$ is bounded in length, and (ii) w_5 encodes a bounded number of uninterrupted list segments. Furthermore, each list segment encoded in w_5 has a bounded set of names for its first element, and either a name or $\top$ or $\perp$ as its last element. Therefore, the only source of unboundedness in w is the length of sequences of $.n$'s in the list segments represented in w_5. Hence, we will assume that each transducer reads $|w_1|w_2|w_3|w_4|$ and remembers the information in this bounded prefix in its finite memory before reading and processing w_5. Similarly, we will assume that when reading a list segment in w_5, each transducer reads the (bounded) set of names representing the first element of the segment, and remembers this in its finite memory before reading the sequence of $.n$'s. Every transducer is also assumed to have two special "sink" states, denoted q_{mem} and q_{err}, with self looping transitions on all symbols of the alphabet. Of these, q_{mem} is an accepting state, while q_{err} is a non-accepting state. A transducer transitions to q_{mem} on reading an input word if it detects creation of garbage, or de-referencing of an uninitialized or **nil**-valued pointer. Such a transition is also accompanied by insertion of a special sequence of symbols, say $\top\top\top$, in the word representation of the next state. Note that the sequence $\top\top\top$ never appears in a word representation of a valid program state. Subsequently, whenever a transducer sees this special sequence in the word representation of the current state, it transitions to q_{mem} and retains the $\top\top\top$ sequence in the word representation of the next state. This ensures that the $\top\top\top$ sequence, one generated, survives repeated applications of the transducer, and manifests itself in the word representation of the final set of reached states. A transducer transitions to q_{err} if it reads an unexpected input. In addition, it transitions to q_{err} if it made an assumption (or guess) about an input word, but subsequently, on reading more of the input word, the assumption was found to be incorrect.

The ability to make a non-deterministic guess and verify it subsequently is particularly useful in our context. To see why this is so, suppose the current program statement is `t3 := x->n` (statement at location `L5` in Example 7.1) and the current program state is $w = |w_1|w_2|w_3|w_4|w_5|$. The transducer must read w and generate the word representation of the next program state, say $w' = |w'_1|w'_2|w'_3|w'_4|w'_5|$. Recall that w'_3 and w'_4 are required to be rank-ordered sequences of program variables names that are uninitialized and set to **nil**, respectively, in the next program state. In order to determine w'_3 and w'_4, the transducer must determine if program variable `t3` is set to an uninitialized value or to **nil** after execution of `t3 := x->n`. This requires knowledge of whether `x->n` is uninitialized or **nil** in the current program state. This information is encoded in sub-word w_5 that represents the uninterrupted list segments in the current program state. Therefore, the transducer must read w_5 before it can generate either w'_3 or w'_4. In addition, the transducer also needs to read w_5 in order to generate w'_5. This is because the uninterrupted list segments in the next program state (encoded by w'_5) are the same as those in the current program state (encoded by w_5), modulo changes effected by the current program statement. Since w'_5 appears to the right of w'_3 and w'_4 in w', it can be generated only after w'_3 and w'_4 have been generated. However, as seen above, generating w'_3 and w'_4 requires reading the whole of w_5 in general. Therefore, the transducer must remember w_5 as it reads w. Unfortunately, the uninterrupted list segments encoded in w_5 are unbounded in general, and a finite state transducer cannot remember unbounded information . One way to circumvent this problem is to have the transducer non-deterministically guess whether `x->n` is uninitialized or **nil** in w_5, generate w'_3 and w'_4 accordingly, remember this guess in its finite memory, and proceed to reading w_5 and generating w'_5. As w_5 is read, if the transducer detects that its guess was incorrect, it must abort the transduction. This is achieved by transitioning to q_{err}.

Transducers for program statements: We now describe how transducers for statements at different program locations are constructed. Using the same notation as before, let $w = |w_1|w_2|w_3|w_4|w_5|$ be the input word read by the transducer and $w' = |w'_1|w'_2|w'_3|w'_4|w'_5|$ be the output word generated by it. We will see how each of the five components of w' are generated. Let us begin with w'_1. The transducer for the statement at location `Li` expects its input to begin with $|C_N L_i|$, i.e. w_1 should be $C_N L_i$. On seeing any input with a different prefix, the transducer simply transitions to q_{err}. Otherwise, the transducer non-deterministically chooses to enter mode C_j $(0 \leq j \leq k)$ for reclaiming heap-shared node name M_j, or remain in mode C_N. In the former case, the transducer changes C_N to C_j and copies the rest of the input word w unchanged to its output. In the latter case, the transducer retains C_N as the first letter of w'_1. It then determines the next program location that would result after executing the statement at location `Li`. For several statements (e.g., all statements except those at `L2` and `L3` in Example 7.1), the next program location can be statically determined, and the transducer replaces L_i with the corresponding next program location in w'_1. For other statements (e.g. those at L2 and L3 in Example 7.1), the next program location has one of two possible values, depending on the truth value of a Boolean expression in the current state. The truth value of the Boolean expression can, of course, be determined from the word representation of the current state, but *only after* a sufficiently large part of the input word has been read. The transducer therefore non-deterministically replaces L_i by one of the two possible next program locations in w'_1, and remembers the corresponding guessed truth value for the Boolean expression in its finite memory. Subsequently, if this guess is found to be incorrect, the transducer transitions to q_{err}.

Having generated w'_1, the transducer must next determine the set of unused names for heap-shared nodes in the next state, in order to generate w'_2. Recalling the constructs in our language (see Table 7.1) and the fact that each program location is associated with at most one primitive statement, it is easy to see that the number of heap-shared nodes in the heap graph can potentially increase only if the primitive statement associated with the current program location is an assignment of the form `PVar->FName := PVar` or `PVar := PVar->FName`. The execution of any other statement either keeps the number of heap-shared nodes unchanged or reduces it by one. For all these other statements, the transducer keeps the set of unused names for heap-shared nodes unchanged in the next state. Note that this may temporarily give rise to a situation wherein the number of heap-shared nodes has reduced by one, but the set of unused names for heap-shared nodes has not changed. Fortunately, this situation is easily remedied in the construction of the overall transducer, since the transducer for individual program statements is non-deterministically combined with transducers for reclaiming heap-shared node names in the overall transducer. Thus, if heap-shared node name M_l was rendered unused by execution of the statement at location L_i, the overall transducer can non-deterministically choose to transition to mode C_l (for reclaiming unused heap-shared node name M_l) after the transducer corresponding to program location L_i has completed its action. We will discuss further about transducers for reclaiming unused heap-shared node names later in this section.

If the statement at the current program location is of the form `PVar->FName := PVar` or `PVar := PVar->FName`, its execution gives rise to a (not necessarily new) heap-shared node unless the right hand side of the assignment evaluates to **nil** or is uninitialized. The statements at locations L5, L6, L7 and L8 in Example 7.1 are examples of such statements. In such cases, the transducer first guesses whether the right hand side of the assignment is **nil** or uninitialized, and remembers this guess in its finite memory. Accordingly, there are two cases to consider.

- If the right-hand side is guessed to be **nil** or uninitialized, the number of heap-shared nodes cannot increase (but can potentially reduce by 1) as a result of executing the current statement. In this case, the transducer keeps the set of unused names for heap-shared nodes unchanged in the next state. The case where the number of heap-shared nodes actually reduces by 1 but the set of unused names is unchanged is eventually taken care of by non-deterministically combining the current transducer with transducers for reclaiming unused names, as discussed above.

- If the right hand side is guessed to be neither **nil** nor uninitialized, execution of the current statement gives rise to a (not necessarily new) heap-shared node, say *nd*, in the heap graph. The transducer can now make one of two non-deterministic choices.
  - If there is at least one name in Σ_M that is outside the set of unused names in the current state (i.e., not in w_2), the transducer can guess that *nd* was already heap-shared earlier and was named M_l, where M_l is non-deterministically chosen from outside the set of unused names. The transducer then remembers M_l as the guessed name for the heap-shared node resulting from execution of the current statement. It also keeps the set of unused names for heap-shared nodes unchanged in the next state.
  - If the set of unused names in the current state, i.e. w_2, is non-empty, the transducer can guess that *nd* is a newly generated heap-shared node. The transducer then removes the first name, say M_l, from w_2, and remembers this as the name for the heap-shared node resulting from execution of the current statement. The set of unused names in the next state is obtained by removing M_l from the corresponding set in the current state.

Once the set of unused names for heap-shared nodes in the next state is determined, it is straightforward to generate w_2'. In all cases, as more of the input word w is read, if any guess made by the transducer is found to be incorrect, the transducer transitions to q_{err}.

In order to generate w_3' and w_4', the transducer must determine the sets of program variables that are uninitialized and set to **nil**, respectively, in the next program state. These sets are potentially changed when the current statement is of the form `PVar := PExp`, `PVar :=` **nil**, `PVar :=` **new** or **free**`(PVar)`. In all other cases, w_3' and w_4' are the same as w_3 and w_4, respectively. If the current statement is of the form `PVar := PVar`, `PVar :=` **nil** or `PVar :=` **new**, the corresponding updations to the sets of uninitialized and **nil**-valued program variables are straightforward, and w_3' and w_4' can be determined after reading w_3 and w_4. For de-allocation statements of the form **free**`(u)`, the sub-word w_4' (encoding the set of **nil**-valued variables) is the same as w_4. However, to determine w_3', we need to guess the set of variables that point to the same memory location as `u`, and are thereby rendered uninitialized by **free**`(u)`. The generation of w_3' in this case is explained later when we discuss generation of w_5'. Finally, if the current statement is of the form `PVar := PVar->FName`, sub-word w_5 of the input may need to be read and remembered, in general, before w_3' and w_4' can be generated. As discussed earlier, this leads to the problem of storing unbounded information in a finite state transducer. To circumvent this problem, the transducer makes a non-deterministic guess about whether the right hand side of the assignment evaluates to an uninitialized value, **nil** or the address of an allocated memory location. It then remembers this guess in its finite memory and generates w'_3 and w'_4 accordingly. By the time the entire input word w has been read, the transducer can determine whether any of its guesses was incorrect. If so, the transducer transitions to q_{err}.

Generating sub-word w'_5 requires determining how the uninterrupted list segments in the current program state (encoded in w_5) are modified by the current program statement. It is not hard to see that only primitive statements, i.e. assignment, memory allocation and de-allocation statements (`AsgnStmt`, `AllocStmt` and `FreeStmt` in Table 7.1), at the current program location can modify the encoding of the heap graph. For all other statements, the encoding of the heap graph in the next program state is the same as that in the current state; in other words, w'_5 is the same as w_5. Let us now look at what happens when each of `AsgnStmt`, `AllocStmt` and `FreeStmt` is executed.

Suppose the current statement is of the form `PVar := PVar`, as exemplified by `t1 := hd` at location `L1` in Example 7.1. In this case, w'_5 is obtained by removing `t1` from the head of any uninterrupted list segment in which it appears in w_5, and by inserting `t1` in the head of any uninterrupted list segment in which `hd` appears in w_5. Next, consider an assignment of the form `PVar := PVar->FName` or `PVar->FName := PVar`, as exemplified by `x->n := t1` and `t1 := t1->n` at locations `L7` and `L8`, respectively, in Example 7.1. Suppose the transducer has guessed that the right hand side of the assignment is neither **nil** nor uninitialized. Let M_l be the guessed name (either already present or an unused name) for the heap-shared node that results from executing the current statement. Let us also assume that all guesses made by the transducer thus far are correct. In the case of `x->n := t1`, sub-word w'_5 is obtained by inserting M_l in the head of the uninterrupted list segment in which `t1` appears in w_5, and by removing M_l from the head of any other list segment in w_5. In addition, the uninterrupted list segment starting from x is made to have only one element with its selector pointing to M_l in w'_5. Similarly, in the case of `t1 := t1->n`, we remove `t1` from the head of any uninterrupted list segment in which it appears in w_5, and putting both $t1$ and M_l at the head of the uninterrupted list segment in w'_5 that starts from the second element of the list originally pointed to by `t1` in w_5. If the current statement is of the form `PVar :=` **nil**, sub-word w'_5 is obtained by simply removing the variable name corresponding to `PVar` from the head of any uninterrupted list segment in which it appears in w_5. If, however, the current statement is of the form `PVar->FName :=` **nil**, say `u->n :=` **nil**, the uninterrupted list segment starting from `u` is made to have a sequence of only one $.n$ selector pointing to $\perp$ in w'_5. For statements of the form `PVar :=` **new**, as exemplified by `t2 := new` at location `L4` in Example 7.1, $t2$ is removed from the head of any uninterrupted list segment in which it appears in w_5, and a separate uninterrupted list segment, $t2.n\top$, is appended at the end of sub-word w'_5. For statements of the form `free(u)`, the transducer first guesses all program variable names and heap-shared node name that appear together with **u** at the head of an uninterrupted list segment in w_5, and remembers this set in its finite memory. All program variable names in this set are removed from the head of the uninterrupted list segment in w_5, and added to the list of uninitialized variables, i.e. w_3'. All heap-shared node names in the above set are also removed from the head of the uninterrupted list segment in w_5, and all list segments that end with any such heap-shared node name are made to end with $\top$. As before, if the transducer subsequently detects that its guess was incorrect, it transitions to q_{err}.

In all cases, if the word representing the current state indicates that an assignment or de-allocation statement de-references a **nil**-valued or uninitialized pointer, the transducer transitions to the control state q_{mem}. In addition, whenever the word representation of the heap graph is changed, if we are left with a list segment without any program variable name or heap-shared node name as the first element of the segment, we can infer that garbage has been created. In such cases too, the transducer transitions to control state q_{mem}.

Transducers for reclaiming heap-shared node names: A transducer for reclaiming the heap-shared node name M_i $(0 \leq i \leq k)$ expects sub-word w_1 of its input to start with C_i. Otherwise, the transducer transitions to q_{err}. Such a transducer always leaves the program location, and sets of uninitialized and **nil**-valued variable names unchanged in the output word. If M_i is already in the set of unused names for heap-shared nodes, i.e. in w_2, the transducer simply changes C_i to C_N in sub-word w_1' and leaves the rest of its input unchanged. If M_i is not in w_2, the transducer assumes that M_i is an unused heap-shared node name and can be reclaimed. This effectively amounts to making one of the following assumptions: (i) M_i does not appear as the head of any uninterrupted list segment in sub-word w_5, or (ii) M_i appears as the sole name at the head of an uninterrupted list segment in w_5, and there is exactly one uninterrupted list segment in w_5 that has the name of its last node as M_i. The transducer non-deterministically chooses one of these cases and remembers its choice in its finite memory. In the first case, the transducer adds M_i to the set of unused names of heap-shared nodes in the next state, i.e. to sub-word w_2', changes the flag C_i to C_N in w_1', and proceeds to replace all occurrences of M_i at the end of uninterrupted list segments in w_5 with $\top$. However, if it encounters an uninterrupted list segment in w_5 that has M_i at its head, the guess made by the transducer was incorrect, and hence it transitions to q_{err}. In the second case, let L_1 be the uninterrupted list segment starting with the sole name M_i in w_5, and let L_2 be the uninterrupted list segment ending with M_i in w_5. The transducer moves one element from the start of L_1 to the end of L_2 in w_5', thereby shortening the list L_1 pointed to by M_i, and lengthening the list L_2 ending with M_i. It also non-deterministically guesses whether the list pointed to by M_i in w_5' has shrunk to length zero, and if so, it adds M_i to the list of unused names of heap-shared nodes in w_2', and replaces C_i by C_N in w_1'. Note that in this case, one application of the transducer may not succeed in reclaiming the name M_i. However, repeated applications of the above transducer indeed reclaims M_i if assumption (ii) mentioned above holds. Of course, if the transducer detects that any of its assumptions/guesses is incorrect, it transitions to control state q_{err}. Additionally, if the list L_1 ends in M_i, we have a garbage cycle and the transducer transitions to control state q_{mem}.

7.5.3. *Computing transitive closures of regular transducers*

Given a heap manipulating program P, let τ represent the overall finite state transducer obtained by non-deterministically combining all the finite state transducers discussed above, i.e. one for the statement (or part of it) at each program location, and one for reclaiming each heap-shared node name in Σ_M. By abuse of notation, we will use τ to also represent the binary relation, R_τ, on words induced by τ, when there is no confusion. Let A_I be a finite state automaton representing a regular set of initial states, say I, of P. The language $\tau^i(I)$ $(i \geq 0)$ represents the set of states reachable from some state in I in i steps, and $\tau^*(I) = \bigcup_{i=0}^{\infty} \tau^i(I)$ represents the set of all states reachable from I. We discuss below approaches to compute $\tau^*(I)$ or over-approximations of it.

It is easy to use a product construction to obtain an automaton representing the set $\tau(I)$. Suppose $A_I = (Q_I, \Sigma_\varepsilon, \Delta_I, Q_{0,I}, F_I)$ and $\tau = (Q_\tau, \Sigma_\varepsilon \times \Sigma_\varepsilon, \Delta_\tau, Q_{0,\tau}, F_\tau)$. To construct an automaton recognizing $\tau(I)$, we first construct a product automaton $A_p = (Q_p, \Sigma_\varepsilon \times \Sigma_\varepsilon, \Delta_p, Q_{0,p}, F_p)$ as follows.

- $Q_p = Q_I \times Q_\tau$
- For every $q_1, q_1' \in Q_I$, $q_2, q_2' \in Q_\tau$ and $\sigma_1, \sigma_2 \in \Sigma_\varepsilon$, $((q_1, q_2), (\sigma_1, \sigma_2), (q_1', q_2')) \in \Delta_p$ iff $(q_1, \sigma_1, q_1') \in \Delta_I$, $(q_2, (\sigma_1, \sigma_2), q_2') \in \Delta_\tau$.
- $Q_{0,p} = Q_{0,I} \times Q_{0,\tau}$
- $F_p = F_I \times F_\tau$

A non-deterministic finite state automaton recognizing $\tau(I)$ is obtained by ignoring the first component of pairs of symbols labeling edges of A_p.

To obtain an automaton recognizing $\tau^2(I) = \tau(\tau(I))$, we can use the same product construction, where an automaton recognizing $\tau(I)$ is first obtained as described above. Alternatively, we can precompute an automaton that induces the binary relation $\tau^2 = \tau \circ \tau$, and then determine $\tau^2(I)$. A non-deterministic finite state automaton that induces τ^2 can be obtained from the automaton that induces τ through a simple product construction. We construct $\tau^2 = (Q_{\tau^2}, \Sigma_\varepsilon \times \Sigma_\varepsilon, \Delta_{\tau^2}, Q_{0,\tau^2}, F_{\tau^2})$, where

- $Q_{\tau^2} = Q_\tau \times Q_\tau$
- For every $q_1, q_2, q_1', q_2' \in Q_\tau$, $\sigma_1, \sigma_2 \in \Sigma_\varepsilon$, $((q_1, q_2), (\sigma_1, \sigma_2), (q_1', q_2')) \in \Delta_{\tau^2}$ iff $\exists \sigma_3 \in \Sigma_\varepsilon.\ (q_1, (\sigma_1, \sigma_3), q_1') \in \Delta_\tau$ and $(q_2, (\sigma_3, \sigma_2), q_2') \in \Delta_\tau$.

- $Q_{0,\tau^2} = Q_{0,\tau} \times Q_{0,\tau}$
- $F_{\tau^2} = F_\tau \times F_\tau$

The above technique can be easily generalized to obtain a non-deterministic finite state automaton inducing τ^i for any given $i > 0$. Once a finite state automaton representation of τ^i is obtained, we can obtain a finite state automaton for $\tau^i(I)$, where I is a regular set of words, through the product construction illustrated above. However, this does not immediately tell us how to compute a finite state automaton representation of $\tau^* = \bigcup_{i=0}^{\infty} T^i$ or of $\tau^*(I)$. If τ is a regular transduction relation, the above constructions show that τ^i and $\tau^i(I)$ is also regular for every $i \geq 0$. However, τ^* and $\tau^*(I)$ may indeed be non-regular, since regular languages are not closed under infinite union. Even if τ^* or $\tau^*(I)$ was regular, a finite state automaton representation of it may not be effectively computable from finite state automata representations of τ and I. A central problem in regular model checking (RMC) concerns computing a regular upper approximation of τ^* or $\tau^*(I)$, for a given regular transduction relation τ and a regular initial set I.

Given finite state automata representing I, τ and a regular set of error states denoted *Bad*, the safety checking problem requires us to determine if $\tau^*(I) \cap Bad = \emptyset$. This can be effectively answered if a finite state automaton representation of $\tau^*(I)$ can be obtained. Depending on τ and I, computing a representation of $\tau^*(I)$ may be significantly simpler than computing a representation of τ^* directly. Several earlier works, e.g. those due to Dams et al. [16], Jonsson et al. [8], Touili [13], Bouajjani et al. [11] and Boigelot et al. [17], have tried to exploit this observation and compute a representation of $\tau^*(I)$ directly. A variety of other techniques have also been developed to compute finite state automata representations of τ^* or $\tau^*(I)$. We outline a few of these below.

Quotienting techniques: In this class of techniques, the product construction outlined above is used to compute finite state automata representations of τ^i for increasing values of i. A suitable equivalence relation $\simeq$ on the states of these automata is then defined based on the history of their creation during the product construction, and the quotient automaton constructed for each i. By defining the equivalence relation appropriately, it is possible to establish equivalence between states of the quotient automata for increasing values of i. Thus, states of different quotient automata can be merged into states of one automaton that over-approximates τ^i for all i. For arbitrary equivalence relations, the language accepted by the resulting automaton is a superset of the language by τ^+. However, it is possible to classify equivalence relations such that certain classes of relations preserve transitive closure under quotienting. In other words, the language accepted by $(\tau/\simeq)^+$ *coincides* with that accepted by τ^+. The reader is referred to the works of Abdulla et al. [9, 10] for details of these special relations. The use of such equivalence relations, whenever possible, provides a promising way of computing $\tau^*(I)$ accurately for special classes of systems.

Abstraction-refinement based techniques: Techniques in this approach can be classified as being either *representation-oriented* or *configuration-oriented.* In representation-oriented abstractions, an equivalence relation $\simeq_r$ of finite index is defined on the set of states of an automaton representation. However, unlike in quotienting techniques, there is no a priori requirement of preservation of transitive closure under quotienting. Therefore, we start with τ (or $\tau(I)$) and compute τ^2 (or $\tau^2(I)$ respectively) as discussed above. The states of the automaton representation of τ^2 (or $\tau^2(I)$) are then quotiented with $\simeq_r$. The language accepted by the resulting automaton is, in general, a superset of that accepted by τ^2 (or $\tau^2(I)$ respectively). The quotiented automaton is then composed with τ to compute an over-approximation of τ^3 (or $\tau^3(I)$ respectively). The states of the resulting automaton are further quotiented with $\simeq_r$, and the process is repeated until a fixed point is reached. Since $\simeq_r$ is an equivalence relation of finite index, convergence of the sequence of automata is guaranteed after a finite number of steps.

In configuration-oriented abstractions, the words (configurations) in the languages $\tau(I)$, $\tau^2(I)$, etc. are abstracted by quotienting them with respect to an equivalence relation $\simeq_c$ of finite index defined on their syntactic structure. Configuration-oriented abstractions are useful for word based state representations in which syntactically different parts of a word represent information of varying importance. For example, in our word based representation of program states, i.e. in $w = |w_1|w_2|w_3|w_4|w_5|$, sub-words w_1, w_2, w_3 and w_4 encode important information pertaining to current program location, **nil**-valued and uninitialized variables, number of heap-shared nodes, etc. Furthermore, these sub-words are bounded and hence represent finite information. Therefore, it may not be desirable to abstract these sub-words in w. On the other hand, long sequences of $.n$'s in the representation of uninterrupted list segments in w_5 are good candidates for abstraction. Bouajjani et al. have proposed and successfully used other interesting configuration-oriented abstractions, like $0 - k$ counter abstractions and closure abstractions, for reasoning about heap manipulating programs [18].

Once we have a regular over-approximation of τ^* or $\tau^*(I)$, we can use it to conservatively check if $\tau^*(I) \cap Bad = \emptyset$. However, since we are working with an over-approximation of $\tau^*(I)$, safety checking may give false alarms. It is therefore necessary to construct a counterexample from an abstract sequence of states from a state in I to a state in Bad, and check if the counterexample is spurious. If the counterexample is not spurious, we have answered the safety checking question negatively. Otherwise, the counterexample can be used to refine the equivalence relation $\simeq_r$ or $\simeq_c$ such that the same counterexample is not generated again by the analysis starting from the refined relation. The reader is referred to [11, 18] for details of refinement techniques for both representation-oriented and configuration-oriented abstractions in RMC.

Extrapolation or widening techniques: In this approach, we compute finite state automata representations of $\tau^i(I)$ for successive values of i, and detect a regular pattern in the sequence. The pattern is then *extrapolated* or *widened* to guess the limit ρ of $\bigcup_{i=0}^{m} \tau^i(I)$ as m approaches ∞. Convergence of the limit can be checked by determining if $I \cup \tau(\rho) \subseteq \rho$. If the check passes, we have computed an over-approximation of $\tau^*(I)$. Otherwise, the guessed limit must be modified to include more states (configurations), and a new regular pattern detected. The reader is referred to the works of Touili [13], Bouajjani et al. [11] and Boigelot et al. [17] for details of various extrapolation techniques. As a particularly appealing example of this technique, Touili [13] showed that if I can be represented as the concatenation of k regular expressions $\rho_1.\rho_2.\cdots\rho_k$, and if $\tau(\rho_1.\rho_2\cdots\rho_k) = \bigcup_{i=1}^{k-1}(\rho_1\cdots\rho_i.\Lambda_i.\rho_{i+1}\cdots\rho_k)$, where the Λ_i are regular expressions, then $\tau^*(I)$ is given by $\rho_1.\Lambda_1^*.\rho_2.\Lambda_2^*\cdots\Lambda_{k-1}^*.\rho_k$.

Regular language inferencing techniques: This approach uses learning techniques, originally developed for inferring regular languages from positive and negative sample sets, to approximate $\tau^*(I)$. The work of Habermehl et al. [14] considers length-preserving transducers, and uses increasingly large complete training sets to infer an automaton representation of $\tau^*(I)$. The increasingly large training sets are obtained by gradually increasing the maximum size i of words in the initial set of states, and by computing the set of all states (words) of size up to i reachable from these initial states. Habermehl et al. use a variant of the Trakhtenbrot-Barzdin algorithm [19] for inferring regular languages for this purpose. Once an approximate automaton for $\tau^*(I)$ has been inferred in this manner, a convergence check similar to the one used for extrapolation techniques can be used to determine if an over-approximation of $\tau^*(I)$ has indeed been reached. It has been shown [14] that if $\tau^*(I)$ is regular, safety checking can always be correctly answered using this technique. Even otherwise, good regular upper approximations of $\tau^*(I)$ can be computed.

Let $UpperApprox(\tau^*(I))$ denote a regular upper approximation of $\tau^*(I)$ obtained using one of the above techniques. We can use $UpperApprox(\tau^*(I))$ to answer interesting questions about the program being analyzed. For example, suppose we wish to determine if execution of the program from an initial state in the regular set I can create garbage, or cause an uninitialized/**nil**-valued pointer to be de-referenced. This can be done by searching for the special sequence $\top\top\top$ in the set of words approximating $\tau^*(I)$. Thus, if $BadMem = \Sigma^*.\{\top\top\top\}.\Sigma^*$, then $UpperApprox(\tau^*(I)) \cap BadMem = \emptyset$ guarantees that no garbage is created, and no uninitialized or **nil**-valued pointer is de-referenced. However, if $UpperApprox(\tau^*(I)) \cap BadMem \neq \emptyset$, we must construct an (abstract) counterexample leading from a state in I to a state in $BadMem$ and check for its spuriousness. If the counterexample is not spurious, we have a concrete way to generate garbage or to de-reference an uninitialized or **nil**-valued pointer starting from a state in I. Otherwise, we must refine or tighten $UpperApprox(\tau^*(I))$ and repeat the analysis.

Consider the program in Example 7.1. By carefully constructing the transducer τ as discussed earlier, and by applying a simple configuration-oriented abstraction technique, we can show that if $I = \{|C_N\, L_1 \,|\, M_0M_1M_2M_3M_4M_5 \,|\, t1\, t2\, t3\,|\cdot$

$|\, hd.n^{+}\bot\, |\} \cup \{|C_N\, L_1\, |\, M_1M_2M_3M_4M_5\, |\, t1\, t2\, t3\, |\, |\, hd.n^{+}.M_0\, |\, xM_0.n^{+}\bot\, |\}$, then $BadMem \cap UpperApprox(\tau^*(I)) = \emptyset$. Thus, regardless of whether the list pointed to by hd contains an element pointed to by x, there are no memory access errors or creation of garbage.

The above discussion assumed that every memory location had a single pointer-valued selector. This was crucial to representing the heap graph as a bounded set of uninterrupted list segments. Recent work by Bouajjani et al. [20] has removed this restriction. Specifically, programs manipulating complex data structures with several pointer-valued selectors and finite-domain non-pointer valued selectors have been analyzed in their work. Regular languages of words no longer suffice to represent the heap graph in such cases. The program state is therefore encoded as a tree backbone annotated with routing expressions [20] to represent arbitrary link structures. The operational semantics of program statements are modeled as tree transducers. Techniques from abstract regular tree model checking are then used to check memory consistency properties and shape invariants. The reader is referred to [20] for a detailed exposition on this topic.

A primary drawback of RMC based approaches for reasoning about heap manipulating programs is the rather indirect way of representing heap graphs and program states as words or extended trees. This, in turn, contributes to the complexity of the transducers. Recently, Abdulla et al. [21] have proposed an alternative technique for symbolic backward reachability analysis of heaps using upward closed sets of heap graphs with respect to a well-quasi ordering on graphs, and using an abstract program semantics that is monotone with respect to this ordering. Their method allows heap graphs to be directly represented as graphs, and the operational semantics is represented directly as relations on graphs. The work presented in [21] considers programs manipulating singly linked lists (like the class of programs we considered), although the general idea can be extended to programs manipulating more complex data structures as well. A detailed exposition on this technique is beyond the scope of the present article. The interested reader is referred to [21] for details.

7.6. An automata based semantics and Hoare-style reasoning

We now present a completely different approach for reasoning about heap manipulating programs. Specifically, we discuss an automata based heap semantics for our programming language, and present a logic for Hoare-style deductive reasoning using this semantics. We show how this technique can be used to check heap related properties of programs, using the program in Example 7.1 as an example. Unlike RMC, the approach outlined in this section has a deductive (theorem-proving) flavour.

There are two predominant paradigms for defining heap semantics of programming languages. In *store based semantics* used by Yorsh et al. [22], Podelski et al. [23], Reps et al. [24], Bouajjani et al. [25], Reynolds [26], Calcagno et al. [27], Distefano et al. [28] and others, the heap is identified as a collection of symbolic memory locations. A *program store* is defined as a mapping from the set of pointer variables and selectors of memory locations to other memory locations. Various formalisms are then used for representing and reasoning about this mapping in a finite way. These include, among others, representation of program stores as logical structures for specialized logics [22, 26] or over formulae that use specially defined heap predicates [23, 24, 28], graph based representations [25], etc. In the alternative *storeless semantics*, originally proposed by Jonkers [29] and subsequently used by Deutsch [30], Bozga [31, 32], Hoare and Jifeng [33] and others, every memory location is identified with the set of paths that lead to the corresponding node in the heap graph. A path in the heap graph is represented by the sequence of edge labels appearing along the path. Thus the heap is identified as a collection of sets of sequences of edge labels, and not as a collection of symbolic memory locations. Different formalisms have been proposed in the literature for representing sets of edge label sequences in a finite way. Regular languages (or finite state automata), and formulae in suitably defined logics have been commonly used for this purpose. Since the focus of this article is on automata based techniques, we discuss below an automata based storeless heap semantics for our programming language. As an aside, we note that reasoning techniques for heap manipulating programs cannot always be partitioned based on whether they use storeless or store based semantics. For example, the work of Rinetzky et al. [34] uses a novel mix of store based and storeless semantics in the framework of TVLA [24] to reason about the effect of procedure calls on data structures in the heap.

7.6.1. *A storeless semantics*

Given a program, let Σ_p and Σ_f denote sets of pointer variables and selectors respectively, as discussed earlier. Let $G_H = (V, E, v_{\mathrm{nil}}, \lambda, \mu)$ be the heap graph at a snapshot of execution of the program. We define an *access path* from a variable $x \in \Sigma_p$ to a node v (possibly v_{nil}) in V as a string $x.\sigma$, where σ is a sequence of selector names appearing as edge labels along a path from $\mu(x)$ to v, if such a path exists in G_H. If no such path exists in G_H, the access path from x to v is undefined.

Let Σ denote $\Sigma_p \cup \Sigma_f$, and $\wp(S)$ denote the powerset of a set S. Adapting the definition of Bozga et al. [31], we define a *storeless structure* Υ as a pair $(S_{\mathrm{nil}}, \Gamma)$, where $S_{\mathrm{nil}} \subseteq \Sigma_p{\cdot}\Sigma_f^*$ and $\Gamma \subseteq \wp(\Sigma_p{\cdot}\Sigma_f^*)$. Furthermore, Γ is either the empty set or a finite set of languages $\{S_1, S_2, \ldots S_n\}$ satisfying the following conditions for all $i, j \in \{1, \ldots n\}$.

- $C1: S_i \neq \emptyset$.
- $C2: i \neq j \Rightarrow S_i \cap S_j = \emptyset$. In addition, $S_i \cap S_{\mathrm{nil}} = \emptyset$.
- $C3: \forall \sigma \in S_i\ (\forall \tau, \theta \in \Sigma^+\ (\sigma = \tau \cdot \theta \Rightarrow \exists k\ ((1 \leq k \leq n) \wedge (\tau \in S_k) \wedge (S_k{\cdot}\{\theta\} \subseteq S_i))))$. A similar property holds for all $\sigma \in S_{\mathrm{nil}}$ as well.

Unlike languages in Γ, there is no non-emptiness requirement on S_{nil}. A storeless structure $\Upsilon = (S_{\text{nil}}, \Gamma)$ with $\Gamma = \{S_1, \ldots S_n\}$ represents n distinct memory locations in the heap and also the **nil** value. Recall that in a heap graph, the **nil** value is represented by a special node v_{nil} with no outgoing edges. Language $S_i \in \Gamma$ may be viewed as the set of access paths in G_H to the ith node (distinct from v_{nil}). Similarly, S_{nil} may be viewed as the set of access paths to v_{nil}. Condition $C1$ requires all nodes other than v_{nil} represented in Υ to have at least one access path. Consequently, *garbage* cannot be represented using this formalism, and we will ignore garbage in the current discussion. Condition $C2$ encodes the requirement that every access path must lead to at most one node. Condition $C3$ states that every prefix τ of an access path σ must itself be an access path to a node represented in Γ. This is also called the *prefix closure* property. Condition $C3$ further encodes the requirement that if multiple access paths reach a node represented by S_k, extending each of these access paths with the same suffix θ must lead us to the same node (represented by S_i or S_{nil} in condition $C3$). This is also called *right regularity*.

Consider a storeless structure $\Upsilon = (S_{\text{nil}}, \Gamma)$, in which $\Gamma = \{S_1 \ldots S_n\}$ represents a set of n nodes $\{v_1 \ldots v_n\}$ in the heap graph G_H. The structure Υ can be represented by an $n+3$ state deterministic finite-state transition system B_Υ. A natural (but not necessarily the only) way to obtain B_Υ is by considering the sub-graph of G_H consisting of nodes $\{v_{\text{nil}}, v_1, \ldots v_n\}$. Specifically, we define a transition system $B_\Upsilon = (Q, \Sigma, q_{init}, \delta)$, where $\Sigma = \Sigma_p \cup \Sigma_f$ as defined earlier, and $Q = \{q_{init}, q_{\text{nil}}, q_{\text{err}}, q_1, \ldots q_n\}$, with $q_{\text{init}}, q_{\text{nil}}$ and q_{err} as distinguished control states. For notational convenience, we will refer to v_{nil}, S_{nil} and q_{nil} as v_0, S_0 and q_0 respectively in the following construction. The transition relation of B_Υ is defined as follows: for every i, j in 0 through n, we let $(q_i, f, q_j) \in \delta$ iff the nodes v_i and v_j in $G_H = (V, E, v_{\text{nil}}, \lambda, \mu)$ are such that $(v_i, v_j) \in E$ and $f \in \lambda((v_i, v_j))$. Furthermore, for every i in 0 through n and $x \in \Sigma_p$, we let $(q_{init}, x, q_i) \in \delta$ iff there exists v_i (represented by S_i in Γ) such that $\mu(x) = v_i$. Finally, for all states q (including q_{err}) and for all $f \in \Sigma_p \cup \Sigma_f$, we let $(q, f, q_{\text{err}}) \in \delta$ iff the above construction does not create any outgoing edge from q labeled f. Right regularity and prefix closure of Υ ensure that for all i in 0 through n, the automaton obtained by letting q_i be the sole accepting state in B_Υ accepts language S_i. The automaton $\mathcal{A}_{\text{err}}$ obtained by letting q_{err} be the sole accepting state accepts all sequences that are *not* valid access paths to nodes represented by Υ.

We now present operational semantics of statements in our programming language with respect to the above storeless representation of the heap. For notational convenience, we will use the following convention:

- The representations of the heap before and after executing a statement are $\Upsilon = (S_{\text{nil}}, \Gamma)$ and $\Upsilon' = (S'_{\text{nil}}, \Gamma')$, respectively.
- If θ denotes a `PExp`, then $\boldsymbol{\Theta}$ denotes the singleton regular language consisting of the access path corresponding to θ. For example, if θ is `u` or `u->n`, then $\boldsymbol{\Theta}$ is $\{u\}$ or $\{u.n\}$, respectively. We will also use $\mathbf{u}$ to denote $\{u\}$ and $\mathbf{u} \cdot \mathbf{n}$ to denote $\{u.n\}$.

- For $L, M \subseteq \Sigma^+$, $L \ominus M$ denotes $L \setminus (M \cdot \Sigma^*)$, i.e. the set of words in L that do not have any prefix in M.
- For $L, X, \subseteq \Sigma^+$, $M \subseteq \Sigma^*$ and $\mathtt{n} \in \Sigma_f$, $\chi^{L,\mathtt{n},M}(X)$ denotes the function $\lambda X.\, X \cup (L{\cdot}\mathbf{n}{\cdot}((M^{-1}L){\cdot}\mathbf{n})^*{\cdot}(M^{-1}X))$. If L, M and X represent sets of access paths to nodes v_L, v_M and v_X respectively in the heap graph, then $\chi^{L,\mathtt{n},M}(X)$ is the augmented set of access paths to v_X after making the n-selector of v_L point to v_M [31]. Note that if $M = \emptyset$, then $\chi^{L,\mathtt{n},M}(X) = X$ for all X.
- If θ denotes a PExp and $\mathcal{S} = \{S_1, S_2, \ldots S_r\}$ denotes a set of mutually disjoint languages, then *FindSet*$(\boldsymbol{\Theta}, \mathcal{S})$ is defined to be $S_i \in \mathcal{S}$ if $\boldsymbol{\Theta} \cap S_i \neq \emptyset$. If $\boldsymbol{\Theta} \cap S_i = \emptyset$ for all $S_i \in \mathcal{S}$, then *FindSet*$(\boldsymbol{\Theta}, \mathcal{S})$ is defined to be $\emptyset$.

With this convention, the storeless operational semantics of primitive statements in our language, i.e., assignment, memory allocation and memory de-allocation statements, is given in Table 7.3. To keep the discussion simple, we have assumed that statements of the form `u := u` or `u := u->n` are not present in programs that we wish to analyze. While this may sound restrictive, every program containing such statements can be translated to a semantically equivalent program without any such statement, as described in Section 7.3.

Table 7.3. Storeless operational semantics of primitive statements.

Notation: θ denotes a PExp, u and v are program variables, n is a selector name

Statement	S'_{nil}	Γ'	Conditions
θ := **nil**	$(S_{\text{nil}} \ominus \boldsymbol{\Theta}) \cup \{\boldsymbol{\Theta}\}$	$\{S_i \ominus \boldsymbol{\Theta} \mid S_i \in \Gamma\} \setminus \{\emptyset\}$	if θ is u->n, $\exists X \in \Gamma,\ u \in X$
u := θ	$S''_{\text{nil}} \cup \mathbf{u} \cdot (\boldsymbol{\Theta}^{-1} S''_{\text{nil}})$, where $S''_{\text{nil}} = S_{\text{nil}} \ominus \mathbf{u}$	$\{S''_j \cup \mathbf{u} \cdot (\boldsymbol{\Theta}^{-1} S''_j) \mid S''_j \in \Gamma''\}$, where $\Gamma'' = \{S_i \ominus \mathbf{u} \mid S_i \in \Gamma\} \setminus \{\emptyset\}$	if θ is v->n, $\exists X \in \Gamma,\ v \in X$
u->n := v	$\chi^{\mathbf{u},\mathtt{n},Y}(S_{\text{nil}} \ominus (\mathbf{u} \cdot \mathbf{n}))$	$\{\chi^{\mathbf{u},\mathtt{n},Y}(S''_j) \mid S''_j \in \Gamma''\}$, where $\Gamma'' = \{S_i \ominus (\mathbf{u} \cdot \mathbf{n}) \mid S_i \in \Gamma\} \setminus \{\emptyset\}$	$Y = \mathit{FindSet}(\mathbf{v}, \{S_{\text{nil}}\} \cup \Gamma)$ $\exists X \in \Gamma,\ u \in X$
u := **new**	$S_{\text{nil}} \ominus \mathbf{u}$	$(\{S_i \ominus \mathbf{u} \mid S_i \in \Gamma\} \cup \{\mathbf{u}\}) \setminus \{\emptyset\}$	
free(u)	$S_{\text{nil}} \ominus X$	$\{S_i \ominus X \mid S_i \in \Gamma\} \setminus \{\emptyset\}$	$\exists X \in \Gamma,\ u \in X$

The last column in Table 7.3 lists necessary and sufficient conditions for the storeless operational semantics to be defined. If these conditions are violated, we say that the operational semantics is undefined. Whenever an assignment is made to **u** (or **u->n**), the membership in S_{nil} or $S_i \in \Gamma$ of all access paths that have u (or $u.n$ respectively) as prefix is invalidated. Therefore, these paths must be removed from all languages in Υ before augmenting the languages with new paths formed as a consequence of the assignment. Similarly, when memory is de-allocated, all paths with a prefix that was an access path to the de-allocated node must be removed from all languages. A formal proof of correctness of the operational semantics involves establishing the following facts for every primitive statement `Stmt`.

(1) If $\Upsilon = (S_{\text{nil}}, \Gamma)$ is a storeless structure (i.e., satisfies all conditions in the definition of storeless structures), then so is $\Upsilon' = (S'_{\text{nil}}, \Gamma')$.
(2) Let $v_i \neq v_{\text{nil}}$ be a node represented by Υ.
   (a) If v_i is neither de-allocated nor rendered garbage by executing `Stmt`, there exists a language in Γ' that contains all and only access paths to v_i after executing `Stmt`.
   (b) If executing `Stmt` de-allocates v_i and if π was an access path to v_i prior to executing `Stmt`, there is no access path with prefix π in any language in Υ'.
(3) S'_{nil} contains all and only access paths to v_{nil} after executing `Stmt`.
(4) If executing `Stmt` allocates a node v'_i, there exists a language in Γ' that contains all and only access paths to v'_i after executing `Stmt`.

We leave the details of the proof as an exercise for the reader.

It is clear from the expressions for S'_{nil} and Γ' in Table 7.3 that if we are given finite state automata representations of S_{nil} and $S_i \in \Gamma$ for $i \in \{1, \ldots n\}$, then finite-state automata representations of S'_{nil} and also of every language in Γ' can be obtained by automata theoretic constructions. Specifically, given a deterministic finite-state transition system B_Υ representing Υ, it is possible to construct a deterministic finite-state transition system B'_Υ representing Υ'.

7.6.2. *A logic for Hoare-style reasoning*

In order to reason about programs using the storeless semantics described above, we choose to use Hoare-style reasoning. The literature contains a rich collection of logics for Hoare-style reasoning using both storeless and store based representations of the heap. Notable among them are separation logic and its variants [26, 28, 35–37], logic of bunched implications [38], logic of reachable patterns (LRP) [22], several transitive closure logics [39], pointer assertion logic (PAL) [40] based on graph types [41], weak alias logic (wAL) [32], L_r [42] and other assertion logics based on monadic second order logic [43]. While separation logic and its variants have arguably received the most attention in recent times, this development has primarily revolved around store based semantics. Since our focus is on automata based storeless semantics, we present below a simplified version of Bozga et al.'s weak alias logic or wAL [32]. We call this *Simplified Alias Logic* or SAL. Both wAL and SAL use storeless representations of the heap as structures for evaluating formulae. SAL however has fewer syntactic constructs than wAL. Implication checking in both logics is undecidable [32]. Nevertheless, decidable fragments with restricted expressiveness can be identified, and practically useful sound (but incomplete) inference systems can be defined. Other logics proposed for storeless representations of the heap are PAL [40], L_r [42] and an assertion logic due to Jensen et al. [43]. Unlike SAL, implication checking in these logics is decidable. While this represents a significant difference and can be very useful for analyzing certain classes of programs, these logics are less expressive than SAL. Our choice of SAL for the current discussion is motivated by the need to express complex heap properties in a logic that is closed under the weakest pre-condition operator. Implication checking is addressed separately either by restricting the logic or by using sound (but incomplete) inference systems.

The logic* SAL*: The syntax and semantics of SAL (adapted from Bozga et al.'s wAL [32]) are given in Tables 7.4a and b. Constants in this logic, shown in bold-

Table 7.4. Syntax and semantics of Simplified Alias Logic.

(a) Syntax of SAL

(Variables & constants)	VC	::=	$X_i, i \in \mathbb{N} \mid \mathbf{c}_{\text{nil}} \mid \mathbf{u}$, for all $\mathrm{u} \in \Sigma_p$
(Selector name sequences)	F	::=	$\mathbf{f} \mid \mathrm{F}.\mathbf{f} \mid (\mathrm{F} + \mathrm{F}) \mid \mathrm{F}^*$, for all $\mathbf{f} \in \Sigma_f$
(Terms)	T	::=	$\mathrm{VC} \mid \mathrm{T} \cdot \mathrm{F} \mid \mathrm{T} \cdot (\mathrm{T}^{-1}\ \mathrm{T}) \mid \mathrm{T} \cup \mathrm{T} \mid \mathrm{T} \cap \mathrm{T} \mid \mathrm{T} \ominus \mathrm{T}$
(Formulae)	φ	::=	$\mathrm{T} = \mathrm{T} \mid \varphi \wedge \varphi \mid \neg\varphi \mid \exists X_i\ \varphi$

(b) Semantics of SAL

Notation:
$V(\varphi)$: Free variables in φ, $\Upsilon = (S_{\text{nil}}, \Gamma)$: A storeless structure, $\nu : V(\varphi) \to \{S_{\text{nil}}\} \cup \Gamma$

$$
\begin{array}{lcl}
[\![X_i]\!]_\nu & = & \nu(X_i),\ i \in \mathbb{N} \\
[\![\mathbf{c}_{\text{nil}}]\!]_\nu & = & S_{\text{nil}} \\
[\![\mathbf{u}]\!]_\nu & = & \{u\},\ \mathrm{u} \in \Sigma_p \\
[\![F.\mathbf{f}]\!]_\nu & = & [\![F]\!]_\nu \cdot \{f\},\ \mathtt{f} \in \Sigma_f \\
[\![(F_1 + F_2)]\!]_\nu & = & [\![F_1]\!]_\nu \cup [\![F_2]\!]_\nu
\end{array}
\qquad
\begin{array}{lcl}
[\![F^*]\!]_\nu & = & ([\![F]\!]_\nu)^* \\
[\![T.F]\!]_\nu & = & [\![T]\!]_\nu \cdot [\![F]\!]_\nu \\
[\![T_1.(T_2^{-1}T_3)]\!]_\nu & = & [\![T_1]\!]_\nu \cdot ([\![T_2]\!]_\nu^{-1} [\![T_3]\!]_\nu) \\
[\![T_1 \cup T_2]\!]_\nu & = & [\![T_1]\!]_\nu \cup [\![T_2]\!]_\nu \\
[\![T_1 \cap T_2]\!]_\nu & = & [\![T_1]\!]_\nu \cap [\![T_2]\!]_\nu \\
[\![T_1 \ominus T_2]\!]_\nu & = & [\![T_1]\!]_\nu \setminus ([\![T_2]\!]_\nu \cdot \Sigma^*),
\end{array}
$$

$$
\begin{array}{lcl}
\nu \models T_1 = T_2 & \text{iff} & [\![T_1]\!]_\nu = [\![T_2]\!]_\nu \\
\nu \models \varphi_1 \wedge \varphi_2 & \text{iff} & \nu \models \varphi_1 \text{ and } \nu \models \varphi_2 \\
\nu \models \neg\varphi & \text{iff} & \nu \not\models \varphi \\
\nu \models \exists X_i\ \varphi & \text{iff} & \exists S \in \{S_{\text{nil}}\} \cup \Gamma,\ \nu \models \varphi[S/X_i]
\end{array}
$$

We say that $\Upsilon \models \varphi$ iff there exists $\nu : V(\varphi) \to \{S_{\text{nil}}\} \cup \Gamma$ such that $\nu \models \varphi$

face, are either $\mathbf{c}_{\text{nil}}$ (denoting the language S_{nil}) or singleton languages consisting of access paths corresponding to pointer variables in the program. Variables are denoted by X_i where i is a natural number. Each variable takes values from the set of regular languages in a storeless structure. Note that this differs from Bozga et al.'s wAL, where free variables can be assigned arbitrary languages in Σ^+ that are neither restricted to be regular, nor required to coincide with one of the languages in the storeless structure over which the formula is evaluated. Terms are formed by applying regular expression operators to variables, constants and sub-terms. Terms denote (possibly empty) subsets of $\Sigma_p \cdot \Sigma_f^*$. Formulae are constructed by applying first-order operators to sub-formulae, where an atomic formula checks language equivalence of two terms ($T_1 = T_2$). We will use the usual shorthand notations $\forall X_i\ \varphi$ and $\varphi_1 \vee \varphi_2$ for $\neg(\exists X_i\ \neg\varphi)$ and $\neg(\neg\varphi_1 \wedge \neg\varphi_2)$ respectively, whenever necessary. Given a storeless structure $\Upsilon = (S_{\text{nil}}, \Gamma)$ and an assignment ν that assigns a language from $\{S_{\text{nil}}\} \cup \Gamma$ to each free variable of φ, we use $[\![T]\!]_\nu$ to denote the regular language obtained by replacing every occurrence of every free variable X in term T with $\nu(X)$. We say that Υ is a model of φ and ν is a satisfying assignment for φ in Υ iff $\nu \models \varphi$, as defined in Table 7.4b. If $\nu \models \varphi$, we will also say that the assignment ν renders φ true.

It can be seen that SAL is expressive enough to describe complex properties of heaps, including some properties of recursive data structures. We list below a few examples to demonstrate the expressiveness of the logic. In each example, we first describe a heap property in English, and then present a shorthand along with a detailed formula in SAL that evaluates to true for a storeless structure iff the heap represented by the structure has the given property.

(1) Term T represents a non-empty set of access paths to a node (possibly v_{nil}) in the heap graph: $\mathit{nempty}(T) \equiv \neg(T = (T \ominus T))$. We will use $\mathit{empty}(T)$ to denote $\neg \mathit{nempty}(T)$.
(2) X has an access path without any prefix in $\mathbf{u}{\cdot}\mathbf{f}^*$: $\mathit{nprefix}(X, \mathbf{u}{\cdot}\mathbf{f}^*) \equiv \mathit{nempty}(X \cap (\Sigma_p{\cdot}\Sigma_f^* \ominus \mathbf{u}{\cdot}\mathbf{f}^*))$.
(3) X can be reached from Y using a sequence of $\mathtt{f}$ selectors: $\mathit{rch}(X, Y, \mathbf{f}) \equiv \mathit{nempty}(X \cap Y.\mathbf{f}^*)$. Note that $\mathit{rch}(X, X, \mathbf{f})$ is true for all non-empty X by definition.
(4) X can be reached from Y using exactly one $\mathtt{f}$ selector: $\mathit{edge}(X, Y, \mathbf{f}) \equiv \mathit{nempty}(X \cap Y.\mathbf{f})$.
(5) X and Y lie on a cycle in the heap graph formed using only $\mathtt{f}$ selectors: $\mathit{cyc}(X, Y, \mathbf{f}) \equiv \neg(X = Y) \wedge \mathit{rch}(X, Y, \mathbf{f}) \wedge \mathit{rch}(Y, X, \mathbf{f})$.
(6) X lies on a lasso or panhandle formed using only $\mathtt{f}$ selectors: $\mathit{lasso}(X, \mathbf{f}) \equiv \exists Y\ (\mathit{rch}(Y, X, \mathbf{f}) \wedge \neg \mathit{rch}(X, Y, \mathbf{f}) \wedge \exists Z\ (\neg(Z = Y) \wedge \mathit{cyc}(Y, Z, \mathbf{f})))$.
(7) X is the root of a tree formed using $\mathtt{f}$ selectors: $\mathit{tree}(X, \mathbf{f}) \equiv \forall Y \forall Z \forall V\ ((\mathit{rch}(Y, X, \mathbf{f}) \wedge \mathit{rch}(Z, X, \mathbf{f}) \wedge \mathit{edge}(V, Y, \mathbf{f}) \wedge \mathit{edge}(V, Z, \mathbf{f})) \Rightarrow (Y = Z))$.

Since terms in SAL are formed by applying regular expression operators, SAL formulae cannot be used to express non-regular properties of the heap, such as those involving unbounded counting. For example, we cannot express "X is the root of a balanced binary tree formed using $\mathtt{f}$ selectors" in SAL.

Example 7.2. As an illustration of how SAL can be used to specify properties of programs, let P be the program in Example 7.1 and let $\varphi \equiv \mathit{rch}(\mathbf{c}_{\text{nil}}, \mathbf{hd}, \mathbf{f})$. The Hoare triple $\{\varphi\}\ P\ \{\varphi\}$ asserts that if program P is started in a state in which $\mathtt{hd}$ points to a **nil**-terminated acyclic list, and if P terminates, then $\mathtt{hd}$ always points to a **nil**-terminated acyclic list after termination of P as well.

Weakest pre-condition calculus: In order to prove the validity of Hoare triples like the one in Example 7.2, we must express the operational semantics of statements using Hoare triples with SAL as the base logic. This involves computing *weakest pre-conditions* [44] of formulae in SAL with respect to primitive statements (i.e., assignment, memory allocation and de-allocation statements) in our language. Let $wp(\mathsf{Stmt}, \varphi)$ denote the weakest pre-condition of φ with respect to primitive statement Stmt. It follows from the definition of wp that $wp(\mathsf{Stmt}, \varphi_1 \wedge \varphi_2) = wp(\mathsf{Stmt}, \varphi_1) \wedge wp(\mathsf{Stmt}, \varphi_2)$. Since the transition relation defined by primitive statements in our language is total (i.e. every state leads to at least one next state after executing a statement) and deterministic (i.e. every state leads to at most one next state after executing a statement), it can be further shown that $wp(\mathsf{Stmt}, \neg\varphi) = \neg wp(\mathsf{Stmt}, \varphi)$ and $wp(\mathsf{Stmt}, \exists X\, \varphi) = \exists X\, wp(\mathsf{Stmt}, \varphi)$. Consequently, the weakest pre-condition of an arbitrary SAL formula φ with respect to a primitive statement can be computed by induction on the structure of φ, and we only need to define weakest pre-conditions of atomic formulae of the form $(T_1 = T_2)$.

Let $V(\varphi)$ denote the set of free variables of a formula φ. For brevity of notation, we will use V for $V(\varphi)$ when φ is clear from the context. For every $\Omega \subseteq V$, let $\langle\varphi\rangle_\Omega$ denote the conjunction $\varphi \wedge \bigwedge_{X \in V \setminus \Omega}(X = \mathbf{c}_{\text{nil}}) \wedge \bigwedge_{Y \in \Omega}(\neg(Y = \mathbf{c}_{\text{nil}}) \wedge \mathit{nempty}(Y))$. Since every satisfying assignment of φ in every model Υ $(= (S_{\text{nil}}, \Gamma))$ sets a (possibly empty) subset of the free variables of φ to S_{nil} and the remaining free variables to languages in Γ, it follows that $\varphi \Leftrightarrow \bigvee_{\Omega \subseteq V} \langle\varphi\rangle_\Omega$. Similarly, for every $\Omega_2 \subseteq \Omega_1 \subseteq V$ and for every pointer expression θ representing a valid access path, let $\langle\varphi\rangle_{\Omega_1,\Omega_2,\Theta}$ denote the conjunction $\varphi \wedge \bigwedge_{X \in V \setminus \Omega_1}(X = \mathbf{c}_{\text{nil}}) \wedge \bigwedge_{Y \in \Omega_1 \setminus \Omega_2}(\neg(Y = \mathbf{c}_{\text{nil}}) \wedge (Y = \mathbf{\Theta})) \wedge \bigwedge_{Z \in \Omega_2}(\neg(Z = \mathbf{c}_{\text{nil}}) \wedge \neg(Z = \mathbf{\Theta}) \wedge \mathit{nempty}(Z))$. By reasoning similar to that above, it can also be shown that $\mathit{empty}(\mathbf{\Theta} \cap \mathbf{c}_{\text{nil}}) \Rightarrow (\varphi \Leftrightarrow \bigvee_{\Omega_2 \subseteq \Omega_1 \subseteq V} \langle\varphi\rangle_{\Omega_1,\Omega_2,\Theta})$.

Let α be a primitive statement in a program written in our language, and let $\Upsilon = (S_{\text{nil}}, \Gamma)$ be a storeless structure representing the program state before execution of α. Furthermore, let φ be a SAL formula such that $\Upsilon \models \varphi$. By definition, there exists an assignment of free variables, say $\nu : V(\varphi) \rightarrow \{S_{\text{nil}}\} \cup \Gamma$, such that $\nu \models \varphi$. For $X \in V(\varphi)$, suppose $\nu(X) = S_i \neq S_{\text{nil}}$. Then X represents the set of access paths to a node, say nd_i, distinct from v_{nil}, in the heap graph prior to execution of α. Suppose nd_i is neither de-allocated nor rendered garbage by executing statement α from the program state Υ. One can then ask: Can the set of access paths to nd_i in the heap graph resulting after execution of α be expressed in terms of X? The operational semantics given in Table 7.3 tells us that this can indeed be done. We will use $\widetilde{X}^\alpha$ to denote the term representing the (potentially new) set of access paths to nd_i after execution of α, where X represented the set of access paths to nd_i before execution of α. Similarly, we will use $\widetilde{\mathbf{c}_{\text{nil}}}^\alpha$ to denote the term representing the set of access paths to v_{nil} after execution of α, where $\mathbf{c}_{\text{nil}}$ represented the set of access paths to v_{nil} prior to execution of α. As an example, if α denotes the statement θ := **nil**, the first row of Table 7.3 tells us that $\widetilde{\mathbf{c}_{\text{nil}}}^\alpha = (\mathbf{c}_{\text{nil}} \ominus \mathbf{\Theta}) \cup \{\mathbf{\Theta}\}$ and $\widetilde{X}^\alpha = X \ominus \mathbf{\Theta}$ for variable $X \neq \mathbf{c}_{\text{nil}}$. For notational clarity, we will henceforth use $\widetilde{X}$ instead of $\widetilde{X}^\alpha$ when α is clear from the context. Given a SAL formula φ and a subset Ω of $V(\varphi)$, we will use $\varphi[\Omega \mapsto \mathbf{c}_{\text{nil}}]$ and $\varphi[\Omega \mapsto \mathbf{\Theta}]$ to denote the formulae obtained from φ by substituting every variable X in Ω with $\mathbf{c}_{\text{nil}}$ and $\mathbf{\Theta}$ respectively. Similarly, we will use $\varphi[\Omega \mapsto \widetilde{\Omega}]$ to denote the formula obtained by substituting every variable X in Ω with $\widetilde{X}$ *and* by substituting every occurrence of $\mathbf{c}_{\text{nil}}$ with $\widetilde{\mathbf{c}_{\text{nil}}}$. Extending the notation, if Ω_1 and Ω_2 are subsets of $V(\varphi)$, we will use $\varphi[\Omega_1 \mapsto \mathbf{c}_{\text{nil}}][\Omega_2 \mapsto \widetilde{\Omega_2}]$ to denote $(\varphi[\Omega_1 \mapsto \mathbf{c}_{\text{nil}}])[\Omega_2 \mapsto \widetilde{\Omega_2}]$. The interpretation of $\varphi[\Omega_1 \mapsto \mathbf{c}_{\text{nil}}][\Omega_2 \mapsto \mathbf{\Theta}][\Omega_3 \mapsto \widetilde{\Omega_3}]$ is similar.

The intuition behind the computation of $wp(\alpha, \varphi)$ can now be explained as follows. As before, let $\Upsilon = (S_{\text{nil}}, \Gamma)$ be a storeless structure representing the program state before execution of α. Let $\Upsilon' = (S'_{\text{nil}}, \Gamma')$ be the corresponding storeless structure after execution of α. Suppose $\Upsilon' \models \varphi$ and $\nu' : V(\varphi) \to \{S'_{\text{nil}}\} \cup \Gamma'$ is a satisfying assignment for φ in Υ'. Let $\Omega \subseteq V(\varphi)$ be the subset of free variables that are assigned languages in Γ' (and not S'_{nil}) by ν', i.e. $\nu'(X) \in \Gamma'$ for all $X \in \Omega$ and $\nu'(Y) = S'_{\text{nil}}$ for all $Y \in V \setminus \Omega$. It follows that $\Upsilon' \models \widehat{\varphi}$, where $\widehat{\varphi} \equiv \varphi[V \setminus \Omega \mapsto \mathbf{c}_{\text{nil}}]$. This is because the assignment $\widehat{\nu} : \Omega \to \Gamma'$, given by $\widehat{\nu}(X) = \nu(X)$ for every $X \in \Omega$, causes $\widehat{\varphi}$ to evaluate to the same truth value that φ evaluates to under the assignment ν', i.e. true. If the execution of α does not allocate any new memory location, every node in the heap graph after execution of α was also present in the heap graph before execution of α. From Table 7.3, we also know how the representations of the sets of access paths to these nodes and to v_{nil} in Υ change to their corresponding representations in Υ', as a result of executing α. It therefore follows that $\Upsilon \models \widehat{\varphi}[\Omega \mapsto \widetilde{\Omega}]$. The corresponding satisfying assignment $\nu : \Omega \mapsto \Gamma$ is such that $\nu(X) = S_i \in \Gamma$ iff $\nu'(X) = S'_i \in \Gamma'$, where S_i and S'_i represent sets of access paths to the same node in the heap graph before and after executing α respectively. Now suppose the formula φ is such that *for all* models $\Upsilon' = (S'_{\text{nil}}, \Gamma')$, every satisfying assignment ν' of φ in Υ' sets all free variables in a subset Ω of $V(\varphi)$ to languages in Γ' and all other free variables to S'_{nil}. We will call such formulae *model-constraining with respect to* Ω. It is easy to see that if φ is model-constraining with respect to Ω, then $wp(\alpha, \varphi)$ is essentially given by $\widehat{\varphi}[\Omega \mapsto \widetilde{\Omega}]$. In reality, $wp(\alpha, \varphi) \equiv \zeta \wedge \widehat{\varphi}[\Omega \mapsto \widetilde{\Omega}]$, where ζ is a SAL formula that asserts conditions to ensure that the execution of α doesn't lead to a memory error. In other words, ζ encodes the conditions listed in Table 7.3 for the operational semantics of primitive statements to be defined. Unfortunately, a general SAL formula φ may not be model-constraining with respect to any $\Omega \subseteq V(\varphi)$. In order to circumvent this problem, we express φ in the equivalent form $\bigvee_{\Omega \subseteq V} \langle\varphi\rangle_\Omega$. Note that for every Ω, the formula $\langle\varphi\rangle_\Omega$ is model-constraining with respect to Ω. Since the wp operator distributes over negation and conjunction (and hence, over disjunction) of formulae with respect to primitive statements in our language, it is now easy to see that $wp(\alpha, \varphi) \equiv \zeta \wedge \bigvee_{\Omega \subseteq V} \left(\langle\varphi\rangle_\Omega[V \setminus \Omega \mapsto \mathbf{c}_{\text{nil}}][\Omega \mapsto \widetilde{\Omega}]\right)$.

The above discussion assumed that the statement α does not allocate a new memory location. However, the same intuition can be generalized even when α

Table 7.5. Weakest pre-conditions and Hoare inference rules for SAL.

(a) Computing weakest pre-conditions for atomic formulae

Notation: $\varphi \equiv (T_1 = T_2)$, V = set of free variables of φ
$\psi_u \equiv \exists X \ (nempty(X \cap \mathbf{u}) \wedge empty(X \cap \mathbf{c}_{\text{nil}}))$

Statement (α)	**wp**(α, φ)
θ := **nil**	$\zeta \wedge \bigvee_{\Omega \subseteq V} \left(\langle \varphi \rangle_{\Omega}[V \setminus \Omega \mapsto \mathbf{c}_{\text{nil}}][\Omega \mapsto \widetilde{\Omega}] \right)$, where $\zeta \equiv \psi_u$ if θ is `u->n`, and True otherwise.
`u := ` θ	$\zeta \wedge \bigvee_{\Omega \subseteq V} \left(\langle \varphi \rangle_{\Omega}[V \setminus \Omega \mapsto \mathbf{c}_{\text{nil}}][\Omega \mapsto \widetilde{\Omega}] \right)$, where $\zeta \equiv \psi_v$ if θ is `v->n`, and True otherwise.
`u->n := v`	$\zeta \wedge \bigvee_{\Omega \subseteq V} \left(\langle \varphi \rangle_{\Omega}[V \setminus \Omega \mapsto \mathbf{c}_{\text{nil}}][\Omega \mapsto \widetilde{\Omega}] \right)$, where $\zeta \equiv \psi_u$
`u :=` **new**	$\bigvee_{\Omega_2 \subseteq \Omega_1 \subseteq V} \left(\langle \varphi \rangle_{\Omega_1, \Omega_2, \mathbf{u}}[V \setminus \Omega_1 \mapsto \mathbf{c}_{\text{nil}}][\Omega_1 \setminus \Omega_2 \mapsto \mathbf{u}][\Omega_2 \mapsto \widetilde{\Omega_2}] \right)$
free`(u)`	$\zeta \wedge \bigvee_{\Omega \subseteq V} \left(\langle \varphi \rangle_{\Omega}[V \setminus \Omega \mapsto \mathbf{c}_{\text{nil}}][\Omega \mapsto \widetilde{\Omega}] \right)$, where $\zeta \equiv \psi_u$

(b) Hoare inference rules

Notation: `[B]`: SAL formula corresponding to Boolean expression `B` in programming language
`[u = v]` $\equiv \exists X \ (nempty(X \cap \mathbf{u}) \wedge nempty(X \cap \mathbf{v}))$
`[IsNil(u)]` $\equiv nempty(\mathbf{c}_{\text{nil}} \cap \mathbf{u})$
`[B1 or B2]` $\equiv [B1] \vee [B2]$
`[not B]` $\equiv \neg[B]$

Inference rules:

$$\frac{}{\{wp(\texttt{Stmt},\varphi)\} \ \texttt{Stmt} \ \{\varphi\}}$$ `Stmt` $\in$ `AsgnStmt`, `AllocStmt` or `FreeStmt`

$$\frac{\{\varphi_1\} \ \texttt{Stmt1} \ \{\varphi_2\} \qquad \{\varphi_2\} \ \texttt{Stmt2} \ \{\varphi_3\}}{\{\varphi_1\} \ \texttt{Stmt1;Stmt2} \ \{\varphi_3\}}$$ Sequential composition

$$\frac{\{\varphi_1\} \ \texttt{Stmt} \ \{\varphi_2\} \qquad \varphi_3 \Rightarrow \varphi_1 \qquad \varphi_2 \Rightarrow \varphi_4}{\{\varphi_3\} \ \texttt{Stmt} \ \{\varphi_4\}}$$ Strengthening pre-condition
Weakening post-condition

$$\frac{\{\varphi_1 \wedge [B]\} \ \texttt{Stmt1} \ \{\varphi_2\} \qquad \{\varphi_1 \wedge \neg[B]\} \ \texttt{Stmt2} \ \{\varphi_2\}}{\{\varphi_1\} \ \texttt{if (B) then Stmt1 else Stmt2} \ \{\varphi_2\}}$$ Conditional branch

$$\frac{\varphi_1 \Rightarrow \varphi_L \qquad \{\varphi_L \wedge [B]\} \ \texttt{Stmt} \ \{\varphi_L\} \qquad \varphi_L \wedge \neg[B] \Rightarrow \varphi_2}{\{\varphi_1\} \ \texttt{while (B) do Stmt} \ \{\varphi_2\}}$$ Looping construct

is a memory allocating primitive statement like **u := new**. The only difference in this case is that we need to express a SAL formula φ as $\bigvee_{\Omega_2 \subseteq \Omega_1 \subseteq V} \langle \varphi \rangle_{\Omega_1, \Omega_2, \mathbf{u}}$. Since $empty(\mathbf{u} \cap \mathbf{c}_{\text{nil}})$ necessarily holds after execution of **u := new**, the formula $\bigvee_{\Omega_2 \subseteq \Omega_1 \subseteq V} \langle \varphi \rangle_{\Omega_1, \Omega_2, \mathbf{u}}$ is equivalent to φ in all program states after execution of **u := new**. Table 7.5a lists $wp(\alpha, \varphi)$ for various primitive statements α in our language and for $\varphi \equiv (T_1 = T_2)$. As discussed earlier, this suffices for computing $wp(\alpha, \varphi)$ for all SAL formulae φ. Table 7.5b gives Hoare inference rules for looping, sequential composition and conditional branching constructs in our programming language. This completes the set of Hoare inference rules for our simple language, with SAL as the base logic.

Decidability issues: In order to prove heap-related properties of programs in our language using Hoare-style reasoning, we must formulate the property as a Hoare triple using SAL as the base logic, and then derive the triple by repeated applications of inference rules in Table 7.5b. Since Hoare logic is relatively complete, if the property holds for the program, there exists a way to derive the triple by repeated applications of inference rules, provided we have an algorithm to check implications in SAL. Implication checking is needed in the rule for weakening pre-conditions and strengthening post-conditions, and also in the rule for looping constructs in Table 7.5b. Given formulas φ and ψ in SAL, $\varphi \Rightarrow \psi$ iff $\varphi \wedge \neg\psi$ is unsatisfiable. In other words, for every storeless structure $\Upsilon = (S_{\text{nil}}, \Gamma)$ and for every assignment $\nu : V(\varphi) \cup V(\psi) \to \{S_{\text{nil}}\} \cup \Gamma$, we have $\nu \not\models (\varphi \wedge \neg\psi)$. Clearly, it suffices to have a satisfiability checker for SAL in order to check implications between SAL formulae that arise in our Hoare-style proofs. Unfortunately, satisfiability checking in SAL is undecidable. The proof of undecidability follows a similar line of reasoning as used by Bozga et al. [32] to show the undecidability of wAL.

Various alternative strategies can, however, be adopted to check satisfiability of subclasses of formulae in practice. A simple strategy that is often used is to work with a set of *sound* (but not *complete*) inference rules in an undecidable logic. Thus, if a formula can be shown to be (un)satisfiable using these rules, the formula is indeed (un)satisfiable. However, there is no guarantee that the satisfiability question for all formulae in the logic can be answered using the set of chosen rules. By carefully choosing the set of rules, it is often possible to use an undecidable logic like SAL quite effectively for proving useful properties of several interesting programs. Example 7.3 below shows an example of such rule schema for SAL. A second strategy is to use a decidable fragment of the logic. An example of this is the logic pAL (propositional Alias Logic) [32], a strict subclass of Bozga et al.'s wAL, but for which implication checking is in NP. A decidable fragment similar to pAL can also be defined for SAL, although the ability to express properties of heaps is reduced (e.g., properties like $rch(X, Y, \mathsf{f})$ are not expressible in pAL). Finally, we can define a notion of *bounded semantics*, in which we only consider storeless structures with at most k languages other than S_{nil}, for a fixed (possibly large) k, to check for satisfiability. Since every storeless structure with k languages can be represented by a deterministic finite-state transition system with $k+3$ states, and since there are finitely many distinct transition systems with $k+3$ states, it follows that satisfiability checking in SAL with bounded semantics is decidable. Note, however, that if a SAL formula φ is found to be unsatisfiable using $k-$bounded semantics, it does not mean that the formula is unsatisfiable. Therefore, if a property is proved by applying Hoare inference rules and by using k-bounded semantics for SAL, then the program satisfies the property as long as the heap contains k or fewer distinct non-garbage memory locations. If, however, the heap grows to contain more than k distinct non-garbage memory locations, a property proved using k-bounded semantics is not guaranteed to hold.

Example 7.3. The Hoare triple in Example 7.2 can be proved using the rules in Table 7.5b, along with the inference rule schema for SAL in Table 7.6. The loop invariant used at location L2 of the program in Example 7.1 is $\varphi_{L2} \equiv ((\mathbf{t1} = \mathbf{c}_{\mathrm{nil}}) \wedge rch(c_{\mathrm{nil}}, \mathbf{hd}, \mathsf{n})) \vee (\neg(\mathbf{t1} = \mathbf{c}_{\mathrm{nil}}) \wedge (rch(\mathbf{t1}, \mathbf{hd}, \mathsf{n}) \wedge rch(\mathbf{c}_{\mathrm{nil}}, \mathbf{t1}, \mathsf{n})))$.

Table 7.6. Sound inference rule schema for SAL.

Rule	Conditions
$\dfrac{nempty(X \cap Y \cdot F_1) \quad nempty(Z \cap X \cdot F_2)}{nempty(Z \cap Y \cdot F_1 \cdot F_2)}$	X: variable or constant, Y, Z: terms F_1, F_2: regular expressions on selector names
$\dfrac{nempty(X)}{nempty(X \cap X \cdot F^*)}$	X: term, F: regular expression on selector names

7.7. A counter automaton based technique

We have seen above two different automata based techniques for reasoning about heap manipulating programs: regular model checking, and Hoare-style reasoning using a logic called SAL that uses automata based storeless representations of the heap as logical structures. In this section, we describe a third technique based on counter automata.

As in Section 7.5, we restrict our attention to a class of heap manipulating programs in which each memory location has a single pointer-valued selector. We have seen earlier that if we ignore garbage, the heap graph of such a program with n variables consists of at most $2n$ uninterrupted list segments. This motivates abstracting such a heap graph by mapping each (unbounded) sequence of selector names in an uninterrupted list segment to an abstract sequence of some fixed size. Unfortunately, such an abstraction does not permit remembering the exact count of nodes that actually existed in the list segment in the original heap graph. An interesting solution to this problem is to associate a counter with every such sequence in the heap graph, and to use the counter to store the count of nodes in the sequence. Bouajjani et al. [25] have used this idea to define a *counter automaton* abstraction of the state transition behaviour of heap manipulating programs. More recently, Abdulla et al. have proposed a technique using graph minors that achieves a similar abstraction [21].

Let $X = \{x_1, \ldots x_n\}$ be a set of counter variables, and let Φ be the set of Presburger logic formulae with free variables in $\{x_i, x_i' \mid x_i \in X\}$. A *counter automaton* with the set X of counter variables is a tuple $\mathcal{A}_c = (Q, X, \Delta)$, where Q is a finite set of control states, and $\Delta \subseteq Q \times \Phi \times Q$ represents the transition relation. A configuration of the counter automaton is a tuple (q, β), where $\beta : X \to \mathbb{N}$

assigns a natural number to each counter variable. The automaton is said to have a transition from (q, β) to (q', β') iff $(q, \varphi, q') \in \Delta$ for some Presburger formula $\varphi \in \Phi$ and the following conditions hold: (i) $\beta'(x_i) = \beta(x_i)$ for every x'_i that is not free in φ, and (ii) φ evaluates to true on substituting $\beta(x_i)$ for all free variables x_i and $\beta'(x_i)$ for all free variables x'_i of φ. A run of $\mathcal{A}_c$ is a sequence of configurations $(q_0, \beta_0), (q_1, \beta_1), \ldots$ such that $\mathcal{A}_c$ has a transition from (q_i, β_i) to (q_{i+1}, β_{i+1}) for every $i \geq 0$.

In order to construct a counter automaton abstraction of the state transition behaviour of a heap manipulating program, we first build a *structural abstraction* of the heap graph. This is done by first defining an *abstract structure* and then establishing a mapping from nodes in the heap graph to nodes in the abstract structure, such that certain technical conditions are met [25]. These conditions ensure that two distinct nodes in the heap graph are mapped to the same node in the abstract structure only if they are not interruptions (see Section 7.5.1 for a definition of "interruptions") and belong to the same uninterrupted list segment. Intuitively, two nodes in the heap graph are mapped to the same node in the abstract structure if they are "internal" to the same uninterrupted list segment. For the class of programs under consideration, this abstraction is similar to the canonical abstraction of Reps et al. [24], in which two nodes "internal" to the same list segment are abstracted into the same *summary* node. We also associate a counter variable with each node in the abstract structure to keep track of the actual count of nodes in the heap graph that have been mapped to it. Furthermore, the abstract structure is constructed in such a way that for every sequence of two abstract nodes connected by an edge, one of the nodes is necessarily pointed to by a program variable, or has an in-degree exceeding 1. Given this condition, it can be shown [25] that the number of different abstract structures representing uninterrupted list segments in the heap graph of a program with a finite number of variables is always finite.

A counter automaton abstraction of the state transition graph of a heap manipulating program is obtained by letting the control states of the automaton be (program location, structural abstraction of heap graph) pairs. Thus, each control state is an abstraction of the program state. Counters associated with nodes in the structural abstraction become counters associated with the control state. Transitions of the counter automaton are guarded by Presburger logic formulae that encode the operational semantics of various primitive program statements. Bouajjani et al. have shown [25] how such Presburger logic formulae can be calculated for assignment, memory allocation and memory de-allocation statements. A transition of the counter automaton corresponds to the execution of a program statement. In general, this can lead to both a change in the counter values as well as change in the shape represented by the abstract structure. The change in counter values allows us to track the lengths of different uninterrupted list segments precisely. Note that a counter automaton abstraction effectively maps a set of memory locations in the heap to a node in the abstract structure. The identity of a memory location is therefore the name of the node in the abstract structure to which it is mapped, and not the set of access paths to this node. In this sense, a counter automaton abstraction uses a store based semantics.

For data-insensitive programs manipulating the heap, it can be shown that a counter automaton abstraction is bisimilar to the state transition graph of the original program. Hence this abstraction preserves all temporal properties of data-insensitive programs. It has been shown by Bouajjani et al. that the counter automaton abstraction also has additional properties that can be used to answer questions about the original program. Although the location reachability problem is undecidable in general for counter automata, these additional properties can be used to prove properties of special classes of programs. The reader is referred to [25] for a detailed exposition on this topic.

7.8. Conclusion

Analysis and formal verification of computer programs is a challenging task, especially for programs that manipulate unbounded structures in the heap. Automata theory provides a rich set of tools and techniques for reasoning about unbounded objects like words, trees, graphs etc. It is therefore not surprising that automata based techniques have attracted the attention of researchers in program analysis and verification. In this article, we surveyed three interesting techniques based on automata and logic for reasoning about programs manipulating the heap. This article is intended to provide an introductory perspective on the use of automata theoretic techniques for analyzing heap manipulating programs. The serious reader is strongly encouraged to refer to the bibliography for further readings.

Acknowledgments

The author thanks the editors of the current volume for their invitation to write this article. The author also thanks the anonymous reviewers and Bhargav Gulavani for their critical comments.

References

[1] J. E. Hopcroft and J. D. Ullman, *Introduction to Automata Theory, Languages and Computation.* (Addison-Wesley, 1979).

[2] W. Landi, Undecidability of static analysis, *Let. Programming Languages and Systems.* **1**(4), 323–337, (1992).

[3] G. Ramalingam, The undecidability of aliasing, *Trans. Programming Languages and Systems.* **16**(5), 1467–1471, (1994).

[4] E. M. Clarke, O. Grumberg, and D. Peled, *Model Checking.* (MIT Press, 2000).

[5] Y. Kesten, O. Maler, M. Marcus, A. Pnueli, and E. Shahar, Symbolic model checking with rich assertional languages, *Theoretical Computer Science.* **256**(1–2), 93–112, (2001).

[6] L. Fribourg. Reachability sets of parametrized rings as regular languages. In *Proc. INFINITY*. Elsevier, (1997).

[7] P. A. Abdulla, B. Jonsson, M. Nilsson, and M. Saksena. A survey of regular model checking. In *Proc. 15th CONCUR*, vol. 3170, pp. 35–48. Springer, (2004).

[8] B. Jonsson and M. Nilsson. Transitive closures of regular relations for verifying infinite-state systems. In *Proc. 6th TACAS*, vol. 1785, *LNCS*, pp. 220–234. Springer, (2000).

[9] P. Abdulla, B. Jonsson, M. Nilsson, and J. d'Orso. Regular model checking made simple and efficient. In *Proc. 13th CONCUR*, vol. 2421, *LNCS*, pp. 116–130. Springer, (2002).

[10] P. A. Abdulla, B. Jonsson, and M. Nilsson. Algorithmic improvements in regular model checking. In *Proc. 15th CAV*, vol. 2725, *LNCS*, pp. 236–248. Springer, (2003).

[11] A. Bouajjani, P. Habermehl, and V. Tomas. Abstract regular model checking. In *Proc. 16th CAV*, vol. 3114, *LNCS*, pp. 372–386. Springer, (2004).

[12] A. Bouajjani, P. Habermehl, P. Moro, and T. Vojnar. Verifying programs with dynamic 1-selector-linked structures in regular model checking. In *Proc. 11th TACAS*, vol. 3440, *LNCS*, pp. 13–29. Springer, (2005).

[13] T. Touili, Regular model checking using widening techniques, *Electronic Notes in Theoretical Computer Science*. **50**(4), (2001).

[14] P. Habermehl and T. Vojnar, Regular model checking using inference of regular languages, *Electronic Notes in Theoretical Computer Science*. **138**(3), 21–36, (2005).

[15] R. Manevich, E. Yahav, G. Ramalingam, and M. Sagiv. Predicate abstraction and canonical abstraction for singly-linked lists. In *Proc. 6th VMCAI*, vol. 3385, *LNCS*, pp. 181–198. Springer, (2005).

[16] D. Dams, Y. Lakhnech, and M. Steffen. Iterating transducers. In *Proc. 13th CAV*, vol. 2102, *LNCS*, pp. 52–53. Springer, (2001).

[17] B. Boigelot, A. Legay, and P. Wolper. Iterating transducers in the large. In *Proc. 15th CAV*, vol. 2725, *LNCS*, pp. 223–235. Springer, (2003).

[18] A. Bouajjani, P. Habermehl, and A. Rogalewicz. Abstract regular tree model checking of complex dynamic data structures. In *Proc. 13th SAS*, vol. 4134, *LNCS*, pp. 52–70. Springer, (2006).

[19] B. A. Trakhtenbrot and Y. A. Barzdin, *Finite Automata: Behaviour and Synthesis*. (North Holland, 1973).

[20] A. Bouajjani, P. Habermehl, A. Rogalewicz, and T. Vojnar. Abstract tree regular model checking of complex dynamic data structures. In *Proc. 13th SAS*, vol. 4134, *LNCS*, pp. 52–70. Springer, (2006).

[21] P. A. Abdulla, A. Bouajjani, J. Cederberg, F. Haziza, and A. Rezine. Monotonic abstraction for programs with dynamic heap graphs. In *Proc. 20th CAV*, vol. 5123, *LNCS*, pp. 341–354. Springer, (2008).

[22] G. Yorsh, A. Rabinovich, M. Sagiv, A. Meyer, and A. Bouajjani, A logic of reachable patterns in linked data structures, *J. Logic and Algebraic Programming*. **73**(1-2), 111–142, (2007).

[23] A. Podelski and T. Wies. Boolean heaps. In *Proc. 12th SAS*, vol. 3672, *LNCS*, pp. 268–283. Springer, (2005).

[24] M. Sagiv, T. Reps, and R. Wilhelm, Parametric shape analysis via 3-valued logic, *Trans. Programming Languages and Systems*. **24**, 2002, (1999).

[25] A. Bouajjani, M. Bozga, P. Habermehl, R. Iosif, and P. Moro. Programs with lists are counter automata. In *Proc. 18th CAV*, vol. 4144, *LNCS*, pp. 517–531. Springer, (2006).

[26] J. C. Reynolds. Separation logic: A logic for shared mutable data structures. In *Proc. 17th LICS*, pp. 55–74. IEEE, (2002).

[27] C. Calcagno. *Semantic and Logical Properties of Stateful Programming.* PhD thesis, University of Genova, (2002).
[28] D. Distefano, P. W. O'Hearn, and H. Yang. A local shape analysis based on separation logic. In *Proc. 12th TACAS*, vol. 3920, *LNCS*, pp. 287–302. Springer, (2006).
[29] H. B. M. Jonkers. Abstract storage structures. In *Algorithmic Languages*, pp. 321–344. North Holland, (1981).
[30] A. Deutsch. Interprocedural may-alias analysis for pointers: Beyond k-limiting. In *Proc. PLDI*, pp. 230–241. ACM, (1994).
[31] M. Bozga, R. Iosif, and Y. Laknech. Storeless semantics and alias logic. In *Proc. PEPM*, pp. 55–65. ACM, (2003).
[32] M. Bozga, R. Iosif, and Y. Lakhnech. On logics of aliasing. In *Proc. 11th SAS*, vol. 3148, *LNCS*, pp. 344–360. Springer, (2004).
[33] C. A. R. Hoare and H. Jifeng. A trace model for pointers and objects. In *Proc. 13th ECOOP*, vol. 1628, *LNCS*, pp. 1–17. Springer, (1999).
[34] N. Rinetzky, J. Bauer, T. Reps, S. Sagiv, and R. Wilhelm. A semantics for procedure local heaps and its abstractions. In *Proc. 32nd POPL*, pp. 296–309. ACM, (2005).
[35] J. Berdine, C. Calcagno, and P. W. O'Hearn. A decidable fragment of separation logic. In *Proc. 24th FSTTCS*, vol. 3328, *LNCS*, pp. 97–109. Springer, (2004).
[36] C. Calcagno, H. Yang, and P. W. O'Hearn. Computability and complexity results for a spatial assertion language for data structures. In *Proc. 21st FSTTCS*, pp. 108–119. Springer, (2001).
[37] C. Calcagno, P. Gardner, and M. Hague. From separation logic to first-order logic. In *Proc. 8th FoSSaCS*, vol. 3441, *LNCS*, pp. 395–409. Springer, (2005).
[38] S. S. Ishtiaq and P. W. O'Hearn. BI as an assertion language for mutable data structures. In *Proc. 28th POPL*, pp. 14–26. ACM, (2001).
[39] N. Immerman, A. Rabinovich, T. Reps, M. Sagiv, and G. Yorsh. The boundary between decidability and undecidability for transitive-closure logics. In *Proc. 18th CSL*, vol. 3210, *LNCS*, pp. 160–174. Springer, (2004).
[40] A. Möller and M. I. Schwartzbach. The pointer assertion logic engine. In *Proc. PLDI*, pp. 221–231. ACM, (2001).
[41] N. Klarlund and M. I. Schwartzbach. Graph types. In *Proc. 20th POPL*, pp. 196–205. ACM, (1993).
[42] M. Benedikt, T. Reps, and M. Sagiv. A decidable logic for describing linked data structures. In *Proc. 8th ESOP*, vol. 1576, *LNCS*, pp. 2–19. Springer, (1999).
[43] J. L. Jensen, M. E. Jörgensen, N. Klarlund, and M. I. Schwartzbach. Automatic verification of pointer programs using monadic second-order logic. In *Proc. PLDI*, pp. 226–236. ACM, (1997).
[44] E. W. Dijkstra. and C. S. Scholten, *Predicate calculus and program semantics.* (Springer, 1990).

Chapter 8

Chop Expressions and Discrete Duration Calculus

S. Ajesh Babu and Paritosh K. Pandya*
Tata Institute of Fundamental Research,
Homi Bhabha Road, Mumbai 400005, India
ajesh@tifr.res.in
pandya@tifr.res.in

Discrete Duration Calculus is a succinct and expressive logic for specifying quantitative timing properties of discrete timed behaviors. We present a conditional equational theory for proving equality and implications between DDC^* formulae. We also investigate the complexities of several decision problems for DDC^*. We introduce a new variant of extended regular expressions called extended chop expressions with tests ($ECET$) for specifying ϵ-free regular languages. Language preserving reductions between DDC^* and $ECET$ giving only linear blowup in size can be formulated. Moreover, $ECET$ satisfy the axioms of Boolean Algebras as well as Kleene Algebras with Tests; these can now be applied to DDC^*. We investigate the complexities of decision problems such as membership, non-emptiness and non-equivalence for extended chop expressions and its subclasses. The algorithmic complexity of synthesis of NFA from extended chop expressions and its subclasses is also investigated. Finally, we formulate the reductions between extended regular expressions and extended chop expressions. Surprisingly, the extended chop expressions are difficult to reduce to extended regular expressions.

8.1. Introduction

Connections between logics and automata have been very influential in formal verification of programs. Especially for reactive systems, various temporal logics have been defined to formally specify the desired behaviors of the system. In spite of their rich vocabulary, these specification notations can be shown to give rise to a regular set of behavior as characterized by a regular language over the observable states of the system. Techniques for algorithmic synthesis of finite state automata recognizing the language of a temporal logic formula, when feasible, are of great value. Exploiting the algorithmic techniques for representation, construction, closure and analysis of finite state automata, it is possible to construct "model checking tools" for automatic verification of reactive programs.

*This work was partially supported by the project 11P-202 at TIFR titled *Laboratory for Construction, Analysis and Verification of Embedded Systems.*

Several important questions about specification and verification can be answered using the automata synthesized from the desired specification.

- Consistency: Is given specification D satisfiable? Note that D may be a conjunction of a finite set of individual properties which must all hold together, or it may be negation of description of bad behaviors. Satisfiability question reduces to checking if $\mathcal{L}(D) \neq \emptyset$. Reachability of final states of formula automaton $A(D)$ can be used to answer this question. The automaton can also be used to find models and counter-models of the specification.
- Membership: Does a given behavior σ satisfy the specification D? Formally, $\sigma \in \mathcal{L}(D)$?
- Equality and Implication: Are two specifications equivalent? Formally, $\mathcal{L}(D_1) = \mathcal{L}(D_2)$? More generally, $\mathcal{L}(D_1) \subseteq \mathcal{L}(D_2)$?
- Model Checking: Given system model A as an automaton, do all its behaviors satisfy the specification D? Formally, $L(A) \subseteq L(D)$?

Specification may be directly given as an automaton. Alternatively, other well established formalisms such as regular expressions or temporal logics can also be used to represent and analyze specifications. We illustrate this approach by taking the example of a rich specification logic called *Discrete Duration Calculus* (DDC^*).

Discrete Duration Calculus (DDC^*) [1–3] is a succinct and highly expressive logic for specifying quantitative timing properties of discrete timed behaviors. Such specifications occur in modelling of synchronous systems (e.g. clocked circuits, Lustre and Esterel Programs) as well as in discrete time approximations of embedded real-time systems. DDC^* is closely related to the Interval Temporal Logic of Moszkowski [4] and the Duration Calculus of Zhou *et al* [5]. It provides novel interval based modalities for describing behaviors and it incorporates constructs for bounded counting. For example, the following formula holds for a behavior σ provided for all fragments (subwords) σ' of σ which have (a) P true in the beginning, (b) Q true at the end, and (c) no occurrences of Q in between, the number of occurrences of states in σ' where R is true is at most 3.

$$\Box(\lceil P \rceil^0 ; [\![\neg Q]\!] ; \lceil Q \rceil^0 \Rightarrow (\textstyle\int R \leq 3)) .$$

Here, the $\Box$ modality ranges over all fragments (observation intervals) of a behavior. Operator ";" is (concatenation-like) fusion of behavior fragments and $[\![\neg Q]\!]$ states invariance of $\neg Q$ over the behavior fragment. Finally, $\int R$ counts the number of occurrences of R within a behavior fragment. A precise definition of the syntax and semantics of DDC^* is given in Section 8.2.

In spite of their high expressive power, DDC^* formulae can be model checked. An automata theoretic decision procedure allows converting a DDC^* formula into a finite state automaton recognizing precisely the models of the formula [1]. This automaton synthesis has been implemented into a tool called DCVALID [1, 2] which permits membership checking, satisfiability checking and model checking of DDC^* properties of synchronous programs written in Esterel, Lustre, Verilog and SMV notations.

In this chapter, we provide a comprehensive account of the complexities of the decision problems such as membership, satisfiability and Language inclusion for the logic DDC^* and its several fragments. The complexity of formula automaton synthesis is also analyzed. Finally, we give a conditional equational theory for proving equivalence and implication between DDC^* formulae. All these investigations are carried out by reducing DDC^* formulae to a new variant of regular expressions called *extended chop expressions with tests* ($ECET$).

Regular expressions are a well studied syntax for specifying regular languages. Regular expressions are based upon the operation of catenation of languages. The Kleene closure L^* denotes words obtained by finitely many (but unbounded) catenations of elements of L. Together with single letters and set theoretic operations such as union, intersection and complementation, we obtain a highly expressive syntax called *extended regular expressions* for specifying regular languages [6].

We consider an alternative theory obtained by replacing the catenation operation by *chop* which is also termed as *fusion* or coalesced product. If $w_1 = a_1a_2 \dots a_n$ and $w_2 = b_1b_2 \dots b_m$, their fusion (or chop) $w_1; w_2$ is the word $a_1a_2 \dots a_nb_2 \dots b_m$ provided $a_n = b_1$ and is undefined otherwise. Thus, the fusion of two words is defined only when the last letter of w_1 coincides with the first letter of w_2 and in the fused word, the last letter a_n is "merged" with the first letter b_1 of the following word. This can be extended to languages in a pointwise manner to give chop (fusion) of languages $L_1; L_2$. The Kleene closure L^* is replaced by iterated chop operator $L^{\otimes}$. In this manner we obtain a novel variant of regular expressions called *chop expressions.* Chop expressions provide a syntax for specifying ϵ-free regular languages.

There are many situations where chop operation is more convenient than catenation: the operation of sequential composition of programs naturally corresponds to fusion (and not catenation) of the state sequences of the components. Thus, in giving automata based semantics of programming languages, especially synchronous languages such as Esterel, the fusion operation occurs naturally [7]. The chop operation is also found in several prominent temporal and program logics and such as the Interval Temporal Logic [4], Duration calculi [1, 5] and the Process Logic [8]. One of our main results is that we can reduce the DDC^* formula to a language equivalent extended chop expression with tests ($ECET$) with only linear blowup in size. This assumes unary encoding of time constants occurring in the formula. Enigmatically, such a reduction to regular expressions seems difficult. Kozen noted similar difficulty in coding the guarded string model of KAT algebras into regular expressions. A language preserving reduction from extended chop expressions with tests ($ECET$) to DDC^* can also be carried out with only linear blowup in size.

In the chapter, we carry out a systematic study of the extended chop expressions with tests (*ECET*) and their relationships with finite state automata as well as extended regular expressions. Extended chop expressions specify ϵ-free regular languages. They share many of the features of extended regular expressions: finite automata can be synthesized from these and the algorithms for decision problems such as membership and non-emptiness of extended regular expressions can be adapted to the extended chop expressions. At the same time there are intriguing points of dissimilarity. We highlight these by formulating reductions between chop expressions and regular expressions in both directions. While regular expressions can be "represented" as chop expressions with only a linear blowup in size, our reverse reductions result in a doubly exponential blowup in size!

We also study a conditional equational axiom system for establishing equivalence between two extended chop expressions. *ECET* satisfy axioms of boolean algebra as well axioms of the Kleene Algebra with Tests (KAT) of Kozen [9]. Due to homomorphic reductions between DDC^* and *ECET* these axioms can now be applied to DDC^* for conditional equational reasoning.

8.2. Discrete Duration Calculus

Let *Pvar* be a finite set of propositional variables representing some observable aspects of system state. Let $VAL(Pvar) \triangleq Pvar \rightarrow \{0,1\}$ be the set of valuations (states) assigning a truth value to each variable. We shall identify behaviors with finite, nonempty sequences of valuations, i.e. elements of $VAL(Pvar)^+$.

Example 8.1. The following picture gives a behavior over variables $\{p, q\}$. Each column vector gives a valuation, and the word is a sequence of such column vectors.

```
p  1 0 1 1 0
q  0 0 0 0 1
```

The above word satisfies the property that p holds initially and q holds at the end but nowhere before that. DDC^* is a logic for formalizing such properties. Each formula specifies a set of such words.

Given a non-empty finite sequence of valuations $\sigma \in VAL^+$, we denote the satisfaction of a DDC^* formula D over σ by $\sigma \models D$.

Syntax of DDC Formulae Let p range over propositional variables *Pvar*, let P, Q range over propositions and D, D_1, D_2 over DDC^* formulae. Propositions are constructed from variables *Pvar* and constants 0, 1 (denoting *true*, *false* respectively) using boolean connectives $\wedge$, $\neg$ etc. as usual.

The syntax of DDC^* is as follows.

$$\begin{aligned}&\lceil P\rceil^0 \mid \lceil\lceil P\rceil \mid D_1; D_2 \mid D_1 \wedge D_2 \mid \neg D \mid D^{\otimes} \mid \\ &\eta\ op\ c \mid \textstyle\int P\ op\ c \qquad \text{where} \quad op \in \{<, \leq, =, \geq, >\}.\end{aligned}$$

Let $\sigma \in VAL(Pvar)^+$ be a behavior. Thus a behavior is finite non-empty sequence of states from VAL. Let $\#\sigma$ denote the length of σ and $\sigma[i]$ the ith element. For example, if $\sigma = \langle v_0, v_1, v_2 \rangle$ then $\#\sigma = 3$ and $\sigma[1] = v_1$. Let $dom(\sigma) = \{0, 1, \ldots, \#\sigma - 1\}$ denote the set of positions within σ. The set of intervals in σ is given by $Intv(\sigma) = \{[b, e] \in dom(\sigma)^2 \mid b \leq e\}$ where each interval $[b, e]$ identifies the subsequence of σ between the positions b and e. Let $\sigma[i : j]$ denote the subword $\sigma[i] \cdot \sigma[i+1] \ldots \sigma[j]$.

Let $\sigma, i \models P$ denote that proposition P evaluates to true at position i in σ. We omit this obvious definition. We inductively define the satisfaction of DDC^* formula D for behavior σ and interval $[b, e] \in Intv(\sigma)$ as follows.

$$\begin{array}{l}
\sigma, [b,e] \models \lceil P \rceil^0 \;\; \textbf{iff} \;\; b = e \text{ and } \sigma, b \models P \\
\sigma, [b,e] \models \lceil\lceil P \rceil \;\; \textbf{iff} \;\; b < e \text{ and } \sigma, i \models P \text{ for all } i : b \leq i < e \\
\sigma, [b,e] \models \neg D \;\; \textbf{iff} \;\; \sigma, [b,e] \not\models D \\
\sigma, [b,e] \models D_1 \wedge D_2 \;\; \textbf{iff} \;\; \sigma, [b,e] \models D_1 \text{ and } \sigma, [b,e] \models D_2 \\
\sigma, [b,e] \models D_1 ; D_2 \;\; \textbf{iff} \;\; \text{for some } m : b \leq m \leq e : \\
\qquad\qquad\qquad \sigma, [b,m] \models D_1 \text{ and } \sigma, [m,e] \models D_2 \,.
\end{array}$$

The operator $D^{\otimes}$ stands for infinite disjunction $\lceil 1 \rceil^0 \vee D \vee D; D \vee D; D; D \vee \ldots$. Formally, its semantics is as follows.

$$\begin{array}{l}
\sigma, [b,e] \models D^{\otimes} \;\; \textbf{iff} \;\; b = e \;\vee\; \exists k \geq 1.\, \exists m_0 \leq m_2 \leq \ldots \leq m_k. \\
\qquad\qquad m_0 = b \wedge m_k = e \wedge \text{for all } 0 \leq i < k, \;\; \sigma, [m_i, m_{i+1}] \models D
\end{array}$$

Entities η and $\int P$ are called *measurements.* Term η denotes the length of the interval whereas $\int P$ denotes the number of times P is true within the interval $[b, e]$ (we treat the interval as being left-closed and right-open). Formally,

$$\begin{array}{l}
eval(\eta, \sigma, [b,e]) \triangleq e - b \\
eval(\int P, \sigma, [b,e]) \triangleq \sum_{i=b}^{e-1} \left\{ \begin{array}{l} 1 \;\; if \;\; \sigma, i \models P \\ 0 \;\; otherwise \end{array} \right\} .
\end{array}$$

Let t range over measurements. A formula $t\ op\ c$ is called a *measurement formula.* Then, $\sigma, [b, e] \models t\ op\ c$ **iff** $eval(t, \sigma, [b, e])\ op\ c$. Finally, a behavior satisfies a formula if the interval spanning the whole behaviors satisfies the formula. Moreover, the language of a formula is the set of behaviors satisfying it.

$$\begin{array}{l}
\sigma \models D \;\; \textbf{iff} \;\; \sigma, [0, \#\sigma - 1] \models D \\
L(D) \triangleq \{\sigma \mid \sigma \models D\}
\end{array}$$

As stated in the introduction, the language of a formula and the automaton recognizing this language play an important role in algorithmic verification of the formula. We will explore the automata based verification of DDC^* in Section 8.8.

Proposition 8.1. $\sigma, [b, e] \models D$ **iff** $\sigma[b : e] \models D$.

Derived Constructs We can also define some derived constructs. These constructs can always be removed from the formula by rewriting them by their definitions. Boolean combinators $\vee, \Rightarrow, \Leftrightarrow$ can be defined using $\wedge, \neg$ as usual. Formula $\top$ denote "true".

- $Pt \triangleq \lceil 1 \rceil^0$ holds for point intervals of the form $[b, b]$. Then, $Ext \triangleq \neg Pt$ defines and non-point intervals and $Unit \triangleq Ext \wedge \neg(Ext; Ext)$ holds for intervals of the form $[b, b+1]$.
- $\Diamond D \triangleq \top; D; \top$ holds provided D holds for some subinterval.
- $\Box D \triangleq \neg\Diamond\neg D$ holds provided D holds for all subintervals.

Example 8.2. A synchronous bus arbiter with n cells has request lines req_1, ..., req_i, ..., req_n and acknowledgment lines $ack_1, \ldots, ack_i, \ldots, ack_n$. At any clock cycle, a subset of the request lines are high. It is the task of the arbiter to set at most one of the corresponding acknowledgment lines high. We formalize the timing properties of the arbiters in logic DDC^*(see [3]).

- There are no spurious acknowledgments.
  $\Box(Pt \;\Rightarrow\; \lceil ack_i \Rightarrow req_i \rceil^0)$.
- *3-cycle response time:* In any observation interval of at least 15 cycles, if req_i is continuously high then at least 3 ack_i are obtained.
  $\Box(\lceil\lceil req_i \rceil \wedge \eta \geq 15 \;\Rightarrow\; \int ack_i \geq 3)$
- *Dead time:* Let $lostcycle \triangleq (\vee_i\, req_i) \;\wedge\; (\wedge_i\, \neg ack_i)$. It holds for any cycle where there are requests but no acknowledgments. The following formula states that there can be at most 5 consecutive lost cycles.
  $\Box(\lceil\lceil lostcycle \rceil \;\Rightarrow\; \eta \leq 5)$.

8.2.1. *Syntactic Subsets of DDC**

- *DDC* consists of DDC^* formulae without the "chop-star" operator $D^{\otimes}$.
- *CoreDDC** formulae are DDC^* formulae without counting constructs $\eta\ op\ c$ and $\int P\ op\ c$.
- *CoreDDC* formulae are $\otimes$-free *CoreDDC** formulae, i.e. $DDC \cap$ *CoreDDC**. Logic *CoreDDC* is called ITL by Lodaya in Chapter 5.

The measurement formulae, although convenient for specifying timing properties of discrete timed behaviors, do not add expressive power to *CoreDDC**. To prove this, we give below a reduction ξ from DDC^* to *CoreDDC**. This reduction was first formulated by Fränzle [10].

In the following equivalences, let t denote a measurement term of the form η or $\int P$ and let c denote a non-negative integer. It is easy to show their validity.

$$\begin{array}{lcl}(t < c) & \Leftrightarrow & \neg(t \geq c)\\ (t \leq c) & \Leftrightarrow & \neg(t > c)\\ (t \geq c) & \Leftrightarrow & (t = c); \top\\ (\eta > c) & \Leftrightarrow & (\eta = c); Ext\\ (\int P > c) & \Leftrightarrow & (\int P = c); (T; \lceil P \rceil^0; Ext)\end{array}$$

By repeatedly applying the first two equivalences (left to right), the $<, \leq$ comparisons can be eliminated; and by applying the last three equivalences, the $>$ and $\geq$ comparisons can be eliminated giving only the formulae with $t = c$ comparisons. This reduction causes only linear blowup in the size of the formula.

Now, we show how to eliminate equality comparisons between measurement terms and integer constants.

$$\begin{array}{lcl}(\eta = 0) & \Leftrightarrow & Pt\\ (\eta = c+1) & \Leftrightarrow & (\eta = c); Unit\\ (\int P = 0) & \Leftrightarrow & (Pt \ \vee \lceil\lceil \neg P \rceil)\\ (\int P = 1) & \Leftrightarrow & (\int P = 0); \lceil P \rceil^0; Unit; (\int P = 0)\\ (\int P = c+1) & \Leftrightarrow & (\int P = c); (\int P = 1)\end{array}$$

By repeatedly applying above equivalences (left to right) we can obtain a *CoreDDC** formula equivalent to any given *DDC** formula. Call this language preserving transformation $\xi : DDC^* \rightarrow CoreDDC^*$. Then, size of $\xi(D)$ is linear in the size of D. This assumes that the integer constants used in the formula are coded in unary notation. Also note that the ξ reduction transforms *DDC* formulae into *CoreDDC* formulae (i.e. it does not introduce additional $\otimes$ operators).

8.3. Chop Expressions and Regular Expressions

Let Σ be a given finite alphabet. Let a range over the elements of Σ. A word over Σ is a finite sequence of letters from Σ. Let ϵ, Σ^*, Σ^+ respectively denote the empty word, the set of all words over Σ and the set of all nonempty words over Σ. We use w, x, y, z to range over words. A language L is a collection of words, i.e. $L \subseteq \Sigma^*$.

8.3.1. *Regular Expressions*

Definition 8.1. Catenation and Kleene Closure

- Given $w_1 = a_1a_2\ldots a_n$ and $w_2 = b_1b_2\ldots b_m$, the catenation $w_1 \cdot w_2$ is the word $a_1a_2\ldots a_nb_1\ldots b_m$ obtained by appending all the letters of w_2 at the end of the letters of w_1.

- Given languages L_1 and L_2 over Σ, their catenation is defined as:

$$L_1 \cdot L_2 \triangleq \{x \cdot y \mid x \in L_1, \text{ and } y \in L_2\}$$

- Let $L^0 \triangleq \{\epsilon\}$. Let $L^{n+1} \triangleq L \cdot L^n$. Then, $L^* \triangleq \bigcup_{n \geq 0} L^n$,

Syntax of Regular Expressions

- Regular Expression (RE): $r ::= \mathbf{0} \mid \epsilon \mid a \mid r \cdot r \mid r + r \mid r^*$.
- Semi-extended Regular Expression (SERE):
  $r ::= \mathbf{0} \mid \epsilon \mid a \mid r \cdot r \mid r + r \mid r^* \mid r \cap r$.
- Extended Regular Expression (ERE):
  $r ::= \mathbf{0} \mid \epsilon \mid a \mid r \cdot r \mid r + r \mid r^* \mid r \cap r \mid \neg r$.
  ($r \cap r$ is optional since it can be expressed in terms of $+$ and $\neg$.)
- Star Free Regular Expression (SFRE): $r ::= \mathbf{0} \mid \epsilon \mid a \mid r \cdot r \mid r + r \mid r \cap r \mid \neg r$.

Each regular expression r denotes a language $\mathcal{L}(r) \subseteq \Sigma^*$. This can be defined by induction on the structure of r. For atomic expressions, $\mathcal{L}(\mathbf{0}) = \emptyset$ (empty set), $\mathcal{L}(a) = \{a\}$ and $\mathcal{L}(\epsilon) = \{\epsilon\}$. The operator $r_1 \cdot r_2$ denotes catenation and the operator r^* denotes kleene closure, i.e. $\mathcal{L}(r_1 \cdot r_2) = \mathcal{L}(r_1) \cdot \mathcal{L}(r_2)$ and $\mathcal{L}(r^*) = \mathcal{L}(r)^*$. Set theoretic operators $+$ and $\cap$ denote union and intersection, and $\mathcal{L}(\neg r) = \Sigma^* - \mathcal{L}(r)$.

8.3.2. *Chop Expressions*

The chop (fusion) of words can be defined as follows.

Definition 8.2 (Chop and chop-star).

- *Let ; be a partial function defined over Σ^+ such that*

$$xa; by = \begin{cases} xay \text{ if } a = b \\ \bot \text{ otherwise} \end{cases}$$

- *Given languages $L_1, L_2 \subseteq \Sigma^+$, let*
  $L_1; L_2 \triangleq \{x; y \mid x \in L_1, y \in L_2, x; y \neq \bot\}$.
- *Given language L over Σ, let $L{\uparrow}^0 \triangleq \Sigma$ and $L{\uparrow}^{n+1} \triangleq L; (L{\uparrow}^n)$. Then, define*
  $L^{\otimes} \triangleq \bigcup_{n \geq 0} L{\uparrow}^n$. *Thus,*
  $L^{\otimes} = \Sigma \cup L \cup (L; L) \cup (L; L; L) \cup \ldots$

Syntax of Chop Expressions

- Chop Expression(*CE*): $c ::= \mathbf{0} \mid \Delta \mid a \mid c; c \mid c + c \mid c^{\otimes}$.
- Semi-extended Chop Expression(*SECE*):
  $c ::= \mathbf{0} \mid \Delta \mid a \mid c; c \mid c + c \mid c^{\otimes} \mid c \cap c$.
- Extended Chop Expression(*ECE*):
  $c ::= \mathbf{0} \mid \Delta \mid a \mid c; c \mid c + c \mid c^{\otimes} \mid \neg c \mid c \cap c$.
- Star Free Chop Expression (*SFCE*): $c ::= \mathbf{0} \mid \Delta \mid a \mid c; c \mid c + c \mid \neg c \mid c \cap c$.

Note that $c \cap c$ is optional in *ECE* and *SFCE* as it can be expressed in terms of $+$ and $\neg$.

Each chop expression c denotes a language $\mathcal{L}(c) \subseteq \Sigma^+$ consisting of non-empty words. This can be defined by induction on the structure of c. The primitive operators $\mathbf{0}$ and a as well as operators $+$ and $\cap$ have the same meaning as in regular expressions. We give the remaining cases: $\mathcal{L}(\Delta) = \Sigma \cdot \Sigma$, $\mathcal{L}(c_1; c_2) = \mathcal{L}(c_1); \mathcal{L}(c_2)$ and $\mathcal{L}(c^{\otimes}) = (\mathcal{L}(c))^{\otimes}$. Moreover, $\mathcal{L}(\neg c) = \Sigma^+ - \mathcal{L}(c)$ (giving only non-empty words).

Definition 8.3. We define *Tests Expressions* (*TE*) as $t ::= \mathbf{0} \mid a \mid t+t \mid t;t \mid {\sim}t$.

We can extend *CE*, *SECE*, *ECE*, *SFCE* respectively to *CET*, *SECET*, *ECET*, *SFCET* by adding test expressions to each of them. For example, *Chop Expressions with Tests*(*CET*) are defined as $c ::= t \mid \Delta \mid c;c \mid c+c \mid c^{\otimes}$ and *Semi-extended Chop Expression with Tests*(*SECET*) are defined as $c ::= t \mid \Delta \mid c;c \mid c+c \mid c^{\otimes} \mid c \cap c$. Test expressions provide a convenient vocabulary akine to propositions for specifying subsets of the alphabets.

Derived expressions: Let $\Sigma = \{a_1, \ldots, a_n\}$. Then,

- $\mathbf{1} \triangleq (a_1 + a_2 + \ldots + a_n)$,
- $\mathit{Ext} \triangleq \neg\mathbf{1}$,
- $c^{\oplus} \triangleq c; (c^{\otimes})$,
- $\top \triangleq \neg\mathbf{0}$,
- ${\sim}c \triangleq (\neg c) \cap \mathbf{1}$.

8.4. Algebraic Laws of Chop Expressions

Let $c_1 \equiv c_2$ be denote $\mathcal{L}(c_1) = \mathcal{L}(c_2)$. Kozen has defined a class of algebraic structures called Kleene Algebras and has given a complete and finite set of axioms for conditional equational reasoning [11, p. 58]. Moreover, he has extended this to structures called Kleene algebras with tests (KAT algebras) and given a complete set of conditional equational axioms for equality of KAT expressions [9]. A conditional equation has the form $P_1 \wedge \ldots \wedge P_{n-1} \Rightarrow P_n$ where each P_i has the form $c_1 \equiv c_2$. We can apply these axioms to the extended chop expressions.

Let *NREG* denote the collection of ϵ-free regular languages over a finite alphabet Σ. We will show later that these are precisely the languages specified by the extended chop expressions. We now outline the algebraic properties of the operators of chop expressions. In these, c ranges over chop expressions and t ranges over test expressions.

(1) *NREG* under the operators $(;, +, \otimes, \mathbf{0}, \mathbf{1})$ forms a *Kleene algebra*. That is, *NREG* with operators $(+, ;, \mathbf{0}, \mathbf{1})$ forms an idempotent semi-ring

$$
\begin{aligned}
&(c_1 + c_2) + c_3 \equiv c_1 + (c_2 + c_3) \\
&\mathbf{0} + c \equiv c + \mathbf{0} \equiv c \\
&c_1 + c_2 \equiv c_2 + c_1 \\
&(c_1; c_2); c_3 \equiv c_1; (c_2; c_3) \\
&\mathbf{1}; c \equiv c; \mathbf{1} \equiv c \\
&c_1; (c_2 + c_3) \equiv c_1; c_2 + c_1; c_3 \\
&(c_2 + c_3); c_1 \equiv c_2; c_1 + c_3; c_1 \\
&\mathbf{0}; c \equiv c; \mathbf{0} \equiv \mathbf{0}
\end{aligned}
$$

and additionally the following axioms (for $\otimes$) also hold. Note that these are conditional (in)equations.

$$
\begin{aligned}
1 + p; p^{\otimes} &\equiv p^{\otimes} & \qquad q + p; r \leq r \;\Rightarrow\; p^{\otimes}; q \leq r \\
1 + p^{\otimes}; p &\equiv p^{\otimes} & \qquad q + r; p \leq r \;\Rightarrow\; q; p^{\otimes} \leq r
\end{aligned}
$$

where the partial order $p \leq q \;\triangleq\; p + q \equiv q$. It is easy to show that the reflexivity, transitivity and anti-symmetry properties hold.

(2) Structure *NREG* under the operators $(\cap, +, \neg, \mathbf{0}, \top)$ forms a *boolean algebra.* We omit listing of the well-known axioms of boolean algebras.

(3) Test expressions t precisely denote elements of 2^{Σ}. It is easy to prove that $t \leq 1$. The key property of tests is that the operators $(;, +, \sim, \mathbf{0}, \mathbf{1})$ over tests (i.e. set 2^{Σ}) form a *finite boolean algebra* which is embedded within *NREG*. The *atoms* of this boolean algebra are the primitive tests a_i where $a_i \in \Sigma$. Hence, $a_i; a_i \equiv a_i$ and $a_i; a_j \equiv \mathbf{0}$ if $i \neq j$.

(4) Moreover, the operations ; and $\cap$ when restricted to the tests are the same. This is given by the axiom

$$c_1 \leq \mathbf{1} \wedge c_2 \leq \mathbf{1} \;\;\Rightarrow\;\; c_1; c_2 \equiv c_1 \cap c_2$$

(5) Some additional properties satisfied by the chop expressions are given below.

$$
\begin{aligned}
&\Delta \;\equiv\; Ext \cap \neg(Ext; Ext) \\
&\top \equiv \Delta^{\otimes} \\
&(t; c) \equiv (t; \top) \cap c \\
&(c; t) \equiv (\top; t) \cap c \\
&\Delta; (c_1 \cap c_2) \equiv (\Delta; c_1) \cap (\Delta; c_2) \\
&(c_1 \cap c_2); \Delta \equiv (c_1; \Delta) \cap (c_1; \Delta) \\
&\neg(t; c) \equiv {\sim}t; \top + t; \neg c \\
&\neg(c; t) \equiv \top; {\sim}t + \neg c; t \\
&\neg(\Delta; c) \equiv \mathbf{1} + \Delta; \neg c \\
&\neg(c; \Delta) \equiv \mathbf{1} + \neg c; \Delta
\end{aligned}
$$

These algebraic laws of *ECET* provide a conditional equational reasoning system for proving the equality of *ECET* expressions. In such reasoning, substitution and cut rules are used apart from the usual properties of equality (and order) [12]. The cut rules states that if $F_1 \wedge e_1' \equiv e_2' \wedge F_2 \Rightarrow e_1 \equiv e_2$ and $G \Rightarrow e_1' \equiv e_2'$ are both derivable then so is $F_1 \wedge G \wedge F_2 \Rightarrow e_1 \equiv e_2$.

Example We show that $a^{\otimes}; a^{\otimes} \equiv a^{\otimes}$.

(1)	$a^{\otimes}; a^{\otimes}$	
$\equiv$	$(1 + a; a^{\otimes}); a^{\otimes}$	axiom for $\otimes$
$\equiv$	$1; a^{\otimes} \ + \ a; a^{\otimes}; a^{\otimes}$	distributivity
$\geq$	$1; a^{\otimes}$	As $(p + q \equiv r \Rightarrow p \leq r)$
$\equiv$	$a^{\otimes}$	
(2)	$1 + a; a^{\otimes} \equiv a^{\otimes}$	axiom
$\Rightarrow$	$a; a^{\otimes} \leq a^{\otimes}$	*(i)*
Also	$a^{\otimes} \leq \ a^{\otimes}$	*(ii)*
Hence	$a^{\otimes} + a; a^{\otimes} \ \leq \ a^{\otimes}$	From (i),(ii) since $p \leq q \wedge r \leq q \Rightarrow p + r \leq q$
$\Rightarrow$	$a^{\otimes}; a^{\otimes} \ \leq a^{\otimes}$	Rule for $\otimes$

Soundness and Completeness Let $\vdash e_1 \equiv e_2$ denote that it is possible to give a proof of $e_1 \equiv e_2$ using a given set of algebraic laws and the substitution rule. Soundness of the algebraic laws ensures that for any e_1, e_2, if $\vdash e_1 \equiv e_2$ then $\mathcal{L}(e_1) = \mathcal{L}(e_2)$. This can be proved inductively by checking that each rule is sound. Completeness of a given set of algebraic laws ensures that for any e_1, e_2, if $\mathcal{L}(e_1) = \mathcal{L}(e_2)$ then $\vdash e_1 \equiv e_2$. Salomaa first gave a complete set of laws for proving equivalence of regular expressions [13]. This was extended to a complete proof system for extended regular expressions [14]. Kozen generalized Regular Expressions to Kleene Algebras with Tests (KAT) and showed that these laws are applicable to regular sets of behaviours arising in diverse situations including sequential program executions [9]. Kozen and Smith showed that the group of axioms (1) and (3) above are complete [15] for Kleene Algebras with Tests. However, Kleene Algebras with Tests do not include the intersection and negation operators of Extended Chop/Regular Expressions.

8.5. Conversions from CE to NFA

We denote a *nondeterministic finite automaton* (NFA) by a 5-tuple $(Q, \Sigma, \delta, s, F)$, where Q is a finite set of states, Σ is a finite input alphabet, $s \in Q$ is the initial state, $F \subseteq Q$ is the set of final states and δ is the transition function mapping $Q \times \Sigma$ to 2^Q. If $\delta(q, a)$ is singleton for each q, a then the automaton is called a *deterministic finite state automaton* (DFA). We refer the reader to [16, p. 19] for the details. In the rest of the paper, we define the size of an NFA as its number of states, i.e. $|Q|$.

Theorem 8.1. *For any given chop expression c, a linear sized NFA N can be constructed such that $\mathcal{L}(c) = \mathcal{L}(N)$.*

Proof. The NFAs for $\mathbf{0}$, a, $r+r$ and Δ are straight forward as in the case of regular expressions (see [16]). Let $N_1 = (Q_1, \Sigma, \delta_1, s_1, F_1)$ and $N_2 = (Q_2, \Sigma, \delta_2, s_2, F_2)$ be the respective NFAs for r_1 and r_2 i.e. $\mathcal{L}(N_1) = \mathcal{L}(r_1)$ and $\mathcal{L}(N_2) = \mathcal{L}(r_2)$. We assume that Q_1 and Q_2 are disjoint.

- Let $N_3 = (Q_1 \cup Q_2, \Sigma, \delta_3, s_1, F_2)$, where $\delta_3 = \delta_1 \cup \delta_2 \cup \{(p, a, q) \mid (p, a, q_f) \in \delta_1 \text{ and } q_f \in F_1 \text{ and } (s_2, a, q) \in \delta_2\}$. Then $\mathcal{L}(N_3) = \mathcal{L}(r_1; r_2)$.
- Let $N_4 = (Q_1 \cup \{f\}, \Sigma, \delta_4, s_1, \{f\})$, where $f \notin Q_1$ and $\delta_4 = \delta_1 \cup \delta'$, and δ' is the smallest set s.t.
  (1) $(s_1, a, f) \in \delta'$ for all $a \in \Sigma$,
  (2) $(p, a, f) \in \delta'$ if $(p, a, q_f) \in \delta_1$ and $q_f \in F_1$,
  (3) $(p, a, q) \in \delta'$ if $(p, a, q_f), (s_1, a, q) \in \delta_1$ and $q_f \in F_1$.
  Then $\mathcal{L}(N_4) = \mathcal{L}(r_1^{\otimes})$.

The size of the resulting automaton is linear in the size of the expression. It can be easily seen that the conversion can be performed in time $O(n^3)$ by an induction on the structure of chop expressions. We omit the details. □

Corollary 8.1. *For any given extended chop expression c, a NFA N (of non-elementary size) can be constructed such that $\mathcal{L}(c) = \mathcal{L}(N)$. If c is a semi-extended chop expression then N is of exponential size.*

Proof. Given $c \in SECE$, there are $O(|c|)$ occurrences of $\cap$ operators. All other operators can be translated with constant blowup in the automaton size. But each $\cap$ operator requires a product construction which can square the automaton size. Hence, the size of the final automaton is at most exponential in $|r|$. If $c \in ECE$, each $\neg$ operator requires determinization of the NFA, which can increase the size of the automaton by an exponent. Hence, the final automaton is of non-elementary size. From Theorem 8.2 and [6], it can be seen that the lower bound for this conversion is also non-elementary. □

Remark 8.1. For any test expression t, we can construct a 3 state DFA $N = (\{s, f, r\}, \Sigma, \delta, s, \{f\})$ with $\delta = \{(s, a, f) | a \in \mathcal{L}(t)\} \cup \{(s, a, r) | a \notin \mathcal{L}(t)\}$, so that $\mathcal{L}(t) = \mathcal{L}(N)$. Hence, the (order of) complexity and the size of the resulting automata in the conversions from *CET* to NFA and *SECET* to NFA are respectively same as those from *CE* to NFA and *SECE* to NFA. This is summarised in Fig. 8.1 below.

Expression	*CE/CET*	*SECE/SECET*	*ECE/ECET*
NFA Size	Linear	EXP	NONELEM

Fig. 8.1. Conversion from (Extended) Chop Expression to NFA.

8.6. Conversions between RE and CE

The chop expressions and regular expressions share many features. We explore the reductions between the two forms of expression.

8.6.1. *Brzozowski derivative*

Definition 8.4 (Brzozowski derivative [17]). *Given a language $\mathcal{L}$, we define the* left derivative *of $\mathcal{L}$ w.r.t an $a \in \Sigma$ as $L_a(\mathcal{L}) = \{w \mid aw \in \mathcal{L}\}$. Similarly the* right derivative *of $\mathcal{L}$ w.r.t an $a \in \Sigma$ is defined as $R_a(\mathcal{L}) = \{w \mid wa \in \mathcal{L}\}$.*

Definition 8.5. Let R be the set of extended regular expressions. We define the function $\lambda : R \to \{\epsilon, \mathbf{0}\}$ such that, for any regular expressions r, $\lambda(r) = \epsilon$ iff $\epsilon \in \mathcal{L}(r)$. It is defined recursively as follows:

$$
\begin{aligned}
\lambda(\epsilon) &= \epsilon \\
\lambda(\mathbf{0}) &= \mathbf{0} \\
\lambda(a) &= \mathbf{0} \text{ (for } a \in \Sigma \text{)} \\
\lambda(r_1 \cdot r_2) &= \begin{cases} \epsilon & \text{if } \lambda(r_1) = \epsilon \text{ and } \lambda(r_2) = \epsilon \\ \mathbf{0} & \text{otherwise} \end{cases} \\
\lambda(r_1 + r_2) &= \begin{cases} \epsilon & \text{if } \lambda(r_1) = \epsilon \text{ or } \lambda(r_2) = \epsilon \\ \mathbf{0} & \text{otherwise} \end{cases} \\
\lambda(r^*) &= \epsilon \\
\lambda(r_1 \cap r_2) &= \begin{cases} \epsilon & \text{if } \lambda(r_1) = \epsilon \text{ and } \lambda(r_2) = \epsilon \\ \mathbf{0} & \text{otherwise} \end{cases} \\
\lambda(\neg r) &= \begin{cases} \epsilon & \text{if } \lambda(r) = \mathbf{0} \\ \mathbf{0} & \text{otherwise} \end{cases}
\end{aligned}
$$

Definition 8.6. The left derivative of (extended) regular expressions is defined follows:

$$
\begin{aligned}
L_a(\mathbf{0}) &= \mathbf{0} \\
L_a(\epsilon) &= \mathbf{0} \\
L_a(a) &= \epsilon \\
L_a(b) &= \mathbf{0} \\
L_a(r_1 + r_2) &= L_a(r_1) + L_a(r_2) \\
L_a(r_1 \cdot r_2) &= L_a(r_1) \cdot r_2 + \lambda(r_1) \cdot L_a(r_2) \\
L_a(r^*) &= L_a(r) \cdot r^* \\
L_a(r_1 \cap r_2) &= L_a(r_1) \cap L_a(r_2) \\
L_a(\neg r) &= \neg L_a(r)
\end{aligned}
$$

Lemma 8.1. *For any (extended) regular expression r, the following holds:*

$$\lambda(r) = \epsilon \text{ iff } \epsilon \in \mathcal{L}(r) \tag{8.1}$$

$$\mathcal{L}(L_a(r)) = L_a(\mathcal{L}(r)) \tag{8.2}$$

$$\mathcal{L}(r) = \lambda(r) + a_1 \cdot L_{a_1}(r) + \ldots + a_n \cdot L_{a_n}(r) \tag{8.3}$$

Proof. Proof of Equation (8.3) is obtained by an induction on the structure of r. Below we give the proof for $\neg$ and $\cap$.

$$\begin{aligned}
\mathcal{L}(L_a(\neg r)) &= \{w \mid aw \in \mathcal{L}(\neg r)\} \text{ (from definition of } L_a \text{)} \\
&= \{w \mid aw \in (\Sigma^* - \mathcal{L}(r))\} \text{ (from definition of } \neg \text{)} \\
&= \{w \mid aw \in \Sigma^*\} - \{w \mid aw \in \mathcal{L}(r)\} \\
&= \{w \mid w \in \Sigma^*\} - \{w \mid aw \in \mathcal{L}(r)\} \\
&= \{w \mid w \in \Sigma^*\} - \{w \mid w \in \mathcal{L}(L_a(r))\} \text{ (from definition of } L_a \text{)} \\
&= \{w \mid w \in (\Sigma^* - \mathcal{L}(L_a(r)))\} \\
&= \mathcal{L}(\neg L_a(r))\} \text{ (from definition of } \neg \text{)} \\
\mathcal{L}(L_a(r_1 \cap r_2)) &= \{w \mid aw \in \mathcal{L}(r_1 \cap r_2)\} \text{ (from definition of } L_a \text{)} \\
&= \{w \mid aw \in \mathcal{L}(r_1) \cap \mathcal{L}(r_2)\} \text{ (from definition of } \cap \text{)} \\
&= \{w \mid aw \in \mathcal{L}(r_1)\} \cap \{w \mid aw \in \mathcal{L}(r_2)\} \\
&= \mathcal{L}(L_a(r_1)) \cap \mathcal{L}(L_a(r_2)) \\
&= \mathcal{L}(L_a(r_1) \cap L_a(r_2)) \text{ (from definition of } \cap \text{)}
\end{aligned}$$

For the complete proof of (8.3), as well as that of (8.1) and (8.2), see [17]. □

Note. For any ERE, r, the size of $L_{a_i}(r)$ is at most $O(|r|^2)$. This can be easily verified from the definition of left derivative.

Similar definitions, laws and complexity hold for the *right derivative* $R_a(r)$ also.

8.6.2. *Conversion from RE to CE*

We give an easy DLOGSPACE computable reduction β from extended regular expressions to extended chop expressions, with a linear blow up in the size of the resulting expression.

Theorem 8.2. *Given an extended regular expression r, we can construct an extended chop expression $\beta(r)$ such that, $\mathcal{L}(\beta(r)) = \mathcal{L}(r) \cdot \Sigma$ and $|\beta(r)|$ is linear in the size of r, where $\beta : ERE \rightarrow ECE$ is a DLOGSPACE computable reduction defined* *inductively as follows:*

$$\begin{aligned}
&\beta(\mathbf{0}) = \mathbf{0}\\
&\beta(\epsilon) = \mathbf{1}\\
&\beta(a) = a; \Delta\\
&\beta(r_1 \cdot r_2) = \beta(r_1); \beta(r_2)\\
&\beta(r_1 + r_2) = \beta(r_1) + \beta(r_2)\\
&\beta(r^*) = (\beta(r))^{\otimes}\\
&\beta(r_1 \cap r_2) = \beta(r_1) \cap \beta(r_2)\\
&\beta(\neg r) = \neg\beta(r)
\end{aligned}$$

Note that under β, the subclass RE maps to CE and SERE maps to SECE.

Proof. By induction on the structure of regular expression r, it can be easily shown that $\mathcal{L}(\beta(r)) = \mathcal{L}(r) \cdot \Sigma$.

$$\begin{aligned}
\mathcal{L}(\beta(r^*)) &= \mathcal{L}(\beta(r)^{\otimes}) \text{ (from definition of } \beta \text{)}\\
&= \Sigma \cup \mathcal{L}(\beta(r)) \cup (\mathcal{L}(\beta(r)); \mathcal{L}(\beta(r))) \cup \ldots \text{ (from definition of } \otimes \text{)}\\
&= \Sigma \cup (\mathcal{L}(r) \cdot \Sigma) \cup ((\mathcal{L}(r) \cdot \Sigma); (\mathcal{L}(r) \cdot \Sigma)) \cup \ldots \text{ (induction hypothesis)}\\
&= \Sigma \cup (\mathcal{L}(r) \cdot \Sigma) \cup (\mathcal{L}(r) \cdot \mathcal{L}(r) \cdot \Sigma) \cup \ldots\\
&= (\mathcal{L}(\epsilon) \cup \mathcal{L}(r) \cup (\mathcal{L}(r) \cdot \mathcal{L}(r)) \cup \ldots) \cdot \Sigma\\
&= \mathcal{L}(r^*) \cdot \Sigma\\
\mathcal{L}(\beta(\neg r)) &= \mathcal{L}(\neg\beta(r)) \text{ (from definition of } \beta \text{)}\\
&= \Sigma^+ - \mathcal{L}(\beta(r)) \text{ (from definition of } \neg \text{)}\\
&= \Sigma^+ - (\mathcal{L}(r) \cdot \Sigma) \text{ (induction hypothesis)}\\
&= (\Sigma^* - \mathcal{L}(r)) \cdot \Sigma = \mathcal{L}(\neg r) \cdot \Sigma
\end{aligned}$$

It should be noted that the β reduction can be carried out in DLOGSPACE (see [18] for the techniques used). □

Lemma 8.2. *For any ϵ-free regular expression r, the chop expression $c = (\beta(R_{a_1}(r)); a_1) + \ldots + (\beta(R_{a_n}(r)); a_n)$ is such that $\mathcal{L}(r) = \mathcal{L}(c)$. The size of c is $O(|r|^2 \times |\Sigma|)$.*

Proof. From the equation for right derivative that corresponds to (8.3) we get

$$\begin{aligned}
\mathcal{L}(r) &= \mathcal{L}(\lambda(r) + R_{a_1}(r) \cdot a_1 + \ldots + R_{a_n}(r) \cdot a_n)\\
&= \mathcal{L}(\mathbf{0} + R_{a_1}(r) \cdot a_1 + \ldots + R_{a_n}(r) \cdot a_n) \quad \text{(since r is } \epsilon \text{ free)}\\
&= \mathcal{L}(R_{a_1}(r) \cdot a_1 + \ldots + R_{a_n}(r) \cdot a_n)
\end{aligned}$$

We show that $\mathcal{L}(R_a(r) \cdot a) = \mathcal{L}((\beta(R_a(r)); a))$. It is easy to see that any word that belong to either $\mathcal{L}(R_a(r) \cdot a)$ or $\mathcal{L}((\beta(R_a(r)); a))$ will be of the form wa.

$$\begin{aligned}
wa \in \mathcal{L}(R_a(r) \cdot a) \;\; &\text{iff} \;\; wa \in \mathcal{L}(R_a(r)) \cdot \Sigma \text{ and } wa \in \mathcal{L}(\Sigma^* \cdot \{a\})\\
&\text{iff} \;\; wa \in (\mathcal{L}(R_a(r)) \cdot \Sigma; \{a\})\\
&\text{iff} \;\; wa \in (\mathcal{L}(\beta(R_a(r))); \mathcal{L}(a)\\
&\text{iff} \;\; wa \in \mathcal{L}(\beta(R_a(r)); a)
\end{aligned}$$

□

For any regular language R, there exists a regular expressions r such that, $R = \mathcal{L}(r)$. Hence by the above lemma, for any ϵ free regular language R, there exists a chop expression c such that $R = \mathcal{L}(c)$

8.6.3. *Conversion from CE to RE*

We now consider the converse question of representing chop expressions by regular expressions. This turns out to be significantly more complex.

Kleene Matrix Algebra Kozen has shown that given a Kleene Algebra K, a square $m \times m$ matrix with elements of K also forms a Kleene Algebra, denoted $M(m, K)$ (see [11, p. 58] [9]). Here, we take the base Kleene algebra K to be the extended regular expressions (ERE).

Definition 8.7. Let P and Q be two $m \times m$ matrices of (extended) regular expressions. We denote by $P+Q$ and $P \star Q$, the matrices of (extended) regular expression obtained respectively, by performing matrix addition and multiplication. Here the addition and multiplication operators on the *elements of the matrices* are respectively the regular expression operators $+$ and $\cdot$. We will denote $P \star Q$ as PQ, whenever there is no ambiguity.

P^* is defined by induction on the size m. If $m = 1$ then $P = [r]$ for some (extended) regular expression r. Therefore $P^* = [r^*]$. For $m > 1$ we break P into four sub-matrices

$$P = \begin{bmatrix} A & | & B \\ - & - & - \\ C & | & D \end{bmatrix}$$

such that A and D are square, say $n \times n$ and $(m-n) \times (m-n)$, respectively. P^* is defined as,

$$P^* = \begin{bmatrix} (A + BD^*C)^* & | & (A + BD^*C)^*BD^* \\ - - - - - - - - & - & - - - - - - - - \\ (D + CA^*B)^*CA^* & | & (D + CA^*B)^* \end{bmatrix}$$

The above definition provides a convenient method of recursively computing the Kleene closure on matrix algebra. We refer the reader to Kozen [11, p. 58] [9] for the details of this beautiful theory. Below, we analyze the complexity of operations of Kleene matrix algebra over extended regular expressions.

Definition 8.8. For a square matrix of extended regular expressions, say A of dimension $m \times m$ with each entry of length at most q, let $size(A)$ denote the tuple (m, q).

Lemma 8.3. *If $size(P) = (m, q_1)$ and $size(Q) = (m, q_2)$ then*

$$size(PQ) = (m, (q_1 + q_2 + 1)m + 1),$$
$$size(P + Q) = (m, q_1 + q_2 + 3),$$
$$size(P^*) \leq (m, 2^{m^2} q_1).$$

Proof. $[P+Q]_{(i,j)}$ is written as $([P]_{(i,j)} + [Q]_{(i,j)})$. While constructing $[PQ]_{(i,j)}$, we ignore the $\cdot$ between $[P]_{(i,k)}$ and $[Q]_{(k,j)}$ and write it as $[P]_{(i,k)}[Q]_{(k,j)}$. That is for example, $(a+b) \cdot (b+c)$ is written as $((a+b)(b+c))$. Hence, the first two are straight forward calculations.
Each entry of P^* is at most the size of each entry of $(A + BD^*C)^*BD^*$. For simplicity assume that $m = 2m'$ and $size(A) = size(D) = (m', q_1)$. We show that $size(P^*) \leq (2m', 2^{(2m')^2} q_1)$

$$\begin{aligned}
size(D^*) &\leq (m', 2^{m'^2} q_1) \text{ (By induction hypothesis)} \\
size(BD^*) &\leq (m', (q_1 + 2^{m'^2} q_1 + 1)m' + 1) \\
size(BD^*C) &\leq (m', ((q_1 + 2^{m'^2} q_1 + 1)m' + 1 + q_1 + 1)m' + 1) \\
&= (m', ((q_1 + 2^{m'^2} q_1 + 1)m' + 2 + q_1)m' + 1) \\
size(A + BD^*C) &\leq (m', ((q_1 + 2^{m'^2} q_1 + 1)m' + 2 + q_1)m' + 1 + q_1 + 3) \\
&= (m', ((q_1 + 2^{m'^2} q_1 + 1)m' + 2 + q_1)m' + q_1 + 4) \\
size((A + BD^*C)^*) &\leq (m', 2^{m'^2}(((q_1 + 2^{m'^2} q_1 + 1)m' + 2 + q_1)m' + q_1 + 4)) \\
size((A + BD^*C)^*BD^*) &\leq (m', (2^{m'^2}(((q_1 + 2^{m'^2} q_1 + 1)m' + 2 + q_1)m' + q_1 + 4) + \\
&\qquad (q_1 + 2^{m'^2} q_1 + 1)m' + 1 + 1)m' + 1) \\
&\leq (m', 2^{4m'^2} q_1) \text{ (conservative upper bound)} \\
\therefore size(P^*) &\leq (2m', 2^{(2m')^2} q_1) = (m, 2^{m^2} q_1)
\end{aligned}$$

□

We now combine the ideas of derivatives and Matrix Kleene Algebra over ERE. This results into Lemma 8.4.

Definition 8.9. For any given (extended) regular expression r, we define

$$\Sigma_v = \begin{bmatrix} \epsilon \; a_1 \; a_2 \ldots a_n \end{bmatrix}$$

$$\Sigma_M = \begin{bmatrix} \epsilon & \mathbf{0} & \mathbf{0} & \ldots & \mathbf{0} \\ \mathbf{0} & a_1 & \mathbf{0} & \ldots & \mathbf{0} \\ \mathbf{0} & \mathbf{0} & a_2 & \ldots & \mathbf{0} \\ \ldots & \ldots & \ldots & \ldots & \ldots \\ \mathbf{0} & \mathbf{0} & \ldots & \ldots & a_n \end{bmatrix}$$

$$\epsilon_v = \begin{bmatrix} \epsilon \, \epsilon \ldots \epsilon \end{bmatrix}$$

$$L_v(r) = \left[\lambda(r)\ L_{a_1}(r) \ldots L_{a_n}(r)\right]$$

$$R_v(r) = \left[\lambda(r)\ R_{a_1}(r) \ldots R_{a_n}(r)\right]$$

$$LR_M(r) = \begin{bmatrix} \lambda(r) & R_{a_1}(r) & \ldots & R_{a_n}(r) \\ L_{a_1}(r) & L_{a_1}(R_{a_1}(r)) & \ldots & L_{a_1}(R_{a_n}(r)) \\ \ldots & \ldots & \ldots & \\ L_{a_n}(r) & L_{a_n}(R_{a_1}(r)) & \ldots & L_{a_n}(R_{a_n}(r)) \end{bmatrix}$$

Lemma 8.4.

- $\Sigma_v \star L_v(r)^T \equiv r \equiv R_v(r) \star \Sigma_v^T$
- $\Sigma_v \star LR_M(r) \star \Sigma_v^T \equiv r$

Proof. Follows from Equation (8.3) (in Lemma 8.1) and Definition 8.9 □

Reducing ECE to ERE We can now formulate our translation from chop expressions to regular expressions.

Theorem 8.3. *Given an extended chop expression c, we can construct an extended regular expression $\alpha(c)$ such that, $\mathcal{L}(\alpha(c)) = \mathcal{L}(c)$ and $|\alpha(c)|$ is $O(2^{(2^{|c|})})$, where $\alpha : ECE \to ERE$ is a reduction defined inductively as follows:.*

$$\begin{aligned}
&\alpha(\mathbf{0}) = \mathbf{0} \\
&\alpha(\Delta) = (a_1 + a_2 + \ldots + a_n) \cdot (a_1 + a_2 + \ldots + a_n) \\
&\alpha(a) = a \\
&\alpha(c_1 + c_2) = \alpha(c_1) + \alpha(c_2) \\
&\alpha(c_1; c_2) = R_v(\alpha(c_1)) \star \Sigma_M \star L_v(\alpha(c_2))^T \\
&\alpha(c^{\otimes}) = \epsilon_v \star (\Sigma_M \star LR_M(\alpha(c)))^* \star (\Sigma_v)^T \\
&\alpha(c_1 \cap c_2) = \alpha(c_1) \cap \alpha(c_2) \\
&\alpha(\neg c) = \neg(\alpha(c) + \epsilon)
\end{aligned}$$

Note that under α, CE maps to RE and SECE maps to SERE.

Proof. All conversions, except that of ; and $\otimes$ are straight forward. $R_v(\alpha(c_1))$ gives a vector of expressions, each of whose language would contain words w such that $w \cdot a_i \in \mathcal{L}(\alpha(c_1))$ and $\mathcal{L}(\alpha(c_1)) = \mathcal{L}(c_1)$ by the induction hypothesis. Similarly, $L_v(\alpha(c_2))$ gives a vector of expressions, each of whose language would contain words w such that $a_i \cdot w \in \mathcal{L}(\alpha(c_2))$ and $\mathcal{L}(\alpha(c_2)) = \mathcal{L}(c_2)$ by the induction hypothesis. Thus, $R_v(\alpha(c_1)) \star \Sigma_M \star L_v(\alpha(c_2))^T$ would have expressions whose language would contain words $x \cdot a_i \cdot y$ such that $x \cdot a_i \in \mathcal{L}(c_1)$ and $a_i \cdot y \in \mathcal{L}(c_2)$, and all such expressions would get constructed in this process. Hence

$\mathcal{L}(\alpha(c_1;c_2)) = \mathcal{L}(R_v(\alpha(c_1)) \star \Sigma_M \star L_v(\alpha(c_2))^T) = \mathcal{L}(c_1;c_2)$.
Let $r' = \alpha(c)$

$$\Sigma_M \cdot LR_M(r') = \begin{bmatrix} \lambda(r') & R_{a_1}(r') & \ldots & R_{a_n}(r') \\ a_1 \cdot L_{a_1}(r') & a_1 \cdot L_{a_1}(R_{a_1}(r')) & \ldots & a_1 \cdot L_{a_1}(R_{a_n}(r')) \\ \ldots & \ldots & \ldots & \\ a_n \cdot L_{a_n}(r') & a_n \cdot L_{a_n}(R_{a_1}(r')) & \ldots & a_n \cdot L_{a_n}(R_{a_n}(r')) \end{bmatrix}$$

A typical entry of this matrix is of the type $a_i \cdot L_{a_i}(R_{a_j}(r'))$. This expression contains all words of the form $a_i \cdot x$ such that $a_i \cdot x \cdot a_j \in \mathcal{L}(r')$ and $\mathcal{L}(r') = \mathcal{L}(\alpha(c)) = \mathcal{L}(c)$ by the induction hypothesis. Thus the corresponding expression in $(\Sigma_M \star LR_M(r')) \star (\Sigma_v)^T$ will contain the word $a_i \cdot x \cdot a_j$. Similarly, $(\Sigma_M \star LR_M(r'))^2$ would have expressions whose language consists of words of the type $a_i \cdot x \cdot a_j \cdot y$ such that $a_i \cdot x \cdot a_j, a_j \cdot y \cdot a_k \in \mathcal{L}(r')$, so that $a_i \cdot x \cdot a_j \cdot y \cdot a_k \in \mathcal{L}(r'); \mathcal{L}(r') = \mathcal{L}(\alpha(c);\alpha(c)) = \mathcal{L}(c;c)$. Thus the reflexive transitive closure o $(\Sigma_M \star LR_M(r'))^\star$ is a matrix whose typical element will contain expressions whose language consists of words of the type $a_i \cdot w$ such that $a_i \cdot w \cdot a_j \in \mathcal{L}(\alpha(c)^\otimes) = \mathcal{L}(c^\otimes)$. □

Complexity: $\otimes$ is the operation with the worst case space complexity and this follows a recurrence equation of the form $T(n+1) = kT(n)^4$, where $k = m^2 2^{m^2}$. Thus for a (extended) chop expression of size n, the equivalent (extended) regular expression is of size at most $O(k^{4^n}) = O((m^2 2^{m^2})^{(4^n)})$, where $m = |\Sigma|$. Therefore the space required is at most $O((m^2 2^{m^2})^{(4^n)}) = O((m^2 2^{m^2})^{(4^n)})$. Hence, this conversion is in double EXPSPACE.

Lemma 8.5. *Given an NFA $N = (Q, \Sigma, \delta, S, F)$, a regular expression r can be constructed such that $\mathcal{L}(N) = \mathcal{L}(r)$ and $|r|$ is $2^{|Q|^{O(1)}}$.*

Proof. Assume without loss of generality that $Q = \{1, \ldots, n\}$. For each $q \in Q$ let X_q denote the set of strings in Σ^* that would be accepted by N, if q were the start state. For any pair of state $i, j \in Q$ let $A_{i,j} = \{a_1, \ldots, a_k\} \subseteq \Sigma$ be the set of letter such that $\forall a \in A_{i,j} : j \in \delta(i,a)$. Further, let $\widehat{A_{i,j}} = a_1 + \cdots + a_k$. Then, X_q satisfies the following system of equations:

$$X_q = \begin{cases} \sum_{i=1}^n \widehat{A_{q,i}} X_i + \phi & \text{if } q \notin F \\ \sum_{i=1}^n \widehat{A_{q,i}} X_i + \epsilon & \text{if } q \in F \end{cases}$$

These equations can be arranged in a single matrix-vector equation of the form

$$X = AX + b$$

where A is an $n \times n$ matrix containing sums of elements of Σ, b is a $n \times 1$ column vector of ϵ and ϕ, and $X = [X_1 \ldots X_n]^T$. The least solution of the above vector equation is $X = A^*b$. For further details, see [11, p. 60]. Hence the regular expression r is obtained by $\sum_{i \in S}[A^*b]_i$. Since $size(A) = (n, |\Sigma|)$, from Lemma 8.3, we know that $size(A^*) = (n, 2^{n^2}|\Sigma|)$. Therefore, size of each entry in $[A^*b]$ is $O(n2^{n^2}|\Sigma|)$ and hence $|r|$ is $O(n^2 2^{n^2}|\Sigma|) = 2^{|Q|^{O(1)}}$. □

Theorem 8.4. *Given a chop expression c, a regular expression r can be constructed, such that,* $\mathcal{L}(c) = \mathcal{L}(r)$ *and* $|(r)|$ *is* $2^{|c|^{O(1)}}$.

Proof. From Theorem 8.1 and Lemma 8.5. □

8.7. Decision Procedures for CE and their Complexity

We consider several decision problems for Chop expressions. These include the membership problem, the non-emptiness problem, and the inequivalence of two chop expressions. These decision problems have been well studied for regular expressions and their decision complexities have been ascertained. Figure 8.2 provides a summary of these results together with their the original sources. Most of these results can be adapted to Chop expressions.

Expression	Membership	Inequivalence	Non-Emptiness
RE	NL Complete [19]	PSPACE Complete [20]	$\geq$ ALOGTIME [21]
			$\leq$ NL
SERE	LOGCFL Complete [22]	EXPSPACE Complete [23]	PSPACE Complete [23]
ERE	P Complete [21]	NONELEM [6]	NONELEM
SFRE	P Complete [21]	NONELEM	NONELEM

Fig. 8.2. Complexity of Decision Problems of (Extended) Regular Expression.

Lemma 8.6. *The membership, inequivalence and non-emptiness problems for chop expressions, and its extensions, are at least as hard as the corresponding decision problems for regular expressions, and its corresponding extensions (as given in table 8.2).*

Proof. By Theorem 8.2 there is a DLOGSPACE reduction from extended regular expressions to extended chop expressions. Under this reduction, the three decision problems for extended regular expressions reduce to their corresponding problems for extended chop expressions (preserving subclasses). The complexity bounds of these problems for regular expressions are given in table 8.2). It is easy to see that under DLOGSPACE reduction, these lower bounds are preserved except the ALOGTIME-hardness of the non-emptiness of regular expressions. The Boolean Sentence Value Problem (BSVP) problem (see Chapter 5) is known to be complete for ALOGTIME [24]. We consider BSVP without negation since there is an ALOGTIME reduction removing negation operator from Boolean Sentences.

Now ALOGTIME hardness of the non-emptiness problem for chop expressions can be shown by the following ALOGTIME reduction from BSVP to *CE*: replace 1 ("true") with a (let $\Sigma = \{a\}$) and 0 ("false") with $\mathbf{0}$ (expression denoting empty language), $\wedge$ with ;, $\vee$ with +. Similar technique was used for showing ALOGTIME hardness of the membership problem for RE [21]. $\square$

We assume that all chop expressions are given in fully parenthesized form. A log sized expression pointer can mark a position within a (extended) chop expression. Such a position uniquely determines the current sub-expression by being at the opening parenthesis of the subexpression. Moreover, using this pointer it is possible to scan the chop expression text to determine the operator (i.e. $+, ;, \otimes$) of the current sub-expression, and to move the pointer to the left or right argument of the operator, or to move the pointer to the following sub-expression. All these operations can be carried out in logspace in the size of the expression. Similarly, a position within the input word can be recorded by a word pointer of size logarithmic in the length of the word.

Lemma 8.7.

(1) The membership problem for CE is in NL.
(2) The membership problem for SECE is in LOGCFL.
(3) The membership problem for ECE and SFCE is in PTIME.

Proof.

(1) We first describe the membership checking procedure for Regular Expressions RE. We can simulate the matching of a given word with given expression e in RE by using a word pointer and an expression pointer. The pointer is moved top-down into leaf sub-expression where the current letter is matched and both the expression pointer and the word pointer are incremented. When an operator such as + is encountered the expression pointer is moved non-deterministically to either the left or the right sub-expression. On finishing the match of a sub-expression the expression pointer jumps to the following sub-expression. Similarly for $\otimes$. The complexity of this procedure is NL. The main difference between chop and regular expressions is illustrated by the following identity: $a; a \equiv a$ but $a; \Delta; a$ is different from a. To handle this, a register of size $\log(|\Sigma|)$ checks at a leaf sub-expression of the form a whether it matches the previous letter (which is stored in a register) as well as the letter at the word pointer. The occurrence of a leaf sub-expression Δ clears the previous letter register and increments the word pointer. We omit the full details which can be found in a separate note. The example below illustrates this procedure. A configuration of the algorithm is a triple giving the word, the expression and the register. The positions of the expression and word pointer are marked by hat symbol above corresponding position. The symbol \$ denotes the end of the

word marker.

$$\langle \hat{b}d\$,\ \hat{(}b;(((a+b);\Delta);d)),\ *\rangle \rightarrow \langle \hat{b}d\$,\ (\hat{b};(((a+b);\Delta);d)),\ *\rangle$$
$$\rightarrow \langle \hat{b}d\$,\ (b;\hat{(}((a+b);\Delta);d)),\ b\rangle \ldots \rightarrow \langle \hat{b}d\$,\ (b;\hat{(}\hat{(}(a+b);\Delta);d)),\ b\rangle$$
$$\rightarrow \langle \hat{b}d\$,\ (a;(((a+\hat{b});\Delta);d)),\ b\rangle \quad \text{(nondeterministically choosing right argument)}$$
$$\rightarrow \langle \hat{b}d\$,\ (a;\hat{(}((a+b);\hat{\Delta});d)),\ b\rangle \rightarrow \langle b\hat{d}\$,\ (a;\hat{(}((a+b);\Delta);\hat{d})),\ *\rangle$$
$$\rightarrow \langle b\hat{d}\$,\ (a;\hat{(}((a+b);\Delta);d)\hat{)},\ d\rangle \rightarrow \langle bd\hat{\$},\ (a;\hat{(}((a+b);\Delta);d))\hat{},\ *\rangle$$

(2) Recall that SECE extend CE with intersection operator. A logCFL algorithm is like the above NL procedure but in addition it can use a stack. Let goal (p_1, p_2, i, j) denote that subexpression between positions p_1 and p_2 must match the subword $w[i, j]$. In order to handle intersection operator $(e_1 \cap e_2)$ (where both sub-expressions e_1, e_2 must match the same subword say $w[i, j]$) corresponding goals are stored in the stack and discharged one by one.

(3) For membership checking of ECE and SFCE, we can follow a dynamic programming method similar to one mentioned in [16, p.75] for ERE. Given an extended chop expression c and a word w, let $n_1 = |c|$ and $n_2 = |w|$. Consider the set of all sub-expressions of c. There are at most $O(n_1)$ of them. Let S be the set of all substrings of w. Thus $|S| \in O(n_2^2)$. Consider the parse tree of c. For each node i of the tree let c_i be the subexpression represented by it. In a bottom up fashion, starting at the leaf nodes of the parse tree we compute $S_{c_i} = \{(k, j) \mid 1 \le k \le j \le n_2 \wedge w(k, j) \in S \wedge w(k, j) \in \mathcal{L}(c_i)\}$.

- $S_a = \{(k, k) \mid 1 \le k \le n_2 \wedge w(k, k) = a\}$
- $S_\Delta = \{(k, k+1) \mid 1 \le k < n_2\}$
- $S_{c_{i_1};c_{i_2}} = \{(k, j) \mid (k, l) \in S_{c_{i_1}} \wedge (l, j) \in S_{c_{i_2}}\}$
- $S_{c_{i_1}+c_{i_2}} = S_{c_{i_1}} \cup S_{c_{i_2}}$, and $S_{c_{i_1} \cap c_{i_2}} = S_{c_{i_1}} \cap S_{c_{i_2}}$, and $S_{\neg c_{i_1}} = S - S_{c_{i_1}}$
- $S_{c_{i_1}^{\otimes}} = RT(S_{c_{i_1}})$, where $RT(S_{c_{i_1}})$ is the reflexive transitive closure of $S_{c_{i_1}}$.

Each of these can be computed in PTIME, and there are at most $O(n_1)$ such operations. Hence the algorithm is in PTIME.

□

Lemma 8.8.

(1) The non-emptiness problem for CE is in NL.
(2) The non-emptiness problem for SECE is in PSPACE.
(3) The non-emptiness problem for ECE and SFCE is NONELEMENTARY.

Proof.

(1-2) Any $e \in CE$ (resp. $SECE$) with non-empty language has a word of size linear (resp. exponential) in size of e. This word can be nondeterministically guessed one letter at a time and the membership checking algorithm can be used to confirm the membership. A register counts the length of the word guessed so far and the algorithm terminates with "no" after linear (resp. exponential) word size is exceeded. The space required for word length register is order log (resp. polynomial) of $|e|$.

(3) Non-emptiness problem for ERE and SFRE is known to be NONELEMENTARY [6]. By Theorems 8.2 and 8.3, there exists elementary conversion between extended chop expressions and extended regular expressions. Hence, non-emptiness problem for ECE and SFCE are also NONELEMENTARY. □

Lemma 8.9.

(1) The inequivalence problem for CE is in PSPACE.
(2) The inequivalence problem for SECE is in EXPSPACE.
(3) The inequivalence problem for ECE and SFCE is NONELEMENTARY.

Proof.

(1) Given a CE of length n, it is easy to build a non-deterministic finite automaton with number of states linear in n (see section 8.5). NFA-INEQUIVALENCE problem is well-known to be in PSPACE [25, p. 315].
(2) The inequivalence problem for SECE has an EXPSPACE upper bound because it is easy to build a non-deterministic finite automaton with 2^n states, which accepts the language described by a SECE of length n (see section 8.5). The NFA-INEQUIVALENCE problem is well-known to be in PSPACE [25, p. 315] with respect to the number of states of the automata.
(3) Inequivalence problem for ERE and SFRE is known to be NONELEMENTARY [6]. By Theorem 8.2 and 8.3, there exists elementary conversion between extended chop expressions and extended regular expressions. Hence, inequivalence problem for ECE and SFCE are also NONELEMENTARY. □

Every ECET can be converted to ECE by by pushing in the $\sim$ using De Morgan's laws and finally replacing any $\sim a_i$ with $(a_1 + \ldots + a_{i-1} + a_{i+1} + \ldots + a_n)$. This reduction can be carried out in DLOGSPACE and the resulting expression would have a blowup by at most $|\Sigma|$ times (which is a constant). Hence, the decision complexities of the membership, non-emptiness and inequivalence problems for *ECE* and its subclasses remains unchanged on addition of test expressions to their syntax.

The table in Figure 8.3 lists the complexities of decision procedures of extended chop expressions and its subclasses.

8.8. From Discrete Duration Calculus to Extended Chop Expressions and Back

In Section 8.2, we gave an interval temporal logic called DDC^* and we gave an effective model (language) preserving transformation ξ from DDC^* to its subset $CoreDDC^*$. In this section, we show that $CoreDDC^*$ and $ECET$ are closely related by giving effective homomorphic conversions between them. This allows the

Expression	Membership	Inequivalence	Non-Emptiness
CE/*CET*	NL Complete	PSPACE Complete	$\geq$ ALOGTIME $\leq$ NL
SECE/*SECET*	LOGCFL Complete	EXPSPACE Complete	PSPACE Complete
ECE	P Complete	NONELEM	NONELEM
SFCE	P Complete	NONELEM	NONELEM

Fig. 8.3. Complexity of Decision Problems of (Extended) Chop Expression.

conditional equational axioms as well as the automaton synthesis techniques for *ECET* to be applied to *CoreDDC** and hence to *DDC**.

A formula D of *CoreDDC** specifies a language over the alphabet $VAL(Pvar)$ (see Section 8.2). Typically, this alphabet turns out to be rather large and syntactically, it is difficult to write expressions directly in terms of its elements. For example, if *Pvar* consists of 50 propositional variables, the alpabet size is 2^{50}.

Let *Extended Chop Expressions with Propositional Tests* (*ECETP*) be a variant of *ECET* with the change that in place of the letters of the alphabet $VAL(Pvar)$, the atomic test expressions have the form $\hat{p}$ where $p \in Pvar$. We do not repeat the full syntax here. The semantics of such an atomic test expression is given below.

$$\mathcal{L}(\hat{p}) = \{s \in VAL(Pvar) \mid s \models p\}$$

Example 8.3. Let $Pvar = \{p, q, r\}$. Then $(\hat{p}; (\hat{q} + {\sim}\hat{r}); \Delta)^{\otimes}; (\hat{p}; \hat{q})$ denotes the collection of words over 3-bit vectors (representing p,q,r) such that the proposition $p \wedge (q \vee \neg r)$ holds for every letter of the word and the last letter satisfies $p \wedge q$.

We now define a language preserving transformation $\gamma : CoreDDC^* \rightarrow ECETP$ as follows. Given proposition P over *Pvar*, let $\hat{P}$ denote the test expression of *ECETP* obtained by substituting in P each p by $\hat{p}$, operator $\wedge$ by ;, operator $\vee$ by + and operator $\neg$ by $\sim$. Further, let

$$\begin{aligned}
&\gamma(\lceil P \rceil^0) \triangleq \hat{P} \\
&\gamma(\lceil\lceil P \rceil) \triangleq Ext \ \wedge \ \neg(\top; ({\sim}\hat{P}); Ext) \\
&\gamma(D_1; D_2) \triangleq \gamma(D_1); \gamma(D_2) \\
&\gamma(D^{\otimes}) \triangleq \gamma(D)^{\otimes} \\
&\gamma(D_1 \wedge D_2) \triangleq \gamma(D_1) \cap \gamma(D_2) \\
&\gamma(\neg D) \triangleq \neg\gamma(D)
\end{aligned}$$

An alternative translation is $\gamma(\lceil\lceil P \rceil) \triangleq \hat{P}; \Delta; (\hat{P}; \Delta)^{\otimes}$.

Lemma 8.10. *For any* $D \in CoreDDC^*$, *we have* $\mathcal{L}(D) = \mathcal{L}(\gamma(D))$

The proof of this lemma is straightforward by induction on the structure of D and Proposition 8.1. We omit this proof.

We now give a language preserving reverse translation ζ : *ECETP* $\to$ *CoreDDC**. We first translate tests to propositions as follows:

$$\begin{aligned}
&\zeta'(\hat{p}) \triangleq p\\
&\zeta'(\mathbf{0}) \triangleq \mathit{false}\\
&\zeta'(\mathbf{1}) \triangleq \mathit{true}\\
&\zeta'(\sim t) \triangleq \neg\zeta(t)\\
&\zeta'(t_1 + t_2) \triangleq \zeta(t_1) \vee \zeta(t_2)\\
&\zeta'(t_1; t_2) \triangleq \zeta(t_1) \wedge \zeta(t_2)
\end{aligned}$$

We now translate chop expressions to *CoreDDC** formulae.

$$\begin{aligned}
&\zeta(t) \triangleq \lceil \zeta'(t) \rceil^0\\
&\zeta(\mathbf{0}) \triangleq \mathit{false}\\
&\zeta(\mathbf{1}) \triangleq Pt\\
&\zeta(\Delta) \triangleq Unit\\
&\zeta(ce_1; ce_2) \triangleq \zeta(ce_1); \zeta(ce_2)\\
&\zeta(ce^{\otimes}) \triangleq \zeta(ce)^{\otimes}\\
&\zeta(ce_1 + ce_2) \triangleq \zeta(ce_1) \vee \zeta(ce_2)\\
&\zeta(ce_1 \cap ce_2) \triangleq \zeta(ce_1) \wedge \zeta(ce_2)\\
&\zeta(\neg ce) \triangleq \neg\zeta(ce)
\end{aligned}$$

Lemma 8.11. *For any $ce \in ECETP$, we have $\mathcal{L}(ce) = \mathcal{L}(\zeta(ce))$.*

The proof of the lemma is by induction on the structure of *ce*. It is omitted here.

Formal Verification of *DDC** The reductions γ and ζ give rise to at most linear blowup in size. Lemmas 8.10 and 8.11 show that *ECETP* and *CoreDDC** are equivalent up to the homomorphisms γ and ζ. Moreover, there is a language preserving reduction ξ from *DDC** to its subset *CoreDDC**. Hence, all the results and practical techniques developed for *ECETP* apply to *DDC**. In particular, the algebraic axioms for proving equality and ordering of chop expressions can be used to prove equivalence and implication between *DDC** formulae.

Example 8.4.

$$\begin{aligned}
&\lceil p \rceil^0; (\lceil\lceil \neg p \rceil \vee Pt); \lceil q \rceil^0 && \text{applying } \gamma\\
\equiv\ &\hat{p}; [(\sim\hat{p}; \Delta; (\sim\hat{p}; \Delta)^{\otimes}) \ + \ \mathbf{1}]; \hat{q}\\
\equiv\ &[\hat{p}; (\sim\hat{p}; \Delta; (\sim\hat{p}; \Delta)^{\otimes}); \hat{q}] \ + \ [\hat{p}; \mathbf{1}; \hat{q}]\\
\equiv\ &\mathbf{0} \ + \ \hat{p}; \hat{q}\\
\equiv\ &\hat{p}; \hat{q} && \text{applying } \zeta\\
\equiv\ &\lceil p \wedge q \rceil^0
\end{aligned}$$

The language preserving reductions ξ : *DDC** $\to$ *CoreDDC** (Section 8.2.1) and γ : *CoreDDC** $\to$ *ECETP* together with the reduction from *ECET* to NFA (Corollary 8.1) give us a method for automatic synthesis of automata for *DDC** formulae.

Theorem 8.5. *For every DDC* formula D over Pvar, we can effectively construct a deterministic finite state automaton A(D) over the alphabet VAL(Pvar) such that* $\mathcal{L}(D) \;=\; \mathcal{L}(A(D))$.

Corollary 8.2. *Satisfiability and validity of DDC* formulae is decidable.*

Proof outline For checking satisfiability of $D \in DDC^*$ we can construct the automaton $A(D)$. A word satisfying the formula can be found by searching for an accepting path within $A(D)$. Such a search can be carried out in time linear in the size (number of nodes + edges) of $A(D)$ by depth-first search.

Example 8.5. The property of Example 8.1 can be stated in DDC^* as formula $\lceil p \rceil^0; \lceil\lceil \neg q \rceil; \lceil q \rceil^0$. On Expansion, this gives us the chop expression $\hat{p}; (Ext \wedge \neg(\top; \hat{q}; Ext)); \hat{q}$. The automaton corresponding to this formula is given below. Each edge is labelled with a column vector giving truth values of variables p, q as in Example 8.1. Letter X is used to denote either 0 or 1.

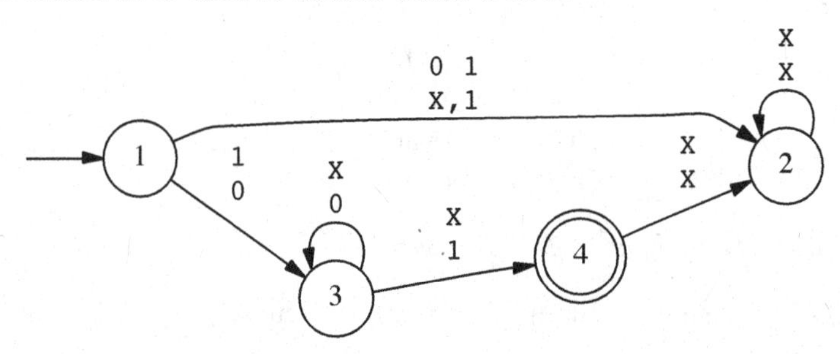

DCVALID The reduction from formulae of DDC^* to finite state automata as outlined in Theorem 8.5 has been implemented into a tool called DCVALID [1], which also checks for the validity of formulae as in corollary 8.2. This tool is built on top of MONA [26]. MONA is a sophisticated and efficient BDD-based implementation of the automata-theoretic decision procedure for monadic logic over finite words [27, 28]. Note that the problem of representing automata with very large alphabet $VAL(Pvar)$ must be overcome. MONA represents the transition function over such large alphabet using multi-terminal BDDs. It also has procedures for determinization, minimization and product of finite automata. The reader is referred to the original paper on MONA [26] for technical details.

An associated tool, called CTLDC, translates the automaton into Esterel, SMV or Verilog module to give a synchronous observer for the property [2]. Using this, DCVALID can model check whether $M \models D$ where M is an Esterel, SMV or Verilog program and D is a DDC^* formula [2]. Using the tools CTLDC and SMV, the properties given in Example 8.2 could be verified for 100 cell arbiter in 3.51 seconds [3]. It should be noted that while the worst-case lower bound on automaton size for *CoreDDC* formulae is NONELEMENTARY, such blowup is rarely observed in practice and we have been able to check validity of some formulae which are over 200 lines long with the tool DCVALID [1].

Acknowledgment

We would like to thank Kamal Lodaya, Deepak D'Souza, Priti Shankar and the anonymous reviewer for their careful reviewing and constructive comments on this article. We thank Nutan Limaye for providing valuable inputs on the complexities of several decision procedures.

References

[1] P. K. Pandya. Specifying and Deciding Quantified Discrete-time Duration Calculus formulae using DCVALID. In *Proc. Real-Time Tools*, (2001).
[2] P. K. Pandya. Model checking CTL*[DC]. In *Proc. 7th TACAS*, vol. 2031, *LNCS*, pp. 559–573. Springer, (2001).
[3] P. K. Pandya, The saga of synchronous bus arbiter: On model checking quantitative timing properties of synchronous programs, *Electr. Notes Theor. Comput. Sci.* **65**(5), (2002).
[4] B. Moszkowski, A temporal logic for multi-level reasoning about hardware, *IEEE Computer.* **18**(2), (1985).
[5] Z. Chaochen, C. A. R. Hoare, and A. P. Ravn, A calculus of durations, *Inf. Process. Lett.* **40**(5), 269–276, (1991).
[6] L. J. Stockmeyer and A. R. Meyer. Word problems requiring exponential time: Preliminary report. In *Proc. 5th STOC*, pp. 1–9. ACM, (1973).
[7] P. K. Pandya, Y. S. Ramakrishna, and R. K. Shyamasundar. A compositional semantics of Esterel in Duration Calculus. In *Proc. AMAST workshop on Real-time Systems: Models and Proofs*. Springer, (1995).
[8] D. Harel, D. Kozen, and R. Parikh, Process logic: Expressiveness, decidability, completeness, *J. Comput. Syst. Sci.* **25**(2), 144–170, (1982).
[9] D. Kozen, Kleene algebra with tests, *Trans. Program. Lang. Syst.* **19**(3), 427–443, (1997).
[10] M. Fränzle. Decidability of duration calculi on restricted model classes. Technical Report ProCoS MF/1, Christian-Albrechts Universitat Kiel, Germany, (1996).
[11] D. Kozen, *Automata and Computability.* (Springer, 1997).
[12] P. Padawitz, *Computing in Horn clause theories.* (Springer, 1988).
[13] A. Salomaa, Two complete axiom systems for the algebra of regular events, *J. ACM.* **13**(1), 158–169, (1966).
[14] A. Salomaa and V. Tixier, Two complete axiom systems for the extended language of regular expressions, *IEEE Trans. on Computers.* **17**, 700–701, (1968).
[15] D. Kozen and F. Smith. Kleene algebra with tests: Completeness and decidability. In *Proc. 10th CSL*, vol. 1258, *LNCS*, pp. 244–259. Springer, (1996).
[16] J. E. Hopcroft and J. D. Ullman, *Introduction to Automata Theory, Languages, and Computation.* (Addison-Wesley, 1979).
[17] J. A. Brzozowski, Derivatives of regular expressions, *J. ACM.* **11**(4), 481–494, (1964).
[18] N. A. Lynch, Log space recognition and translation of parenthesis languages, *J. ACM.* **24**(4), 583–590, (1977).
[19] T. Jiang and B. Ravikumar, A note on the space complexity of some decision problems for finite automata, *Inf. Process. Lett.* **40**(1), 25–31, (1991).
[20] A. R. Meyer and L. J. Stockmeyer. The equivalence problem for regular expressions with squaring requires exponential space. In *Proc. 13th FOCS*, pp. 125–129. IEEE Computer Society, (1972).

[21] H. Petersen. Decision problems for generalized regular expressions. In *Proc. Workshop on Descriptional Complexity of Automata Grammar and Related Structures*, pp. 22–29, (2000).
[22] H. Petersen. The membership problem for regular expressions with intersection is complete in LOGCFL. In *Proc. 19th STACS*, vol. 2285, *LNCS*, pp. 513–522, (2002).
[23] H. Hunt III. The equivalence problem for regular expressions with intersection is not polynomial in tape. Technical Report Report TR73-161, Cornell University, Ithaca, N.Y, (1973).
[24] S. R. Buss. Algorithms for boolean formula evaluation and for tree contraction. In *Arithmetic, Proof Theory and Computational Complexity*, pp. 96–115. Oxford University Press, (1993).
[25] M. Sipser, *Introduction to the theory of computation.* (PWS, 1997).
[26] J. Henriksen, J. Jensen, M. Jørgensen, N. Klarlund, R. Paige, T. Rauhe, and A. Sandholm. Mona: Monadic second-order logic in practice. In *Proc. 17th TACAS*, vol. 6605, *LNCS*, pp. 89–110. Springer, (1995).
[27] J. R. Büchi, Weak second-order arithmetic and finite automata, *Zeitschrift für Mathematische Logik und Grundlagen der Mathematik.* **6**(1-6), 66–92, (1960).
[28] C. C. Elgot, Decision problems of finite automata design and related arithmetics, *Trans. Amer. Math. Soc.* **98**, (1961).

Chapter 9

Automata on Distributed Alphabets

Madhavan Mukund
Chennai Mathematical Institute
H1 SIPCOT IT Park, Padur PO
Siruseri 603103, India
madhavan@cmi.ac.in

Traditional automata theory is an extremely useful abstraction for reasoning about sequential computing devices. For distributed systems, however, there is no clear consensus on how best to incorporate various features such as spatial independence, concurrency and communication into a formal computational model. One appealing and elegant approach is to have a network of automata operating on a distributed alphabet of local actions. Components are assumed to synchronize on overlapping actions and move independently on disjoint actions. We describe two formulations of automata on distributed alphabets, synchronous products and asynchronous automata, that differ in the degree to which distributed choices can be coordinated. We compare the expressiveness of the two models and provide a proof of Zielonka's fundamental theorem connecting regular trace languages to asynchronous automata. Along the way, we describe a distributed time-stamping algorithm that is basic to many interesting constructions involving these automata.

9.1. Introduction

Automata theory provides an extremely useful abstract description of sequential computing devices. A typical computer program manipulates variables. The *state* of the program is the set of values currently assigned to these variables. A computation is a sequence of steps that transforms a program from an initial state to a desired final state.

When we model programs in automata theory, we typically hide the concrete structure of states in terms of the variables used and their values and, instead, assign abstract names to these states. In the same way, we hide the specific nature of the transformations from one state to another and replace them by abstract actions.

In this article, we shift focus from traditional sequential computation to distributed computation. Our aim is to model programs that run on multiple computing devices and have to interact with each other in order to achieve their objective. In this setting, we need to model how programs interact and the way in which they exchange information during these interactions.

There is no clear consensus on how best to incorporate various features such as spatial independence, concurrency and communication into a formal computational model. One appealing and elegant approach is to have a network of automata operating on a distributed alphabet of local actions. The components are assumed to synchronize on overlapping actions and move independently on disjoint actions.

In the most simplistic model, synchronizations serve only to coordinate the actions of independent components and no information is exchanged between components. We call such networks *synchronized products*.

A more elaborate model, proposed by Zielonka [1], is one in which processes share the information in their local states when they synchronize. This facility greatly enhances the computational power of the model. These automata, called *asynchronous automata*, have close connections with Mazurkiewicz trace theory, a language-theoretic formalism for studying concurrent systems [2].

The article begins with a quick introduction to transition systems and automata. We then define direct product automata, which are the building blocks of synchronized products. After establishing some characterization and closure properties of these models, we move on to asynchronous automata and their connection to trace languages. We describe a distributed time-stamping algorithm for these automata that is fundamental for many automata-theoretic results in trace theory. Using this tool we prove Zielonka's theorem that every regular trace language is recognized by an asynchronous automaton whose structure reflects the independence structure of the trace language.

9.2. Transition systems, automata and languages

As a computation evolves, the internal state of the computing device is transformed through a sequence of actions. We model this using labelled transition systems, in which we abstractly represent the various possible states of the system and the moves that the system makes from one state to another, labelled by an appropriate action.

Labelled transition systems Let Σ be a set of *actions*.

- A transition system over Σ is a triple $TS = (Q, \rightarrow, Q_{\mathrm{in}})$ where Q is a set of *states*, $\rightarrow \subseteq Q \times \Sigma \times Q$ is the *transition relation* and $Q_{\mathrm{in}} \subseteq Q$ is the set of *initial states*. We usually write $q \xrightarrow{a} q'$ to denote that $(q, a, q') \in \rightarrow$. As usual, a transition system is *deterministic* if the transition relation $\rightarrow$ satisfies the property that whenever $q \xrightarrow{a} q'$ and $q \xrightarrow{a} q''$, $q' = q''$.
- A (finite-state) automaton over Σ is a quadruple $A = (Q, \rightarrow, Q_{\mathrm{in}}, F)$ where $(Q, \rightarrow, Q_{\mathrm{in}})$ is a transition system with a finite set of states over a finite set of actions Σ, and $F \subseteq Q$ is a set of final states. An automaton is deterministic if the underlying transition system is.

Runs Let $A = (Q, \rightarrow, Q_{\text{in}}, F)$ be an automaton over Σ and let $w = a_1 a_2 \ldots a_n$ be a word in Σ^*. A run of A on w is a sequence of states $q_0 q_1 \ldots q_n$ such that $q_0 \in Q_{\text{in}}$ and for each $i \in [1..n]$, $q_{i-1} \xrightarrow{a_i} q_i$. (For natural numbers $m \leq n$, we write $[m..n]$ to denote the set $\{m, m+1, \ldots, n\}$.) This run is said to be *accepting* if $q_n \in F$.

The automaton A *accepts* or *recognizes* w if it admits at least one accepting run on w. The language of A, $L(A)$ is the set of all words over Σ that A recognizes.

9.3. Direct product automata

A large class of distributed systems can be modelled as networks of local transition systems whose moves are globally synchronized through common actions. To formalize this, we begin with the notion of a distributed alphabet.

Distributed alphabets A distributed alphabet over Σ, or a *distribution* of Σ, is a tuple of nonempty sets $\theta = \langle \Sigma_1, \Sigma_2, \ldots, \Sigma_k \rangle$ such that $\bigcup_{1 \leq i \leq k} \Sigma_i = \Sigma$. For each action $a \in \Sigma$, the *locations* of a with respect to the distribution θ is the set $loc_\theta(a) = \{i \mid a \in \Sigma_i\}$. If θ is clear from the context, we write just $loc(a)$ instead of $loc_\theta(a)$.

Direct product automaton Let $\langle \Sigma_1, \Sigma_2, \ldots, \Sigma_k \rangle$ be a distribution of Σ. For each $i \in [1..k]$, let $A_i = (Q_i, \rightarrow_i, Q^i_{\text{in}}, F_i)$ be an automaton over Σ_i. The *direct product automaton* $(A_1 \parallel A_2 \parallel \cdots \parallel A_k)$ is the automaton $A = (Q, \rightarrow, Q_{\text{in}}, F)$ over $\Sigma = \bigcup_{1 \leq i \leq k} \Sigma_i$, where:

- $Q = Q_1 \times Q_2 \times \cdots \times Q_k$.
- Let $\langle q_1, q_2, \ldots, q_k \rangle, \langle q'_1, q'_2, \ldots, q'_k \rangle \in Q$.
  Then $\langle q_1, q_2, \ldots, q_k \rangle \xrightarrow{a} \langle q'_1, q'_2, \ldots, q'_k \rangle$ if
  - For each $j \in loc(a)$, $q_j \xrightarrow{a}_j q'_j$.
  - For each $j \notin loc(a)$, $q_j = q'_j$.
- $Q_{\text{in}} = Q^1_{\text{in}} \times Q^2_{\text{in}} \times \ldots \times Q^k_{\text{in}}$.
- $F = F_1 \times F_2 \times \ldots \times F_k$.

Direct product language Let $\langle \Sigma_1, \Sigma_2, \ldots, \Sigma_k \rangle$ be a distribution of Σ. $L \subseteq \Sigma^*$ is said to be a direct product language if there is a direct product automaton $A = (A_1 \parallel A_2 \parallel \cdots \parallel A_k)$ such that $L = L(A)$.

Direct product languages can be precisely characterized in terms of their projections onto the local components of the system.

Projections Let $\langle \Sigma_1, \Sigma_2, \ldots, \Sigma_k \rangle$ be a distribution of Σ. For $w \in \Sigma^*$ and $i \in [1..k]$, the projection of w onto Σ_i is denoted $w{\downarrow_{\Sigma_i}}$ and is defined inductively as follows:

- $\varepsilon{\downarrow_{\Sigma_i}} = \varepsilon$, where ε is the empty string.
- $wa{\downarrow_{\Sigma_i}} = \begin{cases} (w{\downarrow_{\Sigma_i}})a & \text{if } a \in \Sigma_i \\ (w{\downarrow_{\Sigma_i}}) & \text{otherwise} \end{cases}$

Shuffle closure The shuffle closure of L with respect to $\langle \Sigma_1, \Sigma_2, \ldots, \Sigma_k \rangle$, *shuffle*$(L, \langle \Sigma_1, \Sigma_2, \ldots, \Sigma_k \rangle)$, is the set

$$\{w \in \Sigma^* \mid \forall i \in [1..k], \exists u_i \in L, w{\downarrow_{\Sigma_i}} = u_i{\downarrow_{\Sigma_i}}\}$$

As usual, we write just *shuffle*(L) if $\langle \Sigma_1, \Sigma_2, \ldots, \Sigma_k \rangle$ is clear from the context.

Proposition 9.1. *Let $\langle \Sigma_1, \Sigma_2, \ldots, \Sigma_k \rangle$ be a distribution of Σ and let $L \subseteq \Sigma^*$ be a regular language. L is a direct product language iff $L =$ shuffle(L).*

Proof Sketch: ($\Rightarrow$) Suppose that L is a direct product language. It is easy to see that $L \subseteq$ *shuffle*(L), so we show that *shuffle*$(L) \subseteq L$. Since L is a direct product language, there exists a direct product automaton $A = (A_1 \parallel A_2 \parallel \cdots \parallel A_k)$ such that $L = L(A)$.

Let $w \in$ *shuffle*(L). For each $i \in [1..k]$, there is a witness $u_i \in L$ such that $w{\downarrow_{\Sigma_i}} = u_i{\downarrow_{\Sigma_i}}$. Since $u_i \in L$, there is an accepting run $q \in Q^i_{\text{in}} \xrightarrow{u_i\downarrow_{\Sigma_i}}_i q_f \in F_i$ in A_i. Since this is true for every i, we can "glue" these runs together and construct an accepting run for A on w, so $w \in L(A) = L$.

($\Leftarrow$) Suppose that $L =$ *shuffle*(L). We prove that L is a direct product language. For $i \in [1..k]$, $L_i = L{\downarrow_{\Sigma_i}}$ is a regular language, since homomorphic images of regular languages are regular. For each $i \in [1..k]$, there exists a deterministic automaton A_i such that $L_i = L(A_i)$. It is then easy to see that $L = L(A_1 \parallel A_2 \parallel \cdots \parallel A_k)$. □

Proposition 9.2. *Direct product languages are not closed under boolean operations.*

Example 9.3.
Let $\theta = \langle \{a\}, \{b\} \rangle$ and let $L = \{ab, ba, aabb, abab, abba, baab, baba, bbaa\}$. Then L is clearly the union of $\{ab, ba\}$ and $\{aabb, abab, abba, baab, baba, bbaa\}$, both of which are direct product languages. However, L is not itself a direct product language because $L \neq$ *shuffle*(L). For instance, $abb \in$ *shuffle*$(L) \setminus L$.

9.4. Synchronized products

We can increase the expressiveness of product automata by removing the restriction that the final states are just the product of the local final states of each component. Instead, we permit an arbitrary subset of global states to be final states.

Synchronized product automaton Let $\langle \Sigma_1, \Sigma_2, \ldots, \Sigma_k \rangle$ be a distribution of Σ. For each $i \in [1..k]$, let $TS_i = (Q_i, \rightarrow_i, Q^i_{\text{in}})$ be a transition system over Σ_i. The *synchronized product automaton* of $(TS_1, TS_2, \ldots, TS_k)$ is an automaton $A = (Q, \rightarrow, Q_{\text{in}}, F)$ over $\Sigma = \bigcup_{1 \leq i \leq k} \Sigma_i$, where:

- $Q = Q_1 \times Q_2 \times \cdots \times Q_k$
- Let $\langle q_1, q_2, \ldots, q_k \rangle, \langle q'_1, q'_2, \ldots, q'_k \rangle \in Q$.
  Then $\langle q_1, q_2, \ldots, q_k \rangle \xrightarrow{a} \langle q'_1, q'_2, \ldots, q'_k \rangle$ if
  - For each $j \in loc(a)$, $q_j \xrightarrow{a}_j q'_j$.
  - For each $j \notin loc(a)$, $q_j = q'_j$.
- $Q_{\text{in}} = Q^1_{\text{in}} \times Q^2_{\text{in}} \times \ldots \times Q^k_{\text{in}}$.
- $F \subseteq Q_1 \times Q_2 \times \ldots \times Q_k$.

Synchronized product language Let $\langle \Sigma_1, \Sigma_2, \ldots, \Sigma_k \rangle$ be a distribution of Σ. $L \subseteq \Sigma^*$ is said to be a synchronized product language if there is a synchronized product automaton A such that $L = L(A)$.

Example 9.4. The language defined in Example 9.3 is a synchronized product language. The synchronized product automaton for this language is shown in Figure 9.1. The set of global final states F is $\{\langle q_1, q'_1 \rangle, \langle q_2, q'_2 \rangle\}$.

Fig. 9.1. A synchronized product automaton for Example 9.3.

Proposition 9.5. *A language is a synchronized product language if and only if it can be written as a finite union of direct product languages.*

Proof Sketch: ($\Rightarrow$) Let $A = (Q, \rightarrow, Q_{\text{in}}, F)$ be a synchronized product such that $\langle TS_1, TS_2, \ldots, TS_k \rangle$ are the component transition systems over $\langle \Sigma_1, \Sigma_2, \ldots, \Sigma_k \rangle$. For each $f = \langle f_1, f_2, \ldots, f_k \rangle \in F$, extend TS_i to an automaton $A^f_i = (TS_i, f_i)$ and construct the direct product $A_f = (A^f_1 \parallel A^f_2 \parallel \cdots \parallel A^f_k)$. Then, $L(A) = \bigcup_{f \in F} L(A_f)$.

($\Leftarrow$) Conversely, let L be a finite union of direct product languages $\{L_i\}_{i \in [1..m]}$, where each L_i is recognized by a direct product $A^i = (A^i_1 \parallel A^i_2 \parallel \cdots \parallel A^i_k)$. For $j \in [1..k]$, let $A^i_j = (Q^i_j, \rightarrow^i_j, Q^{i_j}_{\text{in}}, F_j)$ be the jth component of A^i. We construct a synchronous product $\hat{A} = (\hat{A}_1 \parallel \hat{A}_2 \parallel \cdots \parallel \hat{A}_k)$ as follows. For each component j, we let $\widehat{Q}_j$ be the disjoint union $\biguplus_{i \in [1..m]} Q^i_j$ and define the set of initial states of component j be $\bigcup_{i \in [1..m]} Q^{i_j}_{\text{in}}$. The local transition relations of each component are given by the union $\bigcup_{i\in[1..m]} \rightarrow_j^i$. The crucial point is to define the global set of final states as $(F_1^1 \times F_2^1 \times \cdots \times F_k^1) \cup (F_1^2 \times F_2^2 \times \cdots \times F_k^2) \cup \cdots \cup (F_1^m \times F_2^m \times \cdots \times F_k^m)$. This ensures that the synchronized product accepts only if all components agree on the choice of L_i. □

Proposition 9.6. *Synchronized product languages are closed under boolean operations.*

Proof Sketch. Let L_1 and L_2 be synchronized product languages. Then, by definition, $L_1 = L'_{11} \cup L'_{12} \cup \cdots \cup L'_{1k_1}$ and $L_2 = L''_{21} \cup L''_{22} \cup \cdots \cup L''_{k_2}$, where $\{L'_{11}, L'_{12}, \ldots, L'_{1k_1}\}$ and $\{L''_{21}, L''_{22}, \ldots, L''_{2k_2}\}$ are both sets of direct product languages. It is immediate that $L_1 \cup L_2$ is the union of these two collections, so $L_1 \cup L_2$ is a synchronized product language.

To show closure under complementation, we prove that any synchronized product language is recognized by a *deterministic* synchronized product automaton. If we assume this, we can complement a synchronized product language by exchanging final and non-final states. In other words, if L is a synchronized product language recognized by a deterministic synchronized product automaton $A = (Q, \rightarrow, Q_{\text{in}}, F)$, then $\bar{L}$, the complement of L, is recognized by the automaton $\bar{A} = (Q, \rightarrow, Q_{\text{in}}, Q \setminus F)$.

Let us assume we are working with respect to a distribution $\langle \Sigma_1, \Sigma_2, \ldots, \Sigma_k \rangle$ of Σ. Every synchronized product language L over Σ is a finite union $L_1 \cup L_2 \cup \cdots \cup L_m$ of direct product languages. We establish our claim by induction on m.

If $m = 1$, we have already seen that we can construct the direct product automaton $A = (A_1 \parallel A_2 \parallel \cdots \parallel A_k)$ recognizing L, where for $i \in [1..k]$, A_i is a deterministic automaton recognizing $L{\downarrow_{\Sigma_i}}$.

Now, let $L = L_1 \cup L_2 \cup \ldots \cup L_m$ where L_1 is a direct product language and $L' = L_2 \cup \ldots \cup L_m$ is a synchronized product language. We can assume that L_1 is recognized by a deterministic direct product automaton $A = (Q_A, \rightarrow_A, Q_{\text{in}}^A, F_A) = (A_1 \parallel A_2 \parallel \cdots \parallel A_k)$ and L', by the induction hypothesis, is recognized by a deterministic synchronized product $B = (Q_B, \rightarrow_B, Q_{\text{in}}^B, F_B)$ defined with respect to transition systems $\langle TS_1, TS_2, \ldots, TS_k \rangle$. For each $i \in [1..k]$, let $A_i = (Q_i, \rightarrow_i, Q_{\text{in}}^i, F_i)$ and $TS_i = (\widehat{Q}_i, \Rightarrow_i, \widehat{Q}_{in}^i)$. Define a new deterministic transition system $\widetilde{TS}_i i$ with states $Q_i \times \widehat{Q}_i$, initial states $\{(q_1, q_2) \mid q_1 \in Q_{\text{in}}^i, q_2 \in \widehat{Q}_{in}^i\}$ and transitions of the form $(q_1, q_1') \xrightarrow{a} (q_2, q_2')$ iff $(q_1, a, q_1') \in \rightarrow_i$ and $(q_2, a, q_2') \in \Rightarrow_i$. Clearly, each $\widetilde{TS}_i$ is a deterministic transition system. We now construct a deterministic synchronized product automaton recognizing L from $\langle \widetilde{TS}_1, \widetilde{TS}_2, \ldots, \widetilde{TS}_k \rangle$ by setting $\widetilde{F} = ((F_1 \times \widehat{Q_1}) \times (F_2 \times \widehat{Q_2}) \times \cdots (F_k \times \widehat{Q_k})) \cup \{\langle (q_1, f_1), (q_2, f_2), \ldots, (q_k, f_k) \rangle \mid q_i \in Q_i, \langle f_1, f_2, \ldots, f_k \rangle \in F_B\}$. □

Synchronized product automata are still not as expressive as we would like.

Example 9.7. Let $\theta = \langle \{a,c\}, \{b,c\} \rangle$. Then,

$$L = \Big[\big[\mathit{shuffle}(\{ab\}) + \mathit{shuffle}(\{aabb\})\big].c\Big]^*$$

is not a synchronized product language.

Proof. If L is a synchronized product language, then L can be expressed as a finite union $L_1 \cup L_2 \cup \cdots \cup L_k$ of direct product languages. Let us write 0 for the word *abc* and 1 for word *aabbc*. Consider the following set of $k{+}1$ words of length k with at most one 1: $A_k = \{00\ldots0, 10\ldots0, 010\ldots0, \ldots, 00\ldots01\}$. By the pigeonhole principle, there must be two words $u, v \in A_k$ that belong to the same direct product component L_j, $j \in [1..k]$.

There are two cases to consider.

- Suppose that u and v differ at only one position. Then, without loss of generality, it must be the case that $u = 00\ldots0$ has no 1's. Let v have a 1 at position m, $m \in [1..k]$. Construct a new word $w = (abc)^{m-1}(abbc)(abc)^{k-m}$. It is easy to see that $w \downarrow_{\{a,c\}} = u \downarrow_{\{a,c\}}$ and $w \downarrow_{\{b,c\}} = v \downarrow_{\{b,c\}}$. So, $w \in \mathit{shuffle}(L_j)$ and hence $w \in L_j \subseteq L$ by Proposition 9.1, which is a contradiction.
- Suppose that u and v differ at two positions. Then u has a 1 at position m and v has a 1 at position m' for some $1 \leq m < m' \leq k$. Construct a word $w = (abc)^{m-1}(aabc)(abc)^{m'-m-1}(abbc)(abc)^{k-m'}$. Once again, it is easy to see that $w{\downarrow}_{\{a,c\}} = u{\downarrow}_{\{a,c\}}$ and $w{\downarrow}_{\{b,c\}} = v{\downarrow}_{\{b,c\}}$. So, $w \in \mathit{shuffle}(L_j)$ and hence $w \in L_j \subseteq L$ by Proposition 9.1, which is a contradiction.

□

9.5. Asynchronous automata

To construct a distributed automaton that can recognize the language from Example 9.7, we have to further enhance the structure of distributed automata.

In direct products and synchronized products, when an action a occurs, all components that participate in a must move simultaneously. However, each component is free to choose its local move independent of all other components. In other words, no information is exchanged between components at the time of synchronization.

For instance, if we try to recognize the language of Example 9.7 in our existing model, we have no way of preventing the c-move enabled after one a in the first component from synchronizing with the c-move enabled after two b's in the second component.

To overcome this limitation, Zielonka proposed an enriched definition of the transition relation for each letter a [1]. As usual, let $loc(a)$ denote the components that participate in a. Then, an a-state is a tuple that belongs to the product $\prod_{i \in loc(a)} Q_i$. Let Q_a denote the set of all a-states. We define the a-transition relation Δ_a to be a subset of $Q_a \times Q_a$. In other words, whenever an a occurs, all the components that participate in a share information about their current local states and *jointly* decide on new local states for themselves. These automata are called *asynchronous automata.**

Asynchronous automaton Let $\langle \Sigma_1, \Sigma_2, \ldots, \Sigma_k \rangle$ be a distribution of Σ. For each $i \in [1..k]$, let Q_i be a finite set of states. For each action $a \in \Sigma$, let $\Delta_a \subseteq Q_a \times Q_a$ be the transition relation for a, where $Q_a = \prod_{i \in loc(a)} Q_i$. The *asynchronous automaton* defined by this data is the following.

- $Q = Q_1 \times Q_2 \times \cdots \times Q_k$
- Let $\langle q_1, q_2, \ldots, q_k \rangle, \langle q'_1, q'_2, \ldots, q'_k \rangle \in Q$. Then $\langle q_1, q_2, \ldots, q_k \rangle \xrightarrow{a} \langle q'_1, q'_2, \ldots, q'_k \rangle$ if
  - For $loc(a) = \{i_1, i_2, \ldots, i_j\}$, $(\langle q_{i_1}, q_{i_2}, \ldots, q_{i_j} \rangle, \langle q'_{i_1}, q'_{i_2}, \ldots, q'_{i_j} \rangle) \in \Delta_a$.
  - For each $j \notin loc(a)$, $q_j = q'_j$.
- $Q_{\text{in}} = Q^1_{\text{in}} \times Q^2_{\text{in}} \times \ldots \times Q^k_{\text{in}}$.
- $F \subseteq Q_1 \times Q_2 \times \ldots \times Q_k$.

In other words, though each component has a set of local states as before, the transition relation for each action is *global* within the set of components where it occurs. A synchronized product automaton can be modelled as an asynchronous automaton by setting Δ_a to be the product $\prod_{i \in loc(a)} \xrightarrow{a}_i$.

Here is an asynchronous automaton for the language of Example 9.7.

Example 9.8. The states of components 1 and 2 are $\{q_0, q_1, q_2\}$ and $\{q'_0, q'_1, q'_2\}$, respectively. There is only one initial and one final state, which is $\langle q_0, q'_0 \rangle$ in both cases.

The transition relations are as follows:

- $\Delta_a = \{(q_0, q_1), (q_1, q_2)\}$
- $\Delta_b = \{(q'_0, q'_1), (q'_1, q'_2)\}$
- $\Delta_c = \{(\langle q_1, q'_1 \rangle, \langle q_0, q'_0 \rangle), (\langle q_2, q'_2 \rangle, \langle q_0, q'_0 \rangle)\}$.

In other words, the two components can reset their local states via c to q_0 and q'_0 only if they are jointly in the state $\langle q_1, q'_1 \rangle$ or $\langle q_2, q'_2 \rangle$. There is no move, for instance, from $\langle q_1, q'_2 \rangle$ via c back to $\langle q_0, q'_0 \rangle$.

9.6. Mazurkiewicz traces

A distributed alphabet induces an independence relation on actions—two actions a and b are independent if they occur on disjoint sets of processes. In all three

*The term asynchronous refers to the fact that components process their inputs locally, without reference to a shared global clock. The "exchange of information" at each move is instantaneous, so the communication between automata is actually synchronous!

models of distributed automata that we have considered so far, this means that an occurrence of a cannot enable or disable b, or vice versa.

Instead of deriving independence from a concrete distribution of the alphabet, we can begin with an alphabet equipped with an abstract independence relation. This is the starting point of the theory of traces, initiated by Mazurkiewicz as a language-theoretic formalism for studying concurrent systems [2].

Independence relation An *independence relation* over Σ is a symmetric, irreflexive relation $I \subseteq \Sigma \times \Sigma$. An alphabet equipped with an independence relation (Σ, I) is called a *concurrent alphabet.*

It is clear that the natural independence relation $\mathcal{I}_{loc}$ induced by a distributed alphabet (Σ, loc), $(a, b) \in \mathcal{I}_{loc}$ iff $loc(a) \cap loc(b) = \emptyset$, is irreflexive and symmetric. The following example shows that different distributions may yield the same independence relation.

Example 9.9. Let $\Sigma = \{a, b, c, d\}$. The three distributions $\widetilde{\Sigma} = \langle\{a, c, d\}, \{b, c, d\}\rangle$, $\widetilde{\Sigma}' = \langle\{a, c, d\}, \{b, c\}, \{b, d\}\rangle$ and $\widetilde{\Sigma}'' = \langle\{a, c\}, \{a, d\}, \{b, c\}, \{b, d\}, \{c, d\}\rangle$ all give rise to the independence relation $I = \{(a, b), (b, a)\}$.

Given a concurrent alphabet $(\Sigma, \mathcal{I})$, there are several ways to construct a distributed alphabet (Σ, loc) so that the independence relation $\mathcal{I}_{loc}$ induced by *loc* coincides with $\mathcal{I}$.

We begin by building the dependence graph for $(\Sigma, \mathcal{I})$. Let $\mathcal{D} = (\Sigma \times \Sigma) \setminus \mathcal{I}$. $\mathcal{D}$ is called the *dependence relation.* Construct a graph $G_\mathcal{D} = (V_\mathcal{D}, E_\mathcal{D})$ with $V_\mathcal{D} = \Sigma$ and $(a, b) \in E_\mathcal{D}$ provided $(a, b) \in \mathcal{D}$.

One way to distribute Σ is to create a process p_e for every edge $e \in E_\mathcal{D}$. For each letter a, we then set $loc(a)$ to be the set of processes (edges) incident on the vertex a. Alternately, we can create a process p_C for each maximal clique C in $G_\mathcal{D}$. Then, for each letter a and each clique C, $p_C \in loc(a)$ iff the vertex labelled a belongs to C.

In both cases, it is easy to see that $\mathcal{I}_{loc} = \mathcal{I}$. So, we can go back and forth between a concurrent alphabet $(\Sigma, \mathcal{I})$ and a distributed alphabet (Σ, loc) whose induced independence relation $\mathcal{I}_{loc}$ is $\mathcal{I}$.

Trace equivalence Let $(\Sigma, \mathcal{I})$ be the concurrent alphabet. $\mathcal{I}$ induces a natural *trace equivalence* $\sim$ on Σ^*: two words w and w' are related by $\sim$ iff w' can be obtained from w by a sequence of permutations of adjacent independent letters. More formally, $w \sim w'$ if there is a sequence of words $v_1, v_2, \ldots, v_k$ such that $w = v_1$, $w' = v_k$ and for each $i \in [1..k{-}1]$, there exist words u_i, u_i' and letters a_i, b_i satisfying

$$v_i = u_i a_i b_i u_i', \; v_{i+1} = u_i b_i a_i u_i' \text{ and } (a_i, b_i) \in \mathcal{I}.$$

Proposition 9.10.

- *The equivalance relation* $\sim$ *is a* congruence *on* Σ^* *with respect to concatenation: If* $u \not\sim u'$ *then for any words* w_1 *and* w_2*,* $w_1 u w_2 \sim w_1 u' w_2$*.*
- *Both right and left cancellation preserve* $\sim$*-equivalence:* $wu \sim wu'$ *implies* $u \sim u'$ *and* $uw \sim u'w$ *implies* $u \sim u'$*.*

9.6.1. *Mazurkiewicz traces*

Equivalence classes of words A *(Mazurkiewicz) trace* over $(\Sigma, \mathcal{I})$ is an equivalence class of words with respect to $\sim$. For $w \in \Sigma^*$, we write $[w]$ to denote the trace corresponding to w—$[w] = \{w' \mid w' \sim w\}$.

The intuition is that all words in a trace describe the same underlying computation of a concurrent system. Each word in a trace corresponds to a reordering of independent events. A more direct way to capture this intuition is to represent a trace as a labelled partial order.

Traces as labelled partial orders A *(Mazurkiewicz) trace* over $(\Sigma, \mathcal{I})$ is a labelled partial order $t = (\mathcal{E}, \leq, \lambda)$ where

- $\mathcal{E}$ is a set of *events*
- $\lambda : \mathcal{E} \to \Sigma$ labels each event by a letter
- $\leq$ is a partial order over $\mathcal{E}$ satisfying the following conditions:
  - $(\lambda(e), \lambda(f)) \in D$ implies $e \leq f$ or $f \leq e$
  - $e \lessdot f$ implies $(\lambda(e), \lambda(f)) \in \mathcal{D}$, where

$$\lessdot \,=\, < \setminus <^2 \,=\, \{(e, f) \mid e < f \text{ and } \not\exists g.\ e < g < f\}$$

  is the immediate successor relation in t.

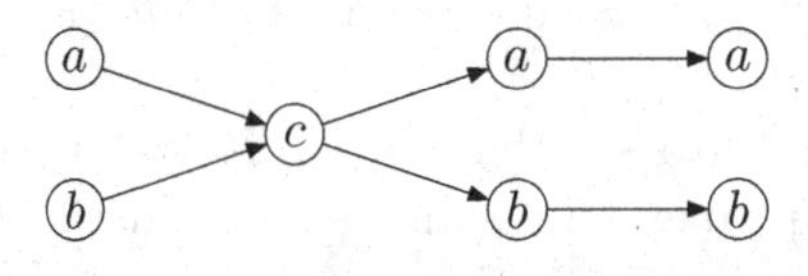

Fig. 9.2. The trace $[bacabba]$ as a labelled partial order.

Example 9.11. Let $\mathcal{P} = \{p, q, r, s\}$ and $\Sigma = \{a, b, c\}$ where $a = \{p, q\}$, $b = \{r, s\}$ and $c = \{q, r, s\}$. Figure 9.2 shows the trace $t = (\mathcal{E}, \leq, \lambda)$ corresponding to the word *bacabba*. The arrows between the events denote the relation $\lessdot$.

If we represent a trace as a labelled partial order, the set of linearizations of this partial order form an equivalence class $[w]$ of words with respect to $\mathcal{I}$. Conversely, it is not difficult to show that each equivalence class $[w]$ generates a unique labelled partial order of which it is the set of linearizations. In fact, the way in which $\leq$ is generated from the dependence relation $\mathcal{D}$ allows us to construct the labelled partial order corresponding to a trace from a single linearization.

Trace languages Traces, like words, form a monoid under concatenation, with the empty trace as the unit. The concatenation operation is easiest to define in terms of the labelled partial order representation—given two traces $t_1 = (\mathcal{E}_1, \leq_1, \lambda_1)$ and $t_2 = (\mathcal{E}_2, \leq_2, \lambda_2)$, the trace $t_i \cdot t_2$ is the trace $t' = (\mathcal{E}', \leq', \lambda')$ where $\mathcal{E}' = \mathcal{E}_1 \cup \mathcal{E}_2$, $\lambda'(e) = \lambda_1(e)$ if $e \in \mathcal{E}_1$ and $\lambda_2(e)$ if $e \in \mathcal{E}_2$, and $\leq'$ is generated by $\lessdot_1 \cup \lessdot_2 \cup \{(x, y) \mid x$ maximal in $\mathcal{E}_1, y$ minimal in $\mathcal{E}_2, (\lambda_1(x), \lambda_2(y)) \in \mathcal{D}\}$.

A *trace language* is a set of traces or, alternatively a subset of the trace monoid. However, we prefer to treat trace languages as string languages which satisfy a closure condition. We say that $L \subseteq \Sigma^*$ is a trace language if L is *$\sim$-consistent*—i.e., for each $w \in \Sigma^*$, w is in L iff every word in $[w]$ is in L. Since traces correspond to equivalence classes of strings, there is a 1-1 correspondence between subsets of the trace monoid and $\sim$-consistent languages over Σ^*.

In the string framework, we say a trace language L is recognizable if it is accepted by a finite-state automaton. Once again, it is not difficult to show that there is a 1-1 correspondence between recognizable subsets of the trace monoid and recognizable $\sim$-consistent languages over Σ^* (see, for instance, [3]).

Henceforth, whenever we use the terms trace language and recognizable trace language, we shall be referring to the definitions in terms of $\sim$-consistent subsets of Σ^* rather than in terms of subsets of the trace monoid.

One of the most fundamental results in trace theory says that every recognizable trace language has a distributed implementation in terms of asynchronous automata.

Theorem 9.12 (Zielonka). *Let L be a recognizable trace language over a concurrent alphabet $(\Sigma, \mathcal{I})$. Then, for* every *distributed alphabet (Σ, loc) such that $\mathcal{I}_{loc} = \mathcal{I}$, we can construct an asynchronous automaton A over (Σ, loc) with $L(A) = L$.*

To prove Zielonka's theorem, we require a distributed timestamping algorithm that is fundamental for many constructions involving asynchronous automata.

9.7. Distributed time-stamping

Let $\mathcal{P} = \{p_1, p_2, \ldots, p_N\}$ be a set of processes which synchronize with each other from time to time and exchange information about themselves and others. The problem we look at is the following: whenever a set $P \subseteq \mathcal{P}$ synchronizes, the processes in P must decide *amongst themselves* which of them has the latest information, direct or indirect, about each process p in the system. We call this the *gossip problem.*

A naïve solution to the gossip problem is to label interactions using a counter whose value increases as time progresses. There are two problems with this approach.

- The counter values (or time-stamps) grow without bound and cannot be generated and stored by finite-state automata.
- Each interaction involves only a subset of processes, so time-stamps have to be generated in a distributed manner.

Our goal is to develop an algorithm to solve the gossip problem that is both finite state and local.

We model the interactions of the processes in $\mathcal{P}$ by a distributed alphabet (Σ, loc) that contains an action X for each nonempty subset $X \subseteq \mathcal{P}$, with the obvious distribution function $loc(X) = X$. A sequence of interactions between the processes is then a trace over (Σ, loc).

Let $t = (\mathcal{E}, \leq, \lambda)$ be a trace representing a computation of our system. For convenience, we assume that t always begins with an initial event $e_\perp$ labelled by $\mathcal{P}$, in which all processes participate. This models the initial exchange of information that is implicit in the fact that all processes agree on a global initial state.

Process-wise ordering Let $t = (\mathcal{E}, \leq, \lambda)$ be a trace. For each event $e \in \mathcal{E}$, we write $p \in e$ to mean that $p \in loc(\lambda(e)) = \lambda(e)$. Each process p orders the events in which it participates. Let us define $\lessdot_p$ to be the strict ordering

$$e \lessdot_p f \stackrel{\text{def}}{=} e < f,\ p \in e \cap f \text{ and for all } e < g < f,\ p \notin g.$$

It is clear that set of all p-events in $\mathcal{E}$ is totally ordered by $\leq_p$, the reflexive, transitive closure of $\lessdot_p$. It is also easy to observe that the overall partial order $\leq$ is generated by $\{\lessdot_p\}_{p \in P}$. If $e \leq f$ we say that e is *below* f. Note that the special event $e_\perp$ is always below every event in t.

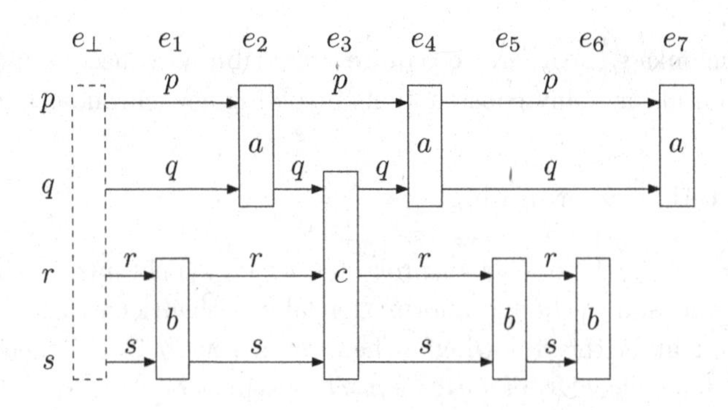

Fig. 9.3. Process-wise ordering in a trace.

Example 9.13. Let $\mathcal{P} = \{p, q, r, s\}$ and $\Sigma = \{a, b, c\}$ where $a = \{p, q\}$, $b = \{r, s\}$ and $c = \{q, r, s\}$. Figure 9.3 has another picture of the trace $t = (\mathcal{E}, \leq, \lambda)$ corresponding to the word *bacabba*. The dashed box corresponds to the event $e_\perp$, which we insert at the beginning for convenience.

In the figure, the labelled arrows between the events denote the relations $\lessdot_p$, $\lessdot_q$, $\lessdot_r$ and $\lessdot_s$. From these, we can compute $<$ and $\leq$. Thus, for example, we have $e_1 \leq e_4$ since $e_1 \lessdot_r e_3 \lessdot_q e_4$, and the events e_5 and e_7 are unordered.

9.7.1. *Ideals*

The main source of difficulty in solving the gossip problem is the fact that the processes in $\mathcal{P}$ need to compute the global information about a trace t while each process only has access to a local, partial view of t. Although partial views of t correspond to subsets of $\mathcal{E}$, not every subset of $\mathcal{E}$ arises from such a partial view. Those subsets of $\mathcal{E}$ which do correspond to partial views of u are called ideals.

Ideals A set of events $I \subseteq \mathcal{E}$ is called an *(order) ideal* if $e \in I$ and $f \leq e$ then $f \in I$.

The requirement that an ideal be downward closed with respect to $\leq$ guarantees that the observation it represents is consistent—whenever an event e has been observed, so have all the events in the computation which necessarily precede e.

Because of our interpretation of $e_\perp$ as an event which takes place *before* the actual computation begins, the minimum possible partial view of a word u is the ideal $\{e_\perp\}$. Henceforth, we assume that every ideal I we consider is non-empty and contains $e_\perp$.

Example 9.14. Let us look once again at Figure 9.3. $\{e_\perp, e_2\}$ is an ideal, but $\{e_\perp, e_2, e_3\}$ is not, since $e_1 \leq e_3$ but $e_1 \notin \{e_\perp, e_2, e_3\}$.

The following observations are immediate.

Proposition 9.15. *Let $t = (\mathcal{E}, \leq, \lambda)$ be a trace.*

- *$\mathcal{E}$ is an ideal.*
- *For any $e \in \mathcal{E}$, the set $\downarrow e = \{f \in \mathcal{E} \mid f \leq e\}$ is an ideal, called the* principal ideal *generated by e. The events in $\downarrow e$ are the only events in $\mathcal{E}$ that are "known" to the processes in e when e occurs.*
- *An ideal I is said to be generated by a set of events X if $I = \bigcup_{e \in X} \downarrow e$. Any ideal I is generated by its maximal events $I_{\max} = \{e \mid e$ is $\leq$-maximal in $I\}$.*
- *If I and J are ideals then $I \cup J$ and $I \cap J$ are ideals.*

Example 9.16. In Figure 9.3, $\{e_\perp, e_1, e_2, e_3, e_5\}$ is the principal ideal $\downarrow e_5$. The ideal $\{e_\perp, e_1, e_2, e_3, e_4, e_5\}$ is generated by $\{e_4, e_5\}$.

Views Let I be an ideal. The maximum p-event in I, $\max_p(I)$, is the last event in I in which p has taken part. In other words, for every $e \in I$, if e is a p-event then $e \leq \max_p(I)$. Since all p-events are totally ordered by $\leq$, and $p \in e_\perp \in I$ for all processes p and for all ideals I, $\max_p(I)$ is well defined.

Let I be an ideal. The *p-view* of I, $\partial_p(I)$ is the set $\downarrow\max_p(I)$. For $P \subseteq \mathcal{P}$, the P-view of I, $\partial_P(I)$, is the *joint view* $\bigcup_{p\in P} \partial_p(I)$.

Example 9.17. In Figure 9.3, let I denote the ideal $\{e_\perp, e_1, e_2, e_3, e_4, e_5, e_6\}$. $\max_q(I) = e_4$ and hence $\partial_q(I) = \{e_\perp, e_1, e_2, e_3, e_4\}$. On the other hand, though $\max_r(I) = e_6$, $\partial_r(I) \neq I$. Rather, $\partial_r(I) = I \setminus \{e_4\}$. The joint view $\partial_{\{q,r\}}(I) = I = \partial_{\mathcal{P}}(I)$.

9.7.2. *Primary and secondary information*

Latest information Let $p, q \in \mathcal{P}$ and I be an ideal. The latest information p has about q in I is $\max_q(\partial_p(I))$, the $\leq$-maximum q-event in $\partial_p(I)$. We denote this event by $latest_{p\to q}(I)$. Observe that $latest_{p\to p}(I) = \max_p(I)$.

Example 9.18. In Figure 9.3, $\max_p(\mathcal{E}) = e_7$ whereas $\max_s(\mathcal{E}) = e_6$. We have $latest_{p\to q}(\mathcal{E}) = e_7$, $latest_{p\to s}(\mathcal{E}) = e_3$ and $latest_{s\to p}(\mathcal{E}) = e_2$.

Primary information Let I be an ideal and $p, q \in \mathcal{P}$. The *primary information* of p after I, $primary_p(I)$, is the set $\{latest_{p\to q}(I)\}_{q\in\mathcal{P}}$. In other words this is the best information that p has about every other process in I. As usual, for $P \subseteq \mathcal{P}$, $primary_P(I) = \bigcup_{p\in P} primary_p(I)$.

More precisely, $primary_p(I)$ is an *indexed* set of events—each event $e = latest_{p\to q}(I)$ in $primary_p(I)$ is represented as a triple (p, q, e). However, we will often ignore the fact that $primary_p(I)$ is an *indexed* set of events and treat it, for convenience, as just a set of events. Thus, for an event $e \in I$, we shall write $e \in primary_p(I)$ to mean that there exists a process $q \in \mathcal{P}$ such that $(p, q, e) \in primary_p(I)$, and so on.

The task at hand is to design a mechanism for processes to compare their primary information when they synchronize. When two processes p and q synchronize, their joint view of the current computation is $\partial_{\{p,q\}}(\mathcal{E})$. At this point, for every other process r, p has in its local information an event $e_r = latest_{p\to r}(\mathcal{E})$ and q has, similarly, an event $e'_r = latest_{q\to r}(\mathcal{E})$. After the current event e, both the p-view and the q-view will correspond to the principal ideal $\downarrow e \subseteq \mathcal{E}$. Assuming that r does not participate in e, $\max_r(\downarrow e)$ will be one of e_r and e'_r. So, p and q have to be able to locally decide whether $e_r \leq e'_r$ or vice versa. If $e_r < e'_r$, then $e_r \in \partial_p(\mathcal{E}) \cap \partial_q(\mathcal{E})$ while $e'_r \in \partial_q(\mathcal{E}) \setminus \partial_p(\mathcal{E})$. Thus, updating primary information is equivalent to computing whether a primary event lies within the intersection $\partial_p(\mathcal{E}) \cap \partial_q(\mathcal{E})$ or outside.

Our first observation is a characterization of the maximal events in the intersection in terms of primary information.

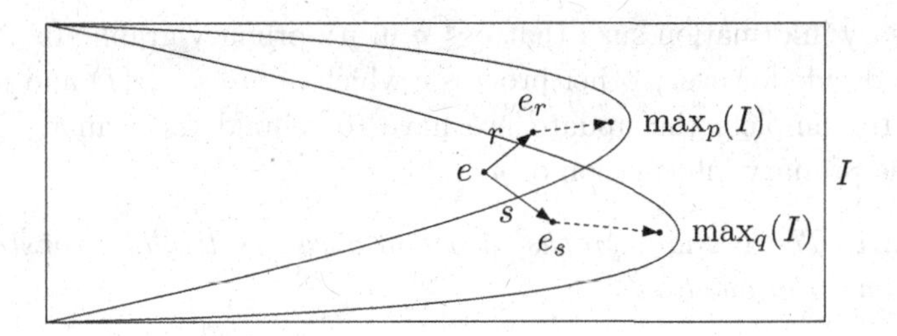

Fig. 9.4. Maximal events in the intersection of two ideals are primary events.

Lemma 9.19. *The maximal elements of $\partial_p(I) \cap \partial_q(I)$ are a subset of $primary_p(I) \cap primary_q(I)$.*

Proof. If $\max_p(I) \leq \max_q(I)$ then $\partial_p(I) = \partial_p(I) \cap \partial_q(I)$ and the only maximal event in this intersection is $e = \max_p(I)$ which is $latest_{p\to p}(I)$ and $latest_{q\to p}(I)$ and hence in $primary_p(I) \cap primary_q(I)$. A symmetric argument holds if $\max_q(I) \leq \max_p(I)$.

The nontrivial case arises when $\max_p(I)$ and $\max_q(I)$ are incomparable. Let e be a maximal event in the intersection. Since $e \leq \max_p(I)$ and $e \leq \max_q(I)$, there is a "path" from e to $\max_p(I)$ in the trace, and another path from e to $\max_q(I)$. So, for some $r, s \in \mathcal{P}$, e, we must have an r event $e_r \in \partial_p(I) \setminus \partial_q(I)$ and an s event $e_s \in \partial_q(I) \setminus \partial_p(I)$ such that $e \lessdot_r e_r \leq \max_p(I)$ and $e \lessdot_s e_s \leq \max_q(I)$ (see Figure 9.4). Notice that e itself is both an r-event and an s-event. Since $e \in \partial_q(I)$ but $e_r \notin \partial_q(I)$ and $e \lessdot_r e_r$, it follows that $e = latest_{q\to r}(I)$. By a symmetric argument, $e = latest_{p\to s}(I)$, so $e \in primary_p(I) \cap primary_q(I)$. □

This characterization yields the following result.

Lemma 9.20. *Let I be an ideal and $p, q, r \in \mathcal{P}$. Let $e = latest_{p\to r}(I)$ and $f = latest_{q\to r}(I)$. Then $e \leq f$ iff there exists $g \in primary_p(I) \cap primary_q(I)$ such that $e \leq g$.*

Proof. Clearly $e \leq f$ iff $e \in \partial_p(I) \cap \partial_q(I)$ iff e is dominated by some maximal element in $\partial_p(I) \cap \partial_q(I)$ iff, by Lemma 9.19, there exists $g \in primary_p(I) \cap primary_q(I)$ such that $e \leq g$. □

To effectively perform the comparison suggested by Lemma 9.20, each process maintains the partial order between events in its primary information.

Primary graph Let I be an ideal and $p \in \mathcal{P}$. The *primary graph* of p in I is the set of primary events together with the partial order between them inherited from the underlying trace.

With primary graphs, it is clear how to perform the comparison indicated in Lemma 9.20. Processes p and q identify an event g that is common to both their sets of primary information such that $e \leq g$ in p's primary graph. In this manner, p and q can decide for every other process r which of $latest_{p \to r}(I)$ and $latest_{q \to r}(I)$ is better. To complete the update, we have to rebuild the primary graph after updating the primary information of p and q.

Lemma 9.21. *The primary graphs of p and q can be locally reconstructed after updating primary infomation.*

Proof. Let e and f be two events in the updated primary information of p and q (recall that both processes have the *same* primary information after synchronization). If both e and f were inherited from the same process, say p, we order them in the new graph if and only if they were ordered in the original primary graph of p.

The interesting case is when e is inherited from p and f from q. In this case, we must have had $e \in \partial_p(I) \setminus \partial_q(I)$ and $f \in \partial_q(I) \setminus \partial_q(I)$. This means that e and f were not ordered in I, so we do not order them in the updated primary graph. □

This procedure generalizes to any arbitrary set $P \subseteq \mathcal{P}$ which synchronizes after a trace t. The processes in P share their primary graphs and compare this information pairwise. Using Lemma 9.20, for each $q \in \mathcal{P} \setminus P$ they decide who has the "latest information" about q and correctly order these events. Each process then comes away with the *same* primary graph, incorporating the best information available among the processes in P.

9.7.3. *Labelling events consistently*

To make Lemma 9.20 effective, we must make the assertions "locally checkable"—for example, if $e = latest_{p \to r}(I)$ and $f = latest_{q \to r}(I)$, processes p and q must be able to decide if there exists an event $g \in \partial_p(I) \cap \partial_q(I)$ between e and f. This can be checked locally provided events in $\mathcal{E}$ are labelled unambiguously.

Since events are processed locally, we must locally assign labels to events in $\mathcal{E}$ so that we can check whether events in the primary graphs of two different processes are equal by comparing their labels. A naïve solution would be for the processes in $loc(a)$ to jointly assign a (sequential) time-stamp to each new occurrence of a, for every letter a. The problem with this approach is that we will need an unbounded set of time-stamps, since u could get arbitrarily large.

Instead we would like a scheme that uses only a finite set of labels to distinguish events. This means that several different occurrences of the same action will eventually get the same label. Since updating primary graphs relies on comparing labels, we must ensure that this reuse of labels does not lead to any confusion.

However, from Lemma 9.20, we know that to compare primary information, we only need to look at the events which are currently in the primary sets of each process. So, it is sufficient if the labels assigned to these sets are consistent across the system—that is, if the same label appears in the current primary information of different processes, then this label does in fact denote the same event in underlying trace.

When a new action a occurs, the processes in $loc(a)$ have to assign it a label that is different from all a-events that are currently in the primary information of all processes. Since the cardinality of $primary_{\mathcal{P}}(\mathcal{E})$ is bounded, such a new label must exist. The catch is to detect which labels are currently in use and which are not.

Unfortunately, the processes in a cannot directly see all the a-events which belong to the primary information of the entire system. An a-event e may be part of the primary information of processes *outside* a—that is, $e \in primary_{\mathcal{P}\setminus a}(\mathcal{E}_u) \setminus primary_a(\mathcal{E}_u)$.

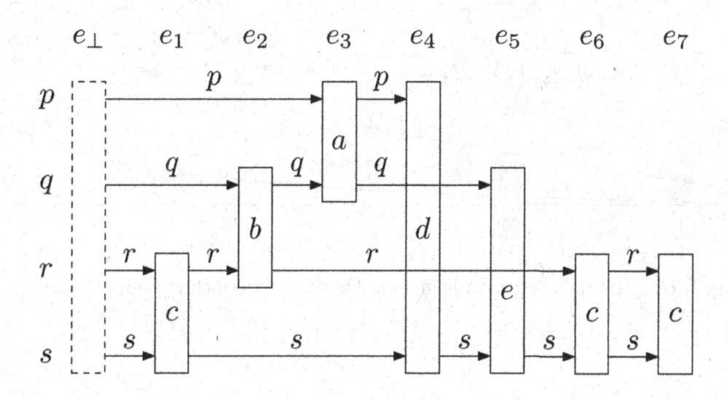

Fig. 9.5. Keeping track of active labels.

Example 9.22. Let $\mathcal{P} = \{p, q, r, s\}$ and $\Sigma = \{a, b, c, d, e\}$ where $a = \{p, q\}$, $b = \{q, r\}$, $c = \{r, s\}$, $d = \{p, s\}$ and $e = \{q, s\}$. Figure 9.5 shows the trace $t = (\mathcal{E}, \leq, \lambda)$ corresponding to the word *cbadecc*.

At the end of this word, $e_2 = latest_{p \to r}(\mathcal{E})$, but $e_2 \notin primary_r(\mathcal{E})$, since $primary_r(\mathcal{E}) = \{(r, p, e_4), (r, q, e_5), (r, r, e_7), (r, s, e_7)\}$.

To enable the processes in a to know about all a-events in $primary_{\mathcal{P}}(\mathcal{E}_u)$, we need to maintain secondary information.

Secondary information The *secondary information* of p after I, $secondary_p(I)$, is the (indexed) set $\bigcup_{q \in \mathcal{P}} primary_q(\downarrow latest_{p \to q}(I))$. In other words, this is the latest information that p has in I about the primary information of q, for each $q \in \mathcal{P}$. Once again, for $P \subseteq \mathcal{P}$, $secondary_P(I) = \bigcup_{p \in P} secondary_p(I)$.

Each event in $secondary_p(I)$ is of the form $latest_{q \to r}(\downarrow latest_{p \to q}(I))$ for some $q, r \in \mathcal{P}$. This is the latest r-event which q knows about up-to the event $latest_{p \to q}(I)$. We abbreviate $latest_{q \to r}(\downarrow latest_{p \to q}(I))$ by $latest_{p \to q \to r}(I)$.

Just as we represented events in $primary_p(I)$ as triples of the form (p, q, e), where $p, q \in \mathcal{P}$ and $e \in I$, we represent each secondary event $e = latest_{p \to q \to r}(I)$ in $secondary_p(I)$ as a quadruple (p, q, r, e). Recall that we often ignore the fact that $primary_p(I)$ and $secondary_p(I)$ are *indexed* sets of events and treat them, for convenience, as just sets of events.

Notice that each primary event $latest_{p \to q}(I)$ is also a secondary event $latest_{p \to p \to q}(I)$ (or, equivalently, $latest_{p \to q \to q}(I)$). So, following our convention that $primary_p(I)$ and $secondary_p(I)$ be treated as sets of events, we write $primary_p(I) \subseteq secondary_p(I)$.

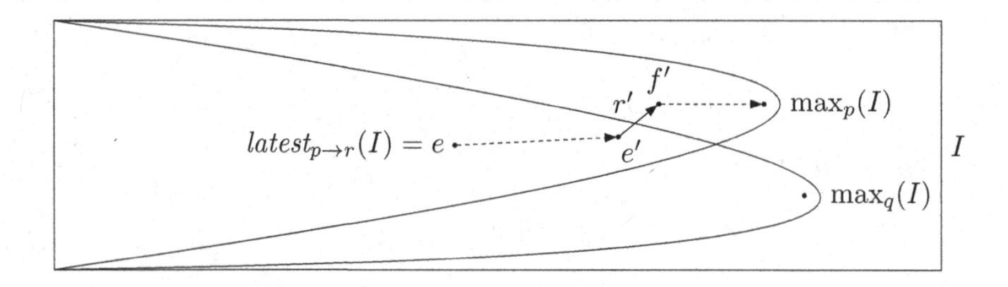

Fig. 9.6. Identifying active labels through secondary information.

Lemma 9.23. *Let I be an ideal and $p \in \mathcal{P}$. If $e \in primary_p(I)$ then for every $q \in e$, $e \in secondary_q(I)$.*

Proof. Let $e = latest_{p \to r}(I)$ for some $r \in \mathcal{P}$ and let $q \in e$. We will show that $e = latest_{q \to r' \to r}(I)$ for some $r' \in \mathcal{P}$. We know that there is a path $e \lessdot f_1 \lessdot \cdots \lessdot \max_p(I)$, since $e \in \partial_p(I)$. This path starts inside $\partial_p(I) \cap \partial_q(I)$.

If this path never leaves $\partial_p(I) \cap \partial_q(I)$ then $\max_p(I) \in \partial_q(I)$. Since $\max_p(I)$ is the maximum p-event in I, it must be the maximum p-event in $\partial_q(I)$. So, $e = \max_r(\max_p(I)) = \max_r(\max_p(\partial_q(I))) = latest_{q \to p \to r}(I)$ and we are done.

If this path does leave $\partial_p(I) \cap \partial_q(I)$, we can find an event e' along the path such that $e \leq e' \lessdot_{r'} f' \leq \max_p(I)$, where $e' \in \partial_p(I) \cap \partial_q(I)$, $f' \in \partial_p(I) \setminus \partial_q(I)$ and $r' \in e' \cap f'$ (see Figure 9.6). It is easy to see that $e' = latest_{q \to r'}(I)$. Since $e = \max_r(\partial_p(I))$, $e \in {\downarrow}e' \subseteq \partial_p(I)$, we have $e = \max_r({\downarrow}e')$. Hence $e = latest_{r' \to r}({\downarrow}e') = latest_{q \to r' \to r}(I)$. □

Corollary 9.24. *Let I be an ideal, $p \in \mathcal{P}$ and e be a p-event in I. If $e \notin secondary_p(I)$ then $e \notin primary_{\mathcal{P}}(I)$.*

So, a process p can keep track of which of its labels are "in use" in the system by maintaining secondary information. Each p-event e initially belongs to $primary_e(I)$, and hence to $secondary_e(I)$, where we also use e to denote the subset of $\mathcal{P}$ that synchronizes at this event. As the computation progresses, e gradually "recedes" into the background and disappears from the primary sets of the system. Eventually, when e disappears from $secondary_p(I)$, p can be sure that e no longer belongs to $primary_{\mathcal{P}}(I)$.

Since $secondary_p(I)$ is a bounded set, p knows that only finitely many of its labels are in use at any given time. So, by using a sufficiently large finite set of labels, each new event can always be assigned an unambiguous label by the processes which take part in the event.

It is easy to see that secondary information can be updated along with primary information. If $latest_{p \to r}(I)$ is better than $latest_{q \to r}(I)$, then all secondary events of the form $latest_{q \to r \to s}(I)$ should also be replaced by the corresponding events $latest_{p \to r \to s}(I)$.

9.7.4. *The "gossip" automaton*

Using our analysis of the primary graph and secondary information maintained by processes, we can now design a deterministic asynchronous automaton to consistently update the primary information of each process whenever a set of processes synchronize.

For $p \in \mathcal{P}$, each local state of p will consist of its primary graph and secondary information, stored as indexed collections or arrays. Each event in these arrays is represented as a pair $\langle P, \ell \rangle$, where P is the subset of processes that synchronized at the event and $\ell \in \mathcal{L}$, a finite set of labels. We shall establish a bound on $|\mathcal{L}|$ shortly.

The initial state is the global state where for all processes p, all entries in these arrays correspond to the initial event $e_\perp$. The event $e_\perp$ is denoted by $\langle \mathcal{P}, \ell_0 \rangle$ for an arbitrary but fixed label $\ell_0 \in \mathcal{L}$.

The local transition functions $\to_a$ modify the local states for processes in a as follows.

(i) When a new a-labelled event e occurs after u, the processes in a assign a label $\langle a, \ell \rangle$ to e which does not appear in $secondary_a(\mathcal{E}_u)$. Corollary 9.24 guarantees that this new label does not appear in $primary_{\mathcal{P}}(\mathcal{E}_u)$.
Let $N = |\mathcal{P}|$. Since each process keeps track of N^2 secondary events and at most N processes can synchronize at an event, there need be only N^3 labels in $\mathcal{L}$. (In fact, in Lemma 9.25 below, we show that it suffices to have N^2 labels in $\mathcal{L}$.)

(ii) The processes participating in e now share their primary graphs and update their primary information about each process $q \notin e$, as described in Lemma 9.20.

The gossip automaton does not "recognize" a set of traces. Rather, it updates the primary graphs and secondary information of all processes appropriately whenever an action is added to the trace, such that at any point in the computation, we can compare the primary information of a set of processes using the information present in their local states.

Lemma 9.25. *In the gossip automaton, the number of local states of each process $p \in \mathcal{P}$ is at most $2^{O(N^2 \log N)}$, where $N = |\mathcal{P}|$.*

Proof. A local state for p consists of its primary graph and secondary information. We estimate how many bits are required to store this.

Recall that for any ideal I, each event in $primary_p(I)$ is also present in $secondary_p(I)$. So it suffices to store just the labels of secondary events. These events are stored in an array with N^2 entries, where each entry is implicitly indexed by a pair from $\mathcal{P} \times \mathcal{P}$. We can store the primary graph as an adjacency matrix with N^2 entries.

Each new event e is assigned a label of the form $\langle P, \ell \rangle$, where P was the set of processes that participated in e and $\ell \in \mathcal{L}$.

We argued earlier that it suffices to have N^3 labels in $\mathcal{L}$. Actually, we can make do with N^2 labels by modifying our transition function slightly. When a letter a is read, instead of immediately labelling the new event, the processes in a first compare and update their primary, secondary information about processes from $\mathcal{P} \setminus a$. These updates concern events which have already been labelled, so the fact that the new event has not yet been labelled is not a problem. Once this is done, all the processes in a will have the same primary and secondary information. At this stage, there are (less than) N^2 distinct labels present in $secondary_a(I)$. So, if $|\mathcal{L}| = N^2$ the processes are guaranteed to find a label they can use for the new event. Regardless of which update strategy we choose, $\ell \in \mathcal{L}$ can be written down using $O(\log N)$ bits.

To write down $P \subseteq \mathcal{P}$, we need, in general, N bits. This component of the label is required to guarantee that all secondary events in the system have distinct labels, since the set $\mathcal{L}$ is common across all processes. However, we do not really need to use all of P in the label for e to ensure this property. If we order $\mathcal{P}$ as $\{p_1, p_2, \ldots, p_N\}$, it suffices to label e by $\langle p_i, \ell \rangle$ where, among the processes in P, p_i has the least index with respect to our ordering of $\mathcal{P}$.

Thus, we can modify our automaton so that the processes label each event by a pair $\langle p, \ell \rangle$, where $p \in \mathcal{P}$ and $\ell \in \mathcal{L}$. This pair can be written down using $O(\log N)$ bits. Overall there are N^2 such pairs in the array of secondary events, so the this can be described using $O(N^2 \log N)$ bits and the primary graph can be represented by $O(N^2)$ bits. Therefore, the number of distinct local states of p is bounded by $2^{O(N^2 \log N)}$. □

9.8. Zielonka's Theorem

It is not difficult to see that any language accepted by an asynchronous automaton over (Σ, loc) is a recognizable trace language over the corresponding concurrent alphabet $(\Sigma, \mathcal{I}_{loc})$. The distributed nature of the automaton guarantees that it is a trace language. To see that the language is recognizable, we note that every asynchronous automaton $\mathcal{A} = (Q, \rightarrow, Q_{\text{in}}, F)$ gives rise to a finite state automaton

$\mathcal{B}$ accepting the same language as $\mathcal{A}$. The states of $\mathcal{B}$ are the global states of $\mathcal{A}$ and the transition relation of $\mathcal{B}$ is given by the global transition relation of $\mathcal{A}$. Since the initial and accepting states of $\mathcal{A}$ are specified as global states, they can directly serve as the initial and final states of $\mathcal{B}$. It is straightforward to verify that $\mathcal{B}$ accepts the same language as $\mathcal{A}$.

On the other hand, the converse is difficult to show. For a given recognizable trace language L over a concurrent alphabet $(\Sigma, \mathcal{I})$, does there exist an asynchronous automaton $\mathcal{A}$ over a distributed alphabet (Σ, loc) such that $\mathcal{A}$ accepts L *and* the independence relation $\mathcal{I}_{loc}$ induced by loc is *exactly* $\mathcal{I}$?

Zielonka's fundamental result is that this is indeed the case [1]. In other words, asynchronous automata accept precisely the set of recognizable trace languages and thus constitute a natural distributed machine model for this class of languages.

Fix a recognizable trace language L over a concurrent alphabet $(\Sigma, \mathcal{I})$, as well as a distribution $loc : \Sigma \to (2^{\mathcal{P}} \setminus \{\emptyset\})$ such that the induced independence relation $\mathcal{I}_{loc}$ is the same as $\mathcal{I}$. We shall construct a deterministic asynchronous automaton $\mathcal{A} = (Q, \to, Q_{\text{in}}, F)$ over (Σ, loc) recognizing L.

Let $\mathcal{B} = (S, \Sigma, \delta, s_0, S_F)$ be the minimum deterministic finite state automaton accepting L, where S denotes the set of states, $\delta : S \times \Sigma \to S$ the transition function, $s_0 \in S$ the initial state and $S_F \subseteq S$ the set of accepting states. As usual, we shall extend δ to a transition function $S \times \Sigma^* \to S$ describing state transitions for input words rather than just single letters. For convenience, we denote this extended transition function also by δ.

The main hurdle in constructing an asynchronous automaton $\mathcal{A}$ from the original DFA $\mathcal{B}$ is the following: On reading an input word u, we must be able to compute whether $\delta(s_0, u) \in S_F$. As we have seen, u is just one linearization of the trace $[u] = (\mathcal{E}_u, \leq_u, \lambda_u)$. After reading u each process in $\mathcal{A}$ only has partial information about $\delta(s_0, [u])$—a process p only "knows about" the events that lie in the p-view $\downarrow\!\max_p(\mathcal{E}_u)$. We have to devise a scheme to recover the state $\delta(s_0, u)$ from the partial information available with each process after reading $[u]$.

Another complication is that processes can only maintain a finite amount of information. So, we need a way of representing arbitrary words in a bounded, finite way. This can be done quite easily—the idea is to record for each word w, its "effect" as dictated by our automaton $\mathcal{B}$.

We first recall a basic fact about recognizable languages.

Definition 9.26. Any language $\hat{L}$ defines a syntactic congruence $\equiv_{\hat{L}}$ on Σ^* as follows:

$$u \equiv_{\hat{L}} u' \stackrel{\text{def}}{=} \forall w_1, w_2 \in \Sigma^*,\ w_1 u w_2 \in \hat{L} \textit{ iff } w_1 u' w_2 \in \hat{L}.$$

It is well known that $\hat{L}$ is recognizable if and only if $\equiv_{\hat{L}}$ is of finite index [4].

Now, consider the relation $\equiv_L$ for the language L we are looking at. We can associate with each word u a function $f_u : S \to S$, where S is the set of states of $\mathcal{B}$, such that $f_u(s) = s'$ iff $\delta(s, u) = s'$. Thus, f_u is a representation of the word u as a "state transformer". Given two words $u, w \in \Sigma^*$, it is easy to see that $f_{uw} = f_w \circ f_u$, where $\circ$ denotes function composition.

Since $\mathcal{B}$ is the minimum DFA recognizing L, it follows that for any words $u, w \in \Sigma^*$, $u \equiv_L w$ if and only if $f_u = f_w$.

Clearly the function $f_w : S \to S$ corresponding to a word w has a bounded representation. So, if we could compute the function f_u corresponding to the input u, we would be able to determine whether $\delta(s_0, u) \in S_F$, since $\delta(s_0, u) = f_u(s_0)$.

However, we still have the original problem arising from the distributed nature of $\mathcal{A}$. Even if each process $p \in \mathcal{P}$ were to maintain the entire p-view of $\mathcal{E}_u$, the only information that we could reasonably hope to extract from the combined view of all the processes is some linearization of the trace $(\mathcal{E}_u, \leq_u, \lambda_u)$. From this labelled partial order, we cannot always recover u uniquely—in general, we can only reconstruct a word $u' \sim u$. Hence, we can at best hope to recover $f_{u'}$ for some $u' \sim u$.

Fortunately, this is not a bottleneck. From the definition of a trace language, it follows that all words that are $\sim$-equivalent are also $\equiv_L$-equivalent.

Proposition 9.27. *Let $\hat{L}$ be a trace language over a concurrent alphabet $(\Sigma, \mathcal{I})$. For any $u, u' \in \Sigma^*$, if $u \sim u'$ then $u \equiv_{\hat{L}} u'$.*

Proof. Suppose $u \sim u'$ but $u \not\equiv_{\hat{L}} u'$. Then, without loss of generality, we can find words w_1 and w_2 such that $w_1 u w_2 \in \hat{L}$ but $w_1 u' w_2 \notin \hat{L}$. Since $w_1 u w_2 \sim w_1 u' w_2$, this contradicts the assumption that $\hat{L}$ is $\sim$-consistent. □

From our previous observation about $\mathcal{B}$, it follows that whenever $u' \sim u$, $u' \equiv_L u$, so $f_{u'} = f_u$. In other words, to determine whether $\delta(s_0, u) \in S_F$, it is sufficient to compute the function $f_{u'}$ corresponding to any word $u' \sim u$. Thus, we can write $f_{[u]}$ to denote the function associated with all linearizations of a trace u.

This, then, is our new goal: for any input word u, we want to compute in $\mathcal{A}$ the function $f_{[u]} : S \to S$ using some representative u' of the trace $[u]$. This still involves finding a scheme to combine the partial views of processes in a sensible way.

We begin by formally defining a projection of a word. A word $u \in \Sigma^*$ of length n can be viewed as a function $u : [1..n] \to \Sigma$ assigning a letter to each position in the word, where $[1..n]$ abbreviates the set $[1..n]$. In the trace $[u] = (\mathcal{E}_u, \leq_u, \lambda_u)$, we henceforth implicitly assume that $\mathcal{E}_u = \{e_\perp, e_1, e_2, \ldots, e_n\}$, where $e_\perp$ is the fictitious initial event that we have assumed for convenience and for $j \in [1..n]$, e_j is the event corresponding to the letter $u(j)$.

Projection of a word Let $u : [1..n] \to \Sigma$ whose corresponding trace is $[u] = (\mathcal{E}_u, \leq_u, \lambda_u)$, and let $X \subseteq \mathcal{E}_u$ where $X \setminus \{e_\perp\} = \{e_{i_1}, e_{i_2}, \ldots, e_{i_k}\}$, with $i_1 < i_2 < \cdots < i_k$. Then $u[X]$, *the projection* of u with respect to X, is the word $u(i_1)u(i_2)\cdots u(i_k)$. If $X \setminus \{e_\perp\} = \emptyset$ then $u[X] = \varepsilon$, the empty string.

Ideals revisited So far, we have implicitly assumed that all ideals are non-empty. However, to construct the asynchronous automaton $\mathcal{A}$ it will be convenient to work with the empty ideal as well. So, henceforth, whenever we encounter an ideal I, unless we explicitly say that I is non-empty we do not rule out the possibility that $I = \emptyset$. Clearly, if $I = \emptyset$, the notions $\max_p(I)$, $primary_p(I)$ and $secondary_p(I)$ are not defined and we shall apply these operators only to non-empty ideals. We also adopt a convention regarding P-views of an ideal. Recall that for $P \subseteq \mathcal{P}$, the P-view $\partial_P(I)$ of a non-empty ideal I is the set of events $\bigcup_{p \in P} \downarrow\max_p(I)$. If $P = \emptyset$, we shall define $\partial_P(I) = \emptyset$.

We begin with the following fact, which is crucial in our construction of $\mathcal{A}$.

Lemma 9.28. *Let u be a word, $[u] = (\mathcal{E}_u, \leq_u, \lambda_u)$ the corresponding trace, and $I, J \subseteq \mathcal{E}_u$ be ideals such that $I \subseteq J$. Then $u[J] \sim u[I]u[J \setminus I]$.*

Proof Sketch: A basic result in trace theory is that $u \sim w$ if and only if $u{\downarrow}_{\{a,b\}} = w{\downarrow}_{\{a,b\}}$ for every pair of dependent letters $(a, b) \in \mathcal{D}$ [5].

Suppose $(a, b) \in \mathcal{D}$ and $u[J]{\downarrow}_{\{a,b\}} \neq (u[I]u[J \setminus I]){\downarrow}_{\{a,b\}}$. Then there must be an occurrence of a and an occurrence of b in $u[J]{\downarrow}_{\{a,b\}}$ which have been transposed in $(u[I]u[J \setminus I]){\downarrow}_{\{a,b\}}$. Assume, as usual, that the events in the trace $[u] = (\mathcal{E}_u, \leq_u, \lambda_u)$ are numbered $\{e_\perp, e_1, e_2, \ldots, e_n\}$ in correspondence with the positions of the letters in u. Let $e_a = e_i$ and $e_b = e_j$ be the events from $\mathcal{E}_J$ corresponding to these occurrences of a and b in u. Without loss of generality, we assume that $i < j$.

It must be the case that $e_a \notin I$ and $e_b \in I$, since the only rearrangement we have performed is to send letters not in $u[I]$ to the right. Since $(a, b) \notin \mathcal{I}$, we can find a process $p \in loc(a) \cap loc(b)$. But then $e_a \leq_p e_b$ and so $e_a \leq e_b$. Since I is an ideal, $e_b \in I$ and $e_a \in \downarrow e_b$, we must have $e_a \in I$ as well, which is a contradiction. □

Corollary 9.29. *Let u be a word, $[u] = (\mathcal{E}_u, \leq_u, \lambda_u)$ the corresponding trace, and $I_1 \subseteq I_2 \subseteq \cdots \subseteq I_k \subseteq \mathcal{E}_u$ a sequence of nested ideals. Then $u[I_k] \sim u[I_1]u[I_2 \setminus I_1] \cdots u[I_k \setminus I_{k-1}]$.*

Proof. Applying Lemma 9.28 once, we get $u[I_k] \sim u[I_{k-1}]u[I_k \setminus I_{k-1}]$. We then apply the lemma to each of $u[I_{k-1}]$, $u[I_{k-2}]$, $\ldots$, $u[I_2]$ in turn to obtain the required expression. □

9.8.1. *Process residues*

Let us return to our problem: We want to compute in $\mathcal{A}$, on any input u, the function $f_{[u]}$ via some $u' \sim u$. We order the processes in $\mathcal{P}$ so that $\mathcal{P} = \{p_1, p_2, \ldots, p_N\}$ and construct subsets $\{Q_j\}_{j \in [1..N]}$, where $Q_1 = \{p_1\}$ and for $j \in [2..N]$, $Q_j = Q_{j-1} \cup \{p_j\}$.

Let $[u] = (\mathcal{E}_u, \leq_u, \lambda_u)$. Construct ideals $I_0, I_1, \ldots, I_N \subseteq \mathcal{E}_u$ where $I_0 = \emptyset$ and for $j \in [1..N]$, $I_j = I_{j-1} \cup \partial_{p_j}(\mathcal{E}_u)$. Clearly $I_j = \partial_{Q_j}(\mathcal{E}_u)$ for $j \in [1..N]$.

Since $\mathcal{E}_u = \partial_{\mathcal{P}}(\mathcal{E}_u) = \partial_{Q_N}(\mathcal{E}_u) = I_N$ and $I_0 \subseteq I_1 \subseteq \cdots \subseteq I_N$, we can write down the following expression based on Corollary 9.29.

$$u = u[I_N] \sim u[I_0]u[I_1 \setminus I_0] \cdots u[I_N \setminus I_{N-1}]$$

For $j \in [2..N]$, $I_j \setminus I_{j-1} = \partial_{Q_j}(\mathcal{E}_u) \setminus \partial_{Q_{j-1}}(\mathcal{E}_u)$ is the same as $\partial_{p_j}(\mathcal{E}_u) \setminus \partial_{Q_{j-1}}(\mathcal{E}_u)$. So, we can rewrite our earlier expression in a more useful form as:

$$\begin{aligned} u &= u[\partial_{Q_N}(\mathcal{E}_u)] \\ &\sim u[\emptyset]u[\partial_{p_1}(\mathcal{E}_u) \setminus \emptyset]u[\partial_{p_2}(\mathcal{E}_u) \setminus \partial_{Q_1}(\mathcal{E}_u)] \cdots u[\partial_{p_N}(\mathcal{E}_u) \setminus \partial_{Q_{N-1}}(\mathcal{E}_u)] \qquad (\diamondsuit) \end{aligned}$$

The word $u[\partial_{p_j}(\mathcal{E}_u) \setminus \partial_{Q_{j-1}}(\mathcal{E}_u)]$ is the portion of u that p_j has seen but which the processes in Q_{j-1} have not seen. This is a special case of what we call a residue.

Residues Let $u \in \Sigma^*$ be a word whose associated trace is $[u] = (\mathcal{E}_u, \leq_u, \lambda_u)$, $I \subseteq \mathcal{E}_u$ an ideal and $p \in \mathcal{P}$ a process. $\mathcal{R}(u, p, I)$ denotes the word $u[\partial_p(\mathcal{E}_u) \setminus I]$ and is called the *residue* u at p with respect to I.

For ideals X and Y, recall that $X \setminus Y = X \setminus (X \cap Y)$, where $X \cap Y$ is also an ideal. So any residue $\mathcal{R}(u, p, I)$ can equivalently be written as $\mathcal{R}(u, p, \partial_p(\mathcal{E}_u) \cap I)$. We will often make use of this fact.

Since $u[\partial_{p_j}(\mathcal{E}_u) \setminus \partial_{Q_{j-1}}(\mathcal{E}_u)]$ can be rewritten as $\mathcal{R}(u, p_j, \partial_{Q_{j-1}}(\mathcal{E}_u))$, we can reformulate $(\diamondsuit)$ as follows:

$$\begin{aligned} u &= u[\partial_{Q_N}(\mathcal{E}_u)] \\ &\sim \mathcal{R}(u, p_1, \emptyset)\mathcal{R}(u, p_2, \partial_{Q_1}(\mathcal{E}_u))\mathcal{R}(u, p_3, \partial_{Q_2}(\mathcal{E}_u)) \ldots \mathcal{R}(u, p_N, \partial_{Q_{N-1}}(\mathcal{E}_u)) \end{aligned}$$

Let us give a special name to residues of this form.

Process residues $\mathcal{R}(u, p, I)$ is a *process residue* if $\mathcal{R}(u, p, I) = \mathcal{R}(u, p, \partial_P(\mathcal{E}_u))$ for some $P \subseteq \mathcal{P}$. We say that $\mathcal{R}(u, p, \partial_P(\mathcal{E}_u))$ is the *P-residue* of u at p.

Notice that $\mathcal{R}(u, p, \emptyset)$ is also a process residue, corresponding to the empty set of processes (by our convention that $\partial_\emptyset(\mathcal{E}_u) = \emptyset$.) Further, $\mathcal{R}(u, p, \emptyset) = u[\partial_p(\mathcal{E}_u)]$, the partial word corresponding to the p-view of $\mathcal{E}_u$.

Example 9.30. Consider our old example—the trace $[bacabba]$ depicted in Figure 9.3. Let $I = \{0, e_1, e_2, e_3\}$. $\partial_s(\mathcal{E}_u) \setminus I = \{e_5, e_6\}$, so $\mathcal{R}(u, s, I) = bb$. Moreover, $\mathcal{R}(u, s, I) = \mathcal{R}(u, s, \partial_p(\mathcal{E}_u))$, so it is the p-residue of u at s.

Suppose that along every input word u, each process p maintains all its P-residues $\mathcal{R}(u, p, \partial_P(\mathcal{E}_u))$, $P \subseteq \mathcal{P}$, as functions from S to S. As we remarked earlier, each of these functions can be represented in a finite, bounded manner. Since each process needs to keep track of only 2^N P-residues, where $N = |\mathcal{P}|$, all these functions can be incorporated into the local state of the process.

Going back to the expression (♦), we can compute the function $f_{u[\partial_{Q_N}(\mathcal{E}_u)\setminus\emptyset]}$ corresponding to $u[\partial_{Q_N}(\mathcal{E}_u) \setminus \emptyset]$ by composing the functions corresponding to the residues $\mathcal{R}(u, p_1, \emptyset)$, $\mathcal{R}(u, p_2, \partial_{Q_1}(\mathcal{E}_u))$, ..., $\mathcal{R}(u, p_N, \partial_{Q_{N-1}}(\mathcal{E}_u))$. Notice that $u[\partial_{Q_N}(\mathcal{E}_u) \setminus \emptyset] = u$. So, we can then decide whether the state $\delta(s_0, u)$ belongs to S_F by applying the function $f_{u[\partial_{Q_N}(\mathcal{E}_u)]}$ to s_0.

Thus, our automaton $\mathcal{A}$ will accept u if $\delta(s_0, u)$ as computed using the process residues corresponding to the expression (♢) lies in S_F. Recall that the accepting states of $\mathcal{A}$ are specified as global states. So, at the end of the word u, we are permitted to observe "externally", as it were, the states of *all* the processes in $\mathcal{A}$ before deciding whether to accept u.

The only hitch now is with computing process residues "on line", as $\mathcal{A}$ reads u. The problem is the following: Let $p \in \mathcal{P}$ and $P \subseteq \mathcal{P}$. If we extend u to ua where $p \notin loc(a)$, it could well happen that $\partial_p(\mathcal{E}_{ua}) \setminus \partial_P(\mathcal{E}_{ua}) \neq \partial_p(\mathcal{E}_u) \setminus \partial_P(\mathcal{E}_u)$, even though $\partial_p(\mathcal{E}_u) = \partial_p(\mathcal{E}_{ua})$.

Example 9.31. Consider the trace [*bacabba*] shown in Figure 9.3. After the subword *bac*, the p-residue at s is bc, corresponding to $\{e_1, e_3\}$. However, when this word is extended to *baca*, the p-residue at s becomes ε, though s does not participate in the final a.

9.8.2. *Primary residues*

Process residues at p can change without p being aware of it. This means that we cannot hope to directly maintain and update process residues locally as $\mathcal{A}$ reads u. To remedy this, we define a new type of residue called a primary residue.

Primary residues Let us call $\mathcal{R}(u, p, I)$ a *primary residue* if I is generated by a subset E of $primary_p(\mathcal{E}_u)$.

Clearly, for $p, q \in \mathcal{P}$, $\mathcal{R}(u, p, \partial_q(\mathcal{E}_u))$, can be rewritten as $\mathcal{R}(u, p, \partial_p(\mathcal{E}_u) \cap \partial_q(\mathcal{E}_u))$. So, by the previous result the q-residue $\mathcal{R}(u, p, \partial_q(\mathcal{E}_u))$ is a primary residue $\mathcal{R}(u, p, {\downarrow}E)$ for some $E \subseteq primary_p(\mathcal{E}_u)$. Further, p can effectively determine the set E given the primary information of both p and q. In fact, it will turn out that *all* process residues can be effectively described in terms of primary residues.

Example 9.32. In the word $u = bacabba$ shown in Figure 9.3, $\mathcal{R}(u, s, \partial_p(\mathcal{E}_u))$ corresponds to the primary residue $\mathcal{R}(u, s, {\downarrow}\{latest_{s \to q}(\mathcal{E}_u)\})$.

Our strategy will now be to maintain primary residues rather than process residues for each process p. The useful property we exploit is that the primary residues at p change *only* when p participates in an event.

Notice that this does not contradict our earlier observation that process residues at p can change independent of p. Even if a synchronization not involving p happens to modify the P-residue at p, the new P-residue remains a primary residue of p, albeit for a different subset of p's primary events.

Further, we show that when p participates in an event, it can recompute its primary residues using *just* the information it receives during the synchronization. At the end of the word u, the expression ($\diamondsuit$) written in terms of process residues that is used to compute $\delta(s_0, u)$ can be effectively rewritten in terms of primary residues. These residues will be available with each process in $\mathcal{P}$, thereby enabling us to calculate $\delta(s_0, u)$.

When discussing how to update primary information in the previous section, we have observed that the maximal events in the intersection of two local views $\partial_p(I) \cap \partial_q(I)$ of an ideal are always primary events for both p and q (Lemma 9.19). We begin with some consequences of this lemma.

Corollary 9.33. *Let $u \in \Sigma^*$ and $p \in \mathcal{P}$.*

(i) For ideals $I, J \subseteq \mathcal{E}_u$, let $\mathcal{R}(u,p,I)$ and $\mathcal{R}(u,p,J)$ be primary residues such that $\mathcal{R}(u,p,I) = \mathcal{R}(u,p,{\downarrow}E_I)$ and $\mathcal{R}(u,p,J) = \mathcal{R}(u,p,{\downarrow}E_J)$ for $E_I, E_J \subseteq primary_p(\mathcal{E}_u)$. Then $\mathcal{R}(u,p,I \cup J)$ is also a primary residue and $\mathcal{R}(u,p,I \cup J) = \mathcal{R}(u,p,{\downarrow}(E_I \cup E_J))$.

(ii) Let $Q \subseteq \mathcal{P}$. Then $\mathcal{R}(u,p,\partial_Q(\mathcal{E}_u))$ is a primary residue $\mathcal{R}(u,p,{\downarrow}E)$ for p. Further, p can effectively compute the set $E \subseteq primary_p(\mathcal{E}_u)$ from the information in $primary_{\{p\}\cup Q}(\mathcal{E}_u)$.

(iii) Let $q, r \in \mathcal{P}$ such that $latest_{p\to r}(\mathcal{E}_u) \leq latest_{q \to r}(\mathcal{E}_u)$. Then $\mathcal{R}(u,p,\partial_r(\partial_q(\mathcal{E}_u)))$ is a primary residue $\mathcal{R}(u,p,{\downarrow}E)$ for p. Further, p can effectively compute the set $E \subseteq primary_p(\mathcal{E}_u)$ from the information in $primary_p(\mathcal{E}_u)$ and $secondary_q(\mathcal{E}_u)$.

Proof.

(i) We can rewrite $\mathcal{R}(u,p,I \cup J)$ as $\mathcal{R}(u,p,\partial_p(\mathcal{E}_u) \cap (I \cup J))$. But $\partial_p(\mathcal{E}_u) \cap (I \cup J) = (\partial_p(\mathcal{E}_u) \cap I) \cup (\partial_p(\mathcal{E}_u) \cap J)$. Since $\mathcal{R}(u,p,I) = \mathcal{R}(u,p,\partial_p(\mathcal{E}_u) \cap I)$, we know that $\partial_p(\mathcal{E}_u) \cap I$ is generated by E_I. Similarly, $\partial_p(\mathcal{E}_u) \cap J$ is generated by E_J. So ${\downarrow}(E_I \cup E_j) = (\partial_p(\mathcal{E}_u) \cap I) \cup (\partial_p(\mathcal{E}_u) \cap J)$. Therefore $E_I \cup E_J$ generates $\partial_p(\mathcal{E}_u) \cap (I \cup J)$ and so the residue $\mathcal{R}(u,p,I \cup J) = \mathcal{R}(u,p,{\downarrow}(E_I \cup E_J))$.

(ii) Let $Q = \{q_1, q_2, \ldots, q_k\}$. We can rewrite $\mathcal{R}(u,p,\partial_Q(\mathcal{E}_u))$ as $\mathcal{R}(u,p,\bigcup_{i\in[1..k]} \partial_{q_i}(\mathcal{E}_u))$. From Lemma 9.19 it follows that for each $i \in [1..k]$, p can compute a set $E_i \subseteq primary_p(\mathcal{E}_u)$ from the information in $primary_{\{p,q_i\}}(\mathcal{E}_u)$ such that $\mathcal{R}(u,p,\partial_{q_i}(\mathcal{E}_u)) = \mathcal{R}(u,p,{\downarrow}E_i)$. From part (i) of this Corollary, it then follows that $\mathcal{R}(u,p,\partial_Q(\mathcal{E}_u)) = \mathcal{R}(u,p,\bigcup_{i\in[1..k]} \partial_{q_i}(\mathcal{E}_u)) = \mathcal{R}(u,p,{\downarrow}E)$ where $E = \bigcup_{i\in[1..k]} E_i$.

(iii) Let $J = \partial_p(\mathcal{E}_u) \cup \partial_r(\partial_q(\mathcal{E}_u))$. J is an ideal. By the construction of J, $\max_p(J) = \max_p(\mathcal{E}_u)$. From the assumption that $latest_{p\to r}(\mathcal{E}_u) \leq latest_{q\to r}(\mathcal{E}_u)$, we have $\max_r(J) = latest_{q\to r}(\mathcal{E}_u)$. So, $\partial_p(J) = \partial_p(\mathcal{E}_u)$ and $\partial_r(J) = \partial_r(\partial_q(\mathcal{E}_u))$. Since $\mathcal{R}(u,p,\partial_r(\partial_q(\mathcal{E}_u))) = \mathcal{R}(u,p,\partial_p(\mathcal{E}_u) \cap (\partial_r(\partial_q(\mathcal{E}_u))) = \mathcal{R}(u,p,\partial_p(J) \cap \partial_r(J))$, it suffices to find a subset $E \subseteq primary_p(\mathcal{E}_u)$ which generates $\partial_p(J) \cap \partial_r(J)$.
By Lemma 9.19, $\partial_p(J) \cap \partial_r(J)$ is generated by $primary_p(J) \cap primary_r(J)$. Since $\max_p(J) = \max_p(\mathcal{E}_u)$, $primary_p(J) = primary_p(\mathcal{E}_u)$.

On the other hand, $primary_r(J) = primary_r(\downarrow latest_{q\to r}(\mathcal{E}_u))$. By definition, this is the set $\{latest_{q\to r\to s}(\mathcal{E}_u)\}_{s\in\mathcal{P}}$.
So the set $E \subseteq primary_p(\mathcal{E}_u)$ generating $\partial_p(J) \cap \partial_r(J)$ is given by $E = primary_p(J) \cap primary_r(J) = primary_p(\mathcal{E}_u) \cap \{latest_{q\to r\to s}(\mathcal{E}_u)\}_{s\in\mathcal{P}}$ and can be computed from $primary_p(\mathcal{E}_u)$ and $secondary_q(\mathcal{E}_u)$.

□

Part (ii) of the preceding Corollary makes explicit our claim that every process residue $\mathcal{R}(u, p, \partial_Q(\mathcal{E}_u))$, $Q \subseteq \mathcal{P}$, can be effectively rewritten as a primary residue $\mathcal{R}(u, p, \downarrow E)$, $E \subseteq primary_p(\mathcal{E}_u)$, based on the information available in $primary_{p\cup\{Q\}}(\mathcal{E}_u)$. In case $Q = \emptyset$, $\mathcal{R}(u, p, \partial_Q(\mathcal{E}_u))$ is given by the primary residue corresponding to $\emptyset \subseteq primary_p(\mathcal{E}_u)$.

9.8.3. *Computing primary residues locally*

We now describe how, while reading a word u, each process p maintains the functions f_w for each primary residue w of u at p.

Initially, at the empty word $u = \varepsilon$, every primary residue from $\{\mathcal{R}(u, p, \downarrow E)\}_{p\in\mathcal{P}, E\subseteq primary_p(\mathcal{E}_u)}$ is just the empty word ε. So, all primary residues are represented by the identity function $Id : S \to S$.

Let $u \in \Sigma^*$ and $a \in \Sigma$. Assume inductively that every $p \in \mathcal{P}$ has computed at the end of u the function f_w for each primary residue $w \in \{\mathcal{R}(u, p, \downarrow E)\}_{E\subseteq primary_p(\mathcal{E}_u)}$. We want to compute for each p the corresponding functions after the word ua, whose associated trace is $(\mathcal{E}_{ua}, \leq_{ua}, \lambda_{ua})$.

For processes not involved in a, these values do not change.

Proposition 9.34. *If $p \notin loc(a)$ then every subset $E \subseteq primary_p(\mathcal{E}_{ua})$ is also a subset of $primary_p(\mathcal{E}_u)$ and the primary residue $\mathcal{R}(ua, p, \downarrow E)$ is the same as the primary residue $\mathcal{R}(u, p, \downarrow E)$.*

Proof. This follows immediately from the fact that $\partial_p(\mathcal{E}_{ua}) = \partial_p(\mathcal{E}_u)$ and $primary_p(\mathcal{E}_{ua}) = primary_p(\mathcal{E}_u)$. □

So, the interesting case is when p participates in a. We show how to calculate all the new primary residues for p using the information available with the processes in $loc(a)$ after u.

Lemma 9.35. *Let $p \in loc(a)$ and $E \subseteq primary_p(\mathcal{E}_{ua})$. The function f_w corresponding to the primary residue $w = \mathcal{R}(ua, p, \downarrow E)$ can be computed from the primary residues at u of the processes in $loc(a)$ using the information in $primary_{loc(a)}(\mathcal{E}_u)$ and $secondary_{loc(a)}(\mathcal{E}_u)$.*

Proof. Let e_a be the event corresponding to the new letter a—that is, $\mathcal{E}_{ua} \setminus \mathcal{E}_u = \{e_a\}$. There are two cases to consider.

Case 1: $(e_a \in E)$
Since $\downarrow E = \downarrow e_a = \partial_p(\mathcal{E}_{ua})$, the residue $\mathcal{R}(ua, p, \downarrow E) = \mathcal{R}(ua, p, \downarrow e_a)$ is the empty word ε. So the corresponding function is just the identity function $Id : S \to S$.

Case 2: $(e_a \notin E)$
We want to compute the function f_w corresponding to the word $w = ua[\partial_p(\mathcal{E}_{ua}) \setminus \downarrow E]$. By Lemma 9.28, we know that

$$ua[\partial_p(\mathcal{E}_{ua})] \sim ua[\downarrow E]ua[\partial_p(\mathcal{E}_{ua}) \setminus \downarrow E]. \tag{9.1}$$

But $ua[\partial_p(\mathcal{E}_{ua})] = u[\partial_{loc(a)}(\mathcal{E}_u)]a$ and so we have

$$ua[\partial_p(\mathcal{E}_{ua})] = u[\partial_{loc(a)}(\mathcal{E}_u)]a \sim u[\downarrow E]u[\partial_{loc(a)}(\mathcal{E}_u) \setminus \downarrow E]a. \tag{9.2}$$

Since $e_a \notin \downarrow E$, $ua[\downarrow E] = u[\downarrow E]$. Thus, cancelling $u[\downarrow E]$ from the right hand sides of (9.1) and (9.2) above, we have $u[\partial_{loc(a)}(\mathcal{E}_u) \setminus \downarrow E]a \sim ua[\partial_p(\mathcal{E}_{ua}) \setminus \downarrow E]$. So, to compute the function f_w, it suffices to compute the function corresponding to $u[\partial_{loc(a)}(\mathcal{E}_u) \setminus \downarrow E]a$.

Let $loc(a) = \{p_1, p_2, \ldots, p_k\}$, where $p = p_1$. Construct sets of processes $\{Q_i\}_{i \in [1..k]}$ such that $Q_1 = \{p_1\}$ and $Q_i = Q_{i-1} \cup \{q_i\}$ for $i \in [2..k]$.

Construct ideals $\{I_j\}_{j \in [0..k]}$ as follows: $I_0 = \downarrow E$ and for $j \in [1..k]$, $I_j = I_{j-1} \cup \mathcal{E}_u|_{p_j}$. Clearly, $I_0 \subseteq I_1 \subseteq \cdots \subseteq I_k \subseteq \mathcal{E}_u$.

By Corollary 9.29, $u[I_k] \sim u[I_0]u[I_1 \setminus I_0] \cdots u[I_k \setminus I_{k-1}]$. Since $u[I_k] = u[\partial_{loc(a)}(\mathcal{E}_u)]$ and $u[I_0] = u[\downarrow E]$, from (9.2) above it follows that the word $u[\partial_{loc(a)}(\mathcal{E}_u) \setminus \downarrow E]a$ which we seek is $\sim$-equivalent to the word $u[I_1 \setminus I_0] \cdots u[I_k \setminus I_{k-1}]a$.

Claim: For each $j \in [1..k]$, $u[I_j \setminus I_{j-1}]$ is a primary residue $\mathcal{R}(u, p_j, \downarrow F_j)$, where $F_j \subseteq primary_{p_j}(\mathcal{E}_u)$. Further, p_j can determine F_j from the information in $primary_{p_j}(\mathcal{E}_u)$ and $secondary_{loc(a)}(\mathcal{E}_u)$.

Assuming the claim, for each word $w_j = u[I_j \setminus I_{j-1}]$, we can find the corresponding function $f_{w_j} : S \to S$ among the primary residues stored by p_j after u. The composite function $f_a \circ f_{w_k} \circ f_{w_{k-1}} \circ \cdots \circ f_{w_1}$ then gives us the function corresponding to the word $u[I_1 \setminus I_0] \cdots u[I_k \setminus I_{k-1}]a$, which is what we need.

Proof of Claim: The way that primary events are updated guarantees that each event $e \in E$ was a primary event in $\mathcal{E}_u$, *before* a occurred, for one of the processes in $loc(a)$; i.e., $E \subseteq primary_{loc(a)}(\mathcal{E}_u)$. For $i \in [1..k]$, let $E_i = E \cap primary_{p_i}(\mathcal{E}_u)$.

First consider $u[I_1 \setminus I_0]$.

Let $E' = E \setminus E_1$. $I_1 \setminus I_0$ is the same as $\partial_{p_1}(\mathcal{E}_u) \setminus \downarrow(E_1 \cup E')$, which is the same as $\partial_{p_1}(\mathcal{E}_u) \setminus (\downarrow E_1 \cup \downarrow E')$. We want to compute $u[I_1 \setminus I_0] = \mathcal{R}(u, p_1, \downarrow E_1 \cup \downarrow E')$.

Each event $e \in E'$ is a primary event of the form $latest_{p_1 \to q_e}(\mathcal{E}_{ua})$ for some $q_e \in \mathcal{P}$. Further, for some $i \in [2..k]$, e was also the primary event $latest_{p_i \to q_e}(\mathcal{E}_u)$ before e_a occurred. Since p_1 has inherited this information from p_i, it must have been the case that $latest_{p_1 \to q_e}(\mathcal{E}_u) \leq latest_{p_i \to q_e}(\mathcal{E}_u)$. So, by part (iii) of Corollary 9.33, the residue $\mathcal{R}(u, p_1, \downarrow e) = \mathcal{R}(u, p_1, \partial_{q_e}(\partial_{p_i}(\mathcal{E}_u)))$ corresponds to a primary residue $\mathcal{R}(u, p_1, \downarrow G_e)$, where p_1 can determine $G_e \subseteq primary_{p_1}(\mathcal{E}_u)$ from $primary_{p_1}(\mathcal{E}_u)$ and $secondary_{p_i}(\mathcal{E}_u)$.

So, by part (i) of Corollary 9.33, $\mathcal{R}(u, p_1, \downarrow E') = \mathcal{R}(u, p_1, \bigcup_{e \in E'} \downarrow e)$ is a primary residue $\mathcal{R}(u, p_1, \downarrow G_1)$ where $G_1 = \bigcup_{e \in E'} G_e$.

$\mathcal{R}(u, p_1, \downarrow E_1)$ is a primary residue since $E_1 \subseteq primary_p(\mathcal{E}_u)$. Applying part (i) of Corollary 9.33 again, $\mathcal{R}(u, p_1, (\downarrow E_1 \cup \downarrow E'))$ corresponds to the primary residue $\mathcal{R}(u, p_1, \downarrow F_1)$, where $F_1 = E_1 \cup G_1$.

Now consider $u[I_j \setminus I_{j-1}]$ for $j \in [2..k]$.

$I_j \setminus I_{j-1}$ is the same as $\partial_{p_j}(\mathcal{E}_u) \setminus (\downarrow E \cup \partial_{Q_{j-1}}(\mathcal{E}_u))$ so we want to compute the residue $\mathcal{R}(u, p_j, \downarrow E \cup \partial_{Q_{j-1}}(\mathcal{E}_u))$.

By a similar argument to the one for $u[I_1 \setminus I_0]$, p_j can compute a set $G_j \subseteq primary_{p_j}(\mathcal{E}_u)$ such that $\mathcal{R}(u, p_j, \downarrow E)$ corresponds to the primary residue $\mathcal{R}(u, p_j, \downarrow G_j)$.

By part (ii) of Corollary 9.33, p_j can compute from $primary_{Q_j}(\mathcal{E}_u)$ a set $H_j \subseteq primary_{p_j}(\mathcal{E}_u)$ such that $\mathcal{R}(u, p_j, \partial_{Q_{j-1}}(\mathcal{E}_u))$ corresponds to the primary residue $\mathcal{R}(u, q, \downarrow H_j)$.

We now use part (i) of Corollary 9.33 to establish that $\mathcal{R}(u, p_j, \downarrow E \cup \partial_{Q_{j-1}}(\mathcal{E})_u)$ corresponds to the primary residue $\mathcal{R}(u, p_j, \downarrow F_j)$, where $F_j = G_j \cup H_j$. □

9.8.4. *An asynchronous automaton for L*

Our analysis of process residues and primary residues immediately yields a deterministic asynchronous automaton $\mathcal{A}$ which accepts the language L. Recall that $\mathcal{B} = (S, \Sigma, \delta, s_0, S_F)$ is the minimal DFA recognizing L.

For $p \in \mathcal{P}$, each local state of p will consist of the following:

- Primary and secondary information for p, as stored by the gossip automaton.
- For each subset E of the primary events of p, a function $f_E : S \to S$ recording the (syntactic congruence class of the) primary residue $\mathcal{R}(u, p, E\downarrow)$ at the end of any word u.

At the initial state, for each process p, all the primary, secondary and tertiary information of p points to the initial event $e_\perp$. For each subset E of primary events, the function f_E is the identity function $Id : S \to S$.

The transition functions $\to_a$ modify the local states of $loc(a)$ as follows:

- Primary, secondary and tertiary information is updated as in the gossip automaton.
- The functions corresponding to primary residues are updated as described in the proof of Lemma 9.35.

Other than comparing primary information, the only operation used in updating the primary residues at p (Lemma 9.35) is function composition. This is easily achieved using the data available in the states of the processes which synchronized.

The final states of $\mathcal{A}$ are those where the value jointly computed from the primary residues in $\mathcal{P}$ yields a state in S_F. More precisely, order the processes as $\mathcal{P} = \{p_1, p_2, \ldots, p_N\}$. Construct subsets of processes $\{Q_i\}_{i \in [1..N]}$ such that $Q_1 = \{p_1\}$ and for $i \in [2..N]$, $Q_i = Q_{i-1} \cup \{p_i\}$.

Let $\vec{v} = \{v_1, v_2, \ldots, v_N\}$ be a global state of $\mathcal{A}$ such that $\mathcal{A}$ is in $\vec{v}$ after reading an input word u.

By Corollary 9.33 (ii), for each $i \in [2..N]$, we can compute from $\{v_1, v_2, \ldots, v_i\}$ a subset E_i of the primary information of p_i such that the Q_{i-1}-residue of p_i is also the primary residue of p_i with respect to E_i. Let f_i denote this primary residue. In addition, from the state v_1, we can extract the function f_1 corresponding to the primary residue $\mathcal{R}(u, p, \emptyset)$.

From the expression ($\diamondsuit$), we know that the composite function $f_N \circ f_{N-1} \circ \cdots \circ f_1$ is exactly the function f_u associated with the input word u leading to the global state $\vec{v}$. So, we put $\vec{v}$ in the set of accepting states F of $\mathcal{A}$ iff $f_N \circ f_{N-1} \circ \cdots \circ f_1(s_0) \in S_F$.

Notice that it does not matter how we order the states in $\vec{v}$ when we try to decide whether $\vec{v} \in F$. We keep track of residues in all processes in a symmetric fashion, and the expression ($\diamondsuit$) holds regardless of how we order $\mathcal{P}$. So, if $\vec{v}$ is a valid (i.e., reachable) global state, the composite function $f_N \circ f_{N-1} \circ \cdots \circ f_1$ which we compute from $\vec{v}$ is always the *same*, no matter how we order $\mathcal{P}$.

From our analysis of residues in this section, we have the following result.

Theorem 9.36. *The language accepted by $\mathcal{A}$ is exactly L.*

The size of $\mathcal{A}$

Proposition 9.37. *Let $M = |S|$ and $N = |\mathcal{P}|$, where S is the set of states of $\mathcal{B}$, the DFA recognizing L, and $\mathcal{P}$ is the set of processes in the corresponding asynchronous automaton $\mathcal{A}$ which we construct to accept L. Then, the number of local states of each process $p \in \mathcal{P}$ is at most $2^{O(2^N M \log M)}$.*

Proof. We estimate the number of bits required to store a local state of a process p.

From Lemma 9.25, we know that the primary and secondary information that we require to keep track of the latest gossip can be stored in $O(N^2 \log N)$ bits.

The new information we store in each local state of p is the collection of primary residues. Each residue, which is a function from S to S, can be written down as an array with M entries, each of $\log M$ bits; i.e., $M \log M$ bits in all. Each primary residue corresponds to a subset of primary events. There are N primary events and so, in general, we need to store 2^N residues. Thus, all the residues can be stored using $2^N M \log M$ bits.

So, the entire state can be written down using $O(2^N M \log M)$ bits, whence the number of distinct local states of p is bounded by $2^{O(2^N M \log M)}$. □

9.9. Discussion

We have characterized the languages recognized by direct product automata as those that are shuffle-closed. On the other hand, we have shown that asynchronous automata accept precisely the class of recognizable trace languages. A very interesting and difficult open problem is to precisely characterize the class of languages recognized by synchronized products. A positive result has been obtained for a very restricted subclass in [6].

The bounded time-stamping algorithm is a fundamental building block for many important constructions involving asynchronous automata. As we have seen, it is at the heart of the proof of Zielonka's theorem. Time-stamping can also be used to define a subset construction for directly determinizing asynchronous automata [7]. Another application of bounded time-stamping is in providing an automata-theoretic decision procedure for TrPTL, a linear time temporal logic with local modalities that is interpreted over traces [8].

Zielonka's theorem was first described in [1]. Subsequently, an equivalent result in terms of an alternative distributed model called *asynchronous cellular automata* was presented in [3]. Our proof is taken from [9] and broadly follows the structure of the original proof in [1], except that the distributed time-stamping function that is an explicit intermediate step in our construction is implicit in the original proof of Zielonka. We believe that clearly separating and identifying the role played by time-stamping helps to make the proof more digestible. Recently, the residue based construction described here has been refined in [10] to show that of the exponentially many primary residues maintained by each process, only polynomially many are actually distinct. This observation eliminates one exponential in the overall complexity of the construction.

Zielonka's theorem can be seen as an algorithm to synthesize a distributed implementation from a sequential specification. Another way to present the same problem is to ask when a global state space can be decomposed into local state spaces such that the product of these local state spaces is isomorphic to the original global state space. This problem can been solved for direct products and asynchronous automata—see [11] for an overview of the results.

However, the following distributed synthesis problem is open: Decompose a global state space into a product of local state spaces that is *bisimilar* to the original state space. A bisimulation [12, 13] is a relation that holds between a pair of transition systems that can simulate each other very faithfully. This is potentially a more useful formulation of the problem since branching time properties are preserved by bisimulation.

Acknowledgments

This material has been used for graduate courses in *Automata and Concurrency* at Chennai Mathematical Institute and the Institute of Mathematical Sciences and the exposition has benefited from the feedback received from the students who attended these courses. The time-stamping algorithm for asynchronous automata and the proof of Zielonka's theorem presented here are both joint work with Milind Sohoni. I thank Namit Chaturvedi and Prateek Karandikar for pointing out some subtle and not-so-subtle errors in the original manuscript.

References

[1] W. Zielonka: Notes on finite asynchronous automata, *R.A.I.R.O.—Inform. Théor. Appl.*, **21** (1987) 99–135.
[2] A. Mazurkiewicz: Basic notions of trace theory, in: J.W. de Bakker, W.-P. de Roever and G. Rozenberg (eds.), *Linear time, branching time and partial order in logics and models for concurrency, LNCS* **354** (1989) 285–363.
[3] R. Cori, Y. Metivier and W. Zielonka: Asynchronous mappings and asynchronous cellular automata, *Inform. and Comput.*, **106** (1993) 159–202.
[4] J. Hopcroft and J.D. Ullman: *Introduction to automata, languages and computation*, Addison-Wesley (1979).
[5] I.J. Aalbersberg and G. Rozenberg: Theory of traces, *Theoret. Comput. Sci.*, **60** (1988) 1–82.
[6] J. Berstel, L. Boasson and M. Latteux: Mixed languages. *Theoret. Comput. Sci.* **332**(1–3) (2005) 179–198.
[7] N. Klarlund, M. Mukund and M. Sohoni: Determinizing asynchronous automata, *Proc. 21st ICALP, LNCS* **820** (1994) 130-141.
[8] P.S. Thiagarajan: TrPTL: A trace based extension of linear time temporal logic, *Proc. 9th IEEE LICS* (1994) 438–447.
[9] M. Mukund and M. Sohoni: Gossiping, asynchronous automata and Zielonka's theorem, *Report TCS-94-2*, Chennai Mathematical Institute, Chennai, India (1994).
[10] B Genest and A Muscholl: Constructing Exponential-Size Deterministic Zielonka Automata. *Proc. 33rd ICALP, LNCS* **4052** (2006) 565–576.
[11] M. Mukund: From global specifications to distributed implementations. in *Synthesis and Control of Discrete Event Systems*, B. Caillaud, P. Darondeau and L. Lavagno (eds), Kluwer (2002) 19–34.
[12] R. Milner: *Communication and Concurrency*, Prentice-Hall (1989).
[13] D. Park: Concurrency and Automata on Infinite Sequences. *Proc. 5th GI-Conference Karlsruhe, Theoret. Comp. Sc.* **104** (1981) 167-183.

Chapter 10

The Theory of Message Sequence Charts

K. Narayan Kumar

Chennai Mathematical Institute
H1 SIPCOT IT Park, Siruseri, India
kumar@cmi.ac.in

Message Sequence Charts or MSCs are a visual formalism used in the specification of systems in many domains including telecommunications, object oriented design and forms a part of the UML language. Consequently, the formal study of MSCs has received considerable attention over the last decade. We survey some of the key results in this area with particular emphasis on the notion of regularity and its relationship to automata, logics and model-checking.

10.1. Message sequence charts

Message Sequence Charts (MSCs) [1] are an appealing visual formalism used in a number of software engineering notational frameworks such as SDL [2] and UML [3]. An MSC is a representation of a single behaviour or run of a system consisting of a collection of processes communicating with each other asynchronously via buffered channels. In this paper we shall restrict ourselves to systems with first-in-first-out (FIFO) channels. Fig. 10.1 shows an MSC involving three processes p, q and r and three messages. An MSC is to be read from top to bottom, the vertical lines denote processes and the arrows represent messages. Formally, MSCs are defined as labelled partial orders.

Let $\mathcal{P} = \{p, q, r, \ldots\}$ be a finite set of processes that communicate with each other through messages via reliable FIFO channels using a finite set of message types $\mathcal{M}$. For $p \in \mathcal{P}$, let $\Sigma_p = \{p!q(m), p?q(m) \mid p \neq q \in \mathcal{P}, m \in \mathcal{M}\}$ be the set of communication actions in which p participates. The action $p!q(m)$ is read as *p sends the message m to q* and the action $p?q(m)$ is read as *p receives the message m from q*. We set $\Sigma = \bigcup_{p \in \mathcal{P}} \Sigma_p$. We also denote the set of *channels* by $Ch = \{(p, q) \in \mathcal{P}^2 \mid p \neq q\}$. Whenever the set of processes $\mathcal{P}$ is clear from the context, we write Σ instead of $\Sigma_{\mathcal{P}}$, etc. Observe that our notation restricts us to a single channel from a process p to any other process q, however this is merely for notational convenience and the results do not change if we permit multiple channels between pairs of processes.

Labelled posets A $\Sigma_{\mathcal{P}}$-labelled poset is a structure $M = (E, \leq, \lambda)$ where $(E, \leq)$ is a partially ordered set and $\lambda : E \to \Sigma_{\mathcal{P}}$ is a labelling function. For $e \in E$, let

$\downarrow e = \{e' \mid e' \leq e\}$. For $p \in \mathcal{P}$ and $a \in \Sigma_{\mathcal{P}}$, we set $E_p = \{e \mid \lambda(e) \in \Sigma_p\}$ and $E_a = \{e \mid \lambda(e) = a\}$, respectively. For $(p, q) \in Ch$, we define the relation $<_{pq}$:

$$e <_{pq} e' \stackrel{\text{def}}{=} \exists m \in \mathcal{M} \text{ such that } \lambda(e) = p!q(m),\ \lambda(e') = q?p(m) \text{ and } |\downarrow e \cap E_{p!q(m)}| = |\downarrow e' \cap E_{q?p(m)}|$$

The relation $e <_{pq} e'$ says that channels are FIFO with respect to *each message*—if $e <_{pq} e'$, the message m read by q at e' is the one sent by p at e.

Finally, for each $p \in \mathcal{P}$, we define the relation $\leq_{pp} = (E_p \times E_p) \cap \leq$, with $<_{pp}$ standing for the largest irreflexive subset of $\leq_{pp}$. We write $Chn(e) = p!q$ if $\lambda(e) = p!q(m)$ and $Chn(e) = p?q$ if $\lambda(e) = p?q(m)$.

Definition 10.1. An MSC over $\mathcal{P}$ is a *finite* $\Sigma_{\mathcal{P}}$-labelled poset $M = (E, \leq, \lambda)$ where:

(1) Each relation $\leq_{pp}$ is a linear (total) order.
(2) If $p \neq q$ then for each $m \in \mathcal{M}$, $|E_{p!q(m)}| = |E_{q?p(m)}|$.
(3) If $e <_{pq} e'$, then $|\downarrow e \cap (\bigcup_{m \in \mathcal{M}} E_{p!q(m)})| = |\downarrow e' \cap (\bigcup_{m \in \mathcal{M}} E_{q?p(m)})|$.
(4) The partial order $\leq$ is the reflexive, transitive closure of $\bigcup_{p,q \in \mathcal{P}} <_{pq}$.

The second condition ensures that every message sent along a channel is received. The third condition says that every channel is FIFO across all messages.

In diagrams, the events of an MSC are presented in *visual order*. The events of each process are arranged in a vertical line and messages are displayed as horizontal or downward-sloping directed edges. Fig. 10.1 shows an example with three processes $\{p, q, r\}$ and six events $\{p_1, p_2, q_1, q_2, r_1, r_2\}$ corresponding to three messages—m_1 from p to q, m_2 from q to r and m_3 from p to r.

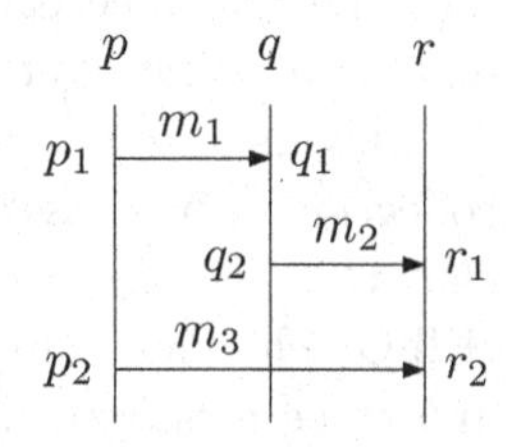

Fig. 10.1. An MSC.

For an MSC $M = (E, \leq, \lambda)$, we let $\text{Lin}(M) = \{\lambda(\pi) \mid \pi \text{ is a linearization of } (E, \leq)\}$. For instance, $p!q(m_1)\, q?p(m_1)\, q!r(m_2)\, p!r(m_3)\, r?q(m_2)\, r?p(m_3)$ is one linearization of the MSC in Fig. 10.1. We write $\text{Lin}(M)$ for the set of linearizations of an MSC M.

Note that under the FIFO assumption an MSC can be reconstructed from any one linearization — the relation $<_{pp}$ is determined by the order of the p events in the linearization while $<_{pq}$ is determined by matching the ith $p!q$ event with the ith $q?p$ event.

Definition 10.2. Let M be an MSC and $B \in \mathbb{N}$. We say that $w \in \text{Lin}(M)$ is *B-bounded* if for every prefix v of w and for every channel $(p, q) \in Ch$, $\sum_{m \in \mathcal{M}} |v{\restriction}\{p!q(m)\}| - \sum_{m \in \mathcal{M}} |v{\restriction}\{q?p(m)\}| \leq B$, where $v{\restriction}\Gamma$ denotes the projection of v on $\Gamma \subseteq \Sigma_{\mathcal{P}}$.

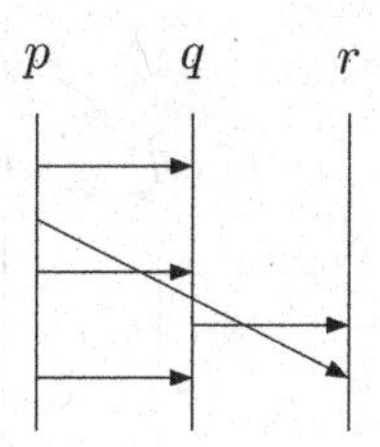

Fig. 10.2. A 3-bounded MSC.

This means that along the sequential execution of M described by w, no channel ever contains more than B-messages. Consider the MSC in Fig. 10.2. The linearization $p!q\ q?p\ p!r\ p!q\ q?p\ q!r\ r?q\ p!q\ q?p\ r?p$ is 1-bounded while the linearization $p!q\ p!r\ p!q\ p!q\ q?p\ q?p\ p!r\ q?p\ r?q\ r?p$ is 3-bounded.

Definition 10.3. We say that M is *universally B-bounded* if every $w \in \mathrm{Lin}(M)$ is B-bounded and that M is *existentially B-bounded* if there is a $w \in \mathrm{Lin}(M)$ which is B-bounded. (We sometimes write B-bounded to mean universally B-bounded.)

The MSC in Fig. 10.2 is universally 3-bounded. The 1-bounded linearization listed earlier shows that this MSC is also existentially 1-bounded. The following proposition (paraphrased) from [4] characterizes universal B-boundedness for MSCs.

Proposition 10.1. *An MSC M is not B-bounded if and only if there are processes p, q and $p!q$ labelled events $e_1 <_{pp} e_2, \ldots <_{pp} e_{B+1}$ such that the $q?p$ labelled event e' with $e_1 <_{pq} e'$ is not below e_{B+1}.*

Optimal linearizations: Given an MSC M the smallest B for which it is B-bounded can be computed in linear time as shown in [5]. In particular, an optimal linearization w.r.t. boundedness can be computed by the following greedy strategy — at each step, extend the linearization, if possible, by a receive event; otherwise extend it by picking from the set of candidate send events the one that minimizes the maximum number of undelivered messages in any channel. Applying this greedy strategy to the MSC in Fig. 10.2 we get $p!q\ q?p\ p!r\ p!q\ q?p\ q!r\ r?q\ r?p\ p!q\ q?p$ which is a 1-bounded linearization.

10.1.1. *Concatenation of MSCs*

The concatenation of M_1 and M_2 written as $M_1 \circ M_2$ denotes the behaviour where each process participates in its events in M_1 and then follows it by participating in its events in M_2. It is not necessary for all processes to complete the events in M_1 before any process enters M_2.

Formally, the (asynchronous) concatenation of MSCs is defined as follows.

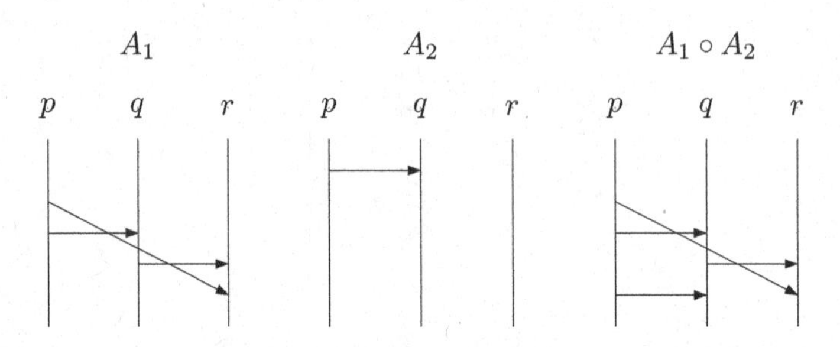

Fig. 10.3. MSC Concatenation.

Definition 10.4. Let $M_1 = (E^1, \leq^1, \lambda_1)$ and $M_2 = (E^2, \leq^2, \lambda_2)$ be a pair of MSCs such that E^1 and E^2 are disjoint. The *(asynchronous) concatenation* of M_1 and M_2 yields the MSC $M_1 \circ M_2 = (E, \leq, \lambda)$ where $E = E^1 \cup E^2$, $\lambda(e) = \lambda_i(e)$ if $e \in E^i$, $i \in \{1, 2\}$, and $<_{pp} = <^1_{pp} \cup <^2_{pp} \cup (E^1_p \times E^2_p)$ and for $(p, q) \in Ch$, $<_{pq} = <^1_{pq} \cup <^2_{pq}$.

MSCs over a given alphabet form a monoid with the empty MSC as the identity (i.e. concatenation is an associative operation). Fig. 10.3 describes the concatenation of two MSCs. In the linearization $p!r\ p!q\ q?p\ q!r\ r?q\ p!q\ q?p\ r?p$ of $A_1 \circ A_2$ observe that the last event of A_1 occurs after both the events of A_2.

We can repeatedly decompose MSCs into the concatenation of smaller (nontrivial) MSCs till we are left with MSCs that cannot be decomposed.

Definition 10.5. An MSC M is said to be an *atom* if it cannot be expressed as the concatenation of two nontrivial MSCs.

The MSCs A_1 and A_2 in Fig. 10.3 are atoms. The MSC M in Fig. 10.2 is the concatenation $A_2 \circ A_1 \circ A_2$. In this case, the decomposition of M into atoms is unique. In general this need not be the case. To understand the exact nature of the decomposition of MSCs into atoms we need some terminology from the theory of traces.

10.1.2. *MSCs and Traces*

A *dependence alphabet* is a pair (Σ, D) where the alphabet Σ is a finite set of actions and the *dependence relation* $D \subseteq \Sigma \times \Sigma$ is reflexive and symmetric. The *independence relation* I is the complement of D. For $A \subseteq \Sigma$, the set of letters independent of A is denoted by $I(A) = \{b \in \Sigma \mid (a, b) \in I$ for all $a \in A\}$ and the set of letters depending on (some action in) A is denoted by $D(A) = \Sigma \setminus I(A)$.

A *Mazurkiewicz trace* is a labelled partial order $t = (V, \leq, \lambda)$ where V is a set of vertices labelled by $\lambda : V \to \Sigma$ and $\leq$ is a partial order over V satisfying the following conditions: For all $x \in V$, the downward set $\downarrow x = \{y \in V \mid y \leq x\}$ is finite, $(\lambda(x), \lambda(y)) \in D$ implies $x \leq y$ or $y \leq x$, and $x \lessdot y$ implies $(\lambda(x), \lambda(y)) \in D$, where $\lessdot \; = \; < \setminus <^2$ is the immediate successor relation in t.

Let $\equiv_I$ be the equivalence relation on Σ^* given by the reflexive transitive closure of the relation $uabv \sim ubav$ whenever $(a, b) \in I$. The set of linearizations of a trace t is an equivalence class of $\equiv_I$. Conversely, it is possible to reconstruct the trace from any given linearization. For a more detailed introduction to Trace theory, the reader is referred to Chapter 9.

We are now in a position to characterize the different decompositions of an MSC into atoms. Let A be any finite collection of atoms over the set of processes $\mathcal{P}$. One can equip A with natural dependence alphabet structure (A, D) where $(a, b) \in D$ if and only if there is a process p that is active in both a and b.

Proposition 10.2. *Let M be an MSC, A be a finite set of atoms that contains the atoms that appear in M and (A, D) be the corresponding dependence alphabet. Then, if $M = a_1 \circ a_2 \circ \ldots a_k$ and $M = a'_1 \circ a'_2 \circ \ldots a'_l$ then $a_1 a_2 \ldots a_k \equiv_I a'_1 a'_2 \ldots a'_l$.*

Thus, the decomposition of an MSC into atoms is unique up to commutations of independent atoms. In particular, the set of atoms that are necessary to decompose an MSC M, denoted by $\mathcal{A}(M)$, is unambiguously defined. We shall write $\mathcal{A}_t(M)$ to denote the trace (or equivalently the linearizations of the trace) over atoms associated with M.

There is a second dependence alphabet associated with MSCs which yields an alternative characterization of B-bounded MSCs. Let $\Sigma_B = \Sigma_{\mathcal{P}} \times \{0, 1, \ldots, B-1\}$ for some natural number B. Let D be the dependence relation given by $(x, i)D(y, j)$ if either x and y occur in the same process or if $Chn(x) = p!q$, $Chn(y) = q?p$ and $i = j$. In a B-bounded MSC, the ith receive on a channel must necessarily occur before the $(i+B)^{th}$ send on the same channel justifying the demand for an ordering between them in the dependency alphabet.

Given an MSC $M = (E, \leq, \lambda)$ we transform it into a Σ_B-labelled partial order $tr(M)$. Let $tr(M) = (E, \leq, \lambda')$ where $\lambda'(e) = (\lambda(e), i)$ where $i = |\{e' \mid e' < e,\ Chn(e') = Chn(e)\}|$. That is, we tag each event by its channel count modulo B. Does this yield a trace w.r.t the alphabet (Σ_B, D)?

There are two kinds of dependencies in the definition of the relation D. A dependency of the first kind, within a process, is also enforced in the MSC. An event labelled $p!q$ is guaranteed to be below the corresponding receive event labelled $q?p$ in the MSC. However, no ordering is necessary between other occurrences of events labelled by these two letters. Thus, the second kind of dependency demanded by D is in general stronger than the ordering in the MSC. Fig. 10.4 describes the trace $tr(M)$ over the

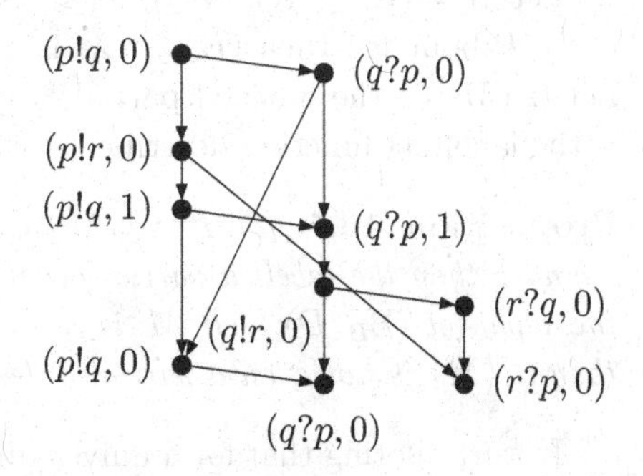

Fig. 10.4. Trace corresponding to MSC of Fig. 10.2.

alphabet $\Sigma_{\mathcal{P}} \times \{0,1\}$ corresponding to the MSC in Fig. 10.2. Observe that there is no ordering between the first $q?p$ and the third $p!q$ in the MSC whilst they are ordered in the trace.

However, for B-bounded MSCS we have the following result.

Proposition 10.3. *[6; 7] Let M be an MSC. Then $tr(M)$ is a trace over the alphabet Σ_B if and only if M is a universally B-bounded MSC.*

Thus, universally B-bounded MSCs are traces over the aforementioned dependence alphabet. This allows us to exploit the well-developed theory of traces in the study of B-bounded MSCs.

Finally, let us examine existentially B-bounded MSCs and their relationship to traces. By definition, an MSC is not existentially B-bounded iff it cannot be linearized into a B-bounded word, and what rules out such a linearization is the following scenario — a $p!q$ event whose associated receive is above the next B $p!q$-labelled events under $\leq$. This can be formalized as follows.

Definition 10.6. Let $(E, \leq, \lambda)$ be a given MSC and B a natural number. The relation $<_{rev}$ on E, defined below, relates the ith receive on a channel with the $(i+B)^{th}$ send on the same channel.

$$e <_{rev} e' \stackrel{\text{def}}{=} \exists f.\, f <_{pq} e \,\&\, \lambda(f) = \lambda(e') \,\&\, |\{e'' | f < e'' \leq e', \lambda(e'') = \lambda(f)\}| = B$$

Existentially B-bounded MSCs are precisely those that do not violate $<_{rev}$.

Proposition 10.4. *[8] Let $M = (E, \leq, \lambda)$ be an MSC and $<_{rev}$ be as defined above. M is existentially B-bounded if and only if $\leq \cup <_{rev}$ is acyclic. Existential B-boundedness of an MSC M can be decided in linear-time.*

In the MSC in Fig. 10.5, with $B = 2$, we find that the $<_{rev}$ edge marked with the dotted line induces a cycle. For $B = 3$, $<_{rev}$ is consistent with the ordering of this MSC implying that this MSC is existentially 3-bounded.

Let $M = (E, \leq, \lambda)$ and $\leq_B = (\leq \cup <_{rev})^*$. If M is existentially B-bounded then $(E, \leq_B, \lambda)$ is a labelled partial order. Let $tr'(M)$ be the labelled partial order $(E, \leq_B, \lambda')$ where λ' is the labelling function described earlier.

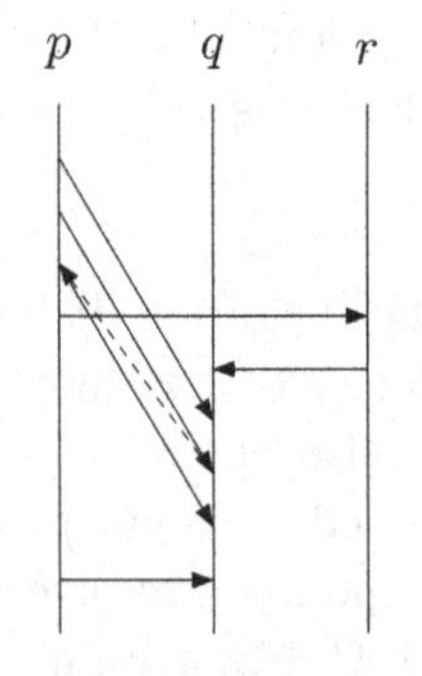

Fig. 10.5. MSC with induced cycle.

Proposition 10.5. *[8] If an MSC M is existentially B-bounded then the labelled partial order $tr'(M)$ is a trace over the alphabet (Σ_B, D). (If M is not existentially B-bounded then $tr'(M)$ is not even a partial order, leave alone a trace.)*

It worth noting that for a universally B-bounded MSC M, $<_{rev} \subseteq \leq$ and therefore $tr'(M) = tr(M)$. Thus, Prop. 10.5 can be thought of as a generalization of Prop. 10.3.

10.1.3. *MSC Languages and regularity*

An *MSC language* is a (finite or infinite) collection of MSCs over a given set of processes $\mathcal{P}$ and messages $\mathcal{M}$. Given the correspondence between MSCs and their linearizations, we may also regard a language of MSCs as a collection of words over $\Sigma_{\mathcal{P}}$ given by the linearizations of the MSCs in the language. In what follows we shall use these two notions interchangeably.

Definition 10.7. [9] A language of MSCs is said to be *regular* if the word language of its linearizations is a regular language.

In any prefix of a linearization of an MSC the number of $p!q$ events is at least as many as the number of $q?p$ events for any pair of processes p and q. We say that a word over $\Sigma_{\mathcal{P}}$ is *proper* if it satisfies this property. In any linearization of an MSC, there are as many $p!q$ events as there are $q?p$ events for every pair of processes p and q. We use *complete* to denote this property. Thus linearizations of MSCs are proper and complete words while their prefixes are proper words.

Let $(Q, \Sigma, \delta, s, F)$ be a deterministic finite automaton accepting a regular MSC language L. We further assume that every state is reachable and that a final state is reachable from every state. Suppose, u and v are two proper words that lead to the same state q from s. Let w be any word that leads from q to some final state. Thus, uw and vw are both complete words. Thus $|\#_{p!q}(v)| - |\#_{q?p}(v)| = |\#_{p!q}(v)| - |\#_{q?p}(v)|$ for each pair of processes p and q (where $\#_a(w)$ denotes the number of a's in the word w). This leads to the following result from [9] which assures us that regular languages are bounded.

Lemma 10.1. *Every regular MSC language L is B-bounded for some B. In particular, B can be chosen to be smaller than the size of the minimal automaton accepting the linearizations of L.*

The converse of this lemma is obviously false — for instance, consider the language $\{(p!q\ q?p)^i\ (r!s\ s?r)^i \mid i \geq 0\}$. We end this section with the definition of finitely generated MSC languages.

Definition 10.8. A language $\mathcal{L}$ of MSCs is said to be finitely generated if $\bigcup_{M \in \mathcal{L}} \mathcal{A}(M)$ is a finite set.

The MSC language given by the complete words $p!q\ (q!p\ p?q)^*\ q?p$ is not finitely generated. As a matter of fact, every word in this language is an atom.

10.2. Message Sequence Graphs

The ITU standard Z.120 describing MSCs also proposes a mechanism to describe collections of MSCs. This mechanism called HMSC (or High-level Message Sequence Charts) or Message Sequence Graphs (MSGs) allows branching, concatenation and iteration.

Definition 10.9. A *Message Sequence Graph* is a structure $\mathcal{G} = (Q, \rightarrow, Q_{in}, F, \Phi)$, where Q is a finite and nonempty set of states, $\rightarrow \subseteq Q \times Q$, $Q_{in} \subseteq Q$ is a set of initial states, $F \subseteq Q$ is a set of final states and Φ labels each state with an MSC.

A *path* π through an MSG $\mathcal{G}$ is a sequence $q_0 \rightarrow q_1 \rightarrow \cdots \rightarrow q_n$ such that $(q_{i-1}, q_i) \in \rightarrow$ for $i \in \{1, 2, \ldots, n\}$. The MSC generated by π is $M(\pi) = M_0 \circ M_1 \circ M_2 \circ \cdots \circ M_n$, where $M_i = \Phi(q_i)$. A path $\pi = q_0 \rightarrow q_1 \rightarrow \cdots \rightarrow q_n$ is a *run* if $q_0 \in Q_{in}$ and $q_n \in F$. The language of MSCs accepted by $\mathcal{G}$ is $L(\mathcal{G}) = \{M(\pi) \mid \pi \text{ is a run through } \mathcal{G}\}$. We say that an MSC language $\mathcal{L}$ is *MSG-definable* if there exists an MSG $\mathcal{G}$ such that $\mathcal{L} = L(\mathcal{G})$.

An example of an MSG is depicted in Fig. 10.6. The initial state is marked $\Rightarrow$ and the final state has a double line. The MSC M corresponding to the path $q_0 \rightarrow q_1 \rightarrow q_0 \rightarrow q_2 \rightarrow q_0$ is also given in the figure.

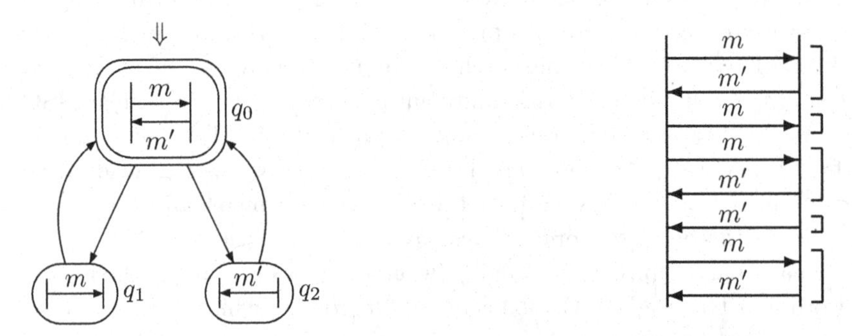

Fig. 10.6. A message sequence graph.

It can be verified that the language generated by the MSG in Fig. 10.6 is a regular language. However, this need not always be the case. There are two reasons why MSGs can generate non-regular languages.

The first reason, as illustrated in Fig. 10.7, is its combination of concurrency and iteration. The language $\mathcal{L}$ generated by this MSG is $\{(M_1 \circ M_2)^n \mid n \geq 0\}$. The events in M_1 and M_2 are completely independent (concurrent) of each other. Thus, we may choose to linearize an MSC of the form $(M_1 \circ M_2)^n$ by first listing all the events involving p and q and then listing the events involving r and s. As a consequence, $\mathcal{L}$ projected to $\{p!q(m), r!s(m)\}^*$ consists of $\sigma \in \{p!q(m), r!s(m)\}^*$ such that $|\sigma\lceil_{p!q(m)}| = |\sigma\lceil_{r!s(m)}|$, which is not a regular string language. Hence $\mathcal{L}$ is not a regular MSC language.

The second reason, as illustrated by the producer-consumer example in Fig. 10.8, is that the buffers can be unbounded. The linearized language of this MSG is $\{w \mid \#_{p!q}w = \#_{q?p}w \ \&\ \forall v \leq w.\ \#_{p!q}v \geq \#_{q?p}v\}$.

However, the language of an MSG is always finitely generated since every MSC in the language can be decomposed using the MSCs that label the nodes of the MSG.

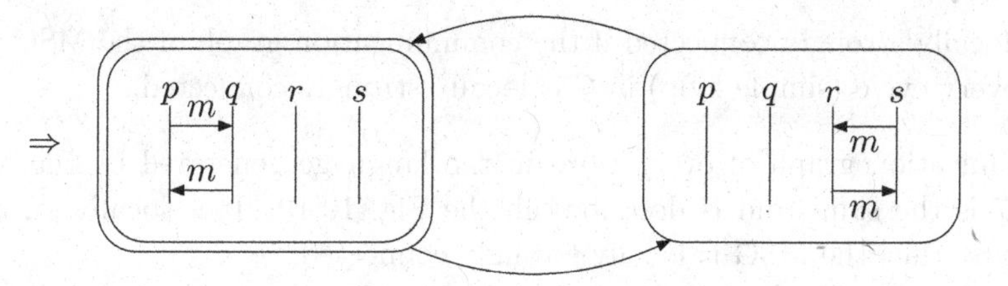

Fig. 10.7. An MSG generating a non-regular language.

Proposition 10.6. *[10] Let $\mathcal{G}$ be a MSG. Let L_N be the set of MSCs that label the nodes of $\mathcal{G}$ and let $\mathcal{L}$ be the language generated by $\mathcal{G}$. Then, $\mathcal{L}$ is a finitely generated MSC language and in particular, $\mathcal{A}(\mathcal{L}) = \mathcal{A}(L_N)$.*

This also means that MSGs are not sufficient to describe every regular MSC language. For example, the language $p!r\ p!q\ q?p\ (q!r\ r?q\ r!q\ q?r)^*\ r?p$ is a regular MSC language that is not finitely generated and hence not MSG definable.

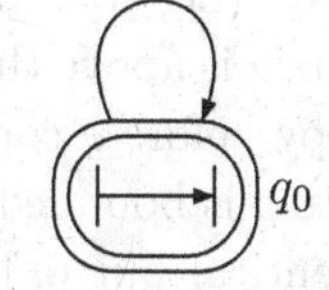

Fig. 10.8. An MSG with unbounded buffer.

10.2.1. *Communication Graph*

The key to understanding the non-regularity of MSGs lies in studying their *communication graphs.*

Definition 10.10. For an MSC $M = (E, \leq, \lambda)$, CG_M, *the communication graph of* M, is the directed graph $(\mathcal{P}, \mapsto)$ where:

- $\mathcal{P}$ is the set of processes of the system.
- $(p, q) \in \mapsto$ iff there exists an $e \in E$ with $\lambda(e) = p!q(m)$.

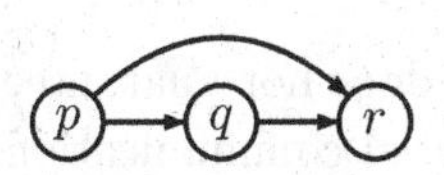

Fig. 10.9. Communication graph of MSC in Fig. 10.2.

The communication graph of the MSC M in Fig. 10.2 is in Fig. 10.9. This graph is not strongly connected. This means that M^*, the iteration of M, is not a bounded language. The reasoning goes as follows: After participating in the events in the first copy of M, the process p can go ahead and participate in its events in the second copy and then the third copy and so on, before q or r participate in any event at all, forcing the channel from p to q to be unbounded. Suppose we modify the MSC M by adding a message from q to p. The iteration of M would still be unbounded — now the processes p and q can participate in all their events in copy one, and then copy two and so on before process r completes any event, forcing the channel from q to r to be unbounded. However the addition of an event from r to p would force the iteration of M to be bounded. But, this also makes the communication graph strongly connected.

Definition 10.11. A (communication) graph is *locally strongly connected* if the graph is the disjoint union of a collection of strongly connected components. An

MSG $\mathcal{G}$ is locally strongly connected if the communication graph of the MSC generated by every cycle (simple loop) in $\mathcal{G}$ is locally strongly connected.

The communication graph of every word in the language generated by the MSG in Fig. 10.7 is the same and is described in the Fig. 10.10. It is locally strongly connected and thus the MSG is locally strongly connected.

Suppose that the communication graph of an MSC M is locally strongly connected and let $X \subseteq \mathcal{P}$ be one of the strongly connected components. Let $p, q \in X$. Since there is a path from q to p in the communication graph of length at most $|X|$, in any MSC of the form $M^{|X|}$ (i.e. $M \circ M \circ \ldots \circ M$, $|X|$ times), there is p event which is above the q events in the first M. Thus, p can at most be in the $(|X|+1)^{st}$ copy before q completes its events in the first copy, ensuring that the channel from p to q is bounded. This argument does not rely on the fact that all the "copies" are identical but only uses the fact that all the copies have the same communication graph. Extending this argument gives

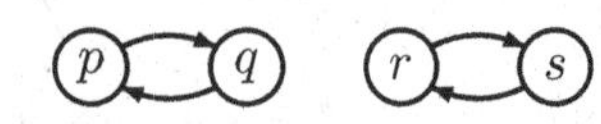

Fig. 10.10. Communication graph of MSG in Fig. 10.7.

Lemma 10.2. *[11; 8] Let $\mathcal{G}$ be an MSG and M be an MSC.*

(1) If the communication graph of M is locally strongly connected then M^* is a B-bounded MSC language for $B \geq |M| \times |\mathcal{P}|$.

(2) If $\mathcal{G}$ is locally strongly connected then $L(\mathcal{G})$ is a B-bounded language for any $B \geq |\mathcal{G}|.|\mathcal{P}|.Max$ where Max is the maximum number of send events in a MSC labelling any one node of $\mathcal{G}$.

(3) If every node in $\mathcal{G}$ is reachable from an initial node and in turn can reach an accepting node and $L(\mathcal{G})$ is a bounded MSC language then $\mathcal{G}$ is locally strongly connected.

Boundedness, a necessary condition for regularity, by itself does not guarantee regularity — the MSG in Fig 10.7 has a locally strongly connected communication graph yet generates a non-regular language (In particular iterating $M_1 \circ M_2$ generates a non-regular language). A further structural restriction on MSGs is needed to rule out the other reason for non-regularity in MSG definable languages — iterations of concurrent behaviours.

Intuitively, the reason for the non-regularity of $(M_1 \circ M_2)^*$ is that $(M_1 \circ M_2)^N = M_1^N \circ M_2^N$, due to the independence of the events in M_1 and M_2, thus implicitly maintaining a counter.

Suppose the communication graph of an MSC M consists of a single nontrivial (i.e. of size at least 2) strongly connected component and a collection of trivial components (corresponding to each process that does not participate in any event in M). If p and q are two processes that participate in M, then in any segment of M^N of the form $M^{|\mathcal{P}|+1}$ there is a p event that depends on a previous q event within the segment and vice versa. Thus all independence is within small segments (of size $M^{|\mathcal{P}|+1}$ or less) and the buffers are also bounded. This is sufficient ([12]) to ensure that M^* is a regular MSC language. This leads to the following definition.

Definition 10.12. An MSC M is said to be *locally synchronized* if its communication graph contains only one nontrivial strongly connected component and a collection of isolated vertices.

An MSG is said to be *locally synchronized* if the communication graph of the MSC generated by any loop in the MSG is locally synchronized. (Locally synchronized MSGs have also been called *bounded* MSGs or *com-connected* MSGs in literature.)

The following result shows that local synchronization is sufficient to guarantee regularity.

Lemma 10.3. *[13; 14] The language of any locally synchronized MSG is regular.*

Note that the definition of locally synchronized MSGs places a demand on all loops in the MSG and not just the cycles. Consider the MSG in Fig 10.11, adapted from [14]. The states q_0 and q_2 are labelled by the empty MSC. The states q_1, q_3 and q_4 are labelled by the MSCs M_1, M_3 and M_4 (described in the figure) respectively. Observe that every cycle in this MSG generates a locally synchronized MSC, however, the language accepted by this MSG is not regular. Every time the processes p and q switch from exchanging the message m to exchanging the message n (or vice versa) the processes r and s exchange a pair of messages. It is easy to derive the non-regularity from this observation.

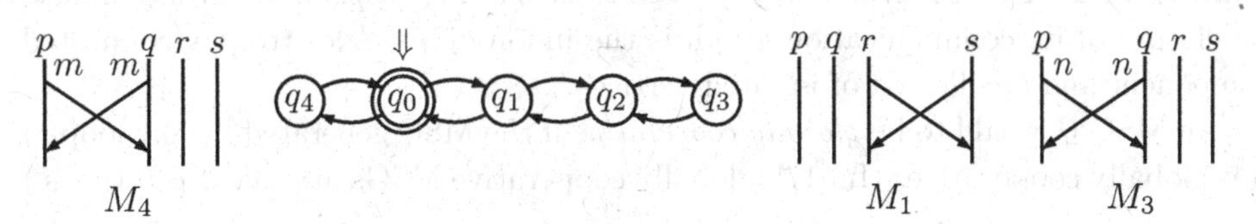

Fig. 10.11. An MSG generating a non-regular language.

Interestingly, if we prohibit the labelling of nodes by the empty MSC, then the definition of a locally synchronized MSG can be weakened.

Proposition 10.7. *Let $\mathcal{G}$ be an MSG in which every node is labelled by a nontrivial MSC. If every cycle in $\mathcal{G}$ describes a locally synchronized MSC then $\mathcal{G}$ is locally synchronized.*

This can be seen as follows: Note that every loop that is not a cycle must properly contain a cycle. Let $p_0p_1 \dots p_k = q_0q_1 \dots q_m = p_{k+m} \dots p_0$ be a loop that contains the cycle $q_0q_1 \dots q_m$. The loop $p_0p_1 \dots p_kp_{k+m+1} \dots p_0$ is smaller and by the induction hypothesis generates a locally synchronized MSC. So does the cycle $q_0q_1 \dots q_m$. Any process that participates in the MSC labelling q_0 (and there is at least one such process) is in the single nontrivial SCC of the communication graph of the smaller loop as well as the single nontrivial SCC of the communication graph of the cycle. Thus, the communication graph of the union of the loop and the cycle is also locally synchronized. Thus,

Lemma 10.4. *Let $\mathcal{G}$ be an MSG in which every node is labelled by a nontrivial MSC. If every cycle in $\mathcal{G}$ generates a locally synchronized MSC then the language of $\mathcal{G}$ is regular.*

Earlier, we remarked that not all regular MSC languages can be described using MSGs. The following result characterizes the collection of MSG definable regular MSC languages.

Lemma 10.5. *[10; 11] A regular MSC language L is definable using MSGs if and only if it is finitely generated. Any such language can also be described using a locally synchronized MSG.*

The proof in [11] (which incidentally defines MSGs as regular expressions constructed using MSCs) exploits the translation from MSCs to traces over the underlying set of atoms and pulls back the corresponding result for traces ([15]).

We have seen that locally strongly connected MSGs are guaranteed to be bounded but place no restrictions on independent iterations. Interestingly, we can also exclude the complications of independent iterations without forcing boundedness (or regularity).

Definition 10.13. [16; 17] An MSC is said to be *globally cooperative* if the symmetric closure of its communication graph is the union of a single strongly connected component and a collected of isolated vertices.

An MSG $\mathcal{G}$ is said to be *globally cooperative* if the MSC generated by any loop in $\mathcal{G}$ is globally cooperative. (In [17], globally cooperative MSGs are called *c-HMSCs.*)

The motivation for this definition comes from trace theory. A word w over a dependence alphabet (Σ, D) is said to be *connected* if the graph $(\Sigma_w, D\downarrow\Sigma_w)$, on the letters that appear in w and the dependency relation restricted to these letters, consists of a single connected component. Fix a dependence alphabet (Σ, D). A finite automaton over Σ in which every loop generates a connected word is said to be *loop connected.* Let the language L be the language of a loop connected finite automaton. Then, the language $\{tr(w) \mid w \in L\}$ of traces represented by words in L is a regular trace language. Equivalently the *trace-closure* of L, $\{w' \mid \exists w \in L.w' \sim w\}$, is a regular language ([18]).

We can easily transform an MSG into an equivalent one where every state is labelled by an atom — simply replace each node labelled by the MSC $A_1 \circ A_2 \ldots \circ A_k$, by a sequence of k nodes each labelled by an atom. (This transformation takes a globally cooperative MSG to a globally cooperative MSG.) Thus, MSGs can be thought of as finite automata over the alphabet of atoms where the states are labelled by letters instead of transitions.

Lemma 10.6. *Let $\mathcal{G}$ be an MSG labelled by atoms and let $\mathcal{A}(\mathcal{G})$ be the set of atoms labelling the nodes of $\mathcal{G}$. Then, $\mathcal{G}$ is globally cooperative if and only if it is loop connected as a finite automaton over the dependency alphabet $(\mathcal{A}(\mathcal{G}), D)$ where aDb whenever there is a process that is active in both a and b.*

This leads to the following regularity theorem for a globally cooperative MSGs. This proves to be a very useful tool in resolving a number of decision problems for globally cooperative MSGs.

Theorem 10.1. *[19] Let $\mathcal{G}$ be a globally cooperative MSG. Then, the language $\{\mathcal{A}_t(M) \mid M \in L(\mathcal{G})\}$ is a regular language and a finite automaton accepting this language with size at most $2^{|\mathcal{G}|\cdot|\mathcal{P}|}$ can be constructed.*

The following result from [19] shows that the if you take away independent iteration what is needed to ensure regularity is the boundedness of channels.

Proposition 10.8. *An MSG $\mathcal{G}$ is locally synchronized if and only if it is globally cooperative and accepts a bounded language.*

10.2.2. *Decision problems*

The language of an MSG is nonempty if and only if there is a path in the MSG from a start state to a final state. Thus, emptiness is decidable.

By Lemma 10.2, to decide whether a MSG accepts a universally bounded language it suffices to check if it is locally strongly connected. Every MSG accepts an existentially bounded language: in particular, if all the MSCs labelling the nodes of an MSG $\mathcal{G}$ are existentially B-bounded, then the language of $\mathcal{G}$ is also existentially B-bounded. The paper by Lohrey and Muscholl [8] establishes a comprehensive collection of the decidability results for a variety boundedness problems for MSGs and MSCs. Most importantly they demonstrate lower bounds for a variety of problems.

Theorem 10.2. *Let $\mathcal{G}$ be an MSG.*

(1) $\mathcal{G}$ is always existentially bounded.
(2) Given $\mathcal{G}$ and B we can check whether $\mathcal{G}$ is existentially B-bounded in linear time ([8]).
(3) Checking whether $\mathcal{G}$ is universally bounded is decidable ([11]). This problem is co-NP complete ([8]).
(4) Given $\mathcal{G}$ and B checking whether $\mathcal{G}$ is universally B-bounded is co-NP complete. This problem is co-NP complete even if B is fixed to the constant 1 ([8]).

There are many factors that contribute to the size of an MSC or an MSG — the number of processes in $\mathcal{P}$, the size of the message alphabet $\mathcal{M}$ and the number of events. Theorem 10.2 holds as it is even if $\mathcal{M}$ is a singleton set ([8]). More often than not, the number of events is likely to be several orders of magnitude larger than the number of processes. Fortunately, if we fix the size of $\mathcal{P}$, all the above problems become efficiently solvable.

Theorem 10.3. *[8] Fix a set $\mathcal{P}$ of processes. The problem of checking whether a MSG over $\mathcal{P}$ is universally bounded (universally B-bounded for a given B) is in NL.*

The following lower-bound for checking structural properties of MSGs is from in [19]. Note that this result uses the assumption that the number of processes is part of the input.

Proposition 10.9. *Deciding whether a given MSG is locally synchronized (globally cooperative) is co-NP complete.*

Local synchronization provides a sufficient condition for regularity. However, there is no hope of obtaining an exact characterization.

Proposition 10.10. *[4; 11] Checking whether an MSG accepts a regular language is an undecidable problem.*

The reason for this undecidability is the following: It is easy to translate a dependency alphabet (Σ, D) into a collection of atoms $\mathcal{A}$ over a set of processes $\mathcal{P}$ by assigning an atom $\mathcal{A}(a)$ for each letter $a \in \Sigma$ in such a way that aDb if and only if $\mathcal{A}(a)$ and $\mathcal{A}(b)$ share an active process. For instance, the MSCs M_1, M_3 and M_4 (from Fig. 10.11) can be used to represent the dependence alphabet $(\{a, b, c\}, (b, c))$. Using this representation we can transform finite automata over Σ into MSGs over $(\mathcal{A}, D)$. The language of this MSG is regular if and only if the trace-closure of the original language is regular. However, checking the regularity of trace-closure is in general an undecidable problem (see for eg. [20; 21]).

Model-checking: A specification describes a collection of behaviours. These could be a set of allowed behaviours that the system should conform to, or a set of disallowed behaviours that the system must avoid. This results in two versions of the model checking problem, the positive and negative model checking problems. In the positive model checking problem the task is to verify that the set of behaviours of the system, L_{sy}, is a subset of the set of behaviours, L_{sp}, described by the specification. In the negative model checking problem, the task is to verify that $L_{sy} \cap L_{sp} = \emptyset$.

Suppose the specification as well as the system are described by MSGs.

Theorem 10.4. *[22; 13] For MSGs the following problems are undecidable:*

(1) Given $\mathcal{G}_1$ and $\mathcal{G}_2$, is $L(\mathcal{G}_1) \subseteq L(\mathcal{G}_2)$?

(2) Given $\mathcal{G}_1$ and $\mathcal{G}_2$, is $L(\mathcal{G}_1) \cap L(\mathcal{G}_2) = \emptyset$?

Once again, translations from the corresponding problems in trace theory suffices to prove the undecidability. As a matter of fact if we rule out independent iterations these problems become decidable.

Theorem 10.5. *[22] For globally cooperative MSGs the following problems are decidable:*

(1) Given $\mathcal{G}_1$ and $\mathcal{G}_2$, is $L(\mathcal{G}_1) \subseteq L(\mathcal{G}_2)$?
(2) Given $\mathcal{G}_1$ and $\mathcal{G}_2$, is $L(\mathcal{G}_1) \cap L(\mathcal{G}_2) = \emptyset$?

The proofs exploit the regular representation via atoms provided by Theorem 10.1 — $L(\mathcal{G}_1) \subseteq L(\mathcal{G}_2)$ if and only if $\mathcal{A}_t(L(\mathcal{G}_1)) \subseteq \mathcal{A}_t(L(\mathcal{G}_2))$ and $L(\mathcal{G}_1) \cap L(\mathcal{G}_2) = \emptyset$ if and only if $\mathcal{A}_t(L(\mathcal{G}_1)) \cap \mathcal{A}_t(L(\mathcal{G}_2)) = \emptyset$. Theorem 10.1 makes globally cooperative MSGs perhaps the most general of the classes of MSGs amenable to algorithmic analysis.

At this point we turn our attention to a regularity property that holds for all MSG definable languages.

Definition 10.14. A set X of linearizations is a *set of representatives* for an MSC language L if $\{M \mid \text{Lin}(M) \cap X \neq \emptyset\} = L$.

Languages that have a regular set of representatives are needless to say interesting. If L is a regular MSC language then $\text{Lin}(L)$ is a regular set of representatives for L. However, as we shall see, the class of languages with regular set of representatives is much larger.

Let $\mathcal{G}$ be an MSG. For each state $q \in \mathcal{G}$, fix a linearization w_q of the MSC labelling q. For any path $\pi = q_1 \to q_2 \to \ldots \to q_k$, let $w(\pi) = w_{q_1} w_{q_2} \ldots w_{q_k}$. Then, $\{w(\pi) \mid \pi \text{ is a run through } \mathcal{G}\}$ is a regular set as well as a set of representatives for $L(\mathcal{G})$.

Proposition 10.11. *[23] Every MSG definable language has a regular set of representatives.*

As such regular representations for L_{sy} and L_{sp} by themselves do not render the model-checking problems effective (as is clear from Theorem 10.4). Even if $L_1 \subseteq L_2$, it is easy to find representative sets X_1 and X_2 respectively in such a way that $X_1 \cap X_2 = \emptyset$. However,

Theorem 10.6. *[24] Suppose L_{sy} is given by a regular set of representatives X_{sy} and L_{sp} is a regular MSC language. Then, the positive and negative model checking problems are decidable. Thus, MSGs can be model-checked w.r.t. regular MSG specifications.*

In proof note that the positive model-checking problem boils down to checking if $X_{sy} \subseteq \text{Lin}(L_{sp})$ which is merely the containment of regular languages, and the negative model-checking involves deciding if $X_{sy} \cap \text{Lin}(L_{sp}) = \emptyset$.

An argument identical to the one used to prove Lemma 10.1 ensures that all the words in any regular representation of an MSC language are B-bounded for some B. Thus,

Proposition 10.12. *If L has a regular set of representatives then L is existentially B-bounded for some B.*

Let $\text{Lin}^B(L)$ be the set of linearizations of L that are B-bounded. Theorem 10.6 can be strengthened as follows: from a regular set of representatives X_{sy} for the system we can derive a bound B such that every word in X_{sy} is B-bounded. Now, $X_{sy} \subseteq \text{Lin}(L_{sp})$ if and only if $X_{sy} \subseteq \text{Lin}^B(L_{sp})$ and $X_{sy} \cap \text{Lin}(L_{sp}) = X_{sy} \cap \text{Lin}^B(L_{sp})$. Thus, it suffices that $\text{Lin}^B(L_{sp})$ be an (effectively constructible) regular set.

Theorem 10.7. *[24] Let L_{sy} be given by a regular set of representatives X_{sy} and let L_{sp} be such that $\text{Lin}^B(L_{sp})$ is effectively regular for some B such that every word in X_{sy} is B-bounded. Then, the positive and negative model-checking problems are decidable.*

Later in this section we shall see that B-bounded linearizations of any globally cooperative MSG language is a regular language. Moreover, as we shall see in Section 10.4 there is another natural class of systems for which B-bounded linearizations are regular for any B. These results, drawn from [24; 25] and [19] show that the model-checking problem for MSGs is decidable for a fairly generous class of specifications.

10.2.3. *Compositional MSGs*

There have been several attempts at extending the definition of MSGs to increase their expressive power. For instance, the *netcharts* model ([26; 27]) attempts at combining the features of MSGs and petri-nets to obtain a model that can generate all regular MSC languages. In this section we consider a natural weakening of the definition of MSCs and MSGs that results in a richer specification language and yet retains most of the useful properties of MSGs.

A compositional MSC is essentially a segment of an MSC, and thus may contain receive events without matching send events and send events without matching receive events. Any association between sends and receives included must satisfy the FIFO assumption. Formally,

Definition 10.15. [28] A CMSC over $\mathcal{P}$ is a *finite* $\Sigma_{\mathcal{P}}$-labelled poset $M = (E, \leq, \lambda, msg)$ (with the notation defined in section 10.1) where

(1) Each relation $\leq_{pp}$ is a linear (total) order.

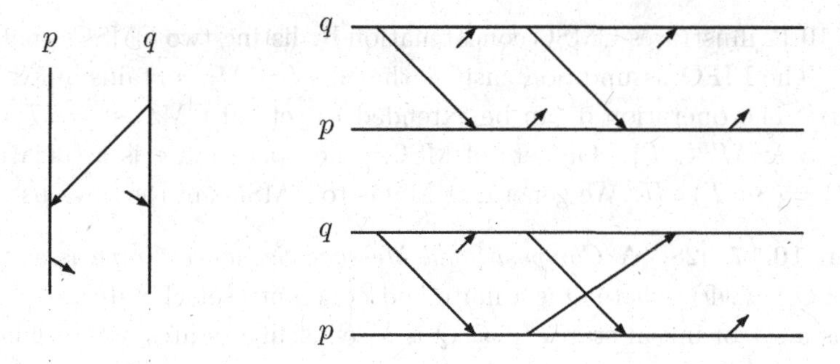

Fig. 10.12. A CMSC M and two elements of $M \circ M$.

(2) *msg* is a partial injective mapping from S to R where
- $S = \{e \in E \mid \lambda(e) = p!q(m) \text{ for some } p, q, m\}$
- $R = \{e \in E \mid \lambda(e) = p?q(m) \text{ for some } p, q, m\}$

satisfying
(a) if $msg(s) = r$ then $s = p!q(m)$ and $r = q?p(m)$ for some p, q, m. We write $s <_{pq} msg(s)$ in that case.
(b) if $s_1 \leq_{pp} s_2$, $Ch(s_1) = Ch(s_2) = p!q$ and $msg(s_1)$ and $msg(s_2)$ are defined then $msg(s_1) \leq_{qq} msg(s_2)$.

(3) $\leq = (\bigcup_{p \in \mathcal{P}} \leq_{pp} \ \cup \ \bigcup_{(p,q) \in Ch} <_{pq})^*$.

In Fig. 10.12 we have a CMSC with one unmatched send and one unmatched receive events.

Every MSC is a CMSC. It is easy to check that a CMSC is an MSC if and only if *msg* is total and onto. Following [25], we define the concatenation of two CMSCs M_1 and M_2 as a set of CMSCs: for each process, the events in M_1 precede its events in M_2, further send events in M_1 may be matched with receive events in M_2, as long as it does not violate the FIFO condition. The result is a set as we are not obliged to match up unmatched sends in M_1 with unmatched receives in M_2 and there may be more than one way to match up unmatched sends in M_1 with unmatched receives in M_2.

Definition 10.16. Let $M_i = (E_i, \leq_i, \lambda_i, msg_i)$, $i = 1, 2$ be CMSCs with $E_1 \cap E_2 = \emptyset$. The concatenation $M_1 \circ M_2$ is the collection of CMSCs of the form $M = (E_1 \cup E_2, \leq, \lambda, msg)$ where

(1) $M\!\downarrow_{E_i} = M_i$ for $i = 1, 2$, where $M\!\downarrow_F$ is the restriction of $\leq$, λ and *msg* to the events in F.
(2) For each $e \in E_2$, if $e \leq e'$ then $e' \in E_2$ (i.e.) send events of E_2 cannot be matched with receive events in E_1.

Figure 10.12 illustrates CMSC concatenation by listing two CMSCs that belong to $M \circ M$. The FIFO assumption ensures that if $M_1 \circ M_2$ contains an MSC then it is unique. The operation $\circ$ can be extended to sets of CMSCs, $S \circ T = \{M \circ M' \mid M \in S \;\&\; M' \in T\}$. On sets of MSCs, the operation $\circ$ is associative, i.e., $S \circ (T \circ U) = (S \circ T) \circ U$. We generalize MSGs to CMSGs in the obvious manner.

Definition 10.17. [28] A *Compositional Message Sequence Graph* is a structure $\mathcal{G} = (Q, \rightarrow, Q_{in}, F, \Phi)$, where Q is a finite and nonempty set of states, $\rightarrow \subseteq Q \times Q$, $Q_{in} \subseteq Q$ is a set of initial states, $F \subseteq Q$ is a set of final states and Φ labels each state with a CMSC.

A path $\pi = q_0 \rightarrow q_1 \rightarrow \cdots \rightarrow q_n$ is an *accepting run* if $q_0 \in Q_{in}$ and $q_n \in F$. The language of MSCs accepted by $\mathcal{G}$, $L(\mathcal{G})$, is the set of MSCs in the set

$$\{\Phi(q_0) \circ \Phi(q_1) \circ \ldots \circ \Phi(q_k) \mid q_0 \rightarrow q_1 \rightarrow \ldots q_k \text{ is an accepting run of } \mathcal{G}\}$$

We say that an CMSC language $\mathcal{L}$ is *CMSG-definable* if there exists an CMSG $\mathcal{G}$ such that $\mathcal{L} = L(\mathcal{G})$.

First of all note that some CMSGs may generate an empty MSC language even though there are paths from the initial state to final states. For instance, a CMSG with a single state that is initial and final, a self-loop, and labelled by the CMSC M from Fig. 10.12 is one such CMSG. Any CMSC generated by this CMSG has unmatched sends (and receives).

Observe that the CMSC $M_1 \circ M^i \circ M_2$, where M is the CMSC from Fig. 10.12, M_1 is the CMSC with just a single $p!q$ event and M_2 is the CMSC with a single $q?p$ event, is an atom for all $i \geq 0$. Thus, CMSG recognizable languages need not be finitely generated. As a matter of fact, as shown in [28], it is quite easy to show that any regular MSC language is CMSG-definable — roughly speaking, we may replace each edge labelled $p!q$ (or $q?p$) in the finite automaton for this language by a node labelled by an MSC with a single $p!q$ (or $q?p$) event. This construction actually implies something stronger:

Proposition 10.13. *If L is an MSC language with a regular set of representatives then L is a CMSG-definable language.*

Interestingly, since every path in this CMSG corresponds to a valid linearization of an MSC in L, every path in this CMSG generates at least one MSC. Checking whether a CMSG generates any MSC or not is undecidable ([28]). Thus, CMSGs are somewhat unrestrained for a specification language.

Definition 10.18. [25] A CMSG $\mathcal{G}$ is said to be *safe* if for any accepting path $q_0 \rightarrow q_1 \rightarrow \ldots \rightarrow q_k$ the set $\Phi(q_0) \circ \Phi(q_1) \circ \ldots \circ \Phi(q_k)$ contains at least one MSC.

Every MSG is a safe CMSG. The property of being safe is decidable and safety gives a sufficient condition to ensure analyzability of CMSGs. Fix a linearization w_q for each node q of a safe CMSG $\mathcal{G}$. It is easy to see that the language

$\{w_{q_0} w_{q_1} \ldots w_{q_k} \mid q_0 \to q_1 \to \ldots q_k$ is an accepting run $\}$ is regular and a set of representatives of the language of $\mathcal{G}$. Combining this with the observation following Proposition 10.13 gives

Proposition 10.14. *[29] An MSC language L has a regular set of representatives if and only if it is the language of a safe CMSG. Thus, every safe CMSG language is existentially B-bounded for some B.*

Safe CMSGs should be as analyzable as the class of MSGs; after all, they enjoy the only regularity property that we have been able to associate with MSGs! To verify this, let us generalize some of the structural restrictions on MSGs to CMSGs.

Definition 10.19. [25] The *communication graph* of a CMSC is the directed graph whose vertices are the elements of $\mathcal{P}$ and there is an edge from p to q whenever there is a $p!q$ event and a $q?p$ event (the two need not necessarily be matched) in the CMSC.

The communication graph of the CMSC M in Fig. 10.12 is the complete directed graph on two vertices. The notions of *locally synchronized* CMSCs and *globally cooperative* CMSCs is defined as before. Finally, a CMSG $\mathcal{G}$ is *locally synchronized* if it is safe and every CMSC generated by every loop in $\mathcal{G}$ is locally synchronized. A CMSG $\mathcal{G}$ is *globally cooperative* if it is safe and every CMSC generated by every loop in $\mathcal{G}$ is globally cooperative.

Theorem 10.8. *[25] Let $\mathcal{G}$ be a globally cooperative CMSG and let B be an integer such that $B \geq |\mathcal{G}|$. Then, the set of B-bounded linearizations of $L(\mathcal{G})$ is a regular language (recognized by a finite automaton whose size is $O(|G|^{Poly(|G|,B,|\mathcal{P}|)})$.)*

The proof relies on the relationship to Mazurkiewicz traces. Combining this with Theorem 10.7 we get

Theorem 10.9. *[25] The positive and negative model checking problems are decidable when L_{sy} is the language of a safe CMSG and L_{sp} is the language of a globally cooperative CMSG.*

The essence of this story, which took some time to develop and culminate in the papers [19; 25], is the following: of the two suspected reasons for the non-analyzability of MSG (CMSG) based specifications, the culprit is independent (concurrent) iterations and eliminating that via a structural restriction (globally cooperative MSGs) delivers a generous decidability result for model-checking.

10.3. Monadic second order logic over MSCs

Monadic second order logic (or MSO) is the logical counter part to automata. Büchi and Elgot ([30; 31]) showed that MSO over finite words has the same expressive power as finite automata, Büchi then extended this to MSO over infinite words and automata over infinite words that bear his name. These connections are not isolated and a host of similar connections have been established between MSO and automata. The most relevant of these results to our context is the one relating regular trace languages and MSO over traces due to Wolfgang Thomas ([32]) and extended to infinite traces by Ebinger and Muscholl ([33]).

Definition 10.20. Fix a set $\mathcal{P}$ of processes and messages. The formulas of the monadic second order logic over MSCs (MSO) are as follows:

$$\varphi ::= a(x) \mid x \in X \mid x \leq y \mid x \leq_{pp} y \mid x \lessdot_p y \mid x <_{pq} y \mid \neg\varphi \mid \varphi \wedge \varphi \mid \exists X.\varphi \mid \exists x.\varphi$$

where $a \in \Sigma_{\mathcal{P}}$ and $(p,q) \in Ch$.

We will also be interested in the fragment of existential monadic second-order formulas (EMSO) which are of the form $\exists X_1 \exists X_2 \ldots \exists X_k.\ \varphi$ where φ is a first order formula. We will also be interested in restricted versions of these logics obtained by permitting only a subset of the 4 relational symbols in the syntax, and this will be made explicit by listing the allowed subset: for e.g. MSO($\lessdot_p, <_{pq}$) to stand for the fragment that does not use $\leq$ and $\leq_{pp}$.

An MSO formula is interpreted over an MSC. The first order variables range over the events in the MSC, second order variables over sets of events and the relational operators have the obvious interpretation: $\leq$ is the ordering on the events of the MSC, $\leq_{pp}$ is the ordering on events in process p, $\lessdot_p$ is the immediate successor relation within a process p, and $<_{pq}$ is the message induced ordering between a send from p and the corresponding receive in q. Observe that the first two relations are ordering relations while the latter two are not. Finally $a(x)$ is true with $x = e$ if $\lambda(e) = a$. The interpretation of the logical operators and quantifiers is as usual. It is quite easy to see that $\leq_{pp}$ and $\lessdot_p$ can be defined using $\leq$, but it turns out that $<_{pq}$ cannot be so defined. Sentences in MSO define languages of MSCs, $L(\varphi) = \{M \mid M \models \varphi\}$ and in this case we say that L is MSO-definable.

Here is a sentence that characterises universally 2-bounded MSCs.

$$\bigwedge_{(p,q)\in Ch} \forall x.\forall y.\forall z.\ (p!q(x)\ \wedge\ p!q(y)\ \wedge\ p!q(z)\ \wedge\ (x < y)\ \wedge (y < z)) \implies$$
$$\exists w.\ (x <_{pq} w)\ \wedge\ (w < z))$$

It asserts that in any sequence of 3 sends, the receive corresponding to the first send must be in the past of the third send (see Prop. 10.1). This can be generalized to describing universally B-bounded MSCs for any fixed B.

It is well-known that the transitive closure of a relation is definable in any monadic second order logic: xR^*y if and only if the smallest set containing x and closed under R also contains y. However, such a definition makes essential use of universal quantification over sets and consequently, such a translation is not always possible in the existential fragment of monadic second order logic. Since $\leq$ and

$\leq_{pp}$ can be defined as transitive closure of $\lessdot_p \cup <_{pq}$ and $\lessdot_p$ respectively, MSO over MSCs is equivalent to MSO($\lessdot_p, <_{pq}$).

In [9; 4] the following characterization theorem is presented.

Theorem 10.10. *A B-bounded language L is regular if and only if* $L = L(\varphi) \cap \{M \mid M \textit{is universally } B\textit{-bounded}\}$ *for some MSO(≤) formula* φ.

The result holds even if the logic is restricted to be EMSO(≤). The proof ([9; 4]) is based on the ideas used in similar results for MSO over traces in [32; 34]. In one direction, given a MSO(≤) formula φ, we construct a formula ψ in MSO over words such that $w \models \psi$ if and only if (i) w is a B-bounded complete and proper word and (ii) the MSC M_w generated by w satisfies φ and (iii) M_w is universally B-bounded. Establishing (i) and (ii) shows that the set of B-bounded linearizations of $L(\varphi)$ is a regular language (which implies Theorem 10.11 below).

Part (i) is easy as the set of B-bounded proper and complete words is a regular language and one may appeal to the Büchi-Elgot theorem. For part (ii), the proof proceeds by defining, in MSO over words, a binary relation $\preceq$ on the positions of the word in such a way that for any B-bounded w and positions i and j in w, $i \preceq j$ if and only if the corresponding events (say e_i and e_j) are ordered under $\leq$ in M_w. Clearly, $e_i \leq_{pp} e_j$ in M_w if and only if $i < j$ and i and j are p events in w. We still have to show that $<_{pq}$ is definable. This makes essential use of the B-boundedness of w. The formula asserts that there is a $k \in \{0, 1, \ldots, B-1\}$ such that i is labelled by $p!q$, the number of positions to its left labelled by $p!q$ is $k(modulo\ B)$, j is to the right of i, it is a labelled by $q?p$ and the number of positions to its left labelled by $q?p$ is $k(modulo\ B)$ and there is no position between i and j labelled by $q?p$ for which the number of positions to its left labelled $q?p$ is $k(modulo\ B)$. Since transitive closure is definable in MSO the result follows. (The transitive closure can be avoided by a slightly more elaborate argument using the fact that one has to hop across processes at most $|\mathcal{P}|$ number of times.) Finally, part (iii) follows from the fact that with $\preceq$ and $\leq$, universally B-boundedness can be defined as explained in the example above.

The other direction is somewhat more involved. Given a regular MSC language and B, use Büchi-Elgot theorem to pick a formula ψ in MSO over words describing the linearizations of this language. Then show that in MSO(≤) one can define a relation $\preceq$ over the MSC that fixes a canonical linearization of the MSC (using techniques from [34]), and then interpret ψ over this linearization.

This result generalizes to MSO over infinite MSCs and regular languages of infinite MSCs as shown by D. Kuske ([7]). As a matter of fact, [7] provides a complete theory of regular languages of infinite MSCs. For the relationship between existentially B-bounded languages and MSO we have the following theorem:

Theorem 10.11. *[29] For any MSO formula* φ *and any* B, $\mathrm{Lin}^B(L(\varphi))$ *is a regular language.*

An immediate consequence is that any existentially B-bounded language described by an MSO formula has a regular set of representatives. As a corollary to the previous two theorems we have

Corollary 10.1. *The problem of checking whether an MSO($\leq$) formula (or a MSO formula) is satisfiable over universally B-bounded MSCs is decidable. Similarly, checking satisfiability over existentially B-bounded MSCs is decidable.*

The natural question then is to ask "Can we model check safe CMSGs (or MSGs) w.r.t. to MSO?". The answer is affirmative. In proof note that, any safe CMSG has a regular set of representatives, Theorem 10.11 implies that $\mathrm{Lin}^B(L(\varphi))$ is regular for any B and any φ in MSO and thus Theorem 10.7 is applicable.

Theorem 10.12. *[23; 29] The problem of deciding whether every MSC generated by a safe CMSG (or MSG) satisfies a MSO formula is decidable.*

B.Bollig and M. Leucker [35] study the expressiveness of MSO and EMSO over MSCs and using techniques from [36] show that

Theorem 10.13.

(1) The monadic quantifier alternation hierarchy of the logic MSO (over MSCs) is infinite. Thus, MSO($\lessdot_p, <_{pq}$) is strictly more expressive than EMSO($\lessdot_p, <_{pq}$).
(2) The logics MSO($\leq$) and EMSO($\lessdot_p, <_{pq}$) are incomparable.

We shall return to the expressive power of MSO over MSCs a little later after we introduce an implementation model for MSCs.

10.4. Message Passing Automata

Safe CMSGs, MSGs and MSO are elegant and expressive languages to describe collections of MSCs. However, they are far removed from an execution model where each process is situated at a different location and there are limitations on what each process actually knows of the global state. The natural execution model for MSCs is that of *message passing automata* (also referred to as *communicating finite-state machines*).

A message passing automaton consists of a collection of finite state processes which communicate with each other by sending messages on FIFO channels. Each transition in a process involves either sending a message to some process or consuming a message from one of its input channels. It is possible to enrich these automata with local moves, however since it has no effect on the results in this section, we work without them. Fig. 10.13 illustrates a message passing automaton implementing a producer-consumer system and an MSC accepted by it. Formally, an MPA is defined as follows:

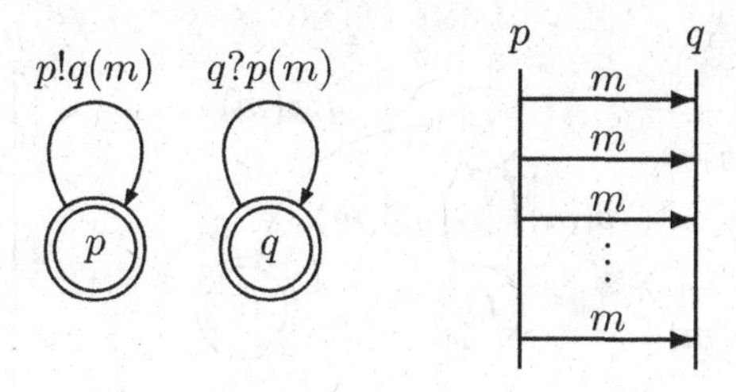

Fig. 10.13. An MPA.

Definition 10.21. [37] Let $\Sigma_{\mathcal{P}}$ be the communication alphabet over the set of processes $\mathcal{P}$ and message alphabet $\mathcal{M}$. A *message-passing automaton (MPA)* over $\Sigma_{\mathcal{P}}$ is a structure $\mathcal{A} = (\{\mathcal{A}_p\}_{p\in\mathcal{P}}, \Delta, s_{in}, F)$ where:

- Δ is a finite alphabet of *auxiliary messages*.
- Each component $\mathcal{A}_p$ is of the form $(S_p, \rightarrow_p)$ where S_p is a finite set of p-local states and $\rightarrow_p \subseteq S_p \times \Sigma_p \times \Delta \times S_p$ is the p-local transition relation.
- $s_{in} \in \prod_{p\in\mathcal{P}} S_p$ is the global initial state.
- $F \subseteq \prod_{p\in\mathcal{P}} S_p$ is the set of global final states.

Observe that our definition allows the *tagging* of each message with auxiliary contents drawn from the set Δ.

The local transition relation $\rightarrow_p$ specifies how the process p sends and receives messages. The transition $(s, p!q(m), x, s')$ says that in state s, p can send the message m to q tagged with auxiliary information x and move to state s'. Similarly, the transition $(s, p?q(m), x, s')$ signifies that at state s, p can receive the message m from q tagged with information x and move to state s'.

A global state of $\mathcal{A}$ is an element of $\prod_{p\in\mathcal{P}} S_p$. For a global state s, s_p denotes the pth component of s. A *configuration* is a pair (s, χ) where s is a global state and $\chi : Ch \rightarrow (\mathcal{M} \times \Delta)^*$ is the *channel state* describing the message queue in each channel c. The *initial configuration* of $\mathcal{A}$ is $(s_{in}, \chi_\varepsilon)$ where $\chi_\varepsilon(c)$ is the empty string ε for every channel c. The set of *final configurations* of $\mathcal{A}$ is $F \times \{\chi_\varepsilon\}$. Observe that in a final configuration all the channels must be empty.

A global move of the automaton involves one of the process depositing a message into a channel (sending a message) or consuming a message from the channel (receiving a message) according to its local transition relation. Suppose, (s, χ) is a configuration and $(s_p, p!q(m), x, s'_p) \in \rightarrow_p$. Then, $(s, \chi) \stackrel{p!q(m)}{\Longrightarrow} (s', \chi')$ where for $r \neq p$, $s_r = s'_r$, for each $r \in \mathcal{P}$, $\chi'((p,q)) = \chi((p,q)) \cdot (m, x)$, and for $c \neq (p,q)$, $\chi'(c) = \chi(c)$. Similarly, if (s, χ) is a configuration and $(s_p, p?q(m), x, s'_p) \in \rightarrow_p$, then there is a global move $(s, \chi) \stackrel{p?q(m)}{\Longrightarrow} (s', \chi')$ where for $r \neq p$, $s_r = s'_r$, for each $r \in \mathcal{P}$, $\chi((q,p)) = (m, x) \cdot \chi'((q,p))$, and for $c \neq (q,p)$, $\chi'(c) = \chi(c)$.

A run of the automaton is a sequence of such global moves and we write $Conf_{\mathcal{A}}$

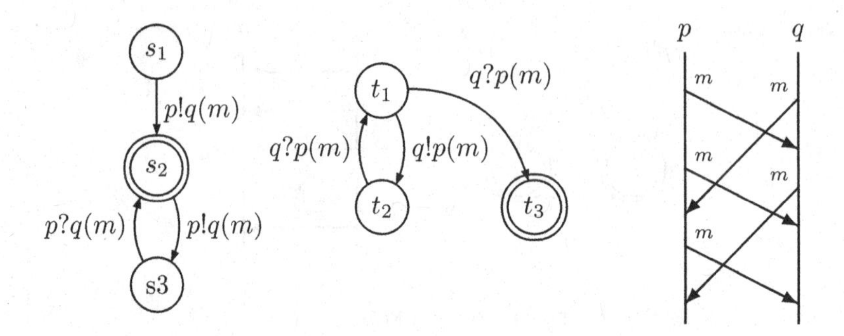

Fig. 10.14. An MPA accepting infinite number of atoms.

for the set of reachable configurations of $\mathcal{A}$. A run is accepting if it ends in a final configuration. For instance

$$((p,q),\varepsilon) \stackrel{p!q(m)}{\Longrightarrow} ((p,q),m) \stackrel{p!q(m)}{\Longrightarrow} ((p,q),mm) \stackrel{q?p(m)}{\Longrightarrow} ((p,q),m) \stackrel{q?p(m)}{\Longrightarrow} ((p,q),\varepsilon)$$

is an accepting run the automaton in Fig. 10.13 on the word $p!q(m)p!q(m)q?p(m)q?p(m)$.

We define $L(\mathcal{A}) = \{\sigma \mid \mathcal{A}$ has an accepting run over $\sigma\}$. Since all channels are empty in the initial and final configurations and a message can be received only if it has already been sent, it is easy to check that any word accepted by an MPA is proper and complete. It is not difficult to see that if $L(\mathcal{A})$ contains one linearization of an MSC M then it contains all the linearizations of the MSC M. As a matter of fact, it is quite easy to define runs of MPAs directly on MSCs as a mapping from events on the MSC to global states of the automaton. Thus $L(\mathcal{A})$ is the set of linearizations of an MSC language. As usual we shall use $L(\mathcal{A})$ to denote the MSC language accepted by $\mathcal{A}$ as well as its linearizations.

Each configuration of an MPA records the messages sent and as yet undelivered. For $B \in \mathbb{N}$, a configuration (s,χ) is *B-bounded* if $|\chi(c)| \leq B$ for every channel $c \in Ch$ and $\mathcal{A}$ is a B-bounded automaton if every reachable configuration $(s,\chi) \in Conf_{\mathcal{A}}$ is B-bounded. Clearly a B-bounded automaton accepts a universally B-bounded language. The global state space of any B-bounded MPA is therefore finite and consequently, every B-bounded MPA accepts a regular MSC language. The converse is also true but we shall get to that a little later. The MPA in Fig. 10.13 accepts a unbounded language.

From any MPA $\mathcal{A}$ and a natural number B we can generate a finite automaton accepting precisely those B-bounded words that are accepted by $\mathcal{A}$ and thus

Proposition 10.15. *For any MPA $\mathcal{A}$ and any natural number B, the language* $\mathrm{Lin}^B(L(\mathcal{A}))$ *is a regular language.*

The language of an MPA need not be finitely generated. Fig. 10.14 describes an MPA and one of the MSCs it accepts. This MPA accepts a regular MSC language

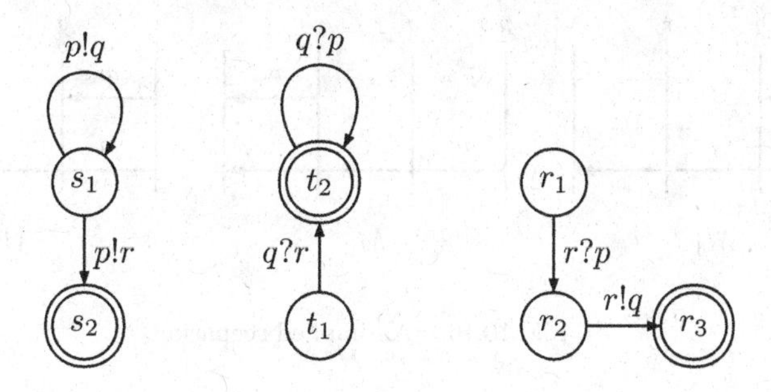

Fig. 10.15. An MPA with an existentially unbounded language.

and every MSC accepted by this automaton is an atom. Thus, there are MPA acceptable languages that cannot be described by MSGs.

Finally, MPAs can also accept languages that are not existentially bounded (for any B), thus MPAs are capable of accepting languages that cannot be described using safe CMSGs. The MPA in Fig. 10.15 accepts the language of MSCs generated by the words $(p!q)^n\ p!r\ r?p\ r!q\ q?r\ (q?p)^n$ and it is easy to verify that this is not an existentially bounded language.

Definition 10.22. An MPA is said have local accepting states if $F = \prod_{p \in \mathcal{P}} F_p$ for some $F_p \subseteq S_p$.

10.4.1. *MPAs without auxiliary messages*

We first examine the power of MPAs without auxiliary message alphabets (i.e.) MPAs whose auxiliary alphabet is singleton.

Definition 10.23. An MSC language L is said to be *weakly realizable* if it is the language of an MPA with a singleton auxiliary message alphabet and with local accepting states.

Consider the language $\{M_1, M_2\}$ of MSCs (from Fig. 10.16). This set is not weakly realizable — Suppose $\mathcal{A}$ is an MPA accepting this language. From the accepting run on M_1, we know that there are runs $p_0 \stackrel{p!q(m)}{\longrightarrow} p_1 \stackrel{p!s(m)}{\longrightarrow} p_2$ and $q_0 \stackrel{q?p(m)}{\longrightarrow} q_1$ for p and q ending in accepting states. Similarly, from the accepting run on M_2, we know that there are runs $r_0 \stackrel{r!s(m)}{\longrightarrow} r_1$ and $s_0 \stackrel{s?r(m)}{\longrightarrow} s_1 \stackrel{s?p(m)}{\longrightarrow} s_2$ ending in accepting states. This means that the MSC M is also accepted, since p and q may behave exactly as they do in accepting M_1 and r and s may behave exactly as they do in accepting M_2 and all four processes end up in an accepting states on M.

The language of an MPA with local accepting states and over a singleton auxiliary message alphabet is merely the shuffle or free product of the local languages

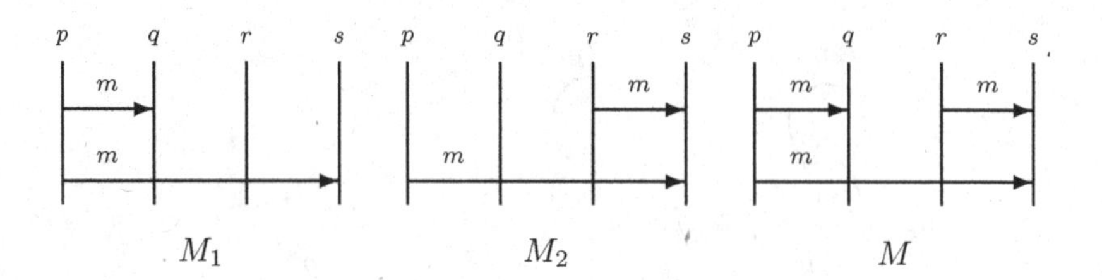

Fig. 10.16. An implied scenario.

of the processes and this is formalized as follows:

Given an MSC M (or any linearization w of M) and a process p, $M\lceil p$ is the word over Σ_p consisting of the projection of M (or equivalently w) to the events in process p. For eg. $M_1\lceil p = p!q(m)p!s(m)$ for the MSC M_1 in Fig. 10.16. For a language L, $L_p = \{M\lceil p \mid M \in L\}$. Finally, given word languages L_p over Σ_p for each $p \in \mathcal{P}$, let $\prod_{p\in\mathcal{P}} L_p = \{M \mid M\lceil p \in L_p\}$ (the usual free product).

Definition 10.24. Given a set L of MSCs its *implied closure* $Imp(L)$ is defined as follows

$$Imp(L) = \{M \mid \forall p \in \mathcal{P}.\ \exists M_p \in L.\ M\lceil p = M_p\lceil p\}$$

If $M \in Imp(L) \setminus L$ then we say that L has *an implied scenario* and that M is an implied scenario of L.

In Fig. 10.16, M is an implied scenario of $\{M_1, M_2\}$. The following characterization (albeit non-effective) is easy to prove.

Proposition 10.16. *[12] If L is weakly realizable then $L = Imp(L)$. Conversely, suppose L is an MSC language and $L\lceil p$ is a regular language for each p, then L is weakly realizable only if $L = Imp(L)$.*

Implied scenarios are of practical interest. Often, a designer specifies a system as a collection of MSCs using say an MSG. The existence of an implied scenario indicates that an implementation by MPAs (w/o auxiliary tagging) would result in behaviours not foreseen by the designer (these might or might not be *bad*). So it would be useful to check if a given MSC language L has any implied scenarios at all, and construct a representation for $Imp(L)$. The implied closure of a B-bounded language may contain MSCs that are not B-bounded. As

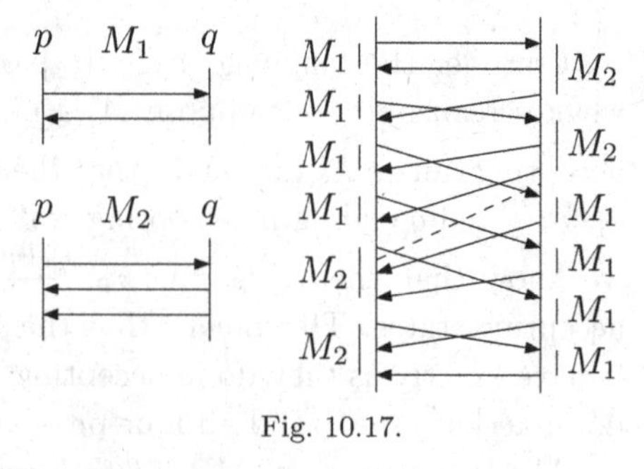

Fig. 10.17.

a matter of fact, the implied closure of a B-bounded regular language need not be bounded at all.

In Fig. 10.17 observe that the two MSCs M_1 and M_2 have complete communication graphs. Therefore, the language $(M_1 + M_2)^*$ is a regular MSG-definable language. On the other hand, for each $k \in \mathbb{N}$, the MSC in which the p-projection matches $M_1^{2k} M_2^k$ and the q-projection matches $M_2^k M_1^{2k}$ has a global cut where the channel (p, q) has capacity $k + 1$. The figure shows the case $k = 2$. The dotted line marks the global cut where the channel (p, q) has maximum capacity. Thus, $Imp(L)$ need not be a regular MSC language when L is a regular MSC language. It gets worse.

Theorem 10.14. *[38] The problem of checking whether $Imp(L) = L$ is undecidable even for regular MSC languages presented as locally synchronized MSGs.*

Let us examine this result a little. Let B be the bound on the channels in L. From the definition of $Imp(L)$ it is not difficult to check that if L is weakly realizable then an MPA $\mathcal{A}$ implementing L can be constructed as follows: For each process p pick a minimal finite automaton (w/o dead or unreachable states) accepting the language $L \upharpoonright p$ as $\mathcal{A}_p$ and set $F = \prod_{p \in \mathcal{P}} F_p$. Thus we have a candidate implementation. Yet, weak realizability is undecidable because it is not possible to check whether the language accepted by $\mathcal{A}$ equals L. By restricting $\mathcal{A}$ to runs where no channel has more than B messages, we have a finite automaton accepting the set of B-bounded words in $Imp(L)$. Thus, checking if there is a B-bounded implied scenario for L is decidable. The difficulty is in finding if there are implied scenarios violating the B-bound. In particular, the existence of a partial run reaching a configuration where some channel has more than B messages does not mean that there are implied scenarios. This is because such a partial run may not be extendable to an accepting run. (However, this means that this partial run has ended in a configuration from where no final configuration is reachable. We shall return to this point a little later.)

Theorem 10.14 has been strengthened [39] to show that this problem is undecidable even when L is a 1-bounded language and undecidability holds even with just 2 processes. These undecidability arguments make essential use of the fact that the channels are FIFO (and therefore requires the message alphabet to have at least two messages.) Restriction to the trivial message alphabet yields the first positive result.

Proposition 10.17. *[17] The problem of checking whether the language of a locally synchronized MSG over a singleton message alphabet is weakly realizable is decidable.*

Since weak realizability is impossible to analyze, it is natural to look for stronger notions of implementability. Alur et al. propose a notion called *safe realizability* which is amenable to algorithmic analysis.

Definition 10.25. A configuration χ of an MPA $\mathcal{A}$ is a *deadlock* if there are no reachable final configurations. An MPA is said to be *deadlock-free* if it has no reachable deadlock configurations. A language is said to be *safely realizable* if it is the language of a deadlock-free MPA with local accepting states and over a singleton auxiliary message alphabet.

It is also possible to characterize safely realizable languages as a closure property akin to Prop. 10.16 (see [12; 40]).

Proposition 10.18. *L is safely realizable if and only if it is weakly realizable and satisfies the following closure property:*

$$\{w \mid \forall p.\exists u_p \in L.\ w{\restriction}p \leq u_p\} \subseteq \{w \mid \exists u \in L. w \leq u\}$$

The closure condition demands that any partial MSC (or proper word) whose projections on every process is consistent with some accepting run of the process must be extendable to an MSC (or proper complete word) in L. Fortunately, safe realizability is analyzable.

Theorem 10.15. *The problem of checking whether a given globally cooperative MSG generates a safely realizable language or not is decidable in EXPSPACE and the problem is EXPSPACE-complete even for locally synchronized MSGs. However, the safe realizability problem for arbitrary MSGs is undecidable.*

We go back to our analysis of why the availability of a candidate implementation does not suffice to ensure decidability of weak implementation. The analysis there ends with a conclusion placed within parenthesis. This conclusion stated in our recently acquired terminology states that the candidate implementation $\mathcal{A}$ has deadlocks or implied scenarios whenever it reaches a configuration violating the B-bound on some channel. Thus, if $\mathcal{A}$ ever reaches a configuration violating the B-bound on some channel, it cannot be a safe implementation of L. Finally, it is not difficult to verify, using Prop. 10.18 that if at all L is safely realizable then $\mathcal{A}$ is such a realization. That completes our sketch of the decidability argument.

The decidability for locally synchronized MSGs appears in [38], that for globally cooperative MSGs in [17] and the exact complexity result as well as the undecidability result is from [40].

10.4.2. *MPAs with auxiliary messages*

In this section we study the expressive power of MPAs with no restriction on the use of auxiliary message contents. We begin by showing that the collection $\{M_1, M_2\}$ from Fig. 10.16 can be implemented using auxiliary messages. An MPA with auxiliary messages drawn from the set $\{1, 2\}$ is described in figure 10.18 The process p signals the process s using the auxiliary information regarding the identity of the MSC (The auxiliary information in all the other messages can be ignored.)

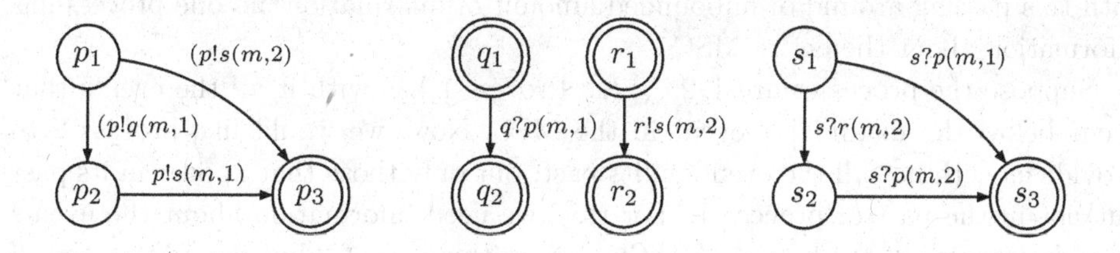

Fig. 10.18. An MPA accepting $\{M_1, M_2\}$.

This example illustrates the use of auxiliary information: it allows a process to convey information about its past (in this case p conveys the information "I sent a message to q" to s by tagging the message with a 1 instead of a 2.) However, since the auxiliary message alphabet is a finite set, it only allows a bounded amount of information about the past to be conveyed in any message. It turns out that this ability to forward bounded amount of information is very powerful.

Theorem 10.16. *[9; 4] An MSC language L over a set of process $\mathcal{P}$ and messages $\mathcal{M}$ is regular if and only if there is an an MPA $\mathcal{A}$, over the same alphabet and with an auxiliary message alphabet Δ, such that $L = L(\mathcal{A})$.*

The proof of this theorem is beyond the scope of this paper. However we provide a brief sketch of the difficulties and main ideas involved.

This theorem is an example of a distributed synthesis theorem—it states that given the global description of a regular MSC language, it is possible to construct a distributed implementation as an MPA. The key ingredients that go into the proof are drawn from the celebrated result of W. Zielonka [41] showing that every regular trace language is recognized by an *asynchronous automaton.*

Let us examine the main difficulty in proving such a result. Suppose the global description is a finite automaton G. We can equip each local process with a copy of G if necessary. Yet, after a sequence of events w, which process is to keep track of the current state of G? Observe that every event takes place only in one process and each process directly observes only the events that it participates in, so it is not possible for any one process to maintain the state of G correctly. For the moment assume that we may tag the messages with unbounded amount of auxiliary information. Then, every process can send the entire history of all the events it has participated in as well as all the events about which it has learnt from others through messages it has received. So, whenever a message is received, the receiving process knows the entire set of events that are below this event in the MSC order. However, an MSC could have up to $|\mathcal{P}|$ maximal events and thus even with this passing around of unbounded amount of information, no one process has information about the entire MSC.

Suppose the processes are $1, 2, \ldots K$. Process 1 has with it all the events that occur below the maximal 1 event in the MSC. Now, we would like process 2 to provide us with not all the events in its past, but only those that appear in its past but not in the past of process 1. For this, we need information about the events in each process $j \in \{2, 3, 4, \ldots, K\}$, that are in the past of 2 but not in the past of 1 (called the 1 residue at 2). With this information we can piece together all the events in the past of 1 and 2 without any ambiguity. Then, we need to obtain from process 3 information about events in $\{3, 4, \ldots, K\}$ that do not appear in the past of 1 and 2 (the $\{1, 2\}$ residue at 3) and so on. Then, the MSC can be reconstructed from the residues available at $1, 2, \ldots K$.

The reduction from unbounded to bounded information hinges on the fact that instead of keeping any partial MSC M (or a linearization w) of the history of a process, we might as well keep the transition function that this word or partial MSC defines on the state space of G.

The ability to program each process to maintain its residues w.r.t. to other processes requires a sophisticated time stamping algorithm from [42] which can perform comparisons such as "is my information about j more recent than i's information about j?". Most importantly this time-stamping algorithm requires each process to maintain only a bounded amount of information and tags each message with only a bounded amount of information. The proof in [43; 4] proceeds along these lines.

An alternative is to use the connection to regular trace languages, then use Zielonka's construction to obtain an asynchronous automaton and then translate back such an automaton into an MPA. This is the structure of the proof in [6]. The above theorem can be strengthened.

Definition 10.26. [43] An MPA is said to deterministic if

- If $(s, p!q, m_1, s_1') \in \longrightarrow_p$ and $(s, p!q, m_2, s_2') \in \longrightarrow_p$ then $m_1 = m_2$ and $s_1' = s_2'$.
- If $(s, p?q, m, s_1') \in \longrightarrow_p$ and $(s, p?q, m, s_2') \in \longrightarrow_p$ then $s_1' = s_2'$.

Determinacy requires that the nature of the message sent from p to q depends only on the local state of the sender p. Note, however, that from the same state, p may have the possibility of sending messages to more than one process. When receiving a message, the new state of the receiving process is fixed uniquely by its current local state and the content of the message. Once again, a process may be willing to receive messages from more than one process in a given state. This definition ensures that a deterministic automaton has at most one run on any $w \in \Sigma_{\mathcal{P}}^*$. Now, we are in a position to state the main characterization theorem for regular MSC languages.

Theorem 10.17. *[4] Let B be any integer and L be a language of universally B-bounded MSCs. Then the following are equivalent:*

(1) L is a regular MSC language
(2) L is definable in MSO($\leq$) (or EMSO($\leq$))
(3) L is the language of a deterministic B-bounded MPA.
(4) L is the language of a B-bounded MPA.

This theorem has been generalized to infinite MSCs by D.Kuske (see [7]). There are couple of other points worth noting: first of all the definition of MPAs uses a global set of final states and the physical realizability of such a global set of acceptance states is debatable. Secondly, the automata constructed in the proof of the above theorem may deadlock, and once again this makes its usability some what limited. In a recent series of papers, N. Baudru and R. Morin [44; 45], have shown that every regular MSC language can be implemented using (nondeterministic) MPAs that are deadlock free and whose acceptance set is local (i.e. $F = \prod_{p \in \mathcal{P}} F_p$ for some collection $(F_p)_{p \in \mathcal{P}}$.)

10.4.3. *Implementing existentially bounded languages*

The characterization in Theorem 10.17 shows that every regular MSC language can be implemented using a deterministic MPA. The corresponding question for MSC languages with a regular set of representatives was solved by Genest, Kuske and Muscholl [24; 25]. An earlier paper [19] set the stage for such a result through a host of results on the virtues of existentially bounded languages including the analyzability of globally cooperative CMSGs (Theorem 10.7) and efficient implementability (as MPAs) for a subclass (locally cooperative MSGs) of globally cooperative CMSGs. The proof of this characterization uses the translation to traces from existentially B-bounded MSCs (described in Section 10.1) to Mazurkiewicz traces.

Theorem 10.18. *[25] Let B be an integer and let L be a language of existentially B-bounded MSCs. Then the following statements are equivalent.*

(1) $\mathrm{Lin}^B(L)$ *is a regular set of representatives for L.*
(2) L is generated by a globally cooperative CMSG.
(3) L is MSO (EMSO) definable.
(4) L is implementable using an MPA.

We have already seen some of the relationships: Theorem. 10.8 shows that one can go from globally cooperative CMSGs to languages with $\mathrm{Lin}^B(L)$ as a regular set of representatives. Theorem 10.11 allows us to move from MSO definable languages to languages with $\mathrm{Lin}^B(L)$ as a regular set of representatives. Prop. 10.13 shows that any language with a regular set of representatives can be translated to a safe CMSG and Proposition 10.14 shows that every safe CMSG generates a language with a regular set of representatives. These relationships are strengthened to restrict the class of CMSGs to globally cooperative CMSGs in [25] using Ochmanski's theorem ([15]). Finally by Prop. 10.15 the set $\text{Lin}^B(L)$ is regular for any MPA.

This leaves the difficult part: a decomposition theorem showing that every existentially B-bounded language has a distributed implementation as an MPA and this involves a fairly complex argument via Mazurkiewicz trace theory and is beyond the scope of this article. The automaton constructed is a nondeterministic automaton and this is unavoidable.

Proposition 10.19. *[5] There are existentially bounded languages recognized by nondeterministic MPAs that cannot be recognized by deterministic MPAs.*

Translating MPAs to EMSO is quite routine, but the converse is not. A remarkable result due to Bollig and Leucker shows the following:

Theorem 10.19. *[35] An MSC language L is EMSO($\lessdot_p, <_{pq}$) definable if and only if it is the language of an MPA.*

In contrast to Theorems 10.17 and 10.18 this theorem applies to arbitrary MSCs. This theorem in combination with Theorem 10.13 shows that

Theorem 10.20. *[35] The class of languages accepted by MPAs is not closed under complementation.*

10.4.4. *Decision Problems*

Every channel is a queue and it is quite easy to simulate counter machines with MPAs. Consequently, general MPAs are too expressive for any sort of analysis:

Theorem 10.21. *[5]*

(1) The language emptiness problem for deterministic MPAs is undecidable.
(2) Given a B checking whether a deterministic MPA accepts a universally B-bounded language is undecidable for every $B > 0$.
(3) Given a deterministic, deadlock-free automaton in which every global state is accepting, checking whether it accepts a universally bounded language is undecidable.
(4) Given a B checking whether a deterministic MPA accepts a existentially B-bounded language is undecidable for every $B > 0$.
(5) Given a deterministic, deadlock-free automaton in which every global state is accepting, checking whether it accepts a existentially bounded language is undecidable.

Items (3) and (5) become decidable for a fixed B, since for a deadlock-free automaton it suffices to check if a configuration with $B+1$ messages in the channel is reachable.

10.5. Conclusion

In this article we have surveyed a selection of the results from the theory of MSCs. These are by no means exhaustive. A number of results that have been omitted here due to lack of space—to name a few, there is a host of decidability (and undecidability) results for the so called "pattern matching" problems on MSCs (see for instance [46; 47]), the related problem of whether a class of MSCs can be implemented by MPAs with additional messages (not merely the addition of auxiliary content to existing messages; see for instance [48]), branching time specification and analysis (see for instance [49–51] and finally there have been some recent results in extending the theory of MSCs with time (see for instance [52–54]).

Acknowledgments

This work was partially supported by *Timed-DISCOVERI*, a project under the Indo-French Networking Project, the ARCUS Ile de France-Inde and the CMI-TCS Academic Alliance.

References

[1] ITU-TS. Recommendation Z.120: *Message Sequence Chart (MSC)*, (1997).

[2] E. Rudolph, P. Graubmann, and J. Grabowski, Tutorial on Message Sequence Charts, *Computer Networks and ISDN Systems.* **28**(12), 1629–1641, (1996).

[3] J. R. G. Booch, I. Jacobson, *Unified Modeling Language User Guide.* (Addison-Wesley, 1997).

[4] J. G. Henriksen, M. Mukund, K. Narayan Kumar, M. A. Sohoni, and P. S. Thiagarajan, A theory of regular MSC languages, *Information and Computation.* **202**(1), 1–38, (2005).

[5] B. Genest, D. Kuske, and A. Muscholl, On Communicating Automata with Bounded Channels, *Fundamenta Informaticae.* **80**(1-3), 147–167, (2007).

[6] D. Kuske. A Further Step towards a Theory of Regular MSC Languages. In *Proc. 19th STACS*, vol. 2285, *LNCS*, pp. 489–500. Springer, (2002).

[7] D. Kuske, Regular sets of infinite message sequence charts, *Information and Computation.* **187**(1), 80–109, (2003).

[8] M. Lohrey and A. Muscholl, Bounded MSC communication, *Information and Computation.* **189**(2), 160–181, (2004).

[9] J. G. Henriksen, M. Mukund, K. Narayan Kumar, and P. S. Thiagarajan. Regular Collections of Message Sequence Charts. In *Proc. 25th MFCS*, vol. 1893, *LNCS*, pp. 405–414. Springer, (2000).

[10] J. G. Henriksen, M. Mukund, K. Narayan Kumar, and P. S. Thiagarajan. On Message Sequence Graphs and Finitely Generated Regular MSC Languages. In *Proc. 27th ICALP*, vol. 1853, *LNCS*, pp. 675–686. Springer, (2000).

[11] R. Morin. On Regular Message Sequence Chart Languages and Relationships to Mazurkiewicz Trace Theory. In *Proc. 4th FoSSaCS*, vol. 2030, *LNCS*, pp. 332–346. Springer, (2001).

[12] R. Alur, K. Etessami, and M. Yannakakis. Inference of message sequence charts. In *Proc. 22nd ICSE*, pp. 304–313. ACM, (2000).

[13] R. Alur and M. Yannakakis. Model Checking of Message Sequence Charts. In *Proc. 10th CONCUR*, vol. 1664, *LNCS*, pp. 114–129. Springer, (1999).
[14] A. Muscholl and D. Peled. Message Sequence Graphs and Decision Problems on Mazurkiewicz Traces. In *Proc. 24th MFCS*, vol. 1672, *LNCS*, pp. 81–91. Springer, (1999).
[15] E. Ochmanski, Regular behaviour of concurrent systems, *Bulletin of the EATCS*. **27**, 56–67, (1985).
[16] B. Genest, A. Muscholl, H. Seidl, and M. Zeitoun. Infinite-State High-Level MSCs: Model-Checking and Realizability. In *Proc. 29th ICALP*, vol. 2380, *LNCS*, pp. 657–668. Springer, (2002).
[17] R. Morin. Recognizable Sets of Message Sequence Charts. In *Proc. 19th STACS*, vol. 2285, *LNCS*, pp. 523–534. Springer, (2002).
[18] M. Clerbout and M. Latteux, Semi-commutations, *Information and Compututation*. **73**(1), 59–74, (1987).
[19] B. Genest, A. Muscholl, H. Seidl, and M. Zeitoun, Infinite-state high-level MSCs: Model-checking and realizability, *J. Comput. Syst. Sci.* **72**(4), 617–647, (2006).
[20] A. Muscholl and H. Petersen, A Note on the Commutative Closure of Star-Free Languages, *Information Processing Letters*. **57**(2), 71–74, (1996).
[21] J. Sakarovitch. The "Last" Decision Problem for Rational Trace Languages. In *LATIN*, vol. 583, *LNCS*, pp. 460–473. Springer, (1992).
[22] A. Muscholl, D. Peled, and Z. Su. Deciding Properties for Message Sequence Charts. In *Proc. 1st FoSSaCS*, vol. 1378, *LNCS*, pp. 226–242. Springer, (1998).
[23] P. Madhusudan. Reasoning about Sequential and Branching Behaviours of Message Sequence Graphs. In *Proc. 28th ICALP*, vol. 2076, *LNCS*, pp. 809–820. Springer, (2001).
[24] B. Genest, A. Muscholl, and D. Kuske. A Kleene Theorem for a Class of Communicating Automata with Effective Algorithms. In *Proc. 8th DLT*, vol. 3340, *LNCS*, pp. 30–48. Springer, (2004).
[25] B. Genest, D. Kuske, and A. Muscholl, A Kleene theorem and model checking algorithms for existentially bounded communicating automata, *Information and Computation*. **204**(6), 920–956, (2006).
[26] M. Mukund, K. Narayan Kumar, and P. S. Thiagarajan. Netcharts: Bridging the gap between HMSCs and executable specifications. In *CONCUR*, vol. 2761, *LNCS*, pp. 293–307. Springer, (2003).
[27] N. Baudru and R. Morin. The Synthesis Problem of Netcharts. In *Proc. 27th ICATPN*, vol. 4024, *LNCS*, pp. 84–104. Springer, (2006).
[28] E. L. Gunter, A. Muscholl, and D. Peled. Compositional Message Sequence Charts. In *Proc. 7th TACAS*, vol. 2031, *LNCS*, pp. 496–511. Springer, (2001).
[29] P. Madhusudan and B. Meenakshi. Beyond Message Sequence Graphs. In *Proc. 21st FSTTCS*, vol. 2245, *LNCS*, pp. 256–267. Springer, (2001).
[30] J. R. Buechi, Weak second-order arithmetic and finite automata, *Z. Math. Logik Grundl. Math.* **6**, 66–92, (1960).
[31] C. C. Elgot, Decision problems of finite automata design and related arithmetics, *Trans. American Mathematical Society*. **98**, 21–52, (1960).
[32] W. Thomas. Automata over infinite objects. In ed. J. van Leeuwen, *Handbook of Theoretical Computer Science, Volume B*, pp. 133–191. Elsevier, (1990).
[33] W. Ebinger and A. Muscholl, Logical Definability on Infinite Traces, *Theor. Comput. Sci.* **154**(1), 67–84, (1996).
[34] P. S. Thiagarajan and I. Walukiewicz. An Expressively Complete Linear Time Temporal Logic for Mazurkiewicz Traces. In *Proc. 12th IEEE LICS*, pp. 183–194, (1997).

[35] B. Bollig and M. Leucker, Message-passing automata are expressively equivalent to EMSO logic, *Theor. Comput. Sci.* **358**(2-3), 150–172, (2006).
[36] O. Matz and W. Thomas. The Monadic Quantifier Alternation Hierarchy over Graphs is Infinite. In *Proc. 12th IEEE LICS*, pp. 236–244, (1997).
[37] D. Brand and P. Zafiropulo, On Communicating Finite-State Machines, *J. ACM.* **30** (2), 323–342, (1983).
[38] R. Alur, K. Etessami, and M. Yannakakis. Realizability and Verification of MSC Graphs. In *Proc. 28th ICALP*, vol. 2076, *LNCS*, pp. 797–808. Springer, (2001).
[39] P. Bhateja, P. Gastin, M. Mukund, and K. Narayan Kumar. Local Testing of Message Sequence Charts Is Difficult. In *Proc. 16th FCT*, vol. 4639, *LNCS*, pp. 76–87. Springer, (2007).
[40] M. Lohrey, Realizability of high-level message sequence charts: closing the gaps, *Theor. Comput. Sci.* **309**(1-3), 529–554, (2003).
[41] W. Zielonka, Notes on Finite Asynchronous Automata, *Informatique Theorique et Applications.* **21**(2), 99–135, (1987).
[42] M. Mukund, K. Narayan Kumar, and M. A. Sohoni, Bounded time-stamping in message-passing systems, *Theor. Comput. Sci.* **290**(1), 221–239, (2003).
[43] M. Mukund, K. Narayan Kumar, and M. A. Sohoni. Synthesizing Distributed Finite-State Systems from MSCs. In *Proc. 11th CONCUR*, vol. 1877, *LNCS*, pp. 521–535. Springer, (2000).
[44] N. Baudru and R. Morin. Safe Implementability of Regular Message Sequence Chart Specifications. In *Proc. 4th SNPD*, pp. 210–217. ACIS, (2003).
[45] N. Baudru and R. Morin. Synthesis of Safe Message-Passing Systems. In *Proc. 27th FSTTCS*, vol. 4855, *LNCS*, pp. 277–289. Springer, (2007).
[46] A. Muscholl. Matching Specifications for Message Sequence Charts. In *Proc. 2nd FoSSaCS*, vol. 1578, *LNCS*, pp. 273–287. Springer, (1999).
[47] B. Genest and A. Muscholl. Pattern Matching and Membership for Hierarchical Message Sequence Charts. In *Proc. 5th LATIN*, vol. 2286, *LNCS*, pp. 326–340. Springer, (2002).
[48] B. Genest. On Implementation of Global Concurrent Systems with Local Asynchronous Controllers. In *Proc. 16th CONCUR*, vol. 3653, *LNCS*, pp. 443–457. Springer, (2005).
[49] W. Damm and D. Harel, LSCs: Breathing Life into Message Sequence Charts, *Formal Methods in System Design.* **19**(1), 45–80, (2001).
[50] D. Harel and R. Marelly, Specifying and executing behavioral requirements: the play-in/play-out approach, *Software and System Modeling.* **2**(2), 82–107, (2003).
[51] D. Harel, H. Kugler, R. Marelly, and A. Pnueli. Smart Play-out of Behavioral Requirements. In *Proc. 4th FMCAD*, vol. 2517, *LNCS*, pp. 378–398. Springer, (2002).
[52] S. Akshay, B. Bollig, and P. Gastin. Automata and Logics for Timed Message Sequence Charts. In *Proc. 27th FSTTCS*, vol. 4855, *LNCS*, pp. 290–302. Springer, (2007).
[53] S. Akshay, M. Mukund, and K. Narayan Kumar. Checking Coverage for Infinite Collections of Timed Scenarios. In *Proc. 18th CONCUR*, vol. 4703, *LNCS*, pp. 181–196. Springer, (2007).
[54] S. Akshay, B. Bollig, P. Gastin, M. Mukund, and K. Narayan Kumar. Distributed Timed Automata with Independently Evolving Clocks. In *Proc. 19th CONCUR*, vol. 5201, *LNCS*, pp. 82–97. Springer, (2008).

Chapter 11

Type Checking of Tree Walking Transducers

Sebastian Maneth

NICTA and University of New South Wales, Sydney, Australia
sebastian.maneth@nicta.com.au

Sylvia Friese and Helmut Seidl

Technische Universität München, Garching, Germany
sylvia.friese@in.tum.de
seidl@in.tum.de

Tree walking transducers are an expressive formalism for reasoning about XSLT-like document transformations. One of the useful properties of tree transducers is decidability of type checking: given a transducer and input and output types, it can be checked statically whether the transducer is type correct, i.e., whether each document adhering to the input type is necessarily transformed into documents adhering to the output type. Here, a "type" means a regular set of trees specified by a finite-state tree automaton. Usually, type checking of tree transducers is extremely expensive; already for simple top-down tree transducers it is known to be EXPTIME-complete. Are there expressive classes of tree transducers for which type checking can be performed in polynomial time? Most of the previous approaches are based on inverse type inference. The approach presented here goes the other direction: it uses forward type inference. This means to infer, given a transducer and an input type, the corresponding set of output trees. In general, this set is not a type, i.e., is not regular. However, its intersection emptiness with a given type can be decided. Using this approach it is shown that type checking can be performed in polynomial time, if (1) the output type is specified by a deterministic tree automaton and (2) the transducer visits every input node only a bounded number of times. If the tree walking transducer is additionally equipped with accumulating call-by-value parameters, then the complexity of type checking also depends (exponentially) on the number of such parameters. For this case a fast approximative type checking algorithm is presented, based on context-free tree grammars. Finally, the approach is generalized from trees to forest walking transducers which additionally support concatenation as a built-in output operation.

11.1. Introduction

The extensible markup language XML is the current standard format for exchanging structured data. Its widespread use has initiated lots of work to support processing of XML on many different levels: customized query languages for XML, such as XQuery, transformation languages like XSLT, and programming language support either in the form of special purpose languages like XDuce, or of binding facilities for mainstream programming languages like JAXB. A central problem in XML processing is the *(static) type checking problem*: given an input and output type and a transformation f, can we statically check whether all outputs generated by f on valid inputs conform to the output type? Since XML types are intrinsically more complex than the types found in conventional programming languages, the type checking problem for XML poses new challenges on the design of type checking algorithms. The excellent survey [1] gives an overview of the different approaches to XML type checking.

In its most general setting, the type checking problem for XML transformations is undecidable. Hence, general solutions are bound to be approximative, but seem to work well for practical XSLT transformations [2]. Another approach is to restrict the types and transformations in such a way that type checking becomes decidable; we then refer to the problem as *exact XML type checking*. For the exact setting, types can be considered as regular or *recognizable* tree languages — thus, capturing the expressive strength of virtually all known type formalisms for XML [3].

Even though the class of transformations for which exact type checking is possible is surprisingly large [4–6], the price to be paid for exactness is also extremely high. The design space for exact type checking comes as a huge "exponential wasteland": even for simple top-down transformations, exact type checking is exponential-time complete [7], and for more complex transformations such as the k-pebble transducers of [5] the problem is non-elementary. For practical considerations, however, one is interested in useful subclasses of transformations for which exact type checking is provably tractable.

In general, we are interested in type checking of transformations formulated through tree walking transducers (2tts) and macro tree walking transducers (2mtts). A 2tt is similar to an attribute grammar which operates on derivation trees, and has trees as semantic domain (with tree top-concatenation as only semantic operation). The 2mtt generalizes the 2tt by adding formal context-parameters to the attributes, i.e., each attribute is seen as a function which can take parameters of type output tree. Such transducers are very expressive and can simulate most features of transformation languages such as XSLT. Given suitable descriptions (types) of admissible inputs and outputs for a 2tt M, type checking M means to test whether all outputs produced by M on admissible inputs are again admissible. Our main result is: if admissible outputs are described by *deterministic* tree automata, then exact type checking can be done in polynomial time for a large class of practically interesting transformations obtained by putting only mild restrictions on the transducers.

Related Work Approximative type checking for XML transformations is typically based on (subclasses of) recognizable tree languages. Using XPath as pattern language, XQuery [8] is a functional language for querying XML documents. It is strongly-typed and type checking is performed via type inference rules computing approximative types for each expression. Approximative type inference is also used in XDuce [9] and its follow-up version CDuce [10]; navigation and deconstruction are based on an extension of the pattern matching mechanism of functional languages with regular expression constructs. Recently, Hosoya et al. proposed a type checking system based on the approximative type inference of [11] for parametric polymorphism for XML [12]. Type variables are interpreted as markings indicating the parameterized subparts. In [2] a sound type checking algorithm is proposed (originally developed for the Java-based language XACT [13]) based on an XSLT flow analysis that determines the possible outcomes of pattern matching operations; for the benefit of better performance the algorithm deals with regular approximations of possible outputs.

Milo et al. [5] propose the k-pebble tree transducer (k-ptt) as a formal model for XML transformations, and show that exact type checking can be done for k-ptts using *inverse* type inference. The latter means to start with an output type O of a transformation f and then to construct the type of the inputs by backwards translating O through f. Each k-pebble tree transducer can be simulated by compositions of $k+1$ stay macro tree transducers (smtts) [4], thus, type checking can be solved in time (iterated) exponential in the number of used pebbles. Intuitively, k-pebble tree transducers for $k = 0$ correspond to our 2tts. In [14] it was shown that inverse type inference for 2tts can be done in exponential time, and can be done for k-fold compositions of 2tts in k-fold exponential time. In [6] it was shown that inverse type inference can be done for a transformation language providing all standard features of most XML transformation languages using a simulation by at most three smtts. Inverse type inference is used in [7, 15] to identify subclasses of top-down XML transformation which have tractable exact type checking. We note that the classes considered there are incomparable to the ones considered in this paper.

11.2. Preliminaries

An XML document can be seen as a sequential representation of an unranked tree. Here is a small example document:

```
<department>
  <employee>
    <data><name>Charles Montgomery Burns</name>...</data>
    <subordinates>
      <employee>
        <data><name>Waylon Smithers</name>...</data>
      </employee>
      <employee> ... </employee> ...
    </subordinates>
  </employee>
  <employee> ... </employee> ...
</department>
```

This example represents a company structure, where each `employee` element has a `data` element, with the personal data of the employee (e.g. the name). Additionally, it may have a **subordinates** element, which is a collection of further **employee** elements. The represented tree has a root node labeled **department**, which has an arbitrary number of children nodes which are labeled **employee**. An **employee**-node has as first child a **data**-node and has possibly a second child labeled **subordinates**. In the following we refer to this tree as t_B.

Formally, an unranked tree over an alphabet Σ consists of a root node labeled by a symbol a from Σ and a forest f, written $\mathsf{a}\langle f\rangle$. A forest (or hedge) is a sequence of an arbitrary number of unranked trees, written $t_1t_2\ldots t_m$. The number m is called the length of the forest. The empty forest, i.e., a forest with length $m = 0$, is denoted by ϵ.

Definition 11.1 (Forests). *Let Σ be a an alphabet (i.e., a finite set). The set $\mathcal{F}_\Sigma$ of forests f over Σ is defined by the grammar rules* $f ::= \epsilon \mid tf, \quad t ::= \mathsf{a}\langle f\rangle$, *where* $\mathsf{a} \in \Sigma$.

Rather than on forests, *tree walking* transducers work on *ranked* trees. There, we assume that a fixed rank is given for every element of Σ, i.e. $\Sigma = \biguplus_{m\in\mathbb{N}} \Sigma^{(m)}$ where $\Sigma^{(m)}$ is the set of all symbols with rank m. We define $\mathit{rank}(\mathsf{a}) = m$ for all symbols $\mathsf{a} \in \Sigma^{(m)}$ for $m \geq 0$. The maximal rank $mr(\Sigma)$ is the smallest number m such that $\Sigma^{(m)} \neq \emptyset$ and $\Sigma^{(m+i)} = \emptyset$ for all $i \geq 1$.

Definition 11.2 (Ranked Trees). *Let Σ be a ranked alphabet. The set $\mathcal{T}_\Sigma$ of ranked trees over Σ is defined by the grammar rules* $t ::= \mathsf{a}(\underbrace{t,\ldots,t}_{m\ times}) \mid \mathsf{b}$, *where* $\mathsf{a} \in \Sigma^{(m)}$ *and* $\mathsf{b} \in \Sigma^{(0)}$.

In the following we use the term 'tree' as a synonym for *ranked tree*. We fix the set $Y = \{y_1, y_2, \ldots\}$ of formal *parameters*. These parameters are of rank 0. For a ranked alphabet Σ, where Σ and Y are disjoint, $\mathcal{T}_\Sigma(Y)$ denotes the set of trees over Σ and Y.

In order to define tree walking transducers on XML documents, we rely on ranked tree representations of forests, e.g., through binary trees. The empty forest then is represented by a leaf with label e (where e is a new symbol that does not appear in the document). The content of an element node a is coded as the left child of a, while the forest of right siblings of a is represented as the right child (this is the well-known "first-child next-sibling" encoding). Accordingly, the ranks of symbols are either zero or two. Figure 11.1 illustrates this relationship between unranked trees and their representation as binary trees. It shows the tree t_B in the left part and its binary encoding t'_B on the right.

The set $\mathcal{N}(t) \subseteq \mathbb{N}^*$ of all *nodes* v in a ranked tree t is defined as $\mathcal{N}(\mathsf{b}) = \{\epsilon\}$ and $\mathcal{N}(\mathsf{a}(t_1,\ldots,t_m)) = \{\epsilon\} \cup \{iv \mid 1 \leq i \leq m, v \in \mathcal{N}(t_i)\}$. where $\mathbb{N}^*$ is the set of strings (including the empty string) over the alphabet of positive natural numbers and ϵ denotes the empty string. The *direction* $\eta(v)$ of a node v indicates whether v is the root of the tree or a particular child, i.e., we define $\eta(\epsilon) = 0$ and $\eta(v'j) = j$. The set $\mathcal{N}(f)$ of nodes of a forest f is defined as $\mathcal{N}(\epsilon) = \{0\}$ and

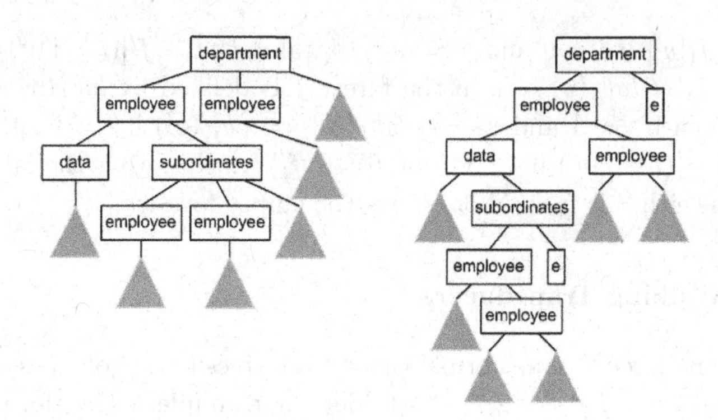

Fig. 11.1. The unranked tree t_B and its binary encoding t'_B.

$\mathcal{N}(\mathtt{a}\langle f_1\rangle f_2) = \{0v' \mid v' \in \mathcal{N}(f_1)\} \cup \{(i+1)v' \mid iv' \in \mathcal{N}(f_2)\}$. For a node v in a forest we define the *direction* $\eta(v)$ which now indicates whether v is at the top-level, has a left sibling or both. Thus, $\eta(0) = 0$, $\eta(i) = 1$ for $i > 0$, $\eta(v'0) = 2$ for $v' \neq \epsilon$ and $\eta(v) = 3$ otherwise.

Note that the definition of nodes of a ranked tree differs from the definition of nodes in a forest consisting of one tree only. Accordingly, also the definitions of *direction* differ. For a ranked tree t and a given node $v \in \mathcal{N}(t)$, $t[v]$ is called the *subtree of t located at v* and is defined as $t[\epsilon] = t$ and $\mathtt{a}(t_1, \ldots, t_m)[iv] = t_i[v]$ for $i = 1, \ldots, m$. For example in the left tree t_B in Figure 11.1 the `subordinates` element is the node 1.2. Here we write 1.2 instead of 12 for the second child of the first child of the root, to distinguish between this node and the twelfth son of the root. With $lab_t(v)$ we refer to the label of the node v in a tree t, or $lab(v)$, if t is given by the context. In the left example $lab_{t_B}(1.2) =$ `subordinates` and in the right tree $lab_{t'_B}(1.2) =$ `employee`.

The height of a tree is recursively defined as $height(\mathtt{b}) = 1$ for $\mathtt{b}$ of rank 0 and $height(\mathtt{a}(t_1, \ldots, t_m)) = 1 + max(height(t_1), \ldots, height(t_m))$ for $\mathtt{a}$ of rank m. The height of a tree is the maximal length of a path from the root to a leaf. If we consider trees on right-hand sides of rules, we have to deal with *state calls.* In this exposition, we will first consider *tree walking transducers* (cf. Section 11.3) which do not support accumulating parameters. For tree walking transducers, state calls are of the form $q(op)$ (op stands for up, $stay$ or $down_i$). For these, we define $height(q(op)) = 1$ for all states q and all operations op. In Section 11.6 we then will add parameters to state calls to obtain *macro tree walking transducers.* Then a state call has the form $q(op, t_1, \ldots, t_n)$ where t_i $(1 \leq i \leq n)$ are ranked trees over $\Sigma \cup Y$ and further state calls ($Y = \{y_1, \ldots, y_k\}$ is a set of parameters). In this case, we define the height recursively by: $height(q(op, t_1, \ldots, t_n)) = 1 + max(height(t_1), \ldots, height(t_k))$. The *size* of a tree t is defined as the number of nodes, i.e., $|t| = |\mathcal{N}(t)|$. Similar notions also apply to forests. In particular, the subforest $f[v]$ at a node v in a forest is defined as f if $i = 0$ and as $f'[i-1]$ if $i > 0$ and $f = tf'$. For $v = iv'$ with

$v' \neq \epsilon$ $f[v] = f_1[v']$ if $i = 0$ and $f = \mathtt{a}\langle f_1\rangle f_2$, and $f[v] = f'[(i-1)v']$ if $i > 0$ and $f = tf'$. The *label* $lab_f(v)$ of v in the forest f is defined by $lab_f(0) = \epsilon$ if $f = \epsilon$, and $lab_f(iv') = \mathtt{a}$ if $i = 1$ and $v' = \epsilon$, $lab_f(iv') = lab_{f_1}(v')$ if $i = 1$ and $v' \neq \epsilon$, and $lab_f(iv') = lab_{f_2}((i-1)v')$ if $i > 0$ and $f = \mathtt{a}\langle f_1\rangle f_2$. Note that the label at a node in a forest thus either is from Σ or equals the empty forest ϵ.

11.3. Tree Walking Transducers

Tree transducers describe transformations τ from trees to sets of trees over a ranked alphabet Σ, i.e., $\tau : \mathcal{T}_\Sigma \rightarrow 2^{\mathcal{T}_\Sigma}$. Consider for example a transformation which translates documents as in the example before into a collection of all employees which are listed under a new root node labeled `staff`. Besides a `name` element, these new `employee` elements now contain an element `boss` if the employee is the subordinate of someone. For our example document, the transformation produces:

```
<staff>
  <employee>
    <data> <name> Charles Montgomery Burns </name> ... </data>
  </employee>
  <employee>
    <data> <name> Waylon Smithers </name> ... </data>
    <boss> <name> Charles Montgomery Burns </name> ... </boss>
  </employee>
  <employee> ...  </employee> ...
</staff>
```

The corresponding tree is referred as s_B and its binary encoding as s'_B. A tree walking transducer starts at the root of the input tree. Depending on the label of the current node, the direction and the state, it produces a tree with leaves which again may contain state calls for nodes of the input tree. These recursively accessed nodes are determined according to the directives specified in the right-hand side of the applied rule: on directive *up*, the father of the current node is processed, on directive $down_i$, the i-th child and on directive *stay* the current node itself. Tree walking transducers can be considered as generalizations of *top-down* tree transducers. While top-down tree transducers are only allowed to move downward in the input tree, tree walking transducers may also stay at the current node or move upward in the tree.

Example 11.1. Using our representation of forests by binary trees (Fig. 11.1), the transformation of our example is realized by a tree walking transducer M_{staff} with the following rules.

1 $q_I(\mathtt{department}) \rightarrow \mathtt{staff}(q(down_1), \mathtt{e})$,
2 $q(\mathtt{employee}) \rightarrow \mathtt{employee}(\mathtt{data}(q_{data}(down_1), q_{boss}(stay)), q_{sub}(down_1))$,
3 $q(\mathtt{e}) \rightarrow q_{up}(up)$,

together with a state q_{data} for copying the personal data

4 $q_{data}(\mathtt{data}) \rightarrow copy(down_1)$,

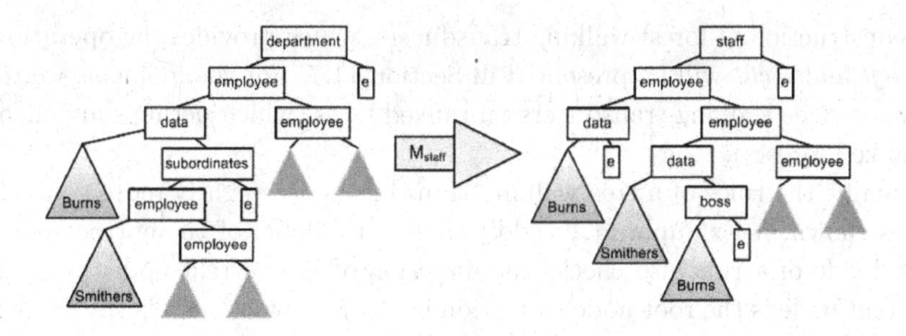

Fig. 11.2. The tree t'_B and its output tree s'_B of the transformation of the 2tt M_{staff}.

as well as a state q_{boss} to find the boss

$$
\begin{array}{lll}
5 & q_{boss}(\texttt{employee}) & \to q_{boss}(up), \\
6 & q_{boss}(\texttt{department}) & \to \texttt{e}, \\
7 & q_{boss}(\texttt{subordinates}) & \to \texttt{boss}(q_{data}(up), \texttt{e}),
\end{array}
$$

and a state q_{sub}, which processes the subordinates

$$
\begin{array}{lll}
8 & q_{sub}(\texttt{data}) & \to q_{sub}(down_2), \\
9 & q_{sub}(\texttt{subordinates}) & \to q(down_1), \\
10 & q_{sub}(\texttt{e}) & \to q_{next}(up).
\end{array}
$$

The state q_{next} searches (in dfs-manner) the next employee

$$
\begin{array}{lll}
11 & q_{next}(\texttt{data}) & \to q_{next}(up), \\
12 & q_{next}(\texttt{employee}) & \to q(down_2),
\end{array}
$$

together with a state q_{up} for going to the boss, if there is no further subordinate

$$
\begin{array}{lll}
13 & q_{up}(\texttt{employee}) & \to q_{up}(up), \\
14 & q_{up}(\texttt{subordinates}) & \to q_{next}(up), \\
15 & q_{up}(\texttt{department}) & \to \texttt{e},
\end{array}
$$

where state *copy* in line 4 is meant to copy the content of `data` (i.e., the left child in the binary representation). The initial state is q_I, which means that we start with state q_I at the root of the tree. The output trees of this transformation are binary representations of the lists of all members of staff. The root, which is labeled `staff`, has a right child with label `e`. The left child of `staff` has label `employee` whose left child is a `data`-node (with the personal data and the boss) and whose right child is a chain of `employee`-nodes. Figure 11.2 illustrates this transformation for the binary example tree t'_B resulting in the tree s'_B. ◁

The example illustrates that the "first-child next-sibling" encoding of forests implies that the *up*-operation of the tree walking transducer may not necessarily access directly the father in the forest representation but may instead reach the *left* sibling – depending whether or not the current node is a left or right child (i.e., has direction 1 or 2). In the example this was no problem: the state q_{boss} simply proceeds upwards in the tree representation until a node with the right label is reached. A direct construction of forest walking transducers, which provides the operations *up*, *down*, *left* and *right* will be presented in Section 11.7. For the moment, we restrict ourselves to tree walking transducers on ranked trees (which perhaps are encodings of unranked forests).

Formally, the rules of a tree walking transducers are slightly more general than the ones shown in Example 11.1: additional to the label of the current node, the left-hand side of a rule also checks the direction of the current node, i.e., whether the current node is the root node (direction is zero), or whether it is the i-th child of its parent node. It is well-known that in the case of tree walking automata (viz. tree walking transducers with output symbols $\{0, 1\}$ of rank zero), such direction tests (or "child number" test) are crucial: without them, the automaton cannot even realize a depth-first left-to-right traversal over the input tree, i.e., it cannot systematically search through every node of the input. For some translations, however, direction tests are not needed (such as our Example 11.1). In that example, we must think of every rule as existing in (at most) three incarnations, for direction zero (root node), direction one (left child), and direction two (right child). For instance, the q-rule for `employee`-nodes (rule number 2 of the example) is needed in the following two incarnations:

$$\begin{array}{ll} \text{2a} & q(\texttt{employee}, 1) \rightarrow \texttt{employee}(\texttt{data}(q_{data}(down_1), q_{boss}(stay)), q_{sub}(down_1)) \\ \text{2b} & q(\texttt{employee}, 2) \rightarrow \texttt{employee}(\texttt{data}(q_{data}(down_1), q_{boss}(stay)), q_{sub}(down_1)) \end{array}$$

Recall from the Preliminaries that the maximal rank of symbols in a ranked alphabet Σ is denoted by $mr(\Sigma)$.

Definition 11.3 (2tt). *A* tree walking transducer M *(*2tt *for short) is a tuple* (Q, Σ, R, Q_0) *where* Q *is a set of states,* Σ *is a ranked alphabet,* $Q_0 \subseteq Q$ *is a set of initial states, and* R *is a finite set of rules. A rule is of the form* $q(\mathtt{a}, \eta) \rightarrow \zeta$ *where* $q \in Q$, $\mathtt{a} \in \Sigma^{(m)}$, $m \geq 0$, $\eta \geq 0$ *and* ζ *is a tree generated by the grammar* $\zeta ::= \mathtt{b}(\underbrace{\zeta, \ldots, \zeta}_{m' \ times}) \mid q'(op)$, *with* $\mathtt{b} \in \Sigma^{(m')}$, $m' \geq 0$, $q' \in Q$, *and* $op \in \{stay, up\} \cup \{down_i \mid 1 \leq i \leq m\}$.

Tree walking transducers are also called 2-way tree transducers, because they generalize to trees the well known concept of 2-way finite state transducer on words (see, e.g., [16]).

Conventionally, tree transducers are defined over two ranked alphabets of input and output symbols. In Definition 11.3 of a 2tt M we only use one alphabet Σ which contains input and output symbols. If we want to distinguish the two, we say that $\mathtt{a} \in \Sigma$ is an *input symbol* if $\mathtt{a}$ appears on the left-hand side of a rule of M; we say that it is an *output symbol* if it appears in the right-hand side of a rule of M. In Example 11.1, `data` is an input and output symbol and `boss` is an output symbol of M_{staff}.

In practice, transducers also have to cope with *unknown* labels in the input such as, e.g., portions of text which then either are ignored or copied into the output. In order to deal with this, we could simply extend our formalism by an extra symbol $\bullet$ of any given rank which serves as a placeholder for unknown labels of this rank. This idea can be extended to placeholders for unknown elements of different atomic types, for instance `String`, `Number` or `Date`. Thus, we can describe the so called "Simple Types" of XML Schema (cf. [17]).

For a right-hand side ζ, we also write $\zeta = s[q_1(op_1), \ldots, q_c(op_c)]$ to refer to all occurrences of state calls in the right-hand side; there $s \in \mathcal{T}_\Sigma(X)$ is a tree which contains exactly one occurrence of the variable x_i for $i = 1, \ldots, c$. Note that s does not contain state calls. For example the right-hand side of the rule in line 2 in the Example 11.1 can be written as $s[q_{data}(down_1), q_{boss}(stay), q_{sub}(down_1)]$ where $s = \texttt{employee}(\texttt{data}(x_1, x_2), x_3)$.

A 2tt is called *deterministic* iff there is at most one state in the set Q_0 and for every triple $(q, \mathtt{a}, \eta)$ of a state, a symbol and a direction there is at most one rule with $q(\mathtt{a}, \eta)$ as left-hand side. The example 2tt M_{staff} is deterministic.

Intuitively, the meaning of the expressions of a right-hand side is as follows: the output can either be an element b whose content is recursively determined, or a recursive call to some state q' on the current input node, on its father or on its i-th subtree. The match patterns in the left-hand side of the rules are restricted to the form "$\mathtt{a}, \eta$", i.e., it is only allowed to check the label of the current input node and its direction. Thus, the transformation of a 2tt M starts at the root node of the input t with one of the initial states. A state q can be applied to an input node v with label $lab(v) = \mathtt{a}$ and direction $\eta = \eta(v)$ if there is a rule with left-hand side $q(\mathtt{a}, \eta)$. The evaluation continues on a child vi of v for each occurrence of a state call $q'(down_i)$, at v itself for each occurrence of a state call $q'(stay)$, and at the parent of v, for each occurrence of a state call $q'(up)$.

Hence, the *meaning* $[\![q]\!]_t$ of a state q of M with respect to an input tree t can be defined as a function from the nodes (of the input tree) to sets of trees, i.e., $[\![q]\!]_t : \mathcal{N}(t) \to 2^{\mathcal{T}_\Sigma}$. The values $[\![q]\!]_t$ for all q are jointly defined as the *least* functions satisfying: $[\![q]\!]_t(v) \supseteq ([\![\zeta]\!]_t(v))$ for rule $q(\mathtt{a}, \eta) \to \zeta$ where v is a node of t with $lab(v) = \mathtt{a}$ and $\eta = \eta(v)$ with

$$[\![\mathtt{b}(\zeta_1, \ldots, \zeta_m)]\!]_t(v) = \{\mathtt{b}(t'_1, \ldots, t'_m) \mid t'_i \in [\![\zeta_i]\!]_t(v)\}$$
$$[\![q'(op)]\!]_t(v) = [\![q']\!]_t([\![op]\!]_t(v)),$$

where op stands for $stay$, up or $down_i$ for $1 \leq i \leq rank(\mathtt{a})$, and $[\![op]\!]_t$ is defined by: $[\![stay]\!]_t(v) = v$, $[\![down_i]\!]_t(v) = vi$, and $[\![up]\!]_t(vi) = v$. The transformation τ_M realized by the 2tt M on an input tree t and sets T of input trees, respectively, is defined by $\tau_M(t) = \bigcup\{[\![q_0]\!]_t(\epsilon) \mid q_0 \in Q_0\}$ and $\tau_M(T) = \bigcup\{\tau_M(t) \mid t \in T\}$. For a deterministic 2tt M the transformation τ_M is a partial function $\tau_M : \mathcal{T}_\Sigma \to \mathcal{T}_\Sigma$. The *domain* of the transducer is the domain of the transformation, i.e., $dom(M) = dom(\tau_M) = \{t \mid \tau_M(t) \neq \emptyset\}$. As usual, the *size* $|M|$ of a 2tt M is the sum of the sizes of all its rules where the size of a rule $q(\mathtt{a}, i) \to \zeta$ is defined as $3 + |\zeta|$. Recall that $|\zeta|$ equals the number of nodes of ζ.

Applying the 2tt M_{staff} from before to t'_B we obtain the tree s'_B. The right-hand sides of rules in a 2tt may be arbitrarily large and contain arbitrarily many state calls. Dealing with such rules increases the complexity of some algorithms on 2tts. Thus, we give a normal form for 2tts where the number of state calls in right-hand sides is bounded by the maximal rank of output symbols. In the particular case where we consider binary representations of forests, the number of state calls in right-hand sides can be restricted to 2.

Lemma 11.1. *For every 2tt M a 2tt M' can be constructed in time $\mathcal{O}(|M|)$ such that (i) $\tau_{M'} = \tau_M$ and (ii) the right-hand side of each rule of M' contains at most k occurrences of states where k is the maximal rank of the output symbols of M.*

Proof. Let $M = (Q, \Sigma, R, Q_0)$. Intuitively, the idea of the construction is to introduce auxiliary states for all proper subtrees which contain more than 1 state call. For a symbol $\mathsf{a} \in \Sigma$ and direction η, let $Z_{\mathsf{a},\eta}$ denote the set of all subterms with more than one state call in right-hand sides of rules for a, η. For each $\zeta \in Z_{\mathsf{a},\eta}$, we introduce a fresh state $q_{\mathsf{a},\eta,\zeta}$. Assume that $\zeta = \mathsf{b}(\zeta_1, \ldots, \zeta_m)$. Then we introduce the new rule $q_{\mathsf{a},\eta,\zeta}(\mathsf{a}, \eta) \to \mathsf{b}(\zeta'_1, \ldots, \zeta'_m)$ where $\zeta'_j = \zeta_j$ if ζ_j contains at most one occurrence of a state, and $\zeta'_j = q_{\mathsf{a},\eta,\zeta_j}(stay)$ otherwise. We construct $M' = (Q', \Sigma, R', Q_0)$ as follows. The set of rules R' of the new transducer consists of all these newly constructed rules. Additionally, we add for every rule $q(\mathsf{a}, \eta) \to \mathsf{b}(\zeta_1, \ldots, \zeta_m)$ of M a new rule $q(\mathsf{a}, \eta) \to \mathsf{b}(\zeta'_1, \ldots, \zeta'_m)$ where for every j, $\zeta'_j = \zeta_j$ if ζ_j contains at most one occurrence of a state, and $\zeta'_j = q_{\mathsf{a},\eta,\zeta_j}(stay)$ otherwise. The set of states Q' contains all states of Q and additionally the new states $q_{\mathsf{a},\eta,\zeta}$ for every symbol $\mathsf{a} \in \Sigma$, direction η and every term $\zeta \in Z_{\mathsf{a},\eta}$.

The resulting transducer M' has a new state at most for every non-leaf node of a right-hand side of a rule in M. Thus, in the worst case, we have at most $|M|$ new states. In the new rules the right-hand side of the original rule of M is split in its subtrees. Thereby, we have $|M'| \in \mathcal{O}(|M|)$. □

In order to describe the behavior of the 2tt $M = (Q, \Sigma, R, Q_0)$ on a fixed input tree t, we are also going to define *runs* of M. A run can itself be described by a ranked tree over the set of rules. Here, the rank of a rule $q(\mathsf{a}, \eta) \to \zeta$ is given by the number of occurrences of calls $q'(op)$ in ζ to states q' in Q.

Definition 11.4 (Run). *Let q denote a state of M and v a node in the input tree t of direction η which is labeled with a. Assume that $r : q(\mathsf{a}, \eta) \to \zeta$ is a rule in R with $\zeta = s[q_1(op_1), \ldots, q_m(op_m)]$. Then the tree $\rho = r(\rho_1, \ldots, \rho_m) \in \mathcal{T}_R$ is a (q, v)-run of the 2tt M on the tree t, if for every $1 \leq i \leq m$, ρ_i is a (q_i, v_i)-run of M on t where v_i is obtained from v by operation op_i. The output $\tau(\rho)$ produced by a run ρ is defined by $\tau(\rho) = s[\tau(\rho_1), \ldots, \tau(\rho_m)]$. A (q_0, ϵ)-run for an initial state q_0 is also called* accepting run *of M on t.*

If M is deterministic, then there exists at most one accepting run on every tree.

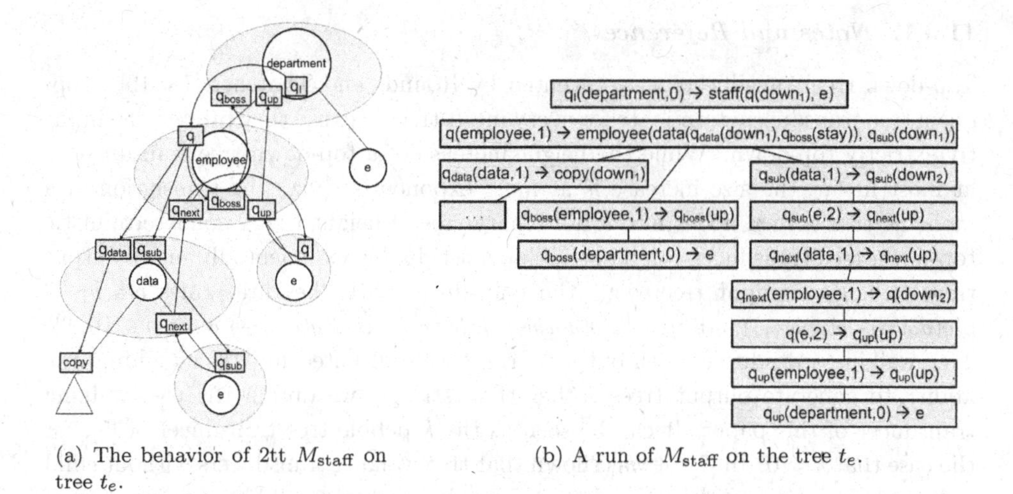

(a) The behavior of 2tt M_{staff} on tree t_e.

(b) A run of M_{staff} on the tree t_e.

Fig. 11.3. The 2tt M_{staff} on the tree t_e.

Example 11.2. Figure 11.3(a) shows the behavior of the (deterministic) example 2tt M_{staff} on the tree $t_e = \texttt{department(employee(data(}\ldots\texttt{, e), e), e)}$ which describes a department with one employee. All states in an oval around a node are applied to this node. The picture includes the dependences of the states. For example, consider the employee node $v = 1$ (with label $lab(v) = \texttt{employee}$). There, we have the state q. In the 2tt, there is just one rule with left-hand side $q(\texttt{employee}, 1)$:

$$\text{2a} \quad q(\texttt{employee}, 1) \rightarrow \texttt{employee(data(}q_{data}(down_1), q_{boss}(stay)\texttt{)}, q_{sub}(down_1)\texttt{)}$$

Thus, we have a $(q, 1)$-run $\rho = r_{2a}(\rho_1, \rho_2, \rho_3)$ where ρ_1, ρ_3 are $(q_{data}, 1.1)$- and $(q_{sub}, 1.1)$-runs, respectively, and ρ_2 is a $(q_{boss}, 1)$-run. This is illustrated by the three arrows starting at q at 1. The Figure 11.3(b) shows a (q_I, ϵ)-run $\rho' = r_1(\rho)$. The state *copy* was not detailed in Example 11.1. Accordingly, the $(copy, 1.1.1)$-run here is not complete. The output $\tau(\rho)$ of this run is the tree $\texttt{staff(employee(data(}\ldots\texttt{, e), e), e)}$. ◁

Accepting runs are another approach to define the semantics of a 2tt. Indeed, this operational semantics of a 2tt coincides with the denotational semantics provided first.

Theorem 11.1. *For a tree t and a 2tt M the following two statements are equivalent. (1) There is an accepting run ρ of M for t with $\tau(\rho) = s$ and (2) $s \in \tau_M(t)$.*

Theorem 11.1 can be proved by fixpoint induction. The denotational view on the semantics of a 2tt allows us to use fixpoint arguments for proving the correctness of constructions, whereas the operational view is better suited for combinatorial arguments.

11.3.1. *Notes and References*

Top-down tree transducers were invented by Rounds and Thatcher [18, 19]. Top-down tree transducers terminate for every input tree, because they process the input tree strictly top-down. While the height increase of a top-down tree transducer is at most linear, the size increase is at most exponential (viz. the translation of a monadic tree with n nodes into a full binary tree of height n). A nondeterministic top-down tree transducer can associate at most double exponentially many output trees to a given input tree; e.g. the transducer with the three rules $q(\mathtt{a}, \eta) \to \mathtt{b}(q(down_1), q(down_1))$, $q(\mathtt{a}, \eta) \to \mathtt{c}(q(down_1), q(down_1))$, and $q(\mathtt{e}, 1) \to \mathtt{e}$ for $\eta \in \{0, 1\}$. Tree walking transducers with output strings were invented in [20]; by adding the ability to generate output trees rather than strings, we obtain the tree walking transducer of this paper. It can be seen as the k-pebble tree transducer of [5], for the case that $k = 0$. In [21] it was shown that tree walking transducers without child number test are not useful: they cannot even check whether all leaves of input trees are labeled by some symbol a. As mentioned in [4], in the total deterministic case the tree walking transducer is essentially the same as the attribute grammar [22]. Similar to the fact that circularity of attribute grammars is decidable, it is possible to change any deterministic tree walking transducer in such a way that all runs are terminating [4]. This is not possible for nondeterministic tree walking transducers, because they can associate infinitely many output trees to a given input tree (viz. the transducer with the two rules $q(\mathtt{a}, 0) \to \mathtt{b}(q(stay))$, and $q(\mathtt{a}, 0) \to \mathtt{e}$). The normal form of Lemma 11.1 is similar to the one for pebble macro tree transducers given in Theorem 16 of [4]. Attribute grammars with tree output are also called "attributed tree transducers" [23]; for total deterministic such transducers (which coincide with our tree walking transducers when they are total deterministic) it is known that the size-to-height relationship of input tree to output tree is linear, and that the number of different output subtrees in an output tree is linear in the size of the corresponding input tree (see, e.g., [24]).

11.4. Type Checking

In this section, we present general techniques for certifying that all outputs produced by a transducer M for trees of a given input type are *well-formed*, i.e., comply with some given output type O. This problem is called *type checking* of the transducer M. Here, a type is just a set of trees, i.e., a *tree language*. Clearly, the tractability of type checking heavily depends on the class of languages used as types, and the class of transformations.

The tree s_B is an example for the output language of the 2tt M_{staff}. Such output trees are binary trees with a root labeled with `staff` and a right-comb of `employee` nodes as left subtree. It is the first-child next-sibling representation of a tree which has a root labeled with `staff` and arbitrary many `employee` nodes as children. A DTD describing this type (not the binary representation) is the following (where `content` stands for further personal data which are not specified here):

```
<!ELEMENT staff (employee)*>
<!ELEMENT employee (data, boss)>
<!ELEMENT data (name, content)>
<!ELEMENT name (#PCDATA)>
<!ELEMENT boss (name, content)>
<!ELEMENT content ... >
```

11.4.1. *Type Checking by Forward Type Inference*

Type checking a transducer M means to verify that all trees produced by M for input trees in the given input type I are necessarily contained in the given output type O. If τ is the transformation induced by the transducer M, we want to check whether or not $\tau(I) \subseteq O$, where $\tau(I) = \{\tau(t) \mid t \in I\}$. If this check succeeds, then we say that M *type checks w.r.t. I and O*. We solve this problem by forward inference, i.e., we determine whether $\tau(I) \cap \overline{O} = \emptyset$, where $\overline{O}$ is the complement of the type O. In order to decide emptiness of this intersection, we proceed in two steps. First, we construct from M a transducer $M_{\overline{O}}$ which produces only those outputs of M which are from $\overline{O}$. This construction is presented in Section 11.5.1. Then we present methods for deciding emptiness of transducers (w.r.t. I).

11.4.2. *Tree Automata*

There are several specification formalisms for XML types, such as DTD, XML Schema, or RELAX NG. For our purpose, the particular type formalisms is not essential, as all of these formalisms can be abstracted by recognizable (or: regular) tree languages. Thus, each type definition can be translated into a finite tree automaton. XML Schema specifications, e.g., can be considered as simple classes of deterministic top-down automata. In Chapter 3 several kinds of tree automata are introduced and basic constructions are presented. Here, we briefly recall crucial definitions of finite tree automata as needed in this chapter. Note that some definitions and notations slightly differ from Chapter 3.

Definition 11.5 (bta). *Let Σ be a ranked alphabet. A* bottom-up finite state tree automaton *A (over Σ),* bta *for short, is a tuple (P, Σ, δ, F) where P is a finite set of states, $F \subseteq P$ is a set of accepting states, and δ is a finite set of transitions $(p, \mathtt{a}, p_1 \dots p_m)$ where $\mathtt{a} \in \Sigma^{(m)}$ and $p, p_1, \dots, p_m \in P$.*

A transition $(p, \mathtt{a}, p_1 \dots p_m)$ denotes that if, for all $1 \le i \le m$, A arrives in state p_i after processing some tree t_i, then it can assign state p to the tree $\mathtt{a}(t_1, \dots, t_m)$. Technically, a p-run ρ of A on a tree $t = \mathtt{a}(t_1, \dots, t_m) \in \mathcal{T}_\Sigma$ is a tree $\rho = r(\rho_1, \dots, \rho_m) \in \mathcal{T}_\delta$ where r is a transition $(p, \mathtt{a}, p_1 \dots p_m) \in \delta$ and ρ_i is a p_i-run of A for t_i. For some applications, it suffices to represent p-runs by trees from $\mathcal{T}_P$ with root p. The tree language $\mathcal{L}(A)$ accepted by A consists of the trees $t \in \mathcal{T}_\Sigma$ by which A can reach an accepting state, i.e. it exists a p-run ρ of A for t with $p \in F$; the latter run is called *accepting run of A on t*. A bottom-up tree automaton $A = (P, \Sigma, \delta, F)$ is *deterministic* (dbta) if for each symbol $\mathtt{a} \in \Sigma^{(m)}$ and every tuple $p_1 \dots p_m$ of states, there is at most one state p with $(p, \mathtt{a}, p_1 \dots p_m) \in \delta$, i.e., δ induces a partial function of type $\Sigma \times P^* \to P$. A bta is called *complete* if there is at least one rule $(p, \mathtt{a}, p_1 \dots p_m) \in \delta$ for all $m \geq 0$, $\mathtt{a} \in \Sigma^{(m)}$, and $p_1, \dots, p_m \in P$.

We may also interpret the transitions of a bta in a *top-down* fashion. Then we obtain the known top-down tree automaton (tta) which starts at the input root node and assigns states to the children of a node, depending on the label of the node and the current state.

Definition 11.6 (dtta). *A bta is called* deterministic top-down *(*dtta *for short), if the set of final states is a singleton set, and the transition relation δ induces a partial function $P \times \Sigma \to P^*$, i.e., for each state $p \in P$ and each symbol $\mathtt{a} \in \Sigma^{(m)}$, there is at most one sequence of states $p_1 \dots p_m \in P^m$ with $(p, \mathtt{a}, p_1 \dots p_m) \in \delta$.*

As usual, the *size* $|A|$ of a finite state tree automaton A is the sum of sizes of all its transitions and the number of states. A transition $(p, \mathtt{a}, p_1 \dots p_m)$ has size $m + 2$. Let BTA, DBTA, and DTTA denote the classes of all languages definable by btas, dbtas, and dttas, respectively. It is known that BTA = DBTA equals the class of regular tree languages, and that DTTA is properly contained in this class.

Example 11.3. Coming back to the transformation from Example 11.1, the set of valid output documents should be lists of staff members, more precisely: `staff` should contain a possibly empty sequence of `employee` elements; Each `employee` element should contain a `data` element and optionally, a `boss` element.

A bta describing (the binary representations of) this set is given by $A_{\text{staff}} = (P, \Sigma, \delta, F)$ where $P = \{r_{\text{staff}}, r_{\text{empl}}, r_{\text{data}}, r_{\text{name}}, r_{\text{boss}}, r_{\mathtt{e}}, \dots\}$ and δ contains the transitions $(r_{\text{staff}}, \mathtt{staff}, r_{\text{empl}}\ r_{\mathtt{e}})$, $(r_{\text{empl}}, \mathtt{employee}, r_{\text{data}}\ r_{\text{empl}})$, $(r_{\text{empl}}, \mathtt{e})$, $(r_{\text{data}}, \mathtt{data}, r_{\text{name}}\ r_{\text{boss}})$, $(r_{\text{boss}}, \mathtt{boss}, r_{\text{name}}\ r_{\mathtt{e}})$, $(r_{\text{boss}}, \mathtt{e})$, $(r_{\text{name}}, \mathtt{name}, r_{\text{content}}\ r_{\mathtt{e}})$, and $(r_{\mathtt{e}}, \mathtt{e})$, where r_{content} is the state characterizing valid personal data of employees. The set of accepting states is, thus, given by $F = \{r_{\text{staff}}\}$. Note that this bta is in fact deterministic top-down. ◁

In the previous example, the bta ran on the first-child next-sibling encoding of forests as binary ranked trees. For convenience, we call such a finite tree automaton also *finite forest automaton* (short: bfa). Again, if it is deterministic or deterministic top-down, then we abbreviate it with dbfa and dtfa, respectively.

11.4.3. *Basic Properties of BTAs*

The approach which we advocate here is called *forward type checking* (cf. Section 11.4.1). Assume that O is the type of all valid output trees. In order to check that the transducer M produces only outputs in O, we construct a transducer which for every input t, only produces those output trees of M which are *not* valid, i.e., which are in $\overline{O}$ (the complement of O). Thus, type correctness for M is reduced to emptiness of the auxiliary transducer $M_{\overline{O}}$. For this idea to work, it is useful to have effective constructions which take the specification of a type and return a specification for its complement. For a *complete* dbta $A = (P, \Sigma, \delta, F)$ this construction is simple: we need to exchange accepting and non-accepting states, i.e., replace F with $P \backslash F$. Since every regular tree language can be accepted by a complete dbta, this construction implies that the complement of a regular tree language is a regular tree language, too. The complement of a type described by a deterministic top-down tree automaton is a regular language as well, but not necessarily in DTTA. The obvious technique for constructing an automaton for the complement therefore is to transform the deterministic top-down automaton into a complete deterministic bottom-up automaton and then apply the complement construction for complete dbtas. This first construction, however, possibly incurs an exponential blow-up in the number of states. Therefore, we approve a different approach: instead of constructing a *deterministic* automaton for the complement, we construct a *non-deterministic* automaton. The latter can be achieved by only moderately increasing the size.

Lemma 11.2. *For a dtta A over the ranked alphabet Σ there is a bta A' over Σ with $\mathcal{L}(A') = \mathcal{T}_\Sigma \backslash \mathcal{L}(A)$ and $|A'| \in \mathcal{O}((|A| + |P| \cdot |\Sigma|) \cdot mr(\Sigma))$, where P is the set of states of A.*

Proof. Intuitively, the automaton A' guesses a path in the input tree to some node where the original automaton A fails. Formally, let $A = (P, \Sigma, \delta, \{p_0\})$ and define $A' = (P', \Sigma, \delta', \{p'_0\})$ with $P' = \{p' \mid p \in P\} \cup \{\bullet\}$ for a new state $\bullet \notin P$. A state p' is meant to generate only trees for which there is *no* p-run of A. The state $\bullet$ describes arbitrary trees, i.e., the language $\mathcal{T}_\Sigma$. The set δ' of transitions of the new bta is defined as follows:

- for every transition $(p, \mathtt{a}, p_1 \dots p_m) \in \delta$ with $m = rank(\mathtt{a}) \geq 1$, and for every $i \in \{1, \dots, m\}$ let $(p', \mathtt{a}, \bullet^{i-1} p'_i \bullet^{m-i}) \in \delta'$
- for every state $p \in P$, $0 \leq m \leq mr(\Sigma)$, and $\mathtt{a} \in \Sigma^{(m)}$, $(p', \mathtt{a}, \bullet^m) \in \delta'$ whenever $\forall p_1, \dots, p_m \in P : (p, \mathtt{a}, p_1 \dots p_m) \notin \delta$
- for every $\mathtt{a} \in \Sigma^{(m)}$, $(\bullet, \mathtt{a}, \bullet^m) \in \delta'$.

For the correctness of the construction, we claim that for every state p of A, and every input tree t, A' has a p'-run on t iff A has no p-run on t. This claim can be proven by induction on the height of input trees.

Now, let $k = mr(\Sigma)$ be the maximal rank of symbols in Σ. For each transition in δ we get at most k new transitions in δ' (one for each successor state). Additionally, we require a new rule of length at most $k + 2$ for each symbol in Σ. Thus, the size of the automaton A' is in $\mathcal{O}((|A| + |\Sigma|) \cdot mr(\Sigma))$. □

Example 11.4. For the dtta $A_{\mathsf{staff}} = (P, \Sigma, \delta, \{r_{\mathsf{staff}}\})$ in Example 11.3, the bta $A'_{\mathsf{staff}} = (P', \Sigma, \delta', \{r'_{\mathsf{staff}}\})$ for the complement has the following transitions for the label `staff`: $(r'_{\text{staff}}, \texttt{staff}, r'_{\text{empl}}\bullet)$, $(r'_{\text{staff}}, \texttt{staff}, \bullet r'_{\text{e}})$, and $(r', \texttt{staff}, \bullet\bullet)$ for all $r' \in P' \setminus \{r'_{\text{staff}}\}$ $\lhd$

11.4.4. *Notes and References*

XML type definition languages such as DTDs [25], XML Schema [26], or RELAX NG [27] are closely related to the regular tree languages [3, 28], that is, to the class of tree languages recognized by finite tree automata.

Tree automata are a well studied formalism in computer science, dating back to the late 1960s. For surveys on tree automata, please see [29–31]. Tree automata inherit most of the good properties of finite automata on strings, such as effective closure under Boolean operations and decidability of emptiness. An important property which will be used later for type checking, is that emptiness of btas can be decided in linear time (see, e.g., Theorem 1.7.4 in [31], Theorem 3.4 in Chapter 3).

Theorem 11.2. *Given a bta A it can be decided in linear time whether or not $\mathcal{L}(A) = \emptyset$.*

Just as in the string case, nondeterministic bottom-up tree automata can be determinized (with a potential and sometimes unavoidable exponential blow up in automaton size). This is not the case for top-down tree automata: the class DTTA of languages accepted by deterministic top-down tree automata is a strict subclass of BTA which does not even contain all finite languages; a famous example of a language not in DTTA is the set $U = \{\texttt{f}(\texttt{a}, \texttt{b}), \texttt{f}(\texttt{b}, \texttt{a})\}$. Note that for a given bta, it is decidable if its language is in DTTA; this is due to the fact that DTTA languages can be characterized by the "path-closed" property [32, 33]; the latter means that the trees in the languages are exactly obtained by combining all paths of the corresponding path language. The language U for instance is not path-closed. Using a similar example, it is easily shown that DTTA is not closed under complementation (and neither under union).

11.5. Type Checking of Tree Walking Transducers

11.5.1. *Intersecting Tree Walking Transducers with Output Types*

In this section we present techniques to type check 2tts against regular tree languages. For a given 2tt we build a second 2tt which produces only output trees in the *complement* of the output type, and otherwise realizes the same transformation as the original 2tt. If the output type is described by a complete dbta $A = (P, \Sigma, \delta, F)$, the complement will be recognized by the complete dbta $\bar{A} = (P, \Sigma, \delta, P \setminus F)$. For a given dtta there exists a bta describing the complement (cf. Lemma 11.2). Thus, it is sufficient to construct a 2tt M_A for a 2tt M and a bta A (which may be the complement automaton of a complete dbta or dtta) with $\tau_{M_A}(t) = \tau_M(t) \cap \mathcal{L}(A)$ for every tree t.

Theorem 11.3. *For every 2tt M and every bta A there is a 2tt M_A with*

$$\tau_{M_A}(t) = \tau_M(t) \cap \mathcal{L}(A)$$

for all $t \in \mathcal{T}_\Sigma$. The size $|M_A|$ of M_A is in $\mathcal{O}(|M| \cdot |A|^{d+1})$, where d is the maximal number of occurrences of states in right-hand sides of M.

Proof. Let $M = (Q, \Sigma, R, Q_0)$ and $A = (P, \Sigma, \delta, F)$. For each state q in Q and all states $p \in P$ we generate new states for M_A of the form $\langle q, p \rangle$. Such a state is meant to generate only trees $t \in \mathcal{T}_\Sigma$ for which there is a run of A starting at the leaves and reaching the root of t in state p. The rules of the new 2tt M_A are $\langle q, p \rangle(\mathtt{a}, \eta) \to \zeta'$ for every rule $q(\mathtt{a}, \eta) \to \zeta$ of M and $\zeta' \in \tau^p[\zeta]$. The sets $\tau^p[.]$ are inductively defined by:

$$\begin{aligned} \tau^p[\mathtt{b}(\zeta_1, \ldots, \zeta_m)] &= \{\mathtt{b}(\zeta_1', \ldots, \zeta_m') \mid (p, \mathtt{b}, p_1' \ldots p_m') \in \delta \wedge \forall i : \zeta_i' \in \tau^{p_i'}[\zeta_i]\} \\ \tau^p[q'(op)] &= \{\langle q', p \rangle(op)\}. \end{aligned}$$

The set of initial states of M_A is $Q_0' = Q_0 \times F$. By fixpoint induction, we verify for every state q, every input tree $t \in \mathcal{T}_\Sigma$, every node $v \in \mathcal{N}(t)$ and every state p that:

$$[\![\langle q, p \rangle]\!]_t(v) = [\![q]\!]_t(v) \cap \{s \in \mathcal{T}_\Sigma \mid \exists \text{ run } \rho \text{ on } s \text{ with } \rho(\epsilon) = p\}$$

For each state in M we have at most $|A|$ new states in M_A. If we have c occurrences of state calls in the right-hand side of a rule r of M, with the state on the left-hand side, we obtain at most $|A|^{c+1}$ new rules for r in M_A. Therefore, the new 2tt is of size $\mathcal{O}(|M| \cdot |A|^{d+1})$ where d is the maximal number of occurrences of state calls in right-hand sides in M. □

Considering only binary trees, we obtain size $\mathcal{O}(|M| \cdot |A|^3)$ for the intersection 2tt (with Lemma 11.1). The last step is to decide whether $\tau_{M_A} \neq \emptyset$. Thereto, we build a bta describing the domain of M_A. This will be done after completing the example.

Example 11.5. Let us try to type check the 2tt $M_{\mathsf{staff}} = (Q, \Sigma, R, Q_0)$ via forward type inference. According to Lemma 11.1, we restrict the maximal number of state calls in right-hand sides to 2. In our example 2tt, the rule $q(\mathtt{employee}, \eta) \to \mathtt{employee}(\mathtt{data}(q_{data}(down_1), q_{boss}(stay)), q_{sub}(down_2))$ has three state calls. We obtain the new rules: $q(\mathtt{employee}, \eta) \to \mathtt{employee}(q'(stay), q_{sub}(stay))$ and $q'(\mathtt{employee}, \eta) \to \mathtt{data}(q_{data}(down_1), q_{boss}(stay))$. According to the proof of Lemma 11.1, the new state q' is $q_{\mathtt{employee}, \mathtt{data}(q_{data}(down_1), q_{boss}(stay))}$. Consider again the dtta $A_{\mathsf{staff}} = (P, \Sigma, \delta, \{r_{\mathsf{staff}}\})$ as output type. The complement bta $A'_{\mathsf{staff}} = (P', \Sigma, \delta', \{r'_{\mathsf{staff}}\})$ is given in Example 11.4. The intersection 2tt $(M_{\mathsf{staff}})_{A'_{\mathsf{staff}}}$ is given by $(Q \times P', \Sigma, R', Q_0 \times \{r'_{\mathsf{staff}}\})$. In what follows we show how the first few rules in R' are constructed from R and δ' as follows. For $q_I(\mathtt{department}, 0) \to \mathtt{staff}(q(down_1), \mathtt{e})$ and r'_{staff} we obtain the rule $\langle q_I, r'_{\mathsf{staff}} \rangle(\mathtt{department}, 0) \to \mathtt{staff}(\langle q, r'_{\mathsf{empl}} \rangle(down_1), \mathtt{e})$. For $q(\mathtt{employee}, \eta) \to \mathtt{employee}(q'(stay), q_{sub}(down_2))$ and r'_{empl} we obtain $\langle q, r'_{\mathsf{empl}} \rangle(\mathtt{employee}, \eta) \to \mathtt{employee}(\langle q', r'_{\mathsf{data}} \rangle(stay), \langle q_{sub}, \bullet \rangle(down_2))$ and

$\langle q, r'_{\text{empl}}\rangle(\texttt{employee}, \eta) \rightarrow \texttt{employee}(\langle q', \bullet\rangle(stay), \langle q_{sub}, r'_{\text{empl}}\rangle(down_2))$. For $q(\texttt{e}, \eta) \rightarrow q_{up}(up)$ and all $r \in P'$ we obtain $\langle q, r\rangle(\texttt{e}, \eta) \rightarrow \langle q_{up}, r\rangle(up)$. ◁

11.5.2. *Deciding Emptiness of 2TTs*

In order to check the emptiness of a tree walking transducer w.r.t. a given input type, we construct a nondeterministic finite state automaton (cf. Section 11.4.2) which then is checked for emptiness. First, for a 2tt M and an input type I we define an *alternating tree walking automaton* M' which ignores the output of the 2tt, but apart from that imitates the behavior of M on trees in I. For M', we then construct a nondeterministic bta $A_{M'}$ accepting all trees t such that $\tau_M(t) \neq \emptyset$. The right-hand sides of transitions of an alternating tree walking automaton are conjunctions. Whereas the empty conjunction, i.e., $\bigwedge \emptyset$, equals **true**.

Definition 11.7 (atwa). *An* alternating tree walking automaton *(*atwa *for short) is a tuple $M = (Q, \Sigma, \delta_M, Q_0)$ where Q is a finite set of states, Σ a ranked input alphabet, $Q_0 \subseteq Q$ a set of initial states, and δ_M a finite set of rules of the form $q(\texttt{a}, \eta) \rightarrow q_1(op_1) \wedge \ldots \wedge q_c(op_c)$ with $c \geq 0$, $q, q_1, \ldots, q_c \in Q$, $\texttt{a} \in \Sigma^{(m)}$, $m \geq 0$, $\eta \geq 0$, and for $1 \leq i \leq c$, $op_i \in \{stay, up\} \cup \{down_j \mid j = 1, \ldots, m\}$.*

A transition $q(\texttt{a}, \eta) \rightarrow H$ is also called q-rule. An atwa traverses a tree like a 2tt, but produces no output. The language $\mathcal{L}(M)$ of an atwa M is defined as the set of all trees for which there exists an accepting run of M.

Definition 11.8 (Run). *For an atwa $M = (Q, \Sigma, \delta_M, Q_0)$, assume that $r : q(\texttt{a}, \eta) \rightarrow H$ is a rule in δ_M with $H = q_1(op_1) \wedge \ldots \wedge q_c(op_c)$. Then the tree $\rho = r(\rho_1, \ldots, \rho_c) \in \mathcal{T}_{\delta_M}$ is a (q, v)-run of the atwa M on the tree t, if v is a node of t with direction η and label $\texttt{a}$ and for all i, ρ_i is a (q_i, v_i)-run of M on t where v_i is obtained from v by operation op_i. A (q_0, ϵ)-run ρ with $q_0 \in Q_0$ is also called* accepting.

A subtree $\rho[w]$ of a run ρ on a node w which is a (q, v)-run for some state q and some node v of t is called (q, v)-*subrun*. The set of rules which are applied to one node v during a run ρ on t is the set $rules_\rho(v) = \{r \mid \exists \rho_1, \ldots, \rho_c : r(\rho_1, \ldots, \rho_c)$ is a (q, v)-subrun of $\rho\}$.

Lemma 11.3. *Assume that M is a 2tt. Then an atwa M' can be constructed in linear time such that for every input tree t, M' has an accepting run for t iff M has an accepting run for t.*

Proof. The atwa M' has the same set of states as M. The rules of M' are obtained from those of M by replacing every right-hand side ζ of M with the conjunction of all $q(op)$ occurring in ζ. Formally, let $M = (Q_M, \Sigma, R, Q_0)$ be a 2tt. Then, the atwa M' is defined by $M' = (Q_M, \Sigma, \delta', Q_0)$ for a set δ' of transitions $\delta' = \{[r] \mid r \in R\}$ where $[q(\texttt{a}, \eta) \rightarrow \zeta] = q(\texttt{a}, \eta) \rightarrow [\zeta]$ and $[\zeta] = q_1(op_1) \wedge \ldots \wedge q_c(op_c)$ for a right-hand side

$\zeta = s[q_1(op_1), \ldots, q_c(op_c)]$. In order to prove the correctness of this construction, we first extend the translation $[.]$ from rules to trees of rules. For $\rho = r(\rho_1, \ldots, \rho_c) \in \mathcal{T}_R$, $[\rho]$ is inductively defined as the tree $[\rho] = [r]([\rho_1], \ldots, [\rho_c])$. Then we claim that ρ is a (q, v)-run of M iff $[\rho]$ is a (q, v)-run of M'. The proof is by structural induction on ρ. Since for every (q, v)-run ρ' of M', some $\rho \in \mathcal{T}_R$ exists with $[\rho] = \rho'$, we conclude that $\tau_M(t) \neq \emptyset$ iff $t \in \mathcal{L}(M')$. $\square$

As an example for this translation of 2tts into atwas, consider the 2tt M_{staff} (Example 11.1). For the second rule, we get: $r_2' : q(\texttt{employee}, \eta) \to q_{data}(down_1) \wedge q_{boss}(stay) \wedge q_{sub}(down_1)$. In order to accept only trees of an input type I, we enlarge an atwa M. Let I be given by a finite state tree automaton (P, Σ, δ_A, F) and $M = (Q, \Sigma, \delta_M, Q_0)$. For every rule $q_0(\mathtt{a}, \eta) \to H$ with $q_0 \in Q_0$ and every $p_f \in F$ we enhance the atwa with the rule $q_0(\mathtt{a}, \eta) \to p_f(stay) \wedge H$. Additionally for every $(p, \mathtt{a}, p_1 \ldots p_m) \in \delta_A$, we add the rule $p(\mathtt{a}, \eta) \to \bigwedge_{i \leq m} p_i(down_i)$.

In an accepting run of an atwa, subruns for the same state and node are interchangeable. Thus, we define *uniform* runs as runs where for each state q and each node v, the (q, v)-subruns are the same.

Definition 11.9 (Uniform Run). *A (q, v)-run ρ of an atwa M on a tree t is called* uniform *if, for every state q, every node v in t, and every two (q, v)-subruns ρ_1, ρ_2 of ρ, $\rho_1 = \rho_2$.*

If an atwa is deterministic, i.e., for each state q, direction η and label $\mathtt{a}$, there is at most one rule of the form $q(\mathtt{a}, \eta) \to H$, then *every* (q', v)-run is uniform. In general, this may not be the case. However, we have:

Lemma 11.4. *For an atwa M and an input tree t the following two statements are equivalent. (1) M has an accepting run for t and (2) M has a uniform accepting run for t.*

Proof. Every uniform accepting run is an accepting run. For the reverse direction, it suffices to prove that for every (q, v)-run ρ of M on t, there is also a uniform (q, v)-run of M on t. For that, we consider the set $B(\rho)$ of all pairs (q', v') for which ρ contains more than one (q', v')-run as a subtree. We proceed by induction on the cardinality of the set $B(\rho)$. If the set $B(\rho)$ is empty, ρ is already uniform. Now assume $B(\rho)$ is non-empty. Then ρ contains a subtree ρ_1 with the following two properties:

(1) ρ_1 is a (q_1, v_1)-run with $(q_1, v_1) \in B$;
(2) all subtrees ρ_1' of ρ_1 are (q', v')-runs with $(q', v') \notin B$.

Then we construct from ρ a tree ρ' by replacing every occurrence of a subtree in ρ which is a (q_1, v_1)-run with ρ_1. Then ρ' is again a (q, v)-run on t, but now the set $B(\rho') \subseteq B(\rho) \backslash \{(q_1, v_1)\}$ contains at least one element less. Thus, by induction hypothesis applied to ρ', there is a (q, v)-run on t which is uniform. $\square$

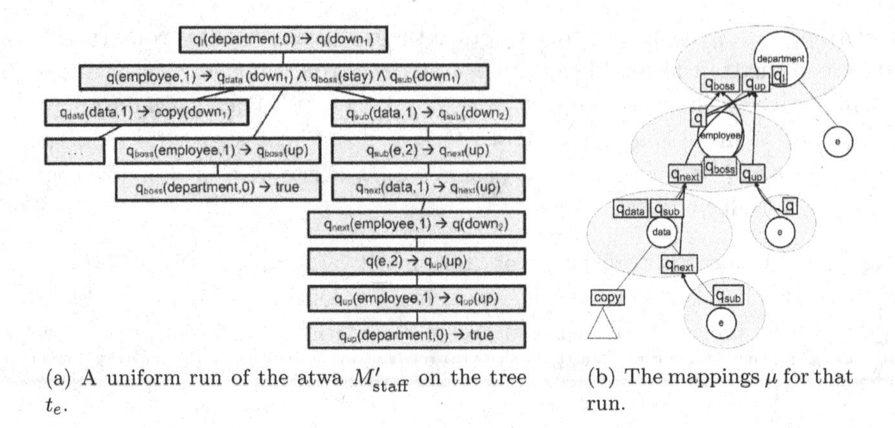

(a) A uniform run of the atwa M'_{staff} on the tree t_e.

(b) The mappings μ for that run.

Fig. 11.4. Behavior of M'_{staff} on t_e.

Figure 11.3(a) shows the behavior of the 2tt M_{staff} on the tree t_e. The corresponding atwa M'_{staff} yields the same behavior. The Figure 11.3(b) shows an accepting run ρ of M_{staff} on t_e. The corresponding run $[\rho]$ of M'_{staff} is illustrated in Figure 11.4(a). This run is uniform. In order to decide emptiness of an atwa M and accordingly of a 2tt, we construct a *nondeterministic* bottom-up finite state tree automaton A_M. In order to accept the domain of M, the bta A_M guesses uniform accepting runs. Since A_M visits each node in the input tree at most once, it *guesses* at every node all transitions which are applied at this node during a uniform run of M.

Technically, the states of the bta A_M consist of guessed directions together with *partial* mappings $\mu : Q \to 2^Q$ of states to sets of states: $\mu(q) = B$ at a given node v means that the (q, v)-run on the input tree will cause calls $q'(up)$ at v only for states q' from B. For a mapping μ we refer to the domain as $\text{dom}(\mu) = \{q \mid \mu(q) \text{ is defined}\}$. Furthermore, to each node v in the input tree we implicitly attach the set $T_v = rules_\rho(v)$ collecting the atwa rules which are applied to the node v in an accepting run ρ of the atwa on the input tree t. Since ρ is uniform, the set $rules_\rho(v)$ contains at most one q-rule for every q.
A pair $\langle \mu, \eta \rangle$ is accepting if $\eta = 0$ and for the partial mapping μ, $q_0 \in \text{dom}(\mu)$ for some accepting state $q_0 \in Q_0$ of the atwa M and $\mu(q) = \emptyset$ for all $q \in \text{dom}(\mu)$.

Now assume that a is a label of arity m, η is a direction and $\mu, \mu_i : Q \to 2^Q$ are partial mappings ($i = 1, \ldots, m$). Then $(\langle \mu, \eta \rangle, \mathsf{a}, \langle \mu_1, 1 \rangle, \ldots, \langle \mu_m, m \rangle)$ is a transition of A_M iff there is a set T of rules of the atwa M with the following properties. Let Q_s, Q_u and $Q_{d,i}$ ($i = 1, \ldots, m$) denote the set of states q for which there is a q-rule in T, the set of states q' with a recursive call $q'(up)$, and the sets of states q' with a recursive call $q'(down_i)$ in some right-hand side of rules in T, respectively. Then the set T of rules should have the following properties:

(1) All rules in T have a left-hand side of the form $q(\mathsf{a}, \eta)$, where $q \in Q$.
(2) Assume $q(\mathsf{a}, \eta) \to q_1(op_1) \wedge \ldots \wedge q_c(op_c) \in T$. Then we have for every $j \leq c$:

- If $op_j = stay$, then T also contains a q_j-rule, i.e., $q_j \in Q_s$;
- If $op_j = down_i$, then $q_j \in \text{dom}(\mu_i)$.

(3) Whenever $q' \in \text{dom}(\mu_i)$ for i, then $\mu_i(q') \subseteq Q_s$.

(4) Consider the following graph G with set of vertices $V = \{q(stay) \mid q \in Q_s\} \cup \{q(down_i) \mid i = 1, \ldots, m, q \in Q_{d,i}\} \cup \{q(up) \mid q \in Q_u\}$ and the following set E of edges:

- If a q-rule in T contains a call $q'(op)$, then $(q(stay), q'(op)) \in E$;
- If $\mu_i(q) = B_i$ is defined, then $(q(down_i), q'(stay)) \in E$ for all $q' \in B_i$.

The resulting directed graph $G = (V, E)$ should be acyclic, and the mapping μ is obtained from G as follows:

- $q \in \text{dom}(\mu)$ iff $q \in Q_s$ and
- $\mu(q) = B$ iff the set B equals the set of all vertices $q'(up)$ which are reachable in G from $q(stay)$.

The size of A_M is exponential in the size of M. We give a detailed example for this construction in Example 11.6. Now, we state the correlation of runs of A_M and of M.

Lemma 11.5. *For a tree t the following statements are equivalent. (1) There is a uniform accepting run of M on t and (2) there is an accepting run of A_M on t.*

Proof. $(1) \Rightarrow (2)$: Let ρ be a uniform accepting run of M on a tree t. For a node v of t with label a and direction η, let T_v denote the set of all atwa rules applied at the root in subruns of ρ starting at v, i.e., $T_v = rules_\rho(v)$. We then construct for every node v of t with label $\mathsf{a} \in \Sigma^{(m)}$ and direction η, a state μ_v and a transition $r_v = (\langle \mu_v, \eta \rangle, \mathsf{a}, \langle \mu_{v1}, 1 \rangle, \ldots, \langle \mu_{vm}, m \rangle)$ of the bta A_M.

The sets T_v allow us to construct a directed graph G_t. The set V_t of vertices of G_t are given by the set of all pairs (q, v) for nodes v of t and states q for which there is a q-rule in T_v. The set E_t of edges consists of:

- all pairs $((q, v), (q', v))$ where the q-rule in T_v contains a call $q'(stay)$;
- all pairs $((q, v), (q', vi))$ where the q-rule in T_v contains a call $q'(down_i)$;
- all pairs $((q, v), (q', v'))$ where the q-rule in T_v contains a call $q'(up)$ and $v = v'i$ for some i.

The graph G_t is acyclic. Moreover since the uniform run ρ is accepting, every vertex in V_t is reachable from some vertex (q_0, ϵ) with $q_0 \in Q_0$.

The graph G_t allows to construct partial mappings μ_v for every node v. $q \in \text{dom}(\mu_v)$ iff (q, v) is a vertex of G_t. Assume (q, v) is a vertex in G_t. We consider two cases. If $v = \epsilon$, then T_ϵ cannot contain any q-rule which has an up-call. In this case, $\eta(v) = 0$, and we set $\mu_\epsilon(q) = \emptyset$. Otherwise, assume that $v = v'i$. Then $\eta(v) = i$ and $q' \in \mu_v(q)$ iff there is an edge $((q_1, v), (q', v'))$ in G_t where (q_1, v) is reachable from (q, v) by a path which contains only vertices (q_2, v_2) referring to nodes v_2 from the subtree at v, i.e., to nodes which have v as a prefix. If no such state q' exists, then $\mu_v(q) = \emptyset$.

It now can be verified for every node v with label $\mathtt{a} \in \Sigma^{(m)}$ and direction η that $(\langle\mu_v, \eta\rangle, \mathtt{a}, \langle\mu_{v1}, 1\rangle, \ldots, \langle\mu_{vm}, m\rangle)$ constitutes a transition of A_M (with T_v as set of rules of the atwa). Since by construction, $\langle\mu_\epsilon, 0\rangle$ is an accepting state of A_M, we have, thus, constructed an accepting run of A_M for t.

(2) $\Rightarrow$ (1) : Let ρ' be an accepting run of A_M on the tree t, and let $\langle\mu_v, \eta\rangle$ and r_v denote the state and transition of A_M attained for the node v in t. We can find sets T_v conforming to the properties of the transition relation of M. These allow to construct a graph G'_t analogously to the graph G_t above. By the definition of the transition relation of A_M, G'_t is acyclic. This allows us to define for every vertex (q, v) in G'_t, the number $h(q, v)$ as the maximal length of a path in G'_t to a leaf, i.e., a vertex with out-degree 0. Using the sets T_v of atwa rules, we now construct for every node v and atwa rule $r : q(\mathtt{a}, \eta) \to H$ from T_v with $H = q_1(op_1) \wedge \ldots \wedge q_c(op_c)$, a tree $\rho[q, v]$ by $\rho[q, v] = H(\rho[q_1, v_1], \ldots, \rho[q_c, v_c])$, where $v_j = [\![op_j]\!](v)$. Note that all these trees are well-defined, since the height of $\rho[q, v]$ precisely equals $h(q, v)$. Moreover, the tree $\rho[q, v]$ is a (q, v)-run of the atwa M on t. Since every (q', v')-subrun of this tree equals the (q', v')-run $\rho[q', v']$, this run is also uniform. In particular, the tree $\rho[q_0, \epsilon]$ constitutes a uniform accepting run of the atwa M. □

By Lemmas 11.3, 11.4, and 11.5, the bta A_M recognizes the domain of the given 2tt M, which gives us Theorem 11.4. Note that this implies, by Theorem 11.2, that also emptiness of M's domain (and hence of M's translation τ_M) can be decided in exponential time.

Theorem 11.4. *Assume M is a 2tt. Then a bta A can be constructed in exponential time such that $\mathcal{L}(A) = dom(M)$. Thus, emptiness for a 2tt can be decided in deterministic exponential time.*

Example 11.6. In this example we consider again the 2tt M_{staff} and its corresponding atwa M'_{staff}. The size of the bta $A_{M'_{\mathsf{staff}}}$ is exponential in the size of the atwa or 2tt. Therefore, we only construct states occurring in a run of $A_{M'_{\mathsf{staff}}}$ on the tree t_e in Figure 11.3(a). In Figure 11.4(a), a uniform accepting run ρ of the atwa is illustrated. The run yields the sets T_v. For instance, for the node labeled with `employee` we get $T_1 = \{q(\mathtt{employee}, 1) \to q_{boss}(stay) \wedge q_{sub}(down_1) \wedge q_{data}(down_1),\ q_{boss}(\mathtt{employee}, 1) \to q_{boss}(up),\ q_{up}(\mathtt{employee}, 1) \to q_{up}(up),\ q_{next}(\mathtt{employee}, 1) \to q(down_2)\}$. The graph G_{t_e} in the proof of Lemma 11.5 is similar to the graph in Figure 11.3(a). There, it spans the tree t_e. To obtain the graph G_{t_e} we have to replace a vertex q, which is located at a node v of t_e, by (q, v). The mappings μ_v are illustrated in Figure 11.4(b). For each state q fixed at a node v, $\mu_v(q)$ is defined. If q has no outgoing edges then $\mu_v(q) = \emptyset$. Otherwise, it is the set of all direct successors of q in this figure. For example

$$\begin{aligned}
\mu_\epsilon &= \{q_I \mapsto \emptyset, q_{up} \mapsto \emptyset, q_{boss} \mapsto \emptyset\} \\
\mu_1 &= \{q \mapsto \{q_{up}, q_{boss}\}, q_{next} \mapsto \{q_{up}\}, q_{boss} \mapsto \{q_{boss}\}, q_{up} \mapsto \{q_{up}\}\} \\
\mu_{1.1} &= \{q_{data} \mapsto \emptyset, q_{sub} \mapsto \{q_{next}\}, q_{next} \mapsto \{q_{next}\}\} \\
\mu_{1.2} &= \{q \mapsto \{q_{up}\}\}
\end{aligned}$$

Note that $q_{up} \notin \mu_{1.1}(q_{next})$ although $(q_{up}, 1)$ is reachable from $(q_{next}, 1.1)$ in G_t. But the path contains $(q, 1.2)$ and 1.1 is not a prefix of 1.2. In order to illustrate a transition of bta $A_{M'_{\text{staff}}}$, consider, e.g., the transition $((\mu_1, 1), \texttt{employee}, \langle\mu_{1.1}, 1\rangle, \langle\mu_{1.2}, 2\rangle) \in \delta_A$ with the set T_1. All rules in T_1 agree in the input label **employee** and the direction 1 (condition 1). Also it contains a q_{boss}-rule for the call $q_{boss}(stay)$. For all states q_{sub}, q_{data} of occurring $down_1$-calls, the mapping $\mu_{1.1}$ is defined. Likewise, for q with a $down_2$-call, $q \in \text{dom}(\mu_{1.2})$. Thus, T_1 also conforms with condition 2. For condition 3, we verify that T_1 has rules both for q_{next} and q_{up}. For the last property, we construct the graph G. The set of vertices is

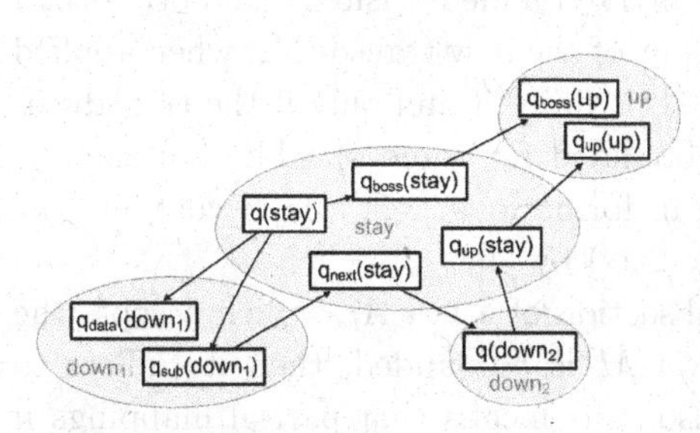

Fig. 11.5. Graph G for transition $((\mu_1, 1), \texttt{employee}, \langle\mu_{1.1}, 1\rangle, \langle\mu_{1.2}, 2\rangle)$.

$$\begin{aligned}
V = \{&q(stay), q_{boss}(stay), q_{next}(stay), q_{up}(stay), \\
&q_{data}(down_1), q_{sub}(down_1), q(down_2), \\
&q_{boss}(up), q_{up}(up)\}
\end{aligned}$$

The edges are illustrated in Figure 11.5. As the last condition requires, the graph G is acyclic, and we can read off the mapping μ_1.

Also we verify, that $\langle\mu_\epsilon, 0\rangle$ is an accepting state. According to Lemma 11.5, the resulting run ρ' is an accepting run of $A_{M'_{\text{staff}}}$ on t. ◁

11.5.3. *Efficient Subcases*

In the previous section, we have provided an algorithm for deciding emptiness of atwas and, thus, also of 2tts which runs in exponential time. This algorithm is indeed worst-case optimal. Not withstanding that, this algorithm allows us to identify subclasses of transducers where emptiness can be decided in polynomial time.

We call a transducer M *b-bounded*, if every accepting run ρ of M has at most b subruns starting at node v. We call M *strictly b-bounded* if every accepting run ρ of M visits each node v in the input tree at most b times, i.e., has at most b subrun *occurrences* starting at node v. The same definitions are also employed for atwas.

Note that the definition of b-boundedness does not exclude that the same node v is traversed arbitrarily often: if so, however, these traversals will be copies of at most b distinct traversals.

Note further that for a given transducer M it is decidable whether or not there exists a b such that M is b-bounded; the same holds for strict b-boundedness. To see this, add for each input symbol a new marked symbol of the same rank. We then consider input trees in which exactly one node is labeled by a marked symbol (this is a regular input tree language). Finally, we change the transducer M in such a way that it produces a specific output tree, for each subrun that starts at the marked input node (resp. for each time the marked node is visited), and other than that does not produce any output. The output of the new transducer, when applied to input trees with exactly one node marked, is finite if and only if the transducer is b-bounded for some b (resp. strictly b-bounded for some b). The finiteness is decidable for a very large class of tree transformations [34], which contains the pebble tree transducers (and hence also the 2tts) by [4].
Consider the construction from the last subsection of a bta A_M which accepts the same language as an atwa M. If the atwa M is b-bounded, then it suffices to consider sets T of atwa rules of size b. Also, this means that partial mappings μ need to be taken into account which are of the form: $B \to 2^{B'}$ for subsets B, B' of states of cardinalities at most b. Note that the number of subsets of size at most b of a set with n elements is bounded by $\frac{1}{b!}(n+1)^b$. Thus, the number of the partial mappings μ can be bounded by: $(n+1)^{2b}\frac{2^{b^2}}{b!}$ if n is the number of states of M. We will not provide an explicit estimation of the number of transitions of the bta since it crucially also depends on other parameters such as the number of rules of M which agree in input symbol and direction (which is typically small). We just note that for b-bounded M, the size of the bta A_M is polynomial in the size of M. The occurring exponent, though, is bounded by $\mathcal{O}(kb^2)$ where k is the maximal arity of an input symbol. Summarizing, we find that emptiness for b-bounded 2tts can be done in polynomial time. Note, however, that neither b-boundedness or strict b-boundedness is preserved by our construction to reduce the number of state calls in right-hand sides. For an efficient method for type-checking, we also require that the bound on the number of visits to every node in the input tree is preserved under the intersection construction. In this respect, we observe:

Lemma 11.6. *If M is strictly b-bounded and A is a bta, then M_A is also strictly b-bounded. If M is just b-bounded, this need not be the case.*

We thus obtain a polynomial-time algorithm for the class of strictly b-bounded 2tts where the number of occurrences of state calls in right-hand sides is also bounded.

11.5.4. *Conclusion*

In this section we presented techniques to type check 2tts against regular tree languages. Our approach is forward type inference. For that purpose, for a given 2tt M we build a second 2tt which produces only output trees in the complement of the output type, and otherwise realizes the same transformation as the original 2tt.

For a bta A describing the complement of the output type, the size of the new 2tt M_A is in $\mathcal{O}(|M| \cdot |A|^{d+1})$ where d is the maximal number of occurrences of states in right-hand sides in M; for binary trees and with Lemma 11.1 it is in $\mathcal{O}(|M| \cdot |A|^3)$.

For this intersection 2tt M_A we build an alternating tree walking automaton M', which imitates the behavior of M_A, but does not produce any output. This construction can be done in linear time. And then, in order to decide emptiness of the atwa M' and accordingly of M_A, we construct a nondeterministic bottom-up finite state tree automaton $A_{M'}$. In general, this construction is exponential in the size of M'. Hence emptiness of a 2tt can be decided in exponential time — a result which has already been known for a long time, see the notes at the end of Section 11.5.5. The general approach, however, allowed us to identify more efficient subclasses. These, we have discussed in Section 11.5.3. If M' is b-bounded, then emptiness can be decided in polynomial time where the exponent of the polynomial only depends on b^2 if the transducer is two-way, i.e., uses *up*-operations. A closer inspection of the construction of a bta from an atwa, though, reveals that the exponent can be reduced to b if the transducer is stay top-down, i.e., uses no *up*-operations (but possibly *stays*). The construction for the intersection, on the other hand, is polynomial in the sizes both of the 2tt and the bta — but may be exponential in the number of occurrences of state calls in right-hand sides. Also, if we start with a b-bounded 2tt M, the construction may not preserve b-boundedness. Instead, the bound on the number of visits to an input node may be increased as much as by a factor of the number of states of the bta. If the 2tt M is *strictly* b-bounded, this property will be retained and also the atwa M' is strictly b-bounded.

As a last step of verifying whether a 2tt type checks w.r.t. input and output types $\mathcal{L}(A_I)$ and $\mathcal{L}(A)$ (for a bta A_I), we construct a bta C with $\mathcal{L}(C) = L(A_{M'}) \cap \mathcal{L}(A_I)$ using the obvious product construction (see, e.g., Section 1.3 in [31]) such that $|C| \in \mathcal{O}(|A_{M'}| \cdot |A_I|)$. According to Theorem 11.2, we can test whether $\mathcal{L}(C) = \emptyset$ (which means that M type checks w.r.t. $\mathcal{L}(A_I)$ and $\mathcal{L}(A)$) in time linear in $|C|$.

Theorem 11.5. *Deciding whether a strictly b-bounded 2tt M type checks w.r.t. regular tree languages I and O, given by btas A_I and A_O, is polynomial in the size of M, A_I, and A_O, but exponential in $b^2 \cdot (d+1)$ where d is the maximal number of occurrences of state calls in right-hand sides. If M has no up-operations, the exponent can be improved to $b \cdot (d+1)$.*

11.5.5. *Notes and References*

Our definition of atwas is equivalent to the alternating two-way finite tree automaton of [35]. Note that alternating tree automata have recently been used in the context of a practical implementation of type checking for tree transducers [36].

The intersection of a 2tt with a given output type (Theorem 11.3) can be seen as a sequential composition of the 2tt with a translation in FTA; the latter is the class of partial identity mappings for regular tree languages. With this in mind, we can, for instance, obtain that top-down tree transducers allow a similar result as the one in Theorem 11.3: by Corollary 2(1) of [37], top-down tree transducers are closed under composition with linear and nondeleting top-down tree transducers; since FTA is included in the latter class, we obtain the desired result for top-down tree transducers. It is an interesting open problem whether a similar composition result holds for 2tts, i.e., whether 2tts are closed under composition with linear and nondeleting 2tts.

The notion of b-boundedness is similar to the notion of finite-copying in tree transducers, see, e.g., [38, 39]. Similar to the results of [40], it probably holds that b-bounded transformations are of linear size increase. A more static version of b-boundedness is the single-use restriction known for attribute grammars [41]. According to [39], it can probably be shown that total, deterministic, strictly b-bounded 2tts are equivalent to single-use restricted attribute grammars. In Section 5 of [42] a similar result as Theorem 11.5 has been shown for stay-macro tree transducers (cf. also the discussion in Section 11.6.4)

Engelfriet et al. show in [43] (Theorem 5), that for every 2tt M (TT in [43]) a regular tree grammar G can be constructed in exponential time such that G generates the domain of τ_M. They refer to the relationship between 2tt and attributed tree transducers explained in [4] and a result of [44] — giving the Theorem 11.4 above. The result of Theorem 11.4 has also been stated in Theorem A.2 of [45] with a proof sketch that uses a game theoretic interpretation of acceptance due to Muller and Schupp. The result also appears as Theorem 1 in [14], where a finite state automaton for the domain of a tree-walking transducer (twt) is constructed in exponential time; Theorem 2 of that paper states that inverse type inference is in k-fold exponential time, for k-fold compositions of twts.

11.6. Macro Tree Walking Transducers

In our running example we have considered the 2tt M_{staff} which lists the members of staff of a department. Although in general, several employees may have the same boss, the transducer spawns for every employee a separate computation to determine the corresponding boss. Conceptually as well as technically, it would be more convenient to determine the boss first, store it in some accumulating parameter and then propagate it to each of his employees. For this reason, we enhance tree transducers with accumulating parameters. A tree transducer with accumulating parameters is also called *macro tree transducer*.

Example 11.7. We omit the state q_{boss} and store the data of the boss in the first parameter y_b. The transformation of the next employee which is not a subordinate of the current, is then stored in the second parameter (y_n). By this construction, we completely omit the states q_{sub}, q_{next}, and q_{up}. The transducer consists of the following rules, for all $\eta \in \{1, 2\}$:

```
1 q_I(department, 0)             → staff(q(down_1, e, e), e)
2 q(employee, η, y_b, y_n)       → employee(data(q_data(down_1), y_b),
3                                    q(down_1, boss(q_data(down_1), e), q(down_2, y_b, y_n)))
4 q(e, η, y_b, y_n)              → y_n
5 q(data, 1, y_b, y_n)           → q(down_2, y_b, y_n),
6 q(subordinates, 2, y_b, y_n)   → q(down_1, y_b, y_n),
7 q_data(data, 1)                → copy(down_1).
```

where state *copy* is meant to copy the content of data (i.e., the left child in the binary representation). ◁

For the formal definition of macro tree walking transducer we assume that every state $q \in Q$ has a fixed rank, i.e., $Q = \biguplus_{n \in \mathbb{N}} Q^{(n)}$ where $Q^{(n)}$ is the set of all states with rank n.

Definition 11.10 (2mtt). *A macro tree walking transducer M (2mtt for short) is a tuple (Q, Σ, R, Q_0) where Q is a set of ranked states, Σ is a ranked alphabet, $Q_0 \subseteq Q^{(1)}$ is a set of initial states, and R is a finite set of rules of the form $q(\mathtt{a}, \eta, y_1, \ldots y_n) \to \zeta$, where $q \in Q^{(n+1)}$, $\mathtt{a} \in \Sigma^{(m)}$, $n, m \geq 0$, $\eta \geq 0$ is a direction and $y_1, \ldots, y_n$ are the accumulating parameters of q. Possible right-hand sides are described by the grammar*

$$\zeta ::= \mathtt{b}(\overbrace{\zeta, \ldots, \zeta}^{m' \ times}) \mid y_j \mid q'(op, \overbrace{\zeta, \ldots, \zeta}^{n' \ times}),$$

with $m', n' \geq 0$, $\mathtt{b} \in \Sigma^{(m')}$, $j \in \{1, \ldots, n\}$, $q' \in Q^{(n'+1)}$, and $op \in \{stay, up\} \cup \{down_\nu \mid 1 \leq \nu \leq m\}$.

In practice, states q may differ in their *rank*, i.e., the numbers of their accumulating parameters. Let $X = \{x_1, x_2, \ldots\}$ denote a countable set of variables of rank (not necessarily 0), and assume that Σ, X and Y are disjoint. For a right-hand side ζ, we write also $\zeta = s[q_1(op_1), \ldots, q_c(op_c)]$ to refer to all occurrences of (maybe nested) state calls in the right-hand side. Here, $s \in \mathcal{T}_{\Sigma \cup X}(Y)$ is a tree which contains each variable $x_1, \ldots, x_c$ exactly once with $\zeta = s[q_1(op_1)/x_1, \ldots, q_c(op_c)/x_c]$ where $s[q_i(op)/x_i]$ denotes the substitution of the state call $q_i(op, s_1, \ldots, s_n)$ for the subtree $x_i(s_1, \ldots, s_n)$ in s where n is the rank of x_i and $n+1$ is the rank of q_i. Note that in s no state call occurs anymore. For example the right-hand side of the rule in Lines 2 and 3 in the Example 11.7 can be written as $s[q_{data}(down_1), q(down_1), q_{data}(down_1), q(down_2)]$ where $s = \mathtt{employee(data(}x_1, y_b\mathtt{)}, x_2\mathtt{(boss(}x_3, \mathtt{e)}, x_4(y_b, y_n)))$.

Intuitively, the meaning of the expressions of a right-hand side is as follows: The produced output is defined analogously to the output of a 2tt up to the accumulating parameters. Here, we consider *call-by-value* parameter passing only. Thus, the expression ζ_j in parameter position j is evaluated first; then the result (which is a tree without state calls) may be copied to the various uses of the formal parameter y_j. This evaluation strategy is also called *inside-out* (IO for short). Note that we slightly abuse Definition 11.10 and use accumulating parameters with names other than $y_1, y_2, \ldots$ (e.g. in Example 11.7 where we use y_b and y_n). Clearly this is without loss of generality, as parameters can easily be renamed according to the definition.

Example 11.8. The rules in the beginning of this section with the initial state q_I form a 2mtt $M_{y,\mathsf{staff}}$. To transform the tree t'_B (the binary representation of our common example tree, see the left tree in Figure 11.2), the 2mtt starts at the root and applies the initial state. Thus, for the first step we get $\mathtt{staff}(q(down_1, \mathtt{e}, \mathtt{e}), \mathtt{e})$ where $down_1$ refers to the node 1. Applying the state q to this `employee`-node, we get several state calls. These state calls are partially nested. Figure 11.6 illustrates the two steps. In the upper picture is the first output with one state call. The lower figure shows the tree after processing $q(1, \mathtt{e}, \mathtt{e})$. There, we get the state call $q(down_1, \mathtt{boss}(q_{data}(down_1), \mathtt{e}), q(down_2, y_b, y_n))$ with nested calls. The first parameter accumulates a tree with root `boss`, whereas the second parameter is a further state call. ◁

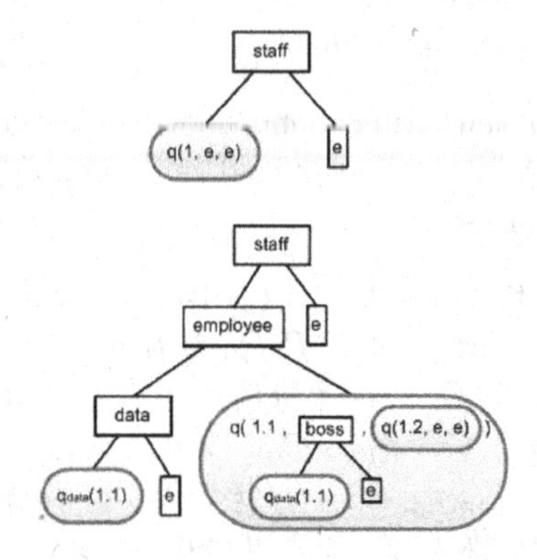

Fig. 11.6. Analyzing $q_I(\epsilon)$ and $q(1, e, e)$.

The order in which nested state calls are evaluated indeed matters. Consider, e.g., a transducer with rules $q_I(\mathtt{a}, 0) \to p(stay, q'(stay))$, $p(\mathtt{a}, 0, y) \to \mathtt{a}$, and $q'(\mathtt{b}, 0) \to \mathtt{b}$. If we evaluate the outermost calls first, the tree $t = \mathtt{a}(t_1, \ldots, t_k)$ will be transformed into $\mathtt{a}$. In this case, the accumulating parameter of p need not to be evaluated. If we start with the innermost calls, there is no rule to evaluate the state call $q'(stay)$ in the right-hand side of the first line. Thus, the output is empty.

We specify the translation induced by a 2mtt using a denotational formulation. Later, we will also consider an operational semantics based on runs. In the denotational semantics, the meaning $[\![q]\!]_t$ of state q of M with n accumulating parameters (with respect to an input tree t) is defined as a mapping from nodes in the input tree to sets of trees with parameters in $Y_n = \{y_1, \ldots, y_n\}$, i.e., $[\![q]\!]_t : \mathcal{N}(t) \to 2^{\mathcal{T}_\Sigma(Y)}$. When we evaluate an innermost call $q(v, s_1, \ldots, s_n)$ during a computation, it suffices to substitute actual parameters s_j for the formal parameters y_j of all terms from $[\![q]\!]_t(v)$ to obtain the set of produced outputs. The values $[\![q]\!]_t$ for all q are jointly defined as the least mappings satisfying: $[\![q]\!]_t(v) \supseteq [\![\zeta]\!]_t$ for rule $q(\mathtt{a}, \eta, \bar{y}) \to \zeta$ where $\bar{y}$ denotes the sequence $y_1, \ldots, y_n$ and v is a node of t with $lab[v] = \mathtt{a}$, $\eta(v) = \eta$ and $[\![\zeta]\!]_t$ is defined by:

$$
\begin{aligned}
[\![y_j]\!]_t &= \{y_j\} \\
[\![\mathtt{b}(\zeta_1, \ldots, \zeta_m)]\!]_t &= \{\mathtt{b}(s_1, \ldots, s_m) \mid s_\nu \in [\![\zeta_\nu]\!]_t\} \\
[\![q'(op, \zeta_1, \ldots, \zeta_{n'})]\!]_t &= \{s[s_1/y_1, \ldots, s_{n'}/y_{n'}] \mid s \in [\![q']\!]_t([\![op]\!]_t(v)), s_\nu \in [\![\zeta_\nu]\!]_t\}
\end{aligned}
$$

Again, op stands for $down_\nu$, $stay$ or up. Recall that the meaning $[\![op]\!]$ is defined by $[\![stay]\!]_t(v) = v$, $[\![down_\nu]\!]_t(v) = v\nu$, and $[\![up]\!]_t(v\nu) = v$. Also, $s[s_1/y_1, \ldots, s_n/y_n]$ denotes the simultaneous substitution of the trees s_j for all occurrences of the variables y_j in the tree s. Note that the call-by-value semantics is reflected in the last equation: the same trees s_j are used for all occurrences of a variable y_j in the tree s corresponding to a potential evaluation of the state q'. The transformation τ_M realized by the 2mtt M on an input tree t and sets T of input trees, respectively, is, thus, defined by $\tau_M(t) = \bigcup\{[\![q_0]\!]_t(\epsilon) \mid q_0 \in Q_0\}$ and $\tau_M(T) = \bigcup\{\tau_M(t) \mid t \in T\}$.

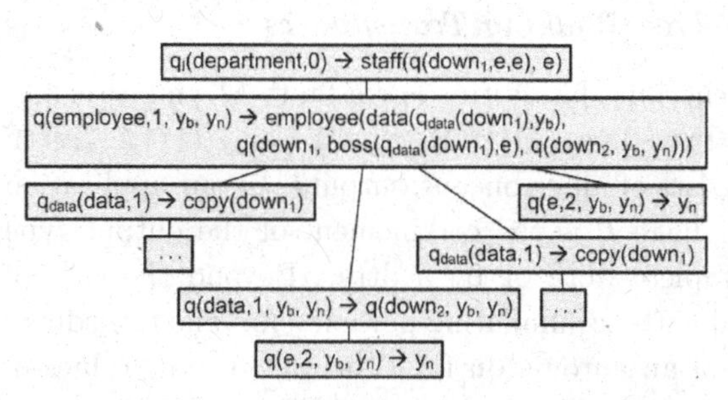

Fig. 11.7. A run of the 2mtt $M_{y,\mathsf{staff}}$ on the tree t_e.

For the operational semantics runs of a 2mtt M on a tree t may be similarly defined as for a 2tt. It is a ranked tree over the set of rules. Here, the rank of a rule $q(\mathsf{a}, \eta, y_1, \ldots, y_n) \to \zeta$ is given by the number of occurrences of recursive calls $q'(op)$ in ζ to states q' in Q. These calls may be nested. Figure 11.7 shows an accepting run ρ of the (deterministic) example 2mtt $M_{y,\mathsf{staff}}$ on the tree t_e, which describes a department with one employee.

The denotational view on the semantics of a 2mtt allows us to use fixpoint arguments for proving the correctness of constructions, whereas the operational view is better suited for more combinatorial arguments. In particular, we can show that the number of occurrences of states in right-hand sides can be restricted to the maximum of the ranks of output symbols and states. We have:

Lemma 11.7. *For every 2mtt M there exists a 2mtt M' with (i) $\tau_{M'} = \tau_M$, (ii) the number of states occurring in each right-hand side is bounded by k, and (iii) $|M'| \in \mathcal{O}(|M| \cdot k^2)$, where k is the maximum of the ranks of output symbols and states of M.*

Proof. The construction proceeds in two phases. In the first phase, we replace every *complicated* call $q'(op, \zeta_1, \ldots, \zeta_n)$ in the right-hand side of a rule $q(\mathsf{a}, \eta, y_1, \ldots, y_{n'}) \to \zeta$ by the simple call $\langle q, op, \zeta_1, \ldots, \zeta_n\rangle(stay, y_1, \ldots, y_{n'})$ for a new state $\langle q, op, \zeta_1, \ldots, \zeta_n\rangle$. Let $[\zeta]$ denote the resulting tree. For the new state $\langle q, op, \zeta_1, \ldots, \zeta_n\rangle$, we introduce the rule $\langle q, op, \zeta_1, \ldots, \zeta_n\rangle(\mathsf{a}, \eta, y_1, \ldots, y_{n'}) \to q(op, \langle\zeta_1, n'\rangle(stay, y_1, \ldots, y_{n'}), \ldots, \langle\zeta_n, n'\rangle(stay, y_1, \ldots, y_{n'}))$ for again fresh states $\langle\zeta_j, n'\rangle$ which are meant to produce the output of ζ_j using n' parameters. For these states, we introduce the rules: $\langle\zeta_j, n'\rangle(\mathsf{a}, \eta, y_1, \ldots, y_{n'}) \to [\zeta_j]$.

As a result of this first transformation phase, we achieve that all right-hand sides either are of the form $q(op, q_1(stay, y_1, \ldots, y_{n'}), \ldots, q_n(stay, y_1, \ldots, y_{n'}))$ or contain only non-nested calls, i.e., calls of the form $q(op, y_1, \ldots, y_{n'})$. In order to restrict the number of calls in right-hand sides of the second type, we essentially proceed as in the proof of Lemma 11.1, i.e., we introduce extra auxiliary states for every proper subtree of right-hand sides of the second kind which contain more than one call.

The resulting transducer has at most one fresh state for every node of a right-hand side while the total sum of sizes of right-hand sides may increase by a factor of k^2 in order to spell out all the auxiliary lists of parameters for the new states.□

11.6.1. *Type Checking Macro Tree Walking Transducers*

As for 2tts we now consider type checking for 2mtts. For a 2mtt M and a regular language $\mathcal{L}$, we again construct a transducer M' with $\tau_{M'}(t) = \tau_M(t) \cap \mathcal{L}$. As in Section 11.5.1, the language $\mathcal{L}$ consists of all erroneous outputs. In our application scenario of type checking, the language $\mathcal{L}$ is the complement of the output type which is either specified by a complete dbta or by a dtta. Beyond the case of 2tts, we now additionally must deal with accumulating parameters. The transducer M' must keep track of the states of an automaton for $\mathcal{L}$ on the current values of the respective parameters. We start with a general construction for deterministic bottom-up automata.

Theorem 11.6. *For every 2mtt M and every dbta A, a 2mtt M_A can be constructed with*

$$\tau_{M_A}(t) = \tau_M(t) \cap \mathcal{L}(A)$$

for all $t \in \mathcal{T}_\Sigma$. The 2mtt M_A is of size $\mathcal{O}(|M| \cdot |A|^{l \cdot (d+1)})$ where l is the maximal rank of a state in M and d is the maximal number of occurrences of states in right-hand sides of M.

Proof. Let $M = (Q, \Sigma, R, Q_0)$ and $A = (P, \Sigma, \delta, F)$. For each state q in $Q^{(n)}$ and all states $p_0, \ldots, p_n \in P$, the 2mtt M_A has a state $\langle q, p_0 p_1 \ldots p_n \rangle$ which is meant to generate all trees s (possibly with variables from $\{y_1, \ldots, y_n\}$) which could be produced by M and for which additionally there is a run of A starting at the leaves y_j with state p_j and reaching the root of s in state p_0. The rules of M_A are: $\langle q, p_0 p_1 \ldots p_n \rangle(\mathtt{a}, \eta, y_1, \ldots, y_n) \to \zeta'$ for every rule $q(\mathtt{a}, \eta, y_1, \ldots, y_n) \to \zeta$ of M and $\zeta' \in \tau^{p_0, p_1, \ldots, p_n}[\zeta]$ where the sets $\tau^{p_0, p_1, \ldots, p_n}[.]$ are inductively defined by:

$$\tau^{p_j, p_1, \ldots, p_n}[y_j] = \{y_j\}$$
$$\tau^{p_0, p_1, \ldots, p_n}[\mathtt{b}(\zeta_1, \ldots, \zeta_m)] = \{\mathtt{b}(\zeta_1', \ldots, \zeta_m') \mid \exists p_1', \ldots, p_m' \in P : (p_0, \mathtt{b}, p_1' \ldots p_m') \in \delta \wedge \forall j : \zeta_j' \in \tau^{p_j', p_1 \ldots p_n}[\zeta_j]\}$$
$$\tau^{p_0, p_1, \ldots, p_n}[q'(op, \zeta_1, \ldots, \zeta_{n'})] = \{\langle q', p_0 p_1' \ldots p_{n'}' \rangle(op, \zeta_1', \ldots, \zeta_{n'}') \mid \exists \zeta_1', \ldots, \zeta_m' \in Z : \forall j : \zeta_j' \in \tau^{p_j', p_1 \ldots p_n}[\zeta_j]\}$$

where Z denotes the set of all subterms of possible right-hand sides of rules of M_A. The set of initial states of M_A is given by $Q_0' = Q_0 \times F$. By fixpoint induction, we verify for every state q of rank $n+1$, every input tree $t \in \mathcal{T}_\Sigma$, every node $v \in \mathcal{N}(t)$ and states $p_0, \ldots, p_n$ of A that:

$$[\![\langle q, p_0, \ldots, p_n\rangle]\!]_t(v) = [\![q]\!]_t(v) \cap \{s \in \mathcal{T}_\Sigma(Y) \mid \delta^*(s, p_1 \ldots p_n) = p_0\} \qquad (*)$$

where $Y = \{y_1, \ldots, y_n\}$ and δ^* is the extension of the transition function of A to trees containing variables from Y, namely, for $\underline{p} = p_1 \ldots p_n$, $\delta^*(y_j, \underline{p}) = p_j$, and $\delta^*(\mathtt{a}(t_1, \ldots, t_m), \underline{p}) = \delta(\mathtt{a}, \delta^*(t_1, \underline{p}) \ldots \delta^*(t_m, \underline{p}))$. The correctness of the construction follows from $(*)$. For each state in M we have at most $|A|^{l+1}$ new states in M_A where l is the maximal rank of states in M. Assume that d is the maximal number of occurrences of states in right-hand sides of rules of M. Then each rule of M gives rise to at most $|A|^{(l+1)\cdot(d+1)}$ new rules in M_A. Therefore, the new 2mtt is of size $\mathcal{O}(|M| \cdot |A|^{(l+1)\cdot(d+1)})$. □

Lemma 11.2 provides a bta describing the complement of a dtta—which, however, is not necessarily deterministic. Theorem 11.6, on the other hand, only holds for deterministic btas. A similar construction is also possible if the bta A is nondeterministic — but then only for transducers which are output-*linear.* Here, we call a 2mtt output-linear if every accumulating parameter occurs at most once in a right-hand side. Nonetheless, we are able to handle complements of output types described by dttas directly. For that, however, we introduce a dedicated construction of an 2mtt M_A.

Theorem 11.7. *For every 2mtt M and every dtta A, a 2mtt $M_{\bar{A}}$ can be constructed with*

$$\tau_{M_{\bar{A}}}(t) = \tau_M(t) \cap \overline{\mathcal{L}(A)}$$

for all $t \in \mathcal{T}_\Sigma$. The 2mtt $M_{\bar{A}}$ is of size $\mathcal{O}(|M| \cdot (h \cdot |A|)^{d+2})$ where $h+1$ is the maximum of the maximal rank of a state in M and the maximal rank of an output symbol, and d is the maximal number of occurrences of state calls in right-hand sides of M.

Proof. Let $M = (Q, \Sigma, R, Q_0)$ and $A = (P, \Sigma, \delta, p_0)$. Our goal is to construct a 2mtt $M_{\bar{A}}$ which simulates the behavior of M while at the same time guessing a path in the output tree which proves non-containment in the set $\mathcal{L}(A)$. For that, the set Q' is defined as: $Q' = Q \cup \{\langle q, p\rangle \mid q \in Q, p \in P\} \cup \{\langle q, p, j, p'\rangle \mid q \in Q^{(n)}, p, p' \in P, j \in \{1, \ldots, n\}\}$. Here, a state $q \in Q$ of $M_{\bar{A}}$ behaves like the state q of M. States $\langle q, p\rangle$ or $\langle q, p, j, p'\rangle$ behave like q in M but additionally make sure that there is a path in the generated output starting from a state p of A at the root which verifies that there is no p-run of A on the output. Thereby, a state $\langle q, p\rangle$ will directly generate the end point of such a path whereas state $\langle q, p, j, p'\rangle$ will only generate a path with p at the root reaching a parameter y_j with state p'. Accordingly, the 2mtt $M_{\bar{A}}$ has the following rules:

$$q(\mathsf{a}, \eta, y_1, \ldots, y_n) \to \zeta$$
$$\langle q, p\rangle(\mathsf{a}, \eta, y_1, \ldots, y_n) \to \zeta' \quad \text{with } \zeta' \in [\zeta]^p$$
$$\langle q, p, j, p'\rangle(\mathsf{a}, \eta, y_1, \ldots, y_n) \to \zeta' \quad \text{with } \zeta' \in [\zeta]^{p,j,p'}$$

for every rule $q(\mathsf{a}, \eta, y_1, \ldots, y_n) \to \zeta$ of M. The sets $[.]^x$ are inductively defined by:

$$\begin{array}{lcl}
[y_i]^p & = & \emptyset \\
[\mathsf{b}(\zeta_1, \ldots, \zeta_m)]^p & = & \{\mathsf{b}(\zeta_1, \ldots, \zeta_m) \mid \forall \bar{p} \in P^m : (p, \mathsf{b}, \bar{p}) \notin \delta\} \\
& \cup & \{\mathsf{b}(\zeta_1, \ldots, \zeta_{\nu-1}, \zeta'_\nu, \zeta_{\nu+1}, \ldots, \zeta_m) \mid \nu \geq 1, (p, \mathsf{b}, p_1 \ldots p_m) \in \delta,\ \zeta'_\nu \in [\zeta_\nu]^{p_\nu}\} \\
[q(op, \zeta_1, \ldots, \zeta_n)]^p & = & \{\langle q, p, \nu, p_\nu\rangle(op, \zeta_1, \ldots, \zeta_{\nu-1}, \zeta'_\nu, \zeta_{\nu+1}, \ldots, \zeta_n) \mid \zeta'_\nu \in [\zeta_\nu]^{p_\nu}\} \\
& \cup & \{\langle q, p\rangle(op, \zeta_1, \ldots, \zeta_n)\}
\end{array}$$

$$\begin{array}{lcl}
[y_j]^{p,j,p'} & = & \{y_j \mid p = p'\} \\
[\mathsf{b}(\zeta_1, \ldots, \zeta_m)]^{p,j,p'} & = & \{\mathsf{b}(\zeta_1, \ldots, \zeta_{\nu-1}, \zeta'_\nu, \zeta_{\nu+1}, \ldots, \zeta_m) \mid \nu \geq 1, (p, \mathsf{b}, p_1 \ldots p_m) \in \delta, \\
& & \qquad \zeta'_\nu \in [\zeta_\nu]^{p_\nu,j,p'}\} \\
[q(op, \zeta_1, \ldots, \zeta_n)]^{p,j,p'} & = & \{\langle q, p, \nu, p_\nu\rangle(op, \zeta_1, \ldots, \zeta_{\nu-1}, \zeta'_\nu, \zeta_{\nu+1}, \ldots, \zeta_n) \mid \zeta'_\nu \in [\zeta_\nu]^{p_\nu,j,p'}\}
\end{array}$$

First, we verify that for every tree $s \in \mathcal{T}_\Sigma(Y)$ the sets $[s]^p$ and $[s]^{p,j',p'}$ either are empty or equal $\{s\}$ where following holds:

(1) $[s]^p = \{s\}$ iff s contains a node $v = i_1 \ldots i_r$ such that there are transitions $(p_0^{(j)}, \mathsf{a}_j, p_1^{(j)}, \ldots, p_{m_j}^{(j)}) \in \delta$ for $j = 1, \ldots, r-1$, such that

   (a) The label of the node $i_1 \ldots i_j$ equals a_j;
   (b) $p_0^{(1)} = p$ and for $j = 1, \ldots, r-2$, $p_0^{(j+1)} = p_{i_j}^{(j)}$;
   (c) there is no $p_{i_r}^{(r-1)}$-transition of A for a_r.

(2) $s \in [s]^{p,j',p'}$ iff s contains a node $v = i_1 \ldots i_{r+1}$ which is labeled with $y_{j'}$ and there are transitions $(p_0^{(j)}, \mathsf{a}_j, p_1^{(j)}, \ldots, p_{m_j}^{(j)}) \in \delta$ for $j = 1, \ldots, r$, such that:

   (a) For $j = 1, \ldots, r$, the label of the node $i_1 \ldots i_j$ equals a_j;
   (b) $p_0^{(1)} = p$ and $p_{i_{r+1}}^{(r)} = p'$; and for $j = 1, \ldots, r-1$, $p_0^{(j+1)} = p_{i_j}^{(j)}$.

Note in particular that by this definition, $s \notin \mathcal{L}(A)$ iff $[s]^{p_0} = \{s\}$ for the initial state p_0 of A. Let us extend the operators $[\,.\,]^p$ and $[\,.\,]^{p,j',p'}$ by:

$$[S]^p = \bigcup\{[s]^p \mid s \in S\} \qquad [S]^{p,j',p'} = \bigcup\{[s]^{p,j',p'} \mid s \in S\}$$

for $S \subseteq \mathcal{T}_\Sigma(Y)$ with $Y = \{y_1, \ldots, y_n\}$. By fixpoint induction, we verify for every state q of rank $n+1$, every $j \in \{1, \ldots, n\}$, every input tree $t \in \mathcal{T}_\Sigma$ and states $p, p' \in P$ that:

$$[\![\langle q, p, j, p'\rangle]\!](t) = [\,[\![q]\!](t)\,]^{p,j,p'}$$
$$[\![\langle q, p\rangle]\!](t) = [\,[\![q]\!](t)\,]^p$$

For each state p and right-hand side ζ of a rule, we assign states of the dtta to the nodes of the tree. Either we immediately hit a node certifying the non-existence of a p-run of the dtta on the output generated from ζ, or we hit an occurrence of a state call $q(op, \ldots)$. If $n \geq 1$ is the rank of q, we have n choices here: either we expect a certificate for the failure of the dtta A inside the evaluation of q or in one of the parameters of q. Overall, we find that every rule of M, thus, gives rise to $(h \cdot |A|)^{d+2}$ rules. Thus, we have $|M_{\bar{A}}| \in \mathcal{O}(|M| \cdot (h \cdot |A|)^{d+2})$.

□

Assume that we have a binary ranked alphabet and that at least one state has an accumulating parameter, i.e., the maximal rank l of states is at least 2. By Lemma 11.7 it is then possible to restrict the number of occurrences of state calls in right-hand sides of the 2mtt to l. This implies that the size of the intersection 2mtt M_A in Theorem 11.6 for a dbta describing the output language is in $\mathcal{O}(|M| \cdot |A|^{l \cdot (l+1)})$. Furthermore, the size of the 2mtt M_A in Theorem 11.7 for a deterministic top-down tree automaton describing the output language is in $\mathcal{O}(|M| \cdot (l \cdot |A|)^{l+1})$.

11.6.2. *Deciding Emptiness of 2MTTs*

To decide emptiness of a 2mtt M, we follow the approach taken for 2tts: we construct an alternating tree walking automaton A_M which is then tested for emptiness. The atwa A_M has the same set of states as M (but they are not ranked anymore now), where the initial states of M and A_M coincide. For every rule $q(\mathtt{a}, \eta, y_1, \ldots, y_n) \to \zeta$ of M, the atwa A_M has a rule $q(\mathtt{a}, \eta) \to q_1(op_1) \wedge \ldots \wedge q_c(op_c)$ if $q_1(op_1, \ldots), \ldots, q_c(op_c, \ldots)$ is the sequence of calls to states of M (possibly nested inside each other), in any order. Since we use 2mtts with call-by-value semantics, M has an accepting run on some input tree t iff A_M has an accepting run on t. Note that this construction is wrong for call-by-need semantics, because M could have an accepting run on a tree $t \notin \mathcal{L}(A_M)$; for instance the 2mtt with the rules $q_I(\mathtt{a}, 0) \to q(stay, q'(stay))$, $\quad q(\mathtt{a}, 0, y) \to \mathtt{a}$ on the tree $\mathtt{a}$.

Theorem 11.8. *For every 2mtt M an atwa A_M can be constructed in polynomial time such that $\mathcal{L}(A_M) = dom(M)$. Thus, it can be decided in deterministic exponential time whether the translation of a 2mtt is empty or not.*

11.6.3. *Input-Linear 2MTTs*

The notions of b-boundedness and strict b-boundedness which we have defined for 2tts stay meaningful also in presence of accumulating parameters. Analogously, we find that emptiness for b-bounded transducers is decidable in polynomial time — independent on the number of accumulating parameters of states.

In order to identify classes of 2mtts where full type-checking is tractable, we therefore take a closer look at the construction for the intersection of 2mtts with (complements of) output types. For simplicity, we first consider 2mtts which are *strictly 1-bounded.* This notion is only meaningful for *top-down* mtts, i.e., 2mtts without operations *up* or *stay*. A top-down transducer M is guaranteed to visit each node of the input tree at most once, if for the same i, the operation $down_i$ does not occur twice in the same right-hand side of M. This property can easily be checked syntactically. Tree transducers satisfying this restriction are called *input-linear.*

Note that input-linearity for a tree transducer implies that the number of state calls in right-hand sides is bounded by the maximal rank of input symbols. Moreover, the output language can be described by rules that are obtained by simply deleting all directives from the transducer's rules. The resulting rules no longer specify a transformation but constitute a *context-free tree grammar* (short: cftg) for generating output trees. As an example of an input-linear mtt consider the following mtt which produces the same output as the common example $M_{y,\mathtt{staff}}$ without the `boss`-subtrees. The transducer only needs one parameter (for the next employee) and has a new state q_{empl}. For $\eta \in \{1, 2\}$, it has the rules:

$$
\begin{array}{lll}
1 & q_I(\mathtt{department}, 0) & \rightarrow \mathtt{staff}(q(down_1, \mathtt{e}), \mathtt{e}) \\
2 & q(\mathtt{employee}, \eta, y_n) & \rightarrow q_{empl}(down_1, q(down_2, y_n)) \\
3 & q_{empl}(\mathtt{data}, 1, y_n) & \rightarrow \mathtt{employee}(\mathtt{data}(q_{data}(down_1), \mathtt{e}), q(down_2, y_n)) \\
4 & q(\mathtt{subordinates}, 2, y_n) & \rightarrow q(down_1, y_n) \\
5 & q(\mathtt{e}, \eta, y_n) & \rightarrow y_n \\
6 & q_{data}(\mathtt{data}, 1) & \rightarrow copy(down_1).
\end{array}
$$

The grammar characterizing its output language looks as follows:

$$
\begin{array}{lll}
1 & q_I & \rightarrow \mathtt{staff}(q(\mathtt{e}), \mathtt{e}) \\
2 & q(y_n) & \rightarrow q_{empl}(q(y_n)) \mid q(y_n) \mid y_n \\
3 & q_{empl}(y_n) & \rightarrow \mathtt{employee}(\mathtt{data}(q_{data}, \mathtt{e}), q(y_n)) \\
4 & q_{data} & \rightarrow copy
\end{array}
$$

where q_I, q, q_{empl}, q_{data}, $copy$ are nonterminals. Selection of rules depending on input symbols and directions now has been replaced with nondeterministic choice.

Context-free tree grammars generalize context-free grammars to trees. Formally, a cftg G can be represented by a tuple (E, Σ, P, E_0) where E is a finite ranked set of function symbols or nonterminals, $E_0 \subseteq E$ is a set of initial symbols of rank 0, Σ is the ranked alphabet of terminal nodes and P is a set of rules of the form $q(y_1, \ldots, y_n) \rightarrow \zeta$ where $q \in E$ is a nonterminal of rank $n \geq 0$. The right-hand side ζ is a tree built up from variables $y_1, \ldots, y_n$ by means of application of nonterminal and terminal symbols. As for 2mtts, inside-out (IO) and outside-in evaluation order for nonterminal symbols must be distinguished. Here, we use the IO or call-by-value evaluation order. The least fixpoint semantics for the cftg G is obtained straightforwardly along the lines for 2mtts — simply by removing the corresponding directive components, i.e., by removing in the last line of the definition of $[\![\zeta]\!]$ for 2mtts the op and $[\![op]\!]_t(v)$. In particular, this semantics assigns to every nonterminal q of rank $n \geq 0$, a set $[\![q]\!] \subseteq \mathcal{T}_\Sigma(Y)$ for $Y = \{y_1, \ldots, y_n\}$. The language generated by G is $\mathcal{L}(G) = \bigcup\{[\![q_0]\!] \mid q_0 \in E_0\}$.

It is easy to see that the output language of an input-linear mtt M can be characterized by a cftg G_M which can be constructed from M in linear time. During this construction every rule $q(\mathtt{a}, \eta, y_1, \ldots, y_n) \rightarrow \zeta$ is rewritten as a production

$q(y_1, \ldots, y_n) \to \zeta'$, where ζ' is obtained from ζ by deleting all occurrences of navigation operators.

The characterization of output languages for input-linear mtts by cftgs is useful because emptiness for (IO-)cftgs is decidable using a similar algorithm as the one for ordinary context-free (word) grammars, and hence can be done in linear time.

Theorem 11.9. *It can be decided in linear time for a cftg G whether or not $\mathcal{L}(G) = \emptyset$.*

Here, we are interested in testing whether a given input-linear mtt M type checks w.r.t. input and output types I and O. Assume that I is given by a (possibly nondeterministic) bta B with only productive states, i.e., for every state p of B there exists a p-run of B on a tree. As a first step, we construct a new input-linear mtt M_B such that M_B's range, i.e., the set $\tau_{M_B}(\mathcal{T}_\Sigma)$ is equal to $\tau_M(I)$. This is done by a straightforward product construction of the bta B and the input-linear mtt M. Note that it may happen that M does not visit a certain subtree t of the input tree. In such a case the checking of t w.r.t. B cannot be done by the new transducer M_B. This does not affect the corresponding output language though. We now construct the intersection 2mtt for M_B and the complement of the output type O. In case, O is given by a dbta, this can be done along the lines of the proof of Theorem 11.6. If O is given by a dtta, we rely on the construction from Theorem 11.7. Since M is input-linear, the intersection 2mtt is again input-linear—meaning that its range can be described by a cftg (thus generating all "illegal outputs" of M w.r.t. I and O). Therefore, Theorem 11.9 gives us:

Theorem 11.10. *Assume M is an input-linear mtt where the ranks of input symbols are bounded, and let I and O denote input and output types for M where I is given by a bta.*

(1) Assume that the output type O is specified with a dbta and the maximal rank of states of M is bounded. Then M can be type checked relative to I and O in polynomial time.
(2) Assume that the output type O is specified with a dtta. Then M can be type checked relative to I and O in polynomial time — even in presence of unbounded ranks of states.

The worst-case complexity bounds for the construction of Theorem 11.10 are exponential in $l \cdot (k+1)$ (for output types given through dbtas) or $k+2$ (for output types given through dttas) where l is the maximal rank of states and k is the maximal rank of an input symbol of M. In practical applications, both k and l may be moderately small. Still, we want to point out that in case of input-linear mtts, the intersection construction can be organized in such a way that only "useful" states are constructed. In order to see this, consider again an input-linear mtt M and a dbta A (representing the incorrect output trees). The idea is to introduce for every

q of M of rank $n+1$, a *Datalog* predicate $q/n+1$. Every rule $q(\mathsf{a}, \eta, y_1, \ldots, y_n) \to \zeta$ of M then gives rise to the *Datalog* implication: $q(Y_0, \ldots, Y_n) \Leftarrow \mathcal{D}[\zeta]_{Y_0}$ where $\mathcal{D}[\zeta]_X$ (X is a variable) is defined by

$$\begin{array}{ll} \mathcal{D}[y_j]_X & = X = Y_j \\ \mathcal{D}[\mathsf{b}(\zeta_1, \ldots, \zeta_m)]_X & = \delta(X, \mathsf{b}, X_1 \ldots X_m) \ \wedge\ \mathcal{D}[\zeta_1]_{X_1} \wedge \ldots \wedge \mathcal{D}[\zeta_m]_{X_m} \\ \mathcal{D}[q'(\zeta_1, \ldots, \zeta_m)]_X & = q'(X, X_1, \ldots, X_m) \ \wedge\ \mathcal{D}[\zeta_1]_{X_1} \wedge \ldots \wedge \mathcal{D}[\zeta_m]_{X_m} \end{array}$$

and the variables $X_1, \ldots, X_m$ in the last two rows are fresh. For subsets $X, X_1, \ldots, X_{m'}$ of the set of states of A, $\delta(X, \mathsf{a}, X_1 \ldots X_{m'})$ denotes the fact that $(x, \mathsf{a}, x_1 \ldots x_{m'}) \in \delta_A$ for all $x \in X$ and $x_j \in X_j$, $j = 1, \ldots, m'$. A bottom-up evaluation of the resulting *Datalog* program computes for every $q/(n+1)$, the set of all tuples $(p_0, \ldots, p_n)$ such that the translation of $\langle q, p_0 \ldots p_n \rangle$ is non-empty. If we additionally want to restrict these predicates only to tuples which may contribute to a terminal derivation of initial nonterminals $\langle q_0, p_f \rangle$, we may top-down query the program with queries $\Leftarrow q_0(p_f)$. Practically, top-down solving organizes the construction such that only useful nonterminals of the intersection grammar are considered. Using this approach, the number of newly constructed nonterminals often will be much smaller than the bounds stated in the theorem. A similar construction is also possible for the intersection of mtts with the complements of dtta languages.

The algorithm for *input-linear* mtts can also be applied to *non-input-linear* 2mtts. Then, the constructed *Datalog* program does no longer precisely characterize the non-empty functions of the intersection 2mtt because dependencies on input subtrees (viz. several transformations of the same input node) have been lost. Accordingly, a *superset* is returned. By means of cftgs, we can express this observation as follows:

Theorem 11.11. *Let G_M be the cftg constructed for a 2mtt M. Then $\tau_M(\mathcal{T}_\Sigma) \subseteq \mathcal{L}(G_M)$.*

Since the cftg still provides a safe *superset* of produced outputs, type checking based on cftgs is sound in the sense that if it does not flag an error, the transformation also will not go wrong. On the other hand, a flagged error may be possibly spurious, i.e., due to the over-approximation of the output language through the cftg.

Consider a top-down transducer M with rule $q_0(\mathsf{a}, 0) \to \mathsf{c}(p(down_1), p(down_1))$, where p realizes the identity using the rules $p(\mathsf{a}, \eta) \to \mathsf{a}(p(down_1))$, $p(\mathsf{b}, \eta) \to \mathsf{b}(p(down_1))$, $p(\mathsf{e}, \eta) \to \mathsf{e}$. In this case, the corresponding approximating cftg G_M is rather coarse: it generates $\mathsf{c}(u, v)$ with $u, v \in \{\mathsf{a}, \mathsf{b}\}^*\mathsf{e}$ (seen as monadic trees). Exact tree copying, however, can be realized through the use of parameters: the transducer with rules $q_0(\mathsf{a}, 0) \to q(down_1, p(down_1))$ and $q(\sigma, \eta, y_1) \to \mathsf{c}(y_1, y_1)$ (for all $\sigma \in \Sigma$) and the same p-rules as M will realize the same translation as M. The cftg for the resulting transducer now does not provide an over-approximation but precisely captures the output language of M.

When approximating the output languages of general 2mtts with cftgs, we no longer can assume that the maximal number d of occurrences of nonterminals in a right-hand side of this grammar is bounded by a small constant. If d turns out to be unacceptably large, we still can apply Lemma 11.7 to limit the maximal number of occurrences of nonterminals in each right-hand side to a number k which is the maximal rank of output symbols and states. This construction, however, introduces *stay*-moves and thus destroys input-linearity.

11.6.4. *Notes and References*

Macro tree transducers [46] are a combination of top-down tree transducers and macro grammars [47]. Macro grammars are just like context-free tree grammars (cftgs), but produce strings (a cftg can be seen as a special macro grammar, because terms are particular strings). Fischer already distinguishes IO and OI for macro grammars, and proves that the corresponding classes of languages are incomparable (which also holds in the tree case). Our normal form of Lemma 11.7 can be seen as a variant of Fischer's IO standard form, which in fact is very similar to Chomsky normal form of context-free grammars: there are exactly 2 or 0 nonterminals in every right-hand side of a grammar in IO standard form. A similar normal form might be possible for 2mtts too, but will cause a larger size increase of the transducer (note that Fischer does not report on grammar sizes, in his constructions). Fischer remarks, just after Corollary 3.1.6, that emptiness of IO macro languages can be reduced to emptiness of context-free languages, by simply dropping all parentheses, commas, argument, and terminal symbols. The resulting context-free (word) grammar generates the empty word if and only if the original language is empty. Since emptiness for a context-free grammar can be decided in linear time (see, e.g., [48]), we obtain a linear time procedure for checking emptiness of IO context-free tree languages, as stated in Theorem 11.9. Context-free tree grammars were considered in [18] and extensively studied in [49].

The fact that output languages of input-linear mtts are IO context-free tree languages is mentioned in Corollary 5.7 of [46] (the class of input-linear mtts is called LMT_{IO} there). A 2mtt without up-moves is called "stay-mtt". Results similar as the ones obtained in this section for 2mtts, have been obtained already in [42] for the restricted case of stay-mtts. For instance, Proposition 3 of that paper is similar to our Theorem 11.7, Theorem 2 of [42] corresponds to our Theorem 11.6, and Theorem 5 of that paper corresponds to our Theorem 11.10. Just before Theorem 11.10, we describe how to incorporate an input type into an input-linear transducer, so that the corresponding output language is preserved. The technical details are exactly as in the proof of Theorem 3.2.1 of [38], where this results was proved for top-down tree transducers. 2mtts are essentially the same as the k-pebble macro tree transducers (k-pmtt) of [4] for the case $k = 0$. For k-pmtts, a normal form similar to our Lemma 11.7 was shown in Theorem 16 of [4].

11.7. Macro Forest Walking Transducers

Conceptually, XML documents are not trees, but forests. Therefore, we extend the concept of tree walking transducers (without or with parameters) to a transformation formalisms of forests. Forests are introduced in Definition 11.1.

Example 11.9. We consider again a transformation from company structures (cf. Section 11.2) to collections of employees which are listed under a new root node labeled `staff` (cf. Section 11.3). In contrast to tree walking transducers, forest transducers do not depend on a ranked alphabet. They can deal with arbitrary many subtrees of nodes. Here, we define a transformation which returns trees of the form $\texttt{staff}\langle f\rangle$ where f is a forest composed of trees of the form $\texttt{employee}(\texttt{data}(\ldots),\texttt{boss}(\ldots))$. The input trees are described by the DTD in Section 11.4. Additionally to the operations *up*, *stay* as in tree walking transducers, forest transducers may use a directive *down* for proceeding to the first child as well as directives *left* and *right* for proceeding to the left or right sibling, respectively.

1	$q_I(\texttt{department}, 0)$	$\to \texttt{staff}\langle q(down, \epsilon)\rangle$
2	$q(\texttt{employee}, \eta, y_b)$	$\to \texttt{employee}\langle\texttt{data}\langle q_{data}(down)\rangle\ y_b\rangle$
3		$q(down, \texttt{boss}\langle q_{data}(down)\rangle)$
4		$q(right, y_b)$
5	$q(\texttt{data}, 2, y_b)$	$\to q(right, y_b)$
6	$q(\texttt{subordinates}, 3, y_b)$	$\to q(down, y_b)$
7	$q(\epsilon, \eta, y_b)$	$\to \epsilon$
8	$q_{data}(\texttt{data}, 2)$	$\to copy(down)$

where state *copy* in line 8 is meant to copy the forest f of a subtree $\texttt{data}\langle f\rangle$. The right-hand side of the second rule is a composition of three forests (line 2-4). The initial state is q_I, which means that we start with state q_I at the root of the first tree of a forest. Here, the transducer walks only the first tree of an input forest. Note also that now rules may be selected depending on the current label of a node in the forest together with its forest direction. Thus, it can check whether the current node is the leftmost node on the top-level (value 0), is on the top-level, but not leftmost (value 1), is leftmost but not on the top-level (value 2) or is neither leftmost nor on the top-level (value 3). ◁

Definition 11.11 (2mft). *A* macro forest walking transducer *(*2mft *for short) is a tuple* $M = (Q, \Sigma, Q_0, R)$, *where* Q *is a finite ranked set of states,* Σ *is a finite alphabet with* $Q \cap \Sigma = \emptyset$, $Q_0 \subseteq Q^{(1)}$ *is the set of initial states and* R *is a finite set of rules of the form:* $q(\epsilon, \eta, y_1, \ldots, y_n) \to \zeta$ *or* $q(\texttt{a}, \eta, y_1, \ldots, y_n) \to \zeta$ *with* $\texttt{a} \in \Sigma$, *direction* $\eta \in \{0, \ldots, 3\}$ *and* $q \in Q^{(n+1)}$ *where the right-hand sides are forests* ζ *of the following form:* $\zeta ::= \epsilon \mid y_j \mid q(op, \zeta_1, \ldots, \zeta_{n'}) \mid \texttt{b}\langle\zeta_1\rangle \mid \zeta_1\,\zeta_2$ *where* $q \in Q^{(n'+1)}$, $\texttt{b} \in \Sigma$, $op \in \{up, stay, down, left, right\}$ *and* $j = 1, \ldots, n$. *Moreover, the right-hand sides for empty input forests* ϵ *must not contain occurrences of the operations down, right or left.*

In case of several rules for the same q, the same direction η and the same symbol a (or ϵ), we also write: $\zeta_1 \mid \ldots \mid \zeta_k$ to list all occurring right-hand sides. In case, that no operation *up* is used, the 2mft is also called *top-down* (short: *1mft* or *mft*). Likewise, if all states are of rank 1, i.e., have no accumulating parameters, the 2mft is an (ordinary) *forest walking* transducer (short: *2ft*). Finally, a 1mft without parameters is also called *forest transducer* (short: *1ft* or *ft*). As for macro tree walking transducers, in practice, states q may differ in their *ranks*, i.e., the numbers of their accumulating parameters plus 1. The set R of rules in Example 11.9 constitute the transducer $M_{y,\mathsf{staff},f} = (Q, \Sigma, Q_0, R)$ with $Q = \{q_I, q, q_{data}, copy\}$ and $Q_0 = \{q_I\}$, which happens to be a 1mft.

A forest transducer behaves similar to a corresponding tree transducer: while walking over the input forest, the transducer chooses rules corresponding to the current states, input symbols and directions at the respective current nodes in the input, and then evaluates the right-hand sides of the rules. Again, we just consider the *inside-out* (IO or call-by-value) strategy for evaluating parameters. There are two significant differences between forest walking and tree walking transducers: First, a forest walking transducer produces output *forests* and therefore, as an extra operation, also supports *concatenation* of output forests. Secondly, the forest transducer has a different set of directions as well as a different set of navigational directives: *up* now means that the transducer moves to its ancestor in the input forest. *left* and *right* now means that the transducer moves to its left or right sibling, whereas *down* means that the transducer moves to its first child. Formally, the semantics of these operations is defined by $[\![up]\!](vi) = v$, $[\![left]\!](vi) = v(i-1)$ if $i > 0$, $[\![right]\!](vi) = v(i+1)$, and $[\![down]\!](v) = v0$. Note that only the operations *down* and *right* have immediate equivalents in commands of a transducer on the *first-child next-sibling* encoding of forests as binary trees where they correspond to the commands $down_1$ and $down_2$, respectively. The *up*-command on the tree, on the other hand, may correspond to the forest commands *left* or *up* — depending on whether the current node is a right or left child. The class of all (forest-walking) macro forest transformations is denoted by (2FMAC) FMAC. Analogously to Lemma 11.7, we find:

Lemma 11.8. *For every 2mft M there exists a 2mft M' with (i) $\tau_{M'} = \tau_M$ (ii) there are at most l occurrences of states on a right-hand side of rules in M' and (iii) $|M'| \in \mathcal{O}(|M| \cdot l^2)$, where l is the maximum of 2 and the maximal rank of states of M.*

Note that it is unfortunately *not* possible to simulate 2mfts by 2mtts which work on (possibly enriched) first-child next-sibling encodings of input and output trees. To see this, consider first the 1mft case. As shown in [50], one can easily construct a 1mft which takes as input a binary tree with m nodes, and outputs a forest consisting of 2^m leaves, i.e., a string of length 2^m. Consider the height increase of the corresponding translation on encoded trees: it is *double*-exponential. However, the height-increase of mtts is at most exponential (see [46]). Now consider the 2mft case. Clearly, a 2mtt can translate a binary tree with m nodes into a monadic tree with 2^m nodes, by doing a depth-first left-to-right traversal, and at each step generating a duplicated state call in a parameter position. As intermediate sentential form the transducer generates $q(\varepsilon, q(\varepsilon, \ldots, q(\varepsilon, \mathtt{e})) \ldots)$ which has 2^m-many occurrences of q; it then replaces $q(\varepsilon, t)$ by $\mathtt{g}(t)$, where $\mathtt{g}$ is an output symbol of rank 1. A 2mft can generate the same sentential form, but can replace $q(\varepsilon, t)$ by tt, i.e., the forest of concatenating two copies of t. In this way, a forest consisting of 2^{2^m}-many $\mathtt{e}$'s is generated. On first-child next-sibling encodings, this corresponds to a tree of height 2^{2^m}. Thus, the translation on encodings has *double*-exponential size-to-height increase. However, it is not difficult to see that the size-to-height increase of 2mtts is at most exponential.

11.7.1. *Intersecting Forest Walking Transducers with Output Types*

In this section, we consider general techniques for intersecting forest walking transducers with output types. Assume that we are given a regular forest language $\mathcal{L}$. Our goal is to construct for a given 2mft M, another 2mft M' which behaves similar to M but produces only outputs in $\mathcal{L}$. If $\mathcal{L}$ describes the set of all invalid outputs, type-checking for M, thus, reduces to checking emptiness of the transformation M'.

In order to provide a general construction for regular $\mathcal{L}$, let us first assume that $\mathcal{L}$ is given as the language defined by a finite forest monoid A, i.e., $\mathcal{L} = \mathcal{L}(A)$. A finite forest monoid (short: ffm) can be considered as a deterministic bottom-up automaton which combines the individual states for the trees t_i in a forest $f = t_1 \ldots t_m$ by means of a monoid operation $\circ$ (compare, e.g., the discussion in [51]). Formally, a *finite forest monoid* consists of a finite monoid G with a neutral element e, a finite subset $F \subseteq G$ of accepting elements, together with a function $up : \Sigma \times G \to G$ mapping a symbol of Σ together with a monoid element for its content to a monoid element representing a forest of length 1. A finite forest monoid accepts a forest f if $up^*(f) \in F$ where $up^*(f_1 f_2) = up^*(f_1) \circ up^*(f_2)$, $up^*(\mathtt{a}\langle f' \rangle) = up(\mathtt{a}, up^*(f'))$, and $up^*(\epsilon) = e$. Given a complete deterministic bottom-up forest automaton $A = (P, \Sigma, \delta, F_A)$, i.e., a dbta operating on the first-child next-sibling representation of forests, we construct a finite forest monoid as follows. Let G be the monoid of functions $P \to P$ where the monoid operation is function composition. In particular, the neutral element of this monoid is the identity function. Moreover, the function up is defined by $up(\mathtt{a}, g) = p \mapsto \delta(\mathtt{a}, g(\delta(\mathtt{e}))\, p)$. Finally, the set of accepting elements is given by $F = \{g \in G \mid g(\delta(\mathtt{e})) \in F_A\}$.

On the other hand every forest monoid G gives rise to a finite tree automaton A_G (running on first-child next-sibling representations) whose set of states is given by the elements of M. The transition function δ of A_G is defined by: $\delta(\mathtt{e}) = e$ and $\delta(\mathtt{a}, g_1\ g_2) = up(\mathtt{a}, g_1) \circ g_2$. Then the set of accepting states simply is given by the accepting elements of G. These constructions show that every recognizable forest language can be recognized by a finite forest monoid and vice versa. Although the ffm for a bottom-up tree automaton generally can be exponentially larger, this need not always be the case.

Example 11.10. For our running example the bta in Example 11.3 is not a complete deterministic bottom-up forest automaton. We get a complete dbfa by adding an extra error state $\bullet$. The new transition function δ' then is defined by: $\delta'(\mathtt{staff}, r_{\text{empl}}\ r_{\text{e}}) = \delta'(\mathtt{staff}, r_{\text{e}}\ r_{\text{e}}) = r_{\text{staff}}$, $\delta'(\mathtt{employee}, r_{\text{data}}\ r_{\text{empl}}) = \delta'(\mathtt{employee}, r_{\text{data}}\ r_{\text{e}}) = r_{\text{empl}}$, $\delta'(\mathtt{data}, r_{\text{name}}\ r_{\text{boss}}) = \delta'(\mathtt{data}, r_{\text{name}}\ r_{\text{e}}) = r_{\text{data}}$, $\delta'(\mathtt{boss}, r_{\text{name}}\ r_{\text{e}}) = r_{\text{boss}}$, $\delta'(\mathtt{name}, r_{\text{content}}\ r_{\text{e}}) = r_{\text{name}}$, $\delta'(\mathtt{e}, \epsilon) = r_{\text{e}}$, and $\delta'(\mathtt{a}, r_1\ r_2) = \bullet$ otherwise. In the corresponding finite forest monoid $A = (G, \Sigma, up, F)$ the monoid G contains the following functions: $g_{\text{empl}} = \{r_{\text{e}} \mapsto r_{\text{empl}}, r_{\text{empl}} \mapsto r_{\text{empl}}\}$, $g_{\text{data}} = \{r_{\text{e}} \mapsto r_{\text{data}}, r_{\text{boss}} \mapsto r_{\text{data}}\}$, $r_{\text{staff}} = \{r_{\text{e}} \mapsto r_{\text{staff}}\}$, $g_{\text{boss}} = \{r_{\text{e}} \mapsto r_{\text{boss}}\}$, $g_{\text{name}} = \{r_{\text{e}} \mapsto r_{\text{name}}\}$, $g_{\text{dataBoss}} = \{r_{\text{e}} \mapsto r_{\text{data}}\}$, $g_{\text{content}} = \{r_{\text{e}} \mapsto r_{\text{content}}\}$, $g_{\bullet} = \emptyset$, $Id = \{r \mapsto r \mid r \in P\}$, where we have omitted all entries $r \mapsto \bullet$. Note that in this example, the forest monoid has only one element more than the underlying finite automaton. Also, the composition table of these functions is given by $Id \circ g = g \circ Id = g$ for all g and furthermore: $g_{\text{data}} \circ g_{\text{boss}} = g_{\text{dataBoss}}$, $g_{\text{empl}} \circ g_{\text{empl}} = g_{\text{empl}}$, and otherwise $g \circ g' = g_{\bullet}$. For the function up, we find: $up(\mathtt{staff}, Id) = g_{\text{staff}}$, $up(\mathtt{staff}, g_{\text{empl}}) = g_{\text{staff}}$, $up(\mathtt{employee}, g_{\text{data}}) = g_{\text{empl}}$, $up(\mathtt{employee}, g_{\text{dataBoss}}) = g_{\text{empl}}$, $up(\mathtt{data}, g_{\text{name}}) = g_{\text{data}}$, $up(\mathtt{boss}, g_{\text{name}}) = g_{\text{boss}}$, $up(\mathtt{name}, g_{\text{content}}) = g_{\text{name}}$, and $up(\mathtt{a}, g) = g_{\bullet}$ otherwise. where the set of accepting functions is given by $F = \{g_{\text{staff}}\}$. $\lhd$

Theorem 11.12. *For every 2mft M and every finite forest monoid A, a 2mft M_A can be constructed such that for all $f \in \mathcal{F}_\Sigma$,*

$$\tau_{M_A}(f) = \tau_M(f) \cap \mathcal{L}(A)$$

The size of M_A is in $\mathcal{O}(|M| \cdot |A|^{l \cdot (d+1)})$ where l is the maximal rank of a state q of M and d is the maximal number of occurrences of states in right-hand sides in M.

Proof. Let $M = (Q, \Sigma, R, Q_0)$ and $A = (G, \Sigma, up, F)$. For each state q in Q with rank $n+1$ and all monoid elements $g_0, \ldots, g_n \in G$, we generate new states for the intersection 2mtt M_A of the form $\langle q, g_0 g_1 \ldots g_n \rangle$. Such a state is meant to generate all forests $f \in \mathcal{F}_\Sigma(\{y_1, \ldots, y_n\})$ for which there is a run of A starting at the leaves y_i with monoid element g_i and reaching the root of f in g_0. The rules of the new 2mft M_A are: $\langle q, g_0 g_1 \ldots g_n \rangle(\mathtt{a}, \eta, y_1, \ldots, y_n) \to \zeta'$ for every rule $q(\mathtt{a}, \eta, y_1, \ldots, y_n) \to \zeta$ of M and $\zeta' \in \tau^{g_0 g_1 \cdots g_n}[\zeta]$, where the sets $\tau^{g_0 g_1 \cdots g_n}[.]$ are inductively defined by:

$$\begin{array}{ll}
\tau^{g_0 g_1 \cdots g_n}[y_j] & = \{y_j \mid g_0 = g_j\} \\
\tau^{g_0 g_1 \cdots g_n}[\mathtt{b}\langle\zeta\rangle] & = \{\mathtt{b}\langle\zeta'\rangle \mid up(\mathtt{b}, g') = g_0 \ \wedge\ \zeta' \in \tau^{g' g_1 \cdots g_n}[\zeta]\} \\
\tau^{g_0 g_1 \cdots g_n}[\epsilon] & = \{\epsilon \mid g_0 = e\} \\
\tau^{g_0 g_1 \cdots g_n}[\zeta_1 \zeta_2] & = \{\zeta_1' \zeta_2' \mid g_0 = g_1' \circ g_2' \ \wedge\ \forall \nu : \zeta_\nu' \in \tau^{g_\nu' g_1 \cdots g_n}[\zeta_\nu]\} \\
\tau^{g_0 g_1 \cdots g_n}[q'(op, \zeta_1, \ldots, \zeta_{n'})] & = \{\langle q', g_0 g_1' \cdots g_{n'}' \rangle(op, \zeta_1', \ldots, \zeta_{n'}') \mid \forall \nu : \zeta_\nu' \in \tau^{g_\nu' g_1 \cdots g_n}[\zeta_\nu]\}
\end{array}$$

The set of initial states of M_A is $Q_0' = Q_0 \times F$. By fixpoint induction, we verify for every state q of rank $n \geq 1$, every input forest $f \in \mathcal{F}_\Sigma$, every node $v \in \mathcal{N}(f)$ and monoid elements $g_0, \ldots, g_n$ that:

$$[\![\langle q, g_0, \ldots, g_n \rangle]\!]_f(v) = [\![q]\!]_f(v) \cap \{f' \in \mathcal{F}_\Sigma(Y) \mid up^*(f', g_1 \ldots g_n) = g_0\} \qquad (*)$$

where $Y = \{y_1, \ldots, y_n\}$ and up^* is the extension of up to forests containing variables from Y, namely, for $\underline{g} = g_1 \ldots g_n$ we have $up^*(y_i, \underline{g}) = g_i$, $up^*(\epsilon, \underline{g}) = e$, $up^*(\mathtt{a}\langle f' \rangle, \underline{g}) = up(\mathtt{a}, up^*(f', \underline{g}))$, $up^*(f_1 f_2, \underline{g}) = up^*(f_1, \underline{g}) \circ up^*(f_2, \underline{g})$. The correctness of the construction follows from $(*)$.

For each state in M we have at most $|A|^l$ new states in M_A, if l is the maximal rank of states in M. If we have d occurrences of states in the right-hand side of a rule r of M, we obtain $|A|^{l \cdot (d+1)}$ new rules for r in M_A. Therefore, the new 2mft is of size $\mathcal{O}(|M| \cdot |A|^{l \cdot (d+1)})$ where l is the maximal rank of a state in M and d bounds the number of occurrences of states in right-hand sides in M. □

Note that this construction differs from the corresponding construction for 2mtts in that we now additionally have to take concatenations of forests into account. It is precisely for this operation, that we rely on the monoid structure of the set G.

Example 11.11. Consider the 2mft M of Example 11.9 and the ffm A in Example 11.10. We get an intersection 2mft with the following rules, for $\eta \in \{2, 3\}$.

1 $\langle q_I, g_{\text{staff}} \rangle(\mathtt{department}, 0) \to \mathtt{staff}\langle\langle q, Id\ Id \rangle(down, \epsilon)\rangle \quad | \quad \mathtt{staff}\langle\langle q, g_{\text{empl}} Id \rangle(down, \epsilon)\rangle$
2 $\langle q, g_{\text{empl}} Id \rangle(\mathtt{employee}, \eta, y_b) \to \mathtt{employee}\langle \mathtt{data}\langle\langle q_{data}, g_{\text{name}} \rangle(down)\rangle\ y_b \rangle$
3 $\qquad \langle q, g_2 g_{\text{boss}} \rangle(down, \mathtt{boss}\langle\langle q_{data}, g_{\text{name}} \rangle(down)\rangle) \quad \langle q, g_3 Id \rangle(right, y_b)$
4 $\langle q, g_{\text{empl}} g_{\text{boss}} \rangle(\mathtt{employee}, \eta, y_b) \to \mathtt{employee}\langle \mathtt{data}\langle\langle q_{data}, g_{\text{name}} \rangle(down)\rangle\ y_b \rangle$
5 $\qquad \langle q, g_2 g_{\text{boss}} \rangle(down, \mathtt{boss}\langle\langle q_{data}, g_{\text{name}} \rangle(down)\rangle) \quad \langle q, g_3 g_{\text{boss}} \rangle(right, y_b)$
6 $\langle q, g_0 g_b \rangle(\mathtt{data}, 2, y_b) \to \langle q, g_0 g_b \rangle(right, y_b)$
7 $\langle q, g_0 g_b \rangle(\mathtt{subordinates}, 3, y_b) \to \langle q, g_0 g_b \rangle(down, y_b)$
8 $\langle q, Id\ g_b \rangle(\epsilon, \eta, y_b) \to \epsilon$
9 $\langle q_{data}, g_{\text{name}} \rangle(\mathtt{data}, 2) \to \langle copy, g_{\text{name}} \rangle(down)$

where for the monoid elements g_2 and g_3 in the third and the forth rules $g_{\text{empl}} = g_{\text{empl}} \circ g_2 \circ g_3$ holds. Thus, g_2 and g_3 are in $\{g_{\text{empl}}, Id\}$. The element g_0 in lines 6 and 7 is either g_{empl} or Id, whereas g_b in lines 6-9 is in $\{g_{\text{boss}}, Id\}$. Additionally there are rules resulting in a state $\langle q, g_\bullet g_b \rangle$ or $\langle q, g_\bullet \rangle$ for a state $q \in Q$. ◁

The draw-back of this general construction, though, is that the (complement of the) output type with which we aim to intersect, first must be represented as a finite forest monoid. In general, this alone may incur an exponential blow-up. If, however, the 2mft is *output-linear*, i.e., uses each parameter at most once, then a much cheaper direct construction is possible. In particular, this cheaper construction applies to 2fts since these transducers have no parameters at all.

Theorem 11.13. *Assume that M is an output-linear 2mft. Then for every (possibly nondeterministic) bfa A, a 2mft M_A can be constructed with*

$$\tau_{M_A}(f) = \tau_M(f) \cap \mathcal{L}(A)$$

for all $f \in \mathcal{F}_\Sigma$. The size of the 2mft $|M_A|$ is in $\mathcal{O}(|M| \cdot |A|^{2l\cdot(d+1)})$ where l is the maximal rank of a state q of M and d is the maximal number of occurrences of states in right-hand sides in M.

Proof. Let $M = (Q, \Sigma, R, Q_0)$ and $A = (P, \Sigma, \delta, \{p_0\})$. The idea for the new 2mft M_A for the intersection is to maintain for every possibly produced output forest f a *pair* of states $\langle p, p' \rangle$ so that the automaton A, when starting in p' to the right of f, possibly may arrive in state p to the left. Accordingly, the set Q' of M_A consists of all states $\langle q, p_0p'_0 \dots p_np'_n \rangle$ where $q \in Q$ is of rank $n+1$, i.e., has n accumulating parameters and $p_i, p'_i \in P$ for all i. Accordingly, the rules of the new 2mft are of the form: $\langle q, p_0p'_0 \dots p_np'_n \rangle(\mathsf{a}, \eta, y_1, \dots, y_n) \;\to\; f'$ with $f' \in \tau^{p_0p'_0\cdots p_np'_n}[f]$ for every rule $q(\mathsf{a}, \eta, y_1, \dots, y_n) \to f$ of M. The sets $\tau^{p_0p'_0\cdots p_np'_n}[.]$ are defined by:

$$
\begin{aligned}
&\tau^{p_jp'_j\; p_1p'_1\cdots p_np'_n}[y_j] &&= \{y_j\}\\
&\tau^{p_0p'_0\; p_1p'_1\cdots p_np'_n}[\mathsf{b}\langle\zeta\rangle] &&= \{\mathsf{b}\langle\zeta'\rangle \mid (p_0, \mathsf{b}, p''_1\, p'_0) \in \delta \wedge (p''_2, \mathsf{e}) \in \delta \wedge \zeta' \in \tau^{p''_1p''_2\; p_1p'_1\cdots p_np'_n}[\zeta]\}\\
&\tau^{p_0p'_0\; p_1p'_1\cdots p_np'_n}[\epsilon] &&= \{\epsilon \mid p_0 = p'_0\}\\
&\tau^{p_0p'_0\; p_1p'_1\cdots p_np'_n}[\zeta_1\zeta_2] &&= \{\zeta'_1\zeta'_2 \mid \exists p :\; \zeta'_1 \in \tau^{p_0p\; p_1p'_1\cdots p_np'_n}[\zeta_1] \;\wedge \zeta'_2 \in \tau^{pp'_0\; p_1p'_1\cdots p_np'_n}[\zeta_2]\}\\
&\tau^{p_0p'_0\; p_1p'_1\cdots p_np'_n}[q'(op, \zeta_1, \dots, \zeta_m)]\\
&\qquad = \{\langle q', p_0p'_0\; p''_1p'''_1 \dots p''_mp'''_m \rangle(op, \zeta'_1, \dots, \zeta'_m) \mid \forall \nu : \zeta'_\nu \in \tau^{p''_\nu p'''_\nu\; p_1p'_1\cdots p_np'_n}[\zeta_\nu]\}.
\end{aligned}
$$

The set of initial states of M_A then consists of all states $\langle q, p_0p' \rangle$ where $q \in Q_0$ and $p_0 \in P$ are accepting states of M and A, respectively, and $(p', \mathsf{e}) \in \delta$. The estimation of the size of the resulting transducer is similar to the case of forest monoids — only that we have to replace the number of monoid elements with the number of *pairs* of states. Thus, the new intersection transducer is of size $\mathcal{O}(|M| \cdot |A|^{2l\cdot(d+1)})$. □

11.7.2. *Deciding Emptiness of 2MFTs*

For deciding emptiness of a forest transducer M, we conceptually follow the approach taken for tree transducers. There, we first constructed an alternating tree walking automaton accepting the domain of M for which in the second step, a nondeterministic tree automaton is constructed. In our case, this would mean that we first formally introduce the concept of alternating *forest walking* automata for which in a separate construction, a nondeterministic forest automaton is constructed. In order to simplify this, we will not intermediately rely on forest walking automata. Instead, we consider for each forest f, an *enriched* first-child next-sibling encoding through binary trees. This means that inside each node of the encoding we additionally record whether or not the current tree node represents a node on the top-level of the forest. Let $\bar{\Sigma} = \{\bar{a} \mid a \in \Sigma\}$ denote a set of new symbols of rank 2. Then the ranked alphabet used by the encoding will be $\Sigma_2 = \Sigma \cup \bar{\Sigma} \cup \{\mathsf{e}, \bar{\mathsf{e}}\}$ where the barred symbols will only occur on the rightmost spine in the tree. A dtta with two states $\{\mathsf{t}, \mathsf{n}\}$ can check whether a tree in $\mathcal{T}_{\Sigma_2}$ is the enriched encoding of a forest or not.

Since the encoding is injective, it suffices for a forest transducer M to construct an atwa M' which defines the set of encodings of the domain of M. Then the set of states of the atwa M' is given by $Q' = \{q'_0, \mathsf{t}, \mathsf{n}\} \cup Q \cup Q^{up}$ where $Q^{up} = \{q^{up} \mid q \in Q\}$ is a set of fresh copies of the states in Q and q'_0 serves as fresh initial state of M'. Assume that $q(\mathsf{a}, \eta, y_1, \ldots, y_n) \to \zeta$ is a rule of M and $q_1(op_1, \ldots), \ldots, q_c(op_c, \ldots)$ is the sequence of recursive calls in ζ. If $\eta \in \{2, 3\}$, i.e., if the rule is not applicable to nodes at the top-level of the input forest, then atwa M' has the rules: $q(\mathsf{a}, \eta - 1) \to q'_1(op'_1) \wedge \ldots \wedge q'_c(op'_c)$ where $q'_j(op'_j) = q_j(stay)$ if $op_j = stay$, $q'_j(op'_j) = q_j(down_1)$ if $op_j = down$, $q'_j(op'_j) = q_j(down_2)$ if $op_j = right$, $q'_j(op'_j) = q_j(up)$ if $op_j = left$ and $\eta = 3$, $q'_j(op'_j) = q_j^{up}(stay)$ if $op_j = up$. For states q^{up}, atwa M' has the rules: $q^{up}(\mathsf{a}, 2) \to q(up)$ and $q^{up}(\mathsf{a}, 3) \to q^{up}(up)$.

If on the other hand $\eta \in \{0, 1\}$, i.e., the rule of the 2mft refers to nodes at the top-level of the input forest, then the atwa M' has the rules: $q(\bar{\mathsf{a}}, 2 \cdot \eta) \to q'_1(op'_1) \wedge \ldots \wedge q'_c(op'_c)$ where $q'_j(op'_j) = q_j(stay)$ if $op_j = stay$, $q'_j(op'_j) = q_j(down_1)$ if $op_j = down$, $q'_j(op'_j) = q_j(down_2)$ if $op_j = right$, $q'_j(op'_j) = q_j(up)$ if $op_j = left$. For every rule $q(\mathsf{a}, 0) \to \zeta$ of M with $q \in Q_0$, we add the rule $q'_0(\bar{\mathsf{a}}, 0) \to \mathsf{t}(stay) \wedge \mathsf{q}(stay)$ where the rules for t and n simulate the computation of a top-down automaton to verify that the input tree is the enriched encoding of a forest. Thus, the rules of the atwa for q'_0 are meant to spawn a subrun which verifies the encoding and to spawn another subrun which simulates an accepting run of the forest transducer on the binary encoding. In particular, the states q^{up} are auxiliary states to implement the operation *up* on the binary representation of the forest. More precisely, the rules for the state q^{up} performs the operation *up* as long as the current node is a right child. If the current node is a left child, a final *up*-operation is executed to arrive at the tree representation of the father node in state q. Overall, we find:

Theorem 11.14. *For every 2mft M, an atwa M' can be constructed in polynomial time such that $\mathcal{L}(M')$ is the set of enriched binary encodings of the set $\{f \mid \tau_M(f) \neq \emptyset\}$. In particular, $\mathcal{L}(M') \neq \emptyset$ iff $\tau_M \neq \emptyset$.*

We thus obtain an exponential algorithm for deciding emptiness of 2mfts which is optimal. Together with our intersection constructions, this algorithm then can be applied also for type-checking 2mft transducers w.r.t. regular input and output types.

In order to arrive at more tractable algorithms or sub-classes, we can apply the same ideas as for 2mtts: in the first place, we can again approximate the set of output forests by means of a context-free forest grammar. A context-free forest grammar for the intersection with a regular forest language specified through a finite forest monoid is polynomial in the size of the grammar and the number of elements in the monoid and exponential only in $l \cdot (d + 1)$ where l is the maximal rank of nonterminals and d is the number of occurrences of nonterminals in right-hand sides. Emptiness for this forest grammar again can be checked in time linear in the size of the grammar. A practical implementation again may construct a *Datalog* program for the sets of useful nonterminals of the grammar. We can also generalize the notions of b-boundedness and strict b-boundedness for 2mfts. While emptiness for b-bounded 2mfts is decidable in polynomial time (where the exponent again depends on b^2), it is only strict b-boundedness which is preserved by our intersection constructions. Here we only state the corresponding result for output types specified through finite forest monoids.

Theorem 11.15. *Assume M is a strictly b-bounded 2mft and I and O are regular forest languages where I and O are given by a finite forest automaton and a finite forest monoid, respectively. Assume further that l is the maximal rank of a state of M and d is the maximal number of occurrences of state calls in right-hand sides. Then M can be type-checked w.r.t. I and O in time polynomial in the sizes of M, the automaton for I and the automaton for O where the exponent linearly depends on $(b+1)^2 \cdot l \cdot (d+1)$.*

11.7.3. *Notes and References*

Top-down macro forest transducers have been introduced by Perst and Seidl in [50]. They are closely related to the top-down transducers of Maneth and Neven [52] (but slightly more general). It was shown in [50] that, even though mfts are more powerful than mtts, they can be type checked with the same complexity bounds, as macro tree transducers. This idea was extended to two-fold compositions of deterministic mtts, in [53]. In [6], a general forest transformation language TL is introduced which captures most features of XML transformation languages such as XSLT. The language TL supports full MSO pattern matching both for the selection of rules applicable at a node in the input tree and for navigation inside the input tree. Thus, 2mfts can be considered as a sub-language of TL where rules are selected depending on the current state and input label only and where navigation is restricted to immediate neighbors in the input forest. The main contribution of that paper is to show how such transformations can be decomposed into three stay macro tree transducers running on the first-child next-sibling encoding of the XML documents in question. The semantics considered there was OI evaluation of nested calls, but similar results can also be proven for IO evaluation, i.e., call-by-value parameter passing as considered here [54].

11.8. Conclusion

In this chapter, we have reviewed basic constructions for tree walking transducers which allow to obtain algorithms for type-checking the transducers w.r.t. regular input and output types. There are three orthogonal variations in which the basic concept of a finite state machine can be made more expressive:

- top-down versus walking
- without parameters versus with parameters;
- on ranked trees versus on unranked forests.

At the very heart of our algorithms for type-checking is to check whether or not a transducer realizes an empty translation. Already for the weakest, i.e., one-way top-down transducers emptiness turns out to be complete for deterministic exponential time. Still, however, we were able to pin-point one major source for the complexity, namely, the number of visits to the same input node. If the transducer visits the same node only constantly often, i.e., is b-bounded for some constant b, then emptiness becomes decidable in polynomial time.

The second ingredient of our algorithm are constructions for computing intersection transducers, i.e., transducers which only produce outputs outside a specified regular set. Here, we considered regular sets as specified by bottom-up deterministic automata (or monoids, in case of forests) or by deterministic top-down automata. The latter construction for forest transducers was at least applicable to *output-linear* transducers, i.e., transducers which use each of their accumulating parameters at most once. Two separate constructions are crucial, since translating top-down deterministic automata into bottom-up automata may incur an extra exponentiation in the number of states. Since these constructions preserve strict b-boundedness we thus overall arrive at a general class of transducers for which type-checking is polynomial.

References

[1] A. Möller and M. I. Schwartzbach. The Design Space of Type Checkers for XML Transformation Languages. In *Proc. 10th ICDT*, vol. 3363, *LNCS*, pp. 17–36. Springer, (2005).
[2] A. Möller, M. Ø. Olesen, and M. I. Schwartzbach, Static validation of XSL transformations, *ACM Trans. Program. Lang. Syst.* **29**(4), (2007).
[3] M. Murata, D. Lee, and M. Mani. Taxonomy of XML Schema Languages using Formal Language Theory. In *Proc. Extreme Markup Languages*, (2001).
[4] J. Engelfriet and S. Maneth, A Comparison of Pebble Tree Transducers with Macro Tree Transducers, *Acta Inf.* **39**, 613–698, (2003).
[5] T. Milo, D. Suciu, and V. Vianu, Typechecking for XML Transformers, *J. Comp. Syst. Sci.* **66**, 66–97, (2003).
[6] S. Maneth, A. Berlea, T. Perst, and H. Seidl. XML Type Checking with Macro Tree Transducers. In *Proc. 24th PODS*, pp. 283–294. ACM, (2005).
[7] W. Martens and F. Neven, On the complexity of typechecking top-down xml transformations, *Theoret. Comput. Sci.* **336**, 153–180, (2005).
[8] S. Boag and D. Chamberlin et.al., editors. XQuery 1.0: An XML Query Language. W3C Working Draft. Available at http://www.w3.org/TR/xquery/, (2003).
[9] H. Hosoya and B. Pierce, XDuce: A Statically Typed XML Processing Language, *ACM Trans. Inter. Tech.* **3**(2), 117–148, (2003).
[10] A. Frisch. Regular Tree Language Recognition with Static Information. In *Proc. 3rd IFIP TCS*. Kluwer, (2004).
[11] H. Hosoya and B. Pierce, Regular expression pattern matching for XML, *Journal of Functional Programming.* **13**(6), 961–1004, (2002).
[12] H. Hosoya, A. Frisch, and G. Castagna. Parametric Polymorphism for XML. In *Proc. 32nd POPL*, pp. 50–62. ACM, (2005).

[13] C. Kirkegaard, A. Möller, and M. Schwartzbach, Static Analysis of XML Transformations in Java, *IEEE Trans. Soft. Eng.* **30**, 181–192, (2004).
[14] J. Engelfriet, The time complexity of typechecking tree-walking tree transducers, *Acta Inf.* **46**(2), 139–154, (2009).
[15] W. Martens and F. Neven. Frontiers of Tractability for Typechecking Simple XML Transformations. In *Proc. 23rd PODS*, pp. 23–34. ACM, (2004).
[16] S. A. Greibach, Hierarchy theorems for two-way finite state transducers, *Acta Inf.* **11**, 80–101, (1978).
[17] D. C. Fallside and P. Walmsley. XML Schema part 0: Primer second edition. W3C recommendation, W3C (October, 2004). http://www.w3.org/TR/xmlschema-0/.
[18] W. Rounds, Mappings and Grammars on Trees, *Math. Systems Theory.* **4**, 257–287, (1970).
[19] J. W. Thatcher. Transformations and translations from the point of view of generalized finite automata theory. In *Proc. 1st STOC*, pp. 129–142. ACM, (1969).
[20] A. V. Aho and J. D. Ullman, Translations on a Context-Free Grammar, *Inform. and Control.* **19**, 439–475, (1971).
[21] T. Kamimura and G. Slutzki, Parallel and two-way automata on directed ordered acyclic graphs, *Inform. and Control.* **49**, 10–51, (1981).
[22] D. E. Knuth, Semantics of context-free languages, *Math. Systems Theory.* **2**(2), 127–145, (1968).
[23] Z. Fülöp, On attributed tree transducers, *Acta Cybern.* **5**, 261–279, (1981).
[24] Z. Fülöp and H. Vogler, *Syntax-Directed Semantics; Formal Models Based on Tree Transducers.* (Springer, 1998).
[25] *Extensible Markup Language (XML) 1.0.* W3C, second edition (6 Oct., 2000). Available at http://www.w3.org/TR/2000/REC-xml-20001006.
[26] D. Fallside, editor. XML Schema. W3C Recommendation, W3C (2 May, 2001). Available at http://www.w3.org/TR/xmlschema-0/.
[27] J. Clark and M. Murata et al. *RelaxNG Specification.* OASIS. Available at http://www.oasis-open.org/committees/relax-ng.
[28] F. Neven, Automata Theory for XML Researchers, *SIGMOD Record.* **31**(3), 39–46, (2002).
[29] F. Gécseg and M. Steinby, *Tree Automata.* (Akadémiai Kiadó, 1984).
[30] F. Gécseg and M. Steinby. Tree languages. In eds. G. Rozenberg and A. Salomaa, *Handbook of Formal Languages, Volume 3*, chapter 1. Springer, (1997).
[31] H. Comon, M. Dauchet, R. Gilleron, F. Jacquemard, C. Löding, D. Lugiez, S. Tison, and M. Tommasi. Tree automata techniques and applications. Available at http://www.grappa.univ-lille3.fr/tata, (2007).
[32] B. Courcelle, A representation of trees by languages II, *Theoret. Comput. Sci.* **7**, 25–55, (1978).
[33] J. Virágh, Deterministic ascending tree automata I, *Acta Cybern.* **5**, 33–42, (1981).
[34] F. Drewes and J. Engelfriet, Decidability of finiteness of ranges of tree transductions, *Inform. and Comput.* **145**, 1–50, (1998).
[35] G. Slutzki, Alternating tree automata, *Theor. Comput. Sci.* **41**(2-3), 305–318, (1985).
[36] A. Frisch and H. Hosoya. Towards practical typechecking for macro tree transducers. In *Proc. 11th DBPL*, vol. 4797, *LNCS*, pp. 246–260. Springer, (2007).
[37] B. S. Baker, Composition of top-down and bottom-up tree transductions, *Inform. and Control.* **41**(2), 186–213, (1979).
[38] J. Engelfriet, G. Rozenberg, and G. Slutzki, Tree transducers, L systems, and two-way machines, *J. Comp. Syst. Sci.* **20**, 150–202, (1980).

[39] J. Engelfriet and S. Maneth, Macro tree transducers, attribute grammars, and MSO definable tree translations, *Inform. and Comput.* **154**, 34–91, (1999).
[40] J. Engelfriet and S. Maneth, Macro tree translations of linear size increase are MSO definable, *SIAM J. Comput.* **32**, 950–1006, (2003).
[41] R. Giegerich, Composition and evaluation of attribute coupled grammars, *Acta Inf.* **25**, 355–423, (1988).
[42] S. Maneth, T. Perst, and H. Seidl. Exact XML type checking in polynomial time. In *Proc. 11th ICDT*, vol. 4353, *LNCS*, pp. 254–268. Springer, (2007).
[43] J. Engelfriet, H. J. Hoogeboom, and B. Samwel. XML transformation by tree-walking transducers with invisible pebbles. In *Proc. 26th PODS*, pp. 63–72. ACM, (2007).
[44] M. Bartha, An algebraic definition of attributed transformations, *Acta Cybern.* **5**, 409–421, (1982).
[45] S. S. Cosmadakis, H. Gaifman, P. C. Kanellakis, and M. Y. Vardi. Decidable optimization problems for database logic programs (preliminary report). In *Proc. 30th STOC*, pp. 477–490. ACM, (1988).
[46] J. Engelfriet and H. Vogler, Macro Tree Transducers, *J. Comp. Syst. Sci.* **31**, 71–146, (1985).
[47] M. J. Fischer. *Grammars with Macro-like Productions.* PhD thesis, Harvard University, Massachusetts, (1968).
[48] J. Hopcroft, R. Motwani, and J. Ullman, *Introduction to Automata Theory, Languages, and Computation.* (Addison-Wesley, 2001), second edition.
[49] J. Engelfriet and E. Schmidt, IO and OI. (I&II), *J. Comp. Syst. Sci.* **15**, 328–353, (1977). and 16:67–99, 1978.
[50] T. Perst and H. Seidl, Macro Forest Transducers, *Inf. Proc. Letters.* **89**, 141–149, (2004).
[51] M. Bojańczyk and I. Walukiewicz. Unranked Tree Algebra. Technical report, University of Warsaw, (2005).
[52] S. Maneth and F. Neven. Structured Document Transformations Based on XSL. In *Proc. 7th DBPL*, vol. 1949, *LNCS*, pp. 80–98. Springer, (1999).
[53] S. Maneth and K. Nakano. XML type checking for macro tree transducers with holes. In *Proc. PLAN-X*, (2008).
[54] T. Perst. *Type Checking XML Transformations.* Dissertation, Technische Universität München, München, (2007).

Chapter 12

Three Case Studies on Verification of Infinite-State Systems

Javier Esparza* and Jörg Kreiker†

Technische Universität München,
Boltzmannstr. 3, 85748 Garching bei München, Germany
**esparza@in.tum.de*
†kreiker@in.tum.de

Most software systems have an infinite number of reachable states due to variables with unbounded data domains, parameters, control structures, time, or communication mechanisms based on buffers or queues. Each of these sources of infinity has been studied using formal models such as timed automata, channel systems, extensions of Petri nets, pushdown processes, broadcast protocols, counter automata, different process algebras, or rewrite systems. This paper is a modest attempt at describing underlying connections by introducing the key concept of symbolic search in a very general framework, along with the conditions for turning it into an effective algorithm.

Papers at conferences usually follow a horizontal approach: they may present a decidability result, leaving an efficient algorithm for further research, or vice versa. Or they may describe an implementation and report on experimental results, without presenting the underlying theory, and without explaining the steps from the system to a formal model. In contrast, we present three case studies following a vertical approach: first, a verification challenge is informally described; then, a formal model of the system is presented; after that, the theory of symbolic search for an adequate class of models is developed, and, finally, the algorithm is executed on a formal model. The concrete case studies are: a mutex protocol modeled as timed automaton, a cache-coherence protocol modeled as a linear automaton, and a skyline plotter modeled as a pushdown automaton.

12.1. Introduction

The automatic verification of systems has made immense progress in the last two decades, in particular thanks to the efforts of the model-checking, program-analysis, and theorem-proving communities. Systems with finitely many states can be automatically verified is, at least in principle, by exhaustively exploring their state space. Symbolic search procedures which use special data structures to compactly store large sets of states have made this a technique with many important applications, especially in the hardware area. However, most software systems have an infinite number of reachable states due to unbounded data structures, timing information, or other factors (see below). While many properties can be decided by analyzing a suitable finite-state abstraction of the system, these abstractions are difficult to find, and may lead to finite state systems far too large for the existing tools. For this reason, since the early 90s much effort has been devoted to identifying classes of infinite-state systems with decidable verification problems.

Infinite state-spaces are often caused by the system exhibiting one or more of the following features:

- **Variables with unbounded data domains** such as integers or dynamic data types such as lists, trees, and DAGs. While the physical constraints of a machine impose a bound on the domain, the bound is too large to be of any practical use.
- **Control structures** like procedure calls and thread creation. Both can generate an unbounded number of processes whose execution has not terminated, and whose local states become part of the global state of the system.
- **Time.** In real-time systems, the current time – be it discrete or continuous – becomes part of the state.
- **Communication mechanisms** based on buffers or queues. Like variables, the physical implementation of a buffer or a queue always has a finite capacity, but the capacity may heavily depend of the details of the implementation, and the bounds are usually too large to be of any use for finite-state verifiers. For these reasons, mathematical models often assume unbounded capacity.
- **Parameters.** A system may contain parameters whose values are left undefined, like for instance the number of processes in a leader election protocol. When these parameters have an infinite domain of possible values, the system becomes in fact an infinite family of systems. For most purposes the family is equivalent to the infinite system consisting of all members of the family running independently of each other.

Each of these five "sources of infinity" has been studied in the literature using a variety of formal models: timed automata, channel systems, extensions of Petri nets, pushdown processes, broadcast protocols, counter automata, different process algebras, and rewrite systems to name but a few. This makes it difficult to perceive research on infinite-state systems as a field on its own. Moreover, due to the space constraints of conferences and to their intended audience, papers on infinite state systems usually follow a "horizontal" approach: a paper in a theoretical conference may present a decidability result, leaving an efficient algorithm for further research, or describe an efficient algorithm, leaving its implementation for further research; a paper in a more applied conference may describe an implementation and report on experimental results, without presenting the underlying theory, and without explaining the steps conducting from the system to a formal model.

This paper is a modest attempt at describing the underlying connections between many of the models and techniques used in the field. It does so by introducing the key concept of symbolic search in a very general framework (Section 12.2), along with the conditions for turning it into an effective algorithm. It then presents three case studies (Sections 12.3, 12.4, and 12.5) following a "vertical" approach: first, a (very small) verification challenge, consisting of a system and a desirable property, is informally described; then, a formal model of the system is presented; then, the theory of symbolic search for an adequate class of models is developed, leading to an effective algorithm for deciding the property; finally, the algorithm is executed on the formal model. The concrete case studies are: a small mutex protocol modeled as timed automaton (Section 12.3), a cache-coherence protocol modeled as a linear automaton (Section 12.4), and a skyline plotter modeled as a pushdown automaton (Section 12.5).

The paper does not contain any novel technical material, and in fact it unashamedly draws from several excellent papers by different colleagues; in particular we mention [1–3]. We hope that it may help researchers from other fields to understand some of the principles of automatic verification in the infinite case, and also provide useful material for undergraduate and graduate courses. In fact, the paper is based on a course given by the first author at the Marktoberdorf Summer School in 2005.

12.2. Symbolic Search for Extended Automata

We introduce symbolic search for *extended automata*, automata whose transitions depend on and update a finite set of variables. In Sections 12.3, 12.4, and 12.5 we will consider three different classes of extended automata.

12.2.1. *Extended Automata*

Loosely speaking, an extended automaton is an automaton whose transitions are guarded by and operate on a set of variables. The variables may have arbitrary types: integers, reals, stacks, queues, etc. More formally, let $X = \{x_1, \dots, x_n\}$ be a finite set of *variables* over sets $V_1, \dots, V_n$ of *values*, and let $\mathcal{V} = V_1 \times \dots \times V_n$ be the set of all possible *valuations* of X.

An *extended automaton* over X is a pair $E = (Q, R)$ where

- Q is a finite set of *control states*, and
- R is a set of *transition rules* or just *rules*. A rule is a tuple $r = (q, g, a, q')$, where
  - $q, q' \in Q$ are the *source* and the *target state* of r, respectively;
  - $g \subseteq \mathcal{V}$ is the *guard* of r; and
  - $a \subseteq \mathcal{V} \times \mathcal{V}$ is the *action* of r.

We often use predicates over the set X to describe guards, and predicates over X and a copy X' to describe actions. The variables in X and X' denote valuations before and after executing the action, respectively. For instance, if $X = \{x_1, x_2\}$ and $V_1 = V_2 = \mathbb{N}$ then we write $x_1 + x_2 > 5$ for the guard $\{(n_1, n_2) \in \mathbb{N} \times \mathbb{N} \mid n_1 + n_2 > 5\}$, and $x_1' = x_2 + 3 \wedge x_2' = 1$ for the action $\{((n_1, n_2), (n_1', n_2')) \mid n_1' = n_2 + 3 \wedge n_2' = 1\}$

An extended automaton is *finite* if $\mathcal{V}$ is finite; otherwise it is *infinite.*

The semantics of extended automata is what one would expect. A *configuration* is a pair $\langle q, \nu \rangle$, where q is a control state and ν is a valuation. If $\nu = (v_1, \ldots, v_n)$, we also write $\langle q, v_1, \ldots, v_n \rangle$. The set of all configurations of an extended automaton E is written $\mathcal{C}_E$ or $\mathcal{C}$, if E is clear from the context. The *transition system* $\mathcal{T}_E$ of an extended automaton E has: all elements of $\mathcal{C}_E$ as nodes, and an edge $\langle q, \nu \rangle \Longrightarrow \langle q', \nu' \rangle$ whenever E has a rule $r = (q, g, a, q')$ such that $\nu \in g$ (we say that ν *satisfies the guard* g, or that $\langle q, \nu \rangle$ *enables* r) and $(\nu, \nu') \in a$. So, loosely speaking, a rule is enabled if the extended automaton is in the source state, and the current valuation satisfies the guard; an enabled rule can occur, and its occurrence leads the automaton to the target state and to a new valuation determined by the execution of the action; note that the action can be nondeterministic.

If $\langle q, \nu \rangle \Longrightarrow \langle q', \nu' \rangle$ then $\langle q', \nu' \rangle$ is an *immediate* r*-successor* of $\langle q, \nu \rangle$, and $\langle q, \nu \rangle$ is an *immediate* r*-predecessor* of $\langle q', \nu' \rangle$. We say that a configuration is an *immediate successor (predecessor)* of another if there exist a rule r in the extended automaton such that the configuration is an immediate r-successor (predecessor) of the other one.

12.2.2. *The Safety Problem*

Given an extended automaton E and a rule r of E, we define the immediate successor functions $post_r^E, post^E : 2^{\mathcal{C}_E} \to 2^{\mathcal{C}_E}$ as follows: c belongs to $post_r^E(C)$ if some immediate r-predecessor of c belongs to C, and c belongs to $post^E(C)$ if some immediate predecessor of c belongs to C. Usually E is clear from the context and we write $post_r(C)$ and $post(C)$. Clearly, if R is the set of rules of E we have $post(C) = \bigcup_{r \in R} post_r(C)$. The reflexive and transitive closure of *post* is denoted by $post^*$, and so we have $post^*(C) = \{c' \in C \mid \exists c \in C.\ c \Rightarrow^* c'\}$. Similarly, we define $pre(C)$ as the set of immediate predecessors of elements in C and pre^* as the reflexive and transitive closure of *pre.* Given two configurations c, c' of an extended automaton E, we say that c' is *reachable* from c, or that c is *backward reachable* from c' if $c' \in post^*(\{c\})$ or, equivalently, if $c \in pre^*(\{c'\})$.

We are interested in the *safety problem* for extended automata, defined as follows: Given an extended automaton E over a set X of variables, a set I of initial configurations, and a set D of "dangerous' configurations, are there $c \in I$ and $c' \in D$ such that c' is reachable from c in E? Notice that we do not require the sets I and D to be finite.

The safety problem is undecidable even for very simple extended automata (notice that the sets I and D might be non-recursive), or for sets I and D containing only one element (Turing machines are a special class of extended automata). In this paper we consider *constrained safety problems* that impose restrictions on

(i) the type and/or the number of variables in the set X,
(ii) the possible combinations of guards and actions,
(iii) the possible sets of initial configurations,
(iv) the possible sets of dangerous configurations.

Some combinations of restrictions correspond to the reachability problem for well-known models of computation. Let us consider two examples.

Example 12.1. Counter machines are extended automata over nonnegative integer variables, i.e., $X = \{x_1, \ldots, x_n\}$ and $V_1 = \ldots = V_n = \mathbb{N}$. There are only two kinds of rules, with the following two combinations of guards and actions: $\mathbf{true}/x'_i = x_i+1$ (where **true** is the empty guard, i.e., the guard indicating that the rule is enabled at any valuation) and $x_i > 0/x'_i = x_i - 1$ (i.e., decrement x_i by 1, but only if $x_i > 0$).

Example 12.2. Pushdown systems are extended automata with a single variable over words, i.e., $X = \{x_1\}$ and $V_1 = \Gamma^*$ for some finite alphabet Γ. There is a single combination of guard and action, which can be represented as γ/w for $\gamma \in \Gamma$ and $w \in \Gamma^*$: the guard γ indicates that the rule is only enabled when $x = \gamma w'$ for some word w'; the action consists of substituting γ by the word w, that is, x becomes ww' after the action.

12.2.3. *Symbolic Search*

When the set of valuations of an extended automaton is finite, the safety problem can be solved by a *forward* or a *backward* search. In forward search, we start at the set I of initial configurations and explore the space of reachable configurations moving "forward", i.e., moving from a configuration to its immediate successors. The search terminates when a dangerous configuration is reached, or when all the reachable configurations have been explored. Backward search starts at the set D, moves from configurations to their immediate predecessors, and terminates when an initial configuration is reached, or when all configurations that can be reached backwards have been explored. Notice that backward search may explore configurations that are not reachable from I.

In conventional search techniques a set of configurations is stored by explicitly storing each of its elements at a different memory address. For large configuration spaces this can be very costly, and for infinite configuration spaces impossible. The idea of *symbolic search* is to directly manipulate possibly infinite *sets of configurations* by means of adequate data structures. For instance, a linear inequation like $x > 5$ can be used as a data structure representing the infinite set of integers $\{6, 7, 8, ...\}$; a regular expression like $(ab)^*$ (or its corresponding finite automaton) can be used to represent the infinite set $\{\epsilon, ab, abab, \ldots\}$ of words. We say that a linear inequation is a *symbolic representation* of its set of solutions and that a finite automaton is a *symbolic representation* of the language it accepts.

The abstract structure of symbolic search is very simple. The forward case is shown in Figure 12.1.

Forward symbolic search

Initialize $C := I$

Iterate $C := C \cup post(C)$ until

$C \cap D \neq \emptyset$; return "reachable", or

a fixed point is reached; return "non-reachable"

Fig. 12.1. The abstract structure of symbolic forward search.

In order to make symbolic search effective for a constrained safety problem, it suffices to find a family $\mathcal{F}$ of sets of configurations satisfying **Conditions (1)–(6)** below. We call the elements of $\mathcal{F}$ *symbolic configurations*:

- **Condition (1).** Every symbolic configuration has a (not necessarily unique) finite symbolic representation.
- **Condition (2).** The set I of initial configurations is a symbolic configuration, i.e., $I \in \mathcal{F}$.
- **Condition (3).** If $C \in \mathcal{F}$, then $C \cup post(C) \in \mathcal{F}$. Moreover, the symbolic representation of $C \cup post(C)$ is effectively computable from the symbolic representation of C.
- **Condition (4).** For every $C \in \mathcal{F}$, the emptiness of $C \cap D$ is decidable (i.e., there is an algorithm that takes the symbolic representations of C and D as input, and decides whether the set $C \cap D$ is empty).
- **Condition (5).** Equality of symbolic configurations is decidable, i.e., there is an algorithm that takes the symbolic representations of two sets of $\mathcal{F}$ as input and decides whether they represent the same set. This is needed in order to check if a fixed point has already been reached: If after executing $C := C \cup post(C)$ the old and new values of C coincide, then the search can stop.
- **Condition (6).** If $post^*(I) \cap D = \emptyset$, then the chain $C_0 \subseteq C_1 \subseteq C_2 \ldots$ of symbolic configurations, where $C_0 = I$ and $C_{i+1} = C_i \cup post(C_i)$, reaches a fixed point after finitely many steps, i.e. there is an index i such that $C_i = C_{i+j}$ for every $j \geq 0$.

Conditions (1)–(5) are needed so that the algorithm can run. Condition (6) guarantees termination. Notice that Condition (5) can be relaxed. It suffices to have an algorithm satisfying the following properties for every $i \geq 0$: if $C_i \neq C_{i+1}$

then the algorithm answers "not equal"; if $C_i = C_{i+1}$ (and so $C_i = C_{i+j}$ for every $j \geq 0$), then there is a $k \geq i$ such that for input C_k, C_{k+1} the algorithm answers "equal". In other words, if two symbolic configurations are not equal, then the algorithm immediately answers "not equal". If they are equal, then the algorithm may not immediately recognize it, but it eventually answers "equal".

The algorithm and the effectivity conditions for symbolic backward search are obtained from those for forward search by exchanging I and D and *pre* and *post*.

Forward and backward search look deceptively symmetric, but in practice they are not. In Section 12.4 we will find a case study in which the set D of dangerous configurations has a certain property not enjoyed by the set I of initial configurations. This property will make backward search from D effective, while forward search from I may not terminate.

12.2.4. *The Powerset Case*

Often the sets belonging to $\mathcal{F}$ are finite unions of sets belonging to another family $\mathcal{B}$. That is, for every $C \in \mathcal{F}$ there are sets $B_1, \ldots, B_m \in \mathcal{B}$ such that $C = B_1 \cup \ldots \cup B_m$. In this case one can symbolically represent C by the set of the symbolic representations of the B_i's (we assume that they have one), and conditions (3) and (4) can be simplified a bit. Since $post(C) = \bigcup_{r \in R} post_r(C) = \bigcup_{r \in R} \bigcup_{1 \leq i \leq m} post_r(B_i)$, Condition (3) can be replaced by: if $B \in \mathcal{B}$, then $post_r(B) \in \mathcal{F}$ for every rule r (this holds in particular if $post_r(B) \in \mathcal{B}$). Condition (4) can be replaced by: for every $B \in \mathcal{B}$, it is decidable whether $B \cap D = \emptyset$.

12.2.5. *Making Symbolic Search Terminate*

In the next sections we present three concrete verification challenges that can be solved using symbolic search. They are examples of three different, and increasingly sophisticated, ways of making Condition (6) hold:

- **Finite $\mathcal{F}$.** If we manage to find a family $\mathcal{F}$ containing only finitely many sets, say k, then Condition (6) is guaranteed to hold, since necessarily $C_k = C_{k+i}$ for every $i \geq 0$. This method will be applied to timed automata in Section 12.3.
- **Finite ascending chains.** If the family $\mathcal{F}$ is infinite but only contains finite ascending chains with respect to set inclusion, then the chain $C_0 \subseteq C_1 \subseteq C_2 \ldots$ of Condition (6) contains a set C_j such that $C_j = C_{j+k}$ for every $k \geq 0$, and so the condition holds. We will use this reasoning in the case of linear automata in Section 12.4.
- **Accelerations.** This is a special case of the well-known widening technique of abstract interpretation [4]. Assume that $\mathcal{F}$ has infinite ascending chains, but we manage to find an operator $\nabla \colon \mathcal{F} \times \mathcal{F} \to \mathcal{F}$ satisfying the following properties:

(a) $C \nabla post(C) \supseteq C \cup post(C)$ for every $C \in \mathcal{F}$, and
(b) the chain $(\widehat{C}_i)_{i \geq 0}$, where $\widehat{C}_0 = C_0$ and $\widehat{C}_{i+1} = \widehat{C}_i \nabla post(C_i)$, reaches a fixed point after finitely many steps.

We call such an operator ∇ a *widening* operator. Consider the modified forward search algorithm in which the line $C := C \cup post(C)$ is replaced by $C := C \nabla post(C)$. The modified algorithm clearly terminates. However, after termination only $C \supseteq post^*(I)$ is guaranteed, and not $C = post^*(I)$, because at some step of the algorithm we can have $C \subseteq post^*(I)$ but $C_i \nabla post(C_i) \not\subseteq post^*(I)$, i.e., the application of ∇ may produce non-reachable states. But assume that ∇ further satisfies

(c) for every $C \in \mathcal{F}$: $C \nabla post(C) \subseteq post^*(C)$

In this case we know that Condition (6) holds for the modified algorithm. We call a modified forward search algorithm with a widening operator satisfying (a), (b) and (c), an *acceleration* of forward search. An instance of this acceleration will be applied to pushdown automata in Section 12.5.

12.3. Timed Automata

As outlined in the Introduction, this section and the next two follow a vertical approach. First we introduce a small but natural verification challenge in an informal way. Then we introduce a class of extended automata and use it to produce a formal model. After that, we define a family of symbolic configurations and show that it satisfies Conditions (1)–(6) of symbolic search. The crucial part is mostly Condition (6), and we shall use increasingly sophisticated techniques to establish it, as described in Section 12.2.5.

12.3.1. *The Case Study: A Small Mutex Algorithm*

Consider the following simple version of the well-known *mutual exclusion* problem. Two processes wish to access a shared resource, and access is granted by entering a critical section. It must be ensured that exactly one process is granted access (we say that this process "wins" and the other "loses"). There is no central process guarding access to the resource, and the processes can only communicate through *one* single shared boolean variable v.

If the variable can be tested and set to a new value *in a single atomic action* then there is a simple solution:

> The initial value of v is set to 1. Both processes execute the following algorithm: if $v = 0$, give up (the process loses); otherwise, *and in the same atomic action*, set v to 0; then enter the critical section (the process wins).

Clearly, the first process to access the variable wins, and the other loses.

If an atomic action can *either* test the value of v *or* set it to a new value, *but not both*, then it is well-known that there is no solution within the limits of a conventional programming language: more variables are needed. However, Michael Fischer observed that there is a very simple solution if one assumes that processes also have access to a clock:

> The initial value of v is arbitrary. The i-th process, with $i \in \{1, 2\}$, executes the following algorithm: at any time before a time unit passes (*strictly* before), the process sets v to i; then it waits for any amount of time exceeding one time unit (*strictly* exceeding), and then it tests whether $v = i$ holds: if $v = i$, then the process enters the critical section (and wins), otherwise it gives up (and loses).

Notice that time is not assumed to be discrete; how much less or more than one time unit the process waits should be irrelevant for the correctness of the algorithm.

We give a formal model of Fischer's protocol as a *timed automaton*, and automatically check that exactly one process accesses the critical section. Timed automata (TA) were introduced in Alur and Dill's seminal paper [5]. Most of the ideas and illustrating examples (except for the worked out case study) are based on the excellent survey paper by Bengtsson and Yi [1], with a number of changes and additions to make the description fit the framework presented in Section 12.2.3.

12.3.2. *Timed Automata*

Timed automata are a special class of extended automata over variables called *clocks*, and with two kinds of rules, called *time-delay* and *state-switch* rules. The formal definition requires some preliminaries.

We call a variable ranging over the non-negative reals a *clock*. Throughout the section we assume a set $\mathcal{X} = \{x_1, \ldots, x_k\}$ of clocks. Consequently, a valuation is a vector of k nonnegative real numbers.

A clock constraint is defined by the following BNF, where $\sim \in \{<, \leq, =, \geq, >\}$, $x, y \in \mathcal{X}$, and n is a non-negative integer:

$$g \ ::= \ \mathbf{true} \mid x \sim n \mid x - y \sim n \mid g \wedge g$$

A constraint is *pure* if it does not contain any atomic formulas of the form $x - y \sim n$. A clock valuation ν may *satisfy* a clock constraint g, written $\nu \models g$. The satisfaction relation is defined as follows:

- $\nu \models \mathbf{true}$ for every valuation ν
- $\nu \models x \sim n$ iff $\nu(x) \sim n$
- $\nu \models x - y \sim n$ iff $\nu(x) - \nu(y) \sim n$
- $\nu \models g_1 \wedge g_2$ iff $\nu \models g_1$ and $\nu \models g_2$

Given a valuation ν of a set $\{x_1, \ldots, x_n\}$ of clocks and a real number δ, we denote by $\nu + \delta$ the valuation given by $(\nu + \delta)(i) = \nu(i) + \delta$ for every $i \in \{1, \ldots, k\}$. The *time-delay action* is the relation $TD = \{(\nu, \nu + \delta) \mid \nu \in \mathcal{V}, \delta > 0\}$. Given a

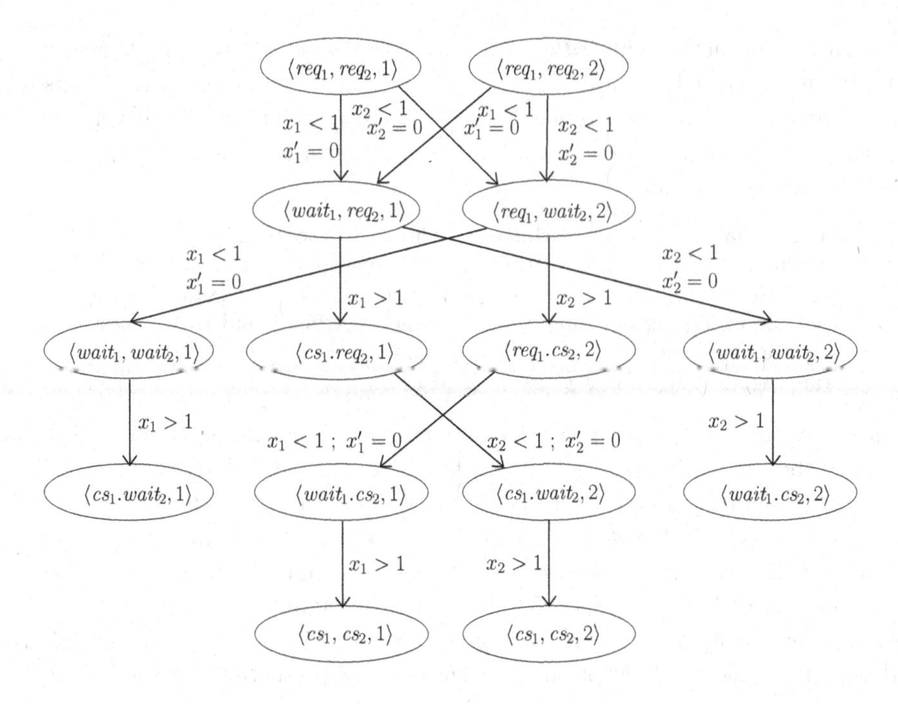

Fig. 12.2. A small mutex algorithm modeled as a timed automaton. Control locations are triples representing the local state of the first process, the second process, and the value of a local variable, which can be either 1 or 2.

subset $Y \subseteq \mathcal{X}$ of clocks and a valuation ν, we denote by ν_Y the valuation given by $\nu_Y(i) = 0$ for every $x_i \in Y$ and $\nu_Y(i) = \nu(i)$ for every $x_i \notin Y$. The *Y-reset action* is the relation $R_Y = \{(\nu, \nu_Y) \mid \nu \in \mathcal{V}\}$.

A *timed automaton* is an extended automaton over $\mathcal{X}$ whose rules satisfy the following conditions:

- For each control state q there is a rule $(q, 2^{\mathcal{V}}, TD, q)$, i.e., a rule that is enabled at every clock valuation, does not change the state, and whose action advances all clocks by a nondeterministically chosen lapse of time δ.
- All other rules are of the form (q, g, R_Y, q') for some *pure* clock constraint g and some $Y \subseteq \mathcal{X}$; i.e., the action consists of resetting the clocks in Y, while keeping the values of those in $\mathcal{X} \setminus Y$.

Rules of the first and second kind are called *time-delay* and *state-switch* rules, respectively. Notice that since the time-delay rules are completely determined by the set of states one does not need to mention them explicitly. For this reason, they are omitted in the usual syntax for timed automata and in the graphical representation.

We can now model our simplified version of Fischer's mutex algorithm as the timed automaton shown in Figure 12.2. A more conventional way of depicting

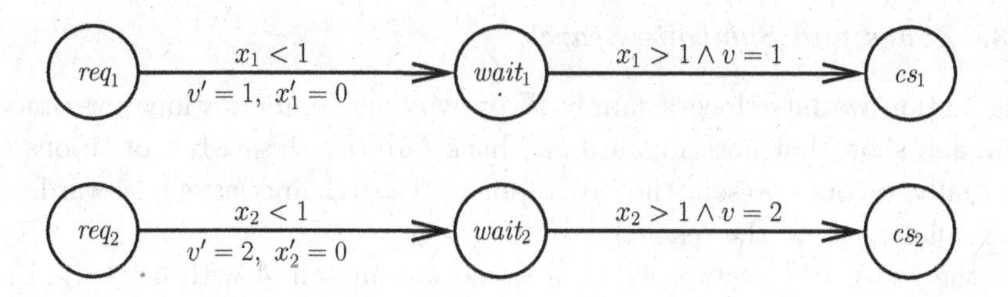

Fig. 12.3. The case study specified more compactly as a network of two automata. The full version of Figure 12.2 results from this one by a product construction.

the protocol in terms of a *network* of two automata is given in Figure 12.3.[a] The automaton has two clocks, x_1 and x_2, one for each process; we denote the action of resetting clock x_i by $x_i' = 0$. We assume that the i-th process ($i = 1, 2$) can be in three states: req_i, previous to setting v to i; $wait_i$, where the process waits before testing if $v = i$ holds, and cs_i, in which the process has reached the critical section. A state of the automaton is a triple (q_1, q_2, v) where q_1 and q_2 are the current states of processes 1 and 2, and v is the current value of the shared variable. A configuration of the automaton consists thus of a state of the automaton and the values of the two clocks. We write it as a five-tuple; for example, $(req_1, req_2, 1, 0.3, 0.3)$ denotes the configuration where process 1 is in state req_1, process 2 is in state req_2, the shared variable has value 1 and both clocks, x_1 and x_2, currently show 0.3 time units. Here are two sample runs of the automaton, where a number on top of a transition indicates the time elapsed.

Sample Run 1:

$$\begin{aligned}(req_1, req_2, 1, 0, 0) &\Longrightarrow (wait_1, req_2, 1, 0, 0)\\ &\overset{2.1}{\Longrightarrow} (wait_1, req_2, 1, 2.1, 2.1)\\ &\Longrightarrow (cs_1, req_2, 1, 2.1, 2.1)\end{aligned}$$

Sample Run 2:

$$\begin{aligned}(req_1, req_2, 1, 0, 0) &\overset{0.7}{\Longrightarrow} (req_1, req_2, 1, 0.7, 0.7)\\ &\Longrightarrow (wait_1, req_2, q, 0, 0.7)\\ &\overset{0.2}{\Longrightarrow} (wait_1, req_2, 1, 0.2, 0.9)\\ &\Longrightarrow (wait_1, wait_2, 2, 0.2, 0)\\ &\overset{1.8}{\Longrightarrow} (wait_1, wait_2, 2, 2, 1.8)\\ &\Longrightarrow (wait_1, cs_2, 2, 2, 1.8)\end{aligned}$$

[a]Loosely speaking, the network consists of two timed automata working in parallel with common clocks. We refrain from giving a formal semantics here.

12.3.3. *Zones and Symbolic Search*

In this section we introduce a family $\mathcal{F}$ of symbolic configurations for timed automata, and show that both forward and backward search satisfy Conditions (1) to (6) (actually, we only present the case of forward search and leave backward search, which is analogous, to the reader).

For the rest of the section we fix a timed automaton $\mathcal{A}$ with a set Q of states over a set $\mathcal{X}$ of clocks, where $k = |\mathcal{X}|$.

Let max be the maximum integer appearing in the guards of the state-switch rules of $\mathcal{A}$. A clock constraint has max as *ceiling* if all the integers appearing in it are smaller than or equal to max. A set of valuations is a *zone* of $\mathcal{A}$ if it is the set of *all* valuations satisfying a clock constraint (pure or not) with ceiling max. We denote the set of all zones by $\mathcal{Z}$. Notice that, while a clock constraint uniquely determines a zone, the same zone may correspond to many clock constraints. Observe further that $\mathcal{Z}$ is finite: while there are infinitely many clock constraints with ceiling max, only finitely many of them are non-equivalent. We call the elements of the set $Q \times \mathcal{Z}$ *indexed zones*. We identify an indexed zone (q, Z) and the set of configurations $\{\langle q, z\rangle \mid z \in Z\}$,

We choose the family $\mathcal{F}$ of symbolic configurations as follows: a set of configurations C is an element of $\mathcal{F}$ if there are indexed zones $(q_1, Z_1), \ldots, (q_m, Z_m)$ such that $C = \bigcup_{i=1}^{m}(q_i, Z_i)$. We say that C is a union of indexed zones, and write $\mathcal{F} = \mathcal{P}(Q \times \mathcal{Z})$, i.e., we choose $\mathcal{F}$ as the powerset of the set of indexed zones.

In the remainder of this section we check that this family satisfies Conditions (1) to (6) for forward search. Since Condition (1) requires the elements of $\mathcal{F}$ to have a finite symbolic representation, we first introduce *Difference Bound Matrices* [6] to represent zones, and describe some of their properties.

Difference Bound Matrices (DBMs) In order to define DBMs, we add an additional reference variable $\mathbf{0}$ with constant value 0 to the set of clocks, and write $\mathcal{X}_0 = \mathcal{X} \cup \{\mathbf{0}\}$. Any clock constraint g, can then be rewritten as a conjunction of constraints of the form $x - y \preceq n$ for $x, y \in \mathcal{X}_0$, $\preceq \in \{<, \leq\}$, and $n \in \mathbb{Z}$. First, every constraint involving only one variable, say $x < 20$, can be rewritten as $x - \mathbf{0} < 20$. Second, teh conjunction of two constraints on the same pair of variables (e.g. $x_1 - x_2 \leq 3$ and $x_1 - x_2 \leq 4$), can be replaced by the intersection of both ($x_1 - x_2 \leq 3$ in our case). Therefore, every constraint is equivalent to another one with one atomic constraint for each pair of clocks from $\mathcal{X}_0$, and so g can be represented as a $(k + 1) \times (k + 1)$ matrix, where each element corresponds to an atomic constraint.

Example 12.3. Consider the zone

$$g_0 = x < 20 \ \wedge \ y \leq 20 \ \wedge \ y - x \leq 10 \ \wedge \ y - x \geq 5 \ \wedge \ z > 5$$

over clocks $\{x, y, z\}$. It can be rewritten using $\mathbf{0}$ and $<, \leq$ only:

$$g_0' = x - \mathbf{0} < 20 \ \wedge \ y - \mathbf{0} \leq 20 \ \wedge \ y - x \leq 10 \ \wedge \ x - y \leq -5 \ \wedge \ \mathbf{0} - z < -5$$

The DBM of g_0' is a 4×4 matrix, whose entries are elements of $(\mathbb{Z} \times \{<, \leq\}) \cup \{\infty\}$. Assuming $\mathbf{0}$ has index 0 and x, y, z have indices 1, 2, and 3, respectively, entry $(-5, \leq)$ in row 2, column 3 denotes $x - y \leq -5$, and entry ∞ in row 4, column 1, means "$z - \mathbf{0} \leq \infty$", that is, z has no upper bound. Entries on the diagonal will always be $(0, \leq)$ to include trivial constraints like $x - x \leq 0$. The matrix, $M(g_0')$, corresponding to g_0' is then:

$$\begin{pmatrix} (0,\leq) & (0,\leq) & (0,\leq) & (-5,<) \\ (20,<) & (0,\leq) & (-5,\leq) & \infty \\ (20,\leq) & (10,\leq) & (0,\leq) & \infty \\ \infty & \infty & \infty & (0,\leq) \end{pmatrix} = M(g_0')$$

Closing a DBM. Different DBMs can represent the same zone. Consider for instance the DBM above, and the one in which the constraint $x - \mathbf{0} < 20$ is replaced by $x - \mathbf{0} \leq 15$. Clearly, the zone of the new DBM is included in the zone of the DBM above. But the contrary also holds: every clock assignment of the DBM above satisfies $x - y \leq -5$ and $y - \mathbf{0} \leq 20$, and so it also satisfies $x - \mathbf{0} \leq 15$; in other words, the constraint $x - \mathbf{0} < 20$ of the original DBM can be tightened without changing the zone.

Fortunately, each nonempty zone has a *unique* DBM that cannot be further tightened. The process of finding it is called *closing* a DBM and it yields a canonical DBM.

For the purpose of this presentation, it shall suffice to say that the canonical form of a DBM can be computed by a graph interpretation of a DBM, where each clock is a node, and a constraint between two clocks is a edge labeled by the constraint. Then closing a DBM amounts to computing all pairs-shortest paths, for example by using an algorithm like Floyd-Warshall [7].

For instance, in the graph representation of $M(g_0')$ there is an edge labeled $(20, <)$ from x to $\mathbf{0}$. However, there is also an edge from x to y labeled $(-5, \leq)$ and an edge1 from y to $\mathbf{0}$ labeled $(20, \leq)$. Going from x to $\mathbf{0}$ via y adds up to $(15, \leq)$ and is in fact a shorter path than the direct one.

We will mostly write down zones as clock constraints or graphically. In particular, for two zones, we will draw some illustrating diagrams as the ones in Figure 12.4.

In the rest of the section we sketch the proof of the following theorem:

Theorem 12.1. *Let $\mathcal{A}$ be a timed automaton, and let I and D be elements of $\mathcal{P}(Q \times \mathcal{Z})$. Forward search satisfies Conditions (1)–(6).*

Condition (1). Since a zone is finitely represented by its associated clock constraint (or DBM), each element of $\mathcal{P}(Q \times \mathcal{Z})$ has a finite representation as a set of pairs (q, g), where q is a state, and g is a clock constraint.

Condition (2). The set I of initial configurations is an element of $\mathcal{P}(Q \times \mathcal{Z})$ by hypothesis. In the case of Fischer's mutex algorithm the singleton set of clock assignments $\{(0,0)\}$ is a zone, because it is the set of solutions of the clock constraint $x_1 = 0 \land x_2 = 0$. Since $I = \{(req_1, req_2, 1, 0, 0), (req_1, req_2, 2, 0, 0)\}$, we have $I \in \mathcal{P}(Q \times \mathcal{Z})$.

Condition (3). We have to show that if C is an element of $\mathcal{P}(Q \times \mathcal{Z})$, then so is $C \cup post(C)$. Since the set of symbolic configurations is the powerset of another set, we can apply the observation to Condition (3) in Section 12.2.4; so it suffices to show that for every indexed zone (q, g) and for every rule r of the timed automaton the set $post_r(\{(q, g)\})$ is the union of indexed zones. In fact, we can show a stronger result: either $post_r(\{(q, g)\})$ is empty (and so the union of the empty set of zones), or a zone (q', g') (and so the union of a set of zones containing only one element). We consider separately the case in which r is a time-delay and a state-switch rule.

- **Time-delay rules**: Consider an indexed zone (q, g). Notice that if g implies a constraint $x - y \leq n$, then so do all clock valuations reachable from g by a time delay (because a delay increases the values of all clocks by the same amount). Furthermore, the valuations reachable through time delays can reach arbitrarily large values for each clock. From this observation it is easy to see that if g is given as a DBM $M(g)$, then the valuations reachable from those satisfying g by means of time delays are represented by the DBM obtained by setting all entries of $M(g)$ in column 0 to ∞. Recall that column 0 contains entries representing $x - \mathbf{0} \leq n$, that is, just upper bounds on individual clocks. From this DBM we can easily get a constraint g' for the set of reachable valuations, and we have $post_r(\{(q, g)\}) = (q, g')$.
- **State-switch rules**: Consider an indexed zone (q, g) and a state-switch rule $r = (q, g'', R_Y, q')$. If $g \land g''$ is unsatisfiable, then $post_r(\{(q, g)\})$ is empty. If $g \land g''$ is satisfiable, let $M(g \land g'')$ be a closed DBM for $g \land g''$ (which can be computed because clock constraints are closed under conjunction). We compute another closed DBM corresponding to resetting the clocks of Y in all clock valuations satisfying $g \land g''$. This DBM is obtained from $M(g \land g'')$ as follows: for every clock $x_i \in Y$, replace the constraints in row i, column 0, and row 0, column i of $M(g)$ by $(0, \leq)$, and remove all other bounds on x_i (i.e., replace all other constants in row i or column i by ∞ or $-\infty$); finally, close the result. From this DBM we extract a clock constraint g', and we have $post_r(\{(q, g)\}) = (q', g')$.

Notice that the operations on DBMs do not introduce any constant larger than the ceiling of the automaton under consideration, and so the result is indeed a zone as we defined it.

Condition (4). We have to show that if $C, D \in \mathcal{P}(Q \times \mathcal{Z})$, then it is decidable whether $C \cap D = \emptyset$. Clearly, it suffices to show that if g and g' are zones, then we can decide whether $g \wedge g'$ is empty. For this, notice that since clock constraints are closed under conjunction we can get a DBM for $g \wedge g'$. For a zone to be empty there must be a pair of clocks such that the upper bound on their difference is smaller than the lower bound. This corresponds to a negative cycle in the graph interpretation, of the DBM, which can be detected using well-known techniques.

Condition (5). The strong version of the condition states that given two symbolic configurations C_1, C_2, it is decidable whether $C_1 = C_2$ holds. While equality of symbolic configurations is in fact decidable, the weaker condition explained in Section 12.2.3 is much easier to prove (and leads to a more efficient search algorithm). Recall the condition: given the chain $C_0 \subseteq C_1 \subseteq C_2 \ldots$ of symbolic configurations, where $C_0 = I$ and $C_{i+1} = C_i \cup post(C_i)$, there is an algorithm satisfying the following properties: if two sets are not equal, then the algorithm answers "not equal"; if there is an index i such that $C_i = C_{i+j}$ for every $j \geq 0$, then there is also $k \geq i$ such that for input C_k, C_{k+1} the algorithm answers "equal". We provide such an algorithm. Let γ_i denote the symbolic representation of C_i as a set of pairs (q, M), where q is a state and M is a closed DBM. Since $post((q, M))$ is a set of zones, we have $\gamma_0 \subseteq \gamma_1 \subseteq \gamma_2 \ldots$. Since the set of indexed zones is finite, there is a γ_k such that $\gamma_k = \gamma_{k+1}$, i.e., γ_k and γ_{k+1} contain exactly the same pairs (q, M). So the algorithm just compares the symbolic representations, and answers 'equal' if they are *syntactically* equal.

Condition (6). The chain $C_0 \subseteq C_1 \subseteq C_2 \ldots$ necessarily reaches a fixed point after finitely many steps, because the set $\mathcal{P}(Q \times \mathcal{Z})$ of symbolic configurations is finite. This is an instance of the finite case of Section 12.2.5.

12.3.4. *Zone Graph of Fischer's Mutex Algorithm*

We have shown in Condition (3) above that if C is an indexed zone, then for every rule r the set $post_r(C)$ is also an indexed zone. One can then define the *zone graph* of a timed automaton: the nodes of the graph are the indexed zones, and there is an edge from (q, g) to (q', g') if $\{(q', g')\} = post_r(\{(q, g)\})$ for some rule r. The reachable part of the zone graph of the timed automaton modelling Fischer's algorithm is shown in Figure 12.5. More precisely, the figure only shows one half of the graph, the other half is completely symmetric with the roles of the processes swapped. We use clock constraints to represent zones. All zones occurring in the zone graph are also given as a two-dimensional point set in Figure 12.4, and it is advisable to digest both figures in parallel.

From the zone graph one can easily see how the symbolic forward search procedure works. The set I is the indexed zone at the top of the figure. After k iterations the set C contains all the indexed zones that can be reached from i in k steps. Since the zone graph does not contain any node in which both processes are in the critical section, mutual exclusion holds.

We shall now *symbolically* follow Sample Run 2 presented at the end of Section 12.3.2. The initial symbolic configuration is a pair of state $\langle req_1, req_2, 1\rangle$ and zone $x_1 = 0 \wedge x_2 = 0$, where both processes are requesting access to the shared resource. The point set representation of this zone consists of a single point in the two-dimensional space and is depicted in Figure 12.4(b).

- From this configuration, we can either let time elapse, as indicated by the dashed arrow, or we could let process 1 move to state $wait_1$, as indicated by the solid arrow. The latter transition does not change the zone part of the symbolic configuration, while the first one results in the zone shown in (c). Graphically, a time-delay action amounts to constructing the shadow of a light source at $(-\infty, -\infty)$. Note that the origin is excluded since at least an arbitrarily small amount of time must elapse. Point exclusion is depicted as a circle.
- While x_1 is still smaller than 1, process 1 may enter the $wait_1$ state, while process 2 remains unaffected. The effect to the zone part of the symbolic configuration, resets x_1 to zero. Note that we apply the reset to the *conjunction* of zone (c) and the guard $x_1 < 1$. So the resulting zone is (d), where x_2 is known to be strictly between 0 and 1.
- From here we let time elapse again and, according to the shadow intuition, end up in a zone that resembles (a) and (g), but is not shown in Figure 12.4. It is obtained by extending (g) to the x_2 axis.
- Now process 2 moves to $wait_2$ thereby setting the value of v to 2. Together with the guard $x_2 < 1$, the previous zone has x_2 reset yielding zone (e).
- Faithful to our behavior so far, we let time elapse again, that is, apply the shadow operation to (e), which results in zone (a).
- As the final transition of this run we let process 2 move to the critical section. The guard $x_2 > 1$ ensures that the shadow only starts at values greater than 1.

12.3.5. *Conclusion and Further Reading*

We have introduced timed automata, and shown that by taking zones as symbolic configurations forward search becomes effective. A slight modification of the arguments shows that backward search is effective as well.

Much of the content of this section is taken from [1], where Bengtsson and Yi describe the efficient implementation of forward and backward search used in UPPAAL, the leading tool for the analysis of timed automata. A survey by Alur and Madhusudan on the decidability and complexity of decision problems for timed automata can be found in [8].

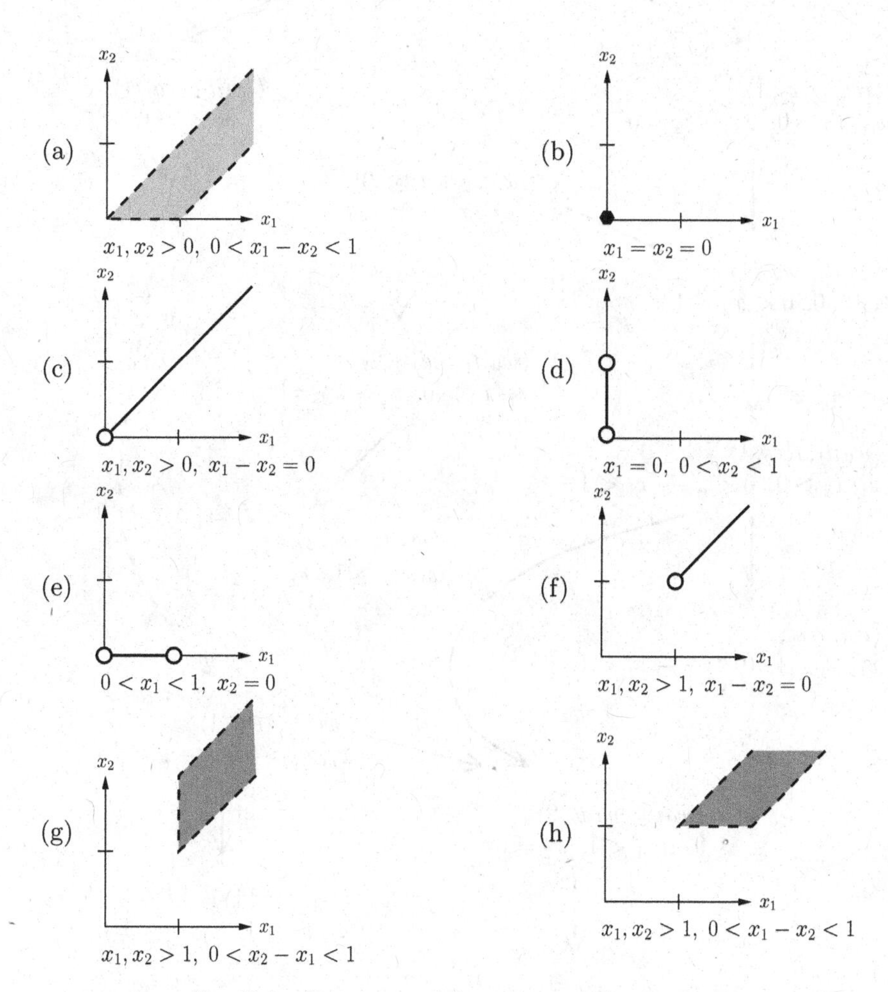

Fig. 12.4. Some zones of the zone graph shown as point sets.

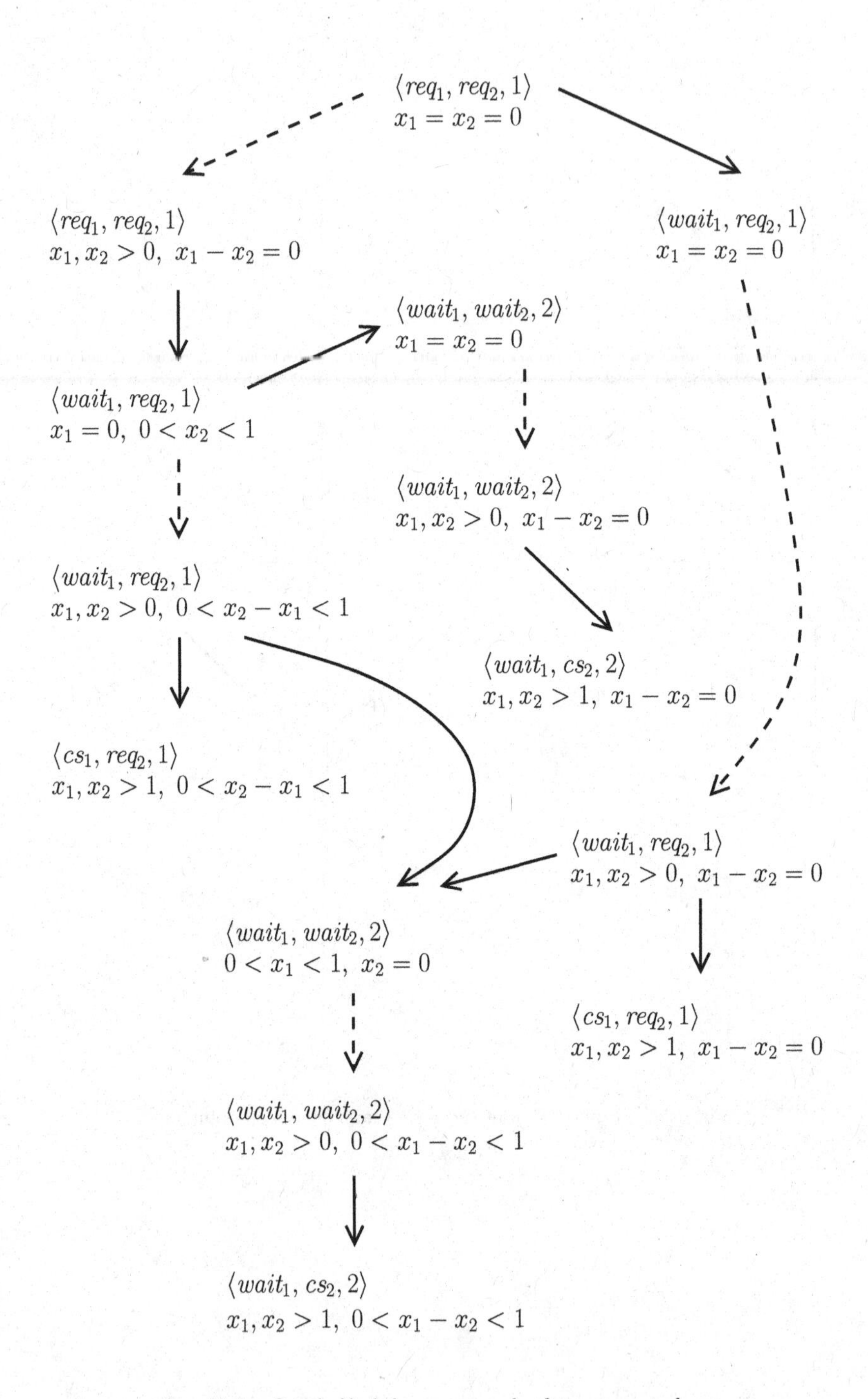

Fig. 12.5. One half of the zone graph of our case study.

12.4. Linear Automata

In this section we model a cache coherence protocol as a linear automaton, an extended automaton over non-negative integer variables whose guards and actions are linear expressions, and verify that it satisfies a property by means of a symbolic backward search. We prove that this search always terminates by showing the absence of infinite ascending chains in our family of symbolic configurations.

12.4.1. *The Case Study: A Cache Coherence Protocol*

Our case study is a simple cache coherence protocol for a multiprocessor system in which each processor has a local cache connected to main memory by a bus.

Recall that a cache memory provides a processor with a copy of a part of the current main memory for fast access during program execution. Local caches on multiprocessor machines improve performance, but introduce the *cache coherence problem*: multiple cached copies of the same block of memory must be kept consistent. The following scenario must be avoided: processor A writes a memory block for which processor B has a local copy in cache; before the cache is updated, processor B reads it; processor B believes it is getting the updated value of the block, while in fact it only gets the old value. We say that a memory block (in main memory or in a cache) is *valid* if it has been updated with the value of the last write access by any processor, and *invalid* otherwise. A cache management protocol is called *coherent* if a read access always provides a valid block.

Coherence is usually guaranteed by means of *write-invalidate* protocols: whenever a processor modifies a cache memory block (a *cache line*), the block address and a bus invalidation signal is sent to all other caches. Receiving this signal invalidates the corresponding cache line and an updated copy of the block must be fetched from main memory. (*Write* protocols send a copy of the new data to all caches sharing the old data). *Snoopy caches* continuously "listen" to the block addresses sent over the bus by other processors, and react when the addresses match their own cache lines (*bus snooping*).

We describe the MESI protocol, a write-invalidate, snoopy protocol [9]. For simplicity, we assume that each processor has one single cache-line, i.e., only one memory block is copied in the cache, and so we speak of a cache instead of a cache line. We model each processor as logically divided into a CPU that sends read and write requests to a *cache-management* unit (CMU), which is also connected to the bus.

In the MESI protocol a CMU is always in one out of four possible states: *modified* (M), *exclusive* (E), *shared* (S), and *invalid* (I), with the following *intended* meaning: in I, the cache is invalid; in S, the cache is valid, and there maybe other valid caches; in E, the last write access was performed by the CMU's own CPU, and so it is valid; moreover, the value has already been transferred to main memory, i.e., the value in main memory is also valid, but no other cache is valid; finally, in state M the last

n most recent writes, with $n > 1$, have been performed by the cache's own CPU; however, the last write has not been transferred to main memory yet; so the cache is valid, but main memory is not, and no other cache is valid.

The CMU can receive read or write requests. A read/write request to a valid cache is a *read/write hit*, otherwise a *read/write miss*. The reactions of the CMU are as follows:

- **Read Hit**. Only possible from states other than *invalid*. The CMU returns the local value of the cache, no coherence action is required.
- **Read Miss**. Only possible in state *invalid*. The CMU goes from *invalid* to *shared*, all caches in states *exclusive* or *modified* go to *shared*.
- **Write Hit**. Only possible in states other than invalid. If the CMU is in state *shared*, it goes to *exclusive* after invalidating the contents of all other caches; if it is in *exclusive* or *modified* it goes to *modified*, no bus transition is required.
- **Write Miss**. Only possible in state *invalid*. The CMU goes from *invalid* to *exclusive* after invalidating all other caches.

The actual property we are interested in is cache coherence. However, proving cache coherence for a protocol is too complex to achieve within the limits of this chapter. So we shall establish and formally prove an important necessary condition for cache coherence: A cache in state M (*modified*) should be the only valid cache.

12.4.2. *Linear Automata*

Linear automata are extended automata over non-negative integer variables, with guards and actions given by linear expressions. The formal definition requires some preliminaries.

Let $\mathbf{x} = (x_1, \ldots, x_n)$ be a tuple of *non-negative* integer variables. A valuation of $\mathbf{x}$ is therefore an element of $\mathbb{N}^n$. A *linear constraint* over $\mathbf{x}$ is inductively defined as follows:

- a linear inequation over $x_1, \ldots, x_n$ with rational coefficients is a linear constraint;
- a boolean combination of linear constraints is a linear constraint.

A *solution* of a linear constraint is a tuple $\mathbf{k} = (k_1, \ldots, k_n) \in \mathbb{N}^n$ such that simultaneously substituting $k_1, \ldots, k_n$ for $x_1, \ldots, x_n$ yields a true expression. The set of solutions of a linear constraint ϕ is denoted by $[\![\phi]\!]$. An example of a linear constraint over the variables x, y, z is $\phi = (x \geq y + 1 \wedge y = 3z)$, and we have $[\![\phi]\!] = \{(3k + 1 + l, 3k, k) \mid k, l \in \mathbb{N}\}$. We often identify a constraint and its set of solutions.

A *linear transformation* is a system of linear equations over two tuples $\mathbf{x} = (x_1, \ldots, x_n)$ and $\mathbf{x}' = (x'_1, \ldots, x'_n)$ of variables of the form

$$x'_1 = f_1(x_1, \ldots, x_n)$$
$$\ldots$$
$$x'_n = f_n(x_1, \ldots, x_n)$$

We sometimes denote a linear transformation by $\mathbf{x}' = \tau(\mathbf{x}$, and write $\tau T(\mathbf{x} = (f_1(\mathbf{x}, \ldots, f_n(\mathbf{x}))$ As for linear constraints, a solution of a linear transformation is a pair $((k_1, \ldots, k_n), (k'_1, \ldots, k'_n)) \in \mathbb{N}^n \times \mathbb{N}^n$ satisfying all equations.

We are now ready to formally define linear automata. A *linear automaton* is an extended automaton over a tuple $\mathbf{x}$ of non-negative integer variables that satisfies the following two conditions: the guards of the rules are linear constraints over $\mathbf{x}$, and the actions of the rules are linear transformations over $\mathbf{x}$ and $\mathbf{x}'$.

Let us now see how the MESI protocol can be formalized as a linear automaton. The automaton has only one state q, and acts on a four-tuple (m, e, s, i) of variables. A configuration $\langle q, (k_m, k_e, k_s, k_i)\rangle$ indicates that the number of CMUs in states M, E, S, and I, is k_m, k_e, k_s and k_i, respectively. Notice that $k_m + k_e + k_s + k_i$ is the total number of CMUs, and does not change during the execution of the protocol. Guards are linear constraints over the tuple (m, e, s, i) of variables, and actions are linear transformations over the tuples $(m, e, s, i), (m', e', s', i')$. The dynamics of the protocol is modelled by the rules of Table 12.1, which correspond to the intended behavior of the protocol after a Read Hit, Read Miss, Write Hit, and Write Miss. For instance, a Write Hit can take place if some cache is in state S, i.e., if $s \geq 1$, and it changes the state of this cache to E, and the state of all other caches to I, i.e., $e' = 1$ and $i' = m + e + s + i - 1$. Notice that, since the automaton only has one state, it is not necessary to indicate the source and the target state of a rule.

Table 12.1. Rules of the MESI protocol.

Event	Guard	Action	Rule Name
Read Hit	$m + e + s \geq 1$	$m' = m$ $e' = e$ $s' = s$ $i' = i$	r_1
Read Miss	$i \geq 1$	$m' = 0$ $e' = 0$ $s' = m + e + s + 1$ $i' = i - 1$	r_2
Write Hit	$m \geq 1$	$m' = m$ $e' = e$ $s' = s$ $i' = i$	r_3
	$e \geq 1$	$m' = m + 1$ $e' = e - 1$ $s' = s$ $i' = i$	r_4
	$s \geq 1$	$m' = 0$ $e' = 1$ $s' = 0$ $i' = m + e + s + i - 1$	r_5
Write Miss	$i \geq 1$	$m' = 0$ $e' = 1$ $s' = 0$ $i' = m + e + s + i - 1$	r_6

The initial configurations of the protocol are those satisfying the constraint $(m = 0 \wedge e = 0 \wedge s = 0)$, i.e., initially all CMUs are in the state *invalid*.

An n-configuration is a configuration whose valuation satisfies the constraint $m + e + s + i = n$. For instance, $\langle q, (2, 0, 2, 0)\rangle$ is a 4-configuration. Since the number of CMUs does not change during the execution of the protocol a transition leaving a n-configuration always leads to another n-configuration. Therefore, for each fixed value of n the set of reachable states of the protocol is finite and can be exhaustively explored. However, this only allows to verify properties of the protocol for one particular instance of the protocol, the one with n participating CMUs. The challenge is to automatically verify that a property holds *for any number of CMUs.*

We can formally state our example property – a cache in state *modified* should be the only valid cache – in terms of linear constraints as follows: The MESI protocol should never reach any configuration satisfying the linear constraint $(m \geq 1 \wedge e+s \geq 1) \vee m \geq 2$. We automatically check that this is the case.

12.4.3. *Lc-sets and Backward Search*

In this section we introduce a family $\mathcal{F}$ such that backward search satisfying conditions (1) to (5), but not condition (6). This means that, while we can apply this version of backward search to linear automata, there are instances for which the algorithm does not terminate. In Section 12.4.4 we will first apply the algorithm to the MESI protocol, and observe that in this case it does terminate. In Section 12.4.5 we will show that this is not a coincidence: we prove that the algorithm always terminates for the subclass of *monotonic linear automata*, of which the MESI protocol is an instance, when the set D of dangerous configurations is *upward closed*, which will also be the case.

All linear automata modeling cache-coherence protocols have only one control state. In what follows, for the sake of simplicity, we only consider this special case, but the reader will have no difficulty in extending the results to the case of several states.

When the automaton has one state, a configuration is completely determined by its valuation. So we write ν instead of $\langle q, \nu\rangle$ and speak of "the configuration ν". Also, instead of $\langle q, \nu\rangle \Longrightarrow \langle q, \nu\rangle$ we just write $\nu \Longrightarrow \nu'$.

A set of configurations C is *linearly constrained*, or an *lc-set* for short, if there is a linear constraint ϕ such that $C = [\![\phi]\!]$. We choose the lc-sets as symbolic configurations, i.e., $\mathcal{F}$ contains all lc-sets. In the rest of this section we prove the following result:

Theorem 12.2. *Let $\mathcal{A}$ be a linear automaton, and let I and D be lc-sets. Symbolic backward search with lc-sets satisfies conditions (1)–(5), but not condition (6).*

Condition (1). An lc-set C is finitely represented by the linear constraint ϕ such that $C = [\![\phi]\!]$.

Condition (2). The set D is an lc-set by hypothesis. Notice that in the case of the MESI protocol we have $D = (m \geq 1 \wedge e + s \geq 1) \vee m \geq 2$, and so D is an lc-set.

Condition (3). We have to show that if C is an lc-set, then $C \cup pre(C)$ is also an lc-set. Since linear constraints are closed under disjunction, the union of two lc-sets is also an lc-set. So it suffices to show that $pre_r(C)$ is an lc-set for every rule r. We need a bit of notation.

Let ϕ be a linear constraint over a tuple $\mathbf{x} = (x_1, \ldots, x_n)$ of variables, and let $\mathbf{x}' = \tau(\mathbf{x})$ be a linear transformation . We write $\phi[\mathbf{x}/\tau(\mathbf{x})]$ to denote the constraint obtained by syntactically replacing every occurrence of x_i in ϕ by the expression $f_i(x_1, \ldots, x_n)$ of τ (i.e., the right-hand-side of the equation for x'_i in τ).

For example, let $\phi_0 = (e \leq m + 1 \wedge s = 3m + 2i)$, and let $\tau(\mathbf{x})$ be the transformation of rule r_4 of Table 12.1. Then

$$\begin{aligned} \phi_0[\mathbf{x}/\tau(\mathbf{x})] &= (e \leq m + 1 \wedge s = 3m + 2i)[e/e - 1, m/m + 1] \\ &= (e - 1 \leq m + 1 + 1 \wedge s = 3(m + 1) + 2i) \end{aligned}$$

which can be simplified to $(e \leq m + 3 \wedge s = 3m + 2i + 3)$.

We can now prove the following lemma, which shows how to compute a linear constraint for $pre_r(C)$ given a linear constraint for C.

Lemma 12.1. *Let ϕ and τ be the linear constraint and the linear transformation corresponding to the guard and the action of a rule r. Let ψ be a constraint representing an lc-set C (i.e., $C = [\![\psi]\!]$). We have:*

$$pre_r(C) = [\![\phi \wedge \psi[\mathbf{x}/\tau(\mathbf{x})]]\!] \,.$$

Proof. Observe first that $[\![\psi[\mathbf{x}/\tau(\mathbf{x}]\!]$ is the weakest precondition of $[\![\psi]\!]$ under the simultaneous assignment $\mathbf{x} := \tau(\mathbf{x})$. This follows immediately from the well-known weakest-precondition rule for simultaneous assignment. So $[\![\psi[\mathbf{x}/\tau(\mathbf{x})]]\!]$ contains all states that are transformed into states of C by τ. However, for such a state to belong to $pre_r(C)$ it must also satisfy the guard ϕ, and the lemma follows. □

For example, consider rule r_4 of the MESI-protocol of Table 12.1, and let $C = [\![\phi_0]\!]$ with ϕ_0 as above. We have

$$\begin{aligned} pre_{r_4}(C) &= pre_{r_3}(e \leq m + 1 \wedge s = 3m + 1/2i) \\ &= e \geq 1 \ \wedge \ e \leq m + 3 \wedge s = 3m + 1/2i + 3 \end{aligned}$$

With the help of this lemma we can easily establish **Condition (3)**: if C is an lc-set of configurations, then so is $C \ \cup \ pre(C)$. Since linear constraints are closed under disjunction, it suffices to show that $pre_r(C)$ is an lc-set for every rule r. By Lemma 12.1, it suffices to show that $\phi \wedge \phi[\mathbf{x}/\tau(\mathbf{x})]$ is a linear constraint, where ϕ and τ are the guard and the transformation of the rule r, respectively. Since ϕ is linear, we only have to show that $\phi[\mathbf{x}/\tau(\mathbf{x})]$ is linear. This follows from the fact that τ is a linear transformation. Let $x'_i = f_i(\mathbf{x})$ the the i-th inequation of τ. For every $1 \leq i \leq n$, we have to replace every occurrence of x_i in ϕ by $f_i(\mathbf{x})$. Since $f_i(\mathbf{x})$ is linear, the result of the substitution is again a linear constraint.

Condition (4). For every lc-set C the emptiness of $C \cap I$ is decidable. Since we assume that both C and I are lc-sets represented by linear constraints, say ϕ_C and ϕ_I, we have $C \cap I = [\![\phi_C \wedge \phi_I]\!]$, and so it suffices to prove the decidability of the satisfiability problem for linear constraints. This follows easily from the following observations:

- Every linear constraint is equivalent (has the same solutions) to a linear constraint without negations.
  To prove this, observe first that negations can be pushed inwards through conjunctions and disjunctions. Then, observe that the negation of a linear (in)equation is equivalent to a disjunction of linear (in)equations. For instance $\neg(x \leq y)$ is equivalent to $(x > y')$, and $\neg(x = y)$ is equivalent to $(x < y) \vee (y < x)$.
- Every constraint is equivalent to a constraint in disjunctive normal form without negations.
- Every disjunct of a constraint in disjunctive normal form without negations is a system of Diophantine equations and inequations.

So the satisfiability problem reduces to deciding if a system of linear Diophantine equations and inequations has a solution. In turn, this problem can be reduced to Integer Linear Programming, which is known to be solvable in nondeterministic polynomial time (see for instance [10]). Specific algorithms for Diophantine equations and inequations also exist, see for instance [11]. The algorithm corresponding to this sketch of a decidability proof is not efficient, but the issue of efficiency is beyond the scope of this paper, and we refer the reader to the literature.

Condition (5). The equality of lc-sets is decidable. Given two constraints ϕ_1, ϕ_2, we have $[\![\phi_1]\!] = [\![\phi_2]\!]$ if and only if $[\![(\phi_1 \wedge \neg\phi_2) \vee (\neg\phi_2 \wedge \phi_1)]\!] = \emptyset$, which we have shown to be decidable when proving that Condition (4) holds.

It remains to show that **Condition (6)** fails. Consider the linear automaton acting on two variables x, y, and having one single rule with $x \leq 1$ as guard and $x' = x - 1, y' = y + 1$ as action. Let $D = (x = 0)$. Then $pre^*(D) = \mathbf{true}$, but $pre^i(D) = (x \leq i)$, and so backward search does not terminate.

12.4.4. *Backward Search for the MESI Protocol*

We apply backward search with lc-sets to the MESI protocol. We start with:

$$D = (m \geq 1 \wedge e + s \geq 1) \vee m \geq 2\,.$$

We compute $pre(D)$ using the procedure sketched in the proof of **Condition (3)**. Notice that for our reachability problem it is not necessary to consider rules r_1 and r_3 of Table 12.1, because their actions are the identity transformation.

$$\begin{aligned}
& pre(D) \\
= \; & pre_{r_2}(D) \cup pre_{r_4}(D) \cup pre_{r_5}(D) \cup pre_{r_6}(D) \\
= \; & \quad (i \geq 1 \wedge D[m/0, e/0, s/m+e+s+1, i/i-1]) \\
& \vee (e \geq 1 \wedge D[m/m+1, e/e-1]) \\
& \vee (s \geq 1 \wedge D[m/0, e/1, s/0, i/m+e+s+i-1]) \\
& \vee (i \geq 1 \wedge D[m/0, e/1, s/0, i/m+e+s+i-1]) \\
= \; & \quad i \geq 1 \wedge ((0 \geq 1) \wedge 0+m+e+s+1 \geq 1) \vee 0 \geq 2) \\
& \vee e \geq 1 \wedge ((m+1 \geq 1 \wedge e-1+s \geq 1) \vee m+1 \geq 2) \\
& \vee s \geq 1 \wedge ((0 \geq 1) \wedge 1+0 \geq 1) \vee 0 \geq 2) \\
& \vee i \geq 1 \wedge ((0 \geq 1) \wedge 1+0 \geq 1) \vee 0 \geq 2) \\
= \; & e \geq 1 \wedge (e+s \geq 2 \vee m \geq 1) \\
= \; & (e \geq 1 \wedge e+s \geq 2) \vee (e \geq 1 \wedge m \geq 1)
\end{aligned}$$

(Note that, since we are reasoning about non-negative integers only, it is always possible to drop constraints of the form $x \geq 0$.)

We can simplify $D \cup pre(D)$ by getting rid of $e \geq 1 \wedge m \geq 1$, which is implied by the first disjunct of D, to obtain

$$D \cup pre(D) = (m \geq 1 \wedge e+s \geq 1) \vee m \geq 2 \vee (e \geq 1 \wedge e+s \geq 2).$$

It is easy to see that $D \subset D \cup pre(D)$ holds – for instance because the configuration $(0,1,1,0) \in pre(D) \setminus D$. So we proceed with the search by computing $pre^2(D)$, this time without so much detail as in the previous case:

$$\begin{aligned}
& pre^2(D) \\
& = pre_{r_4}(pre(D)) \\
& = e \geq 1 \wedge (e-1 \geq 1 \wedge (m+1 \geq 1 \vee e-1+s \geq 2)) \\
& = e \geq 2 \wedge (m \geq 0 \vee e+s \geq 3) \\
& = e \geq 2
\end{aligned}$$

Since $[\![e \geq 2]\!] \subset [\![e \geq 1 \wedge e+s \geq 2]\!]$, we obtain $pre^2(D) \subset pre(D)$, which implies $D \cup pre(D) = D \cup pre(D) \cup pre^2(D)$. So backward search has reached a fixed point, and we can conclude

$$pre^*(D) = D \cup pre(D) = (m \geq 1 \wedge e+s \geq 1) \vee m \geq 2 \vee (e \geq 1 \wedge e+s \geq 2).$$

Recall that the set I of initial configurations of the MESI protocol is given by $I = (m = 0 \wedge e = 0 \wedge s = 0)$. So $I \cap pre^*(D) = \emptyset$, and so the MESI protocol satisfies that a cache in state *modified* is the only valid cache.

12.4.5. *Monotonic Linear Automata and Upward-Closed Sets*

Termination of the algorithm for the MESI protocol is not just luck. We show that backward search always terminates for *monotonic linear automata*, of which the MESI protocol is an instance, when the set of dangerous configurations is *upward closed*, which is also the case.

A linear automaton is *monotonic* if the following condition holds: for all configurations $\nu_1, \nu_1', \nu_2 \in \mathbb{N}^n$, if $\nu_2 \geq \nu_1$ and $\nu_1 \Longrightarrow \nu_1'$ (i.e., ν_1' is reachable from ν_1 in one step), then there exists $\nu_2' \geq \nu_1'$ such that $\nu_2 \Longrightarrow \nu_2'$.

The automaton of the MESI protocol is monotonic: Assume $\nu_2 \geq \nu_1$ and $\nu_1 \Longrightarrow \nu_1'$. Then there exists a rule r of the automaton with a guard ϕ and a linear transformation τ such that $\nu_1 \in [\![\phi]\!]$ and $\nu_1' = \tau(\nu_1)$. Since $\nu_2 \geq \nu_1$ and ϕ is of the form $q \geq 1$ for some $q \in \{m, e, s, i\}$ (see Table 12.1), we have $\nu_2 \in [\![\phi]\!]$. So $\nu_2 \Longrightarrow \nu_2'$ for $\nu_2' = \tau(\nu_2)$. Since $\nu_2 = \nu_1 + \delta$ for some non-negative vector δ and τ is linear, we have $\nu_2' = \tau(\nu_1 + \delta) = \tau(\nu_1) + \tau(\delta)$. For the linear transformations of Table 12.1 the vector $\tau(\delta)$ is also non-negative, and so $\nu_2' \geq \nu_1'$.

We study symbolic backward on monotonic linear automata with a new class of symbolic configurations, the upward-closed sets.

Given $\nu, \nu' \in \mathbb{N}^n$, we say $\nu \leq \nu'$ if $\nu(i) \leq \nu'(i)$, for every $i \in [1..n]$, where $\nu(i)$ and $\nu'(i)$ denote the i-th component of ν and ν', respectively; we say $\nu \lneq \nu'$ if $\nu \leq \nu'$ and $\nu(i) < \nu'(i)$ for some $i \in [1..n]$. A set of configurations C of a linear automaton is *upward-closed* if $\nu \in C$ and $\nu' \geq \nu$ implies $\nu' \in C$. A linear constraint ϕ is *upward closed* if $[\![\phi]\!]$ is upward closed.

Observe that the linear constraint $\phi = (m \geq 1 \wedge s + e \geq 1) \vee m \geq 2$ is upward closed, and so the set $D = [\![\phi]\!]$ of dangerous configurations of the MESI protocol is upward closed.

At first glance, the relation between the upward-closed and the lc-sets is not clear. We show that upward-closed sets are a special class of lc-sets. For this we need the following well-known result, a variant of Dickson's lemma. For the sake of completeness, we sketch a proof.

Lemma 12.2. *Any set $C \subseteq \mathbb{N}^n$ has finitely many minimal elements with respect to the partial order $\leq$.*

Proof. Assume there exists $C \subseteq \mathbb{N}^n$ such that the set $M \subseteq C$ of minimal elements of C is infinite. We prove by induction on n that M contains two elements ν, ν' such that $\nu \lneq \nu'$, contradicting the assumption that ν' is a minimal element of C. In fact, we prove a stronger statement: M contains an infinite chain $\nu_1 \lneq \nu_2 \lneq \nu_3 \ldots$. For the base case $n = 1$, let ν_1 be a minimal element of M, ν_2 a minimal element of $M \setminus \{\nu_1\}$, ν_3 a minimal element of $M \setminus \{\nu_1, \nu_2\}$ etc. Since for $n = 1$ the order $\leq$ is total, we have $\nu_1 \lneq \nu_2 \lneq \nu_3 \ldots$. For the induction step, assume $n \geq 1$. Given a configuration $\nu \in \mathbb{N}^n$, let $\nu' \in \mathbb{N}^{(n-1)}$ denote the projection of ν onto the first $(n-1)$ components, and let $l \in \mathbb{N}$ denote the value of ν's last component, i.e., ν is obtained by adding l to ν' as last component. By induction hypothesis, M contains configurations $\nu_1, \nu_2, \nu_3, \ldots$ whose projections satisfy $\nu_1' \lneq \nu_2' \lneq \nu_3' \ldots$. Choose an index i_1 such that $l_{i_1} \leq l_j$ for every $j \in \mathbb{N}$; then an index i_2 such that $l_{i_2} \leq l_j$ for every $j \in \mathbb{N} \setminus \{1, 2, \ldots, i_1\}$; then an index i_3 such that $l_{i_3} \leq l_j$ for every $j \in \mathbb{N} \setminus \{1, 2, \ldots, i_2\}$etc. This produces an infinite sequence of indices $i_1 < i_2 < i_3 \ldots$ such that $l_{i_1} \leq l_{i_2} \leq l_{i_3} \ldots$. Since $\nu_{i_1}' \lneq \nu_{i_2}' \lneq \nu_{i_3}' \ldots$ also holds, we get $\nu_{i_1} \lneq \nu_{i_2} \lneq \nu_{i_3} \ldots$. □

Now, let C be an arbitrary upward-closed set. By Lemma 12.2, C has finitely many minimal elements $\nu_1, \ldots, \nu_k$. Let $\nu_i = (k_i^1, \ldots, k_i^n)$; then $C = [\![\phi_1 \vee \ldots \vee \phi_k]\!]$, where $\phi_i = x_1 \geq k_i^1 \wedge \ldots \wedge x_n \geq k_i^n$. So C is an lc-set.

In the rest of this section we sketch the proof of the following result:

Theorem 12.3. *Backward search satisfies Conditions (1)–(6) for each monotonic linear automaton $\mathcal{A}$, upward closed set D, and lc-set I.*

Condition (1). Since upward-closed sets are linearly constrained, they have a finite symbolic representation as the set of solutions of a constraint. Moreover, since an upward-closed set is completely determined by its set of minimal elements, and by Lemma 12.2 this set is finite, an upward-closed set can also be finitely represented by its set of minimal elements.

Condition (2) holds by hypothesis. In the case of the MESI protocol we have $D = (m \geq 1 \wedge e + s \geq 1) \vee m \geq 2$, which is equal to the upward-closed set with $\{(1,1,0,0),(1,0,1,0),(2,0,0,0)\}$ as set of minimal elements.

For **Condition (3)** we have to show that, if C is an upward-closed set of configurations of a monotonic linear automaton, then *pre*(C) is also upward closed. This amounts to proving that $\nu_1 \in pre(C)$ and $\nu_2 \geq \nu_1$ imply $\nu_2 \in pre(C)$. Since $\nu_1 \in pre(C)$, there exists $\nu_1' \in C$ such that $\nu_1 \Longrightarrow \nu_1'$ and $\nu_1' \in C$. Since the automaton is monotonic and $\nu_2 \geq \nu_1$, there exists $\nu_2' \geq \nu_1'$ such that $\nu_2 \Longrightarrow \nu_2'$. Since C is upward-closed, $\nu_1' \in C$, and $\nu_2' \geq \nu_1'$, we have $\nu_2' \in C$, and so $\nu_2 \in pre(C)$. Note that the union of two upwards closed sets is upwards closed, and so $C \cup pre(C)$ is upwards closed if C is.

We have not yet shown that a symbolic representation of *pre*(C) can be effectively computed from the symbolic representation of C, but this holds for the representation as linear constraint because of Lemma 12.1. It is not difficult to see that it also holds for the representation by the set of minimal elements. We leave the design of an algorithm that, given the minimal elements of C as input, yields the minimal elements of *pre*(C) as output, as an exercise for the reader.

Conditions (4) and (5) follow immediately from Lemma 12.2 and the fact that the same conditions hold for lc-sets.

Finally, we prove that **Condition (6)** also holds, namely that every infinite chain $U_1 \subseteq U_2 \subseteq U_3 \ldots$ of upward-closed sets contains an element U_k such that $U_k = \bigcup_{i \geq 1} U_i$. By Lemma 12.2, the set M of minimal elements of $\bigcup_{i \geq 1} U_i$ is finite, and so there exists an index k such that $M \subseteq U_k$. Let M_k denote the set of minimal elements of U_k. Since U_k is upward-closed and $M \subseteq U_k$, we necessarily have $M = M_k$. It follows $U_k = \bigcup_{i \geq i} U_i$, which concludes the proof of Theorem 12.3.

Therefore, the fact that our backward search computation terminated for the MESI protocol was not a stroke of luck: for monotonic protocols, and if the set D is upward-closed, the computation is guaranteed to terminate.

12.4.6. *Conclusion and Further Reading*

Delzanno has used backward search to automatically prove properties of other cache-coherence protocols [2]. Some of them can be modelled by monotonic automata, others cannot. Remarkably, backward search terminates in all cases

The termination of backward search has been shown by proving that the domain of symbolic configurations satisfies the ascending chain condition. In turn, this result is based on the existence of an order $\leq$ on the set $\mathcal{V}$ of valuations ($\mathbb{N}^n$ in our case) satisfying the following two fundamental properties:

- the extended automaton is monotonic w.r.t. $\leq$, and
- every subset of $\mathcal{V}$ has finitely many minimal elements w.r.t. $\leq$; an order satisfying this property is called a *well-order*.

These properties turn out to hold not only for monotonic linear automata, but for other classes of extended automata working on other data structures, like lossy channel systems, a class of automata working on tuples of words, or timed Petri nets. Historically, the approach was first applied by Abdulla and Jonsson to lossy channel systems [12]. In a series of papers, with different co-authors, Abdulla and Jonsson have vastly generalized and extended the original idea (see for instance [13]); Finkel and Schnoebelen have also contributed very substantially [14].

12.5. Pushdown Automata

In this section we model a small recursive sequential program as a pushdown automaton. Pushdown automata were already shown to be an instance of extended automata in Example 12.2. We prove a simple property of the program by means of an *accelerated* forward symbolic search.

12.5.1. *The Case Study: Skylines*

Consider the C-program of Figure 12.6, taken from [3]. The question marks in the conditionals mean nondeterministic choice, which is not a primitive in C but could be easily simulated. The program may be understood as a controller for a plotter or for printing on the screen. The procedures `up`, `right`, and `down` are supposed to draw a line of unit length, starting at the current position of the pen or the cursor, in the direction indicated by the name of the procedure; the code for this is omitted, for our purposes we can just assume that they return immediately.

The program is supposed to draw *skylines* like the one shown on the left of Figure 12.7. In the course of this section we will prove that the program never draws *degenerate skylines* like the one on the right of Figure 12.7. A skyline is degenerate if at some point the program draws a line going up, and immediately after a line going down. In a non-degenerate skyline, the program always moves right at least once between any pair of up and down moves.

```
   main() {
0:     s();
1: }

   void s() {
0:     if (?) {
1:         up();
2:         m();
3:         down();
       }
4: }
```

```
   void m() {
0:     if (?) {
1:         s();
2:         right();
3:         if (?)
4:             m();
       } else {
5:         up();
6:         m();
7:         down();
       }
8: }
```

Fig. 12.6. A C program drawing skylines. Procedures for actual drawing – up, right, and down – are left unspecified.

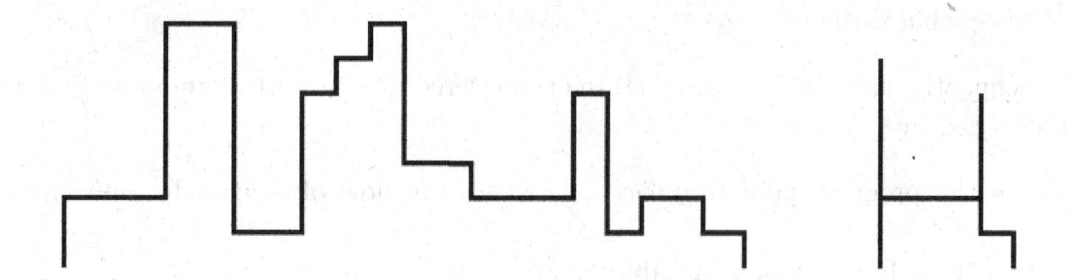

Fig. 12.7. A skyline and a degenerate skyline.

12.5.2. *Pushdown Automata*

We model sequential programs with possibly recursive procedures as pushdown automata. Recall that, as explained in Example 12.2, we define a pushdown automaton as an extended automaton over a single variable, the *stack*, whose values are words over an alphabet Γ of *stack symbols*. The words are also called *stack contents*. A configuration of a PDA is therefore a pair $\langle q, v\rangle$, where q is a state and $v \in \Gamma^*$. The guards of the rules check if the topmost symbol of the current stack content is equal to a fixed symbol; the actions replace the topmost symbol by a fixed word. So a rule is determined by its source and target states, the fixed symbol that the guard compares with the topmost stack symbol, and the word that the action replaces the top symbol with.

We formally define a PDA using a notation that slightly differs from the one we have used for extended automata. The reason is that we wish the notation to resemble the classical one for pushdown automata. A PDA is a triple (Q, Γ, Δ), where Q is the set of states, often called *control states*, Γ is the stack alphabet, and $\Delta \subseteq (Q \times \Gamma) \times (Q \times \Gamma^*)$ is the set of rules. We write $\langle q, \gamma \rangle \longrightarrow \langle q', v \rangle$, if $(q, \gamma, q', v) \in \Delta$. Without loss of generality we assume $|v| \leq 2$ for all $\langle q, \gamma \rangle \longrightarrow \langle q', v \rangle$ in Δ.

A configuration of a PDA is a pair of a control state and the current stack content (the current value of the stack variable in terms of extended automata). The PDA can move from configuration $\langle q, \gamma w \rangle$ to configuration $\langle q', vw \rangle$ by means of a rule $\langle q, \gamma \rangle \longrightarrow \langle q', v \rangle$ that replaces the topmost stack symbol γ with v. Formally, $\langle q, \gamma w \rangle \Longrightarrow \langle q', vw \rangle$ if there exists a rule $\langle q, \gamma \rangle \longrightarrow \langle q', v \rangle$ such that $w = \gamma v'$ and $w' = vv'$.

Sequential Programs as PDA We shall now discuss how a sequential, procedural program may be encoded as a PDA. Such programs are mainly determined by

- control flow: assignments, conditionals, loops, and procedure calls, possibly with parameters and return values;
- local variables of each procedure; and
- global variables.

Consequently, any given runtime state encountered during a program execution is determined by

- the program pointer indicating, where the flow of control has currently arrived;
- the values of global variables; and
- the values of local variables of the each procedure that has not yet returned.

Note that there is a tight correspondence between the *stack* of activation records and the *stack* of a PDA. Indeed, we shall interpret a configuration $\langle q, \gamma v \rangle$ such that γ encodes the current activation record and v encodes the pending records. Following these lines, γ represents the program pointer within the active procedure as well as the content of the local variables of this procedure. Pending procedures, pending return address and pending local variable contents are saved in v. Control state q has a more global flavor and will be used to hold global variables. Since we are dealing with finite sets of stack symbols, we must restrict ourselves to values of finite domain data types when saving variable contents.

The correspondence between program statements and PDA rules is as follows:

- A simple statement, like an assignment, only influences the topmost activation record and, potentially, the value of global variables. So it is encoded by rules of the form $\langle q, \gamma \rangle \longrightarrow \langle q', \gamma' \rangle$.
- Procedure calls push a new record on top of the call stack. The global variables and the current activation record do not change. So procedure calls can be modeled by rules of the form $\langle q, \gamma \rangle \longrightarrow \langle q, \gamma' \gamma \rangle$.

- When a procedure returns, it simply pops the topmost record from the stack. This can be modeled by a rule of the form $\langle q, \gamma \rangle \longrightarrow \langle q, \epsilon \rangle$.

Skylines – Formally. We model the skyline program as a PDA. Since the program does not have any global variables, the PDA will only needs a single control state, which we call q_s. Also since there are not any local variables either, the stack symbols correspond to the possible program points. We denote the program points belonging to procedure p by $p_0, p_1, \ldots$, with p_0 as the initial point. Note that the actual drawing procedures, which are not explicitly modeled, only have a single program point each: up_0, $down_0$, and $right_0$, respectively. So we can model the skyline program by a PDA

$$A_s = (\{q_s\},\ \Gamma_s,\ \Delta_s)$$

over the set Γ_s of stack symbols given by

$$\Gamma_s = \{main_0, main_1, s_0, \ldots, s_5, m_0, \ldots, m_8, up_0, down_0, right_0\}$$

Notice how there is exactly one stack symbol for each control point of the skyline program in Figure 12.6. The set Δ_s of rules is shown in Figure 12.8. For instance, the rule $\langle q_s, m_2 \rangle \longrightarrow \langle q_s, right_0 m_3 \rangle$ models the call to procedure `right()` at program point 2 of procedure `main()`. The rules marked with `if (true)` and `if (false)` model the conditionals. The marks `(*)` and `(**)` are used later.

Non-degenerate skylines. Recall that we are interested in proving the absence of a call to `up` immediately followed by a call to `down`. In terms of our encoding we like to prove the absence of transitions like

$$\langle q_s, v \rangle \Longrightarrow \langle q_s, up_0 v' \rangle \Longrightarrow \langle q_s, v' \rangle \Longrightarrow \langle q_s, down_0 v'' \rangle$$

Our property is thus a property of a *sequence* of configurations instead of a property of a *single* configuration. For this reason symbolic reachability of configurations cannot directly prove this property. However, there is a generally useful trick to circumvent this problem: monitors. A monitor is a finite automaton running in parallel with the pushdown system. Our monitor has two states, q_{up} and $\overline{q_{up}}$. The monitor will be in state q_{up} if the most recent drawing action (`up`, `down`, or `right`) encountered has been `up`. So the monitor moves to state q_{up} whenever up_0 is pushed onto the stack, stays there during all subsequent `up` calls, and moves to $\overline{q_{up}}$ when it sees a call to `down` or `right`.

Running the monitor in parallel with the skyline PDA is achieved by constructing the product of the monitor and the PDA, which is simply the product of a finite automaton and a PDA. So we consider the "monitored" skyline PDA

$$A_s' = (\{q_{up}, \overline{q_{up}}\},\ \Gamma_s,\ \Delta_s')$$

where Δ_s' is obtained from Δ_s by

Procedure `main`

$\langle q_s, main_0\rangle \longrightarrow \langle q_s, s_0 main_1\rangle$
$\langle q_s, main_1\rangle \longrightarrow \langle q_s, \epsilon\rangle$

Procedure `up`

$\langle q_s, up_0\rangle \longrightarrow \langle q_s, \epsilon\rangle$

Procedure `down`

$\langle q_s, down_0\rangle \longrightarrow \langle q_s, \epsilon\rangle$

Procedure `right`

$\langle q_s, right_0\rangle \longrightarrow \langle q_s, \epsilon\rangle$

Procedure `s`

$\langle q_s, s_0\rangle \longrightarrow \langle q_s, s_1\rangle$	`if (true)`
$\langle q_s, s_0\rangle \longrightarrow \langle q_s, s_4\rangle$	`if (false)`
$\langle q_s, s_1\rangle \longrightarrow \langle q_s, up_0 s_2\rangle$	$(\star)$
$\langle q_s, s_2\rangle \longrightarrow \langle q_s, m_0 s_3\rangle$	
$\langle q_s, s_3\rangle \longrightarrow \langle q_s, down_0 s_4\rangle$	$(\star\star)$
$\langle q_s, s_4\rangle \longrightarrow \langle q_s, \epsilon\rangle$	

Procedure `m`

$\langle q_s, m_0\rangle \longrightarrow \langle q_s, m_1\rangle$	`if (true)`
$\langle q_s, m_0\rangle \longrightarrow \langle q_s, m_5\rangle$	`if (false)`
$\langle q_s, m_1\rangle \longrightarrow \langle q_s, s_0 m_2\rangle$	
$\langle q_s, m_2\rangle \longrightarrow \langle q_s, right_0 m_3\rangle$	$(\star\star)$
$\langle q_s, m_3\rangle \longrightarrow \langle q_s, m_4\rangle$	`if (true)`
$\langle q_s, m_3\rangle \longrightarrow \langle q_s, m_8\rangle$	`if (false)`
$\langle q_s, m_4\rangle \longrightarrow \langle q_s, m_0 m_8\rangle$	
$\langle q_s, m_5\rangle \longrightarrow \langle q_s, up_0 m_6\rangle$	$(\star)$
$\langle q_s, m_6\rangle \longrightarrow \langle q_s, m_0 m_7\rangle$	
$\langle q_s, m_7\rangle \longrightarrow \langle q_s, down_0 m_8\rangle$	$(\star\star)$
$\langle q_s, m_8\rangle \longrightarrow \langle q_s, \epsilon\rangle$	

Fig. 12.8. The encoding of the skyline program as a PDA. Program points belonging to procedure p are written p_i for naturals i. They correspond to the program labels of the C program of Figure 12.6.

- replacing each rule $\langle q_s, \gamma\rangle \longrightarrow \langle q_s, up_0\gamma'\rangle$ marked by $(\star)$ in Figure 12.8 by two rules $\langle q_{up}, \gamma\rangle \longrightarrow \langle q_{up}, up_0\gamma'\rangle$ and $\langle \overline{q_{up}}, \gamma\rangle \longrightarrow \langle q_{up}, up_0\gamma'\rangle$;
- replacing each rule $\langle q_s, \gamma\rangle \longrightarrow \langle q_s, \gamma'\gamma''\rangle$ marked $(\star\star)$ in Figure 12.8 by two rules $\langle q_{up}, \gamma\rangle \longrightarrow \langle \overline{q_{up}}, \gamma'\gamma''\rangle$ and $\langle \overline{q_{up}}, \gamma\rangle \longrightarrow \langle \overline{q_{up}}, \gamma'\gamma''\rangle$; and
- replacing each unmarked rule $\langle q_s, \gamma\rangle \longrightarrow \langle q_s, w\rangle$ in Figure 12.8 by two rules $\langle q_{up}, \gamma\rangle \longrightarrow \langle q_{up}, w\rangle$ and $\langle \overline{q_{up}}, \gamma\rangle \longrightarrow \langle \overline{q_{up}}, w\rangle$.

We assume that the skyline program starts with a call to `main` in control location $\overline{q_{up}}$; that is, no drawing action, and in particular no upwards drawing action, has yet been observed. Our set I_s of initial configurations can thus be written as:

$$I_s = \{\langle \overline{q_{up}}, main_0\rangle\}$$

Now, the skyline program produces degenerate configurations if A'_s can reach any of the configurations of the set

$$D_s = \{\langle q_{up}, down_0 w\rangle \mid w \in \Gamma_s^*\}$$

a configuration in D_s is reached if an `up` drawing action is immediately followed by a `down` drawing action, regardless of the history represented by w in the definition of D_s. We can automatically check whether the program produces degenerate configurations by answering the question

$$post^*_{A'_s}(I_s) \cap D_S \stackrel{?}{=} \emptyset$$

In Section 12.5.3, we introduce *regular configurations* as our set of symbolic configurations, and show that they satisfy **Conditions (1)–(6)**. The symbolic representation of the set $post^*_{A'_s}(I_s)$ is computed and discussed in Section 12.5.4.

12.5.3. *Regular Configurations*

The number of reachable configurations of a PDA may be infinite, due to the possibility of unbounded stacks. In particular, the skyline program may go up forever, stacking more and more unfinished calls to procedure s. So we must find a finite symbolic representation of possibly infinite sets of configurations.

We fix a PDA $A = (Q, \Gamma, \Delta)$. Since stack contents are words, sets of stack contents are languages over the alphabet Γ. Since *regular* languages can be finitely represented by finite automata, we choose as our family $\mathcal{F}$ of symbolic configurations the *regular sets of configurations.* A set C of configurations of A is *regular* if the set $\{w \in \Gamma^* \mid \langle q, w\rangle \in C\}$ is regular for all $q \in Q$.

We show that regular configurations satisfy **Conditions (1)–(5)** for forward search, but not **Condition (6)**. We then show that it is possible to accelerate the search so that it terminates.

Theorem 12.4. *Let I and D be regular sets of configurations. Symbolic forward search with regular sets satisfies Conditions (1)–(5), but not Condition (6).*

Condition (1). We need to show that every regular set of configurations has a finite symbolic representation. As the possible stack contents for each control location are known to be regular languages, one can compactly represent regular sets of configurations as a non-deterministic finite automaton with ϵ-transitions (NFAs). Accepting a configuration is achieved by making all control locations of a given PDA, A, initial states of the NFA. A configuration $\langle q, w\rangle$ is accepted by the NFA if it accepts w starting in state q. It is straightforward that every regular set of configurations of A has such a finite symbolic representation as an NFA. Formally, we shall call an NFA for accepting regular sets of configurations of A an A-automaton. It is defined as follows:

Let $A = (Q, \Gamma, \Delta)$ be a PDA. An A-*automaton* is a finite automaton $B = (P, \Gamma, Q, F, \delta)$, where P is a set of states containing Q, i.e., $P \supseteq Q$ holds, Γ is an alphabet, Q is a set of initial states, $F \subseteq P$ is a set of final states, and $\delta \subseteq P \times (\Gamma \cup \{\epsilon\}) \times (P \setminus Q)$ is a transition relation. B *accepts* a configuration $\langle q, w\rangle$ of A if some path of B leading from q to some final state q' is labeled by w; in this case we write $q \overset{w}{\rightsquigarrow}{}^* q'$. Notice that δ denotes the transition relation of an A-automaton, while Δ denotes the set of rules of the PDA.

Our definition of δ forbids transitions in B leading into some initial state. This is important for the correctness of the forward search algorithm presented below. Observe that every regular language is accepted by some NFA satisfying this condition.

Example 12.4. Figure 12.9 shows two A'_s-automata. The one on the left accepts the set I_s, and the one on the right the set D_s.

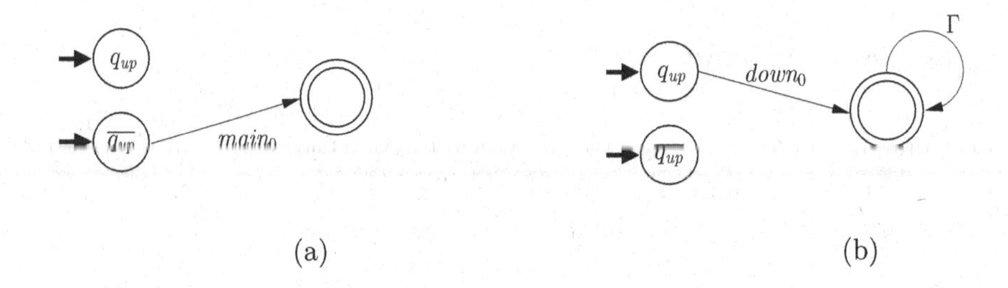

Fig. 12.9. Automata accepting the initial and dangerous configurations of the skyline PDA. Initial states are marked by big arrows, final states by double circles. If a set of stack symbols is attached to an arc, it denotes a set of transitions, one for each element of the set.

Condition (2). The set I_s is regular by hypothesis. The set $I_s = \{\langle \overline{q_{up}}, main_0 \rangle\}$ of our case study is accepted by the automaton on the left of Figure 12.9.

Condition (3). We have to show that if C is a regular set of configurations of PDA A, then also $C \cup post_A(C)$ is regular and efficiently computable from C's symbolic representation.

Let B be an A-automaton accepting a regular set C of configurations of A. We construct an A-automaton accepting $C \cup post_A(C)$. First, as we are interested in computing the *union* $C \cup post_A(C)$, we can safely keep all states and transitions of B. We add more states and transitions to also accept the configurations of $post_A(C)$.

The configurations of $post_A(C)$ are obtained by applying a rule to the configurations of C. The PDA A can have three types of rules: rules removing the topmost stack symbol, rules replacing it with one other symbol, and rules replacing it with a word of length 2. Figure 12.10 illustrates how to add new states and transitions to deal with each of these three cases. The second row of the Figure shows the three kinds of rules, and the third the transitions and states added. If in B we can go from q to a state p by reading the symbol γ (possibly together with an arbitrary number of ϵ's), and we have a rule $\langle q, \gamma \rangle \longrightarrow \langle q', \epsilon \rangle$ in Δ, then $post_A(C)$ must accept any configuration $\langle q', w \rangle$ such that $\langle p, \gamma w \rangle$ is accepted by B. Therefore, we add an ϵ-transition (q', ϵ, p) to the transition relation δ of B. This will exactly achieve this goal. The other two kinds of actions are treated similarly .

We obtain the whole of $C \cup post_A(C)$ by applying this construction to all rules in Δ and to all matchings of $q \overset{\gamma}{\leadsto}^* p$ in B. Note that we must indeed require initial states in B not to have incoming transitions. The reason is essentially the same as the reason for introducing a new initial state in the union construction of two

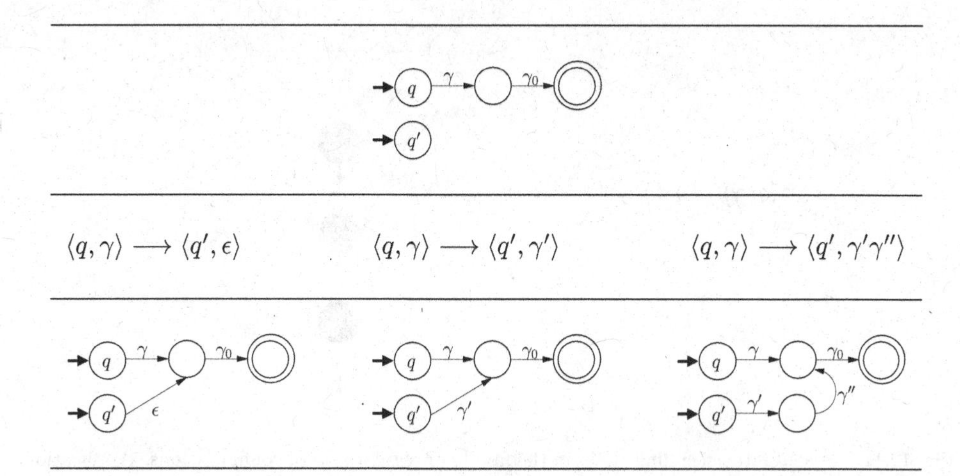

Fig. 12.10. Construction of $C \cup post_A(C)$. The first row shows part of an automaton accepting C. The second row shows the three possible type of PDA rules. The third row shows an automaton accepting $C \cup post_r(C)$ for each rule r.

NFAs. If, for instance, q' in Figure 12.10 had a self-loop, it would be easy to see that this may add non-reachable configurations to $C \cup post(C)$.

Having defined the notion of A-automata makes the proof of **Condition (4).** The emptiness of $C \cap D$ can be easily checked using standard automata-theoretic techniques.

Condition (5). Checking $C_1 = C_2$ is decidable, because equality of languages accepted by NFA is decidable. Checking equality is known to be computationally expensive, but once we present the acceleration in the next section it will become clear that in our special case the procedure can be extremely simplified.

To check that **Condition (6)** does not hold, consider the PDAwith only one rule shown on the left of Figure 12.11 and the A-automaton shown on the right (ignore the black states for the moment), accepting only the configurations $\langle q, \gamma \rangle$. Each application of the procedure described in the proof of Condition (3) adds a new state to the automaton, and so forward search computes an infinite sequence of A-automata, each of them accepting exactly one word more than the previous one. So we have $C_0 \subset C_1 \subset C_2 \ldots$, and forward search does not terminate. This concludes the proof of Theorem 12.4.

If we look at the example of Figure 12.11 in more detail, we observe that $post^*(\langle q, \gamma \rangle) = \langle p, \gamma\gamma'^* \rangle$ is a regular set. This set is recognized by an automaton where we keep only the first black state in Figure 12.11, and add a self-loop to it, labeled by γ'. So, while the fixed point is a member of the family $\mathcal{F}$, forward search never reaches it, it keeps constructing better and better approximations to it, but never getting there. In the next section we show how to deal with this problem.

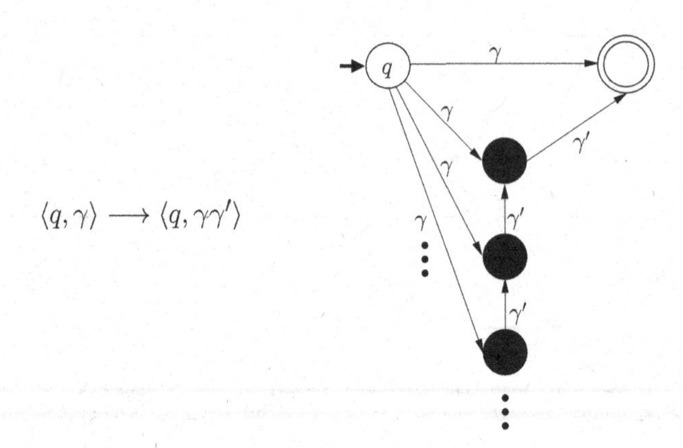

Fig. 12.11. An infinite ascending chain in the family of regular sets of configurations. Application of *post* for the rule to the left keeps on generating a new state (a black one). However, $post^* = \langle p, \gamma\gamma'^* \rangle$ is regular.

Termination by Acceleration. Figure 12.11 shows that forward search can compute an infinite ascending chain of regular sets of configurations. We resort to an *acceleration* ∇, as defined in Section 12.2.5.

Instead of computing $C \cup post_A(C)$ we will compute $C \nabla post_A(C)$ such that $post^*_A(C) \supseteq C \nabla post_A(C) \supseteq C \cup post_A(C)$. The key to defining ∇ is to *re-use states.* As hinted above, we only introduce *one* of the black states in Figure 12.11. More generally, we only introduce one state for each pair $\langle q', \gamma' \rangle$ such that Δ contains at least one rule of the form $\langle q, \gamma \rangle \longrightarrow \langle q', \gamma'\gamma'' \rangle$ (we also call the state $\langle q', \gamma' \rangle$). These states will then be "re-used".

Below we present the algorithm for computing an A-automaton accepting $C \nabla post_A(C)$ from an A-automaton B accepting C. Again, we assume that B has no initial states with incoming transitions. The algorithm adds states and transitions to B according to the following saturation rules:

(1) If A has a rule $\langle q, \gamma \rangle \longrightarrow \langle q', \epsilon \rangle$ and B has a transition (q, γ, p), then add a transition (q', ϵ, p) to B.
(2) If A has a rule $\langle q, \gamma \rangle \longrightarrow \langle q', \gamma' \rangle$ and B has a transition (q, γ, p), then add a transition (q', γ', p) to B.
(3) If A has a rule $\langle q, \gamma \rangle \longrightarrow \langle q', \gamma'\gamma'' \rangle$ and B has a transition (q, γ, p), then

- if B does not yet contain a state $\langle q', \gamma' \rangle$, then add such a state to B; and
- add transitions $(q', \gamma', \langle q', \gamma' \rangle)$ and $(\langle q', \gamma' \rangle, \gamma'', p)$ to B.

The accelerated forward search replaces the line $C := C \cup post_A(C)$ by $C := C \nabla post_A(C)$, where the automaton for $C \nabla post_A(C)$ is computing using the algorithm we have just sketched. Let us apply the accelerated search to the example of Figure 12.11. The first iteration adds a new state $\langle q, \gamma \rangle$, which corresponds to the first black state in the Figure, and a transition $(q, \gamma, \langle q, \gamma \rangle)$. In the second iteration, because of this transition and the third saturation rule, the algorithm adds no new states (the state $\langle q, \gamma \rangle$ has already been added), but a new transition $(\langle q', \gamma' \rangle, \gamma', \langle q', \gamma' \rangle)$: this is the self-loop mentioned above. In the third iteration, the algorithm does not add any new state or transition, and so the accelerated forward search terminates. The search has "jumped" to the limit $post^*(\langle q, \gamma \rangle)$ in only two steps.

The accelerated search *always terminates.* The number of states that can be added is bounded, for instance by the number of rules of the PDA, and so the number of transitions is also bounded. So the search eventually reaches a point at which the saturation rules cannot add any new state or transition.

The accelerated search must also satisfy $post^*_A(C) \supseteq C \nabla post_A(C)$ and $C \nabla post_A(C) \supseteq C \cup post_A(C)$. The second of this conditions is easy to prove: each word accepted by the A-automaton for $C \cup post_A(C)$ is also accepted by the A-automaton for $C \nabla post_A(C)$. The other condition is non-trivial, but not difficult. We refer the reader to the correctness proof in [3].[b] In the same reference, a detailed complexity analysis of the accelerated search is carried out. The search needs polynomial time in both the size of PDA and the size of the A-automaton accepting the set I of initial configurations.

12.5.4. *Forward Search for the Skyline Program*

Initial and Dangerous Configurations. The symbolic representations of I_s and D_s as A'_s-automata are shown in Figure 12.9, which proves both sets regular. Actually, initial configurations are typically quite simple and regular, and the same holds for many interesting sets of dangerous configurations, applying the monitor trick if necessary. However, we must mention that there are interesting properties that cannot be tested by a monitor, if the monitor has to be a finite automaton. For instance, the skyline program satisfies that in any execution the number of up and down moves is equal. However, no finite automaton can monitor this property, because, loosely speaking, it would then be an automaton accepting the language $\{\mathtt{up}^n\mathtt{down}^n \mid n \geq 0\}$. If we allow general PDAs as monitors, then the approach cannot be applied, because the intersection of two PDAs may not be equivalent to a PDA.

Let us now illustrate the first few iterations of the accelerated search, starting from the symbolic representation of I_s in Figure 12.9. The only rule "matching" I_s is $\langle \overline{q_{up}}, main_0 \rangle \longrightarrow \langle \overline{q_{up}}, s_0 main_1 \rangle$, because in I_s we have $\overline{q_{up}} \rightsquigarrow^* q$ reading $main_0$. Therefore, according to saturation rule (3) we add the transitions $(\overline{q_{up}}, s_0, \langle \overline{q_{up}}, s_0 \rangle)$ and $(\langle \overline{q_{up}}, s_0 \rangle, main_1, q)$ to I_s and obtain (an automaton for) I^1_s depicted in Figure 12.12.

I^1_s accepts all configurations reachable within one step: $\langle \overline{q_{up}}, main_0 \rangle$ and $\langle \overline{q_{up}}, s_0 main_1 \rangle$.

[b] For presentation reasons our notation differs; in particular, we use Q and P for the states of A and B, respectively, while in [3] it is the other way round.

Fig. 12.12. Automata representing the set I_s^1 and I_s^2 of skyline configurations reachable within one, respectively two, steps from the initial configuration.

Now, two further rules match in I_s^1: $\langle\overline{q_{up}}, s_0\rangle \longrightarrow \langle\overline{q_{up}}, s_1\rangle$ and $\langle\overline{q_{up}}, s_0\rangle \longrightarrow \langle\overline{q_{up}}, s_4\rangle$. Applying saturation rule (2) we introduce s_1 and s_4 labeled transitions in parallel to s_0 to arrive at I_s^2 also shown in Figure 12.12. New reachable configurations are thus $\langle\overline{q_{up}}, s_1 main_1\rangle$ and $\langle\overline{q_{up}}, s_4 main_1\rangle$. The first is reached when the non-deterministic choice in s evaluates to true, and the second when the same choice yields false.

Before we comment on the final outcome of the $post^*_{A'_s}(I_s)$ computation depicted in Figures 12.13 and 12.14, note that we will add $(\overline{q_{up}}, \epsilon, \langle\overline{q_{up}}, s_0\rangle)$ due to rule $\langle\overline{q_{up}}, s_4\rangle \longrightarrow \langle\overline{q_{up}}, \epsilon\rangle$, that is, procedure s returning. Thanks to this new transition the configuration $\langle\overline{q_{up}}, main_1\rangle$ becomes reachable. This, in turn, allows us to add an ϵ-transition from $\overline{q_{up}}$ to q indicating program termination.

Finally, observe that rule $\langle\overline{q_{up}}, s_1\rangle \longrightarrow \langle q_{up}, up_0 s_2\rangle$ matches in I_s^2. Using saturation rule (2) we add $(q_{up}, s_0, \langle q_{up}, s_0\rangle)$ and $(\langle q_{up}, s_0\rangle, s_2, \langle\overline{q_{up}}, s_2\rangle)$ to I_s^2. We hereby obtain the first reachable configuration with control location q_{up}: $\langle q_{up}, up_0 s_2 main_1\rangle$, which corresponds to entering s, taking the true branch of the choice, and calling up.

The automaton accepting $post^*_{A'_s}(I_s)$ is computed as delineated above. It does not fit in one figure, and so it is shown in Figures 12.13 and 12.14. Figure 12.13 shows the reachable configurations of the form $\langle q_{up}, \cdot\rangle$ and Figure 12.13 those of the form $\langle\overline{q_{up}}, \cdot\rangle$. The complete automaton obtained by merging the vertical "backbones" of the two automata.

In order to decide $post^*_{A'_s}(I_s) \cap D_S \stackrel{?}{=} \emptyset$ one can construct the product of the automata for D_s and $post^*_{A'_s}(I_s)$, and check for emptiness. But in our case this is not necessary. It suffices to check whether $post^*_{A'_s}(I_s)$ accepts anything starting in q_{up} with a subsequent $down_0$ labeled transition. To the skyline programmer's luck, it is easy to see that this is not the case, which means that a plotter driven by this program will not draw degenerate skylines.

We conclude this section by inviting the interested reader to undertake the instructive exercise of picking a reachable configuration from Figures 12.13 and 12.14 and finding an execution of A'_s that reaches it. As an example, Figure 12.15 shows an execution leading to the configuration $\langle q_{up}, up_0 m_6 m_8 s_3 main_1\rangle$. Each plotter move—that is, each line segment—is annotated with the configuration at

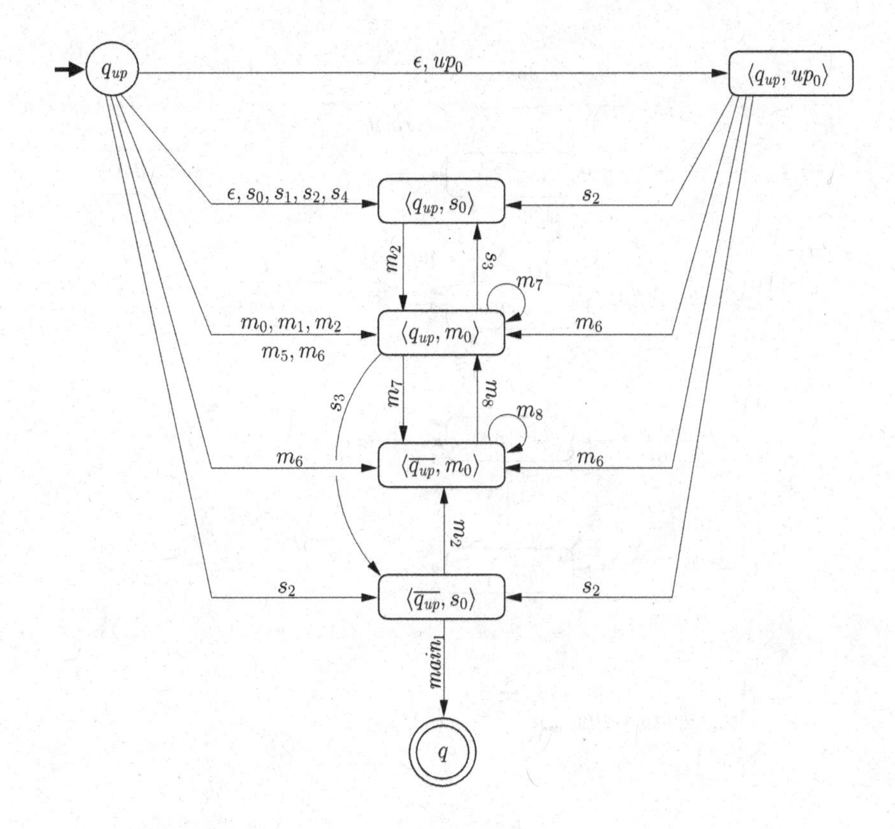

Fig. 12.13. An automaton representing the configurations of $post^*_{A'_s}(I_s)$ that are of the form $\langle q_{up}, \cdot\rangle$. Note that none of the dangerous configurations can be accepted by this automaton, because there is no accepting path starting with $down_0$.

which it was drawn. One can also check that all these configurations are accepted by one of the automata in Figures 12.13 and 12.14.

12.5.5. *Conclusion and Further Reading*

Modelling sequential procedural programs by pushdown automata is the basis of the Moped model checker developed by Schwoon [3, 15]. Suwimonteerabuth et al. have implemented a Java front end for this model checker, called jMoped [16, 17]. Recursive state machines are a model of computation equivalent to pushdown systems; model-checking algorithms for them have been proposed by several authors (see for instance [18, 19]).

Backward search with regular sets of configurations can also be accelerated to guarantee termination, and the algorithm is even slightly simpler. The algorithm and its proof can be found in Chapter 16 of this book. The saturation algorithms were presented in [20]. Efficient algorithms for both forward and backward search can be found in [21].

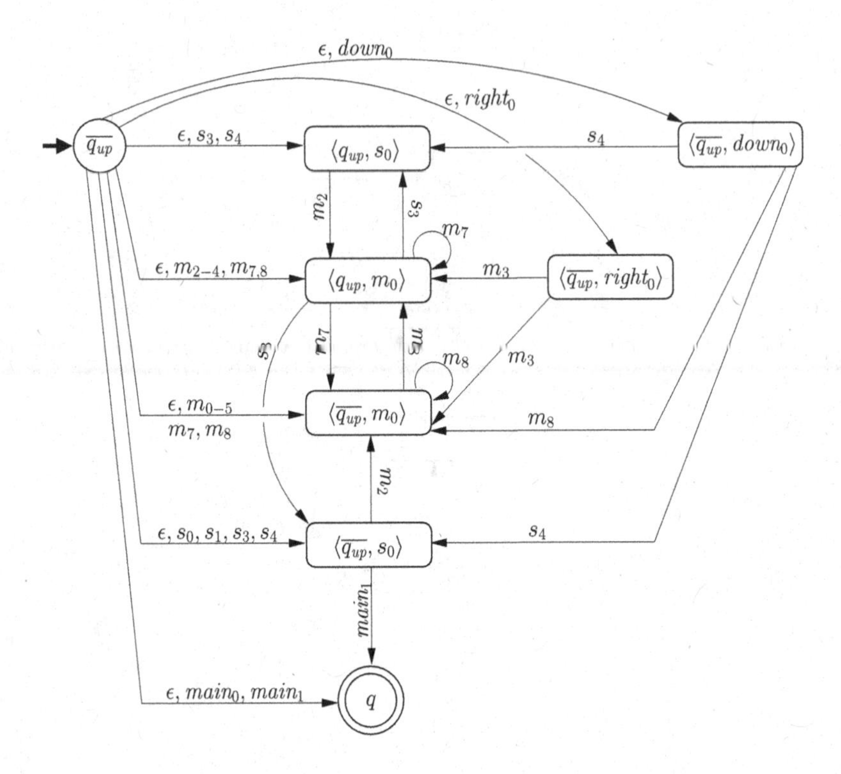

Fig. 12.14. An automaton representing the configurations of $post^*_{A'_s}(I_s)$ that are of the form $\langle \overline{q_{up}}, \cdot\rangle$.

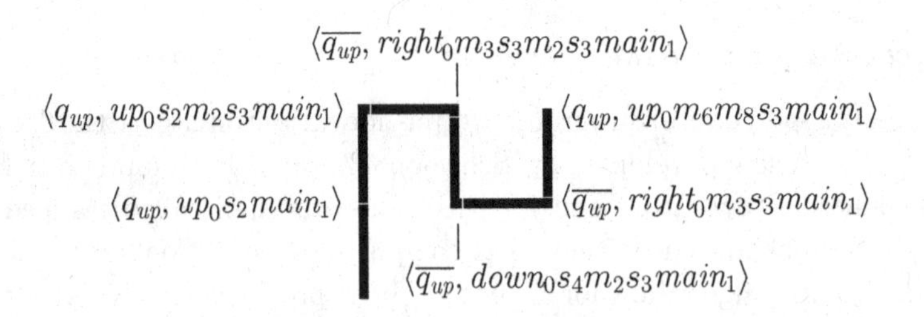

Fig. 12.15. A sample skyline and potential program configurations after each drawing step. By checking how the automata in Figures 12.13 and 12.14 accept the respective configurations, one gains quite some insight into how the program works.

12.6. Conclusion

We have presented an introduction to the verification of systems with an infinite state space. We have argued that many different sources of infinity can be modelled within the framework of extended automata and symbolic search, and have considered three different case studies involving real-time, parametric systems, and control structures. The key issue of symbolic search is finding an adequate class of symbolic configurations, or, in other words, an adequate data structure for representing infinite sets. We have presented several data structures: constraints, sets of minimal elements, and finite automata. We have also addressed the problem of guaranteeing termination of the search, and introduced three different strategies for achieving the goal.

References

[1] J. Bengtsson and W. Yi. Timed automata: Semantics, algorithms and tools. In *Lectures on Concurrency and Petri Nets*, vol. 3098, *LNCS*, pp. 87–124. Springer, (2003).
[2] G. Delzanno, Constraint-based verification of parameterized cache coherence protocols, *Formal Methods in System Design.* **23**(3), 257–301, (2003).
[3] S. Schwoon. *Model-Checking Pushdown Systems.* PhD thesis, Technische Universität München, (2002).
[4] P. Cousot and R. Cousot. Abstract interpretation: A unified lattice model for static analysis of programs by construction or approximation of fixpoints. In *Proc. 4th POPL*, pp. 238–252, (1977).
[5] R. Alur and D. L. Dill, A theory of timed automata, *Theor. Comput. Sci.* **126**(2), 183–235, (1994).
[6] D. L. Dill. Timing assumptions and verification of finite-state concurrent systems. In *Proc. Workshop on Automatic Verification Methods for Finite State Systems*, vol. 407, *LNCS*, pp. 197–212. Springer, (1990).
[7] R. W. Floyd, Algorithm 97: Shortest path, *Commun. ACM.* **5**(6), 345, (1962).
[8] R. Alur and P. Madhusudan. Decision problems for timed automata: A survey. In *Proc. SFM-RT*, vol. 3185, *LNCS*, pp. 1–24. Springer, (2004).
[9] J. Handy, *The Cache Memory Book.* (Academic Press, 1993).
[10] M. R. Garey and D. S. Johnson, *A guide to the theory of NP-completeness.* (W.H Freeman, 1979).
[11] E. Contejean and H. Devie, An efficient incremental algorithm for solving systems of linear diophantine equations, *Inf. Comput.* **113**(1), 143–172, (1994).
[12] P. A. Abdulla and B. Jonsson, Undecidable verification problems for programs with unreliable channels, *Inf. Comput.* **130**(1), 71–90, (1996).
[13] P. A. Abdulla, K. Cerans, B. Jonsson, and Y.-K. Tsay, Algorithmic analysis of programs with well quasi-ordered domains, *Inf. Comput.* **160**(1-2), 109–127, (2000).
[14] A. Finkel and P. Schnoebelen, Well-structured transition systems everywhere!, *Theor. Comput. Sci.* **256**(1-2), 63–92, (2001).
[15] J. Esparza and S. Schwoon. A BDD-based model checker for recursive programs. In *Proc. 13th CAV*, vol. 2102, *LNCS*, pp. 324–336. Springer, (2001).
[16] D. Suwimonteerabuth, S. Schwoon, and J. Esparza. jMoped: A Java bytecode checker based on Moped. In *Proc. 11th TACAS*, vol. 3440, *LNCS*, pp. 541–545. Springer, (2005).

[17] D. Suwimonteerabuth, F. Berger, S. Schwoon, and J. Esparza. jMoped: A test environment for Java programs. In *CAV*, vol. 4590, *LNCS*, pp. 164–167. Springer, (2007).

[18] R. Alur, K. Etessami, and M. Yannakakis. Analysis of recursive state machines. In *CAV*, vol. 2102, *LNCS*, pp. 207–220. Springer, (2001).

[19] M. Benedikt, P. Godefroid, and T. W. Reps. Model checking of unrestricted hierarchical state machines. In *ICALP*, vol. 2076, *LNCS*, pp. 652–666. Springer, (2001).

[20] A. Bouajjani, J. Esparza, and O. Maler. Reachability analysis of pushdown automata: Application to model checking. In *Proceedings of CONCUR'97*, vol. 1243, *LNCS*. Springer, (1997).

[21] J. Esparza, D. Hansel, P. Rossmanith, and S. Schwoon. Efficient algorithms for model checking pushdown systems. In *CAV*, vol. 1855, *LNCS*, (2000).

Chapter 13

Introduction to Hybrid Automata

Madhu Gopinathan* and Pavithra Prabhakar†

*Computer Science and Automation,
Indian Institute of Science, Bangalore, India
gmadhu@csa.iisc.ernet.in

†Department of Computer Science
University of Illinois at Urbana-Champaign
pprabha2@illinois.edu

A dynamical system is a system whose state changes with time. A hybrid system is a dynamical system whose state evolves either by continuous evolution specified by the current mode of the system or by a discrete jump that may change the mode of continuous evolution. In this chapter, we introduce hybrid automata for modeling hybrid systems. For general hybrid automata, the reachability problem, which is central to the analysis of models, is undecidable. There has been extensive research on finding subclasses of hybrid automata which are expressive enough to model real-life systems and at the same time are amenable to analysis. We focus on two such subclasses called rectangular hybrid automata and o-minimal hybrid automata.

13.1. Introduction

Hybrid systems can be thought of as finite state machines which interact with systems whose behaviour is governed by a set of differential equations. The state of such a system changes either by discrete jumps or by continuous evolution. For example, consider how the speed of a car equipped with gears changes: there are discrete jumps (shifting of gears) and continuous evolution of speed according to the acceleration determined by the selected gear. In this chapter, we describe *hybrid automata* [1], a formal model that combines finite state automata with real valued variables whose evolution is governed by sets of differential equations.

Hybrid systems are often used in safety critical applications. Therefore, it is important to analyze whether a system implementation conforms to a given safety specification. A hybrid automaton model of a hybrid system defines the valid behaviours of that system. In the reachability problem, the issue is to check whether a bad state (according to the specification) is reachable by a valid behaviour from an initial state of the hybrid system. The reachability problem is undecidable in the general case. We review results that show how restrictions on the general definition lead to the decidability of the reachability problem for rectangular [2] and o-minimal [3] hybrid automata.

This chapter is organized as follows. First, we give a brief overview of dynamical systems. Then we present some examples of hybrid automata followed by the formal definitions. This is followed by a discussion of the reachability problems for rectangular and o-minimal hybrid automata.

13.2. Dynamical Systems

A dynamical system is a system whose state changes with time. We first give an example of a system whose behaviour is governed by a set of differential equations, i.e. its state evolves purely by continuous evolution.

Example 1. Consider a pendulum attached to a rod of negligible weight and moving under gravity (see Fig. 13.1). For simplicity, let us assume that the rod has unit length and the friction is negligible. Let θ denote the angle that the pendulum makes with the vertical (positive counterclockwise). There is a downward gravitational force mg acting on the pendulum. This force has two components: the first component $mg\cos\theta$ contributes to the tension on the rod and the second component $mg\sin\theta$ contributes to the swinging of the pendulum.

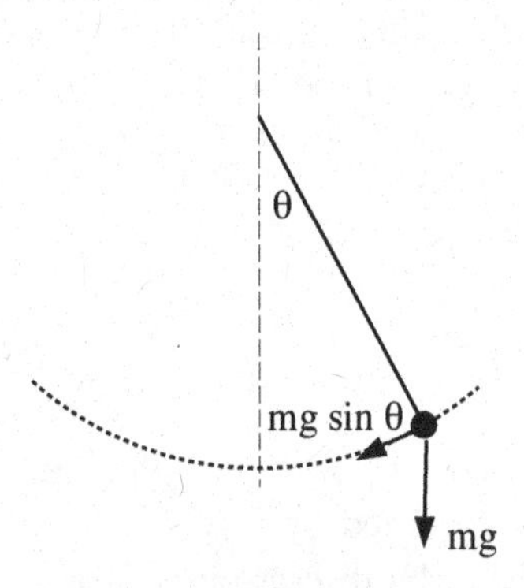

Fig. 13.1. Pendulum.

Using Newton's law, we can write the following non-linear, second order, ordinary differential equation (ODE) that governs the motion of the pendulum ($\dot{\theta}$ denotes the first derivative $\frac{d\theta}{dt}$ and $\ddot{\theta}$ denotes the second derivative $\frac{d^2\theta}{dt^2}$):

$$-mg\sin\theta(t) = m\ddot{\theta}(t) \tag{13.1}$$

Note that when θ is positive, the force that contributes to swinging seeks to decrease θ and is therefore negative. Assuming that at time $t = 0$, the pendulum starts at some initial position θ_0 with some initial velocity $\dot{\theta}_0$, a function $\theta(t) : \mathbb{R}_{\geq 0} \to \mathbb{R}$ is a solution to the ODE if it satisfies the following system of equations:

$$\theta(0) = \theta_0$$
$$\dot{\theta}(0) = \dot{\theta}_0$$
$$\ddot{\theta}(t) + mg \sin \theta(t) = 0$$

Henceforth, we assume that the mass $m = 1$ and the units of time and distance are chosen such that $g = 1$. A solution for equation (13.1) given $\theta_0 = 1$ and $\dot{\theta}_0 = 0.1$ is shown in Fig. 13.2 where θ is shown in bold and $\dot{\theta}$ in dashed lines. The solution curves are called trajectories of the system.

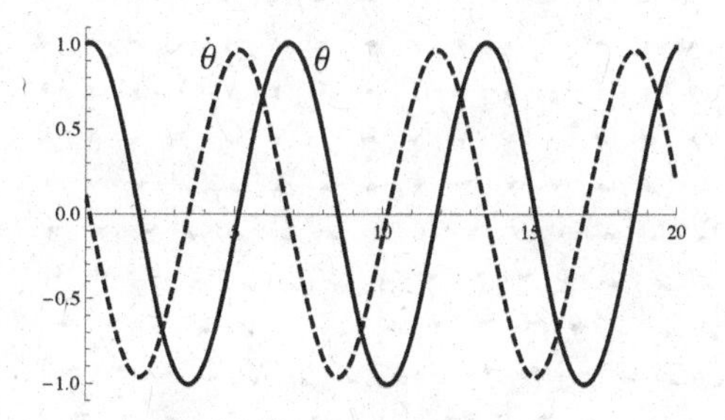

Fig. 13.2. Trajectory for equation (13.1).

Instead of studying a higher-order differential equation such as (13.1), we use the state space formalism [4]. The state space refers to the space whose axes are the state variables. The state of the system is represented as a vector in that space. The size of the state vector is called the *dimension* of the system.

We can now study a system of first-order differential equations of the form:

$$\dot{\mathbf{x}} = f(\mathbf{x}) \tag{13.2}$$

where $\mathbf{x}$ is the state of the system. Let $x_1, \ldots, x_n$ denote the elements of a state vector of an n-dimensional system, where $n \geq 1$. Let the state space $\mathbf{X}$ be $\mathbb{R}^n$. Therefore, each element x_i assumes a value from $\mathbb{R}$. Equation 13.2 stands for the following set of functions which govern the evolution of the state.

$$\dot{x_1} = f_1(x_1, \cdots, x_n, t)$$
$$\vdots$$
$$\dot{x_n} = f_n(x_1, \cdots, x_n, t)$$

In our example, the state of the pendulum at time t can be described using a vector (x_1, x_2) where x_1, x_2 stand for $\theta(t)$ and $\dot{\theta}(t)$ respectively. The state space equations corresponding to (13.1) are as follows:

$$\begin{bmatrix} \dot{x_1} \\ \dot{x_2} \end{bmatrix} = \begin{bmatrix} x_2 \\ -\sin x_1 \end{bmatrix}.$$

The expression f in equation 13.2 induces a function $f : \mathbb{R}^n \to \mathbb{R}^n$ that assigns a "direction" vector to each point in the state space and is called a *vector field*. Fig. 13.3 shows the vector field corresponding to equation (13.1).

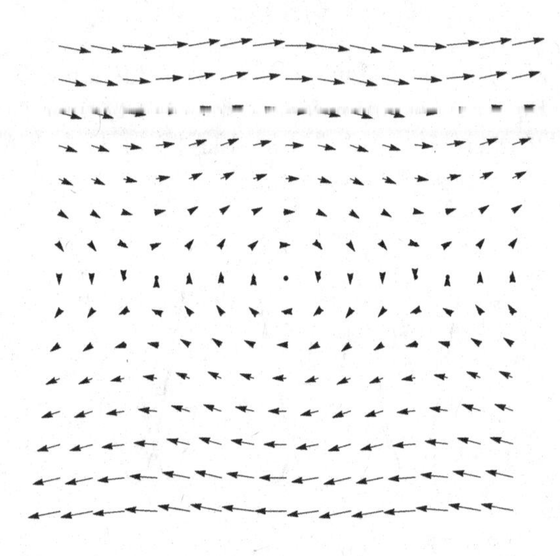

Fig. 13.3. Vector field for equation (13.1).

13.3. Hybrid Systems

A hybrid system is a dynamical system with two types of dynamics: discrete jumps and continuous evolution. Our second example illustrates such a system.

Example 2. Consider the model of a bouncing ball shown in Fig. 13.4. There is a single control state (also called mode) and the continuous state is represented using a vector (p, v) where p denotes the vertical position above the ground and v denotes the vertical velocity which we take to be positive in the upward direction. The state space equations are as follows:

$$\begin{bmatrix} \dot{p} \\ \dot{v} \end{bmatrix} = \begin{bmatrix} v \\ -g \end{bmatrix}.$$

Here g denotes the acceleration due to gravity. These equations hold only when $p \geq 0$. This condition is called the *invariant* of the state. The ball bounces when $p = 0$ and $v \leq 0$ and its velocity v is reset to $-cv$ where c is the coefficient of restitution, a value between 0 and 1, which indicates the bounciness of the ball. The discrete jump from the state to itself is annotated with a *guard* $p = 0 \wedge v \leq 0$. If the guard condition is satisfied, then v is reset to $-cv$ following the jump. The sourceless arrow indicates the initial state while its label denotes the initial condition.

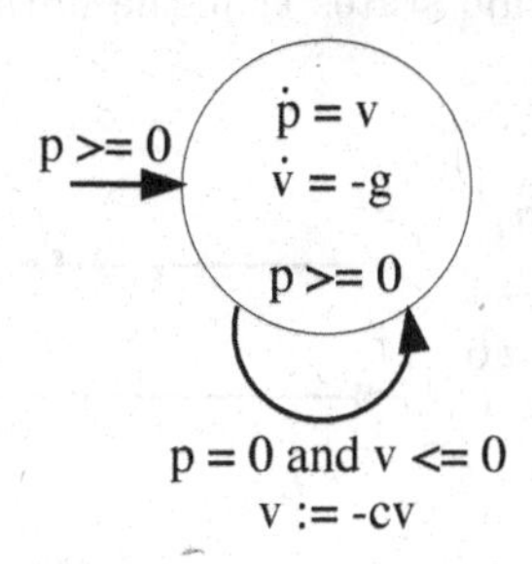

Fig. 13.4. Model of a bouncing ball.

Figure 13.5 shows the trajectory of a ball dropped from an initial position 1 meter above the ground. Its state evolves continuously until it hits the ground ($p = 0$). Then the velocity is reset to $-cv$ (in this example, $c = 0.8$). The state of the ball again evolves continuously until another bounce occurs.

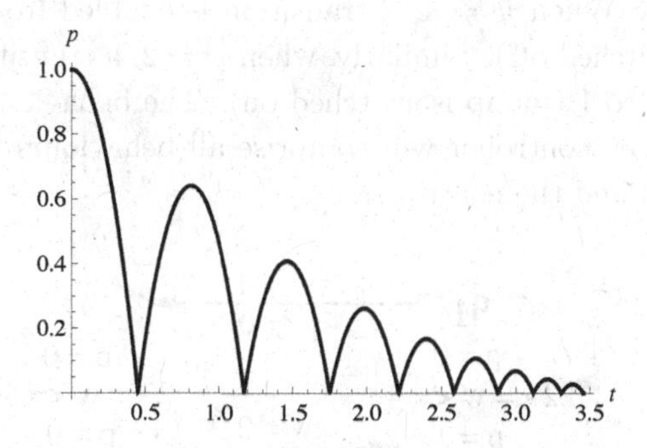

Fig. 13.5. Trajectory of a bouncing ball.

13.4. Hybrid Control

Examples 1 and 2 did not involve the use of a controller. We now look at another example which involves the control of a hybrid system.

Example 3. Consider a water tank with a pump connected to it. If the pump is switched on, then water rises at the rate of 1 cm/sec and if it is switched off, then water falls at the rate of 1 cm/sec. Figure 13.6 shows a hybrid automaton model for this water tank. The variable $\mathbf{w}$ denotes the water level in the tank, which is always between 0 and 6 cm. The variable p denotes the status of the pump (on $=$ 1, off $=$ 0). The water tank model assumes that the pump will be switched off when the water level reaches 6 cm and it will be switched on when the water level reaches 0. This is modeled using the invariants $w < 6 \wedge p = 1$ in state b_1 and $w > 0 \wedge p = 0$ in state b_2. The initial condition states that the initial control state is b_1 and the water level is at 2 cm.

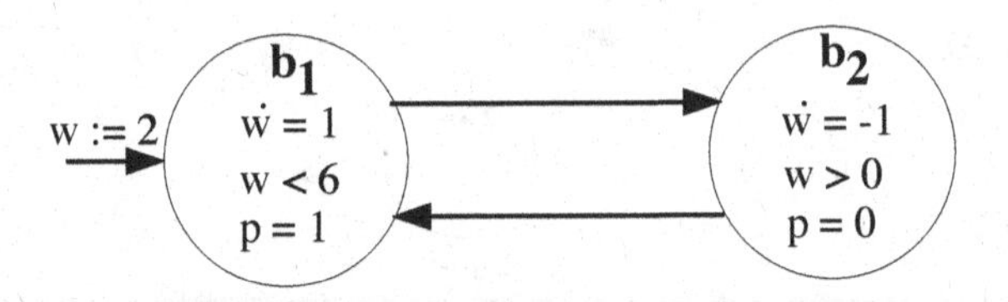

Fig. 13.6. Model of a water tank.

Figure 13.7 shows a controller which controls the water level in the tank by switching the pump on or off. This controller has been designed with the goal of always keeping the water level in the tank between 2 cm and 4 cm. It has two states q_1 and q_2. The initial condition states that the initial control state is q_1 and the pump is switched on. The invariant in q_1 is $2 \leq w \leq 4 \ \wedge p = 1$ and that in q_2 is $2 \leq w \leq 4 \ \wedge p = 0$. When $w = 4$, a transition is enabled from q_1 to q_2 and p is set to 0 (pump is switched off). Similarly, when $w = 2$, a transition is enabled from q_2 to q_1 and p is set to 1 (pump is switched on). The behaviour of the water tank controlled by the given controller will comprise all behaviours that are allowed by both the tank model and the controller.

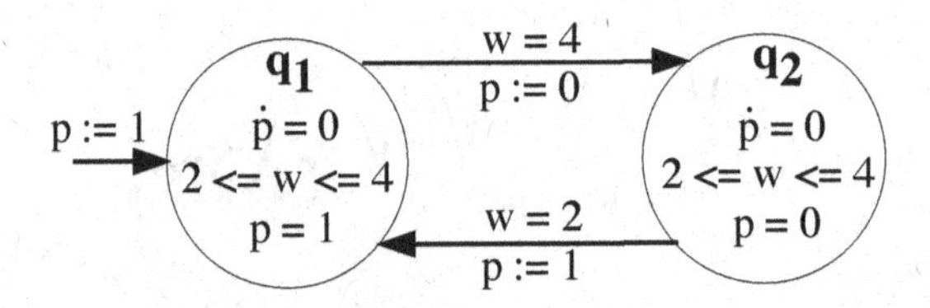

Fig. 13.7. A controller for the water tank.

The state space equations for state (b_1, q_1) when the pump is switched on are

$$\begin{bmatrix} \dot{w} \\ \dot{p} \end{bmatrix} = \begin{bmatrix} 1 \\ 0 \end{bmatrix}$$

and the equations for state (b_2, q_2) when the pump is switched off are

$$\begin{bmatrix} \dot{w} \\ \dot{p} \end{bmatrix} = \begin{bmatrix} -1 \\ 0 \end{bmatrix}.$$

Figure 13.8 shows an example of a behaviour of the controlled system where the water level rises from 2 cm to 4 cm as the pump is switched on and then the water level falls from 4 cm to 2 cm as the pump is switched off.

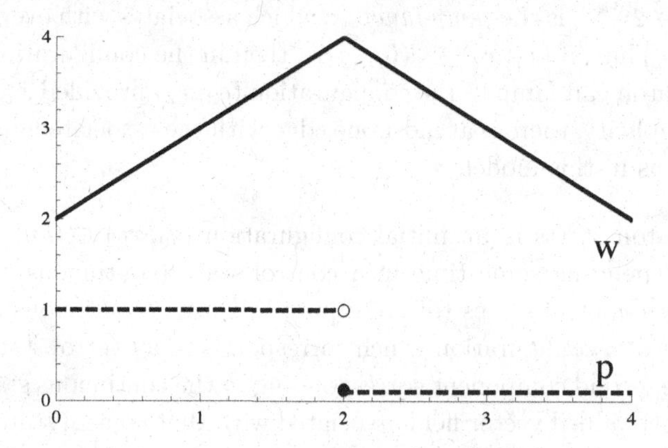

Fig. 13.8. Behaviour of controlled water tank over variables $\{p, w\}$.

We have presented a brief overview of hybrid systems above. For a more detailed introduction, the reader is referred to [1, 5].

13.5. Hybrid Automata

Our definition of a hybrid automaton is adapted from the definitions of hybrid automata in the literature [1]. We first define the components of a hybrid automaton and give an informal description of what the components stand for. The way in which the components are used will be described formally when we define the semantics of a hybrid automaton.

Definition 13.1. A *hybrid automaton* is a tuple

$$\mathcal{H} = (Q, X, init, F, I, E, G, J)$$

where

- Q is a finite set of *control states*;
- X is the continuous state space, a subset of $\mathbb{R}^n$, defined by the state variables $x_1, \ldots, x_n$. A *configuration* of $\mathcal{H}$ is a pair $(q, x) \in Q \times X$;
- $init \subseteq Q \times X$ is the set of initial configurations;
- $F : Q \to (X \to 2^{\mathbb{R}^n})$ is the *flow function* which associates with every control state a set-valued vector field which constrains the time derivative of the continuous flow;
- $I : Q \to 2^X$ is the *invariant function* which associates with every control state an invariant set;
- $E \subseteq Q \times Q$ is an *edge relation*;

- $G : E \to 2^X$ is the *guard function* which associates a guard condition with every edge;
- $J : E \to 2^{X \times X}$ is the *jump function* which associates with every edge a set of jumps. Thus, if $(x_1, x_2) \in J((q_1, q_2))$, then in the configuration (q_1, x_1), the automaton can jump to the configuration (q_2, x_2) provided $x_1 \in G((q_1, q_2))$. For simplicity, there is at most one edge with the associated guard and jump functions in this model.

The automaton starts in an initial configuration $(q_0, x_0) \in \mathit{init}$. It then alternates between spending some time in a control state (continuous evolution) and jumping between control states (discrete jump). During continuous evolution, the first component of a configuration which corresponds to a control state remains the same, while the second component corresponding to the continuous state evolves according to the set valued vector field associated with that control state. In addition, at all points of time in a control state, the continuous part of the state must satisfy the invariant associated with that control state. The automaton can jump via an edge e from one control state q to another control state q' if its current configuration (q, x) satisfies the guard associated with the edge e. The resulting configuration will be (q', x') where x' is determined by the jump function associated with the edge e in that $(x, x') \in J(e)$.

For the water tank automaton in Example 3 (Fig. 13.6), the control states are $\mathbf{b}_1$ and $\mathbf{b}_2$. The state space is $\mathbb{R}^2$. The set of initial configurations *init* is $\{(\mathbf{b}_1, (w = 2, p = 1))\}$. The flow function is given by $F(\mathbf{b}_1)(x) = \{(1, 0)\}$ and $F(\mathbf{b}_2)(x) = \{(-1, 0)\}$ for all continuous states x. The invariant function of $\mathbf{b}_1$ requires $w < 6$ and $p = 1$ and that of $\mathbf{b}_2$ requires $w > 0$ and $p = 0$. The edges between $\mathbf{b}_1$ and $\mathbf{b}_2$ are always enabled, i.e. $G(e) = \mathit{true}$ representing the set of all possible continuous states for both the edges. The jump functions associated with these edges do not modify the continuous state.

We define the semantics of a hybrid automaton in the continuous time setting as follows. A *transition system* is a structure of the form $\mathcal{S} = (Q, Q^0, \Sigma, \to)$, where Q is a set of states, $Q^0 \subseteq Q$ is the set of initial states, and $\to \subseteq Q \times \Sigma \times Q$ is the transition relation. We will denote the triple $(q, a, q') \in \to$ by $q \to_a q'$.

Definition 13.2. The transition system corresponding to $\mathcal{H}$, denoted $[\![\mathcal{H}]\!]$, is defined as $[\![\mathcal{H}]\!] = (Q \times X, \mathit{init}, \Sigma, \to)$ where

- The configurations of $\mathcal{H}$, i.e. $S = Q \times X$, are the states of $[\![\mathcal{H}]\!]$;
- The initial configurations of $\mathcal{H}$ are the initial states of $[\![\mathcal{H}]\!]$;
- $\Sigma = \{\mu\} \cup \mathbb{R}_{\geq 0}$, where μ is a symbol denoting a discrete jump;
- $\to \subseteq (Q \times X) \times (\mu \cup \mathbb{R}_{\geq 0}) \times (Q \times X)$ is the set of transitions. There are two kinds of transitions namely discrete and continuous.
  - *Discrete or μ transition.* $((q, x), \mu, (q', x')) \in \to$ if there exists $(q, q') \in E$ such that $x \in I(q)$, $x' \in I(q')$, $(x, x') \in J((q, q'))$, and $x \in G((q, q')$;

– *Continuous transition.* $((q, x), t, (q', x')) \in \rightarrow$ if $t \geq 0$, $q = q'$ and there exists a continuous function $f : [0, t] \rightarrow X$ such that $f(0) = x$, $f(t) = x'$, and for all $t' \in (0, t)$, $\dot{f}(t') \in F(q)(f(t'))$ and $f(t') \in I(q)$.

A *trajectory* of $\mathcal{H}$ starting from an initial configuration (q_0, x_0) of $\mathcal{H}$ is a finite sequence of configurations $(q_0, x_0) \rightarrow (q_1, x_1) \rightarrow \cdots \rightarrow (q_n, x_n)$ such that between each pair of consecutive configurations, there is either a discrete transition or a continuous transition.

The reachability problem for a hybrid automaton $\mathcal{H}$ asks whether a given configuration (q, x) is reachable from an initial configuration in the induced transition system $[\![H]\!]$. The control state reachability problem for $\mathcal{H}$ is the following: given a control state q of $\mathcal{H}$, does there exist a configuration of the form (q, x) that is reachable from an initial state in the transition system $[\![\mathcal{H}]\!]$.

The hybrid automaton model described above is very general, in the sense that it does not constrain the form of the functions F, I, G and J. Also nothing is said about how to represent these. Even if we assume that these functions have a finite representation (for example, linear or polynomial functions), it turns out that for this general class, the reachability problem is undecidable [1, 6].

13.6. Undecidability of reachability

We now show that reachability is undecidable even for a very simple class of hybrid automata called 2-rate timed automata.

Definition 13.3. A 2-rate timed automaton $\mathcal{R}$ is a hybrid automaton

$$\mathcal{R} = (Q, X, \mathit{init}, F, I, E, G, J)$$

which satisfies the following conditions:

- For all control states $q \in Q$, the set $\mathit{init}(q)$ is such that all state variables are initialized to 0;
- For all control states $q \in Q$, F_q is such that all variables evolve either at the rate of 1 or 2;
- For every edge $e \in E$, the set $G(e)$ is defined by guards of the form $x_i \leq k$ or $k \leq x_i$ or $x_i = x_j$ where x_i, x_j are state variables and k is an integer constant;
- During a jump, each state variable x_i is either unchanged or is reset to 0.

Theorem 13.1. *Reachability is undecidable for 2-rate timed automata [1].*

Proof Sketch. Let C be a 2-counter machine with 2 counters c_1, c_2, each capable of holding a non-negative integer and states $q_1, \cdots, q_k$, with q_1 as the initial state and q_k as the halting state. The transitions of C are of the form $q_i \overset{op}{\rightarrow} q_j$ where op is one of following: increment a counter, decrement a counter if it is not zero, test if a counter is zero and then go to one of two states based on the result of the test. A configuration of $\mathcal{C}$ is of the form (q_i, v_1, v_2) where v_1, v_2 represent the values of c_1, c_2 respectively. The halting problem for 2-counter machines asks if there exists an algorithm to determine whether any given counter machine can reach a halting configuration $(q_k, (\_, \_))$ from its initial configuration $(q_1, (c_1 = 0, c_2 = 0))$.

We now show that a 2-rate timed automaton can simulate a nondeterministic 2-counter machine. Then the result above follows from the fact that the halting problem for 2-counter machines is undecidable [7]. Let $\mathcal{M}$ be a 2-rate timed automaton. The control state of $\mathcal{M}$ encodes the state of $\mathcal{C}$. Let x_1, x_2, y, u and v be state variables initialized to 0. All variables except v continuously evolve at the rate of 1 and v evolves at the rate of 2. The variables x_1 and x_2 respectively encode the values of the counters c_1 and c_2. If the value of c_i is n, then the value of x_i is $\frac{1}{2^n}$. The variable y is used to mark intervals of length 1: whenever y is 1, it is reset to 0. The i-th configuration of $\mathcal{C}$ is encoded by the state of $\mathcal{M}$ at time i.

We now consider three cases: at time $(i+1)$, the counter value encoded by x (where x stands for x_1 or x_2) at time i (a) remains unchanged, (b) is incremented and (c) is decremented. Suppose that the value of x at time i is $\frac{1}{2^n}$. In the first case, for x to have the value $\frac{1}{2^n}$ at time $(i+1)$, x must be reset to 0 when x becomes 1 at time $(i+1) - \frac{1}{2^n}$. In the second case, for x to have the value $\frac{1}{2^{n+1}}$ at time $(i+1)$, x has to be reset to 0 at time $(i+1) - \frac{1}{2^{n+1}}$. In order to do this, reset u when $x = 1$, then nondeterministically reset both x and v simultaneously and test $u = v$ at time $(i+1)$. A successful test at time $(i+1)$ ensures that v has the value $\frac{1}{2^n}$ and therefore x has the value $\frac{1}{2^{n+1}}$. In the third case, for x to have the value $\frac{1}{2^{n-1}}$ at time $(i+1)$, x has to be reset to 0 at time $(i+1) - \frac{1}{2^{n-1}}$. In order to do this, nondeterministically reset u, then reset x and v simultaneously at time $i - \frac{1}{2^n}$ and test $u = v$ at time i. A successful test ensures that u has the value $\frac{1}{2^{n-1}}$ at time i. If x is reset when u becomes 1, then at time $(i+1)$, x has the value $\frac{1}{2^{n-1}}$. Figure 13.9 illustrates the three cases for $n = 1$. The thick lines show the evolution of x and the dashed lines show the time points at which the variables are reset.

Thus, the executions of $\mathcal{M}$ correspond to the executions of $\mathcal{C}$ and the halting problem of $\mathcal{M}$ reduces to the problem of whether a halting state $(q_k, (\_, \_, \_, \_, \_))$ of $\mathcal{M}$ is reachable from its initial state $(q_1, (x_1 = 0, x_2 = 0, y = 0, u = 0, v = 0))$. □

There has been a significant amount of research on finding subclasses [6, 8–12] which are tractable and are a good approximation of real-life systems. We now discuss two of these subclasses of hybrid automata.

13.7. Rectangular Hybrid Automata

Rectangular hybrid automata [2, 6] is an interesting subclass of hybrid automata because they can be used for conservatively approximating sets of arbitrary hybrid trajectories [13]. Thus, rectangular automata could be used to create abstractions of complex hybrid systems which can then be verified [14]. The reachability problem

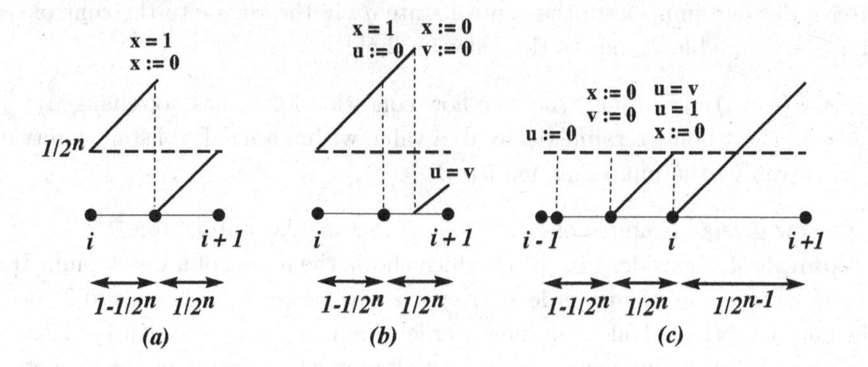

Fig. 13.9. Simulating a 2-counter machine.

of *initialized* rectangular automata (to be defined below) is decidable [6].

A *rectangular* set $R \subseteq \mathbb{R}^n$ is of the form $R_1 \times \cdots \times R_n$ where each R_i is an interval (bounded or unbounded), i.e. R is a product of n intervals of the real line. For example, Fig. 13.10 shows a rectangular set in $\mathbb{R}^2$.

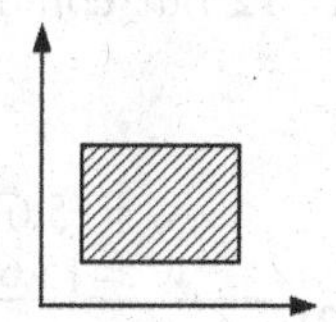

Fig. 13.10. Rectangular set in $\mathbb{R}^2$.

Definition 13.4. A rectangular hybrid automaton $\mathcal{R}$ is a hybrid automaton

$$\mathcal{R} = (Q, X, init, F, I, E, G, J)$$

which satisfies the following conditions:

- For all control states $q \in Q$, the sets $init(q)$ and $I(q)$ are rectangular;
- For all control states $q \in Q$, there is a rectangular set F_q such that for all $x \in X$, $F(q)(x) = F_q$;
- For every edge $e \in E$, the set $G(e)$ is rectangular; $J(e) = (X, Y)$ such that Y is a rectangular set $Y_1 \times \cdots \times Y_n$ and during a jump via e, a state variable x_i is either not reset or is set nondeterministically to a value within a fixed, constant interval Y_i $(1 \leq i \leq n)$.

In a rectangular hybrid automaton, in every control state, the derivative of each variable x_i always lies between two fixed bounds $[l_i, u_i]$. These bounds may vary from one control state to another. A rectangular automaton is said to be *initialized* if after a discrete jump from the control state q via the edge e to the control state q', for every variable x_i, one of the following holds:

- $F(q)(x_i) = F(q')(x_i)$ (i.e. the flow condition for x_i has not changed)
- x_i is set nondeterministically to a value within a fixed, constant interval as given by the jump function $J(e)$.

We now give an example of an initialized rectangular automaton.

Example 4. Consider Fig. 13.11 which shows the model of a water tank. If the pump is switched on, the water level rises at a rate between 1 cm/sec and 2 cm/sec. If the pump is switched off, then the water level falls at a rate between 1 cm/sec and 2 cm/sec. Also, the pump can be switched off only when the water level is between 5 and 6 cm/sec and it can be switched on again only when the water level is between 0 and 1 cm/sec. This automaton is an initialized rectangular automaton while the water tank automaton of Example 3 (Fig. 13.6) is a rectangular automaton that is not initialized as the jump from b_1 to b_2 does not initialize w even though the flow condition of w changes from 1 to -1. Figure 13.12 shows an example trajectory where the water level rises at the rate of 2 cm/sec for a second and then falls at the rate of 1 cm/sec during the next second. The water level need not change at a constant rate as shown in Figure 13.12 but could change at any rate that satisfies the flow condition.

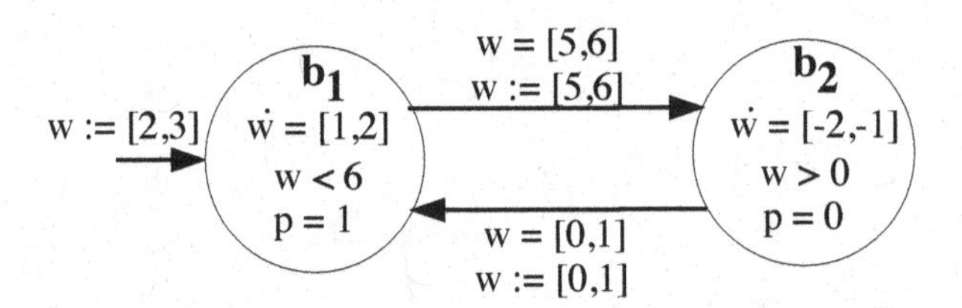

Fig. 13.11. Model of a water tank.

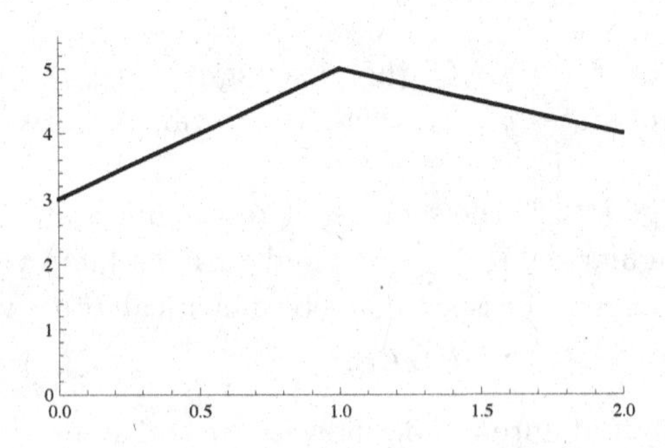

Fig. 13.12. Behaviour of controlled water tank over variables $\{p, w\}$.

The reachability problem for initialized rectangular automata is decidable [2, 6, 15]. This result is obtained by first translating an initialized rectangular automaton $\mathcal{R}$ to an initialized multirate automaton $\mathcal{M}_{\mathcal{R}}$ in which each variable evolves according to a constant, rational slope, which may be different in different control states. Then the multirate automaton $\mathcal{M}_{\mathcal{R}}$ is translated into a timed automaton $\mathcal{T}_{\mathcal{R}}$ for which the reachability problem is known to be decidable (see Chapter 4 in this volume). If either the initialization condition is removed or if rectangularity is violated (for example, with a guard condition such as $x_1 = x_2$), then the reachability problem becomes undecidable [6].

We now sketch the key ideas behind this two-step translation. Let $\mathcal{R}$ be an initialized rectangular automaton of n variables. Consider a variable x_i with $\dot{x}_i = [l, u]$ in $\mathcal{R}$. In the corresponding initialized multirate automaton $\mathcal{M}_{\mathcal{R}}$, each variable x_i is replaced by two variables y_{2i-1} and y_{2i} such that $\dot{y}_{2i-1} = l$ and $\dot{y}_{2i} = u$. Let $h_{\mathcal{M}} : Q \times \mathbb{R}^{2n} \to Q \times \mathbb{R}^n$ be a function which maps the state space of $\mathcal{M}_{\mathcal{R}}$ to that of $\mathcal{R}$ defined by

$$h_{\mathcal{M}}((q, y)) = \{(q, x) \mid y_{2i-1} \leq x_i \leq y_{2i}\}.$$

The variable y_{2i-1} tracks the least possible value of x_i and the variable y_{2i} tracks the greatest possible value of x_i. The jump relation of $\mathcal{M}_{\mathcal{R}}$ is constructed such that this continues to hold even after a jump. Let $x_u \in [k, m]$ be the guard condition and $x_i \in [k', m']$ be the reset condition associated with an edge (q, q') in $\mathcal{R}$. If the intervals $[y_{2i-1}, y_{2i}]$ and $[k, m]$ do not intersect, i.e. when $y_{2i-1} > m \vee y_{2i} < k$, the jump is not enabled. Therefore, the guard condition in $\mathcal{M}_{\mathcal{R}}$ is rewritten as $y_{2i-1} \leq m \wedge y_{2i} \geq k$ and the reset condition is rewritten as $y_{2i-1} := k', y_{2i} := m'$.

For example, while translating the rectangular automaton $\mathcal{R}$ in Fig. 13.13(a), two variables y_1 and y_2 are introduced in the corresponding multirate automaton $\mathcal{M}_{\mathcal{R}}$ shown in Fig. 13.13(b). The guard and reset conditions of the edges are rewritten as discussed above. It is shown in [6] that

- for every execution of any initialized rectangular automaton $\mathcal{R}$, a corresponding execution of $\mathcal{M}_{\mathcal{R}}$ can be constructed;
- for every finite execution of $\mathcal{M}_{\mathcal{R}}$, starting from the end of the execution, a corresponding execution of $\mathcal{R}$ can be constructed.

The initialized multirate automaton $\mathcal{M}_{\mathcal{R}}$ is further translated into a timed automaton $\mathcal{T}_{\mathcal{R}}$ by rescaling the state space. Let $h_{\mathcal{T}} : Q \times \mathbb{R}^{2n} \to Q \times \mathbb{R}^{2n}$ be a function which maps the state space of $\mathcal{T}_{\mathcal{R}}$ to that of $\mathcal{M}_{\mathcal{R}}$ defined by

$$h_{\mathcal{T}}((q, (y_1, \ldots, y_{2n}))) = (q, (l_1 \cdot y_1, \ldots, l_{2n} \cdot y_{2n}))$$

where $l_i = \dot{y}_i$ if $\dot{y}_i \neq 0$ and $l_i = 1$ otherwise. For example, consider the states q_2, q_3 and the edge e_3 in Fig. 13.13 (b). In the corresponding timed automaton $\mathcal{T}_{\mathcal{R}}$, all the states have $\dot{y}_1 = 1$ and $\dot{y}_2 = 1$. The guard and reset conditions of the edge e_3 are modified as $y_1 \leq 5 \wedge y_2 \geq \frac{5}{2}$ and $y_1 := \frac{-5}{2} \wedge y_2 := -5$ respectively.

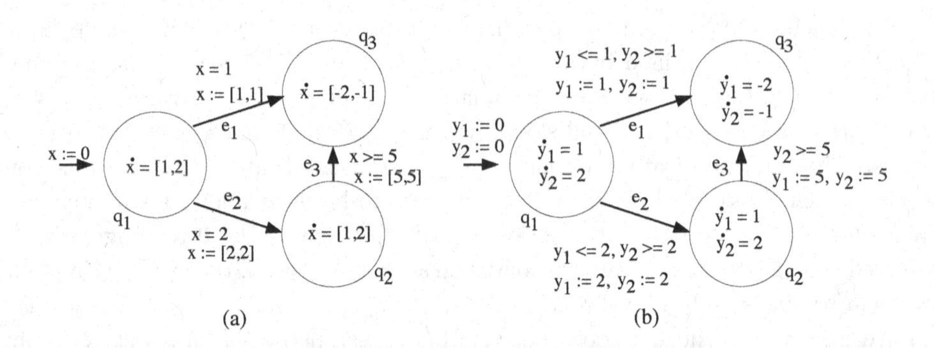

Fig. 13.13. Translation of a rectangular automaton $\mathcal{R}$ (a) to a multirate automaton $\mathcal{M}_{\mathcal{R}}$ (b).

We say that a set of configurations A of the automaton $\mathcal{R}$ is "representable" as a set of regions if there exists a set of regions B of $\mathcal{T}_{\mathcal{R}}$ such that $h_{\mathcal{M}}(h_{\mathcal{T}}(B)) = A$. Let $Reach_{\mathcal{H}}(I)$ denote the set of states that can be reached from the set of states I using a trajectory of the automaton $\mathcal{H}$.

Proposition 13.1. *Let A be a set of configurations of the initialized rectangular automaton $\mathcal{R}$ which is representable as a set of regions B of the timed automaton $\mathcal{T}_{\mathcal{R}}$. Then $Reach_{\mathcal{R}}(A)$ is representable as the set of regions $Reach_{\mathcal{T}_{\mathcal{R}}}(B)$.*

We note that the set of initial configurations of $\mathcal{R}$ associated with a particular control state is assumed to be rectangular and hence the set of initial configurations I is clearly representable as a set of regions, say J, of $\mathcal{T}_{\mathcal{R}}$. Thus the set of configurations of $\mathcal{R}$ reachable from I can be computed in terms of the set of regions that are reachable from the set of regions J in the region automaton corresponding to the timed automaton $\mathcal{T}_{\mathcal{R}}$.

13.8. O-minimal Hybrid Automata

O-minimal hybrid systems are a class of hybrid systems which allow a rich class of continuous dynamics, and yet are amenable to analysis. They are based on o-minimal theories whose domain is the set of real numbers. In this section, we will present a proof of the decidability of o-minimal hybrid systems with respect to various properties. An important approach in establishing decidability results for hybrid systems is to construct a *bisimulation* which is a finite quotient of the original infinite state system. Bisimulation preserves various properties between systems, such as, finite state reachability and properties specified by various branching time logics such as CTL^* and μ-calculus. We will use such an approach to show decidability results for o-minimal hybrid systems. We begin by introducing some of the preliminary notions related to bisimulations, and then go on to the definition of o-minimal hybrid systems.

Preliminaries. Recall that a binary relation R on a set A is a subset of $A \times A$. We will sometimes write aRb to denote $(a, b) \in R$. For a binary relation R on a set A we define the inverse of R, denoted R^{-1}, to be the relation $\{(b, a) \mid (a, b) \in R\}$. An equivalence relation R on a set A partitions A into equivalence classes of the form $[a]_R = \{b \in A \mid aRb\}$. A partition Π of a set A defines a natural equivalence relation $\equiv_\Pi$ on A, given by $a \equiv_\Pi b$ iff a and b belong to the same partition in Π. In the sequel we will use the partition Π to mean both the partition, as well as the equivalence relation associated with it. We say that an equivalence relation is of finite index if the number of equivalence classes is finite. Finally, we will say an equivalence relation R_1 *refines* another equivalence relation R_2 iff $R_1 \subseteq R_2$.

Let $\mathcal{S}_1 = (Q_1, Q_1^0, \Sigma, \rightarrow_1)$ and $\mathcal{S}_2 = (Q_2, Q_2^0, \Sigma, \rightarrow_2)$ be two transition systems. A *bisimulation relation* between the transition systems $\mathcal{S}_1$ and $\mathcal{S}_2$ is a binary relation $R \subseteq Q_1 \times Q_2$ such that if $(q_1, q_2) \in R$ and $q_1 \rightarrow_a q_1'$ then there exists q_2' such that $q_2 \rightarrow_a q_2'$ and $(q_1', q_2') \in R$; and if $(q_1, q_2) \in R$ and $q_2 \rightarrow_a q_2'$, then there exists q_1' such that $q_1 \rightarrow_a q_1'$ and $(q_1', q_2') \in R$. If $\mathcal{S}_1 = \mathcal{S}_2 = \mathcal{S}$, then we call R a bisimulation relation on $\mathcal{S}$. A state $q_1 \in Q$ is said to be *bisimilar* to $q_2 \in Q$, written $q_1 \cong q_2$, if there exists a bisimulation R on $\mathcal{S} = (Q, Q_0, \Sigma, \rightarrow)$ such that $(q_1, q_2) \in R$. The bismilarity relation $\cong$ on Q is the set of all $(q_1, q_2) \in Q \times Q$ such that q_1 is bisimilar to q_2. The bisimilarity relation $\cong$ on Q can be seen to be an equivalence relation and a bisimulation relation for $\mathcal{S}$ [16].

If a bisimulation relation R for the transition system $\mathcal{S}$ above is an equivalence relation, then we can construct a quotient transition system $\mathcal{S}/_R = (Q', Q'^0, \Sigma, \rightarrow')$ as follows.

- Q' is the set of equivalence classes of R.
- Q_0' is the set of equivalence classes of the elements in Q_0.
- $q_1' \rightarrow_a' q_2'$ iff there exist q_1 and q_2 such that $q_1' = [q_1]_R$ and $q_2' = [q_2]_R$ and $q_1 \rightarrow_a q_2$.

We have the following proposition which relates $\mathcal{S}$ and $\mathcal{S}/_R$.

Proposition 13.2. *Let $\mathcal{S} = (Q, Q_0, \Sigma, \rightarrow)$ be a transition system. Let R be an equivalence relation which refines the partition $\{Q_0, Q \backslash Q_0\}$ and is a bisimulation on $\mathcal{S}$. Then the relation $R' = \{(q, [q]_R) \mid q \in Q\}$ is a bisimulation between $\mathcal{S}$ and $\mathcal{S}/_R$.*

In particular, the above proposition says that a state q' of $\mathcal{S}/_R$ is reachable iff there exists some $q \in q'$ which is reachable in $\mathcal{S}$.

Next, we give a quick overview of first-order logic. More information can be found in any standard textbook on mathematical logic, for example [17].

A *language* or *vocabulary* L is a disjoint union of relation symbols L^r and constant symbols L^c. Each relation symbol has an associated arity which is a natural number. The formulas over a language L and a set of variables *Var* are defined inductively as:

$$\varphi ::= R(t_1, \cdots, t_n) \mid t_1 = t_2 \mid \neg\varphi \mid \varphi \vee \varphi \mid \exists x \varphi,$$

where $t_i \in Var \cup L^c$, and $R \in L^r$. We don't have function symbols explicitly but they can be modelled as relations by their graphs. Thus a k-ary function f can be modelled as the $k+1$-ary relation

$$R_f = \{(a_1, \ldots, a_k, f(a_1, \ldots, a_k)) \mid a_i \in A\}.$$

A *sentence* is a closed formula, that is, each occurrence of a variable x in the formula is in the scope of a quantifier $\exists x$.

The formulas of L are interpreted over *structures* which give interpretations to the symbols in the language. A *structure* $\mathcal{A}$ for L (or L-structure) is a triple $(A; (R^{\mathcal{A}})_{R \in L^r}; (c^{\mathcal{A}})_{c \in L^c})$ consisting of a non-empty set A, the underlying set of $\mathcal{A}$; for each relation $R \in L^r$ of arity m a set $R^{\mathcal{A}} \subseteq A^m$, the interpretation of R in $\mathcal{A}$; and for each constant $c \in L^c$ an element $c^{\mathcal{A}} \in A$, the interpretation of c in $\mathcal{A}$. Given a structure $\mathcal{A}$ for L, a formula φ over L and Var, and a valuation v which maps each variable x in Var to an element in the domain A, we can define when $\mathcal{A}$ and v satisfy φ, denoted $\mathcal{A}, v \models \varphi$, inductively. Let v' be a function from $Var \cup L^c$ to A such that for $x \in Var$, $v'(x) = v(x)$ and for $c \in L^c$, $v'(c) = c^{\mathcal{A}}$. Also, let $v[x \mapsto a]$ be the function identical to v except that it maps the variable x to a.

- $\mathcal{A}, v \models R(t_1, \cdots, t_n)$ iff $R^{\mathcal{A}}(v'(t_1), \cdots, v'(t_n))$.
- $\mathcal{A}, v \models t_1 = t_2$ iff $v'(t_1) = v'(t_2)$.
- $\mathcal{A}, v \models \neg\varphi$ iff $\mathcal{A}, v \not\models \varphi$.
- $\mathcal{A}, v \models \varphi_1 \vee \varphi_2$ iff $\mathcal{A}, v \models \varphi_1$ or $\mathcal{A}, v \models \varphi_2$.
- $\mathcal{A}, v \models \exists x \varphi$ if there exists $a \in A$ such that $\mathcal{A}, v[x \mapsto a] \models \varphi$.

Example 13.1. The language $L_{Ri} = \{0, 1, \leq, +, \cdot\}$ of ordered semi-rings has the constant symbols 0 and 1, a binary symbol $\leq$ and ternary relation symbols $+$ and $\cdot$. Let $(\mathbb{R}, 0, 1, \leq, +, \cdot)$ be the L_{Ri} structure, where $\mathbb{R}$ is the set of reals, constants 0 and 1 are interpreted as the real numbers 0 and 1, and $\leq$, $+$ and $\cdot$ are relations denoting the standard total ordering, addition and multiplication ($+$ and $\cdot$ denote the graphs of the corresponding functions) on the reals, respectively. Note that we use $\leq, +$ and $\cdot$ to denote both the symbol and its interpretations.

Returning to the logical language L above, let us fix a structure $\mathcal{A}$ for L as above. Let us use the shorthand $v[x_1 \mapsto a_1, \cdots, x_k \mapsto a_k]$ to denote the valuation which maps every variable $x \notin \{x_1, \cdots, x_k\}$ to $v(x)$ and maps each x_i to a_i. A k-ary relation $S \subseteq A^k$, is said to be *definable* in the structure $\mathcal{A}$ if there is a first-order formula $\varphi(x_1, x_2, \ldots x_k)$ over L, such that $S = \{(a_1, \ldots, a_k) \mid \mathcal{A}, v[x_1 \mapsto a_1, \cdots, x_k \mapsto a_k] \models \varphi\}$. Hence, a set $S \subseteq A$ is definable in $\mathcal{A}$ if there exists a formula $\varphi(x)$ over L such that $S = \{a \mid \mathcal{A}, v[x \mapsto a] \models \varphi\}$. For example, the set $\{x \in \mathbb{R} \mid x \leq \sqrt{2}\}$ is definable in $(\mathbb{R}, 0, 1, \leq, +, \cdot)$ by the formula $(x \cdot x) \leq (1 + 1)$ (we use $\cdot$ and $+$ as functions here for readability, but we could use the formula

$\forall z \forall z'(\neg(\cdot(x,x,z) \wedge +(1,1,z')) \vee \leq (z,z'))$ instead). A k-ary function f will be said to be definable if its graph is definable. Note that the definable sets are clearly closed under boolean operations.

A *theory* *Th*($\mathcal{A}$) of a structure $\mathcal{A}$ is the set of all sentences that hold in $\mathcal{A}$, i.e., $\{\varphi \mid \mathcal{A} \models \varphi\}$. We will use $(\mathbb{R}, \leq, +, \cdot)$ to denote the theory of reals, that is, the set of sentences true in the L_{Ri} structure $(\mathbb{R}, 0, 1, \leq, +, \cdot)$ defined in Example 13.1. *Th*($\mathcal{A}$) is said to be *decidable* if there is an effective procedure to decide membership in the set *Th*($\mathcal{A}$). One of the consequences of *Th*($\mathcal{A}$) being decidable is that it is also decidable to check the emptiness of a definable relation, and whether two definable relations are equal.

A binary relation $\leq$ on a set A is said to be a total order if $\leq$ is reflexive, transitive and anti-symmetric. The set A is said to be totally ordered by the ordering relation $\leq$. An *interval* is a subset of a totally ordered set defined using one or two bounds as follows: $\{x \mid a \sim_1 x \sim_2 b\}$, $\{x \mid x \sim a\}$, and $\{x \mid a \sim x\}$, where $\sim_1, \sim_2, \sim \in \{<, \leq\}$. Trivially, $\{x \mid a \leq x \leq b\}$ with $a = b$, is an interval consisting of a single point. We write $\mathcal{A} = (A, \leq, \ldots)$ to convey that the L-structure $\mathcal{A}$ has an ordering relation $\leq$ along with other elements in its structure.

Definition 13.5. A totally ordered structure $\mathcal{A} = (A, \leq, \ldots)$ is *o-minimal* (order-minimal) if every definable subset of A is a finite union of intervals [18].

If $\mathcal{A}$ is an o-minimal structure, then we will call *Th*($\mathcal{A}$) an o-minimal theory. Examples of o-minimal theories include $(\mathbb{R}, \leq, +, -, \cdot)$ and $(\mathbb{R}, \leq, +, -, \cdot, \mathbf{exp})$, where "$-$" and $\mathbf{exp}$ are the subtraction and exponentiation operations on reals, respectively. Additional examples can be found in [18, 19]. The theory $(\mathbb{R}, \leq, +, -, \cdot)$ is known to be decidable [20].

Next we state a property of o-minimal structures which will be used later (see [18], Corollary 3.6, p.60).

Theorem 13.2. *(Uniform Finiteness) Let $Y \subseteq \mathbb{R}^m \times \mathbb{R}^n$ be definable in an o-minimal theory, we denote by Y_a the fibre $\{z \in \mathbb{R}^n \mid (a, z) \in Y\}$. Then there is a number $N_Y \in \mathbb{N}$ such that for each $a \in \mathbb{R}^m$, the set Y_a has at most N_Y definable connected components.*

O-minimal hybrid systems. An *o-minimal* hybrid system is a hybrid system of the form

$$\mathcal{H} = (Q, X, \mathit{init}, F', I, E, G, J)$$

where all the components except F' are similar to those in the definition of a hybrid system with the following additional constraints. There exists an *o-minimal* structure $\mathcal{A} = (\mathbb{R}, \leq, \ldots)$ such that

- the sets *init* and X are definable in $\mathcal{A}$.
- the sets $I(q)$, $G(e)$ and $J(e)$ are definable in $\mathcal{A}$ for every $q \in Q$ and $e \in E$.
- $J \colon E \to 2^{X \times X}$ maps every $e \in E$ to a set $X_1 \times X_2$ where $X_1, X_2 \subseteq X$.

- $F' : Q \times X \to (\mathbb{R}_{\geq 0} \to X)$ such that $F'(q,x)(0) = x$, F' is a continuous function, i.e, for each q, the function f_q defined by $f_q(x,t) = F'(q,x)(t)$ is continuous, and F' is definable in $\mathcal{A}$. Here the function F' is interpreted as a solution of the vector field instead of the vector field itself. The induced transition system $[\![\mathcal{H}]\!]$ is defined as for a hybrid automaton in Definition 13.2, except that $((q,v),t,(q',v')) \in\to$ if $t \geq 0$, $q = q'$, $F'(q,v)(t) = v'$, and for all $t' \in [0,t]$, $F'(q,v)(t') \in I(q)$.

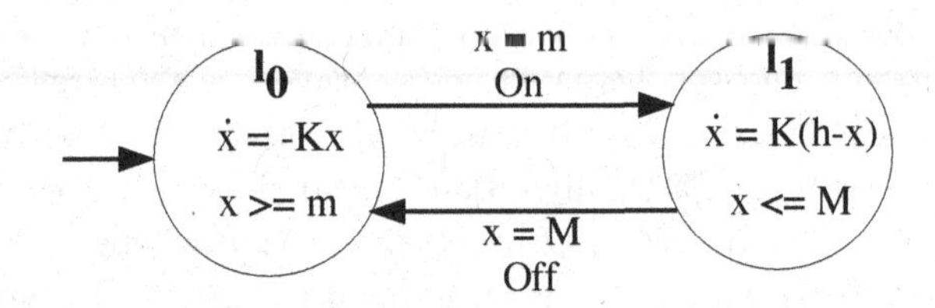

Fig. 13.14. Example of an o-minimal hybrid system.

Figure 13.14 gives an example of a thermostat represented as an o-minimal hybrid system [21]. The temperature of a room has to be kept between m and M degrees. The room is equipped with a thermostat which senses the temperature and turns a heater on and off. The temperature is governed by differential equations. Let us denote the temperature by the variable x. When the heater is on, the temperature decreases according to the function $x(t) = \Theta e^{-Kt}$; when the heater is on, the temperature increases according to the function $x(t) = \Theta e^{-Kt} + h(1 - e^{-Kt})$ where t is the time, Θ the initial temperature, h and K are parameters for the heater and the room. Note that the resets of the hybrid automaton are of the required form. For example the reset corresponding to the edge from l_0 to l_1 is given by the set $\mathbb{R} \times \{m\}$.

Also the water tank example given in Figure 13.11 is an o-minimal hybrid system.

It is shown in [3, 21] that verification of properties like control state reachability is decidable for the class of o-minimal systems with a decidable o-minimal theory. This is shown by exhibiting a finite bisimulation quotient of the original transition system in which the exact time is abstracted away. Hence one can construct a finite state system on which a model-checking algorithm can be run and the result can then be lifted to the verification problem of the original system. This approach is similar to the region automaton for timed automata (see Chapter 4 in this volume). So we consider a time-abstract semantics of a hybrid system $\mathcal{H}$, denoted $[\![\mathcal{H}]\!]_\tau$, which is similar to $[\![\mathcal{H}]\!]$ except that every transition labeled with $t \in \mathbb{R}$ is replaced by the symbol τ. For simplicity of the proof, we will assume the following consistency condition on the flow function: $F'(q,x)(t+t') = F'(q, F'(q,x)(t))(t')$ for all x, t and t'. The above property states that the future of the flow after reaching a point y is independent of how it reached y. Given a subset $S \subseteq Q \times X$, let $S|_q$ denote the set $\{(q,x) \mid (q,x) \in S\}$, that is the restriction of S to the discrete location q.

Theorem 13.3. *Let $\mathcal{H} = (Q, X, init, F, I, E, G, J)$ be an o-minimal hybrid system on the o-minimal structure $\mathcal{A}$. Let its time abstract transition system be $[\![\mathcal{H}]\!]_\tau = (S, init, \{\mu, \tau\}, \rightarrow)$. Given any $\mathcal{A}$-definable finite partition Π of S, there exists a definable equivalence relation $\sim$ on S of finite index which refines the partition Π and is a bisimulation relation on $[\![\mathcal{H}]\!]_\tau$. Further, the bisimulation quotient $[\![\mathcal{H}]\!]_\tau/_\sim$ is constructible if Th($\mathcal{A}$) is decidable.*

Proof. (*Sketch.*) The proof proceeds in two steps. In the first step, a partition of the state space of $[\![\mathcal{H}]\!]_\tau$ is defined which is a bisimulation when only μ transitions are considered. The partition defined has the property that any refinement of the partition is also a bisimulation with respect to μ transition. Then a refinement of the above defined partition is constructed which is a bisimulation with respect to the τ transitions.

Given a discrete location q, consider a partition $\equiv_q$ of $S|_q$ which refines the initial continuous state space, invariants, guards and reset associated with the location in the following sense. For all $s_1 = (q, x_1), s_2 = (q, x_2) \in S|_q$, $s_1 \equiv_q s_2$ implies

- $s_1 \Pi s_2$.
- $s_1 \in init|_q$ iff $s_2 \in init|_q$,
- $x_1 \in I(q)$ iff $x_2 \in I(q)$,
- for every edge $e = (q, q')$ out of q, $x_1 \in G(e)$ iff $x_2 \in G(e)$ and for the jump $J(e) = X_1 \times X_2$ associated with the edge e, $x_1 \in X_1$ iff $x_2 \in X_1$.

Note that given that the various components of the hybrid system $\mathcal{H}$ are o-minimally definable, we can define the maximal relation $\equiv_q$ satisfying the above conditions. Further $\equiv_q$ is of finite index, and $\Pi_0 = \bigcup_q \equiv_q$ is an equivalence relation of finite index on S. And if $s_1 \Pi_0 s_2$ and $(s_1, \mu, s_1') \in\rightarrow$, then there exists s_2' such that $(s_2, \mu, s_2') \in\rightarrow$. This is because, if $s_1 = (q, x_1)$, $s_1' = (q', x_1')$ and $(s_1, \mu, s_1') \in\rightarrow$, there is an edge $e = (q, q')$ such that $x_1 \in G(e)$, $J(e) = X_1 \times X_2$, $x_1 \in X_1$ and $x_1' \in X_2$. Then $((q, x_2), \mu, (q', x_1')) \in\rightarrow$, since $(q, x_1)\Pi_0(q, x_2)$ implies $(q, x_1) \equiv_q (q, x_2)$ which along with $x_1 \in G(e)$ and $x_1 \in X_1$ implies $x_2 \in G(e)$ and $x_2 \in X_1$. Therefore for $s_2' = s_1'$, $(s_2, \mu, s_2') \in\rightarrow$. In this case we say that Π_0 is a bisimulation on S with respect to the μ transitions. Further Π_0 is a refinement of Π. Observe that any refinement of Π_0 is also a bisimulation on S with respect to μ transitions.

Next we refine each $\equiv_q$ to obtain a finite definable $\equiv_q'$ which is a bisimulation on $S|_q$ with respect to τ transitions. Then $\sim = \bigcup_q \equiv_q'$ is a bisimulation on $[\![\mathcal{H}]\!]_\tau$ which refines Π, is definable and is of finite index.

It remains to show that we can find a finite refinement of $\equiv_q$ which is a bisimulation with respect to τ transitions. Let us fix a location q. Let $\mathcal{P} = \{P_1, \cdots, P_k\}$ be the partitions of $\equiv_q$, and let $S_q = \{q\} \times X$. Let $\varphi_i(x,t)$ be a formula which says that the trajectory starting from state (q, x) at time t is in a state (q, y) and $(q, y) \in P_i$. Note that such a formula exists, since each of the P_is is definable. Let $Y_i = \{(x, t) \,|\, t \geq 0, \varphi_i(x, t)\}$. Since Y_i is a definable subset of $\mathbb{R}^n \times \mathbb{R}$, from Theorem 13.2 we have that there is a bound N_i such that for every x, the times at which the trajectory starting from (q, x) is in P_i forms a finite union of intervals of time, and the number of intervals is bounded by N_i. Hence there is a number N $(= \sum_i N_i)$ such that the sequence of partitions a trajectory starting from any (q, x) visits can be represented as a finite word of length at most N. With every $x \in X$, let us associate a word w_x which denotes the sequence of partitions of $\mathcal{P}$ visited by the trajectory starting from (q, x). Figure 13.15 gives a partition of the state space into parts A and B. The word corresponding to the trajectory f is A,

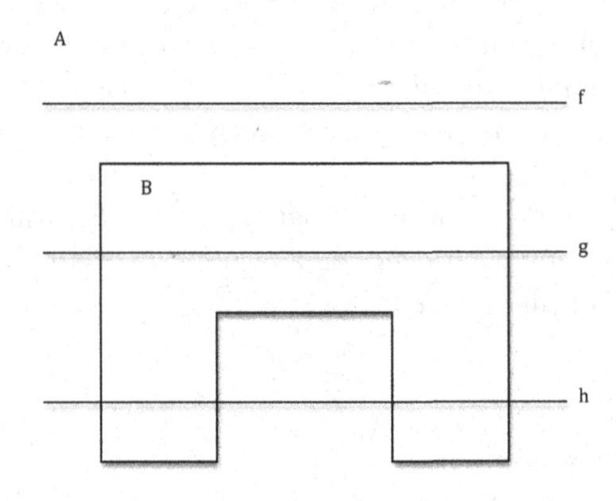

Fig. 13.15. Encoding a trajectory as a word.

that corresponding to g is ABA and the word corresponding to h is $ABABA$.

Let $\Omega = \{w_x \,|\, x \in X\}$. Note that Ω is finite and if the underlying o-minimal theory is decidable, then Ω can be computed. Define an equivalence relation $\equiv'_q$ on S_q whose equivalence classes are given by $\{S_w \,|\, w \in \Omega\}$ where $S_w = \{(q, x) \in S_q \,|\, w_x \in \Omega\}$. Note that $\equiv'_q$ refines $\equiv_q$ since $(q, x_1) \equiv'_q (q, x_2)$ implies that the first symbols of w_{x_1} and w_{x_2} are the same, which implies that the partitions to which (q, x_1) and (q, x_2) belong to are the same. Further $\equiv'_q$ has finite index and is definable in the underlying o-minimal theory. Finally, $\equiv'_q$ is a bisimulation with respect to τ transitions. To see the last remark, note that if $s_1 \equiv'_q s_2$ and $(s_1, \tau, s'_1) \in \rightarrow$, then w_{y_1} is a suffix of w_{x_1} where $s_1 = (q, x_1)$ and $s'_1 = (q, y_1)$. But then $w_{x_2} = w_{x_1}$, where $s_2 = (q, x_2)$. The trajectory starting from (q, x_2) reaches a state (q, y_2) such that $w_{y_2} = w_{y_1}$, since w_{y_1} is also a suffix of w_{x_2}. Hence there is a $s'_2 = (q, y_2)$ such that $(s_2, \tau, s'_2) \in \rightarrow$. This proves the first part of the theorem.

If the underlying o-minimal theory is decidable, we can check if there is a μ or τ transition between the sets of states defined by any two partitions. That is, given two definable sets P and Q we can construct a sentence in the language of $\mathcal{A}$ which is true if and only if there is a μ (or τ) transition from P to Q. Since the theory of $\mathcal{A}$ is decidable we can check for the satisfaction of this formula. Hence we can compute the bisimulation quotient $[\![\mathcal{H}]\!]_\tau \backslash \sim$.

Alternately, to compute a finite bisimulation quotient of $[\![\mathcal{H}]\!]_\tau$, we run the bisimulation algorithm [22] on the initial partition. If the theory of the underlying o-minimal structure is decidable, then each step in the bisimulation algorithm is computable (because of the closure of the o-minimally definable sets under boolean operations and the decidability of their emptiness). Further, the bisimulation algorithm terminates since $[\![\mathcal{H}]\!]_\tau$ has a finite bisimulation. However, it might not terminate with the partition $\sim$. □

More details can be found in [3, 21].

Corollary 13.1. *Let $\mathcal{A}$ be an o-minimal structure such that Th($\mathcal{A}$) is decidable. Then the control state reachability problem is decidable for the class of o-minimal hybrid systems definable in $\mathcal{A}$.*

Proof. Let $\mathcal{H}$ be an o-minimal hybrid system definable in $\mathcal{A}$. Consider the following partition $\Pi = \{init, S \backslash init\}$ of the state space of $[\![\mathcal{H}]\!]_\tau = (S, init, \{\mu, \tau\}, \rightarrow)$. Then $\sim$ given by Theorem 13.3 is a bisimulation of $[\![\mathcal{H}]\!]_\tau$ which refines Π. Hence from Proposition 13.2, $[\![\mathcal{H}]\!]_\tau$ is bisimilar to $[\![\mathcal{H}]\!]_\tau \backslash \sim$. First note that all the states in a particular partition of $\sim$ all have the same location. Hence a location q of $\mathcal{H}$ is reachable iff there exists some x such that (q, x) is reachable in $[\![\mathcal{H}]\!]_\tau$ iff there exists some partition of $\sim$ corresponding to q which is reachable in $[\![\mathcal{H}]\!]_\tau \backslash \sim$. Since $[\![\mathcal{H}]\!]_\tau \backslash \sim$ is a finite state system, we can compute whether some state of $[\![\mathcal{H}]\!]_\tau \backslash \sim$ corresponding to a location q of $\mathcal{H}$ is reachable. □

One of the important implications of the above theorem is that the verification of various properties which are bisimulation invariant, like properties expressed in CTL or mu-calculus, can be solved algorithmically for o-minimal systems defined in decidable theories.

It turns out that the jumps in the o-minimal hybrid systems are a limiting factor in that they do not allow the continuous values of a variable to be carried across locations. Note that jumps are of the form $X_1 \times X_2$, so a continuous value belonging to X_1 before the jump can be reset to any value in X_2. Other models have been proposed which relax this condition, for example, see [23, 24].

References

[1] R. Alur, C. Courcoubetis, N. Halbwachs, T. A. Henzinger, P.-H. Ho, X. Nicollin, A. Olivero, J. Sifakis, and S. Yovine, The algorithmic analysis of hybrid systems, *Theor. Comput. Sci.* **138**(1), 3–34, (1995).

[2] A. Puri and P. Varaiya. Decidability of hybrid systems with rectangular differential inclusion. In *Proc. 6th CAV*, pp. 95–104, (1994).

[3] G. Lafferriere, G. Pappas, and S. Sastry, O-minimal hybrid systems, *Mathematics of Control, Systems and Signals.* **13**(1), 1–21, (2000).

[4] E. D. Sontag, *Mathematical Control Theory: Deterministic Finite Dimensional Systems.* (Springer, 1998).

[5] J. Lygeros. Lecture notes on hybrid systems.

[6] T. A. Henzinger, P. W. Kopke, A. Puri, and P. Varaiya, What's decidable about hybrid automata?, *J. Comput. Syst. Sci.* **57**(1), 94–124, (1998).

[7] M. L. Minsky, *Computation: finite and infinite machines.* (Prentice-Hall, 1967).

[8] R. Alur and D. L. Dill, A theory of timed automata, *Theor. Comput. Sci.* **126**(2), 183–235, (1994).

[9] E. Asarin, O. Maler, and A. Pnueli, Reachability analysis of dynamical systems having piecewise constant derivatives, *Theor. Comput. Sci.* **138**(1), 35–65, (1995).

[10] E. Asarin, G. Schneider, and S. Yovine. On the decidability of the reachability problem for planar differential inclusions. In *Proc. 4th HSCC*, vol. 2034, *LNCS*, pp. 89–104. Springer, (2001).

[11] E.Asarin, G.Schneider, and S.Yovine, Algorithmic analysis of polygonal hybrid systems, part I: Reachability, *Theor. Comput. Sci.* **379**(1-2), 231–265, (2007).

[12] P. Prabhakar, V. Vladimerou, M. Viswanathan, and G. E. Dullerud. A decidable class of planar linear hybrid systems. In *Proc. 11th HSCC*, vol. 4981, *LNCS*, pp. 401–414. Springer, (2008).

[13] A. Puri, P. Varaiya, and V. Borkar. ϵ-approximation of differential inclusions. In *Proc. 34th CDC*, vol. 3, pp. 2892–2897. IEEE (Dec, 1995).

[14] R. Alur, T. A. Henzinger, G. Lafferriere, and G. J. Pappas, Discrete abstractions of hybrid systems, *Proceedings of the IEEE.* **88**(7), 971–984 (Jul, 2000).

[15] A. Puri and P. Varaiya. Decidable hybrid systems. In *Proc. Workshop on Hybrid Systems III*, vol. 1066, *LNCS*, pp. 413–423. Springer, (1996).

[16] R. Milner, *Communication and Concurrency.* (Prentice-Hall, Inc, 1989).

[17] H. D. Ebbinghaus, J. Flum, and W. Thomas, *Mathematical Logic.* (Springer, 1996).

[18] L. van den Dries, *Tame Topology and O-minimal Structures.* (Cambridge Univesity Press, 1998).

[19] L. van den Dries and C. Miller, On the real exponential field with restricted analytic functions, *Israel Journal of Mathematics.* (85), 19–56, (1994).

[20] A. Tarski, *A Decision Method for Elementary Algebra and Geometry.* (University of California Press, 1951), 2nd edition.

[21] T. Brihaye and C. Michaux, On the expressiveness and decidability of o-minimal hybrid systems, *J. Complexity.* **21**(4), 447–478, (2005).

[22] R. Paige and R. E. Tarjan, Three partition refinement algorithms, *SIAM J. Comput.* **16**(6), 973–989, (1987).

[23] R. Gentilini. Reachability problems on extended o-minimal hybrid automata. In *Proc. 3rd FORMATS*, vol. 3829, *LNCS*, pp. 162–176. Springer, (2005).

[24] V. Vladimerou, P. Prabhakar, M. Viswanathan, and G. E. Dullerud. STORMED hybrid systems. In *Proc. 35th ICALP (Part 2)*, vol. 5126, *LNCS*, pp. 136–147. Springer, (2008).

Chapter 14

The Discrete Time Behaviour of Restricted Linear Hybrid Automata

Manindra Agrawal[1*], Frank Stephan[2†], P. S. Thiagarajan[3] and Shaofa Yang[4‡]

[1]*Department of Computer Science and Engineering, Indian Institute of Technology, Kanpur, India*
manindra@cse.iitk.ac.in

[2]*School of Computing and Department of Mathematics, National University of Singapore (NUS), Singapore*
fstephan@comp.nus.edu.sg

[3]*School of Computing, NUS, Singapore*
thiagu@comp.nus.edu.sg

[4]*UNU-IIST, Macao*
ysf@iist.unu.edu

We summarize results from [1–3] on the discrete time behaviour of a class of restricted linear hybrid automata. Specifically, we show the regularity of the discrete time behaviour of hybrid automata in which the rates of continuous variables are governed by linear operators in a diagonal form and in which the values of the continuous variables can be observed only with finite precision. Crucially, we do not demand—as is usually done—that the values of the continuous variables be reset during mode changes. We can cope with polynomial guards and we can tolerate bounded delays both in sampling the values of the continuous variables and in effecting changes in their rates required by mode switchings. We also show that if the rates are governed by diagonalizable linear operators with rational eigenvalues *and* there is no delay in effecting rate changes, the discrete time behaviour of the hybrid automaton is recursive. However, the control state reachability problem in this setting is undecidable.

14.1. Introduction

Hybrid automata are models of control systems which consist of digital components interacting with continuously evolving physical attributes. Such control systems

---

*Part of this work was done when the author was a Distinguished Visiting Professor at NUS.
†Supported in part by NUS grant R252–000–212–112.
‡This work was done when the author was a *Ph. D.* student at NUS.

appear in domains such as avionics, automotive electronics and industrial process control. Hybrid automata can be used to provide effective methods for modeling and verifying the correct functioning of such systems.

A hybrid automaton is a finite transition system where each state (termed mode or control state) is labelled with a differential equation specifying the evolution rates of a fixed set of real-valued variables $x_1, x_2, \ldots, x_n$. Each transition is labelled with a constraint (called a guard) on the values of $x_1, x_2, \ldots, x_n$. A transition can be taken only when its guard is satisfied. The semantics of a hybrid automaton is given by an infinite state transition system where each state consists of the current mode and the current values of $x_1, x_2, \ldots, x_n$. There are two types of transitions, one signifying the passage of time during which the continuous variables evolve at a mode and the other—usually instantaneous—consisting of a change of modes.

We study the behaviour of hybrid automata in which the rate functions associated with the modes are restricted linear differential equations. We show that if the values of the continuous variables can be observed only with finite precision, then the discrete time behaviour of a large class of hybrid automata is regular. Further, these behaviours can be effectively computed. The key feature of our setting is that we do *not* demand that the value of a continuous variable be reset during a mode switch. The point is, this reset assumption—often a crippling one from the modeling point of view—is often made to obtain tractable classes of hybrid automata [4, 5]. Our results suggest that focusing on discrete time semantics and the realistic assumption of finite precision leads to effective analysis methods for hybrid automata.

In the related literature, one often assumes that the rates are piecewise constant. This is so especially in settings where one does not make the reset assumption and yet obtains positive verification results [6–8]. Even here, since the mode changes can take place over continuous time (a transition may be taken any time its guard is satisfied), basic verification problems often become undecidable [4, 9]. In contrast, it was shown in [10] that one can go much further in the positive direction for piecewise constant rate automata, if one defines their behaviour using a discrete time semantics. As argued in [10], if the hybrid automaton models the closed loop system consisting of a digital controller interacting with a continuous plant, then the discrete time semantics is the natural one; the controller will observe via sensors, the states of the plant and effect, via actuators, changes in the plant dynamics at discrete time points determined by its internal clock. In [1] it was shown that, in this setting, one can in fact tolerate bounded delays both in the observation of the plant states and in effecting changes in the plant dynamics.

Both in [10] and [1], the transition guards were required to be rectangular; conjunctions of simple linear inequalities involving just one variable. In [2], it was shown that one can cope with much more expressive guards—essentially all effectively computable guards—if one further assumes that the values of the continuous variables can be observed only with finite precision. In many settings including the one where the hybrid automaton models a digital controller interacting with a continuous plant, finite precision is a natural assumption.

In [3] we showed that the combination of discrete time semantics and finite precision can not only allow more expressive guards but can also take us beyond piecewise constant rates. One of our main results here is that under finite precision, the discrete time behaviour of a hybrid automaton is regular and effectively computable even when the rate of a continuous variable x_i in each control state q is governed by an equation of the form $dx_i/dt = c_q \cdot x_i(t)$. Further, we can cope with arbitrary computable guards. We can also tolerate bounded delays in sampling the values of the continuous variables and in effecting changes in their rates required by mode switchings. Again, it is a crucial feature that we do not demand resetting of the values of the continuous variables during mode changes. Our regularity result implies that a variety of model checking problems can be effectively solved for this class of hybrid automata. For instance, it follows easily that the control state reachability problem in this setting is decidable. One can also effectively determine whether the discrete time behaviors exhibited by such a hybrid automaton satisfy properties asserted in various temporal logics [11].

We also showed that the discrete time behaviours of hybrid automata in a much richer setting are *recursive.* That is, for such a hybrid automaton $\mathcal{A}$, one can effectively determine whether a given finite sequence of control states is a discrete time behaviour of $\mathcal{A}$. Specifically, the rates of continuous variables at the control state q are governed by a linear differential operator represented by a diagonalizable ([12]) matrix A_q with rational eigenvalues. Further, we allow polynomial guards but do not permit delays in effecting rates changes. A consequence of this positive result is that one can effectively solve a variety of bounded model checking problems [13] in this rich setting. However, it turns out that the control state reachability problem is undecidable for this class of automata; this is so, even if the guards are restricted to be rectangular.

The proofs of the above two results seem to suggest that one can hope to go much further if update delays are allowed. This will prevent the hybrid automaton from retaining an unbounded amount of information as its dynamics evolves. However, we do not know at present how to take advantage of this observation since we lack suitable techniques for tracking rational approximations of exponential terms with *real* exponents. In this connection, the fundamental theory presented in [14] may turn to be important. We also feel that the techniques presented in [15, 16] will turn out to be useful even though they are developed under a regime where continuous variables are reset during mode changes.

In the next two sections, we define our hybrid automata and develop their discrete time semantics. Section 14.4 provides a brief summary of the results on these hybrid automata in [1–3]. In Section 14.5, we present the results in the setting of finite precision. Namely, we show that the discrete time behaviour of a class of hybrid automata is regular where the evolution rates are governed by linear operators in diagonal form. We also show that the discrete time behaviour of a class of hybrid automata is recursive where the rates matrices are diagonalizable with rational eigenvalues. In Section 14.6, we discuss the results under the assumption of perfect precision. We show that the results in 14.5 also hold in the perfect-precision setting provided that every transition guard is rectangular. We conclude in Section 14.7. In our presentation we emphasize the main constructions and proof ideas. The technical details can be found in [17].

14.2. Preliminaries

Through the rest of this chapter, we fix a positive integer n and one function symbol x_i for each i in $\{1, 2, \ldots, n\}$. We will often refer to the x_i's as "continuous" variables and will view each x_i as a function (of time) $x_i : \mathbb{R}_{\geq 0} \rightarrow \mathbb{R}$. As usual, $\mathbb{R}$ is the set of reals and $\mathbb{R}_{\geq 0}$, the set of non-negative reals. We let $\mathbb{Q}$ denote the set of rationals.

The transitions of the hybrid automaton will have associated guards. A *polynomial constraint* is an inequality of the form $p(x_1, x_2, \ldots, x_n) \leq 0$ or $p(x_1, x_2, \ldots, x_n) < 0$ where $p(x_1, x_2, \ldots, x_n)$ is a polynomial over $x_1, x_2, \ldots, x_n$ with integer coefficients. A *polynomial guard* is a finite conjunction of polynomial constraints. We let *Grd* denote the set of polynomial guards. Unless otherwise stated, by a guard we will mean a polynomial guard.

A *valuation* V is just a member of $\mathbb{R}^n$. It prescribes the value $V(i)$ to the variable x_i. The notion of a valuation satisfying a guard is defined in the obvious way.

A *lazy finite-precision linear hybrid automaton* is a structure $\mathcal{A} = (Q, q_{in}, V_{in}, Delay, \epsilon, \{\rho_q\}_{q \in Q}, \{\gamma_{min}, \gamma_{max}\}, \longrightarrow)$ where:

- Q is a finite set of *control states* with q, q' ranging over Q.
- $q_{in} \in Q$ is the initial control state.
- $V_{in} \in \mathbb{Q}^n$ is the initial valuation.
- $Delay = \{\delta^0_{ob}, \delta^1_{ob}, \delta^0_{up}, \delta^1_{up}\} \subseteq \mathbb{Q}$ is the set of *delay parameters* such that $0 \leq \delta^0_{up} \leq \delta^1_{up} < \delta^0_{ob} \leq \delta^1_{ob} \leq 1$.
- ϵ, a positive rational, is the *precision of measurement.*
- $\{\rho_q\}_{q \in Q}$ is a family of rate functions associated with the control states. In the general case, ρ_q will be of the form $\dot{x} = A_q x + b_q$ where A_q is an $n \times n$ matrix with rational entries and $b_q \in \mathbb{Q}^n$. For each i in $\{1, 2, \ldots, n\}$ this specifies the rate function of x_i as the differential equation $dx_i/dt = \sum_{j=1}^{n} A_q(i,j) \cdot x_j(t) + b_q(i)$ where $A_q(i,j)$ is the (i,j)-th entry of A_q.
- $\gamma_{min}, \gamma_{max} \in \mathbb{Q}$ are *range parameters* such that $0 < \gamma_{min} < \gamma_{max}$.
- $\longrightarrow \subseteq Q \times Grd \times Q$ is a transition relation such that $q \neq q'$ for every (q, g, q') in $\longrightarrow$.

Figure 14.1 displays an example of such a hybrid automaton. The control states $q1, q2$ are represented by two circles. The rate functions of $q1, q2$ are inscribed in their respective circles. The initial state $q1$ is indicated with a pointing arrow.

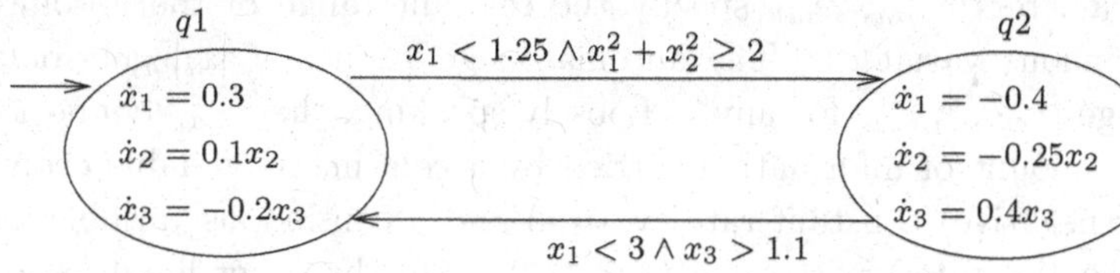

q1: $\dot{x}_1 = 0.3$, $\dot{x}_2 = 0.1x_2$, $\dot{x}_3 = -0.2x_3$

q2: $\dot{x}_1 = -0.4$, $\dot{x}_2 = -0.25x_2$, $\dot{x}_3 = 0.4x_3$

q1 → q2: $x_1 < 1.25 \wedge x_1^2 + x_2^2 \geq 2$

q2 → q1: $x_1 < 3 \wedge x_3 > 1.1$

$\delta_{ob}^0 = 0.8,\ \delta_{ob}^1 = 0.9,\ \delta_{up}^0 = 0.1,\ \delta_{up}^1 = 0.2$
$V_{in}(1) = V_{in}(2) = V_{in}(3) = 1$
$\epsilon = 0.1$
$\gamma_{min} = 0.5,\ \gamma_{max} = 4$

Fig. 14.1. A lazy finite-precision linear hybrid automaton.

In what follows, for convenience we will often say "finite-precision hybrid automaton" instead of "lazy finite-precision linear hybrid automaton". We are interested in the discrete time behaviour of our hybrid automata assuming that a suitable granularity of time has been fixed. At each time instant T_k, the automaton receives a measurement regarding the current values of the x_i's. However, the value of x_i that is observed at time T_k is the value that held at some time $t \in [T_{k-1} + \delta_{ob}^0, T_{k-1} + \delta_{ob}^1]$. Further, the value is observed with a precision of ϵ. More specifically, any value of x_i in the half-open interval $[(m - 1/2)\epsilon, (m + 1/2)\epsilon)$ is reported as $m\epsilon$ where m is an integer. For a real number v, we will denote this rounded-off value relative to ϵ as $\langle v \rangle_\epsilon$ and often just write $\langle v \rangle$. More sophisticated rounding-off functions can also be considered (see [2]) but for ease of presentation, we shall not do so here.

If at T_k, the automaton is in control state q and the observed n-tuple of values $(\langle v_1 \rangle, \langle v_2 \rangle, \ldots, \langle v_n \rangle)$ satisfies the guard g with (q, g, q') being a transition, then the automaton may perform this transition instantaneously and move to the control state q'. As a result, the x_i's will cease to evolve according to the rate function ρ_q and instead start evolving according to the rate function $\rho_{q'}$. However, for each x_i, this change in the rate of evolution of each x_i will not kick in at T_k but at some time $t \in [T_k + \delta_{up}^0, T_k + \delta_{up}^1]$. In this sense, both the sensing of the x_i's and the rate changes associated with mode switching take place in a lazy fashion but with bounded delays. We expect $\delta_{ob}^0, \delta_{ob}^1$ to be close to 1 and $\delta_{up}^0, \delta_{up}^1$ to be close to 0 while both $\delta_{ob}^1 - \delta_{ob}^0$ and $\delta_{up}^1 - \delta_{up}^0$ to be small compared to 1. The restriction we have imposed on the delay parameters is mainly for minimizing the notational overhead. More relaxed conditions can be imposed without affecting the main results.

In the idealized setting, the value observed at T_k is the value that holds at exactly T_k ($\delta_{ob}^0 = 1 = \delta_{ob}^1$) and the change in rates due to mode switching would kick in immediately ($\delta_{up}^0 = 0 = \delta_{up}^1$).

The parameters $\gamma_{min}, \gamma_{max}$ specify the relevant range of the absolute values of the continuous variables. The automaton gets stuck if $|x_i|$ gets outside the allowed range $[\gamma_{min}, \gamma_{max}]$ for any i. Loosely speaking, the γ_{max} bound is used to restrict the amount of information carried by a continuous variable evolving at a (positive or negative) constant rate ($\dot{x}_i = c$) and a continuous variable increasing at an exponential rate ($\dot{x}_i = c \cdot x_i(t)$, $c > 0$). On the other hand, γ_{min} is used to restrict the amount of information carried by a continuous variable decreasing at an exponential rate ($\dot{x}_i = c \cdot x_i(t)$, $c < 0$). We note that our setting is quite different from the classical continuous setting. Hence the standard control objective of driving a system variable to 0 is not relevant here and thus does not pose a serious limitation.

We define a lazy *perfect-precision* differential hybrid automaton in the obvious way; the parameter ϵ is dropped and $\langle v \rangle$ is set to v for every real number v. For convenience, we shall refer to a lazy perfect-precision linear hybrid automaton as a perfect-precision hybrid automaton. For further convenience, a finite-precision hybrid automaton will often be referred to as just a hybrid automaton.

14.3. The Transition System Semantics

In what follows, we define the behaviour of one of our hybrid automaton in terms of an associated transition system. The behaviour of a perfect-precision hybrid automaton will be an easy adaptation of this notion.

Through the rest of this section, we fix a hybrid automaton $\mathcal{A}$ and assume its associated notations and terminology as defined in the previous section. A *configuration* is a triple (q, V, q') where q, q' are control states and V is a valuation. q is the current control state, q' is the control state that held at the previous time instant and V captures the *actual* values of the variables at the current time instant. The valuation V is said to be *feasible* if $\gamma_{min} \leq |V(i)| \leq \gamma_{max}$ for every i in $\{1, 2, \ldots, n\}$. The configuration (q, V, q') is *feasible* iff V is a feasible valuation. The initial configuration is (q_{in}, V_{in}, q_{in}) and is assumed to be feasible. We let $Conf_{\mathcal{A}}$ denote the set of configurations. We assume that the unit of time has been fixed at some suitable level of granularity and that the rate functions $\{\rho_q\}_{q \in Q}$ have been scaled accordingly.

Suppose the automaton $\mathcal{A}$ is in the configuration (q_k, V_k, q'_k) at time T_k. Then one unit of time will be allowed to pass and at time instant T_{k+1}, the automaton $\mathcal{A}$ will make an instantaneous move by executing a transition or the silent action τ and move to a configuration $(q_{k+1}, V_{k+1}, q'_{k+1})$. The silent action τ will be used to record that no mode change has taken place during this move. The action μ will be used to record that a transition has been taken and as a result, a mode change has taken place. As is common, we will collapse the unit-time-passage followed by an instantaneous transition into one "time-abstract" transition labelled by τ or μ. We wish to formalize the transition relation $\Longrightarrow \subseteq Conf_{\mathcal{A}} \times \{\tau, \mu\} \times Conf_{\mathcal{A}}$. For doing so, we note that given a matrix $A \in \mathbb{Q}^{n \times n}$, a vector $b \in \mathbb{Q}^n$, a positive real T

and a valuation V, we can find a unique family of curves (see [12]) $\{x_i\}_{1\leq i\leq n}$ with $x_i : [0,T] \to \mathbb{R}$ such that for every i we have $x_i(0) = V(i)$ and for every $t \in [0,T]$ we have $dx_i/dt = \sum_{j=1}^{n} A_q(i,j) \cdot x_j(t) + b_q(i)$. In what follows, we shall denote the valuation $(x_1(T), x_2(T), \ldots, x_n(T))$ thus obtained as $Val(A,b,T,V)$ without explicitly displaying the curves x_i's.

Let (q,V,q'), $(q1,V1,q1')$ be in $Conf_{\mathcal{A}}$. Suppose there exist reals t_i^{up}, $i = 1,2,\ldots,n$, in $[\delta_{up}^0, \delta_{up}^1]$ such that $V1$ is related to V as follows: Let $t_{\pi_1}^{up} \leq t_{\pi_2}^{up} \leq \cdots \leq t_{\pi_n}^{up}$ with $\pi_1, \pi_2, \ldots, \pi_n$ being a permutation of the indices $1,2,\ldots,n$. Then there exist valuations U_i, $i = 1,2,\ldots,n$, such that $U_1 = Val(A_{q'}, b_{q'}, t_{\pi_1}^{up}, V)$; $U_{i+1} = Val(A_i, b_i, t_{\pi_{i+1}}^{up} - t_{\pi_i}^{up}, U_i)$ for $i = 1,2,\ldots,n-1$; and $V1 = Val(A_n, b_n, 1 - t_{\pi_n}^{up}, U_n)$, where for $i = 1,2,\ldots,n$, the matrix $A_i \in \mathbb{Q}^{n\times n}$ and the vector $b_i \in \mathbb{Q}^n$ are given by: if $j \in \{\pi_1, \pi_2, \ldots, \pi_i\}$, then the j-th row of A_i (b_i) equals the j-th row of A_q (b_q); otherwise the j-th row of A_i (b_i) equals the j-th row of $A_{q'}$ ($b_{q'}$).

The intuition is that at time T_{k+1} the continuous variables have valuation $V1$ while at time T_k, the continuous variables have valuation V and $\mathcal{A}$ resides at control state q. Further, at time T_{k-1}, the automaton was at control state q'. For each i, the real number $T_k + t_i^{up}$ is the time at which x_i ceases to evolve at the rate $dx_i/dt = \sum_{j=1}^{n} A_{q'}(i,j) \cdot x_j + b_{q'}(i)$ and starts to evolve at the rate $dx_i/dt = \sum_{j=1}^{n} A_q(i,j) \cdot x_j + b_q(i)$.

Now we state the condition that $\Longrightarrow$ must fulfil. Let (q,V,q'), $(q1,V1,q1')$ be in $Conf_{\mathcal{A}}$. Suppose there exist reals t_i^{up}, $i = 1,2,\ldots,n$, in $[\delta_{up}^0, \delta_{up}^1]$ such that $V1$ is related to V as dictated above.

(i): Suppose $q1 = q1' = q$. Then $(q,V,q') \xRightarrow{\tau} (q1,V1,q1')$.

(ii): Suppose $q1' = q$ and there exists a transition $(q,g,q1)$ in $\longrightarrow$ and reals t_i^{ob}, $i = 1,2,\ldots,n$, in $[\delta_{ob}^0, \delta_{ob}^1]$ such that $(\langle w_1\rangle, \langle w_2\rangle, \ldots, \langle w_n\rangle)$ satisfies g, where w_i is the i-th component of the valuation $Val(A_n, b_n, t_i^{ob} - t_{\pi_n}^{up}, U_n)$ for $i = 1,2,\ldots,n$. Then $(q,V,q') \xRightarrow{\mu}_{\mathcal{A}} (q1,V1,q1')$.

As might be expected, the real $T_k + t_i^{ob}$ is the time at which the value of x_i was observed for each $i = 1,2,\ldots,n$. We illustrate the transitions satisfying condition (i),(ii) in figure 14.2(i) and figure 14.2(ii), respectively. The time point Δ is $T_k + \delta_{up}^0$, the shaded triangle represents $T_k + \delta_{up}^1$, $\Box$ is $T_k + \delta_{ob}^0$, and the shaded box corresponds to $T_k + \delta_{ob}^1$. For each $i = 1,2,\ldots,n$, Δ_i is $T_k + t_{\pi_i}^{up}$. The time points $\Box_1$, $\Box_2$, …, $\Box_n$ form a permutation of $T_k + t_1^{ob}$, $T_k + t_2^{ob}$, …, $T_k + t_n^{ob}$.

Basically, there are four possible transition types depending on whether $q = q'$ and whether τ or μ is the action label. For convenience, we have collapsed these four possibilities into two cases according to τ or μ being the action label, and in each case have handled the subcases $q = q'$ and $q \neq q'$ simultaneously.

Now define the transition system $TS_{\mathcal{A}} = (RC_{\mathcal{A}}, (q_{in}, V_{in}, q_{in}), \{\tau, \mu\}, \Longrightarrow_{\mathcal{A}})$ via:

- $RC_{\mathcal{A}}$, the set of *reachable configurations* of $\mathcal{A}$ is the least subset of $Conf_{\mathcal{A}}$ that contains the initial configuration (q_{in}, V_{in}, q_{in}) and satisfies:

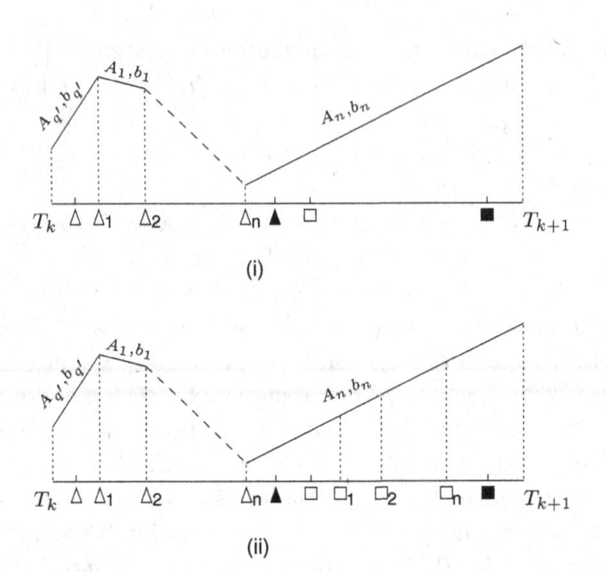

Fig. 14.2. Conditions on transition relation.

Suppose (q, V, q') is in $RC_{\mathcal{A}}$ and is a feasible configuration. Suppose further, $(q, V, q') \stackrel{\alpha}{\Longrightarrow} (q1, V1, q1')$ for some $\alpha \in \{\tau, \mu\}$. Then $(q1, V1, q1') \in RC_{\mathcal{A}}$.

- $\Longrightarrow_{\mathcal{A}}$ is $\Longrightarrow$ restricted to $RC_{\mathcal{A}} \times \{\tau, \mu\} \times RC_{\mathcal{A}}$.

We note that a reachable configuration can be the source of a transition in $TS_{\mathcal{A}}$ only if it is feasible. Thus infeasible reachable configurations will be deadlocked in $TS_{\mathcal{A}}$.

To illustrate the definition of $TS_{\mathcal{A}}$, we describe a few transitions in $TS_{\mathcal{A}}$ for the finite-precision hybrid automaton shown in figure 14.1. At time T_0, the automaton assumes the initial configuration $C_{in} = (q1, (1, 1, 1), q1)$. At time T_1, the observed values of x_1, x_2, x_3 are reported. Suppose that the observed value of x_1 (respectively, x_2,x_3) was the ϵ-precision measurement of its value held at time 0.82 (respectively, 0.89, 0.81). Then the observed values of x_1,x_2,x_3 are w_1, w_2, w_3 where

$$w_1 = \langle 1 + 0.3 \times 0.82 \rangle = 1.2,$$
$$w_2 = \langle e^{0.1 \times 0.89} \rangle = 1.1,$$
$$w_3 = \langle e^{-0.2 \times 0.81} \rangle = 0.9.$$

It follows that (w_1, w_2, w_3) satisfies the guard $x_1 < 1.25 \wedge x_1^2 + x_2^2 \geq 2$ and thus at time T_1 the automaton *may* (instantaneously) switch to control state $q2$. Hence, we have the transition $C_{in} \stackrel{\mu}{\Longrightarrow}_{\mathcal{A}} C1$, where $C1$ is the configuration $(q2, (1.3, e^{0.1}, e^{-0.2}), q1)$. However, the automaton is also allowed to remain at control state $q1$ at time T_1. That is, we also have the transition $C_{in} \stackrel{\tau}{\Longrightarrow}_{\mathcal{A}} C1'$, where $C1'$ is the configuration $(q1, (1.3, e^{0.1}, e^{-0.2}), q1)$. Now, continuing from $C1$, suppose that the rate updates of x_1, x_2, x_3 kick in respectively at times $T_1 + 0.14$, $T_1 + 0.12$,

$T_1 + 0.18$. Then the values of x_1, x_2, x_3 at time T_2 are v_1, v_2, v_3, where

$$v_1 = 1.3 + 0.3 \times 0.14 + (-0.4) \times 0.86 = 0.998,$$
$$v_2 = e^{0.1} \cdot e^{0.1 \times 0.12 + (-0.25) \times (1-0.12)} = e^{-0.108},$$
$$v_3 = e^{-0.2} \cdot e^{-0.2 \times 0.18 + 0.4 \times (1-0.18)} = e^{0.092}.$$

It follows that we have the transition $C1 \stackrel{\tau}{\Longrightarrow}_{\mathcal{A}} C2$, where $C2$ is the configuration $(q2, (v_1, v_2, v_3), q2)$. Other transitions of $TS_{\mathcal{A}}$ can be similarly computed.

A *run* of $TS_{\mathcal{A}}$ is a finite sequence of the form

$$\sigma = (q_0, V_0, q_0') \, \alpha_0 \, (q_1, V_1, q_1') \, \alpha_1 \, (q_2, V_2, q_2') \ldots (q_\ell, V_\ell, q_\ell')$$

where (q_0, V_0, q_0') is the initial configuration and $(q_k, V_k, q_k') \stackrel{\alpha_k}{\Longrightarrow}_{\mathcal{A}} (q_{k+1}, V_{k+1}, q_{k+1}')$ for $k = 0, 1, \ldots, \ell - 1$. The *state sequence* induced by the run σ above is the sequence $q_0 q_1 \ldots q_\ell$. We define the state sequence language of $\mathcal{A}$ denoted $\mathcal{L}(\mathcal{A})$ to be the set of state sequences induced by runs of $TS_{\mathcal{A}}$.

Lastly, we remark that the above transition system $TS_{\mathcal{A}}$ and associated definitions apply in exactly the same way to a perfect-precision hybrid automaton, except that $\langle v \rangle$ is set to v for every real number v.

14.4. Summary of the Results

In [2], it is shown that, if each continuous variable evolves at (possibly different) constant rates in all the control states (each ρ_q takes the form $\dot{x} = b_q$), then the discrete time behaviour of a finite-precision hybrid automaton is regular and can be effectively computed. This result is generalized in [3], where we showed that for the case that each A_q is a diagnonal matrix, the control state sequence language generated by a finite-precision hybrid automaton is regular and can be effectively computed. In fact this holds provided every continuous variable either evolves at constant rates in all the control states or at exponential rates in all the control states.

In [3], we also consider the more general case that each A_q is a diagonalizable matrix and show that, with the additional restriction that there are no delays associated with rates update ($\delta_{up}^0 = 0 = \delta_{up}^1$), the control state sequence language generated by a finite-precision hybrid automaton is recursive. However, the control state reachability problem in this setting (with the restriction of no update delays) is undecidable.

The setting of perfect precision was studied in [1]. There it was shown that if every continuous variable evolves at constant rates in all the control states, then the control state sequence language generated by a perfect-precision hybrid automaton is regular and can be effectively computed, provided that every transition guard is *rectangular*. A *rectangular guard* is a finite conjunction of inequalities of the forms $x_i < c$, $x_i \leq c$, $x_i > c$, $x_i \geq c$, where $i \in \{1, 2, \ldots, n\}$, and c is a rational number.

For more general rate functions, it can be shown that each of the three main results of [3] described above also holds for perfect-precision hybrid automata provided every transition guard is rectangular.

14.5. The Finite-Precision Setting

In this section, we present in a unified way the results from [2, 3] on finite-precision hybrid automata.

14.5.1. *The Constant Rates Case*

It was shown in [2] that if every continuous variable evolves at (possibly different) constant rates in all the control states, then the language of state sequences of a finite-precision hybrid automaton is regular. More precisely:

Theorem 14.1 ([2]). *Let $\mathcal{A}$ be a hybrid automaton such that for each q, ρ_q takes the form $\dot{x} = b_q$ (that is, each A_q is the zero matrix). Then $\mathcal{L}(\mathcal{A})$ is a regular subset of $Q^\star$. Further, a finite state automaton accepting $\mathcal{L}(\mathcal{A})$ can be effectively computed from a given presentation of $\mathcal{A}$.*

Proof Sketch: The proof consists of two major steps. The first one is to quotient the set of reachable configurations $RC_{\mathcal{A}}$ into a *finite number* of equivalence classes using a suitably chosen equivalence relation $\approx$. The crucial property required of $\approx$ is that it should be a congruence with respect to the transition relation of $TS_{\mathcal{A}}$. In other words, if $(q1, V1, q1') \approx (q2, V2, q2')$ and $(q1, V1, q1') \stackrel{\alpha}{\Longrightarrow}_{\mathcal{A}} (q3, V3, q3')$, then we require that there exists a configuration $(q4, V4, q4')$ such that $(q2, V2, q2') \stackrel{\alpha}{\Longrightarrow}_{\mathcal{A}} (q4, V4, q4')$ *and* $(q3, V3, q3') \approx (q4, V4, q4')$. The second step is to show that we can effectively compute these equivalence classes and a transition relation over them such that the resulting finite state automaton generates the language of control state sequences.

To define the equivalence relation $\approx$, we shall partition the real line into a *finite* number of intervals and assert that $(q1, V1, q1') \approx (q2, V2, q2')$ iff $q1 = q2$ and $V1(i)$ falls into the same interval as $V2(i)$ for every i, and $q1' = q2'$. More precisely, we shall firstly construct a *finite* set Θ of rational numbers. Secondly, list the members of Θ in *increasing* order as $\theta_1, \theta_2, \ldots, \theta_m$ where $m = |\Theta|$, and divide the real line into finitely many open intervals $\{(-\infty, \theta_1), (\theta_1, \theta_2), \ldots, (\theta_{m-1}, \theta_m), (\theta_m, \infty)\}$, and finitely many singleton intervals $[\theta_j, \theta_j]$, $j = 1, 2, \ldots, m$.

We now describe Θ. Pick Δ to be the largest positive rational number that *integrally* divides every number in the set of rational numbers $\{\delta^0_{ob}, \delta^1_{ob}, \delta^0_{up}, \delta^1_{up}, 1\}$. Define Γ to be the largest rational which *integrally* divides every number in the finite set of rational numbers $\{b_q(i) \cdot \Delta \mid q \in Q, i \in \{1, 2, \ldots, n\}\} \bigcup \{\gamma_{min}, \gamma_{max}\} \bigcup \{\epsilon/2\}$. Let $\mathbb{Z}$ denote the set of integers. Now Θ is defined to be the *finite* set of rational numbers $\{h\Gamma \in [-\gamma_{max}, \gamma_{max}] \mid h \in \mathbb{Z}\}$. In other words, Θ contains integral multiples of Γ in the interval $[-\gamma_{max}, \gamma_{max}]$.

Next we show that the equivalence relation $\approx$ defined above is indeed a congruence relation. Suppose $(q1, V1, q1') \approx (q2, V2, q2')$ and $(q1, V1, q1') \stackrel{\alpha}{\Longrightarrow}_{\mathcal{A}} (q3, V3, q3')$, where $\alpha \in \{\mu, \tau\}$. We need to argue that there exists a reachable configuration $(q4, V4, q4')$ such that $(q2, V2, q2') \stackrel{\alpha}{\Longrightarrow}_{\mathcal{A}} (q4, V4, q4')$ and $(q4, V4, q4') \approx (q3, V3, q3')$. For brevity, we consider just the case $\alpha = \mu$.

Firstly, it is easy to see that $(q2, V2, q2')$ is feasible from the facts that $V2(i)$ falls into the same interval as $V1(i)$ for every i, and that $\gamma_{min}, -\gamma_{min}, \gamma_{max}, -\gamma_{max}$ are members of Θ.

Secondly, suppose that in the transition $(q1, V1, q1') \stackrel{\alpha}{\Longrightarrow}_{\mathcal{A}} (q3, V3, q3')$, for each i, the rate update of x_i (from $b_{q1'}(i)$ to $b_{q1}(i)$) kicks in at time $T_k + t_i^{up}$ and the value of x_i was observed at time $T_k + t_i^{ob}$, where T_k is the time at which $\mathcal{A}$ assumes the configuration $(q1, V1, q1')$. Let w_i be the (actual) value of x_i at time $T_k + t_i^{ob}$ for each i and $(q1, g, q3)$ be a transition of $\mathcal{A}$ where $(\langle w_1 \rangle, \ldots, \langle w_n \rangle)$ satisfies g. We show that there exist reals $\hat{t}_i^{up}$ in $[\delta_{up}^0, \delta_{up}^1]$, $\hat{t}_i^{ob}$ in $[\delta_{ob}^0, \delta_{ob}^1]$, $i = 1, 2, \ldots, n$, which fulfil the condition detailed as follows. Suppose $\mathcal{A}$ assumes configuration $(q2, V2, q2')$ at time $\widehat{T}_k$, and the rate update of x_i (from $b_{q2'}(i)$ to $b_{q2}(i)$) kicks in at time $\widehat{T}_k + \hat{t}_i^{up}$, and the value of x_i is observed at time $\widehat{T}_k + \hat{t}_i^{ob}$. Let $\hat{w}_i$ be the actual value of x_i at time $\widehat{T}_k + \hat{t}_i^{ob}$ for each i. Let $V4$ be the valuation of $x_1, x_2, \ldots, x_n$ at time $\widehat{T}_{k+1}$. Then for each i, $\hat{w}_i$ falls into the same interval (as defined above via Θ) as w_i. And $V4(i)$ falls into the same interval as $V3(i)$.

One can then easily show that $\langle \hat{w}_i \rangle = \langle w_i \rangle$ for each i, and conclude that the configuration $(q4, V4, q4')$ with $q4 = q3$, $q4' = q3'$, possesses the desired property.

We now fix an i and show the existence of $\hat{t}_i^{up}$, $\hat{t}_i^{ob}$. The argument applies to any i in $\{1, 2, \ldots, n\}$. Assume that $b_{q1'}(i) > b_{q1}(i) > 0$, $V3(i)$ falls into the open interval (θ, θ') (defined by Θ) and w_i into the open interval (ϑ, ϑ'), where $\theta, \theta', \vartheta, \vartheta' \in \Theta$. Other cases can be similarly handled.

It follows from the definition of $TS_{\mathcal{A}}$ that $\Phi(V1(i))$ holds where $\Phi(v)$ is the condition

$$
\begin{aligned}
\exists t^{up} \in \mathbb{R}. \exists t^{ob} \in \mathbb{R}.\ & \delta_{up}^0 \le t^{up} \le \delta_{up}^1 \\
& \bigwedge\ \theta < v + b_{q1'}(i) \cdot t^{up} + b_{q1}(i) \cdot (1 - t^{up}) < \theta' \\
& \bigwedge\ \delta_{ob}^0 \le t^{ob} \le \delta_{ob}^1 \\
& \bigwedge\ \vartheta < v + b_{q1'}(i) \cdot t^{up} + b_{q1}(i) \cdot (t^{ob} - t^{up}) < \vartheta' .
\end{aligned}
$$

It is not difficult to see that the condition $\Phi(v)$ is in fact equivalent to $\eta < v < \eta'$, where η is the larger of $\theta - b_{q1'}(i) \cdot \delta_{up}^1 - b_{q1}(i) \cdot (1 - \delta_{up}^1)$ and $\vartheta - b_{q1'}(i) \cdot \delta_{up}^1 - b_{q1}(i) \cdot (\delta_{ob}^1 - \delta_{up}^1)$. On the other hand, η' is the smaller of $\theta' - b_{q1'}(i) \cdot \delta_{up}^0 - b_{q1}(i) \cdot (1 - \delta_{up}^0)$ and $\vartheta' - b_{q1'}(i) \cdot \delta_{up}^0 - b_{q1}(i) \cdot (\delta_{ob}^0 - \delta_{up}^0)$. Thus, both η, η' are members of Θ (if they fall into the interval $[-\gamma_{max}, \gamma_{max}]$). Consequently, $\Phi(V2(i))$ also holds and this establishes the existence of reals $\hat{t}_i^{up}$,$\hat{t}_i^{ob}$. □

We wish to point out that in [2], theorem 14.1 was shown for the slightly more general case where γ_{min} is allowed to be zero. As remarked earlier, the bound

γ_{min} is used to bound the amount of information carried by a continuous variable decreasing at an exponential rate. Thus, the bound γ_{min} does not play a role in theorem 14.1.

14.5.2. *The Exponential Rates Case*

In [3], we generalize theorem 14.1 to the case that each A_q is a diagonal matrix but where every continuous variable either evolves at constant rates in all the modes or at exponential rates in all the modes. More precisely, we have:

Theorem 14.2 **([3]).** *Let $\mathcal{A}$ be a finite-precision hybrid automaton such that A_q is a diagonal matrix for every control state q. Suppose there exists a* fixed *partition $\{DIF, CON\}$ of the indices $\{1, 2, \ldots, n\}$ such that for each control state q, ρ_q is such that $\dot{x}_i = A_q(i,i) \cdot x_i$ if $i \in DIF$ and $\dot{x}_i = b_q(i)$ if $i \in CON$. Then $\mathcal{L}(\mathcal{A})$ is a regular subset of $Q^\star$. Further, a finite state automaton accepting $\mathcal{L}(\mathcal{A})$ can be effectively computed from $\mathcal{A}$.*

Proof Sketch: The basic strategy is to extend the proof of theorem 14.5, to deal with continuous variables evolving at exponential rates. The simple but key observation that enables this is, in the (natural) logarithmic scale, exponential rates get represented as *constant* rates. However, this logarithmic representation brings a new complication. As in the proof of theorem 14.5, due to the guards and the bounds $\gamma_{min}, \gamma_{max}$, one need to handle inequalities involving comparisions of x_i with γ_{min}, γ_{max}, $(m + 1/2)\epsilon$, where m is an integer. Thus, to partition the real line into finitely many intervals, one need to take into account *irrational* numbers of the forms $\ln \gamma_{min}$, $\ln \gamma_{max}$, $\ln((m + 1/2)\epsilon)$.

More precisely, we shall define a finite set Θ_{con} of rational numbers relevant to variables x_i with $i \in CON$, and a finite set Θ_{dif} of real numbers relevant to variables x_i with $i \in DIF$. As in the proof of theorem 14.5, we list members of Θ_{con} in increasing order and divide the real line into finitely many open intervals and finitely many singleton intervals. Do the same for Θ_{dif}.

Now we define the equivalence relation $\approx$ over configurations via: $(q1, V1, q1') \approx (q2, V2, q2')$ iff $q1 = q2$, $V1(i)$,$V2(i)$ fall into the same interval defined by Θ_{con} for $i \in CON$, $\ln V1(i)$,$\ln V2(i)$ fall into the same interval defined by Θ_{dif} for $i \in DIF$, and $q2 = q2'$. For notational convenience, we assume $V_{in}(i) > 0$ for every $i \in DIF$. The key consequence of this assumption is that in any reachable configuration, the value of x_i for $i \in DIF$ will be positive.

We describe $\Theta_{con}, \Theta_{dif}$. Pick Δ as in the proof of theorem 14.5. Define Γ to be the largest rational which *integrally* divides every number in the finite set of rational numbers $\{A_q(i,i) \cdot \Delta \mid q \in Q, i \in DIF\} \bigcup \{b_q(j) \cdot \Delta \mid q \in Q, j \in CON\} \bigcup \{\gamma_{min}, \gamma_{max}\} \bigcup \{\epsilon/2\}$. Define Θ_{con} to be the *finite* set of rational numbers $\{h\Gamma \in [-\gamma_{max}, \gamma_{max}] \mid h \in \mathbb{Z}\}$. Let Θ_{IR} be the set of irrational numbers $\{\ln((m + 1/2)\epsilon) \mid m \in \mathbb{Z}, \langle\gamma_{min}\rangle \leq m\epsilon \leq \langle\gamma_{max}\rangle\} \bigcup \{\ln \gamma_{min}, \ln \gamma_{max}\}$. Define Θ_{dif} to be the *finite* set of real numbers $\{h\Gamma \in [\ln \gamma_{min}, \ln \gamma_{max}] \mid h \in \mathbb{Z}\} \bigcup \{\ell\Gamma + \theta \in [\ln \gamma_{min}, \ln \gamma_{max}] \mid$

$\ell \in \mathbb{Z}$, $\theta \in \Theta_{IR}\}$. In other words, Θ_{dif} contains rational numbers of the form $h\Gamma$ in the interval $[\ln \gamma_{min}, \ln \gamma_{max}]$ where h is a (positive) integer, and irrational numbers of the form $\ell\Gamma + \theta$ in the interval $[\ln \gamma_{min}, \ln \gamma_{max}]$ where ℓ is an integer (that can be positive, zero or negative) and θ is a member of Θ_{IR}.

Similar to the proof of theorem 14.1, one can then argue that the equivalence relation $\approx$ defined above is a congruence relation with respect to the transition relation of $TS_{\mathcal{A}}$. Further, one can effectively compute these equivalence classes and a transition relation over them such that the resulting finite state automaton generates the language of control state sequences. □

14.5.3. *Diagonalizable Rate Matrices*

Here we describe the investigation in [3] for the more general case that each A_q is *diagonalizable* [12] and has n distinct rational eigenvalues. Recall that the $n \times n$ matrix A is diagonalizable in case there is a basis of eigenvectors $\{f_1, f_2, \ldots, f_n\}$ so that under the associated coordinate transformation, A can be represented as the diagonal matrix $diag(\lambda_1, \lambda_2, \ldots, \lambda_n)$ with the λ_i's being the eigenvalues of A. The demand on each A_q to have n distinct rational eigenvalues is natural given the concern for effective computations.

We further restrict ourselves to the case where there is no delay associated with the update of rates of the continuous variables ($\delta^0_{up} = 0 = \delta^1_{up}$). This is due to the fact at present we don't know how to deal with differential equations of the form $\dot{x} = Ax + b$. One will have to deal with such equations if update delays are present ($\delta^0_{up} < \delta^1_{up}$). This is due to the fact that the rate changes of the continuous variables may kick in at *different* times in the interval $[T_k + \delta^0_{up}, T_k + \delta^1_{up}]$.

With the above restrictions in place, [3] showed that the control state sequence language generated by a finite-precision hybrid automaton is recursive. However, [3] also showed that the control state reachability problem is undecidable in this setting. We now state and explain these two results.

Theorem 14.3. *Suppose $\mathcal{A}$ is a finite-precision hybrid automaton such that $\delta^0_{up} = 0 = \delta^1_{up}$ and for every control state q, A_q is a diagonalizable matrix having n distinct rational eigenvalues. Then $\mathcal{L}(\mathcal{A})$ is a recursive subset of $Q^\star$.*

Proof Sketch: First we note that the first order theory of the reals augmented with the *constant* e is decidable. Next, we show that, given a control state sequence $q_0q_1 \ldots q_\ell$, one can construct a sentence $\Phi_{q_0q_1\ldots q_\ell}$ in this augmented structure such that $\Phi_{q_0q_1\ldots q_\ell}$ is true iff $q_0q_1 \ldots q_\ell$ is in $\mathcal{L}(\mathcal{A})$.

To see that the above augmented theory is decidable, we observe that one can effectively determine whether $p(e) < 0$ for any given polynomial $p(e)$ with integer coefficients, by approximating e sufficiently with the power series expansion of e.

As for the construction of $\Phi_{q_0q_1\ldots q_\ell}$, it suffices to show that, given control states $q, q', q1, q1'$ and $\alpha \in \{\tau, \mu\}$, one can construct (in the above augmented structure) a formula $\Phi_{q,q',q1,q1',\alpha}(V, V1)$ with free variables $V(i)$, $V1(i)$, $i = 1, 2, \ldots, n$, which asserts $(q, V, q') \overset{\alpha}{\Longrightarrow}_{\mathcal{A}} (q1, V1, q1')$. For brevity, we consider just the case $\alpha = \mu$.

We now fix $q, q', q1, q1'$ and describe the formula $\Phi_{q,q',q1,q1',\mu}(V, V1)$. It follows from definition of $TS_{\mathcal{A}}$ that $(q, V, q') \overset{\alpha}{\Longrightarrow}_{\mathcal{A}} (q1, V1, q1')$ iff $V1 = Val(A_q, b_q, 1, V)$ and there exist reals t_i^{ob}, $i = 1, 2, \ldots, n$, in $[\delta_{ob}^0, \delta_{ob}^1]$, and a transition $(q, g, q1)$ of $\mathcal{A}$ such that $(\langle w_1 \rangle, \langle w_2 \rangle, \ldots, \langle w_n \rangle)$ satisfies g, where w_i is the i-th component of $Val(A_q, b_q, t_i^{ob}, V)$ for each i. In fact, from [12] it is easy to see that for a real $T \in [0, 1]$, $Val(A_q, b_q, T, V) = H(e^T)$ where $H : \mathbb{R} \to \mathbb{R}^n$ is given by

$$H(u) = F\, diag(u^{\lambda_1}, u^{\lambda_2}, \ldots, u^{\lambda_n})\, F^{-1}(V + A_q^{-1} b_q) - A_q^{-1} b_q \,.$$

Here, λ_1, λ_2, ..., λ_n are the n distinct rational eigenvalues of A_q. And F is the matrix in $\mathbb{Q}^{n \times n}$ whose i-th column is the eigenvector corresponding to λ_i, for $i = 1, 2, \ldots, n$. Clearly we can effectively compute λ_1, λ_2, ..., λ_n and the matrix F.

Now the construction of $\Phi_{q,q',q1,q1',\mu}(V, V1)$ follows from two observations. Firstly, it is easy to see that there exist reals t_i^{ob}, $i = 1, 2, \ldots, n$, satisfying the dictated condition iff there exist reals $u_i \in [e^{\delta_{ob}^0}, e^{\delta_{ob}^1}]$, $i = 1, 2, \ldots, n$, such that w_i is the i-th component of $H(u_i)$ for each i. Secondly, if $(\langle w_1 \rangle, \langle w_2 \rangle, \ldots, \langle w_n \rangle)$ satisfies g, then $(\langle w_1 \rangle, \langle w_2 \rangle, \ldots, \langle w_n \rangle)$ is one of the *finitely many* valuations $(v_1, v_2, \ldots, v_n)$ which satisfies g and moreover fulfils the following condition: for each i, $v_i = m_i \epsilon$ where m_i is an integer with $\langle -\gamma_{max} \rangle \leq m_i \epsilon \leq \langle \gamma_{max} \rangle$. □

A consequence of theorem 14.3 is that a number of bounded model checking problems involving finite-precision hybrid automata can in principle be effectively solved.

Theorem 14.4. *There is no effective procedure which can, given a finite-precision hybrid automaton $\mathcal{A}$ satisfying the restrictions stated in theorem 14.3 and a control state q_f of $\mathcal{A}$, determine whether q_f is reachable in $\mathcal{A}$. In other words, whether there exists a reachable configuration (q, V, q') of $\mathcal{A}$ such that $q = q_f$.*

Proof Sketch: The proof is to reduce the halting problem of two-counter automata ([18]) to the control state reachability problem of the class of hybrid automata stated in the theorem. Given a two-counter automaton $\mathcal{C}$, one can construct a finite-precision hybrid automaton $\mathcal{A}$ satisfying the restrictions dictated in the theorem. Further, a designated control state q_f of $\mathcal{A}$ is reachable in $TS_{\mathcal{A}}$ iff the halting state of $\mathcal{C}$ is reachable.

We describe the main ingredients for constructing $\mathcal{A}$. Naturally, the finitely many states of $\mathcal{C}$ will be encapsulated in the control states of $\mathcal{A}$. The automaton $\mathcal{A}$ will have three continuous variables x_1, x_2, x_3. Variables x_1, x_2 will respectively represent values of the two counters of $\mathcal{C}$, while x_3 will be used to enforce that values of x_1, x_2 are updated appropriately. More precisely, a counter having value h will be represented by the corresponding variable taking the value $1 + e^{-1} + e^{-2} + \cdots + e^{-h}$. In particular, a counter with value zero will be represented by the corresponding continuous variable taking the value 1.

Incrementation or decrementation of a counter is simulated by letting the corresponding continuous variable evolve for two time units with suitable rates. For instance, to increment a counter represented by x_1, one lets x_1 evolve at rate $\dot{x}_1 = -x_1$ for *exactly* one time unit, and then at rate $\dot{x}_1 = 1$ for *exactly* one more time unit. We need to enforce that x_1 "stays" for exactly one time unit in each of these two steps. This is achieved by setting suitable rates on x_3 and suitable (rectangular) guards involving x_3 so that those configurations of $\mathcal{A}$ in which x_1 "stays" beyond one time unit in one of these two steps will *not* lead to a configuration of the form (q, V, q') where q is the designated state q_f. □

14.6. The Perfect-Precision Setting

In this section, we discuss the results in [1, 3] regarding perfect-precision hybrid automata. Firstly, in [1] it is shown that if each continuous variable always evolves at (possibly different) constant rates in all control states and each transition guard is rectangular, then the control state sequence language of a perfect-precision hybrid automaton is regular.

Theorem 14.5 ([1]). *Let $\mathcal{A}$ be a perfect-precision hybrid automaton such that for each q, ρ_q takes the form $\dot{x} = b_q$ (that is, each A_q is the zero matrix) Further, each transition guard is rectangular. Then $\mathcal{L}(\mathcal{A})$ is a regular subset of $Q^\star$. Further, a finite state automaton accepting $\mathcal{L}(\mathcal{A})$ can be effectively computed from $\mathcal{A}$.*

Proof Sketch: The proof is similar to that of theorem 14.1. Pick Δ as in the proof of theorem 14.1. Let C be the set of rational numbers c for which some transition guard of $\mathcal{A}$ contains an inequality in one of the forms $x_i < c$, $x_i \leq c$, $x_i > c$, $x_i \geq c$. Define Γ to be the largest rational which *integrally* divides every number in the finite set of rational numbers $\{b_q(i) \cdot \Delta \mid q \in Q, i \in \{1, 2, \dots, n\}\} \bigcup C \bigcup \{\gamma_{min}, \gamma_{max}\}$. Define Θ to be the *finite* set of rational numbers $\{h\Gamma \in [-\gamma_{max}, \gamma_{max}] \mid h \in \mathbb{Z}\}$.

Now define the equivalence relation $\approx$ over configurations in the same way as in the proof of theorem 14.1. One can then argue that $\approx$ is a congruence relation with respect to the transition relation of $TS_{\mathcal{A}}$. Further, one can effectively compute these equivalence classes and a transition relation over them such that the resulting finite state automaton generates the language of control state sequences. □

We also showed in [3] that all of theorem 14.2, 14.3, 14.4 hold if the combination of finite precision and polynomial guards is replaced by that of perfect precision and rectangular guards (that is, every transition guard is rectangular). We explain here only the intuition behind this observation, and refer to [17] for detailed proofs. As noted from the proof of theorem 14.3, for a valuation V with $\gamma_{min} \leq |V(i)| \leq \gamma_{max}$ and a polynomial guard g, if $(\langle V(1)\rangle, \langle V(2)\rangle, \dots, \langle V(n)\rangle)$ satisfies g, then $(\langle V(1)\rangle, \langle V(2)\rangle, \dots, \langle V(n)\rangle)$ is one of the *finitely many* valuations $(v_1, v_2, \dots, v_n)$ which satisfies g and moreover fulfils the following condition: for each i, $v_i = m_i\epsilon$

where m_i is an integer with $\langle -\gamma_{max} \rangle \leq m_i\epsilon \leq \langle \gamma_{max} \rangle$. Hence, in our approaches to finite precision and polynomial guards as in the proofs of theorem 14.2, 14.3, 14.4, we are dealing with rectangular constraints involving comparisons of *actual values* of continuous variables with rational numbers of the form $m\epsilon$ where $m \in \mathbb{Z}$. This is the key observation which enables us to adapt the proofs of theorem 14.2, 14.3, 14.4, to the setting of perfect precision and rectangular guards.

14.7. Discussion

We have shown here that the twin features of discrete time and finite precision allows to deal with hybrid automata whose dynamics are governed by restricted linear differential operators and whose transitions have polynomial guards. Indeed, their control state sequence languages will be regular and these state sequence languages can be effectively constructed although we do not impose the reset restriction.

Our results seem to suggest that once observational and update delays are included to further reduce the expressive power of these automata, one may be able to handle much richer continuous dynamics. The key obstacle here is our lack of techniques for constructing effective rational approximations of the continuous dynamics. Here, the mathematical foundations provided in [14] and the logical underpinnings developed in [15, 16] may provide a sound basis.

References

[1] M. Agrawal and P. Thiagarajan. Lazy rectangular hybrid automata. In *Proc. 7th HSCC*, vol. 2993, *LNCS*, pp. 1–15. Springer, (2004).
[2] M. Agrawal and P. Thiagarajan. The discrete time behaviour of lazy linear hybrid automata. In *Proc. 8th HSCC*, vol. 3414, *LNCS*, pp. 55–69. Springer, (2005).
[3] M. Agrawal, F. Stephan, P. Thiagarajan, and S. Yang. Behavioural approximations for restricted linear differential hybrid automata. In *Proc. 9th HSCC*, vol. 3927, *LNCS*, pp. 4–18. Springer, (2006).
[4] R. Alur, T. Henzinger, G. Lafferriere, and G. Pappas, Discrete abstractions of hybrid systems, *Proc. of the IEEE.* **88**, 971–984, (2000).
[5] A. Casagrande, C. Piazza, and B. Mishra. Semi-algebraic constant reset hybrid automata - SACoRe. In *Proc. 44th IEEE Conf. on Decision and Control and European Control Conf.*, pp. 678–683, (2005).
[6] V. Gupta, T. Henzinger, and R. Jagadeesan. Robust timed automata. In *Proc. Workshop on Hybrid and Real-Time Systems*, vol. 1201, *LNCS*, pp. 331–345. Springer, (1997).
[7] T. Henzinger, P. Kopke, A. Puri, and P. Varaiya, What's decidable about hybrid automata?, *J. of Comp. and Sys. Sci.* **57**, 94–124, (1998).
[8] Y. Kesten, A. Pnueli, J. Sifakis, and S. Yovine. Integration graphs: A class of decidable hybrid systems. In *Proc. Hybrid Systems*, vol. 736, *LNCS*, pp. 179–208. Springer, (1993).
[9] T. Henzinger and P. Kopke. State equivalences for rectangular hybrid automata. In *Proc. 7th CONCUR*, vol. 1119, *LNCS*, pp. 530–545. Springer, (1996).

[10] T. Henzinger. The theory of hybrid automata. In *11th LICS*, pp. 278–292. IEEE Press, (1996).
[11] E. Emerson. Temporal and modal logics. In *Handbook of Theoretical Comp. Sci., Vol. B*, pp. 997–1072. Elsevier, (1990).
[12] M. Hirsch and S. Smale, *Differential Equations, Dynamical Systems and Linear Algebra.* (Academic Press, 1974).
[13] A. Biere, A. Cimatti, E. Clarke, O. Strichman, and Y. Zhu, *Bounded Model Checking.* Advances in Computers 58, (Academic Press, 2003).
[14] A. Baker, *Transcendental Number Theory.* (Cambridge University Press, 1979).
[15] G. Lafferriere, G. Pappas, and S. Sastry, O-minimal hybrid systems, *Math. Control Signals Systems.* **13**, 1–21, (2000).
[16] G. Lafferriere, G. Pappas, and S. Yovine, Symbolic reachability computation for families of linear vector fields, *J. Symbolic Computation.* **32**, 231–253, (2001).
[17] M. Agrawal, F. Stephan, P. Thiagarajan, and S. Yang. The discrete time behaviour of restricted linear hybrid automata. Technical Report, School of Computing, NUS, Singapore, Available at `https://dl.comp.nus.edu.sg/dspace/handle/1900.100/12`, (2008).
[18] J. Hopcroft and J. Ullman, *Introduction to Automata Theory, Languages and Computation.* (Addison-Wesley, 1979).

Part III

Automata and Logic

Chapter 15

Specification and Verification using Temporal Logics*

Stéphane Demri* and Paul Gastin†
LSV, ENS Cachan, CNRS, INRIA Saclay, France
* *demri@lsv.ens-cachan.fr*
† *gastin@lsv.ens-cachan.fr*

This chapter illustrates two aspects of automata theory related to linear-time temporal logic LTL used for the verification of computer systems. First, we present a translation from LTL formulae to Büchi automata. The aim is to design an elementary translation which is reasonably efficient and produces small automata so that it can be easily taught and used by hand on real examples. Our translation is in the spirit of the classical tableau constructions but is optimized in several ways. Secondly, we recall how temporal operators can be defined from regular languages and we explain why adding even a single operator definable by a context-free language can lead to undecidability.

Keywords: temporal logic, model-checking, Büchi automaton, temporal operator, context-free language.

15.1. Introduction

Temporal logics as specification languages. Temporal logics (TL) are modal logics [1] designed to specify temporal relations between events occurring over time. They first appear as a branch of logic dedicated to reasoning about time, see e.g. [2]. The introduction of TL for reasoning about program behaviours is due to [3]. Among the desirable properties of formal specification languages, the temporal logics have an underlying flow of time in its models, define mathematically the correctness of computer systems, express properties without ambiguity and are useful for carrying out formal proofs. Moreover, compared with the mathematical formulae, temporal logic notation is often clearer and simpler. This is a popular formalism to express properties for various types of systems (concurrent programs, operating systems, network communication protocols, programs with pointers, etc.). An early success of the use of TL has been the verification of finite-state programs with TL specifications, see e.g. [4, 5].

*This work has been partially supported by projects ARCUS Île de France-Inde, ANR-06-SETIN-003 DOTS, and P2R MODISTE-COVER/Timed-DISCOVERI.

Automata-based approach. Model-checking [4, 6] is one of the most used methods for checking temporal properties of computer systems. However, there are different ways to develop theory of model-checking and one of them is dedicated to the construction of automata from temporal logic formulae. In that way, an instance of a model-checking problem is reduced to a nonemptiness check of some automaton, typically recognizing infinite words [7]. This refines the automata-based approach developed by R. Büchi for the monadic second-order theory of $\langle \mathbb{N}, < \rangle$ in [8] (for which nonelementary bounds are obtained if the translation is applied directly). This approach has been successfully developed in [9] for linear-time temporal logics and recent developements for branching-time temporal logics with best complexity upper bounds can be found in [10], see also a similar approach for description logics, program logics or modal logics [1].

On the difficulty of presenting simple automata. The translations from temporal formulae into automata providing best complexity upper bounds, often require on-the-fly algorithms and are not always extremely intuitive. For instance, on-the-fly algorithms for turning specifications into automata and doing the emptiness check can be found in [11–13]. In order to explain the principle of such translations, it is essential to be able to show how to build simple automata for simple formulae. For instance, the temporal formula $\mathsf{X}^n p$ stating that the propositional variable p holds at the n-th next step, may lead to an exponential-size automaton in n when maximally consistent sets of formulae are states of the automata even though $\mathsf{X}^n p$ has a linear-size automaton. This gain in simplification is of course crucial for practical purposes but it is also important to have simple constructions that can be easily taught. That is why we share the pedagogical motivations from [14] and we believe that it is essential to be able to present automata constructions that produce simple automata from simple formulae.

Our contribution. This chapter presents two aspects of automata theory for LTL model-checking and it can be viewed as a follow-up to [9, 15, 16]. First, we present a translation from LTL formulae to generalized Büchi automata such that simple formulae produce simple automata. We believe this translation can be easily taught and used by hand on real examples. So, Section 15.2 recalls standard definitions about the temporal logics LTL and CTL* whereas Section 15.3 provides the core of the translation in a self-contained manner. A nice feature of the construction is the use of target transition-based Büchi automata (BA) which allows us to obtain concise automata: the acceptance condition is a conjunction of constraints stating that some transitions are repeated infinitely often. This type of acceptance condition has already been advocated in [12, 14, 17–20]. Secondly, we consider richer temporal logics by adding either path quantifiers (Section 15.4.1) or language-based temporal operators (regular or context-free languages) following the approach introduced in [21]. We recall the main expressive power and complexity issues related to

Extended Temporal Logic ETL [21] (Section 15.4.2). Finally, we show that model-checking for propositional calculus augmented with a simple context-free language recognized by a visibly pushdown automaton (VPA) [22] is highly undecidable (Section 15.4). It is worth observing that Propositional Dynamic Logic with programs [23] (PDL) augmented with visibly pushdown automata has been recently shown decidable [24] and the class of VPA shares very nice features (closure under Boolean operations for instance [22]).

15.2. Temporal logics

15.2.1. *Modalities about executions*

The languages of TL contain precisely modalities having a temporal interpretation. A modality is usually defined as a syntactic object (term) that modifies the relationships between a predicate and a subject. For example, in the sentence "Tomorrow, it will rain", the term "Tomorrow" is a temporal modality. TL makes use of different types of modalities and we recall below some of them interpreted over runs (a.k.a. executions or ω-sequences). The temporal modalities (a.k.a. temporal combinators) allow one to speak about the sequencing of states along an execution, rather than about the states taken individually. The simplest temporal combinators are X ("neXt"), F ("sometimes") and G ("always"). Below, we shall freely use the Boolean operators $\neg$ (negation), $\vee$ (disjunction), $\wedge$ (conjunction) and $\rightarrow$ (material implication).

- Whereas φ states a property of the current state, $\mathsf{X}\,\varphi$ states that the next state (X for "neXt") satisfies φ. For example, $\varphi \vee \mathsf{X}\,\varphi$ states that φ is satisfied now or in the next state.

$\mathsf{X}\,p$: next-time p

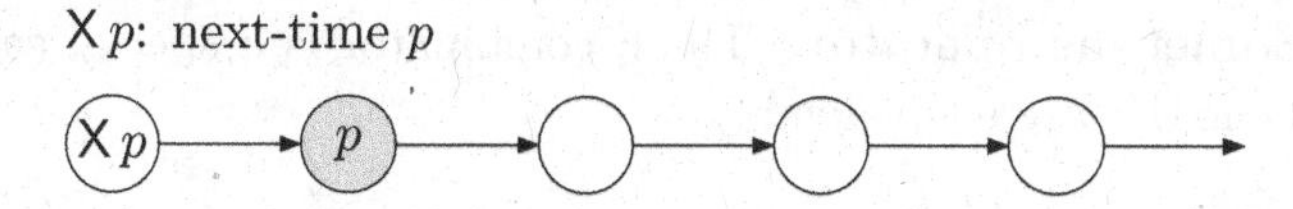

- $\mathsf{F}\,p$ announces that a future state (F for "Future") satisfies φ without specifying which state, and $\mathsf{G}\,\varphi$ that all the future states satisfy φ. These two combinators can be read informally as "φ will hold some day" and "φ will always be". Foundations of the modal approach to temporal logic can be found in [25] in which are introduced the combinators F and G.

$\mathsf{F}\,p$: sometimes p

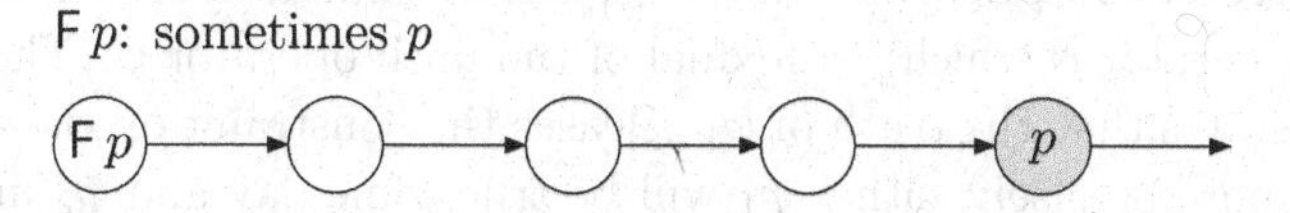

Duality The operator G is the dual of F: whatever the formula φ may be, if φ is always satisfied, then it is not true that $\neg\varphi$ will some day be satisfied, and conversely. Hence $\mathsf{G}\,\varphi$ and $\neg\,\mathsf{F}\,\neg\varphi$ are equivalent.

$\mathsf{G}\,p$: always p

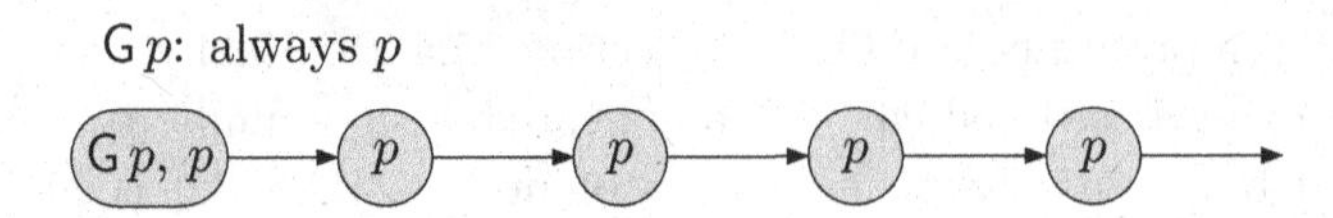

By way of example, the expression `alert` $\rightarrow$ F `halt` means that if we (currently) are in a state of alert, then we will (later) be in a halt state.

Past-time operators Likewise, the past operators F^{-1} ("sometime in the past") and G^{-1} ("always in the past") are also introduced in [25].

$\mathsf{F}^{-1}\,p$: sometime in the past p

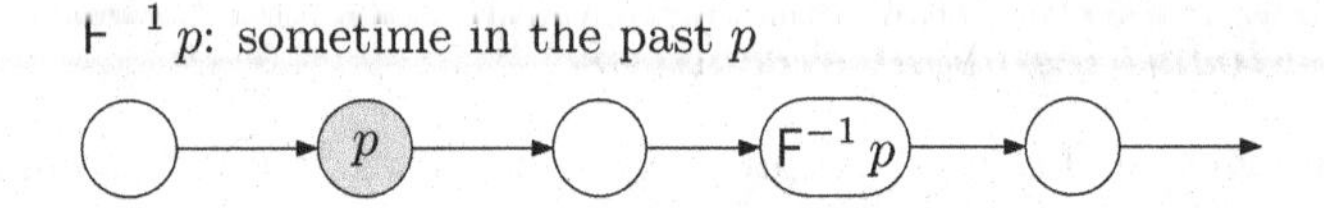

- The U combinator (for "Until") is richer and more complicated than the combinator F. $\varphi_1 \mathsf{U} \varphi_2$ states that φ_1 is true until φ_2 is true. More precisely: φ_2 will be true some day, and φ_1 will hold in the meantime.

  $p \,\mathsf{U}\, q$: p until q

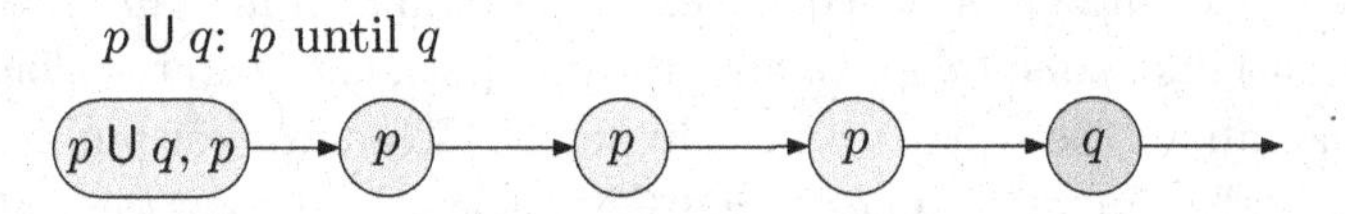

The example $\mathsf{G}($`alert` $\rightarrow$ F `halt`$)$ can be refined with the statement that "starting from a state of alert, the alarm remains activated until the halt state is eventually reached":

$$\mathsf{G}(\texttt{alert} \rightarrow (\texttt{alarm}\ \mathsf{U}\ \texttt{halt})).$$

Sometime operator The F combinator is a special case of U: $\mathsf{F}\,\varphi$ and $\texttt{true}\ \mathsf{U}\ \varphi$ are equivalent.

Weak until There exists also a "weak until", denoted W. The statement $\varphi_1 \mathsf{W} \varphi_2$ still expresses "$\varphi_1 \mathsf{U} \varphi_2$", but without the inevitable occurrence of φ_2 and if φ_2 never occurs, then φ_1 remains true forever. So, $\varphi_1 \mathsf{W} \varphi_2$ is equivalent to $\mathsf{G}\,\varphi_1 \vee (\varphi_1 \mathsf{U} \varphi_2)$.

Release operator In the sequel, we shall also use the so-called "release" operator R which is the dual of the until operator U. The formula $\varphi_1 \mathsf{R} \varphi_2$ states that the truth of φ_1 releases the constraint on the satisfaction of φ_2, more precisely, either φ_1 will be true some day and φ_2 must hold between the current state and that day, or φ_2 must be true in all future states.

We provide below other examples of properties that can be expressed thanks to these temporal modalities.

(safety) $\mathsf{G}(\mathtt{halt} \rightarrow \mathsf{F}^{-1}\,\mathtt{alert})$,
(liveness) $\mathsf{G}(p \rightarrow \mathsf{F}\,q)$,
(total correctness) $(\mathtt{init} \wedge p) \rightarrow \mathsf{F}(\mathtt{end} \wedge q)$,
(strong fairness) $\mathsf{G}\,\mathsf{F}\,\mathtt{enabled} \rightarrow \mathsf{G}\,\mathsf{F}\,\mathtt{executed}$.

15.2.2. *Linear-time temporal logic* LTL

As far as we know, linear-time temporal logic LTL in the form presented herein has been first considered in [26] based on the early works [3, 27]. Indeed, the strict until operator in [27] can express the temporal operators from LTL (without past-time operators) and in [3] temporal logics are advocated for the formal verification of programs. However, the version of LTL with explicitly the next-time and until operators first appeared in [26]. Actually, the next-time operator has been introduced in [28] in order to define LTL restricted to the next-time and sometime operators (see also a similar language in [29]). Nowadays, LTL is one of the most used logical formalisms to specify the behaviours of computer systems in view of formal verification. It has been also the basis for numerous specification languages such as PSL [30]. Moreover, it is used as a specification language in tools such as SPIN [31] and SMV [32]. LTL formulae are built from the following abstract grammar:

$$\varphi ::= \overbrace{\bot \mid \top \mid p \mid \neg\varphi \mid \varphi\wedge\psi \mid \varphi\vee\psi}^{\text{propositional calculus}} \mid \overbrace{\mathsf{X}\varphi \mid \mathsf{F}\varphi \mid \mathsf{G}\varphi \mid \varphi\,\mathsf{U}\,\psi \mid \varphi\,\mathsf{R}\,\psi}^{\text{temporal extension}}$$

where p ranges over a countably infinite set AP of propositional variables. Elements of AP are obtained by abstracting properties, for instance p may mean "$x = 0$". Given a set of temporal operators $\mathcal{O} \subseteq \{\mathsf{X}, \mathsf{F}, \mathsf{G}, \mathsf{U}, \mathsf{R}\}$, we write LTL($\mathcal{O}$) to denote the restriction of LTL to formulae with temporal connectives from $\mathcal{O}$. We write sub(φ) to denote the set of subformulae of the formula φ and $|\varphi|$ to denote the size of the formula φ viewed as a string of characters. LTL models are program runs, i.e., executions viewed as ω-sequences. The reason why the models are infinite objects (instead of finite sequences to encode finite runs) is mainly due to the possibility to specify limit behaviours such as fair behaviours. So, a structure (or model) for LTL is an infinite sequence $u : \mathbb{N} \rightarrow 2^{\mathrm{AP}}$, i.e., an infinite word of $(2^{\mathrm{AP}})^{\omega}$. Here are the five first states of an LTL model:

$$\{p\} \longrightarrow \{q\} \longrightarrow \{\} \longrightarrow \{p, q, r\} \longrightarrow \{q\} \longrightarrow$$

Given a structure u, a position $i \in \mathbb{N}$ and a formula φ, we define inductively the satisfaction relation $\models$ as follows:

- always $u, i \models \top$ and never $u, i \models \bot$,
- $u, i \models p \stackrel{\text{def}}{\Leftrightarrow} p \in u(i)$, for every $p \in \mathrm{AP}$,
- $u, i \models \neg\varphi \stackrel{\text{def}}{\Leftrightarrow} u, i \not\models \varphi$,
- $u, i \models \varphi_1 \wedge \varphi_2 \stackrel{\text{def}}{\Leftrightarrow} u, i \models \varphi_1$ and $u, i \models \varphi_2$,

- $u, i \models \varphi_1 \vee \varphi_2 \stackrel{\text{def}}{\Leftrightarrow} u, i \models \varphi_1$ or $u, i \models \varphi_2$,
- $u, i \models \mathsf{X}\varphi \stackrel{\text{def}}{\Leftrightarrow} u, i+1 \models \varphi$,
- $u, i \models \mathsf{F}\varphi \stackrel{\text{def}}{\Leftrightarrow}$ there is $j \geq i$ such that $u, j \models \varphi$,
- $u, i \models \mathsf{G}\varphi \stackrel{\text{def}}{\Leftrightarrow}$ for all $j \geq i$, we have $u, j \models \varphi$,
- $u, i \models \varphi_1 \mathsf{U} \varphi_2 \stackrel{\text{def}}{\Leftrightarrow}$ there is $j \geq i$ such that $u, j \models \varphi_2$ and $u, k \models \varphi_1$ for all $i \leq k < j$,
- $u, i \models \varphi_1 \mathsf{R} \varphi_2 \stackrel{\text{def}}{\Leftrightarrow} u, j \models \varphi_2$ for all $j \geq i$, or there is $j \geq i$ such that $u, j \models \varphi_1$ and $u, k \models \varphi_2$ for all $i \leq k \leq j$.

We say that two formulae φ and ψ are *equivalent* whenever for all models u and positions i, we have $u, i \models \varphi$ if and only if $u, i \models \psi$. In that case, we write $\varphi \equiv \psi$. Roughly speaking, φ and ψ state equivalent properties over the class of ω-sequences indexed by propositional valuations. For instance, $\mathsf{F}\varphi$ is equivalent to $\top \mathsf{U} \varphi$ and $\mathsf{G}\varphi$ is equivalent to $\bot \mathsf{R} \varphi$. Consequently, it is clear that our set of connectives is not minimal in terms of expressive power but it provides handy notations. Moreover, we shall use the following abbreviations: $\varphi_1 \rightarrow \varphi_2$ for $\neg\varphi_1 \vee \varphi_2$ and $\mathsf{F}^{\infty} \varphi$ for $\mathsf{G}\mathsf{F}\varphi$ ("φ holds infinitely often"). Finally, one can check that G is the dual of F (since $\mathsf{G}p \equiv \neg \mathsf{F} \neg p$) and R is the dual of U (since $\varphi_1 \mathsf{R} \varphi_2 \equiv \neg(\neg\varphi_1 \mathsf{U} \neg\varphi_2)$).

We write $u \models \varphi$ instead of $u, 0 \models \varphi$. The standard automata-based approach for LTL, see e.g. [9], considers the models for a formula φ as a language $\mathcal{L}(\varphi)$ over the alphabet $2^{\mathrm{AP}(\varphi)}$ where $\mathrm{AP}(\varphi)$ denotes the set of propositional variables occurring in φ (these are the only relevant ones for the satisfaction of φ):

$$\mathcal{L}(\varphi) = \{u \in (2^{\mathrm{AP}(\varphi)})^{\omega} \mid u \models \varphi\}.$$

We say that φ is satisfiable if $\mathcal{L}(\varphi)$ is non-empty. Similarly, φ is valid if $\mathcal{L}(\neg\varphi)$ is empty. The satisfiability problem for LTL, denoted by SAT(LTL), is defined as follows:

input: an LTL formula φ,
output: 1 if $u \models \varphi$ for some infinite word $u \in (2^{\mathrm{AP}(\varphi)})^{\omega}$; 0 otherwise.

The validity problem VAL(LTL) is defined similarly. In order to be precise, it is worth observing that herein we consider *initial* satisfiability since the formula φ holds at the position 0. Since LTL (and the other temporal logics considered in this chapter) does not deal with past-time operators, *initial* satisfiability is equivalent to satisfiability (satisfaction at some position, not necessarily 0).

Let us now consider the model-checking problem. A Kripke structure $\mathcal{M} = \langle W, R, \lambda \rangle$ is a triple such that

- W is a non-empty set of states,
- R is a binary relation on W (accessibility relation, one-step relation),
- λ is a labeling $\lambda : W \rightarrow 2^{\mathrm{AP}}$.

$\mathcal{M}$ is simply a directed graph for which each node is labeled by a propositional interpretation (labeled transition system). A path in $\mathcal{M}$ is a sequence

$\sigma = s_0 s_1 s_2 \cdots$ (finite or infinite) such that $(s_i, s_{i+1}) \in R$ for every $i \geq 0$. We write Paths$(\mathcal{M}, s_0)$ to denote the set of infinite paths of $\mathcal{M}$ starting at state s_0. We also write λPaths$(\mathcal{M}, s_0)$ to denote the set of *labels* of infinite paths starting at s_0: $\lambda\text{Paths}(\mathcal{M}, s_0) = \{\lambda(s_0)\lambda(s_1)\lambda(s_2)\cdots \mid s_0 s_1 s_2 \cdots \in \text{Paths}(\mathcal{M}, s_0)\}$.

The (*universal*) model-checking problem for LTL, denoted by $\text{MC}^{\forall}(\text{LTL})$, is defined as follows:

input: an LTL formula φ, a finite and total† Kripke structure $\mathcal{M}$ and $s_0 \in W$,
output: 1 if $u \models \varphi$ for all $u \in \lambda\text{Paths}(\mathcal{M}, s_0)$ (written $\mathcal{M}, s_0 \models_{\forall} \varphi$); 0 otherwise.

Without any loss of generality, in the above statement we can assume that the codomain of the labeling λ is restricted to AP(φ). The size of $\langle W, R, \lambda \rangle$ is defined by $\text{card}(W) + \text{card}(R) + \Sigma_{w \in W}\text{card}(\lambda(w))$. It is easy to check that $\mathcal{M}, s_0 \models_{\forall} \varphi$ if and only if $\lambda\text{Paths}(\mathcal{M}, s_0) \cap \mathcal{L}(\neg\varphi) = \emptyset$.

There is a dual definition, called *existential* model checking and denoted by $\text{MC}^{\exists}(\text{LTL})$, where an existential quantification on paths is considered. We write $\mathcal{M}, s_0 \models_{\exists} \varphi$ if $u \models \varphi$ for some $u \in \lambda\text{Paths}(\mathcal{M}, s_0)$. Similarly, $\mathcal{M}, s_0 \models_{\exists} \varphi$ if and only if $\lambda\text{Paths}(\mathcal{M}, s_0) \cap \mathcal{L}(\varphi) \neq \emptyset$.

We present below a Kripke structure in which ON and OFF are propositional variables and we identify them with states where they hold respectively.

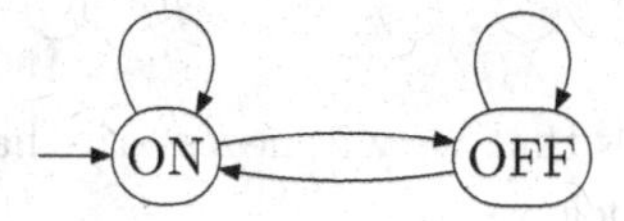

We leave to the reader to check that the properties below hold:

- $\mathcal{M}, \text{ON} \models_{\exists} \mathsf{F}^{\infty}\,\text{ON} \wedge \mathsf{F}^{\infty}\,\text{OFF}$,
- $\mathcal{M}, \text{ON} \models_{\exists} \neg\,\mathsf{F}^{\infty}\,\text{OFF}$,
- $\mathcal{M}, \text{ON} \models_{\exists} \mathsf{G}(\text{ON} \rightarrow \mathsf{XX}\ \text{OFF})$.

15.2.3. *Branching-time temporal logic* CTL*

The language introduced so far can only state properties along one execution. It is also often desirable to express the branching aspect of the behavior: many futures are possible starting from a given state. For instance, consider the property φ which is informally defined as: "whenever we are in a state where p holds, it is possible to reach a state where q holds". This natural property cannot be expressed in LTL. Indeed, with the models $\mathcal{M}_1$ and $\mathcal{M}_2$ of Figure 15.1, we have $\lambda\text{Paths}(\mathcal{M}_1) = \lambda\text{Paths}(\mathcal{M}_2)$. Hence $\mathcal{M}_1$ and $\mathcal{M}_2$ satisfy the same LTL formulae. But indeed, $\mathcal{M}_1$ satisfies φ whereas $\mathcal{M}_2$ does not.

The logic CTL* introduces special purpose quantifiers, A (compare with $\forall$ in first-order logic) and E (compare with $\exists$ in first-order logic), which allow to quantify over the set of executions. These are called *path quantifiers*. The expression $\mathsf{A}\,\varphi$

† $\forall x \in W, \exists y \in W,\ \langle x, y \rangle \in R$.

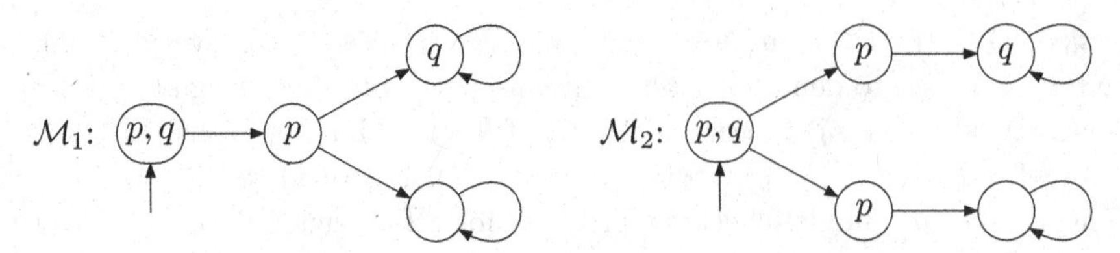

Fig. 15.1. Two models undistinguishable for LTL.

states that all executions out of the current state satisfy property φ. Dually, $\mathsf{E}\,\varphi$ states that from the current state, there exists an execution satisfying φ. Our "natural" property above can be written $\mathsf{A}\,\mathsf{G}(p \rightarrow \mathsf{E}\,\mathsf{F}\,q)$. It is worth observing that A and E quantify over paths whereas G and F quantify over positions along a path. The expression $\mathsf{E}\,\mathsf{F}\,\varphi$ states that it is possible (by following a suitable execution) to have φ some day, which is illustrated below.

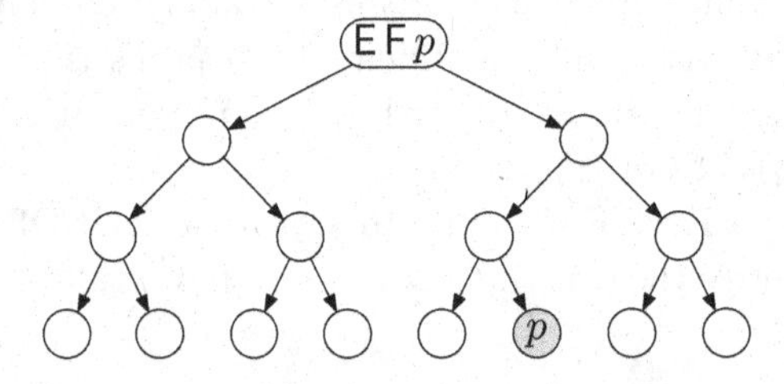

The expression $\mathsf{A}\,\mathsf{F}\,\varphi$ states that we will necessarily have φ some day, regardless of the chosen execution (see below).

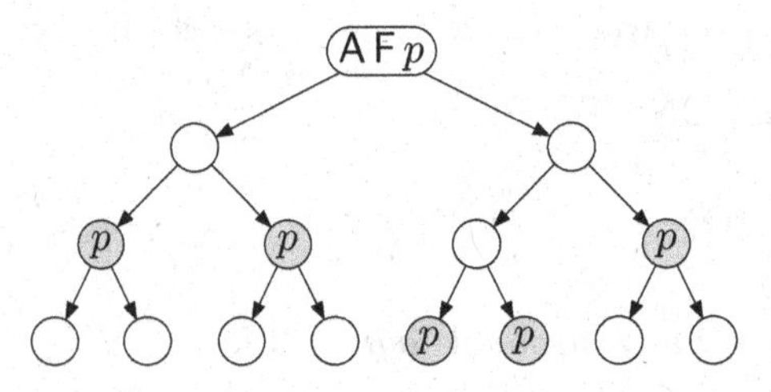

The expression $\mathsf{A}\,\mathsf{G}\,\varphi$ states that φ holds in all future states including now (see below).

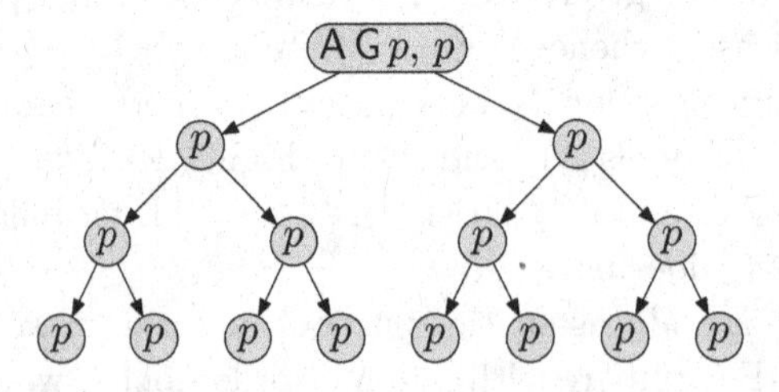

The expression $\mathsf{E}\,\mathsf{G}\,\varphi$ states that there is an execution on which φ always holds (see below).

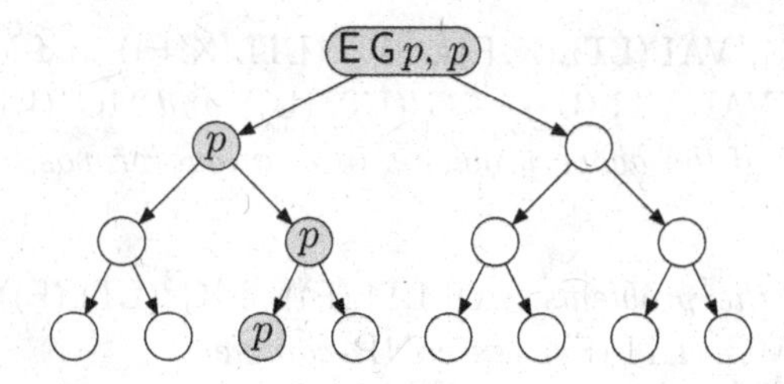

"Branching-time logics" refers to logics that have the ability to freely quantify over paths. Standard examples of branching-time temporal logics include Computation Tree Logic CTL [33], CTL* [34] and the modal μ-calculus. We define below the logic CTL* (that is more expressive than both LTL and CTL) for which the model-checking problem can be solved easily by using a subroutine solving the model-checking problem for LTL. Hence, even though the object of this chapter is not especially dedicated to branching-time logics, we explain how CTL* model-checking can be dealt with. CTL* formulae are built from the following abstract grammar:

$$\varphi ::= p \mid \neg\varphi \mid \varphi_1 \wedge \varphi_2 \mid \mathsf{E}\varphi \mid \mathsf{A}\varphi \mid \mathsf{X}\varphi \mid \varphi_1 \mathsf{U} \varphi_2$$

where p ranges over AP. CTL* models are total Kripke models. Let $\sigma = s_0 s_1 \ldots$ be an infinite path in $\mathcal{M}$, $i \geq 0$ and φ be a formula. The satisfaction relation $\sigma, i \models \varphi$ is defined inductively as follows (we omit the clauses for Boolean connectives):

- $\sigma, i \models p \overset{\text{def}}{\Leftrightarrow} p \in \lambda(s_i)$,
- $\sigma, i \models \mathsf{X}\varphi \overset{\text{def}}{\Leftrightarrow} \sigma, i+1 \models \varphi$,
- $\sigma, i \models \varphi_1 \mathsf{U} \varphi_2 \overset{\text{def}}{\Leftrightarrow}$ there is $j \geq i$ such that $\sigma, j \models \varphi_2$ and $\sigma, k \models \varphi_1$ for all $i \leq k < j$,
- $\sigma, i \models \mathsf{E}\varphi \overset{\text{def}}{\Leftrightarrow}$ there is an infinite path $\sigma' = s'_0 s'_1 \ldots$ such that $s'_0 = s_i$ and $\sigma', 0 \models \varphi$,
- $\sigma, i \models \mathsf{A}\varphi \overset{\text{def}}{\Leftrightarrow}$ for every infinite path $\sigma' = s'_0 s'_1 \ldots$ such that $s'_0 = s_i$, we have $\sigma', 0 \models \varphi$.

The model-checking problem for CTL*, denoted by $\mathrm{MC}^{\forall}(\mathrm{CTL}^*)$, is defined as follows:

input: a CTL* formula, a finite and total Kripke model $\mathcal{M} = \langle W, R, \lambda \rangle$ and $s \in W$;
output: 1 if $\sigma, 0 \models \varphi$ for all infinite paths $\sigma \in \mathrm{Paths}(\mathcal{M}, s)$ starting from s; 0 otherwise.

15.2.4. *Complexity issues*

Let us recall a few complexity results.

Theorem 15.1. *[34–37] The following problems are* PSPACE*-complete.*

(i) SAT(LTL), VAL(LTL), $\mathrm{MC}^{\exists}(\mathrm{LTL})$ *and* $\mathrm{MC}^{\forall}(\mathrm{LTL})$.

(ii) SAT(LTL(X, F)), VAL(LTL(X, F)), $\mathrm{MC}^{\exists}$(LTL(X, F)) *and* $\mathrm{MC}^{\forall}$(LTL(X, F)).
(iii) SAT(LTL(U)), VAL(LTL(U)), $\mathrm{MC}^{\exists}$(LTL(U)) *and* $\mathrm{MC}^{\forall}$(LTL(U)).
(iv) The restriction of the above problems to a unique propositional variable.
(v) MC(CTL*).

On the other hand, the problems SAT(LTL(F)), $\mathrm{MC}^{\exists}$(LTL(F)) *are* NP*-complete and* VAL(LTL(F)), $\mathrm{MC}^{\forall}$(LTL(F)) *are* coNP*-complete.*

The treatment in Section 15.3 will establish that SAT(LTL), VAL(LTL), $\mathrm{MC}^{\forall}$(LTL) and $\mathrm{MC}^{\exists}$(LTL) are in PSPACE.

The Computation Tree Logic CTL [33] is a strict fragment of CTL* for which model-checking can be solved in polynomial-time. We briefly recall that CTL formulae are defined by the grammar below:

$$\varphi ::= p \mid \neg\varphi \mid \varphi_1 \wedge \varphi_2 \mid \mathsf{E}\varphi_1 \,\mathsf{U}\, \varphi_2 \mid \mathsf{A}\varphi_1 \,\mathsf{U}\, \varphi_2 \mid \mathsf{EX}\varphi \mid \mathsf{AX}\varphi$$

CTL model-checking is PTIME-complete and the complexity function is bilinear in the size of the formula and in the size of the Kripke structure [38] (see also the survey paper [37]) whereas CTL satisfiability is EXPTIME-complete [33]. By contrast, the satisfiability problem for CTL* is much more complex: 2EXPTIME-complete (upper bound from [39] and lower bound from [40]).

15.3. From LTL formulae to Büchi automata

In this section, we explain how to translate an LTL formula φ into an automaton $\mathcal{A}_\varphi$ such that the language recognized by $\mathcal{A}_\varphi$ is precisely $\mathcal{L}(\varphi)$. However, in order to be of practical use, the translation process can be divided in four stages (at least):

(1) preprocessing the LTL formula using simple logical equivalences,
(2) translation of φ into a generalized Büchi automaton $\mathcal{A}_\varphi$ (the core of the construction),
(3) simplification and optimization of $\mathcal{A}_\varphi$,
(4) translation of $\mathcal{A}_\varphi$ into a Büchi automaton.

Indeed, it is legitimate to aim at building Büchi automata as simple as possible, even though we know that in the worst case the translation has an exponential blow-up. This section is mainly dedicated to step (2) with the construction of simple automata. Considerations about steps (1) and (3) can be found in Sections 15.3.3 and 15.3.6, respectively.

15.3.1. *Automata-based approach*

The construction of $\mathcal{A}_\varphi$ from the formula φ remains the core for the decision procedures of the satisfiability problem and the model checking problem for LTL specifications. Indeed, the (initial) satisfiability problem amounts to checking the Büchi automaton $\mathcal{A}_\varphi$ for emptiness. To solve the model-checking problem, one constructs first the product $\mathcal{B} = \mathcal{M} \times \mathcal{A}_{\neg\varphi}$ of the model $\mathcal{M}$ with the automaton $\mathcal{A}_{\neg\varphi}$ so that successful runs of $\mathcal{B}$ correspond to infinite runs of $\mathcal{M}$ satisfying the formula $\neg\varphi$. Therefore, $\mathcal{L}(\mathcal{B}) = \emptyset$ if and only if $\mathcal{M} \models_{\forall} \varphi$ and the model-checking problem is again reduced to the emptiness problem for a Büchi automaton.

Note that checking nonemptiness of a Büchi automaton can be done efficiently (NLOGSPACE or linear time, see e.g. [41, Theorem 12]) since it reduces to several reachability questions in the underlying graph of the automaton: we have to find a reachable accepting state with a loop around it. Since both the satisfiability problem and the model-checking problem for LTL specifications are PSPACE-complete, we cannot avoid an exponential blow-up in the worst case when constructing a Büchi automaton $\mathcal{A}_\varphi$ associated with an LTL formula φ. Fortunately, in most practical cases, we can construct a *small* Büchi automaton $\mathcal{A}_\varphi$.

It is therefore very important to have *good* constructions for the Büchi automaton $\mathcal{A}_\varphi$ even though there are several interpretations of *good*. It is indeed important to obtain a *small* automaton and several techniques have been developed to reduce the size of the resulting automaton [42]. On the other hand, it is also important to have a quick construction. Some constructions, such as the tableau construction [15], may take an exponential time even if the resulting *reduced* automaton is small. The problem with the most efficient constructions [17] is that they are involved, technical and based on more elaborate structures such as alternating automata. Herein we are interested in a *good* translation from a pedagogical point of view. Our construction is a middle term between tableau constructions and more elaborate constructions based on alternating automata. As a result, it will be efficient, it will produce small automata, and it will be possible to translate non trivial LTL formulae by hand. However, we admit that it is neither the most efficient nor the one that produces the smallest automata.

15.3.2. *Büchi automata in a nutshell*

We first recall the definition of Büchi automata (BA) and some useful generalizations; a self-contained introduction to the theory of finite-state automata for infinite words can be found in Chapter 2 . A BA is a tuple $\mathcal{A} = (Q, \Sigma, I, T, F)$ where Q is a finite set of states, Σ is the alphabet, $I \subseteq Q$ is the set of initial states, $T \subseteq Q \times \Sigma \times Q$ is the set of transitions, and $F \subseteq Q$ is the set of accepting (repeated, final) states.

A *run* of $\mathcal{A}$ is a sequence $\rho = s_0, a_0, s_1, a_1, s_2, \ldots$ such that $(s_i, a_i, s_{i+1}) \in T$ is a transition for all $i \geq 0$. The run ρ is *successful* if $s_0 \in I$ is initial and some state of F is repeated infinitely often in ρ: $\inf(\rho) \cap F \neq \emptyset$ where we let $\inf(\rho) = \{s \in Q \mid \forall i,\ \exists j > i,\ s = s_j\}$. The label of ρ is the word $u = a_0 a_1 \cdots \in \Sigma^\omega$. The automaton $\mathcal{A}$ accepts the language $\mathcal{L}(\mathcal{A})$ of words $u \in \Sigma^\omega$ such that there exists a successful run of $\mathcal{A}$ on the word u, i.e., with label u. For instance, the automaton in Figure 15.2 accepts those words over $\{a, b\}$ having infinitely many a's (the initial states are marked with an incoming arrow and the repeated states are doubly circled).

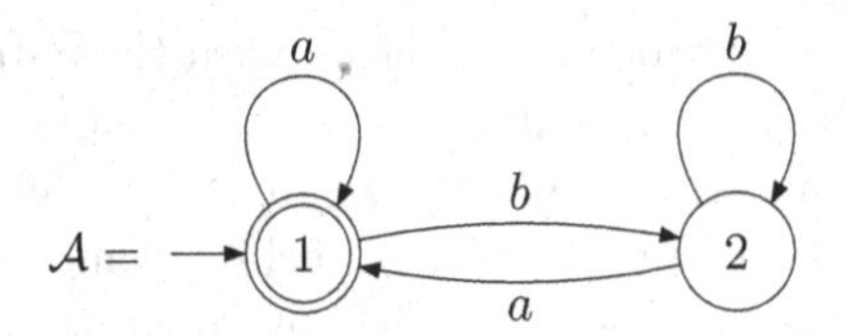

Fig. 15.2. $\mathcal{L}(\mathcal{A}) = \{u \in \{a,b\}^\omega \mid |u|_a = \omega\}$.

When dealing with models for an LTL formula φ, the words are over the alphabet $\Sigma = 2^{\mathsf{AP}(\varphi)}$ where $\mathsf{AP}(\varphi)$ is the set of propositional variables occurring in φ. A letter $a \in \Sigma$ is read as a propositional valuation for which exactly the propositional variables in a hold. We take advantage of this natural interpretation for defining sets of letters: given a propositional formula ψ, we let $\Sigma_\psi = \{a \in \Sigma \mid a \models \psi\}$ for which "$\models$" refers to the satisfaction relation from propositional calculus. For instance, we have $\Sigma_p = \{a \in \Sigma \mid p \in a\}$, $\Sigma_{\neg p} = \Sigma \setminus \Sigma_p$, $\Sigma_{p\wedge q} = \Sigma_p \cap \Sigma_q$, $\Sigma_{p\vee q} = \Sigma_p \cup \Sigma_q$ and $\Sigma_{p\wedge\neg q} = \Sigma_p \setminus \Sigma_q$. In general, a transition between two states s, s' will be enabled for all letters satisfying some propositional formula ψ. We use $s \xrightarrow{\Sigma_\psi} s'$ as a concise representation of the set of transitions $\{s \xrightarrow{a} s' \mid a \in \Sigma_\psi\}$.

Several examples of Büchi automata corresponding to LTL formulae are given in Figure 15.3. In these automata, transitions are labeled with subsets of Σ meaning that all letters in the subset are allowed for the transition. In some cases, the automaton associated with a formula is deterministic, that is for all $s \in Q$ and $a \in \Sigma$, $\{s' \mid \langle s, a, s'\rangle \in T\}$ has at most one state. Although, determinism is a very desirable property, it is not always possible. For instance, the automaton for $\mathsf{G}\,\mathsf{F}\,p$ is deterministic whereas the automaton for its negation $\neg\,\mathsf{G}\,\mathsf{F}\,p \equiv \mathsf{F}\,\mathsf{G}\,\neg p$ must be nondeterministic. This is an easy example showing that deterministic BA are not closed under complement.

By contrast, Büchi automata are closed under union, intersection and complement, which corresponds to the Boolean operations on formulae. It is also easy to construct an automaton for $\mathsf{X}\,\varphi$ from an automaton for φ, see for instance the automaton for $\mathsf{X}\,\mathsf{X}\,p$ in Figure 15.3. Finally, one can construct an automaton for $\varphi\,\mathsf{U}\,\psi$ from automata for φ and ψ. Hence, we have a modular construction of $\mathcal{A}_\varphi$ for any LTL formula φ. But both negation and until yield an exponential blowup. Hence this modular construction is non-elementary and useless in practice, see also [43, 44].

Now we introduce a generalization of the acceptance condition of Büchi automata. First, it will fall on transitions instead of states, as considered also in [18]. Second it will allow conjunctions of classical Büchi conditions. Formally, a generalized Büchi automaton (GBA) is a tuple $\mathcal{A} = (Q, \Sigma, I, T, T_1, \ldots, T_n)$ where Q, Σ, I, T are as above and the acceptance condition which deals with transitions is given by the sets $T_i \subseteq T$ for $1 \le i \le n$. For a run ρ of $\mathcal{A}$, we denote by $\inf_T(\rho)$ the set of transitions that occur infinitely often in ρ and the run is *successful* if $\inf_T(\rho) \cap T_i \neq \emptyset$ for each $1 \le i \le n$. Often, we simply write $\inf(\rho)$ instead of $\inf_T(\rho)$. For instance,

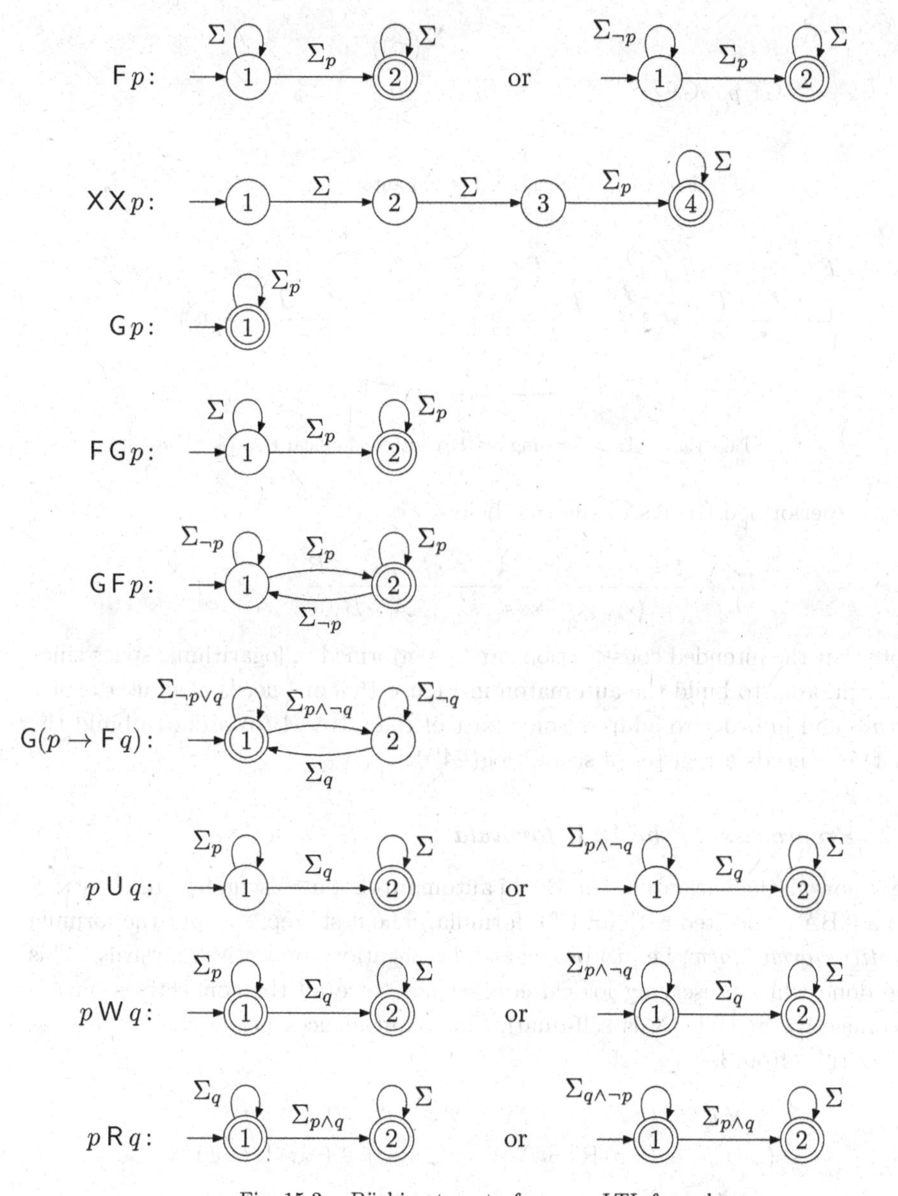

Fig. 15.3. Büchi automata for some LTL formulae.

a GBA is given in Figure 15.4 where we require both that transitions with short dashes and transitions with long dashes are repeated infinitely often.

Each GBA $\mathcal{A}$ can be easily translated into a classical BA (preserving the language of accepted ω-words). Indeed, it is sufficient to perform a synchronized product $\mathcal{A} \otimes \mathcal{B}$ with the automaton $\mathcal{B}$ in Figure 15.5 where the accepting Büchi states in the product are those containing the state n in the second component. Synchro-

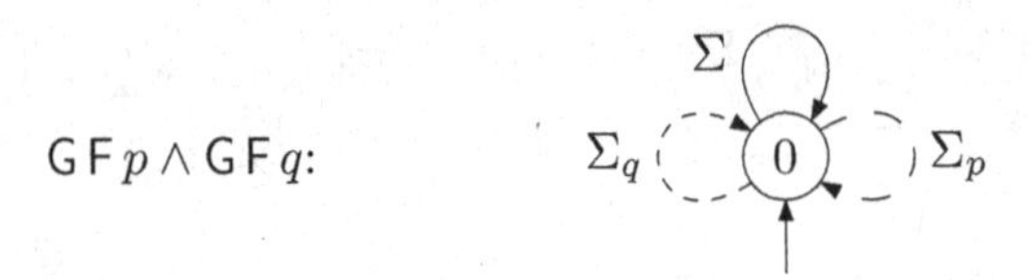

Fig. 15.4. Generalized Büchi automaton.

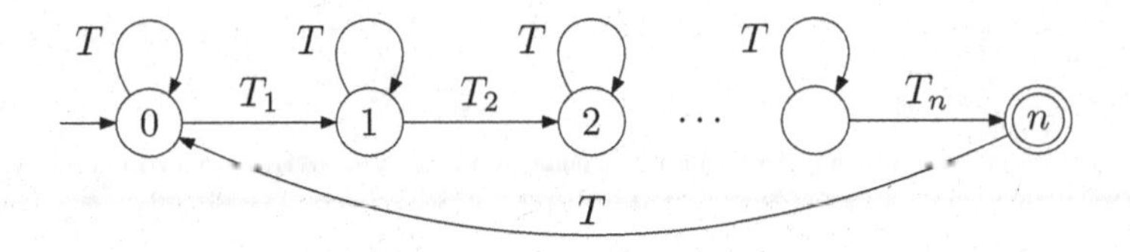

Fig. 15.5. Transforming a GBA into a classical BA.

nization is performed thanks to the rule below:

$$\frac{t = s_1 \xrightarrow{a} s_1' \in \mathcal{A} \qquad s_2 \xrightarrow{t} s_2' \in \mathcal{B}}{\langle s_1, s_2 \rangle \xrightarrow{a} \langle s_1', s_2' \rangle \in \mathcal{A} \otimes \mathcal{B}}$$

Note that the intended construction can be performed in logarithmic space since typically in order to build the automaton in Figure 15.5 one needs a counter of size $\mathcal{O}(\log(n))$ and in order to address some part of the GBA $\mathcal{A}$ (in order to build the product) one needs a register of size $\mathcal{O}(\log(|\mathcal{A}|))$.

15.3.3. *Preprocessing the* LTL *formula*

We have now all the background on Büchi automata that are useful for our construction of a GBA associated with an LTL formula. The first step is to put the formula in *negative normal form*, i.e., to propagate the negation connectives inwards. This can be done while preserving logical equivalence since all the connectives have a dual connective in LTL (X is self-dual). The equivalences below can be read as rewriting rules from left to right:

$$\begin{array}{cc} \neg(\varphi \vee \psi) \equiv (\neg\varphi) \wedge (\neg\psi) & \neg(\varphi \wedge \psi) \equiv (\neg\varphi) \vee (\neg\psi) \\ \neg(\varphi \mathsf{U} \psi) \equiv (\neg\varphi) \mathsf{R} (\neg\psi) & \neg(\varphi \mathsf{R} \psi) \equiv (\neg\varphi) \mathsf{U} (\neg\psi) \\ \neg \mathsf{X} \varphi \equiv \mathsf{X} \neg\varphi & \neg\neg\varphi \equiv \varphi \end{array}$$

Formally, an LTL formula is in *negative normal form* (NNF) if it follows the syntax given by

$$\varphi ::= \top \mid \bot \mid p \mid \neg p \mid \varphi \vee \varphi \mid \varphi \wedge \varphi \mid \mathsf{X}\varphi \mid \varphi \mathsf{U} \varphi \mid \varphi \mathsf{R} \varphi$$

where p ranges over atomic propositions in AP.

In the following, a *temporal* formula is defined as either a literal (i.e., a propositional variable or its negation) or a formula in NNF with outermost connective among $\{\mathsf{X}, \mathsf{U}, \mathsf{R}\}$. Therefore, any LTL formula in NNF is a positive Boolean combination of temporal formulae. Note that, translating an arbitrary LTL formula in NNF does not increase the number of temporal subformulae. This is important since the size of the GBA $\mathcal{A}_\varphi$ that we will construct depends on the number of temporal subformulae of φ. Therefore, before starting the construction, it is useful to rewrite the formula in order to reduce the number of temporal subformulae. Several rewriting rules are presented in [45, 46] and we only give below some examples which again should be applied from left to right.

$$(\mathsf{X}\,\varphi) \wedge (\mathsf{X}\,\psi) \equiv \mathsf{X}(\varphi \wedge \psi) \qquad (\mathsf{X}\,\varphi)\;\mathsf{U}\;(\mathsf{X}\,\psi) \equiv \mathsf{X}(\varphi\;\mathsf{U}\;\psi)$$

$$(\varphi\;\mathsf{R}\;\psi_1) \wedge (\varphi\;\mathsf{R}\;\psi_2) \equiv \varphi\;\mathsf{R}\;(\psi_1 \wedge \psi_2) \qquad (\varphi_1\;\mathsf{R}\;\psi) \vee (\varphi_2\;\mathsf{R}\;\psi) \equiv (\varphi_1 \vee \varphi_2)\;\mathsf{R}\;\psi$$

$$(\mathsf{G}\,\varphi) \wedge (\mathsf{G}\,\psi) \equiv \mathsf{G}(\varphi \wedge \psi) \qquad \mathsf{G}\,\mathsf{F}\,\varphi \vee \mathsf{G}\,\mathsf{F}\,\psi \equiv \mathsf{G}\,\mathsf{F}(\varphi \vee \psi)$$

It is worth noting that the above simplification rules are useful in practice. By contrast, writing a formula in NNF remains a step that mainly eases the presentation of the forthcoming construction. Indeed, propagating the negation connectives inwards can be performed symbolically by storing subformulae of the initial formula augmented with polarities in $\{0, 1\}$.

15.3.4. *Building simple automata*

We start now the description of the core of our construction. A state Z of an automaton $\mathcal{A}_\varphi$ will be a subset of $\mathrm{sub}(\varphi)$, the set of subformulae of φ. We say that a set Z of formulae is *consistent* if it does not contain $\bot$ or a pair $\{\psi, \neg\psi\}$ for some formula ψ (since our formulae are in NNF, ψ could only be a propositional variable). We often need the conjunction of formulae in Z which will be written $\bigwedge Z = \bigwedge_{\psi \in Z} \psi$. Note that $\bigwedge \emptyset = \top$. The formulae in Z are viewed as *obligations*, i.e., if a run ρ on a word u starts from Z and satisfies the acceptance condition then $u \models \bigwedge Z$, i.e., $u \models \psi$ for all $\psi \in Z$. More precisely, if we denote by $\mathcal{A}_\varphi^Z$ the automaton $\mathcal{A}_\varphi$ where Z is the unique initial state, then our construction will guarantee that

$$\mathcal{L}(\mathcal{A}_\varphi^Z) = \{u \in \Sigma^\omega \mid u \models \bigwedge Z\}$$

Therefore, the unique initial state of $\mathcal{A}_\varphi$ will be the singleton set $\{\varphi\}$.

We say that a set Z of LTL formulae in NNF is *reduced* if all formulae in Z are either literals or formulae with outermost connective X. Given a *consistent* and *reduced* set Z, we write $\mathrm{next}(Z)$ to denote the set $\{\psi \mid \mathsf{X}\,\psi \in Z\}$ and Σ_Z to denote the set of letters which satisfy all literals in Z:

$$\Sigma_Z = \bigcap_{p \in Z} \Sigma_p \cap \bigcap_{\neg p \in Z} \Sigma_{\neg p}$$

Equivalently, Σ_Z is the set of letters $a \in \Sigma$ such that for every $p \in \mathrm{AP}(\varphi)$, $p \in Z$ implies $p \in a$ and $\neg p \in Z$ implies $p \notin a$. From a consistent and reduced set Z, the automaton is ready to perform any transition of the form $Z \xrightarrow{a} \mathrm{next}(Z)$ with $a \in \Sigma_Z$.

Table 15.1. Reduction rules.

If $\psi = \psi_1 \wedge \psi_2$:	$Y \xrightarrow{\varepsilon} (Y \setminus \{\psi\}) \cup \{\psi_1, \psi_2\}$
If $\psi = \psi_1 \vee \psi_2$:	$Y \xrightarrow{\varepsilon} (Y \setminus \{\psi\}) \cup \{\psi_1\}$ $Y \xrightarrow{\varepsilon} (Y \setminus \{\psi\}) \cup \{\psi_2\}$
If $\psi = \psi_1 \mathsf{R} \psi_2$:	$Y \xrightarrow{\varepsilon} (Y \setminus \{\psi\}) \cup \{\psi_1, \psi_2\}$ $Y \xrightarrow{\varepsilon} (Y \setminus \{\psi\}) \cup \{\psi_2, \mathsf{X}\,\psi\}$
If $\psi = \mathsf{G}\,\psi_2$:	$Y \xrightarrow{\varepsilon} (Y \setminus \{\psi\}) \cup \{\psi_2, \mathsf{X}\,\psi\}$
If $\psi = \psi_1 \mathsf{U} \psi_2$:	$Y \xrightarrow{\varepsilon} (Y \setminus \{\psi\}) \cup \{\psi_2\}$ $Y \xrightarrow[!\psi]{\varepsilon} (Y \setminus \{\psi\}) \cup \{\psi_1, \mathsf{X}\,\psi\}$
If $\psi = \mathsf{F}\,\psi_2$:	$Y \xrightarrow{\varepsilon} (Y \setminus \{\psi\}) \cup \{\psi_2\}$ $Y \xrightarrow[!\psi]{\varepsilon} (Y \setminus \{\psi\}) \cup \{\mathsf{X}\,\psi\}$

For instance, for every $a \in \Sigma$, $\emptyset \xrightarrow{a} \emptyset$ is a transition which can be interpreted as: when no obligations have to be satisfied, any letter can be read and there are still no obligations. Note that $\Sigma_Z \neq \emptyset$ since Z is consistent, but $\text{next}(Z)$ is not reduced in general. We will use ε-transitions to reduce arbitrary sets of formulae in reduced sets so that the semantics of the automaton is preserved. These transitions are handy but they will not belong to the final GBA $\mathcal{A}_\varphi$. So let Y be a set of formulae which is not reduced and choose some $\psi \in Y$ *maximal* among the non-reduced formulae in Y (here maximal is for the *subformula* ordering). Depending on the form of ψ, the ε-transitions allowing to reduce ψ are presented in Table 15.1. The rules for G and F can indeed be derived from those for R and U, they are included for convenience. Indeed, we only introduce transitions between *consistent* sets. While $\xrightarrow{\varepsilon}$ denotes the one-step reduction relation, as usual, we write $\xrightarrow[*]{\varepsilon}$ to denote the reflexive and transitive closure of $\xrightarrow{\varepsilon}$.

When ψ is a conjunction or a G formula then we introduce only one ε-transition $Y \xrightarrow{\varepsilon} Y_1$ and $\bigwedge Y \equiv \bigwedge Y_1$. In the other cases, we introduce two ε-transitions $Y \xrightarrow{\varepsilon} Y_1$ and $Y \xrightarrow{\varepsilon} Y_2$ and $\bigwedge Y \equiv \bigwedge Y_1 \vee \bigwedge Y_2$.

We introduce these ε-transitions iteratively until all states have been reduced. The construction terminates since each step removes a *maximal* non-reduced formula and introduces only strictly smaller non-reduced formulae (note that $\mathsf{X}\,\alpha$ is not smaller than α but is reduced).

Finally, note the mark $!\psi$ on the second transitions for U and F. It denotes the fact that the *eventuality* ψ_2 has been postponed. The marked transitions will be used to define the acceptance condition of the GBA in such a way that all eventualities are satisfied along an accepting run.

In Figure 15.6 we show the ε-transitions that are introduced when we start with a singleton set $\{\varphi\}$ with $\varphi = \mathsf{G}(p \to \mathsf{F}\,q) \equiv \bot\,\mathsf{R}\,(\neg p \vee (\top\,\mathsf{U}\,q))$. Note again the mark $!\mathsf{F}\,q$ on the last transition.

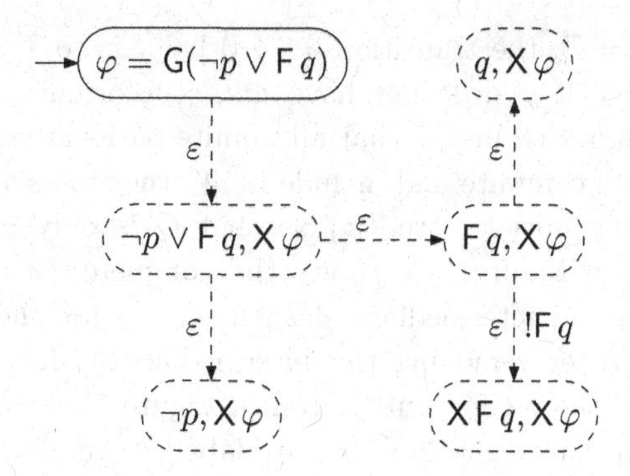

Fig. 15.6. Reduction of a state to reduced states.

An *until formula* α of φ (in NNF) is a subformula of φ with outermost connective either U or F. The set of until formulae of φ is denoted by $\mathsf{U}(\varphi)$. For each subset Y of formulae in NNF we define

$$\mathrm{Red}(Y) = \{Z \text{ consistent and reduced} \mid \text{there is a path } Y \xrightarrow[*]{\varepsilon} Z\}$$

and for each $\alpha \in \mathsf{U}(\varphi)$ we also define

$$\mathrm{Red}_\alpha(Y) = \{Z \text{ consistent and reduced} \mid \text{there is a path } Y \xrightarrow[*]{\varepsilon} Z \\ \text{without using an edge marked with } !\alpha\}$$

Thanks to the nice properties of the reduction rules, we obtain the equivalence

$$\bigwedge Y \equiv \bigvee_{Z \in \mathrm{Red}(Y)} \bigwedge Z$$

Consequently, by using the reductions from Figure 15.6 we obtain

$$\mathrm{Red}(\{\varphi\}) = \{\{\neg p, \mathsf{X}\varphi\}, \{q, \mathsf{X}\varphi\}, \{\mathsf{X}\,\mathsf{F}\, q, \mathsf{X}\varphi\}\}$$
$$\mathrm{Red}_{\mathsf{F}\,q}(\{\varphi\}) = \{\{\neg p, \mathsf{X}\varphi\}, \{q, \mathsf{X}\varphi\}\}$$

Observe that in $\mathrm{Red}_{\mathsf{F}\,q}(\{\varphi\})$, the subscript $\mathsf{F}\,q$ refers to an *absence* of ε-transitions marked by $!\mathsf{F}\,q$ along the reduction path. This is the case when we do not have the obligation $\mathsf{F}\,q$ or if the eventuality $\mathsf{F}\,q$ is satisfied now by imposing the obligation q. By contrast, an ε-transition $\{\mathsf{F}\,q, \mathsf{X}\varphi\} \xrightarrow[!\mathsf{F}\,q]{\varepsilon} \{\mathsf{X}\,\mathsf{F}\,q, \mathsf{X}\varphi\}$ with mark $!\mathsf{F}\,q$ indicates that the eventuality $\mathsf{F}\,q$ is *not* satisfied now. We hope this is not too confusing.

We give now the formal definition of $\mathcal{A}_\varphi = (Q, \Sigma, I, T, (T_\alpha)_{\alpha \in \mathsf{U}(\varphi)})$. The set of states is $Q = 2^{\mathrm{sub}(\varphi)}$ and the initial state is the singleton $I = \{\varphi\}$. The set of transitions is defined as follows:

$$T = \{Y \xrightarrow{a} \mathrm{next}(Z) \mid Y \in Q,\ a \in \Sigma_Z \text{ and } Z \in \mathrm{Red}(Y)\}$$

For each $\alpha \in \mathsf{U}(\varphi)$, we define the acceptance set T_α:

$$T_\alpha = \{Y \xrightarrow{a} \mathrm{next}(Z) \mid Y \in Q,\ a \in \Sigma_Z \text{ and } Z \in \mathrm{Red}_\alpha(Y)\}$$

Since $\emptyset \in Q$ and $\Sigma_\emptyset = \Sigma$, the transition $\emptyset \xrightarrow{\Sigma} \emptyset$ belongs to T and to T_α for each $\alpha \in \mathsf{U}(\varphi)$. Note that, if φ does not have *until* subformulae then there are no acceptance conditions, which means that all infinite paths are successful.

In practice, we only compute and include in $\mathcal{A}_\varphi$ the states and transitions that are reachable from the initial state $\{\varphi\}$ so that Q is only a subset of $2^{\mathrm{sub}(\varphi)}$. The first automaton in Figure 15.7 shows the complete construction, including the ε-transitions and the intermediary dashed states, for the response formula $\varphi = \mathsf{G}(p \to \mathsf{F}\, q)$. After removing the intermediary dashed states and the ε-transitions, we obtain the second automaton in Figure 15.7 where the transitions from the unique acceptance condition $T_{\mathsf{F}\, q}$ are labelled with $\mathsf{F}\, q$. As $\Sigma_{\neg p} \subseteq \Sigma$, the loop labeled $\Sigma_{\neg p}$ on the second state is redundant‡. Similarly, the transitions labeled $\Sigma_{\neg p \wedge q}$ is redundant. We obtain the third GBA $\mathcal{A}_\varphi$ in Figure 15.7. It is then easy to check that $\mathcal{L}(\mathcal{A}_\varphi) = \{u \in \Sigma^\omega \mid u \models \varphi\}$.

15.3.5. *Correctness*

More examples will conclude this section. Let us first show the correctness of the construction. The main result is stated in Theorem 15.2.

Theorem 15.2. *The automaton $\mathcal{A}_\varphi$ accepts precisely the models of φ, i.e.,*

$$\mathcal{L}(\mathcal{A}_\varphi) = \mathcal{L}(\varphi) = \{u \in \Sigma^\omega \mid u \models \varphi\}$$

In order to be precise, φ satisfies much more models by considering the larger set of propositional variables that do not appear in φ. Indeed such propositional variables are simply irrelevant for the satisfaction of φ. The proof of this theorem requires several lemmas and propositions. The first lemma is trivial.

Lemma 15.3. *Let Z be a* consistent *and* reduced *set of formulae in NNF. Let $u = a_0 a_1 a_2 \cdots \in \Sigma^\omega$ and $n \geq 0$. Then $u, n \models \bigwedge Z$ if and only if $u, n{+}1 \models \bigwedge \mathrm{next}(Z)$ and $a_n \in \Sigma_Z$.*

Using the equivalence $\bigwedge Y \equiv \bigvee_{Z \in \mathrm{Red}(Y)} \bigwedge Z$ we prove now:

Lemma 15.4. *Let Y be a subset of formulae in NNF and let $u \in \Sigma^\omega$ be an infinite word. If $u \models \bigwedge Y$ then there is $Z \in \mathrm{Red}(Y)$ such that $u \models \bigwedge Z$ and for every $\alpha = \alpha_1 \mathsf{U}\, \alpha_2 \in \mathsf{U}(\varphi)$, if $u \models \alpha_2$, then $Z \in \mathrm{Red}_\alpha(Y)$.*

Proof. Consider again the reduction rules presented in Table 15.1. At each step, either we have a single ε-transition $Y \xrightarrow{\varepsilon} Y_1$ and $\bigwedge Y \equiv \bigwedge Y_1$ or we have two ε-transitions $Y \xrightarrow{\varepsilon} Y_1$ and $Y \xrightarrow{\varepsilon} Y_2$ and $\bigwedge Y \equiv \bigwedge Y_1 \vee \bigwedge Y_2$. So there is a reduction path from Y to some $Z \in \mathrm{Red}(Y)$ such that $u \models \bigwedge Z$ and whenever we reduce an until formula $\alpha = \alpha_1 \mathsf{U}\, \alpha_2$ with $u \models \alpha_2$ we take the first reduction $Y' \xrightarrow{\varepsilon} Y' \setminus \{\alpha\} \cup \{\alpha_2\}$.

‡Actually this corresponds to a set of transitions which is contained in the set of transitions described by the loop labeled Σ.

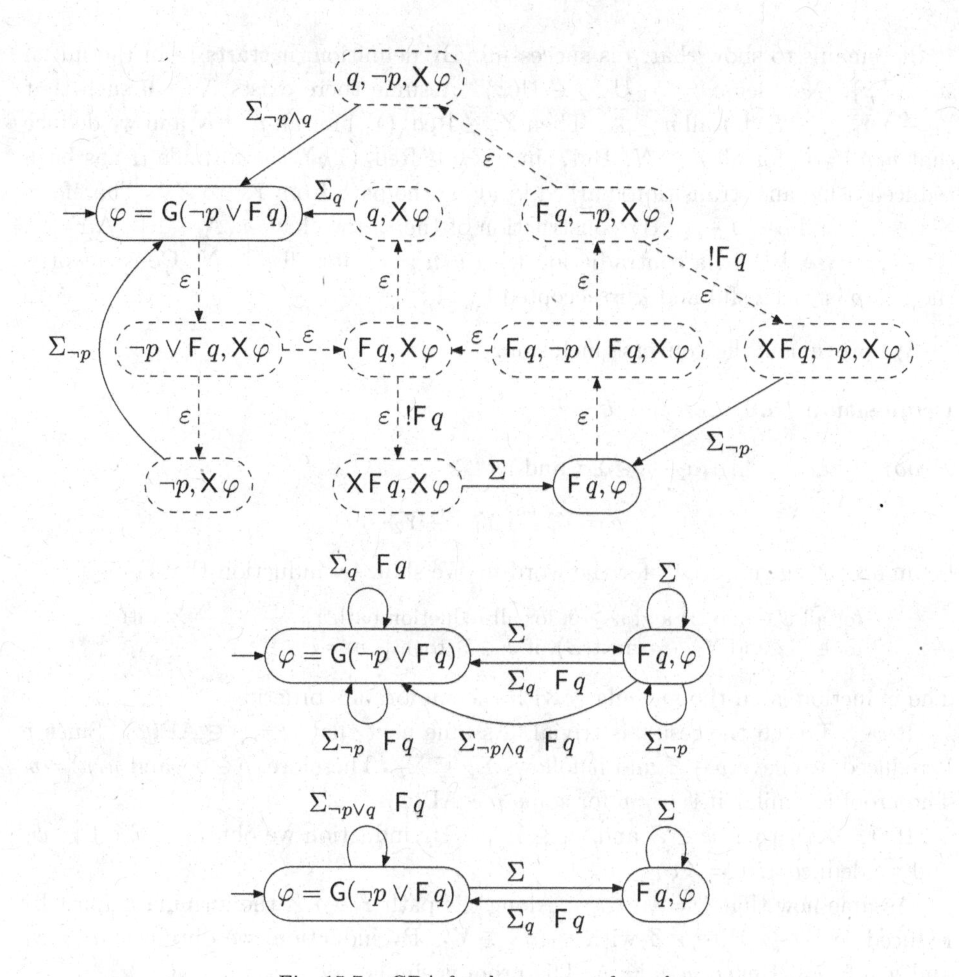

Fig. 15.7. GBA for the response formula.

Now, let $\alpha = \alpha_1 \mathsf{U}\, \alpha_2 \in \mathsf{U}(\varphi)$ be such that $u \models \alpha_2$. Either α is never reduced along this path and indeed $Z \in \mathrm{Red}_\alpha(Y)$ or α is reduced and by the hypothesis above we took the unmarked ε-transition. Hence $Z \in \mathrm{Red}_\alpha(Y)$. $\square$

Proposition 15.5. $\mathcal{L}(\varphi) \subseteq \mathcal{L}(\mathcal{A}_\varphi)$.

Proof. Let $u = a_0 a_1 a_2 \cdots \in \Sigma^\omega$ be such that $u \models \varphi$. By induction, we build

$$\rho = Y_0 \xrightarrow{a_0} Y_1 \xrightarrow{a_1} Y_2 \cdots$$

of $\mathcal{A}_\varphi$ such that for all $n \geq 0$ we have $u, n \models \bigwedge Y_n$ and there is some $Z_n \in \mathrm{Red}(Y_n)$ with $a_n \in \Sigma_{Z_n}$ and $Y_{n+1} = \mathrm{next}(Z_n)$. We start with $Y_0 = \{\varphi\}$. Assume now that $u, n \models \bigwedge Y_n$ for some $n \geq 0$. By Lemma 15.4, there is $Z_n \in \mathrm{Red}(Y_n)$ such that $u, n \models \bigwedge Z_n$ and for all until subformulae $\alpha = \alpha_1 \mathsf{U}\, \alpha_2 \in \mathsf{U}(\varphi)$, if $u, n \models \alpha_2$ then $Z_n \in \mathrm{Red}_\alpha(Y_n)$. Then we define $Y_{n+1} = \mathrm{next}(Z_n)$. Since $u, n \models \bigwedge Z_n$, Lemma 15.3 implies $a_n \in \Sigma_{Z_n}$ and $u, n+1 \models \bigwedge Y_{n+1}$. Therefore, ρ is a run for u in $\mathcal{A}_\varphi$.

It remains to show that ρ is successful. By definition, it starts from the initial state $\{\varphi\}$. Now let $\alpha = \alpha_1 \mathsf{U} \alpha_2 \in \mathsf{U}(\varphi)$. Assume there exists $N \geq 0$ such that $Y_n \xrightarrow{a_n} Y_{n+1} \notin T_\alpha$ for all $n \geq N$. Then $Z_n \notin \mathrm{Red}_\alpha(Y_n)$ for all $n \geq N$ and we deduce that $u, n \not\models \alpha_2$ for all $n \geq N$. But, since $Z_N \notin \mathrm{Red}_\alpha(Y_N)$, the formula α has been reduced using an ε-transition marked $!\alpha$ along the path from Y_N to Z_N. Therefore, $\mathsf{X}\alpha \in Z_N$ and $\alpha \in Y_{N+1}$. By construction of the run we have $u, N+1 \models \bigwedge Y_{N+1}$. Hence, $u, N+1 \models \alpha$, a contradiction with $u, n \not\models \alpha_2$ for all $n \geq N$. Consequently, the run ρ is successful and u is accepted by $\mathcal{A}_\varphi$. □

We prove now the converse inclusion.

Proposition 15.6. $\mathcal{L}(\mathcal{A}_\varphi) \subseteq \mathcal{L}(\varphi)$.

Proof. Let $u = a_0 a_1 a_2 \cdots \in \Sigma^\omega$ and let

$$\rho = Y_0 \xrightarrow{a_0} Y_1 \xrightarrow{a_1} Y_2 \cdots$$

be an accepting run of $\mathcal{A}_\varphi$ for the word u. We show by induction that

for all $\psi \in \mathrm{sub}(\varphi)$ and $n \geq 0$, for all reduction path $Y_n \xrightarrow[*]{\varepsilon} Y \xrightarrow[*]{\varepsilon} Z$ with $a_n \in \Sigma_Z$ and $Y_{n+1} = \mathrm{next}(Z)$, if $\psi \in Y$ then $u, n \models \psi$.

The induction is on the formula ψ with the subformula ordering.

If $\psi = \top$ then the result is trivial. Assume next that $\psi = p \in \mathrm{AP}(\varphi)$. Since p is reduced, we have $p \in Z$ and it follows $\Sigma_Z \subseteq \Sigma_p$. Therefore, $p \in a_n$ and $u, n \models p$. The proof is similar if $\psi = \neg p$ for some $p \in \mathrm{AP}(\varphi)$.

If $\psi = \mathsf{X}\psi_1$ then $\psi \in Z$ and $\psi_1 \in Y_{n+1}$. By induction we obtain $u, n+1 \models \psi_1$ and we deduce $u, n \models \mathsf{X}\psi_1 = \psi$.

Assume now that $\psi = \psi_1 \wedge \psi_2$. Along the path $Y \xrightarrow[*]{\varepsilon} Z$ the formula ψ must be reduced so $Y \xrightarrow[*]{\varepsilon} Y' \xrightarrow[*]{\varepsilon} Z$ with $\psi_1, \psi_2 \in Y'$. By induction, we obtain $u, n \models \psi_1$ and $u, n \models \psi_2$. Hence, $u, n \models \psi$. The proof is similar for $\psi = \psi_1 \vee \psi_2$.

Assume next that $\psi = \psi_1 \mathsf{U} \psi_2$. Along the path $Y \xrightarrow[*]{\varepsilon} Z$ the formula ψ must be reduced so $Y \xrightarrow[*]{\varepsilon} Y' \xrightarrow{\varepsilon} Y'' \xrightarrow[*]{\varepsilon} Z$ with either $Y'' = Y' \setminus \{\psi\} \cup \{\psi_2\}$ or $Y'' = Y' \setminus \{\psi\} \cup \{\psi_1, \mathsf{X}\psi\}$. In the first case, we obtain by induction $u, n \models \psi_2$ and therefore $u, n \models \psi$. In the second case, we obtain by induction $u, n \models \psi_1$. Since $\mathsf{X}\psi$ is reduced we get $\mathsf{X}\psi \in Z$ and $\psi \in \mathrm{next}(Z) = Y_{n+1}$.

Let $k > n$ be minimal such that $Y_k \xrightarrow{a_k} Y_{k+1} \in T_\psi$ (such a value k exists since ρ is accepting). We first show by induction that $u, i \models \psi_1$ and $\psi \in Y_{i+1}$ for all $n \leq i < k$. Recall that $u, n \models \psi_1$ and $\psi \in Y_{n+1}$. So let $n < i < k$ be such that $\psi \in Y_i$. Let $Z' \in \mathrm{Red}(Y_i)$ be such that $a_i \in \Sigma_{Z'}$ and $Y_{i+1} = \mathrm{next}(Z')$. Since k is minimal we know that $Z' \notin \mathrm{Red}_\psi(Y_i)$. Hence, along any reduction path from Y_i to Z' we must use a step $Y' \xrightarrow[!\psi]{\varepsilon} Y' \setminus \{\psi\} \cup \{\psi_1, \mathsf{X}\psi\}$. By induction on the formula we obtain $u, i \models \psi_1$. Also, since $\mathsf{X}\psi$ is reduced, we have $\mathsf{X}\psi \in Z'$ and $\psi \in \mathrm{next}(Z') = Y_{i+1}$.

Second, we show that $u, k \models \psi_2$. Since $Y_k \xrightarrow{a_k} Y_{k+1} \in T_\psi$, we find some $Z' \in \mathrm{Red}_\psi(Y_k)$ such that $a_k \in \Sigma_{Z'}$ and $Y_{k+1} = \mathrm{next}(Z')$. Since $\psi \in Y_k$, along some

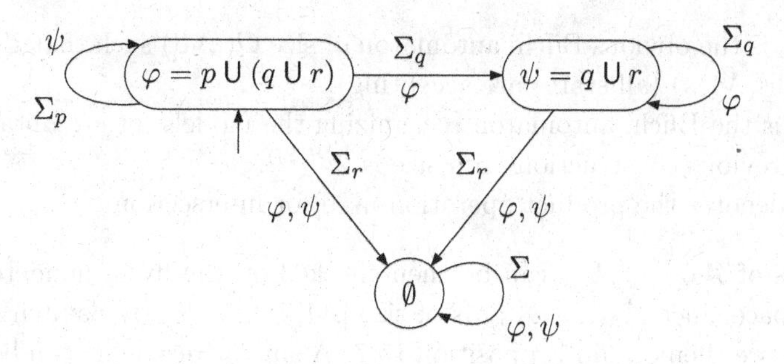

Fig. 15.8. GBA for nested until.

reduction path from Y_k to Z' we use a step $Y' \xrightarrow{\varepsilon} Y' \setminus \{\psi\} \cup \{\psi_2\}$. By induction we obtain $u, k \models \psi_2$. Finally, we have shown $u, n \models \psi_1 \,\mathsf{U}\, \psi_2 = \psi$.

The last case is when $\psi = \psi_1 \,\mathsf{R}\, \psi_2$. Along the path $Y \xrightarrow[*]{\varepsilon} Z$ the formula ψ must be reduced so $Y \xrightarrow[*]{\varepsilon} Y' \xrightarrow{\varepsilon} Y'' \xrightarrow[*]{\varepsilon} Z$ with either $Y'' = Y' \setminus \{\psi\} \cup \{\psi_1, \psi_2\}$ or $Y'' = Y' \setminus \{\psi\} \cup \{\psi_2, \mathsf{X}\,\psi\}$. In the first case, we obtain by induction $u, n \models \psi_1$ and $u, n \models \psi_2$. Hence, $u, n \models \psi$ and we are done. In the second case, we obtain by induction $u, n \models \psi_2$ and we get also $\psi \in Y_{n+1}$. Continuing with the same reasoning, we deduce easily that either $u, n \models \mathsf{G}\,\psi_2$ or $u, n \models \psi_2 \,\mathsf{U}\, (\psi_1 \wedge \psi_2)$. □

We have proved the correctness of our construction. We give now more examples and discuss simplifications that may be applied during the construction. First, consider $\varphi = p \,\mathsf{U}\, (q \,\mathsf{U}\, r)$. Here we have two until formulae, φ itself and $\psi = q \,\mathsf{U}\, r$, hence the GBA will have two acceptance sets T_φ and T_ψ. We can easily check that

$$\begin{aligned}
\mathrm{Red}(\{\varphi\}) &= \{\{p, \mathsf{X}\,\varphi\}, \{q, \mathsf{X}\,\psi\}, \{r\}\} \\
\mathrm{Red}_\varphi(\{\varphi\}) &= \{\{q, \mathsf{X}\,\psi\}, \{r\}\} \\
\mathrm{Red}_\psi(\{\varphi\}) &= \{\{p, \mathsf{X}\,\varphi\}, \{r\}\}
\end{aligned}$$

Hence, starting from the initial state $\{\varphi\}$, the construction introduces two new states $\{\psi\}$ and $\emptyset$. We compute

$$\begin{aligned}
\mathrm{Red}(\{\psi\}) &= \{\{q, \mathsf{X}\,\psi\}, \{r\}\} \\
\mathrm{Red}_\varphi(\{\psi\}) &= \{\{q, \mathsf{X}\,\psi\}, \{r\}\} \\
\mathrm{Red}_\psi(\{\psi\}) &= \{\{r\}\}
\end{aligned}$$

There are no new states, so the construction terminates and we obtain the GBA of Figure 15.8 where the transitions from T_φ and T_ψ are marked φ and ψ respectively.

The polynomial space upper bound for LTL model-checking can be then stated as follows.

Proposition 15.7. *[41] Given a finite and total Kripke structure $\mathcal{M}$, a state s in $\mathcal{M}$ and an LTL formula φ, it is possible to check in space polynomial in $|\varphi| + \log |\mathcal{M}|$ whether $\mathcal{M}, s \models_\forall \varphi$ and $\mathcal{M}, s \models_\exists \varphi$.*

Indeed, $\mathcal{M}, s \models_\forall \varphi$ holds if and only if $\mathrm{L}(\mathcal{A}_{\mathcal{M},s} \otimes \mathcal{A}_{\neg\varphi}) = \emptyset$ where

- $\mathcal{A}_{\mathcal{M},s}$ is the obvious Büchi automaton of size $\mathcal{O}(|\mathcal{M}|)$ such that $\mathcal{L}(\mathcal{A}_{\mathcal{M},s}) = \lambda\text{Paths}(\mathcal{M}, s)$ (all states are accepting),
- $\mathcal{A}_{\neg\varphi}$ is the Büchi automaton recognizing the models for $\neg\varphi$ obtained with the previous constructions. Its size is $2^{\mathcal{O}(|\varphi|)}$.
- "$\otimes$" denotes the product operation used for intersection.

Nonemptiness of $\mathcal{A}_{\mathcal{M},s} \otimes \mathcal{A}_{\neg\varphi}$ can be then checked on the fly in nondeterministic polynomial space since $\mathcal{A}_{\mathcal{M},s} \otimes \mathcal{A}_{\neg\varphi}$ is of size $|\mathcal{M}| \times 2^{\mathcal{O}(|\varphi|)}$. By Savitch's theorem (see e.g. [47]), we then obtain Proposition 15.7. A similar reasoning can be done for existential model checking since $\mathcal{M}, s \models_{\exists} \varphi$ holds if and only if $\text{L}(\mathcal{A}_{\mathcal{M},s} \otimes \mathcal{A}_{\varphi}) \neq \emptyset$. Furthermore, the properties about the construction of $\mathcal{A}_{\varphi}$ allow also to get the polynomial space upper bound for satisfiability and validity.

Proposition 15.8. *[35] Checking whether an LTL formula φ is satisfiable (or valid) can be done in space polynomial in $|\varphi|$.*

15.3.6. *On the fly simplifications of the GBA*

One optimization was already included in the contruction: when reducing a set Y of formulae, we start with *maximal* formulae. This strategy produces fewer ε-transitions and fewer states in the set $\text{Red}(Y)$. For instance, assume that $Y = \{\varphi, \mathsf{G}\,\varphi\}$. If we reduce first the maximal formula $\mathsf{G}\,\varphi$ we obtain $Y' = \{\varphi, \mathsf{X}\,\mathsf{G}\,\varphi\}$ and it remains to reduce φ. We obtain the sets $Z \cup \{\mathsf{X}\,\mathsf{G}\,\varphi\}$ for $Z \in \text{Red}(\{\varphi\})$. If instead we start by reducing φ we obtain the sets $Z \cup \{\mathsf{G}\,\varphi\}$ for $Z \in \text{Red}(\{\varphi\}$. But then $\mathsf{G}\,\varphi$ has to be reduced so that we obtain the sets $Z \cup \{\varphi, \mathsf{X}\,\mathsf{G}\,\varphi\}$ and φ has to be reduced again. We obtain finally sets $Z_1 \cup Z_2 \cup \{\mathsf{X}\,\mathsf{G}\,\varphi\}$ for $Z_1, Z_2 \in \text{Red}(\{\varphi\})$.

Next, during the reduction, we may replace a set Y by Y' provided they are *equivalent*, i.e., $\bigwedge Y \equiv \bigwedge Y'$. Checking equivalence is as hard as constructing the automaton, so we only use easy syntactic equivalences. For instance, we may use the following rules:

$$\begin{array}{ll} \text{If } \psi = \psi_1 \vee \psi_2 \text{ and } \psi_1 \in Y \text{ or } \psi_2 \in Y: & Y \xrightarrow{\varepsilon} Y \setminus \{\psi\} \\ \text{If } \psi = \psi_1 \,\mathsf{U}\, \psi_2 \text{ and } \psi_2 \in Y: & Y \xrightarrow{\varepsilon} Y \setminus \{\psi\} \\ \text{If } \psi = \psi_1 \,\mathsf{R}\, \psi_2 \text{ and } \psi_1 \in Y: & Y \xrightarrow{\varepsilon} Y \setminus \{\psi\} \cup \{\psi_2\} \end{array}$$

We explain now an easy and useful simplification of the constructed GBA: when two states have the same outgoing transitions, then they can be merged. More precisely, two states s_1 and s_2 of a GBA $\mathcal{A} = (Q, \Sigma, I, T, T_1, \ldots, T_n)$ have the same outgoing transitions if for all $a \in \Sigma$ and $s \in Q$, we have

$$\begin{array}{lll} & (s_1, a, s) \in T \iff (s_2, a, s) \in T & \\ \text{and} & (s_1, a, s) \in T_i \iff (s_2, a, s) \in T_i & \text{for all } 1 \leq i \leq n. \end{array}$$

In this case, the two states s_1 and s_2 can be merged without changing the accepted language. When merging these states, we redirect all transitions to either s_1 or s_2 to the new merged state $s_{1,2}$.

With our construction, we have an easy sufficient condition ensuring that two states Y and Y' have the same outgoing transitions:

$$\begin{cases} \mathrm{Red}(Y) = \mathrm{Red}(Y') & \text{and} \\ \mathrm{Red}_\alpha(Y) = \mathrm{Red}_\alpha(Y') & \text{for all } \alpha \in \mathsf{U}(\varphi) \end{cases} \tag{15.1}$$

For instance, consider the formula $\varphi = \mathsf{G}\,\mathsf{F}\,p$. We have

$$\begin{aligned} \mathrm{Red}(\{\varphi\}) &= \{\{p, \mathsf{X}\,\varphi\}, \{\mathsf{X}\,\mathsf{F}\,p, \mathsf{X}\,\varphi\}\} \\ \mathrm{Red}_{\mathsf{F}\,p}(\{\varphi\}) &= \{\{p, \mathsf{X}\,\varphi\}\} \end{aligned}$$

Hence, from the initial state $\{\varphi\}$, we reach a new state $\{\mathsf{F}\,p, \varphi\}$. We can easily check that the two states satisfy (15.1) hence they can be merged and the resulting GBA has only one state and two transitions:

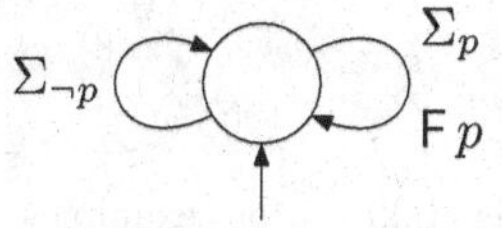

Similarly, if we consider $\varphi = \mathsf{G}\,\mathsf{F}\,p \wedge \mathsf{G}\,\mathsf{F}\,q$. We have

$$\begin{aligned} \mathrm{Red}(\{\varphi\}) = \{&\{p, q, \mathsf{X}\,\mathsf{G}\,\mathsf{F}\,p, \mathsf{X}\,\mathsf{G}\,\mathsf{F}\,q\}, \{p, \mathsf{X}\,\mathsf{F}\,q, \mathsf{X}\,\mathsf{G}\,\mathsf{F}\,p, \mathsf{X}\,\mathsf{G}\,\mathsf{F}\,q\}, \\ &\{q, \mathsf{X}\,\mathsf{F}\,p, \mathsf{X}\,\mathsf{G}\,\mathsf{F}\,p, \mathsf{X}\,\mathsf{G}\,\mathsf{F}\,q\}, \{\mathsf{X}\,\mathsf{F}\,p, \mathsf{X}\,\mathsf{F}\,q, \mathsf{X}\,\mathsf{G}\,\mathsf{F}\,p, \mathsf{X}\,\mathsf{G}\,\mathsf{F}\,q\}\} \\ \mathrm{Red}_{\mathsf{F}\,p}(\{\varphi\}) = \{&\{p, q, \mathsf{X}\,\mathsf{G}\,\mathsf{F}\,p, \mathsf{X}\,\mathsf{G}\,\mathsf{F}\,q\}, \{p, \mathsf{X}\,\mathsf{F}\,q, \mathsf{X}\,\mathsf{G}\,\mathsf{F}\,p, \mathsf{X}\,\mathsf{G}\,\mathsf{F}\,q\}\} \\ \mathrm{Red}_{\mathsf{F}\,q}(\{\varphi\}) = \{&\{p, q, \mathsf{X}\,\mathsf{G}\,\mathsf{F}\,p, \mathsf{X}\,\mathsf{G}\,\mathsf{F}\,q\}, \{q, \mathsf{X}\,\mathsf{F}\,p, \mathsf{X}\,\mathsf{G}\,\mathsf{F}\,p, \mathsf{X}\,\mathsf{G}\,\mathsf{F}\,q\}\} \end{aligned}$$

Hence, a direct application of the construction produces an automaton with 4 new states. However, we can easily check with (15.1) that all states have the same outgoing transitions. Hence, again, the resulting GBA has only one state and 4 transitions which are marked $\mathsf{F}\,p$ or $\mathsf{F}\,q$ if they belong to $T_{\mathsf{F}\,p}$ or $T_{\mathsf{F}\,q}$:

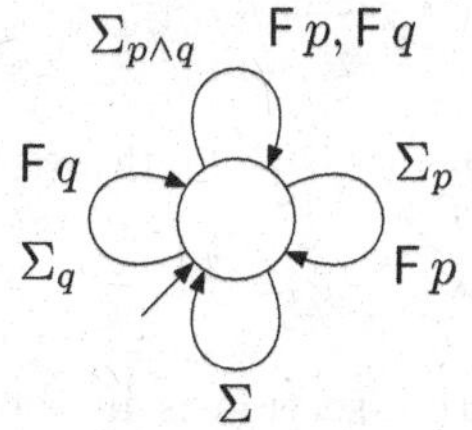

More examples are given in Figure 15.9.

Other optimizations can be found in [14, 16], as well as simplifications of Büchi automata generated from LTL formulae in [17].

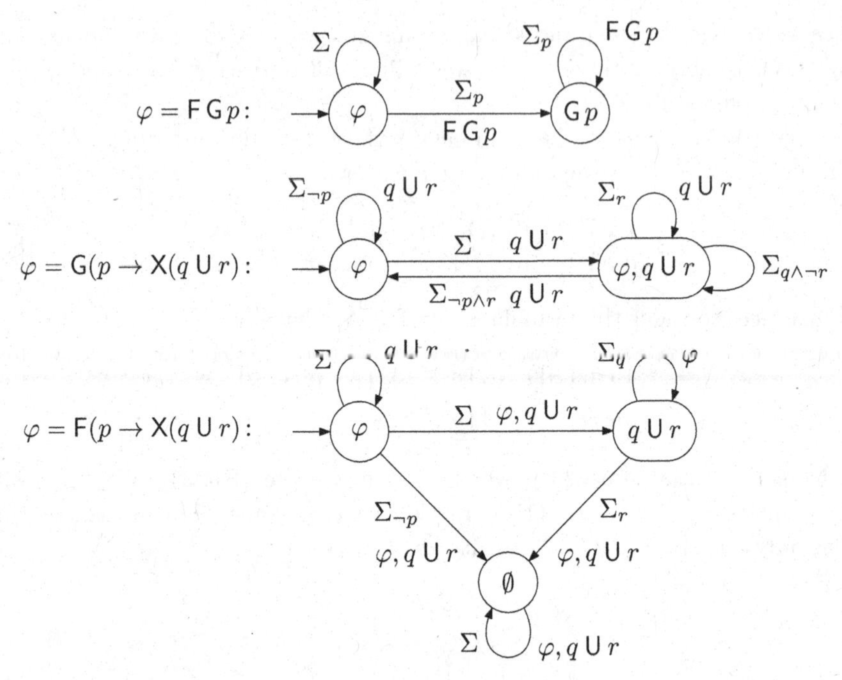

Fig. 15.9. More examples.

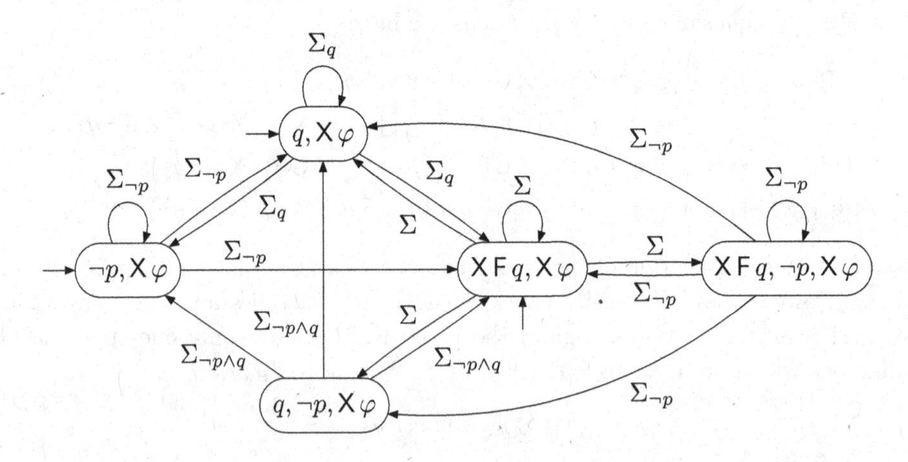

Fig. 15.10. Another GBA for the response formula.

15.3.7. *Related work*

The construction presented in this section is in the same vein as those presented in [12, 14, 16, 48], see also [20]. For instance, the states of the automata in [16] correspond to our reduced sets and the transitions are constructed by applying first the *next* step and then the *reduction* phase. For the response formula $\mathsf{G}(p \to \mathsf{F}\,q)$ there are 5 reduced sets (see Figure 15.7) and we would get the automaton of

Figure 15.10. Note that, except possibly for the initial state, each of our states is the next state of some reduced set. But we usually have several reduced sets having the same next set. Hence, our construction cannot yield more states (still apart possibly for the initial state) and usually yield fewer states than the construction of [16] as in the example above. Another difference is that [16] uses acceptance conditions based on states whereas we use transition-based acceptance.

Moreover, tableau methods for LTL, see e.g. [15], contain decomposition rules similar to the reduction rules and contain an additional step to check global conditions about eventuality formulae (in $\mathsf{U}(\varphi)$). Our construction is therefore similar to such methods since the expansion of a state (or branch in the tableau terminology) is done on demand and the verification of eventualities is simply performed by transition-based acceptance conditions. Hence, the two phases of the method in [15] apply also herein and in the worst-case we also obtain exponential-size automata. It is worth recalling that a *one-pass* tableaux calculus for LTL is presented in [49] by using additional control structures (no step to check global conditions). Finally, there is another solution to encode the second phase of the method in [15], which is to translate LTL model-checking into CTL model-checking with fairness conditions [50].

Helpful bibliographical remarks can also be found at the end of [48, Chapter 5] as well as in [20].

15.4. Extensions

In this section, we consider three extensions for developments made in Section 15.3. Firstly, we show how a procedure solving LTL model-checking can be used to solve CTL* model-checking. Secondly, we present an extension of LTL with temporal operators defined from finite-state automata. Thirdly, we show that adding a single temporal operator defined by a context-free language leads to undecidability.

15.4.1. *Model-checking for branching-time* CTL*

Even though CTL* is a branching-time logic, $\mathrm{MC}^{\forall}(\mathrm{CTL}^*)$ can be solved by using as subroutine the algorithm for LTL model-checking with a simple renaming technique.

Proposition 15.9. *[34]* $\mathrm{MC}^{\forall}(\mathrm{CTL}^*)$ *is* PSPACE*-complete.*

Proof. Since $\mathrm{MC}^{\forall}(\mathrm{LTL})$ is a subproblem of $\mathrm{MC}^{\forall}(\mathrm{CTL}^*)$, the PSPACE-hardness is immediate. In order to show that $\mathrm{MC}^{\forall}(\mathrm{CTL}^*)$ is in PSPACE, we use known techniques for LTL plus renaming.

For each quantifier $\mathcal{Q} \in \{\exists, \forall\}$, we write $\mathrm{MC}^{\mathcal{Q}}_{\mathrm{LTL}}(\mathcal{M}, s, \varphi)$ to denote the function that returns `true` if and only if $\mathcal{M}, s \models_{\mathcal{Q}} \varphi$. We have seen that these functions can be computed in polynomial space in $|\varphi| + \log(|\mathcal{M}|)$ (Proposition 15.7). In order to establish the PSPACE upper bound, here is an algorithm based on formulae renaming using only polynomial space:

$\mathrm{MC}^{\forall}_{\mathrm{CTL}^*}(\mathcal{M} = \langle W, R, \lambda\rangle, s \in W, \varphi)$

- If E, A do not occur in φ, then return $\mathrm{MC}^{\forall}_{\mathrm{LTL}}(\mathcal{M}, s, \varphi)$.
- Otherwise φ contains a subformula $\mathcal{Q}\psi$ where E, A do not occur in ψ and $\mathcal{Q} \in \{\mathsf{E}, \mathsf{A}\}$. This means that the formula ψ belongs to LTL. Let $\mathcal{Q}'$ be "$\forall$" if $\mathcal{Q} = \mathsf{A}$, "$\exists$" otherwise. Let $p_{\mathcal{Q}\psi}$ be a new propositional variable. We define λ' an extension of λ for every $s' \in W$ by:

$$\lambda'(s') = \begin{cases} \lambda(s') \cup \{p_{\mathcal{Q}\psi}\} & \text{if } \mathrm{MC}^{\mathcal{Q}'}_{\mathrm{LTL}}(\mathcal{M}, s', \psi) \\ \lambda(s') & \text{otherwise.} \end{cases}$$

  Return $\mathrm{MC}^{\forall}_{\mathrm{CTL}^*}(\langle W, R, \lambda'\rangle, s, \varphi[\mathcal{Q}\psi \leftarrow p_{\mathcal{Q}\psi}])$ where $\varphi[\mathcal{Q}\psi \leftarrow p_{\mathcal{Q}\psi}]$ is obtained from φ by replacing every occurrence of $\mathcal{Q}\psi$ by $p_{\mathcal{Q}\psi}$.

Since in $\mathrm{MC}^{\forall}_{\mathrm{CTL}^*}(\mathcal{M}, s, \varphi)$, the recursion depth is at most $|\varphi|$, we can show that $\mathrm{MC}^{\forall}_{\mathrm{CTL}^*}$ uses only polynomial space since $\mathrm{MC}^{\exists}_{\mathrm{LTL}}$ and $\mathrm{MC}^{\forall}_{\mathrm{LTL}}$ require only polynomial space. The soundness of the algorithm is not very difficult to show. □

15.4.2. *Automata-based temporal operators*

We have seen how to build a GBA $\mathcal{A}_\varphi$ such that $\mathcal{L}(\mathcal{A}_\varphi)$ is equal to the set of models for φ (in LTL). However, it is known that LTL is strictly less expressive than Büchi automata [21]. It is not always easy to figure out whether a given Büchi automaton on the alphabet $2^{\mathrm{AP}(\varphi)}$ corresponds to an LTL formula where $\mathrm{AP}(\varphi)$ denotes the set of propositional variables occurring in φ. For instance, is there an LTL formula, counterpart of the automaton presented below?

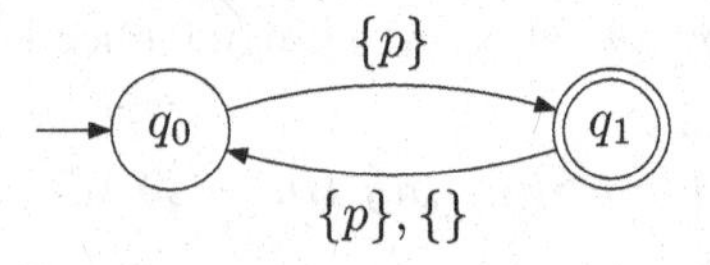

We invite the reader to analyze why none of the formulae below is adequate:

(1) $p \wedge \mathsf{X}\neg p \wedge \mathsf{G}(p \leftrightarrow \mathsf{X}\mathsf{X}p)$,
(2) $p \wedge \mathsf{G}(p \rightarrow \mathsf{X}\mathsf{X}p)$,
(3) $q \wedge \mathsf{X}\neg q \wedge \mathsf{G}(q \leftrightarrow \mathsf{X}\mathsf{X}q) \wedge \mathsf{G}(q \rightarrow p)$.

For instance, formula (1) defines a unique model over the two-letters alphabet $2^{\{p\}}$. Note that formula (3) requires that p holds at all even positions but uses an extra propositional variable which is not allowed by the alphabet of the automaton. Actually, there exist quite simple properties that cannot be expressed with LTL.

To check whether an ω-regular language L can be expressed in LTL, one may compute the *syntactic monoid* of L from the automaton which recognizes L and check that this monoid is *aperiodic*, see e.g. the survey chapter [51]. This procedure can be applied to prove the following result (even though the proof in [21] is different).

Proposition 15.10. *[21] There is no* LTL *formula φ built over the unique propositional variable p such that $\mathcal{L}(\varphi)$ is exactly the set of* LTL *models such that p holds on every even position (on odd positions, p may be true or not).*

That is why, in [21], an extension of LTL has been introduced by adding temporal operators defined with finite-state automata. Alternatively, right-linear grammars can also be used to define regular languages. Let $\mathcal{A} = \langle \Sigma, S, S_0, \rho, F \rangle$ be a finite-state automaton with letters from a linearly ordered alphabet Σ, say with the ordering $a_1 < \ldots < a_k$. Assume that we already have defined the formulae $\varphi_1, \ldots, \varphi_k$. Then, $\mathcal{A}(\varphi_1, \ldots, \varphi_k)$ is a new formula in the *Extended Temporal Logic* (ETL). The relation $u, i \models \mathcal{A}(\varphi_1, \ldots, \varphi_k)$ holds when a finite pattern induced from $\mathcal{L}(\mathcal{A})$ exists from position i. There is a correspondence between the letters $a_1, \ldots, a_k$ and the arguments $\varphi_1, \ldots, \varphi_k$. More precisely $u, i \models \mathcal{A}(\varphi_1, \ldots, \varphi_k)$ if there is a finite word $a_{i_1} a_{i_2} \ldots a_{i_n} \in \mathcal{L}(\mathcal{A})$ such that for every $1 \leq j \leq n$, we have $u, i + (j-1) \models \varphi_{i_j}$.

Note that, if $S_0 \cap F \neq \emptyset$, then $\varepsilon \in \mathcal{L}(\mathcal{A})$ and $\mathcal{A}(\varphi_1, \ldots, \varphi_k)$ is equivalent to $\top$. Observe also that in the condition above, the index of the k-th letter (with $k \in \{1, \ldots, n\}$) determines which argument must hold at the $(k-1)$-th next position. We present below a model for the ETL formula $\mathcal{A}(p, q)$ with $\mathcal{L}(\mathcal{A}) = \{ab^i a \mid i \geq 0\}$ and $a < b$.

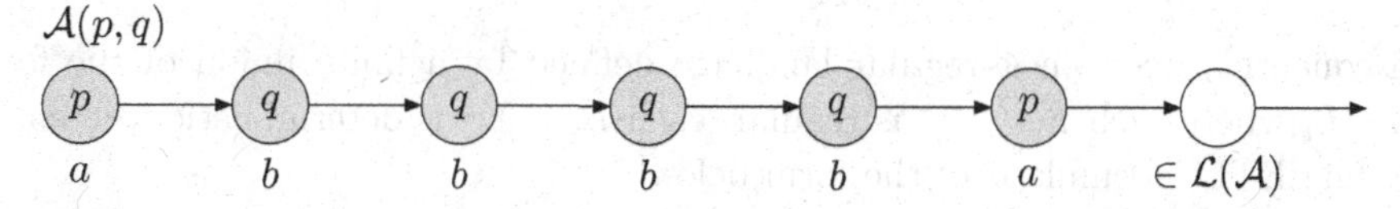

By way of example, the formula $\varphi \mathsf{U} \psi$ is equivalent to $\mathcal{B}(\varphi, \psi)$ with $\mathcal{L}(\mathcal{B}) = a^* b$ and $a < b$. Similarly, the *weakness* of LTL described in Proposition 15.10 can be fixed within ETL: the formula $\neg\mathcal{A}(\top, \neg p)$ with $\mathcal{L}(\mathcal{A}) = (a^2)^* b$ holds exactly in models such that the propositional variable p holds on every even position.

Formally, the syntax of ETL allows the propositional variables, the boolean connectives and temporal modalities of the form $\mathcal{A}(\varphi_1, \ldots, \varphi_k)$. Note that it does not include the temporal operators from LTL since they can be expressed with automata.

In order to illustrate the expressive power of ETL, it is sufficient to consider the fragment ETL^- defined by the syntax below:

$$\varphi ::= \top \mid \neg\varphi \mid \varphi \vee \varphi \mid K \cdot \varphi$$

where $K \subseteq \Sigma^*$ is a regular language of finite words over some finite alphabet $\Sigma \subseteq 2^{\text{AP}}$ such that each letter $a \in \Sigma$ is a finite set. The semantics is that $w \models K \cdot \varphi$ if we can write $w = uv$ with $u \in K$ and $v \models \varphi$. To show that $K \cdot \varphi$ can be expressed in ETL, consider an automaton $\mathcal{A}$ for $K \cdot \#$ where $\#$ is a new letter (larger than all letters from Σ). For each $a \in \Sigma$, let $\varphi_a = \bigwedge_{p \in a} p \wedge \bigwedge_{p \notin a} \neg p$. Then, $K \cdot \varphi$ is expressed by $\mathcal{A}((\varphi_a)_{a \in \Sigma}, \varphi)$ where φ_a is substituted for $a \in \Sigma$ and φ is substituted for the trailing #.

Lemma 15.11. *For any ω-regular language L over a finite alphabet $\Sigma \subseteq 2^{\mathrm{AP}}$ made of finite letters, there is a formula φ in* ETL$^-$ *such that $L = \mathcal{L}(\varphi)$.*

Proof. First, observe that any ω-regular language L over Σ can be written as a finite union of languages of the form $K \cdot L$ where $K \subseteq \Sigma^*$ is regular and $L \subseteq \Sigma^\omega$ is deterministic, i.e., recognized by a deterministic Büchi automaton. This can be easily derived from a deterministic Muller automaton recognizing L.

So consider a deterministic and complete Büchi automaton $\mathcal{A} = (Q, \Sigma, \{i\}, \delta, F)$. For each $s \in Q$, we define

$$M^s = \{u \in \Sigma^* \mid \delta(i, u) = s\}$$
$$N^s = \{v \in \Sigma^* \mid \delta(s, v) \in F\}.$$

Then, we can show

$$\overline{\mathcal{L}(\mathcal{A})} = \bigcup_{s \in Q} M^s \cdot \overline{N^s \cdot \Sigma^\omega}$$

We deduce immediately that $\mathcal{L}(\mathcal{A})$ can be expressed with the formula

$$\neg \bigvee_{s \in Q} M^s \cdot \neg(N^s \cdot \top)$$

Consequently, given an ω-regular language defined by a finite union of the form $\bigcup K_i \cdot L_i$ where each $K_i \subseteq \Sigma^*$ is regular, each $L_i \subseteq \Sigma^\omega$ is deterministic, the corresponding ETL$^-$ formula is of the form below:

$$\bigvee_i K_i \cdot \left(\neg \bigvee_{s \in Q_i} M_i^s \cdot \neg(N_i^s \cdot \top)\right)$$

□

As a corollary, any ω-regular language can be defined by an expression obtained from the grammar $\mathcal{L} ::= \Sigma^\omega \mid \mathcal{L} \cup \mathcal{L} \mid \overline{\mathcal{L}} \mid K \cdot \mathcal{L}$, where K ranges over the regular languages in Σ^*.

Even though ETL formulae are seldom used in specification languages, its main theoretical assets rest on its high expressive power and on the relatively low complexity of satisfiability/model-checking problems as stated below.

Proposition 15.12. *[9]*

(I) MC$^\forall$(ETL), MC$^\exists$(ETL) *and* SAT(ETL) *are* PSPACE-*complete.*
(II) ETL *has the same expressive power as Büchi automata.*

An automata-based construction for ETL formulae can be found in [21], leading to Proposition 15.12(I). Proposition 15.12(II) is a corollary of Lemma 15.11.

Proposition 15.12(I) entails that ETL model-checking is not more difficult than LTL model-checking in the worst case, modulo logarithmic space many-one reductions. This is quite surprising in view of the expressive power of ETL – Proposition 15.12(II). Hence, the class of languages defined by ETL formulae is equal to the class of languages defined by

- Büchi automata (Proposition 15.12(II)),
- formulae from monadic second-order theory for $\langle \omega, < \rangle$ (S1S) and ω-regular expressions (finite unions of sets UV^ω with regular $U, V \subseteq \Sigma^*$), see e.g. [52, Chapter III],
- formulae from LTL with second-order quantification. In such an extension of LTL, we allow formulae of the form $\forall p \,.\, \varphi$ with $u, i \models \forall p \,.\, \varphi$ if for every u' such that u and u' agree on all propositional variables different from p, we have $u', i \models \varphi$, see e.g. [53].
- formulae from LTL with fixed-point operators [54].

So, ETL is a powerful extension of LTL but the above equivalences do not mean that all the above formalisms have the same conciseness. Actually, we know the following complexity results:

- the nonemptiness problem for Büchi automata is NLogSpace-complete,
- $MC^{\forall}$(ETL), $MC^{\exists}$(ETL) and SAT(ETL) are PSpace-complete,
- satisfiability for LTL with fixed-point operators is PSpace-complete [54],
- satisfiability for S1S is non-elementary (time complexity is not bounded by any tower of exponential of fixed height) [43].

So, S1S is the most concise language for describing ω-regular languages.

15.4.3. *Context-free extensions*

It is possible to extend the definition of ETL by replacing formulae of the form $\mathcal{A}(\varphi_1, \ldots, \varphi_n)$ by formulae of the form $\mathrm{L}(\varphi_1, \ldots, \varphi_n)$ where L is a language of finite words specified within a fixed formalism. The language L is again viewed as a set of patterns, not necessarily regular. For a class $\mathcal{C}$ of languages, we write PC[$\mathcal{C}$] to denote the extension the propositional calculus with formulae of the form $\mathrm{L}(\varphi_1, \ldots, \varphi_n)$ for some $\mathrm{L} \in \mathcal{C}$. Obviously, ETL is precisely equivalent to PC[REG] where REG is the class of regular languages represented by finite-state automata. We have seen that ETL is decidable and it is natural to wonder whether PC[CF] is also decidable where CF is the class of context-free languages (represented by context-free grammars).

15.4.3.1. *Undecidability of* PC[CF]

Since numerous problems for context-free languages are undecidable, it is not very surprising to get the following result.

Proposition 15.13. SAT(PC[CF]) *is undecidable.*

Before presenting the proof, let us recall that the next operator X and the until operator U (and the derived operators F and G) can be defined as operators obtained from finite-state automata. Hence, in the proof below, we use them freely.

Proof. We show that the validity problem for PC[CF] is undecidable, which entails the undecidability of SAT(PC[CF]) since PC[CF] is closed under negations. We reduce the universality problem for context-free grammars (see e.g. the textbook [55]) into the validity problem. Let G be a context-free grammar over the terminal alphabet $\Sigma = \{a_1, \ldots, a_n\}$. We write G^+ to denote the CF grammar over the terminal alphabet $\Sigma^+ = \{a_1, \ldots, a_n, a_{n+1}\}$ such that $\mathcal{L}(G^+) = \mathcal{L}(G) \cdot \{a_{n+1}\}$. The letter a_{n+1} is simply an end marker. G^+ can be effectively computed from G.

Let UNI be the formula

$$(\neg p_{n+1} \,\mathsf{U}\, (p_{n+1} \wedge \mathsf{X}\,\mathsf{G}\,\neg p_{n+1})) \wedge \mathsf{G} \bigvee_{1 \leq i \leq n+1} \left(p_i \wedge \bigwedge_{1 \leq j \leq n+1, j \neq i} \neg p_j \right)$$

The structures satisfying UNI are precisely those for which exactly one variable from $p_1, \ldots, p_{n+1}$ holds at each state and, p_{n+1} holds at a unique state of the model. Hence, we characterize structures that can be naturally viewed as finite words, possibly in $\mathcal{L}(G^+)$. We show that $\mathcal{L}(G) = \Sigma^*$ if and only if UNI $\rightarrow \mathcal{L}(G^+)$ is valid.

Indeed, if $\mathcal{L}(G) \neq \Sigma^*$, say $a_{i_1} a_{i_2} \cdots a_{i_\ell} \notin \mathcal{L}(G)$. Let u be the model $\{p_{i_1}\} \cdot \{p_{i_2}\} \cdots \{p_{i_\ell}\} \cdot \{p_{n+1}\} \cdot \{p_1\}^\omega$. We have $u \models \text{UNI} \wedge \neg\mathcal{L}(G^+)(p_1, \ldots, p_{n+1})$. So UNI $\rightarrow \mathcal{L}(G^+)$ is not valid. Conversely, it is easy to show that $\mathcal{L}(G) = \Sigma^*$ implies UNI $\rightarrow \mathcal{L}(G^+)$ is valid since every structure satisfying the formula UNI corresponds to a word in Σ^*. □

Proposition 15.13 is interesting, but after all, it rests on the fact that PC[CF] can easily encode universality for context-free grammars. It would be more interesting to establish that undecidability still holds for a very small fragment of CF, which is the subject of the next section.

15.4.3.2. *When a single context-free language leads to high undecidability*

The main result of this section is the (high) undecidability of the model-checking problem for PC[L$_1$] for the context-free language $\mathrm{L}_1 = \{a_1^k \cdot a_2 \cdot a_1^k \cdot a_3 \mid k \geq 0\}$. We start by introducing the auxiliary language $\mathrm{L}_0 = \{a_1^k \cdot a_2 \cdot a_1^{k-1} \cdot a_3 \mid k \geq 1\} = a_1 \cdot \mathrm{L}_1$. The next operator X, the eventuality operator F and the temporal operator defined from L$_0$ are definable in PC[L$_1$] thanks to the following equivalences:

- $\mathsf{X}\varphi \equiv \mathrm{L}_1(\perp, \top, \varphi)$,
- $\mathsf{F}\varphi \equiv \mathrm{L}_1(\top, \varphi, \top)$,
- $\mathrm{L}_0(\varphi_1, \varphi_2, \varphi_3) \equiv \varphi_1 \wedge \mathsf{X}\,\mathrm{L}_1(\varphi_1, \varphi_2, \varphi_3)$ since $\mathrm{L}_0 = a_1 \cdot \mathrm{L}_1$.

In the sequel, we therefore freely use these operators in PC[L_1] as well as the dual operator G.

Proposition 15.14. *Satisfiability for* PC[L_1] *is undecidable.*

Proof. We reduce the recurring domino problem DOMREC [56] to satisfiability for PC[L_1]. Let Sides $= \{left, right, up, down\}$ and recall that a domino game is a structure Dom $= \langle C, D, \gamma \rangle$ where C is a finite set of colours, D is a finite set of dominoes, and $\gamma : D \times \text{Sides} \to C$ is a map that assigns a color to each side of the dominoes. Dom can tile $\mathbb{N} \times \mathbb{N}$ if and only if there is a map $f : \mathbb{N} \times \mathbb{N} \to D$ that satisfies the color constraints. This means that only domino sides with identical colors can be adjacent (no rotation of dominoes is allowed). The problem DOMREC, known to be Σ_1^1-complete [56], takes as input a domino game Dom with a distinguished color c and asks whether Dom can pave $\mathbb{N} \times \mathbb{N}$ where the color c occurs infinitely often in the first column.

Let Dom $= \langle C, D, \gamma \rangle$ be a domino game with $C = \{1, \ldots, n\}$, $D = \{1, \ldots, m\}$, and $c = 1$. So we have an instance of DOMREC. Let us explain the syntactic ressources we shall need. We use the following propositional variables:

- *in* is a propositional variable that holds when the state encodes a position in $\mathbb{N}^2$. Indeed, there are states in the model that do not correspond to positions in $\mathbb{N}^2$. In order to facilitate the presentation, we also introduce *out* that is equivalent to the negation of *in*.
- For every $j \in D$, we introduce the variable d_j with intended meaning that "the position in $\mathbb{N}^2$ associated with the current state is occupied by a domino of type j".
- For every $i \in C$, we use the variables up_i, $down_i$, $left_i$, $right_i$. For instance, up_1 holds whenever the domino on the position associated with the current state has color 1 on its top.

Every state encoding a position in $\mathbb{N}^2$ is occupied by a unique domino:

$$\mathsf{G}\left(in \to \bigvee_{1 \leq j \leq m} \left(d_j \wedge \bigwedge_{1 \leq k \leq m, k \neq j} \neg d_k \right) \right)$$

Propositional variables for colours are compatible with the definition of domino types:

$$\bigwedge_{1 \leq j \leq m} \bigwedge_{s \in \text{Sides}} \mathsf{G}\left(in \wedge d_j \to s_{\gamma(j,s)} \wedge \bigwedge_{1 \leq k \leq n, k \neq \gamma(j,s)} \neg s_k \right)$$

We write PAVE to denote the conjunction of the above formulae. Now, we shall define the states of the model that correspond to positions in $\mathbb{N}^2$. We write SNAKE to denote the conjunction of the following formulae:

- $\mathsf{G}(in \leftrightarrow \neg out)$,

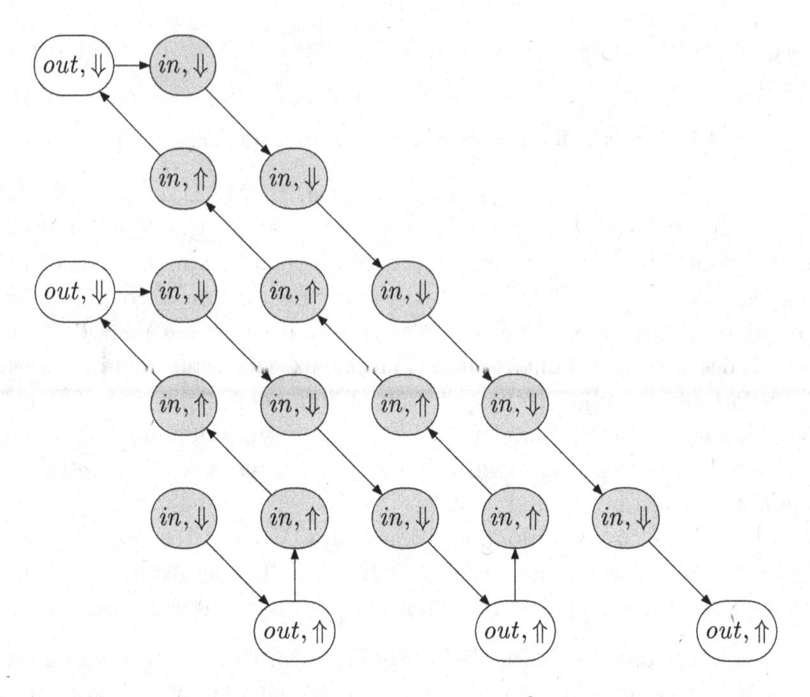

Fig. 15.11. The path in $\mathbb{N}^2$.

- $in \wedge \mathsf{X}\, out \wedge \mathsf{XX}\, in \wedge \mathsf{XXX}\, in \wedge \mathsf{XXXX}\, out$,
- $\mathsf{G}(out \rightarrow \mathsf{X}\, \mathrm{L}_1(in, out, in \wedge \mathsf{X}\, out))$.

The only structure (built over $\{in, out\}$) satisfying SNAKE is:

$$\{in\} \cdot \{out\} \cdot \{in\}^2 \cdot \{out\} \cdot \{in\}^3 \cdot \{out\} \cdot \{in\}^4 \ldots$$

This sequence makes reference to the path presented in Figure 15.11, where the set of grey nodes encodes $\mathbb{N}^2$. The difficulty of the proof is not to design a path through $\mathbb{N}^2$ but rather to define a path on which it is easy to access to neighbours (right or up). Moreover, we need to identify the positions encoding the first column of $\mathbb{N}^2$. Consequently, a bit more work is required.

For every state *in*, we need to remember if it occurs in a sequence of *in* that is upward or downward (alternately, this encodes the parity of i for each block $\{out\} \cdot \{in\}^i \cdot \{out\}$). Indeed, this criterion is relevant to access to (right or up) neighbours. We introduce the variables $\Uparrow$ and $\Downarrow$. The former one $\Uparrow$ is only introduced to facilitate the presentation. We write DIRECTION to denote the conjunction of the following formulae:

- $\mathsf{G}(\Uparrow \leftrightarrow \neg \Downarrow)$,
- $\Downarrow \wedge \mathsf{X} \Uparrow$,
- $\mathsf{G}(in \wedge \mathsf{X}\, in \wedge \Uparrow \rightarrow \mathsf{X} \Uparrow)$ ("we stay on upward sequence"),
- $\mathsf{G}(in \wedge \mathsf{X}\, in \wedge \Downarrow \rightarrow \mathsf{X} \Downarrow)$ ("we stay on downward sequence"),

- $\mathsf{G}(in \wedge \mathsf{X}\, out \wedge \Uparrow \rightarrow (\mathsf{X} \Downarrow \wedge \mathsf{X}\mathsf{X} \Downarrow))$ ("we pass from upward to downward sequence"),
- $\mathsf{G}(in \wedge \mathsf{X}\, out \wedge \Downarrow \rightarrow (\mathsf{X} \Uparrow \wedge \mathsf{X}\mathsf{X} \Uparrow))$ ("we pass from downward to upward sequence").

The only structure (built over $\{in, out, \Uparrow, \Downarrow\}$) satisfying SNAKE $\wedge$ DIRECTION is

$$\{in, \Downarrow\} \cdot \{out, \Uparrow\} \cdot \{in, \Uparrow\}^2 \cdot \{out, \Downarrow\} \cdot \{in, \Downarrow\}^3 \cdot \{out, \Uparrow\} \cdot \{in, \Uparrow\}^4 \cdots$$

This structure encodes the path through $\mathbb{N}^2$ described in Figure 15.11. The path allows to access to adjacent states as follows:

- in a state $\{in, \Uparrow\}$, we access to the up and right neighbours with the help of L_0 and L_1, respectively,
- in a state $\{in, \Downarrow\}$, we access to the up and right neighbours with the help of L_1 and L_0, respectively,

We write CONSTRAINTS to denote the conjunction of following formulae that express color constraints for adjacent dominoes:

- $\mathsf{G}(in \wedge \Uparrow \rightarrow (\bigwedge_{1 \leq i \leq n} right_i \rightarrow L_1(in, out, left_i)))$,
- $\mathsf{G}(in \wedge \Uparrow \rightarrow (\bigwedge_{1 \leq i \leq n} up_i \rightarrow L_0(in, out, down_i)))$,
- $\mathsf{G}(in \wedge \Downarrow \rightarrow (\bigwedge_{1 \leq i \leq n} right_i \rightarrow L_0(in, out, left_i)))$,
- $\mathsf{G}(in \wedge \Downarrow \rightarrow (\bigwedge_{1 \leq i \leq n} up_i \rightarrow L_1(in, out, down_i)))$.

We write REC to denote the formula that states that colour 1 occurs infinitely often in the first column:

$$\mathsf{GF}\left(\mathsf{X}(out \wedge \Downarrow) \wedge \bigvee_{s \in \mathsf{Sides}} s_1\right) \vee \mathsf{GF}\left(out \wedge \Downarrow \wedge \mathsf{X} \bigvee_{s \in \mathsf{Sides}} s_1\right)$$

The domino game Dom can pave $\mathbb{N}^2$ with colour 1 occurring infinitely often on the first column if and only if PAVE$\wedge$SNAKE$\wedge$DIRECTION$\wedge$CONSTRAINTS$\wedge$REC is satisfiable in PC[L_1]. □

Satisfiability for PC[L_1] is therefore Σ^1_1-hard and is not recursively enumerable. The proof of Proposition 15.14 is inspired from the undecidability of propositional dynamic logic (PDL) augmented with the context-free language $\{a_1^k a_2 a_1^k \mid k \geq 0\}$ (see for example [23, chapter 9] for more details). For formal verification, the following result is more meaningful since it illustrates the strength of adding a single context-free language.

Corollary 15.15. *The model-checking problem for* PC[L_1] *is undecidable.*

Proof. It is indeed simple to reduce satisfiability for PC[L_1] to model-checking for PC[L_1]. Let φ be a formula built over the propositional variables $p_1, \ldots, p_n$. We write $\mathcal{M}_n = \langle W, R, \lambda \rangle$ to denote the complete Kripke structure such that $W = 2^{\{p_1, \ldots, p_n\}}$, $R = W \times W$ and λ is the identity. Then φ is valid if and only if for all $s \in W$, we have $\mathcal{M}_n, s \models_\forall \varphi$. Since PC[$L_1$] is closed under negations, by Proposition 15.14, the validity problem for PC[L_1] is also undecidable. Hence, the model-checking problem for PC[L_1] is undecidable. □

Surprisingly, a similar undecidability result holds for a specific context-free language that can be recognized by a visibly pushdown automaton (VPA) [22] as shown below. For instance, L_1 cannot be recognized by a VPA. Let L_2 be the context-free language $\{a_1^k a_2 a_3^k a_4 \mid k \geq 0\}$ built over the alphabet $\Sigma = \{a_1, \ldots, a_4\}$. This language can be easily defined by a VPA. A very nice feature of the class of VPA is that it defines context-free languages that are closed under Boolean operations. Moreover, PDL generalized to programs defined by VPA is still decidable [24] (generalizing for instance [57]). Unlike this extension of PDL, we have the following undecidability results.

Corollary 15.16. *Satisfiability and model-checking for* PC[L_2] *are undecidable.*

Indeed, the temporal operator defined with the language L_2 can easily express the temporal operator defined with L_1 since $L_1(\varphi_1, \varphi_2, \varphi_3) \equiv L_2(\varphi_1, \varphi_2, \varphi_1, \varphi_3)$. By contrast, the characterization of decidable positive fragments of PC[CF] (without negation) is still open (the above undecidability proofs use negation in an essential way).

15.5. Concluding remarks

In this chapter, we have presented two distinct aspects of the use of automata theory for the verification of computer systems. A translation from LTL formulae into Büchi automata has been defined with the main advantage to produce simple automata for simple formulae. This follows the automata-based approach advocated in [9] with simplicity and pedagogical requirements from [12, 14]. We believe that this is an adequate translation to be taught to students. The second use of automata is related to automata-based temporal operators generalizing the more standard temporal connectives such as the next operator or the until operator. After recalling the standard results about the operators defined from regular languages, we have explained how a restricted addition of operators defined from context-free languages can easily lead to undecidability. For instance, we show why model-checking for propositional calculus augmented with a simple CF language recognized by a VPA [22] is highly undecidable.

References

[1] P. Blackburn, M. de Rijke, and Y. Venema, *Modal Logic.* (Cambridge University Press, 2001).
[2] N. Rescher and A. Urquhart, *Temporal Logic.* (Springer, 1971).
[3] A. Pnueli. The temporal logic of programs. In *Proc. 18th FOCS*, pp. 46–57. IEEE, (1977).

[4] J.-P. Queille and J. Sifakis. Specification and verification of concurrent systems in CESAR. In *Proc. 23rd FOCS*, pp. 337–351. IEEE, (1982).
[5] E. M. Clarke, E. A. Emerson, and A. P. Sistla. Automatic verification of finite-state concurrent systems using temporal logic specifications: A practical approach. In *Proc. 10th POPL*, pp. 117–126. ACM, (1983).
[6] E. Clarke, O. Grumberg, and D. Peled, *Model Checking*. (MIT Press, 2000).
[7] W. Thomas. Automata on infinite objects. In *Handbook of Theoretical Computer Science, Volume B, Formal models and semantics*, pp. 133–191. Elsevier, (1990).
[8] J. R. Büchi. On a decision method in restricted second-order arithmetic. In *International Congress on Logic, Method and Philosophical Science'60*, pp. 1–11, (1962).
[9] M. Vardi and P. Wolper, Reasoning about infinite computations, *Information and Computation*. **115**, 1–37, (1994).
[10] D. Harel, O. Kupferman, and M. Vardi, On the complexity of verifying concurrent transition systems, *Information and Computation*. **173**(2), 143–161, (2002).
[11] C. Courcoubetis, M. Vardi, P. Wolper, and M. Yannakakis. Memory efficient algorithms for the verification of temporal properties. In *Proc. 2nd CAV*, vol. 531, *LNCS*, pp. 233–242. Springer, (1990).
[12] R. Gerth, D. Peled, M. Vardi, and P. Wolper. Simple on-the-fly automatic verification of linear temporal logic. In *Proc. 15th PSTV*, vol. 38, *IFIP Conference Proceedings*, pp. 3–18. Chapman & Hall, (1995).
[13] M. Daniele, F. Giunchiglia, and M. Vardi. Improved automata generation for linear temporal logic. In *Proc. 11th CAV*, vol. 1633, *LNCS*, pp. 249–260. Springer, (1998).
[14] J.-M. Couvreur. On-the-fly verification of linear temporal logic. In *Proc. Formal Methods*, vol. 1708, *LNCS*, pp. 253–271. Springer, (1999).
[15] P. Wolper, The tableau method for temporal logic: An overview, *Logique et Analyse*. **110–111**, 119–136, (1985).
[16] P. Wolper. Constructing automata from temporal logic formulas: A tutorial. In *European Educational Forum: School on Formal Methods and Performance Analysis*, vol. 2090, *LNCS*, pp. 261–277. Springer, (2000).
[17] P. Gastin and D. Oddoux. Fast LTL to Büchi automata translation. In *Proc. 13th CAV*, vol. 2102, *LNCS*, pp. 53–65. Springer, (2001).
[18] D. Giannakopoulou and F. Lerda. From states to transitions: improving translation of LTL formulae to Büchi automata. In *Proc. 22nd FORTE*, vol. 2529, *LNCS*, pp. 308–326. Springer, (2002).
[19] J. Couvreur, A. Duret-Lutz, and D. Poitrenaud. On-the-fly emptiness checks for generalized Büchi automata. In *Proc 12th SPIN*, vol. 3639, *LNCS*, pp. 169–184. Springer, (2005).
[20] H. Tauriainen. *Automata and Linear Temporal Logic: Translations with Transition-based Acceptance*. PhD thesis, Helsinki University of Technology, (2006).
[21] P. Wolper, Temporal logic can be more expressive, *Information and Computation*. **56**, 72–99, (1983).
[22] R. Alur and P. Madhusudan. Visibly pushdown languages. In *Proc. 36th STOC*, pp. 202–211. ACM, (2004).
[23] D. Harel, D. Kozen, and J. Tiuryn, *Dynamic Logic*. (MIT Press, 2000).
[24] C. Löding and O. Serre. Propositional dynamic logic with recursive programs. In *Proc. 9th FOSSACS*, vol. 3921, *LNCS*, pp. 292–306. Springer, (2006).
[25] A. Prior, *Past, Present and Future*. (Oxford University Press, 1967).
[26] D. Gabbay, A. Pnueli, S. Shelah, and J. Stavi. On the temporal analysis of fairness. In *Proc. 7th POPL*, pp. 163–173. ACM, (1980).

[27] J. Kamp. *Tense Logic and the Theory of Linear Order.* PhD thesis, UCLA, USA, (1968).
[28] Z. Manna and A. Pnueli. The modal logic of programs. In *Proc. 6th ICALP*, vol. 71, *LNCS*, pp. 385–409. Springer, (1979).
[29] A. Pnueli. The temporal semantics of concurrent programs. In *Proc. Semantics of Concurrent Computation*, vol. 70, *LNCS*, pp. 1–20. Springer, (1979).
[30] C. Eisner and D. Fisman, *A Practical Introduction to PSL.* (Springer, 2006).
[31] G. Holzmann, The model checker SPIN, *IEEE Transactions on Software Engineering.* **23**(5), 279–295, (1997).
[32] K. McMillan, *Symbolic Model Checking.* (Kluwer Academic Publishers, 1993).
[33] E. Clarke and A. Emerson. Design and synthesis of synchronization skeletons using branching time temporal logic. In *Proc. Worshop on Logic of Programs*, vol. 131, *LNCS*, pp. 52–71. Springer, (1981).
[34] A. Emerson and J. Halpern, "Sometimes" and "Not Never" revisited: on branching versus linear time temporal logic, *J. ACM.* **33**, 151–178, (1986).
[35] A. Sistla and E. Clarke, The complexity of propositional linear temporal logic, *J. ACM.* **32**(3), 733–749, (1985).
[36] S. Demri and P. Schnoebelen, The complexity of propositional linear temporal logics in simple cases, *Information and Computation.* **174**(1), 84–103, (2002).
[37] P. Schnoebelen. The complexity of temporal logic model checking. In *Advances in Modal Logic, vol. 4, selected papers from 4th Conf. Advances in Modal Logic*, pp. 437–459. King's College Publication, (2003).
[38] E. M. Clarke, E. A. Emerson, and A. P. Sistla, Automatic verification of finite-state concurrent systems using temporal logic specifications, *ACM Trans. Programming Languages and Systems.* **8**(2), 244–263, (1986).
[39] E. Emerson and C. Jutla, The complexity of tree automata and logics of programs, *SIAM J. Computing.* **29**(1), 132–158, (2000).
[40] M. Vardi and L. Stockmeyer. Improved upper and lower bounds for modal logics of programs. In *Proc. 17th STOC*, pp. 240–251. ACM, (1985).
[41] M. Vardi. Nontraditional applications of automata theory. In *2nd TACS*, vol. 789, *LNCS*, pp. 575–597. Springer, (1994).
[42] K. Etessami, T. Wilke, and R. Schuller, Fair simulation relations, parity games, and state space reduction for Büchi automata, *SIAM J. Computing.* **34**(5), 11591175, (2005).
[43] A. Meyer. Weak second order theory of successor is not elementary-recursive. Technical Report MAC TM-38, MIT, (1973).
[44] L. Stockmeyer. *The Complexity of Decision Problems in Automata Theory and Logic.* PhD thesis, Department of Electrical Engineering, MIT, (1974).
[45] K. Etessami and G. Holzmann. Optimizing Büchi automata. In *Proc. 11th CONCUR*, vol. 1877, *LNCS*, pp. 153–167. Springer, (2000).
[46] F. Somenzi and R. Bloem. Efficient Büchi automata from LTL formulae. In *Proc. 12th CAV*, vol. 1855, *LNCS*, pp. 248–263. Springer, (2000).
[47] C. Papadimitriou, *Computational Complexity.* (Addison-Wesley, 1994).
[48] Z. Manna and A. Pnueli, *Temporal Verification of Reative Systems: Safety.* (Springer, 1995).
[49] S. Schwendimann. A new one-pass tableau calculus for PLTL. In *TABLEAUX'98*, vol. 1397, *LNAI*, pp. 277–291. Springer, (1998).
[50] E. Clarke, O. Grumberg, and K. Hamaguchi. Another look at LTL model checking. In *Proc. 6th CAV*, vol. 818, *LNCS*, pp. 415–427. Springer, (1994).
[51] V. Diekert and P. Gastin. First-order definable languages. In *Logic and Automata:*

History and Perspectives, vol. 2, *Texts in Logic and Games*, pp. 261–306. Amsterdam University Press, (2008).

[52] H. Straubing, *Finite Automata, Formal Logic, and Circuit Complexity.* Progress in Theoretical Computer Science, (Birkhäuser, 1994).

[53] P. Wolper, M. Vardi, and A. Sistla. Reasoning about infinite computation paths. In *Proc. 24th FOCS*, pp. 185–194. IEEE, (1983).

[54] M. Vardi. A temporal fixpoint calculus. In *Proc. 15th POPL*, pp. 250–259. ACM, (1988).

[55] D. Kozen, *Automata and Computability.* (Springer, 1997).

[56] D. Harel, Recurring dominoes: making the highly undecidable highly understandable, *Annals of Discrete Mathematics.* **24**, 51–72, (1985).

[57] D. Harel and E. Singerman, More on nonregular PDL: Finite models and Fibonacci-like programs, *Information and Computation.* **128**, 109–118, (1996).

Chapter 16

Finite Automata and the Analysis of Infinite Transition Systems

Wolfgang Thomas
Lehrstuhl Informatik 7, RWTH Aachen University, Germany
thomas@automata.rwth-aachen.de

In this tutorial, we present basic concepts and results from automata theory for the description and analysis of infinite transition systems. We introduce and discuss the classes of rational, automatic, and prefix-recognizable graphs and in each case address the question whether over such graphs the model-checking problem (with respect to natural logics) is decidable. Then we treat two different extensions of prefix-recognizable graphs, namely the graphs of the "Caucal hierarchy" and the graphs presented by ground tree rewriting systems, again with an analysis of their suitability for model-checking. This application of automata theoretic ideas helps to clarify the balance between the expressiveness of frameworks for the specification of models and the possibility to automatize verification.

16.1. Introduction

The analysis of infinite transition systems is a fundament in infinite-state system verification and at the same time one of the most promising application domains of automata theory. This tutorial aims at an overview on some central ideas and topics currently studied in this field.

The set-up of algorithmic verification is built on two pillars: transition systems as models of "systems" (programs, protocols, control units), and specifications given by logical formulas that express some desired behaviour. The model-checking problem is the question "Given a transition graph G and a formula φ, does G satisfy φ?". As logical frameworks we consider mainly classical logics like first-order or monadic second-order logic. Since first-order logic is too weak to express reachability properties (which are a central objective in verification), we have to include constructs that allow to cover reachability. For example, we consider FO-logic with a signature that is expanded by the transitive closure E^* of the edge relation E. Monadic second-order logic is a much more powerful system (in which E^* is definable from E). It is even more expressive than the branching time logics CTL and CTL*.

On the side of the transition graphs, there are numerous methods to obtain finite presentations. (Such presentations are needed when infinite structures – in our case: graphs – occur as instances of algorithmic problems.) For example, one can use grammars or equation systems as generators of structures, as done in the work of Courcelle [1]. In the present paper we pursue a different track and consider presentations of infinite structures in terms of finite automata. In this approach, the domain of a structure is described as a regular set of words (or trees), and the relations of the structure are defined by automata of different types that accept tuples of words (or tuples of trees). There are several kinds of automata for the definition of relations, leading to different types of relational structures.

The first part of this paper is concerned with three fundamental classes of transition graphs, namely the rational, the automatic, and the prefix-recognizable graphs (and the pushdown graphs as a special case of the latter). These classes of graphs are cornerstones in an automata based theory of infinite models. We shall see that the first two classes are too extensive to allow algorithmic solutions for interesting problems in verification, while the third is very well-behaved – as seen in the decidability of the model-checking problem for monadic second-order logic.

In the subsequent two sections of the paper we consider two proper extensions of the class of prefix-recognizable graphs. The first extension is based on an idea of Caucal [2] to generate a much larger class of models where the model-checking problem with respect to monadic second-order logic is still decidable: One applies the two model transformations "monadic second-order interpretation" and "unfolding" in alternation, starting from the finite trees. We introduce the resulting "Caucal hierarchy" of graphs and illustrate its large range by some examples.

The second extension is motivated by the fact that very natural types of infinite graphs are not located in the Caucal hierarchy. A prominent example is the infinite $(\mathbb{N} \times \mathbb{N})$-grid; the associated model-checking problem with respect to monadic second-order logic is undecidable. We introduce "ground tree rewriting graphs" that contain the infinite grid as a special case but nevertheless permit a solution of the model-checking problem for first-order logic expanded by the reachability predicate. For the analysis of these graphs we use automata over finite trees rather than over finite words.

In the final section we address complementary issues: First we note connections between the "internal" presentation of graphs (as it is used for the rational and automatic graphs) and the "external" presentation in terms of transformations of given graphs. Then we briefly discuss the problem of linking automata theoretic presentations to structural properties of graphs. Finally, we sketch connections to formal language theory; here an infinite transition graph is used as an infinite automaton, and the relation between the presentation of such graphs and the form of the accepted languages is studied.

The application of automata theory to verification as outlined in this chapter is only one method among many others. Let us mention an alternative approach that is found, for example, in the analysis of Petri nets or lossy channel systems [3]. In these cases the reachability problem can be treated (and solved) using certain monotonicity properties of the reachability relation. A general development of this method is found in the theory of "well-structured transition systems" (see [4]).

Our exposition assumes knowledge of basic automata theory and logic. In several cases we only give proofs in an informal style and have to refer to the literature for details.

16.2. Technical Preliminaries

16.2.1. *Transition Systems*

We consider structures in the format of edge-labelled and vertex-labelled transition graphs

$$G = (V, (E_a)_{a \in \Sigma}, (P_b)_{b \in \Sigma'})$$

with two finite alphabets Σ, Σ' for labelling edges, respectively vertices. V is the (at most countable) set of vertices (in applications: "states"), $E_a \subseteq V \times V$ (for a symbol $a \in \Sigma$) is the set of a-labelled edges, and $P_b \subseteq V$ (for $b \in \Sigma'$) is the set of b-labelled vertices (in applications representing a state property). We write E for the union of the E_a. As special cases, we allow Σ and Σ' to be empty. In the first case we have a structure $(V, E, (P_b)_{b \in \Sigma'})$, in the second case a structure $(V, (E_a)_{a \in \Sigma'})$, and if both label alphabets are empty we consider directed graphs (V, E).

More generally, one can consider relational structures $\mathcal{A} = (A, R_1^{\mathcal{A}}, \ldots, R_k^{\mathcal{A}})$, where the $R_i^{\mathcal{A}}$ are relations of possibly different arities over A, say $R_i^{\mathcal{A}}$ of arity n_i. In the sequel we stay with transition graphs for ease of notation and for their significance in verification.

As examples of transition graphs we mention the following:

- *Kripke structures*, which are graphs of the form $G = (V, E, (P_b)_{b \in \Sigma'})$, where each P_b collects states which satisfy certain atomic propositions,
- the ordering $(\mathbb{N}, <)$ of the natural numbers,
- the *binary tree* $T_2 = (\{0,1\}^*, S_0, S_1)$ where $S_i = \{(w, wi) \mid w \in \{0,1\}^*\}$ (analogously, the n-ary tree is $T_n := (\{0, \ldots, n-1\}^*, S_0^n, \ldots, S_{n-1}^n)$).

16.2.2. *Logics*

First-order logic FO over the signature with the symbols E_a, P_b is built up from variables $x, y, \ldots$ and atomic formulas $x = y$, $E_a(x, y)$, $P_b(x)$ where x, y are first-order variables, using the standard propositional connectives $\neg, \wedge, \vee, \rightarrow, \leftrightarrow$ and the quantifiers $\exists, \forall$.

The reachability relation over G is the relation E^* defined by

$$E^*(u, v) \quad \Leftrightarrow \quad \exists v_0 \ldots v_k \in V(v_0 = u \;\; \wedge \;\; \forall i < k : (v_i, v_{i+1}) \in E \;\; \wedge v_k = v)$$

It is well-known that E^* is not FO-definable (see, e.g., [5]). We call FO(R) the logic obtained from FO by adjoining a symbol for the reachability relation E^*

to the signature. A slightly stronger variant is FO(Reg) which involves regular expressions r over the edge label alphabet. Rather than E^* we then use (symbols for) the relations E_r where $E_r(u, v)$ holds if there is a path from u to v whose edge label sequence yields a word in the language defined by the regular expression r.

Monadic second-order logic MSO is obtained by adjoining variables $X, Y, \ldots$ for sets of elements (of the universe V under consideration) and atomic formulas $X(y)$ (meaning that the element y is in the set X) as well as quantifiers over set variables. We note that MSO encompasses FO(R), since we can express $E^*(x, y)$ by the formula saying that each set which contains x and is closed under E must contain y.

We use the standard notations; e.g. $G \models \varphi[v]$ indicates that G satisfies the formula $\varphi(x)$ with the element v as interpretation of x. Given a formula $\varphi(x_1, \ldots, x_n)$, the relation defined by it in G is

$$\varphi^G = \{(v_1, \ldots, v_n) \in V^n \mid G \models \varphi[v_1, \ldots, v_n]\}.$$

The model-checking problem "Does the transition system G satisfy the sentence φ?" comes in two forms, the "uniform" version where an instance is a pair (G, φ), and a "non-uniform" one where G is considered fixed and the instance is φ. In the latter case (when G is fixed), we consider the (FO- or FO(R)- or MSO-) *theory* of G, i.e., the respective set of sentences which are true in G. In all the cases discussed in this paper, we can obtain decidability of a uniform model-checking problem from decidability of the associated non-uniform version (either by an explicit proof or by an analysis of the given proof for the non-uniform version).

16.3. Rational Graphs

In this section we discuss a first type of infinite transition graph that is presented in terms of finite automata. The idea is to use words over some alphabet as names of vertices, regular languages for vertex properties, and automaton-definable relations over words for the edge relations. For the latter, we consider the definition of word relations in terms of regular expressions over word-tuples, or equivalently in terms of "transducers", i.e., nondeterministic automata that asynchronously scan a given tuple of input words.

A relation $R \subseteq \Gamma^* \times \Gamma^*$ is *rational* if it can be defined by a regular expression starting from the atomic expressions $\emptyset$ (denoting the empty relation) and (u, v) for words u, v (denoting the relation $\{(u, v)\}$) by means of the operations union, concatenation (applied componentwise), and iteration of concatenation (Kleene star). An alternative characterization of these relations is given by nondeterministic automaton that work one-way from left to right, but asynchronously, on the two components of an input $(w_1, w_2) \in \Gamma^* \times \Gamma^*$ (see [6] or [7]). A transition of such an automaton is simply a triple $(p, u/v, q)$ with states p, q and words u, v. A pair (w_1, w_2) is accepted if for some successful path with label sequence $u_1/v_1, \ldots, u_k/v_k$

we have $w_1 = u_1 \dots u_k$ and $w_2 = v_1 \dots v_k$. The generalization of the definition to n-ary relations for $n > 2$ is obvious.

Example 16.1. Consider the suffix relation $\{(w_1, w_2) \mid w_1 \text{ is a suffix of } w_2\}$. A corresponding automaton (nondeterministic transducer) would progress with its reading head on the second component w_2 until it guesses that the suffix w_1 starts; this, in turn, can be checked by moving the two reading heads on the two components simultaneously, comparing w_1 letter by letter with the remaining suffix of w_2.

A *rational transition graph* (or just *rational graph*) has the form $G = (V, (E_a)_{a\in\Sigma}, (P_b)_{b\in\Sigma'})$ where V and the sets P_b are regular sets of words over an auxiliary alphabet Γ and where each $E_a \subseteq \Gamma^* \times \Gamma^*$ is a rational relation.

Clearly, each rational graph is recursive in the sense that the edge relations and the vertex properties are decidable. However, very simple properties of rational graphs may be undecidable.

Proposition 16.1. *For each instance $(\overline{u}, \overline{v})$ of PCP (Post's Correspondence Problem) one can construct a rational graph $G_{(\overline{u},\overline{v})}$ such that $(\overline{u}, \overline{v})$ has a solution (i.e., an index sequence $i_1, \dots, i_k$ exists such that $u_{i_1} \dots u_{i_k} = v_{i_1} \dots v_{i_k}$) iff $G_{(\overline{u},\overline{v})}$ has a loop edge from some vertex to itself.*

Proof. Given a PCP-instance $(\overline{u}, \overline{v}) = ((u_1, \dots, u_m), (v_1, \dots, v_m))$ over an alphabet Γ, we specify a rational graph $G_{(\overline{u},\overline{v})} = (V, E)$ as follows. The vertex set V is Γ^*. The edge set E consists of the pairs of words of the form $(u_{i_1} \dots u_{i_k}, v_{i_1} \dots v_{i_k})$ where $i_1, \dots, i_k \in \{1, \dots, m\}$ and $k \geq 1$. Clearly, an asynchronously progressing nondeterministic automaton can check whether a word pair (w_1, w_2) belongs to E; basically the automaton has to guess successively the indices $i_1, \dots, i_k$ and at the same time to check whether w_1 starts with u_{i_1} and w_2 starts with v_{i_1}, whether w_1 continues by u_{i_2} and w_2 by v_{i_2}, etc. So the graph $G_{(\overline{u},\overline{v})}$ is rational. Clearly, in this graph there is an edge from some vertex w back to the same vertex w iff the PCP-instance $(\overline{u}, \overline{v})$ has a solution (namely by the word w). □

The existence of a loop edge (w, w) is expressible by the first-order formula $\exists x\ E(x, x)$. Hence we obtain that the uniform model checking-problem is undecidable over rational graphs (Morvan [8]):

Theorem 16.1. *There is no algorithm which, given a presentation of a rational graph G and a first-order sentence φ, decides whether $G \models \varphi$.*

Let us now construct a *single* rational graph with an undecidable first-order theory (following [9]); so also the non-uniform model-checking problem can be undecidable for a rational graph.

Theorem 16.2. *There is a rational graph G with an undecidable first-order theory.*

Proof. We use a Turing machine M that accepts a (recursively enumerable but) non-recursive language. We encode its undecidable halting problem (for different input words x) into a family of PCP-instances.

For simplicity of exposition, we refer here to the standard construction of the undecidability of PCP as one finds it in textbooks (see [10, Section 8.5]): A Turing machine M with input word x is converted into a PCP-instance $((u_1, \ldots, u_m), (v_1, \ldots, v_m))$ over an alphabet A whose letters are the states and tape letters of M and a symbol # (for the separation between M-configurations in M-computations). If the input word is $x = a_1 \ldots a_n$, then u_1 is set to be the initial configuration word $c(x) := \#q_0a_1 \ldots a_n$ of M; furthermore we always have $v_1 = \#$, and $u_2, \ldots, u_m, v_2, \ldots, v_m$ only depend on M. Then the standard construction (of [10]) ensures the following:

M halts on input x iff the PCP-instance $((c(x), u_2, \ldots, u_m), (\#, v_2, \ldots, v_m))$ has a special solution. Here a special solution is given by an index sequence $(i_2, \ldots, i_k)$ such that $c(x)u_{i_2} \ldots u_{i_k} = \#v_{i_2} \ldots v_{i_k}$.

Let G be the graph as defined from these PCP-instances as above: The vertices are the words over A, and we have a single edge relation E with $(w_1, w_2) \in E$ iff there are indices $i_2, \ldots, i_k$ and a word x such that $w_1 = c(x)u_{i_2} \ldots u_{i_k}$ and $w_2 = \#v_{i_2} \ldots v_{i_k}$. Clearly G is rational, and we have an edge from a word w back to itself if it is induced by a special solution of some PCP-instance $((c(x), u_2, \ldots, u_m), (\#, v_2, \ldots, v_m))$.

In order to address the input words x explicitly in the graph, we add further vertices and edge relations E_a for $a \in A$. A $c(x)$-labelled path via the new vertices will lead to a vertex of G with prefix $c(x)$; if the latter vertex has an edge back to itself, then a special solution for the PCP-instance $((c(x), u_2, \ldots, u_m), (\#, v_2, \ldots, v_m))$ can be inferred. The new vertices are words over a copy $\underline{A}$ of the alphabet A (consisting of the underlined versions of the A-letters). For any word $c(x)$ we shall add the vertices which arise from the underlined versions of the proper prefixes of $c(x)$, and we introduce an E_a-edge from any such underlined word $\underline{w}$ to $\underline{wa}$ (including the case $\underline{w} = \underline{\varepsilon}$). There are also edges to non-underlined words: We have an E_a-edge from $\underline{w}$ to any non-underlined word which has wa as a prefix. Call the resulting graph G'. It is easy to see that G' is rational.

By construction of G', the PCP-instance $((c(x), u_2, \ldots, u_m), (\#, v_2, \ldots, v_m))$ has a special solution iff there is a path in G', labelled with the word $c(x)$, from the vertex $\underline{\varepsilon}$ to a vertex which has an edge back to itself.

Note that the vertex $\underline{\varepsilon}$ is definable as the only one with outgoing E_a-edges but without any ingoing E_a-edge. Thus the above condition is formalizable by a first-order sentence φ_x, using variables for the $|c(x)| + 1$ vertices of the desired path. Altogether we obtain that the Turing machine M halts on input x iff $G' \models \varphi_x$. □

This result shows that rational graphs in general are much too complex for decidability results even regarding a weak logic like FO; hence they do not play an interesting role in algorithmic approaches to verification. On the other hand, the rational word relations underlying these graphs constitute a beautiful chapter of automata theory; for a recent exposition see [7].

16.4. Automatic Graphs

In *automatic (or synchronized rational) relations* a more restricted processing of an input (w_1, w_2) by an automaton is required than in the asynchronous mode as mentioned for nondeterministic transducers: We now require that an automaton scans a pair (w_1, w_2) of words strictly in parallel letter by letter. Thus one can assume that the automaton reads letters from $\Gamma \times \Gamma$ for word pairs over Γ. In order to cover the case that w_1, w_2 are of different length, one assumes that the shorter word is prolonged by dummy symbols \$ to achieve equal length. Let $[w_1, w_2]$ be the word over the alphabet $(\Gamma \times \Gamma) \cup ((\Gamma \cup \{\$\}) \times \Gamma) \cup (\Gamma \times (\Gamma \cup \{\$\}))$ associated with (w_1, w_2). A relation $R \subseteq \Gamma^* \times \Gamma^*$ induces the language $L_R = \{[w_1, w_2] \mid (w_1, w_2) \in R\}$. The relation R is called *automatic* if the associated language L_R is regular. Again, the generalization to n-ary relations for $n > 2$ is obvious.

From this definition it is clear that the automatic relations share many good properties which are familiar from the theory of regular word languages. For example, one can transform a nondeterministic automaton (that recognizes a word relation in the synchronous mode) to an equivalent deterministic one, a fact which does not hold for the asynchronous transducers.

A graph $(V, (E_a)_{a \in \Sigma}, (P_b)_{b \in \Sigma'})$ is called *automatic* if V and each $P_b \subseteq V$ are regular languages over an alphabet Γ and each edge relation $E_a \subseteq \Gamma^* \times \Gamma^*$ is automatic.

Example 16.2. The infinite two-dimensional grid $G_2 := (\mathbb{N} \times \mathbb{N}, E_a, E_b)$ (with E_a-edges $((i,j),(i,j+1))$ and E_b-edges $((i,j),(i+1,j))$) is automatic: It can be obtained using the words in X^*Y^* as vertices, whence the edge relations become $E_a = \{(X^iY^j, X^iY^{j+1}) \mid i,j \geq 0\}$ and $E_b = \{(X^iY^j, X^{i+1}Y^j) \mid i,j \geq 0\}$, which both are clearly automatic.

Example 16.3. Consider the transition graph over $\Gamma = \{X_0, X, Y\}$ where there is an a-edge from X_0 to X and from X^i to X^{i+1} (for $i \geq 1$), a b-edge from X^iY^j to $X^{i-1}Y^{j+1}$ (for $i \geq 1, j \geq 0$), and a c-edge from Y^{i+1} to Y^i (for $i \geq 0$). We obtain the automatic graph of Figure 16.1. (This graph also has a natural meaning as "infinite automaton", using the vertex X_0 as "initial state" and the vertex ε as "final state". The accepted language is the context-sensitive language of the words $a^ib^ic^i$ with $i > 0$. We return to this aspect in the last section of the paper.)

Example 16.4. Let $T_2' = (\{0,1\}^*, S_0, S_1, \leq, \mathrm{EquLev})$ be the expansion of the binary tree $T_2 = (\{0,1\}^*, S_0, S_1)$ by the prefix relation $\leq = \{(u,v) \in \{0,1\}^* \mid u$ is a prefix of $v\}$ and the "equal level relation" $\mathrm{EquLev} = \{(u,v) \in \{0,1\}^* \mid |u| = |v|\}$. Clearly T_2' is automatic.

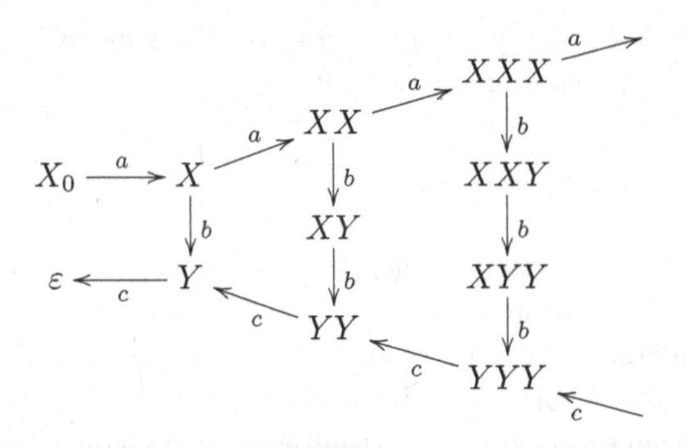

Fig. 16.1. An automatic graph.

In the literature, the automatic relations appear also under several other names, among them “regular”, “sequential”, and “synchronized rational”.

We give another example which illustrates the power of automatic relations.

Example 16.5. Given a Turing machine M with state set Q and tape alphabet Γ, we consider the graph G_M with vertex set $V_M = \Gamma^* Q \Gamma^*$, considered as the set of M-configurations. By an appropriate treatment of the blank symbol, we can assume that the length difference between two successive M-configurations is at most 1; thus it is easy to see that the relation E_M of word pairs which consist of successive M-configurations is automatic. So the configuration graph $G_M = (V_M, E_M)$ is automatic.

The relation that contains the pairs of successive Turing machine configurations can as well be described in terms of an infix rewriting system: For example, the effect of a Turing machine instruction that requires, in state p with letter a on the work cell, to print b, move to the right, and go into state q, is captured by the infix rewriting rule $pa \to bq$. Extending Example 16.5, we see that in general a graph (with a regular set of vertices) whose edge relation is defined by a finite infix rewriting system is also automatic.

Let us show that first-order properties of automatic graphs are decidable:

Theorem 16.3. *The FO-theory of an automatic graph is decidable.*

Proof. Let $G = (V, (E_a)_{a \in \Sigma}, (P_b)_{b \in \Sigma'})$ be a graph with an automatic presentation over Γ. We verify inductively over FO-formulas $\varphi(x_1, \ldots, x_n)$ that the following relation is automatic:

$$R_\varphi := \varphi^G = \{(w_1, \ldots, w_n) \mid G \models \varphi[w_1, \ldots, w_n]\}$$

For the atomic formulas, this is clear by the automatic presentation of G. In the induction step, the Boolean connectives are easy due to the closure of regular sets under Boolean operations. (Note that the complement is applied with respect to the set of words $[w_1, w_2]$, i.e. the words where the letter \$ may occur only in one component, and only at the end.) For the step of existential quantification, assume – as a typical case – that the binary relation R is recognized by the finite automaton $\mathcal{A}$, say with final state set F. We have to verify that also

$$S = \{w_1 \in \Gamma^* \mid \exists w_2 : (w_1, w_2) \in R\}$$

is automatic (i.e. in this unary case: a regular language).

The automaton checking S is obtained from $\mathcal{A}$ by a projection of the input letters to the first components and by an extension of F to a set F'. A state is included in F' if some (possibly empty) sequence of letters $(\$, a)$ leads to F. This covers the case that the component w_2 is longer than w_1.

If this inductive construction is applied to an FO-sentence φ (i.e., a formula without free variables), the final result is a finite automaton with unlabelled edges, such that a successful run (a path from the initial to some final state) exists iff the sentence φ is true in G. Since the existence of a successful run can be decided, we obtain the claim of the Theorem. □

An analogous argument shows that Presburger arithmetic, the FO-theory of the structure $(\mathbb{N}, +)$, is decidable (see [11]). For this purpose, one codes an n-tuple of natural numbers by the n-tuple of the reversed binary representations. The atomic formula $x_1 + x_2 = x_3$ defines a ternary relation over $\{0,1\}^*$ which is automatic, since the usual check that an addition of binary numbers is correct can be done by a finite automaton. For the logical connectives one proceeds as in the proof above. For an analysis of the complexity bounds of this decision procedure see [12]. An introduction to applications in verification is given in [13].

If we extend the logic FO by including the reachability relation E^*, then the above-mentioned decidability result fails.

Theorem 16.4. *There is an automatic graph $G = (V, E)$ such that the relation E^* is undecidable.*

Proof. As in Example 16.5, we take the automatic configuration graph G_M of a Turing machine M. We consider a Turing machine M that accepts an undecidable (but of course recursively enumerable) language $L(M)$. So the vertices are configuration words in $\Gamma^* Q \Gamma^*$ (where Γ is the tape alphabet of M and Q is its set of states). Assume that the machine M halts in a unique configuration, say q_s with a stop state q_s and a blank tape inscription. Then M accepts the input word w iff in G_M from the configuration $q_0 w$ the configuration q_s can be reached. Since $L(M)$ is undecidable, we obtain the claim of the theorem. □

This small result is one of the main obstacles in developing algorithmic solutions of the model-checking problem over infinite systems: The automatic graphs are a very natural framework for modelling interesting infinite systems, but most applications of model-checking involve some kind of reachability analysis; so the undecidability phenomenon of the theorem above enters. Current research tries to find good restrictions or variants of the class of automatic graphs where the reachability problem is still solvable.

Let us also look at a more ambitious problem than reachability: decidability of the monadic second-order theory of a given graph. Here we get undecidability already for automatic graphs with a much simpler transition structure than that of the graph G_M of the previous theorem. The most prominent example is the infinite two-dimensional grid (introduced as an automatic graph in Example 16.2). Note that the reachability problem over the grid (say from a given vertex to another given vertex) is decidable.

Theorem 16.5. *The monadic second-order theory of the infinite two-dimensional grid G_2 is undecidable.*

Proof. The idea is to code the computations of Turing machines in a more uniform way than in the previous result. Instead of coding a Turing machine configuration by a single vertex and capturing the Turing machine steps directly by the edge relation, we now use a whole row of the grid for coding a configuration (by an appropriate coloring of its vertices with tape symbols and a Turing machine state). A computation of a Turing machine, say with m states and n tape symbols, is thus represented by a sequence of colored rows (using $m+n$ colors), i.e., by a coloring of the grid. (We can assume that even a halting computation generates a coloring of the whole grid, by repeating the final configuration ad infinitum.) In this view, the horizontal edge relation is used to progress in space, while the vertical one allows to progress in time. A given Turing machine M halts on the empty tape iff there is a coloring of the grid with $m + n$ colors which

- represents the initial configuration (on the empty tape) in the first row,
- respects the transition table of M between any two successive rows,
- contains a vertex which is colored by a halting state.

Such a coloring corresponds to a partition of the vertex set $\mathbb{N} \times \mathbb{N}$ of the grid into $m + n$ sets. One can express the existence of the coloring by saying "there exist sets $X_1, \ldots, X_{m+n}$ which define a partition and satisfy the requirements of the three items above". In this way one obtains effectively an MSO-sentence φ_M such that M halts on the empty tape iff $G_2 \models \varphi_M$. □

16.5. Prefix Rewriting and Pushdown Systems

The undecidability of the reachability problem over automatic graphs (Theorem 16.4) is no surprise to a reader who knows the undecidability of the word problem for Semi-Thue systems, i.e. infix rewriting systems. Following Example 16.5, we remarked that infix rewriting systems induce automatic graphs.

As observed already by Büchi in 1964, the situation changes when we use prefix rewriting instead. Büchi showed that the words which are generated from a fixed word w by a finite prefix rewriting system form an effectively constructible regular language L. As an application one obtains the well-known fact that the reachable configurations of a pushdown automaton constitute a regular set. As a second application we note an elegant solution of the reachability problem over prefix rewriting systems: In order to decide whether from the word w one can reach the word v in finitely many steps, one computes a finite automaton recognizing the "reachability language" L mentioned above, and then checks whether this automaton accepts v.

In the first part of this section we introduce two types of graphs based on the idea of prefix rewriting. The first (and more restricted) version is the notion of pushdown graph, with edges corresponding to moves of a pushdown automaton. The second allows to capture infinitely many instances of prefix rewriting in a single rule; the graphs obtained in this way are called prefix-recognizable.

In a second part we present the solution of the reachability problem as indicated above. There are two approaches to this problem, "forward search" as mentioned above, or "backward search" starting from a target vertex or a set T of target vertices. We shall pursue the second approach.

In a third part we treat a much stronger result than decidability of the reachability problem over pushdown graphs and prefix-recognizable graphs. We sketch the proof that even the MSO-theory of any such graph is decidable. As starting point we use Rabin's Theorem on the decidability of the MSO-theory of the binary tree T_2 [14].

16.5.1. *Definitions*

A graph $G = (V, (E_a)_{a\in\Sigma})$ is called *pushdown graph* (over the label alphabet Σ) if it is the transition graph of the reachable global states of an ε-free pushdown automaton. Here a pushdown automaton is of the form $\mathcal{P} = (P, \Sigma, \Gamma, p_0, Z_0, \Delta)$, where P is the finite set of control states, Σ the input alphabet, Γ the stack alphabet, p_0 the initial control state, $Z_0 \in \Gamma$ the initial stack symbol, and $\Delta \subseteq P \times \Sigma \times \Gamma \times \Gamma^* \times P$ the transition relation. (A transition $\tau = (p, a, \gamma, v, q)$ proceeds from state p to q while processing input letter a and replacing the top stack symbol γ by the word v; note that we consider "real-time" automata without ε-transitions.) A configuration (sometimes also called global state) of the automaton is given by a control state and a stack content, i.e., by a word from $P\Gamma^*$. The graph $G = (V, (E_a)_{a\in\Sigma})$ is now specified as follows:

- V is the set of configurations in $P\Gamma^*$ which are reachable (via finitely many applications of transitions of Δ) from the initial configuration p_0Z_0.
- E_a is the set of all pairs $(p\gamma w, qvw)$ from $V \times V$ for which there is a transition (p, a, γ, v, q) in Δ.

Then the edge relation E coincides with the one-step derivation relation $p_1w_1 \vdash p_2w_2$ over V, and the transitive closure E^* with the derivability relation $\vdash^*$.

A more general class of graphs, which includes the case of vertices of infinite degree, consists of the "prefix-recognizable graphs" (introduced by Caucal [15]). These graphs are defined in terms of prefix-rewriting systems in which "control states" (as they occur in pushdown automata) are no longer used and where a word on the top of the stack (rather than a single letter) may be rewritten. Thus, a rewriting step can be specified by a triple (u_1, a, u_2), describing a transition from a word $u_1 w$ via letter a to the word $u_2 w$. The feature of infinite degree is introduced by allowing generalized rewriting rules of the form $U_1 \xrightarrow{a} U_2$ with regular sets U_1, U_2 of words. Such a rule leads to the (in general infinite) set of rewrite triples (u_1, a, u_2) with $u_1 \in U_1$ and $u_2 \in U_2$. A graph $G = (V, (E_a)_{a \in \Sigma})$ is called *prefix-recognizable* if for some finite system $\mathcal{S}$ of such generalized prefix rewriting rules $U_1 \xrightarrow{a} U_2$ over an alphabet Γ, we have

- $V \subseteq \Gamma^*$ is a regular set,
- E_a consists of the pairs $(u_1 w, u_2 w)$ where $u_1 \in U_1$, $u_2 \in U_2$ for some rule $U_1 \xrightarrow{a} U_2$ from $\mathcal{S}$, and $w \in \Gamma^*$.

Example 16.6. The structure $(\mathbb{N}, \mathrm{Succ}, <)$ is prefix recognizable. We write the structure as $(\mathbb{N}, E_a, E_b)$ and represent numbers by sequences over the one-letter alphabet with the symbol $|$ only. So $V = |^*$, and the two relations E_a, E_b are defined by the prefix rewriting rules $\varepsilon \xrightarrow{a} |$ and $\varepsilon \xrightarrow{b} |^+$.

The prefix-recognizable graphs coincide with the pushdown graphs when ε-rules are added to pushdown automata and edges are defined in terms of transitions in the composed relation $\xrightarrow{\varepsilon}^* \circ \xrightarrow{a} \circ \xrightarrow{\varepsilon}^*$.

Before turning to a closer analysis of pushdown graphs and prefix-recognizable graphs, let us settle the inclusion relations between the four classes of graphs introduced so far.

Theorem 16.6. *The pushdown graphs, prefix-recognizable graphs, automatic graphs, and rational graphs constitute, in this order, a strictly increasing inclusion chain of graph classes.*

Proof. For the proof, we first note that the prefix-recognizable graphs are clearly a generalization of the pushdown graphs and that the rational graphs generalize the automatic ones. To verify that a prefix-recognizable graph is automatic, we first proceed to an isomorphic graph which results from reversing the words under consideration, at the same time using suffix rewriting rules instead of prefix rewriting ones. Given this format of the edge relations, we can verify that it is automatic: Consider a word pair (wu_1, wu_2) which results from the application of a suffix rewriting rule $U_1 \xrightarrow{a} U_2$, with regular U_1, U_2 and $u_1 \in U_1$, $u_2 \in U_2$. A nondeterministic automaton can easily check this property of the word pair by scanning the two components simultaneously letter by letter, guessing when the common prefix w of the two components is passed, and then verifying (again proceeding letter by letter) that the remainder u_1 of the first component is in U_1 and the remainder u_2 of the second component is in U_2.

The strictness of the inclusions may be seen as follows. The property of having bounded degree separates the pushdown graphs from the prefix-recognizable ones (see Example 16.6). To distinguish the other graph classes, one may use logical decidability results. It will be shown in Section 16.5.3 that the monadic second-order theory of a prefix-recognizable graph is decidable, which fails for some automatic graphs (Theorem 16.5). Furthermore, the first-order theory of an automatic graph is decidable (Theorem 16.3), which fails in general for the rational graphs (Theorem 16.2). □

The next two subsections show two decidability results on transition systems that are generated in terms of prefix rewriting. First we show that reachability over pushdown systems is decidable, then that the MSO-theory of a prefix-recognizable graph is decidable. The second result is of course much stronger, both regarding the class of graphs and the class of properties addressed. However, it seems useful to present the weaker result (on mere reachability) since the proof method is important and leads to a polynomial-time procedure.

16.5.2. *Reachability over Pushdown Graphs*

In this section it is convenient to consider unlabelled pushdown graphs rather than pushdown automata; so we abstract from the input alphabet, the initial state, and the initial stack symbol. We work with pushdown systems in the format $\mathcal{P} = (P, \Gamma, \Delta)$ where P is the set of control states, Γ the stack alphabet, and $\Delta \subseteq P \times \Gamma \times \Gamma^* \times P$ the finite set of transitions. For a set $T \subseteq P\Gamma^*$ of "target configurations" let

$$\text{pre}^*(T) = \{pv \in P\Gamma^* \mid \exists qw \in T : pv \vdash^* qw\}$$

We show the following fundamental result which (in different terminology) goes back to Büchi [16]:

Theorem 16.7. *Given a pushdown automaton $\mathcal{P} = (P, \Sigma, \Gamma, p_0, Z_0, \Delta)$ and a finite automaton recognizing a set $T \subseteq P\Gamma^*$, one can compute a finite automaton recognizing* $\text{pre}^*(T)$.

We can then decide the reachability of a configuration p_2w_2 from p_1w_1 by setting $T = \{p_2w_2\}$ and checking whether the automaton recognizing $\text{pre}^*(T)$ accepts p_1w_1.

The transformation of a given automaton $\mathcal{A}$ which recognizes T into the desired automaton $\mathcal{A}'$ recognizing *pre*$^*(T)$ works by a simple process of "saturation", which involves adding more and more transitions but leaves the set of states unmodified. This construction, which improves the original one by Büchi regarding efficiency, appears in several sources, among them [17], [18], and [19]; we follow the latter. It is convenient to work with P as the set of initial states of $\mathcal{A}$; so a configuration pw of the pushdown automaton is scanned by $\mathcal{A}$ starting from state p and then processing the letters of w. This use of P as the set of initial states of $\mathcal{A}$ motivates the term P-automaton in the literature. The P-automata we use for specifying T do not have transitions into P; we call them *normalized.*

The saturation procedure is based on the following idea: Suppose a pushdown transition allows to rewrite the configuration $p\gamma w$ into qvw, and that the latter one is accepted by $\mathcal{A}$. Then the configuration $p\gamma w$ should also be accepted. If $\mathcal{A}$ accepts qvw by a run starting in state q and reaching, say, state r after processing v, we enable the acceptance of $p\gamma w$ by adding a direct transition from p via γ to r. The saturation algorithm performs such insertions of transitions as long as possible.

Saturation Algorithm:
Input: P-automaton $\mathcal{A}$, pushdown system $\mathcal{P} = (P, \Gamma, \Delta)$
$\mathcal{A}_0 := \mathcal{A}$, $i := 0$
REPEAT:
  IF $pa \to qv \in \Delta$ and $\mathcal{A}_i : q \xrightarrow{v} r$ THEN
    add (p, a, r) to $\mathcal{A}_i$ and obtain $\mathcal{A}_{i+1}$
    $i := i + 1$
UNTIL no transition can be added
$\mathcal{A}' := \mathcal{A}_i$
Output: $\mathcal{A}'$

As an example consider $\mathcal{P} = (P, \Gamma, \Delta)$ with $P = \{p_0, p_1, p_2\}$, $\Gamma = \{a, b, c\}$, $\Delta = \{(p_0a \to p_1ba), (p_1b \to p_2ca), (p_2c \to p_0b), (p_0b \to p_0)\}$ and $T = \{p_0aa\}$. The P-automaton for T is the following:

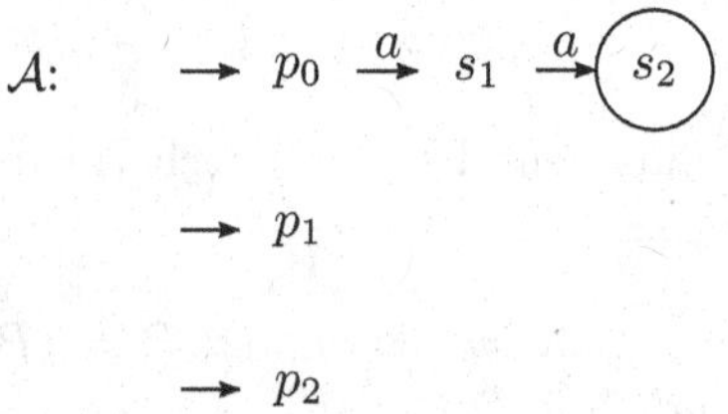

$\mathcal{A}$: $\to p_0 \xrightarrow{a} s_1 \xrightarrow{a} (s_2)$

$\to p_1$

$\to p_2$

Execution of the saturation algorithm introduces edges as indicated in the following figure. Insertion of $p_0 \xrightarrow{b} p_0$ is based on the rule $p_0b \to p_0$ and $\mathcal{A}_0(= \mathcal{A}) : p_0 \xrightarrow{\varepsilon} p_0$, insertion of $p_2 \xrightarrow{c} p_0$ on the rule $p_2c \to p_0b$ and $\mathcal{A}_1 : p_0 \xrightarrow{b} p_0$, insertion of $p_1 \xrightarrow{b} s_1$ on the rule $p_1b \to p_2ca$ and $\mathcal{A}_2 : p_2 \xrightarrow{ca} s_1$, insertion of $p_0 \xrightarrow{a} s_2$ on the rule $p_0a \to p_1ba$ and $\mathcal{A}_3 : p_1 \xrightarrow{ba} s_2$, and insertion of $p_1 \xrightarrow{b} s_2$ on the rule $p_1b \to p_2ca$ and $\mathcal{A}_4 : p_2 \xrightarrow{ca} s_2$. The final result is the following:

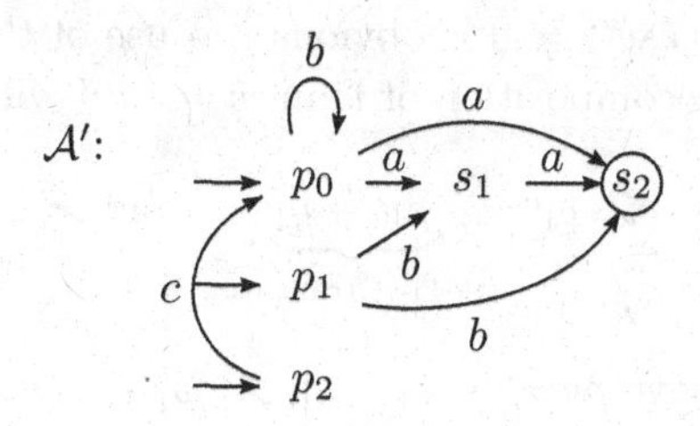

So for $T = \{p_0aa\}$ we extract the following result.

$$pre^*(T) = p_0b^*(a+aa) + p_1b + p_1ba + p_2cb^*(a+aa)$$

Proposition 16.2. *The Saturation Algorithm terminates and gives, for an input automaton $\mathcal{A}$ recognizing T, as output an automaton $\mathcal{A}'$ recognizing* $\mathrm{pre}^*(T)$.

Proof. Termination of the algorithm is clear since new transitions (p, a, q) can be added only finitely often to the given automaton.

Next we have to show:

$$pw \in \mathrm{pre}^*(T) \Leftrightarrow \mathcal{A}' : p \xrightarrow{w} F$$

For the direction from left to right we use induction over the number $n \geq 0$ of steps to get to T and show: $pw \to^n ru \in T \Rightarrow \mathcal{A}' : p \xrightarrow{w} F$.

The case $n = 0$ is obvious. In the induction step assume $pw \to^{n+1} ru$ and $ru \in T$. We have to show that $\mathcal{A}'$ accepts pw. Consider the decomposition of the step sequence to $ru \in T$: $paw' \to p'vw' \to^n ru$ with $w = aw'$ and a pushdown transition $pa \to p'v$. The induction assumption gives $\mathcal{A}' : p' \xrightarrow{vw'} F$. So, there exists an $\mathcal{A}'$-state q with $\mathcal{A}' : p' \xrightarrow{v} q \xrightarrow{w'} F$. Consequently, the saturation algorithm produces the transition $(p, a, q) \in \Delta_{\mathcal{A}'}$, and pw is accepted by $\mathcal{A}'$.

For the direction from right to left we show

$$\mathcal{A}' : p \xrightarrow{w} q \implies \exists p'w' \in P\Gamma^* \text{ such that } \mathcal{A} : p' \xrightarrow{w'} q \wedge pw \vdash^* p'w'$$

For $q \in F$ (the final state-set of $\mathcal{A}$) we obtain the claim; note that $\mathcal{A} : p' \xrightarrow{w'} q$ says that $p'w' \in T$.

We denote by $\mathcal{A}_i$ the P-automaton which originates from $\mathcal{A}$ after i insertions of new transitions by the saturation algorithm. We show inductively over i:

$$\text{If } \mathcal{A}_i : p \xrightarrow{w} q, \text{ then } \exists p'w' \in P\Gamma^* \text{ such that } \mathcal{A} : p' \xrightarrow{w'} q \wedge pw \vdash^* p'w'$$

The case $i = 0$ obvious. For the induction claim assume that $\mathcal{A}_{i+1} : p \xrightarrow{w} q$. Consider an accepting run $\mathcal{A}_{i+1} : p \xrightarrow{w} q$. Let j be the number of applications of the $(i+1)$-st transition that was added by the algorithm. We prove the claim inductively over j. The case $j = 0$ is obvious (no use of the $(i+1)$-st transition). For $j+1$, consider the decomposition of w in $w = uau'$ with

$$\mathcal{A}_i : p \xrightarrow{u} p_1, \quad \mathcal{A}_{i+1} : \underbrace{p_1 \xrightarrow{a} q_1}_{(i+1)\text{-st transition}} \quad \text{and } \mathcal{A}_{i+1} : q_1 \xrightarrow{u'} q$$

By induction (on i) we have $pu \vdash^* p_1' u_1$ with $\mathcal{A} : p_1' \xrightarrow{u_1} p_1$. Since $\mathcal{A}$ is normalized, its initial state p_1 has no ingoing transitions, hence $u_1 = \varepsilon$ and $p_1 = p_1'$; thus $pu \vdash^* p_1$.

The saturation algorithm adds (p_1, a, q_1) to $\mathcal{A}_i$. So, there are p_2 and a pushdown rule $p_1 a \to p_2 v$ with $\mathcal{A}_i : p_2 \xrightarrow{v} q_1$.

Finally, in the run on u', the $(i+1)$-st transition is used $\leq j$ times, so by induction assumption on j, we know for the run $\mathcal{A}_{i+1} : p_2 \xrightarrow{v} q_1 \xrightarrow{u'} q$ that there is $p'w'$ with $\mathcal{A} : p' \xrightarrow{w'} q$ and $p_2 vu' \vdash^* p'w'$.

Altogether we have $pw = puau' \vdash^* p_1 au' \vdash p_2 vu' \vdash^* p'w' (\in T)$. □

It is easily seen that the number of iterations of the saturation algorithm is bounded by the number $|Q|^2 \cdot |\Sigma|$ of possible transitions, and that each iteration only costs polynomial time; hence the saturation algorithm is polynomial.

Our treatment of the reachability problem was based on the idea of backward search: From a regular target set T we worked backwards and obtained the regular set $\text{pre}^*(T)$. In an analogous way one can work forward, then proceeding from a set C of configurations to $\text{post}^*(C)$, the set of configurations that are reachable from configurations in C. For discussion of this approach and applications in verification we refer the reader to the chapter [20] of this handbook.

The idea of the saturation algorithm has been transferred to many related problems, for example for solving reachability problems over higher-order pushdown graphs [21], for checking "recurrent reachability" over pushdown graphs [19], for two-player reachability games played on pushdown graphs [22], and for reachability over transition graphs associated with tree rewriting systems (see [23; 24] and Section 16.7 below).

16.5.3. *The MSO-Theory of Pushdown Graphs*

The aim of this section is to show that the MSO-theory of a prefix-recognizable graph is decidable. The starting point is a deep and difficult decidability result, "Rabin's Tree Theorem", which we use here without proof. A self-contained exposition is in [25].

Theorem 16.8. (Rabin [14])
The MSO-theory of the infinite binary tree T_2 is decidable.

In order to proceed from the binary tree to prefix-recognizable graphs we apply the method of interpretation. The idea is to describe (using MSO-formulas) a structure $\mathcal{A}$ in another structure $\mathcal{B}$ whose MSO-theory is known to be decidable. Once such a description is possible, one can derive that also the MSO-theory of $\mathcal{A}$ is decidable. In our case, the structure $\mathcal{A}$ is a prefix-recognizable graph and $\mathcal{B}$ the binary tree T_2.

Let us first illustrate the idea of MSO-interpretation by showing that the MSO-theory of the n-branching tree T_n is decidable also for $n > 2$. As typical example consider $T_3 = (\{0,1,2\}^*, S_0^3; S_1^3, S_2^3)$. We obtain a copy of T_3 in T_2 by considering only the T_2-vertices in the set $T = (10 + 110 + 1110)^*$. A word in this set has the form $1^{i_1}0 \ldots 1^{i_m}0$ with $i_1, \ldots, i_m \in \{1,2,3\}$; and we take it as a representation of the element $(i_1 - 1) \ldots (i_m - 1)$ of T_3.

The following MSO-formula $\varphi(x)$ (written in abbreviated suggestive form, using successor functions rather than successor relations) defines the set T in T_2:

$$\forall Y[Y(\varepsilon) \wedge \forall y(Y(y) \rightarrow (Y(y10) \wedge Y(y110) \wedge Y(y1110))) \rightarrow Y(x)]$$

It says that x is in the closure of ε under 10-, 110-, and 1110-successors. The relation $\{(w, w10) | w \in \{0,1\}^*\}$ is defined by the following formula:

$$\psi_0(x,y) := \exists z(S_1(x,z) \wedge S_0(z,y))$$

With the analogous formulas ψ_1, ψ_2 for the other successor relations, we see that the structure with universe φ^{T_2} and the relations $\psi_i^{T_2}$ restricted to φ^{T_2} is isomorphic to T_3.

In general, an MSO-interpretation of a structure $\mathcal{A}$ in a structure $\mathcal{B}$ is given by a "domain formula" $\varphi(x)$ and, for each relation $R^{\mathcal{A}}$ of $\mathcal{A}$, say of arity m, an MSO-formula $\psi(x_1, \ldots, x_m)$, such that $\mathcal{A}$ with the relations $R^{\mathcal{A}}$ is isomorphic to the structure with universe $\varphi^{\mathcal{B}}$ and the relations $\psi^{\mathcal{B}}$ restricted to $\varphi^{\mathcal{B}}$.

Then for an MSO-sentence χ (in the signature of $\mathcal{A}$) one can construct a sentence χ' (in the signature of $\mathcal{B}$) such that $\mathcal{A} \models \chi$ iff $\mathcal{B} \models \chi'$. In order to obtain χ' from χ, one replaces every atomic formula $R(x_1, \ldots, x_m)$ by the corresponding formula $\psi(x_1, \ldots, x_m)$ and one relativizes all quantifications to $\varphi(x)$. As a consequence, we note the following:

Proposition 16.3. *If $\mathcal{A}$ is MSO-interpretable in $\mathcal{B}$ and the MSO-theory of $\mathcal{B}$ is decidable, then so is the MSO-theory of $\mathcal{A}$.*

As a second example of MSO-interpretation, consider a pushdown automaton $\mathcal{A}$ with stack alphabet $\{0, \ldots, k-1\}$ and states $q_1, \ldots, q_m$. Let $G_{\mathcal{A}} = (V_{\mathcal{A}}, E_{\mathcal{A}})$ be its configuration graph. Choosing $n = \max\{k, m\}$, we can exhibit an MSO-interpretation of $G_{\mathcal{A}}$ in T_n: Just represent configuration $q_j i_1 \ldots i_r$ by the vertex $i_r \ldots i_1 j$ of T_n. For example, the configuration $(i, 001)$ is represented by the tree node $100i$. Applying the pushdown rule $(i, 0, 11, j)$ we obtain the new tree node $1011j$. The application of this rule thus corresponds to a step from a tree node $u0i$

to $u11j$. So the one-step relation of the transition $\tau = (i, 0, 11, j)$ is described by the formula (in short notation, again using successor functions rather than successor relations)

$$\varphi_\tau(x, y) = \exists z(x = z0i \wedge y = z11j)$$

The transition relation of the configuration graph is thus defined by

$$\bigvee_{\tau \in \Delta} \varphi_\tau,$$

and the domain of the configuration graph is easily defined as the closure of the initial configuration under the transition relation.

Hence we obtain the following result of Muller, Schupp [26]:

Proposition 16.4. *The MSO-theory of a pushdown graph is decidable.*

By an easy generalization of the proof we obtain the corresponding statement for the prefix-recognizable graphs. The difference to the proof above is just a refinement of the formula φ_τ expressing the one-step derivation relation between configurations induced by a transition τ. Instead of describing a single move from one word to another, say from wap to $wbbq$, we have to describe all admissible moves from words wu to words wv where $u \in U, v \in V$ for a prefix-rewriting rule $U \to V$. (Since we deal with the representation of configurations as tree nodes, where the changes occur in the suffix rather than the prefix, we assume that we have reversed the words in U, V in order to match our coding.)

Suppose the sets U, V are recognized by the finite automata $\mathcal{A}_U, \mathcal{A}_V$ with state sets Q_U, Q_V, respectively. In order to describe the application of the rule $\tau = (U \to V)$, we write down a formula $\varphi_\tau(x, y)$ that expresses the following:

> there are z, u, v s.t. $x = zu$, $y = zv$ and on the path segment from z to $x = zu$, from z to zv, respectively, the automaton $\mathcal{A}_U$, respectively $\mathcal{A}_V$ has an accepting run.

The existence claims on the accepting runs are easily formalizable using quantifications over sets. Let us consider the case of $\mathcal{A}_U$, where $Q_U = \{1, \ldots, k\}$ and, for example, 1 is the initial and k the only final state. We express that there are k subsets $X_1, \ldots, X_k$ that form a partition of the path segment $\{z, \ldots, zu\}$, where the set X_i is intended to contain those vertices where state i is visited in the run. The property of being a successful run for these sets X_i is captured by three clauses, namely that the vertex z belongs to X_1, zu belongs to X_k (since k was the only final state), and that for any vertex s on the path from z to (and excluding) zu, a disjunction over the $\mathcal{A}_U$-transitions $\tau = (i, a, j)$ holds. Such a disjunction member for (i, a, j) expresses that $s \in X_i$, the next vertex of the path to zu is the node sa, and $sa \in X_j$.

The domain of the configuration graph is defined as for the case of pushdown graphs. Hence we have proved the following result, using again the interpretation in a suitable tree T_n.

Theorem 16.9. (Caucal [15]) *The monadic second-order theory of a prefix-recognizable graph is decidable.*

16.6. Unfoldings and the Caucal Hierarchy

The decidability of the MSO-theory of pushdown (and prefix recognizable) graphs can be generalized in two directions, in order to cover more general types of models. First, one tries to widen the class of graphs such that the decidability result on the MSO-theory still holds. This approach is pursued in the present section.

In view of Theorem 16.5, using this appoach we shall not be able to handle simple models such as the infinite grid G_2. In the next section we thus restrict the logic under consideration to the fragment FO(R) of MSO-logic and present a class of graphs that includes G_2 and allows to show decidability of the model-checking problem with respect to FO(R).

In the previous section we considered interpretations as a method to generate a model "within" a given one, via defining formulas. A more "expansive" way of model construction is the unfolding of a graph $(V, (E_a)_{a\in\Sigma}, (P_b)_{b\in\Sigma'})$ from a given vertex v_0, which yields a tree $T_G(v_0) = (V', (E'_a)_{a\in\Sigma}, (P'_b)_{b\in\Sigma'})$: V' consists of the vertices $u_0a_1u_1\ldots a_ru_r$ with $u_0 = v_0$, $(u_i, u_{i+1}) \in E_{a_{i+1}}$ for $i < r$, E'_a contains the pairs $(u_0a_1u_1\ldots a_ru_r, u_0a_1u_1\ldots a_ru_rau)$ with $(u_r, u) \in E_a$, and P'_b the vertices $u_0a_1u_1\ldots a_ru_r$ with $u_r \in P_b$. The unfolding operation has no effect in bisimulation invariant logics, but is highly nontrivial for MSO-logic. Consider, for example, the singleton graph G_0 over $\{v_0\}$ with a 0-labelled and a 1-labelled edge from v_0 to v_0. Its unfolding is the infinite binary tree T_2. While checking MSO-formulas over G_0 is trivial, this is a deep result for T_2. A powerful result going back to Muchnik 1985 implies that unravelling preserves decidability of the MSO-theory.

Theorem 16.10. (Muchnik 1985, Courcelle and Walukiewicz [27])
If the MSO-theory of G is decidable and v_0 is an MSO-definable vertex of G, then the MSO-theory of $T_G(v_0)$ is decidable.

The result holds also for a slightly more general construction ("tree iteration") which can also be applied to relational structures other than graphs. We cannot go into details here; a good presentation is given in [28].

MSO-interpretations and unfoldings are two operations which preserve decidability of MSO model-checking. Caucal [2] studied the structures generated by applying *both* operations, alternating between unfoldings and interpretations. He introduced the following hierarchy $(\mathcal{G}_n)$ of graphs, together with a hierarchy $(\mathcal{T}_n)$ of trees:

- $\mathcal{T}_0$ = the class of finite trees
- $\mathcal{G}_n$ = the class of graphs which are MSO-interpretable in a tree of $\mathcal{T}_n$
- $\mathcal{T}_{n+1}$ = the class of unfoldings of graphs in $\mathcal{G}_n$

By the results of the preceding sections (and the fact that a finite structure has a decidable MSO-theory), each structure in the Caucal hierarchy has a decidable MSO-theory. By a hierarchy result of Damm [29] on higher-order recursion schemes, the hierarchy is strictly increasing (for a new and transparent poof see [30]).

In Caucal's orginal paper [2], a different formalism of interpretation (via "inverse rational substitutions") is used instead of MSO-interpretations. We work with the latter to keep the presentation more uniform; the equivalence between the two approaches has been established by Carayol and Wöhrle [31]. Referring to yet another characterization (see also [31]) in terms of higher-order pushdown systems (that are derived from pushdown automata with nested stacks), one also speaks of the "pushdown hierarchy".

Let us take a look at some structures which occur in this hierarchy (following [32]). It is clear that $\mathcal{G}_0$ is the class of finite graphs, while $\mathcal{T}_1$ contains the so-called regular trees (alternatively defined as the infinite trees which have only finitely many non-isomorphic subtrees). Figure 16.2 (upper half) shows a finite graph and its unfolding as a regular tree.

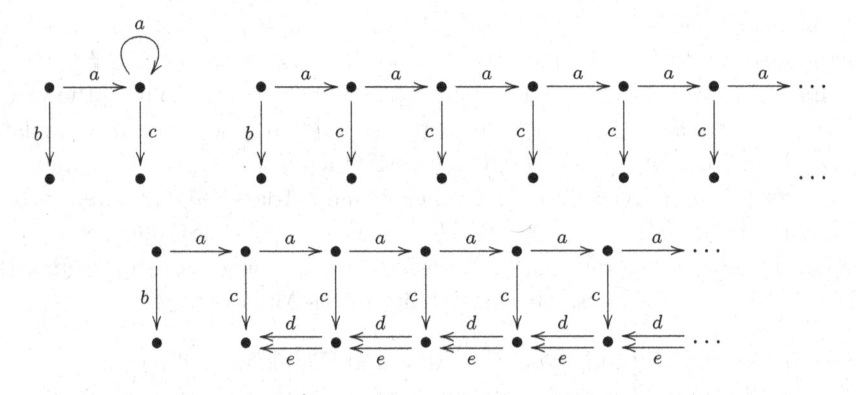

Fig. 16.2. A graph, its unfolding, and a pushdown graph.

By an MSO-interpretation we can obtain the pushdown graph of Figure 16.2 in the class $\mathcal{G}_1$; the domain formula and the formulas defining E_a, E_b, E_c are trivial, while

$$\psi_d(x, y) = \psi_e(x, y) = \exists z \exists z' (E_a(z, z') \wedge E_c(z, y) \wedge E_c(z', x)).$$

Let us apply the unfolding operation again, from the only vertex without incoming edges. We obtain the "algebraic tree" of Figure 16.3, belonging to $\mathcal{T}_2$ (for the moment one should ignore the dashed line).

As a next step, let us apply an MSO-interpretation to this tree which will produce a graph (V, E, P) in the class $\mathcal{G}_2$ (where E is the edge relation and P a unary predicate). Referring to Figure 16.3, V is the set of leaves located along the dashed line, E contains the pairs which are successive vertices along the dashed line, and

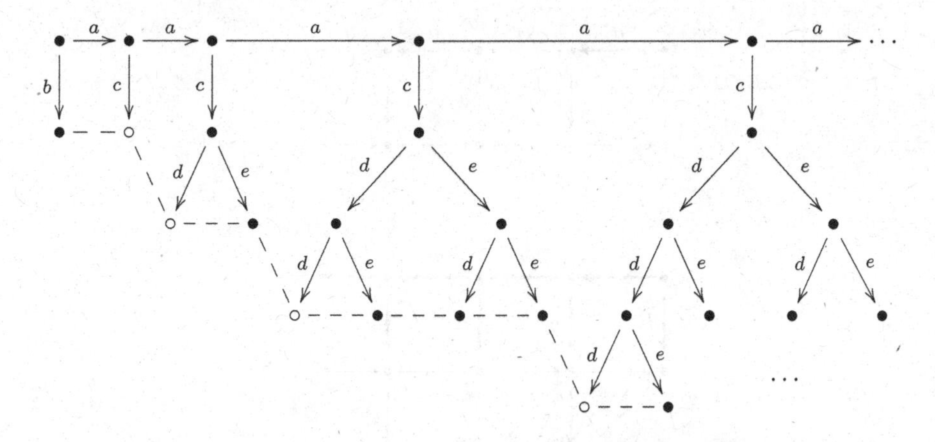

Fig. 16.3. Unfolding of the pushdown graph of Figure 16.2.

P contains the special vertices drawn as non-filled circles. This structure is isomorphic to the structure $(\mathbb{N}, \text{Succ}, P_2)$ with the successor relation Succ and predicate P_2 containing the powers of 2.

To prepare a corresponding MSO-interpretation, we use formulas such as $E_{d^*}(x, y)$ which expresses: "Each set which contains x and is closed under E_d-successors contains y".

As domain formula we use the formula $\varphi(x)$ saying that from x there is no outgoing edge.

The required edge relation E is defined by

$$\psi(x, y) = \varphi(x) \wedge \varphi(y) \wedge \exists z \exists z'(\psi_1(x, y, z, z') \vee \psi_2(x, y, z, z') \vee \psi_3(x, y, z))$$

where

- $\psi_1(x, y, z, z') \;=\; E_a(z, z') \wedge E_b(z, x) \wedge E_c(z', y)$
- $\psi_2(x, y, z, z') \;=\; E_a(z, z') \wedge E_{ce^*}(z, x) \wedge E_{cd^*}(z', y)$
- $\psi_3(x, y, z) \;=\; E_{de^*}(z, x) \wedge E_{ed^*}(z, y)$

Finally we define P by the formula $\chi(x) = \varphi(x) \wedge \exists z \exists z'(E_c(z, z') \wedge E_{d^*}(z', x))$.

We infer that the MSO-theory of $(\mathbb{N}, \text{Succ}, P_2)$ is decidable, a result first proved by Elgot and Rabin in 1966 with a different approach.

Let us discuss another interesting structure of this kind, namely the structure $(\mathbb{N}, \text{Succ}, \text{Fac})$ where Fac is the set of factorial numbers. We start from a simpler pushdown graph than the one used above (see upper part of Figure 16.4) and consider its unfolding, which is the comb structure indicated by the thick arrows of the lower part of Figure 16.4.

We number the vertices of the first horizontal line by $0, 1, 2 \ldots$ and call the vertices of the respective column below to be of "level 0", "level 1", "level 2" etc. Now we use the simple MSO-interpretation which takes all tree nodes as domain

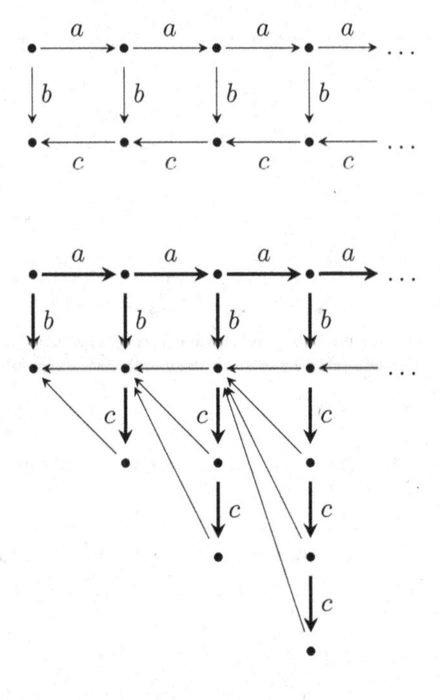

Fig. 16.4. Preparing for the factorial predicate.

and introduces for $n \geq 0$ a new edge from any vertex of level $n+1$ to the first vertex of level n. This introduces the thin edges in Figure 16.4. It is easy to write down a defining MSO-formula. Note that the top vertex of each level plays a special role since it is the target of an edge labelled b, while the remaining ones are targets of edges labelled c.

Consider the tree obtained from this graph by unfolding. It has subtrees consisting of a single branch off level 0, 2 branches off level 1, $3 \cdot 2$ branches off level 2, and generally $(n+1)!$ branches off level n. Via the top-to-bottom order of the c-labelled edges, these branches are arranged from left to right in a natural (and MSO-definable) order. To capture the structure $(\mathbb{N}, \mathrm{Succ}, \mathrm{Fac})$, we apply an interpretation which (for $n \geq 1$) cancels the branches starting at the b-edge target of level n (and leaves only the branches off the targets of c-edges). As a result, $(n+1)! - n!$ branches off level n remain for $n \geq 1$, while there is one branch off level 0. Numbering these remaining branches, the $n!$-th branch appears as first branch off level n. Note that we traverse this first branch off a given level by disallowing c-edges after the first c-edge. So a tree shape similar to Figure 16.3 emerges, now for the factorial predicate. Summing up, we have generated the structure $(\mathbb{N}, \mathrm{Succ}, \mathrm{Fac})$ as a graph in $\mathcal{G}_3$.

There are interesting structures $(\mathbb{N}, \mathrm{Succ}, P)$ (with unary predicate P) for which the decidability of the MSO-theory is unsettled. An example is given by the prime number predicate Prime. If the MSO-theory of $(\mathbb{N}, \mathrm{Succ}, \mathrm{Prime})$ were decidable, one could invoke the decision procedure to solve the (open) twin prime problem (asking whether there are infinitely many pairs of primes with distance 2). On the other hand, an undecidability proof will be difficult since the standard approach (via interpretation of first-order arithmetic in the MSO-theory of $(\mathbb{N}, \mathrm{Succ}, \mathrm{Prime})$) will not work (cf. [33]). However, we know of an expansion $(\mathbb{N}, \mathrm{Succ}, P_0)$ whose MSO-theory is decidable but which does not occur in Caucal's hierarchy. One takes P_0 to consist of the hyperexponentials of 2, i.e. the numbers $2, 2^2, 2^{2^2}$ and so on (see [31]).

So far we have considered expansions of the successor structure of the natural numbers by unary predicates. Only very few (and somehow artificial) examples of binary relations R are known such that the MSO-theory of $(\mathbb{N}, \mathrm{Succ}, R)$ is decidable. Let us mention a unary function (considered as a binary relation): the *flip function*. It associates 0 to 0 and for each nonzero n the number which arises from the binary expansion of n by modifying the least significant 1-bit to 0 (see Figure 16.5).

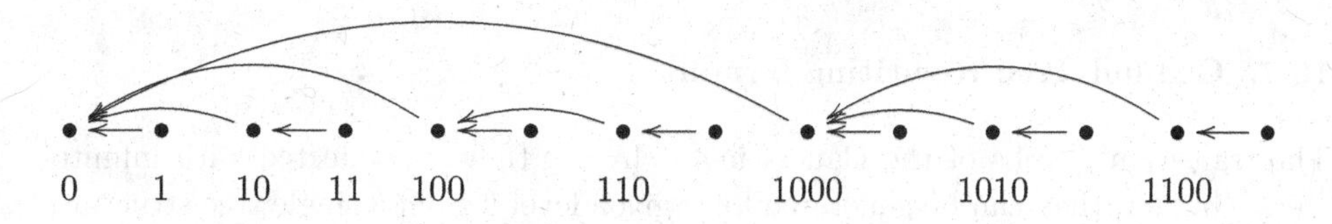

Fig. 16.5. The flip function.

It is easy to see that the structure $(\mathbb{N}, \mathrm{Succ}, \mathrm{Flip})$ can be obtained from the algebraic tree of Figure 16.3 by an MSO-interpretation. A flip-edge will connect vertex u to the last leaf vertex v which is reachable by a d^*-path from an ancestor of u; if such a path does not exist, an edge to the target of the b-edge (representing number 0) is taken.

The graphs in the Caucal hierarchy supply a vast universe of structures which has not been understood very well on the higher levels (say from level 3 onwards). Many interesting questions arise, for example the problem whether one can compute the lowest level on which a given structure that belongs to the hierarchy occurs.

Let us finally discuss the relation of the Caucal hierarchy to the class of automatic structures. The grid G_2 shows that there are automatic graphs outside the Caucal hierarchy (just note that the MSO-theory of G_2 is undecidable; cf. Theorem 16.5). For the converse we use an example of Kuske [34]: The ordinal ordering $(\omega^\omega, <)$ is not automatic (see [35]) but, as we now see, occurs in the Caucal hierarchy. Invoking Cantor's normal form (see, e.g., [36, IV.2.14]), we represent $(\omega^\omega, <)$ as the set of vectors $(k_n, \ldots, k_0)$ of natural numbers (where $k_n > 0, n \geq 0$) with the order by length and the lexicographical order for vectors of same length, preceded by the vector (0). To present this ordering, we start with the graph of

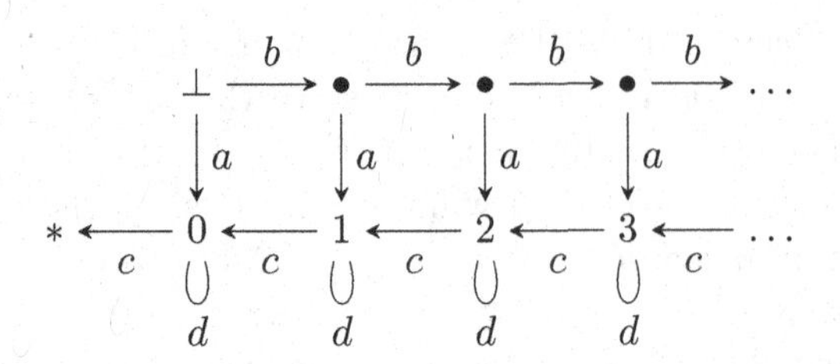

Fig. 16.6. Preparing for the model $(\omega^\omega, <)$.

Figure 16.6 (which belongs to $\mathcal{G}_1$). Its unfolding from $\perp$ yields a tree with paths labelled by words $b^n\ a\ d^{k_n}\ c\ d^{k_{n-1}}\ c \ldots c\ d^{k_0}\ c$. We select the paths with $k_n > 0$ (i.e., where a d-edge follows the a-edge); they correspond to the vectors $v = (k_n, \ldots, k_0)$ with $k_n > 0$. We obtain also $v = (0)$ by adding the path $\perp \xrightarrow{a} 0 \xrightarrow{c} *$. The $*$-labelled leaves of these paths with their left-to-right order (induced by the order $a < b < c < d$ of the edge labels) thus give a copy of $(\omega^\omega, <)$ as a graph in $\mathcal{G}_2$.

16.7. Ground Tree Rewriting Graphs

The transition graphs of the Caucal hierarchy are tightly connected with infinite trees – in fact, they can be generated for a given level k from a single tree structure via MSO-interpretations. For many purposes of verification the graphs in the Caucal hierarchy are too restricted (except for applications in the implementation of higher-order recursion).

A more flexible kind of model is generated when the idea of prefix-rewriting is generalized in a different direction, proceeding from word rewriting to tree rewriting (which we identify here with term rewriting). Instead of modifying the prefix of a word by applying a prefix-rewriting rule, we may rewrite a subtree of a given tree, precisely as it is done in ground term rewriting. We shall speak of "ground tree rewriting". So a rule $t \to t'$ applied to some tree s allows to replace one occurrence of subtree t of s by t'. To fix state properties, we refer to the well-known concept of regular sets of trees, defined by finite tree automata (see the capter [37] of this volume for an introduction).

A *ground tree rewriting graph (GTRG)* $G = (V, (E_a)_{a\in\Sigma}, (P_b)_{b\in\Sigma'})$ has a vertex set V consisting of finite trees. The subsets $P_b \subseteq V$ are given by regular tree languages, and each edge relation E_a is defined by a finite ground tree rewriting system. Usually one restricts V to contain only trees which are reachable from some regular set of initial trees via the edge relations E_a.

The concept is best introduced by an example. Consider the graph generated from the tree $f(c,d)$ by applying the rules $c \to g(c)$ and $d \to g(d)$ which produce the trees $f(g^i(c), g^j(d))$ in one-to-one correspondence with the elements (i,j) of $\mathbb{N} \times \mathbb{N}$ (see Figure 16.7).

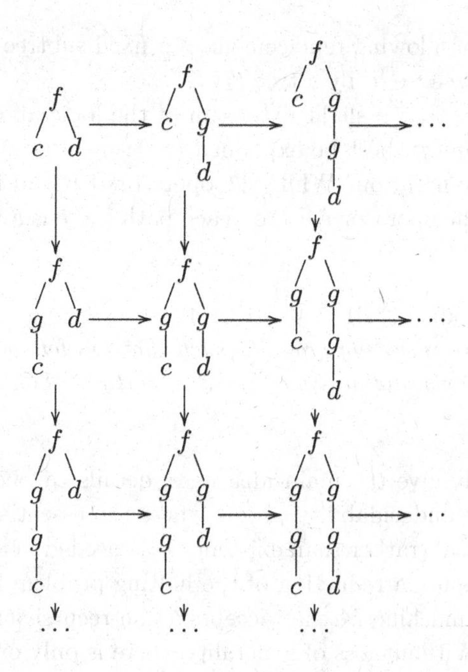

Fig. 16.7. The grid as a ground tree rewriting graph.

We thus see that the infinite $\mathbb{N} \times \mathbb{N}$-grid G_2 is a GTRG. Hence the MSO-theory of a GTRG can be undecidable. (Since G_2 is automatic, we know that the FO-theory of G_2 is decidable.)

However, for interesting properties beyond FO-logic the model-checking problem is still decidable. It is possible to combine the techniques of Section 2 (on automatic graphs) and of Section 3 (saturation algorithm), now applied over the domain of finite trees rather than words. Since the methodology does not change, we only state the result. In the second claim of the theorem below we refer to operators of the logic CTL*, namely

- $\mathrm{EX}_a\varphi$ for "there is an a-labelled edge to a successor state satisfying φ",
- $\mathrm{EF}\varphi$ for "there is a finite path to a state satisfying φ",
- $\mathrm{EGF}\varphi$ for "there is an infinite path with infinitely many occurrences of states satisfying φ".

Theorem 16.11. (Dauchet, Tison [23], Löding [24])
Over a ground tree rewriting graph, the model-checking problem is decidable for the logic FO(R), *and also for the branching-time logic with atomic formulas for regular state properties (specified by tree automata), the Boolean connectives, and the operators* EX_a, EF, *and* EGF.

As for the step from pushdown graphs to prefix-recognizable graphs, it is possible to generalize the rewriting rules to the format $T \to T'$ with regular tree languages

T, T'. Here, instead of allowing replacement of a fixed subtree by another one, one may replace any subtree $t \in T$ by a tree $t' \in T'$.

We now shall note that a slight extension of the logic above leads to undecidability. This extension can best be explained in terms of branching time temporal operators in CTL-like notation: While the operators EF and EGF preserve decidability, this fails for the operator AF ("on each path there is a vertex with a certain property").

Theorem 16.12. (Löding [24])
There is a ground tree rewriting graph G such that the following problem is undecidable: Given a vertex v and a regular set T of vertices of G, does every path from v through G reach T?

Proof. We can only give the main idea here; details can be found in [24]. The method is typical for undecidability proofs where the essential logical operator to be exploited is universal (rather than existential, as needed in a direct coding of the halting problem). We use a reduction of the halting problem for Turing machines, considering a Turing machine M that accepts a non-recursive (but recursively enumerable) language. Without loss of generality, there is only one accepting configuration c_{acc}. We represent a Turing machine configuration $c = a_1 \dots a_k \; q \; b_\ell \dots b_1$ by a tree t_c with two branches: From the top node with label $\bullet$, we have a unary left-hand branch whose nodes are labelled $X, a_1, \dots, a_k$, and a unary right-hand branch with labels $X, b_1, \dots, b_\ell, q$. So the left-hand branch ends with the symbol that is left to the current work cell of the Turing machine, and the right-hand branch ends with the symbol in the work cell and the current state of the Turing machine. Let t_{acc} be the tree coding the configuration c_{acc}.

The task is to set up ground rewriting rules that simulate steps of the Turing machine M. The main problem for a correct update of a tree t_c, coding a Turing machine configuration c, is the fact that one has to use several rewriting steps, independently on the left-hand and on the right-hand branch, to simulate a change of c. Without giving details, let G_M be the ground tree rewriting graph given by these rewriting rules.

One cannot eliminate the possibility that rewriting steps carried out on the left-hand branch and on the right-hand branch do not correspond to a correct transformation (according to a Turing machine step). The main idea is now to specify a regular set R of "admissible" trees which collects all trees generated during "correct" updates according to Turing machine steps. One can fix R in such a way that any application of rules that does not conform to a Turing machine step will eventually lead outside the set R. Let us call T_{error} the complement of R; clearly this tree language is regular. Let $T = T_{\text{error}} \cup \{t_{\text{acc}}\}$, which is again a regular set.

Given this, the claim of the theorem follows easily: For each input word w of M, M will accept w iff in the graph G_M, each path from the tree coding the configuration $q_0 w$ will meet T. □

Theorem 16.12 extends to several other variants of the reachability problem where the universal quantifier enters. We mention three such variants (see also [24]): For example, instead of the CTL modality AF (expressing termination) one may consider the CTL modality EU (where $\mathrm{E}(\varphi \mathrm{~U~} \psi)$ means that there exists a path to a vertex v satisfying ψ such that for all vertices of the path up to v, φ is true). Also we obtain undecidability for *regular reachability* over ground tree rewriting graphs; here we consider the extension FO(Reg) of FO (see Section 16.2), where for each regular expression r we allow the atomic formula $E_r(x, y)$, meaning that there is a path from x to y whose edge label sequence satisfies r. Finally, the undecidability result holds also for *alternating reachability*: Here one assumes that from vertex v two players, called 1 and 2, build up a path by choosing successive edges in alternation; the target set T is said to be "reachable" from v if Player 2 has a strategy to guarantee a visit to a vertex of T.

So Theorem 16.12 and the subsequent remarks indicate rather severe limitations for showing decidability of generalized reachability properties over ground tree rewriting graphs.

The class of ground tree rewriting graphs and the Caucal hierarchy are two incompatible extensions of the class of pushdown graphs. The grid G_2 is an example of a ground tree rewriting graph that does not belong to the Caucal hierarchy. On the other hand, by [38], ground tree rewriting graphs of bounded tree-width are isomorphic to pushdown graphs. So a tree on the second level of the Caucal hierarchy cannot be presented as a ground tree rewriting graph.

16.8. Completing the Picture

16.8.1. *Internal vs. External Presentations*

We have discussed four basic types of infinite transition graphs: the rational, automatic, prefix-recognizable, and the ground tree rewriting graphs. As specialization of the prefix-recognizable graphs we considered the pushdown graphs, and as a generalization of prefix-recognizable graphs the graphs of the Caucal hierarchy.

For the definition of these structures, two approaches were pursued:

- the internal presentation in terms of automaton definable sets and relations of words, respectively trees,
- the external presentation by means of model transformations (such as interpretations or unfoldings), starting from certain fundamental structures (in our case, finite trees or the structure T_2).

It can be shown that in many cases the two approaches can be merged. In [45] it is shown that a transition graph is automatic iff is can be obtained by a FO-interpretation from the binary tree structure $T_2' = (\{0,1\}^*, S_0, S_1, \mathrm{EquLev})$ where EquLev is the "equal level predicate". A corresponding result for prefix-recognizable graphs and MSO-interpretations in the (standard) binary tree T_2 was shown by Blumensath [44] (see also Chapter 15 of [49]). There are analogous results on rational graphs ([8]), on the graphs of the Caucal hierarchy (in terms of the so-called higher-order pushdown graphs; see, e.g., [31]), and on the ground tree rewriting graphs ([48]).

The combination of both views (internal and external) is necessary for developing a nice algorithmic theory of infinite structures. Usually, the internal description is helpful in devising efficient algorithmic solutions, and the external presentation gives a convenient way of generating models without entering too much into "details of implementation". In classical mathematics, these two views are standard and complement each other. For example, if we specify a vector space by a basis (and the rule that linear combinations over the basis generate the elements of the space), we give an internal representation. If we take all linear maps over some vector space to construct a new vector space, we are building an external presentation.

16.8.2. *Structural Characterizations*

In order to separate classes of graphs as introduced in this chapter, "structural characterizations" would be useful that do not involve a reference to the presentations. We mention a master example of such a characterization, due to Muller and Schupp, that is concerned with pushdown graphs.

Let $G = (V, (E_a)_{a \in \Sigma})$ be a graph of bounded degree and with designated "origin" vertex v_0. Let V_n be the set of vertices whose distance to v_0 is at most n (via paths formed from edges as well as reversed edges). Define G_n to be the subgraph of G induced by the vertex set $V \setminus V_n$, calling its vertices in $V_{n+1} \setminus V_n$ the "boundary vertices". The *ends* of G are the connected components (using edges in both directions) of the graphs G_n with $n \geq 0$. In [26], Muller and Schupp established a beautiful characterization of pushdown graphs in terms of the isomorphism types of their ends (where an end isomorphism is assumed to respect the vertex property of being a boundary vertex):

Theorem 16.13. (Muller, Schupp [26])
A transition graph G of bounded degree is a pushdown graph iff the number of distinct isomorphism types of its ends is finite.

As an application, we see directly (i.e., without resorting to (un-) decidability results on model-checking) that the infinite $(\mathbb{N} \times \mathbb{N})$-grid is not a pushdown graph. The ends G_n exclude all vertices from the origin up to distance n. The vertices of distance precisely n form a counter-diagonal from vertex $(0, n)$ to vertex $(n, 0)$. This counter-diagonal shows in particular that no two graphs G_m, G_n for $m \neq n$ are isomorphic.

A second structural characterization of pushdown graphs in terms of ground tree rewriting graphs is due to Löding [38] (and was already mentioned at the end of Section 16.7): A ground tree rewriting graph is of bounded tree-width iff it is isomorphic to a pushdown graph.

For many graph classes discussed in this chapter, elegant structural characterizations are still missing.

16.8.3. *Recognized Languages*

A transition graph $G = (V, (E_a)_{a\in\Sigma}, I, F)$ with unary predicates $I, F \subseteq V$ (of "initial" and "final" vertices) may be used as an acceptor of words in the obvious way: A word is accepted if it occurs as a labelling of a path from a vertex in I to a vertex in F.

If V is finite, we obtain the usual model of nondeterministic finite automata (here with several initial states), which yields the regular languages as corresponding class of languages. It is not surprising that the pushdown graphs (and, as it is easily verified, also the prefix-recognizable graphs) yield precisely the context-free languages:

Theorem 16.14. (Muller-Schupp [26], Caucal [15])
A language L is context-free iff L is recognized by a pushdown graph (with regular sets of initial and final states) iff L is recognized by a prefix-recognizable graph (with regular sets of initial and final states).

This track of research was continued by surprising results regarding the rational and automatic graphs:

Theorem 16.15. (Morvan-Stirling [39], Rispal [50])
A language L is context-sensitive iff L is recognized by an automatic graph (with regular sets of initial and final states) iff L is recognized by a rational graph (with regular sets of initial and final states).

For an exposition of this theorem as well as several variants we recommend [40].

The graphs of the Caucal hierarchy also correspond to known language classes which have been introduced in terms of "higher-order pushdown automata". For instance, the languages recognized by Caucal graphs of level 2 coincide with the "indexed languages" introduced in the 1960's by Aho [41]. It is an open problem to provide a corresponding description for the languages recognized by ground tree rewriting graphs.

16.9. Retrospective and Outlook

In this chapter we gave an introduction to fundamental classes of infinite transition graphs defined in terms of automata, with some emphasis on the question which types of model-checking problems can be solved algorithmically.

Let us summarize some central ideas:

- The reduction of the Post Correspondence Problem and of the Halting Problem for Turing machines to simple questions about rational and automatic graphs,
- the decidability of the FO-theory of an automatic graph using an inductive construction of automata for definable relations,
- the reachability analysis for pushdown systems using the saturation algorithm,
- the method of interpretations, used to show that the MSO-theory of a prefix-recognizable graph is decidable, and the combination of interpretations and unfoldings for building up the Caucal hierarchy,
- the role of the infinite grid, as a structure with an undecidable MSO-theory but – as a ground tree rewriting graph – sharing still some decidability properties,
- the undecidability of properties over ground tree rewriting graphs that involve universal path quantification.

The subject of finitely presented infinite structures using automata theoretic ideas is fastly developing. Many tracks of research are pursued. We mention just a few:

- The application of grammars for the generation of infinite graphs (see [42]),
- the systematic study of *all* possible automatic / prefix recognizable presentations of a structure and their relation; in particular the influence of presentations on the efficiency of algorithms,
- the consideration of more transformations for the generation of models, for example different kinds of products or variants of the unfolding operation (for example, using sets rather than sequences as elements of the new model); see e.g. [43],
- the generation of more general structures than graphs (e.g., hypergraphs),
- better insight into the gap between FO and MSO (by interesting intermediate logics), and similarly between automatic and pushdown graphs (by interesting intermediate types of graphs),
- a merge of the theory of infinite transition systems with other sources of infinity, especially arithmetical constraints over infnite domains such as $\mathbb{N}$ and $\mathbb{R}$.

Acknowledgments

Many thanks are due to Christof Löding for his remarks on a preliminary version of this paper and to the two anonymous referees for their very careful reading and helpful suggestions.

References

[1] B. Courcelle, The expression of graph properties and graph transformations in monadic second-order logic, in *Handbook of Graph Grammars*, pp. 313-400, World Scientific, (1997).
[2] D. Caucal. On infinite graphs having a decidable monadic theory. In *Proc. 27th MFCS*, vol. 2420, *LNCS*, pp. 165-176. Springer, (2002).
[3] P.A. Abdulla and B. Jonsson, Verifying programs with unreliable channels, *Inform. and Comput.* **127**, 91–101, (1996).
[4] A. Finkel and Ph. Schnoebelen, Well-structured transition systems everywhere!, *Theor. Comput. Sci.* **256**, 63–92, (2001).
[5] H. D. Ebbinghaus, J. Flum and W. Thomas, *Mathematical Logic.* (Springer 1994).
[6] J. Berstel, *Transductions and Context-Free Languages.* (Teubner Verlag, 1979).
[7] J. Sakarovitch, *Eléments de Théorie des Automates*, Vuibert, Paris 2003. Engl. transl. *Elements of Automata Theory*, Cambridge Univ. Press, to appear.
[8] C. Morvan, On rational graphs. In *Proc. 3rd FoSSaCS*, vol. 1784, *LNCS*, pp. 252–261. Springer, (2000).
[9] W. Thomas, A short introduction to infinite automata. In *Proc. 5th DLT*, vol. 2295, *LNCS*, pp. 130-144. Springer. (2002).
[10] J. E. Hopcroft and J. D. Ullman, *Introduction to Automata Theory, Languages, and Computation.* (Addison-Wesley, 1979).
[11] J.R. Büchi, Weak second-order arithmetic and finite automata, *Z. Math. Logik Grundl. Math.* **6**, 66–92, (1960).
[12] F. Klaedtke, Bounds on the automata size for Presburger arithmetic, *Trans. Comput. Log.* **9**(2), ACM, (2008).
[13] B. Boigelot and P. Wolper. Representing arithmetic constraints with automata: An overview. In *Proc. 18th Logic Programming*, vol. 2401, *LNCS*, pp. 1–19. Springer, (2002).
[14] M. O. Rabin, Decidability of second-order theories and automata on infinite trees, *Trans. Amer. Math. Soc.* **141**, 1–35, (1969).
[15] D. Caucal. On infinite transition graphs having a decidable monadic theory. In *Proc. 23rd ICALP*, vol. 1099, *LNCS*, pp. 194–205. Springer, (1996). Full version in: *Theor. Comput. Sci.* **290**, 79–115, (2003).
[16] J. R. Büchi, Regular canonical systems, *Archiv Math. Logik und Grundlagenforschung* **6**, 91–111, (1964).
[17] R. V. Book and F. Otto, *String-Rewriting Systems.* (Springer, 1993).
[18] J. L. Coquidé, M. Dauchet, R. Gilleron and S. Vágvölgyi, Bottom-up tree pushdown automata: Classification and connection with rewrite systems, *Theor. Comput. Sci.* **127**, 69–98, (1994).
[19] J. Esparza, D. Hansel, P. Rossmanith and S. Schwoon. Efficient algorithms for model-checking pushdown systems. In *Proc. 12th CAV*, vol. 1855, *LNCS*, pp. 232–247. Springer, (2000).
[20] J. Esparza and J. Kreiker, Three case studies on verification of infinite-state systems, this volume.
[21] M. Hague and L. Ong. Symbolic backwards-reachability analysis for higher-order pushdown systems. In *Proc. 11th FoSSaCS*, vol. 4423, *LNCS*, pp. 213–227. Springer, (2007).
[22] T. Cachat, Symbolic strategy synthesis for games on pushdown graphs. In *Proc. 29th ICALP*, vol. 2380, *LNCS*, pp. 704–715. Springer, (2002).

[23] M. Dauchet and S. Tison, The theory of ground rewrite systems is decidable, *Proc. 5th LICS*, pp. 242–248. IEEE, (1990).
[24] C. Löding, Reachability problems on regular ground tree rewriting graphs. *Theory of Computing Systems* **39**, 347–383, (2006).
[25] W. Thomas, Languages, automata, and logic, in: *Handbook of Formal Languages vol. 3*, pp. 389–455. Springer, (1997).
[26] D. Muller and P. Schupp, The theory of ends, pushdown automata, and second-order logic, *Theor. Comput. Sci.* **37**, 51–75, (1985).
[27] B. Courcelle and I. Walukiewicz, Monadic second-order logic, graph coverings and unfoldings of transition systems, *Ann. Pure Appl. Logic* **92** 51–65, (1998).
[28] D. Berwanger and A. Blumensath. The monadic theory of tree-like structures. In *Proc. Automata, Logics, and Infinite Games*, vol. 2500, *LNCS*, pp. 285–302. Springer, (2002).
[29] W. Damm, The IO- and OI-hierarchies, *Theor. Comput. Sci.* **20**, 95–207, (1982).
[30] A. Blumensath, On the structure of graphs in the Caucal hierarchy, *Theor. Comput. Sci.* **400**, 19–45, (2008).
[31] A. Carayol and S. Wöhrle. The Caucal hierarchy of infinite graphs in terms of logic and higher-order pushdown automata. In *Proc. 23rd FSTTCS*, vol. 2914, *LNCS*, pp. 112–123. Springer, (2003).
[32] W. Thomas. Constructing infinite graphs with a decidable MSO-theory. In *Proc. 28th MFCS*, vol. 2747, *LNCS*, pp. 113–124. Springer, (2003).
[33] W. Thomas, The theory of successor with an extra predicate, *Math. Ann.* **237**, 121–132, (1978).
[34] D. Kuske, personal communication, 2009.
[35] B. Khoussainov, S. Rubin and F. Stephan, Automatic linear orders and trees, *Trans. Comput. Logic* **6**(4), ACM, 2005.
[36] A. Levy, *Basic Set Theory.* (Springer, 1979).
[37] C. Löding, Basics on tree automata, this volume.
[38] C. Löding, Ground tree rewriting graphs of bounded tree width, In *Proc. 19th STACS*, vol. 2285, *LNCS*, pp. 559-570. Springer, (2002).
[39] C. Morvan and C. Stirling. Rational graphs trace context-sensitive languages. In *Proc. 26th MFCS*, vol. 2136, *LNCS*, pp. 548–559. Springer, (2001).
[40] A. Carayol and A. Meyer, Context-Sensitive languages, rational graphs and determinism, *Logical Methods in Computer Science* **2**, (2006).
[41] A. V. Aho, Indexed grammars — an extension of context-free grammars, *J. ACM* **15**, 647–671, (1968).
[42] D. Caucal, Deterministic graph grammars, in *Logic and Automata*, pp. 169–250. Amsterdam Univ. Press, (2008).
[43] A. Blumensath, Th. Colcombet and C. Löding, Logical theories and compatible operations, in: *Logic and Automata*, pp. 73–106, Amsterdam Univ. Press, (2008).
[44] A. Blumensath, Prefix-recognisable graphs and monadic second-order logic, Tech. Rep. AIB-2001-06, RWTH Aachen, 2001.
[45] A. Blumensath and E. Grädel. Automatic structures. In *Proc. 15th LICS*, pp. 51–62. IEEE, (2000).
[46] A. Blumensath and E. Grädel, Finite presentations of infinite structures: Automata and interpretations, *Theory of Computing Systems* **37**, 641–674, (2004).
[47] D. Caucal, On the regular structure of prefix rewriting, *Theor. Comput. Sci.* **106**(1), 61–86, (1992).
[48] Th. Colcombet. On families of graphs having a decidable first order theory with reachability. *Proc. 29th ICALP*, vol. 2380, *LNCS*, pp. 98–109. Springer, (2002).

[49] E. Grädel, W. Thomas, and Th. Wilke (Eds.), *Proc. Automata, Logics, and Infinite Games*, vol. 2500, *LNCS*, Springer, (2002).

[50] C. Rispal, The synchronized graphs trace the context-sensitive languages, *Electr. Notes Theor. Comput. Sci.* **68**, (2002).

Chapter 17

Automata over Infinite Alphabets

Amaldev Manuel and R. Ramanujam
Institute of Mathematical Sciences, C.I.T campus,
Taramani, Chennai - 600113
amal@imsc.res.in
jam@imsc.res.in

In many contexts such as validation of XML data, software model checking and parametrized verification, the systems studied are naturally abstracted as finite state automata, but whose input alphabet is infinite. The use of such automata for verification requires their non-emptiness problems to be decidable. However, ensuring this is non-trivial, since the space of configurations of such automata is infinite. We describe some recent attempts in this direction, and suggest that an entire theoretical framework awaits development.

17.1. Motivation

The theory of finite state automata over (finite) words is an area that is rich in concepts and results, offering interesting connections between computability theory, algebra, logic and complexity theory. Moreover, finite state automata provide an excellent abstraction for many real world applications, such as string matching in lexical analysis [1, 2], model checking finite state systems [3] etc.

Considering that finite state machines have only bounded memory, it is *a priori* reasonable that their input alphabet is finite. If the input alphabet were infinite, it is hardly clear how such a machine can tell infinitely many elements apart. And yet, there are many good reasons to consider mechanisms that achieve precisely this.

Abstract considerations first: consider the set of all finite sequences of natural numbers (given in binary) separated by hashes. A word of this language, for example, is 100#11#1101#100#10101. Now consider the subset L containing all sequences with some number repeating in it. It is easily seen that L is not regular, it is not even context-free. The problem with L has little to do with the representation of the input sequence. If we were given a bound on the numbers occurring in any sequence, we could easily build a finite state automaton recognizing L. The difficulty arises precisely because we don't have such a bound or because we have 'unbounded data'. It is not difficult to find instances of languages like L occurring naturally in the computing world. For example consider the sequences of all *nonces* used in a security protocol run. Ideally this language should be $\overline{L}$. The question is how to recognize such languages, and whether there is any hope of describing regular collections of this sort.

Note that we could simply take the set of binary numbers as the alphabet in the example above: $D = \{\#, 0, 1, 10, 11, \ldots\}$). Now, $L = \{w = b_0\#b_1\#\ldots b_n \mid w \in D^*, \exists i, j.b_i = b_j\}$. Note further that D itself is a regular language over the alphabet $\{\#, 0, 1\}$.

There are more concrete considerations that lead to infinite alphabets as well, arising from two strands of computation theory: one from attempts to extend classical model checking techniques to *infinite state systems*, and the other is the realm of *databases*. Systems like software programs, protocols (communication, cryptography, ...), web services and alike are typically infinite state, with many different sources of unbounded data: program data, recursion, parameters, time, communication media, etc. Thus, model checking techniques are confronted with infinite alphabets. In databases, the XML standard format of *semi-structured data* consists of labelled trees whose nodes carry data values. The trees are constrained by schemes describing the tree structure, and restrictions on data values are specified through data constraints. Here again we have either trees or paths in trees whose nodes are labelled by elements of an infinite alphabet.

Building theoretical foundations for studies of such systems leads us to the question of how far we can extend finite state methods and techniques to infinite state systems. The attractiveness of finite state machines can mainly be attributed to the easiness of several decision problems on them. They are robust, in the sense of invariance under nondeterminism, alternation etc. and characterizations by a plurality of formalisms such as Kleene expressions, monadic second order logic, and finite semigroups. Regular languages are logically well behaved (closed under boolean operations, homomorphisms, projections, and so on). What we would like to do is to introduce mechanisms for unbounded data in finite state machines in such a way that we can retain as many of these nice properties as possible.

In the last decade, there have been several answers to this question. We make no attempt at presenting a comprehensive account of all these, but point to some interesting automata theory that has been developed in this direction. Again, while many theorems can be discussed, we concentrate only on one question, that of emptiness checking, guided by concerns of system verification referred to above.

The material discussed here covers the work of several researchers, and much of it can be found in [4–8]. Our own contribution is limited to the material in 17.5.

17.2. Languages of data words

Notation: Let $k > 0$; we use $[k]$ to denote the set $\{1, 2, \ldots k\}$. When we say $[k]_0$, we mean the set $\{0\} \cup [k]$. By $\mathbb{N}$ we mean the set of natural numbers $\{0, 1, \ldots\}$. When $f : A \to B$, $(a, b) \in (A \times B)$, by $f \oplus (a, b)$, we mean the function $f' : A \to B$, where $f'(a') = f(a')$ for all $a' \in A, a' \neq a$, and $f'(a) = b$.

Before we consider automaton mechanisms, we discuss languages over infinite alphabets. We will look only at languages of **words** but it is easily seen that similar notions can be defined for languages of *trees*, whose nodes are labelled from an infinite alphabet. We will use the terminology of database theory, and refer to languages over infinite alphabets as **data languages**. However, it should be noted that at least in the context of database theory, data trees (as in XML) are more natural than data words, but as it turns out, the questions discussed here happen to be considerably harder for tree languages than for word languages.

Customarily, the infinite alphabet is split into two parts: it is of the form $\Sigma \times D$, where Σ is a finite set, and D is a countably infinite set. Usually, Σ is called the *letter alphabet* and D is called the *data alphabet.* Elements of D are referred to as *data values.* We use letters a, b etc to denote elements of Σ and use d, d' to denote elements of D.

The letter alphabet is a way to provide the data values 'contexts'. In the case of XML, Σ consists of tags, and D consists of data values. Consider the XML description: <name> ''Tagore''</name>: the tag <name> can occur along with different strings; so also, the string ''Tagore'' can occur as the value associated with different tags. As another example, consider a system of unbounded processes with states $\{b, w\}$ for 'busy' and 'wait'. When we work with the traces of such a system, each observation records the state of a process denoted by its process identifier (a number). A word in this case will be, for example, $(b, d_1)(w, d_2)(w, d_1)(b, d_2)$.

A **data word** w is an element of $(\Sigma \times D)^*$. A collection of data words $L \subseteq (\Sigma \times D)^*$ is called a *data language.* In this article, by default, we refer to data words simply as words and data languages as languages. As usual, by $|w|$ we denote the length of w.

Let $w = (a_1, d_1)(a_2, d_2) \dots (a_n, d_n)$ be a data word. The *string projection* of w, denoted as $str(w) = a_1 a_2 \dots a_n$, the projection of w to its Σ components. Let $i \in [n] = |w|$. The **data class** of d_i in w is the set $\{j \in [n] \mid d_i = d_j\}$. A subset of $[n]$ is called a data class of w if it is the data class of some d_i, $i \in [n]$. Note that the set of data classes of w form a partition of $[|w|]$.

We introduce some example data languages which we will keep referring to in the course of our discussion; these are over the alphabet $\Sigma = \{a, b\}, D = \mathbb{N}$.

- $L_{\exists n}$ is the set of all words in $(\Sigma \times D)^*$ in which at least n distinct data values occur.
- $L_{<n}$ is the set of all data words in which every data value occurs at most n times.
- $L_{a^*b^*}$ is the set of all data words whose string projections are in the set a^*b^*.
- $L_{\forall a \exists b}$ is the set of all data words where every data value occurring under a occurs under b also.
- $L_{fd(a)}$ is the collection of all data words in which all the data values in context a are distinct. ($fd(a)$ stands for functional dependency on a.)

Let $\cdot$ denote concatenation on data words. For $L \subseteq (\Sigma \times D)^*$, consider the Myhill-Nerode equivalence on $(\Sigma \times D)^*$ induced by L: $w_1 \sim_L w_2$ iff $\forall w. w_1 \cdot w \in L \Leftrightarrow w_2 \cdot w \in L$. L is said to be regular when $\sim_L$ is of finite index. A classical theorem of automata theory equates the class of regular languages with those recognized by finite state automata, in the context of languages over finite alphabets.

It is easily seen that $\sim_{L_{fd(a)}}$ is not of finite index, since each singleton data word (a, d) is distinguished from (a, d'), for $d \neq d'$. Hence we cannot expect a classical finite state automaton to accept $L_{fd(a)}$; we need to look for another device, perhaps an infinite state machine.

Indeed, for most data languages, the associated equivalence relation is of infinite index. Is there a notion of *recognizability* that can be defined meaningfully over such languages and yet corresponds (in some way) to finite memory? This is the central question addressed in this article.

17.3. Formulating an automaton mechanism

The notion of regularity on languages over infinite alphabets can be approached using different mechanisms: *descriptive* ones like logics or rational expressions, or *operational* ones like automata models. There are two reasons for our discussing only automata here: one, machine models are closer to our algorithmic intuition about language behaviour, and enable us to compare the computational power of different machines; two, there are relatively fewer results to discuss on the descriptive side.

The first challenge in formulating an automaton mechanism is the question of 'finite representability'. It is essential for a machine model that the automaton is presented in a finite fashion. In particular, we need implicit finite representations of the data alphabet. An immediate implication is that we need algorithms that work with such implicit representations. Towards this, from now on, *we consider only data alphabets D in which membership and equality are decidable.*

Automata for words over finite alphabets are usually presented as working on a read-only finite tape, with a *tape head* under finite state control. One detail which is often taken for granted is the complexity of the tape head. Since we can recognize a finite language (which is the alphabet!) by a constant-sized circuit the computing power of the tapehead is far inferior to that of the automaton.

In the case of infinite alphabets, the situation is different, and our assumption about decidable membership and equality in D makes sense when we consider the complexity of the tape head. For example, if we consider the alphabet as the encodings of all halting Turing machines, the tapehead has to be a Σ_1^0 machine, which is obviously hard to conceive of as a machine model relevant to software verification. Therefore, we see that our assumption needs tightening and we should require the membership and equality checking in the alphabet to be *computationally feasible.* In fact, we should also ensure that the language accepted by the automaton, when restricted to a finite subset of the infinite alphabet, remains regular.

One obvious way of implementing finite presentations is by insisting that the automaton uses only *finitely many* data values in its transition relation. This seems reasonable, since any property of data that we wish to specify can refer explicitly only to finitely many data values. However, when the only allowed operation on data is checking for equality of data values, such an assumption becomes drastic: it is easily seen that having infinite data alphabets is superfluous in such automata. Every such machine is equivalent to a finite state machine over a finite alphabet.

Thus we note that infinite alphabets naturally lead us to infinite state systems, whose space of configurations is infinite. The theory of computation is rich in such models: pushdown systems, Petri nets, vector addition systems, Turing machines etc. In particular, we are led to models in which we equip the automaton with some additional mechanism to enable it to have infinitely many configurations.

This takes us to a striking idea from the 1960's: "automata theory is the study of memory structures". These are structures that allow us to fold infinitely many actions into finitely many instructions for manipulating memory, which can be part of the automaton definition. These are storage mechanisms which provide specific tools for manipulating and accessing data. Obvious memory mechanisms are *registers* (which act like scratch pads, for memorizing specific data values encountered), *stacks*, *queues* etc.

One obvious memory structure is the input tape, which can be 'upgraded' to an unbounded sequential read-write memory. But then it is easily noted that a finite state machine equipped with such a tape is Turing-complete. On the other hand, if the tape is read-only, the machine accepts all datawords whose string projections belong to the letter language (subset of Σ^*) defined by the underlying automaton. Clearly this machine is also not very interesting. We therefore look for structures that keep us in between: those with infinitely many configurations, but for which reachability is yet decidable. Note that such ambition is not unrealistic, since Petri nets and pushdown systems are systems of this kind.

17.4. Automata with registers

The simplest form of memory is a finite random access read-write storage device, traditionally called *register*. In register automata [4], the machine is equipped with finitely many registers, each of which can be used to store one data value. Every automaton transition includes access to the registers, reading them before the transition and writing to them after the transition. The new state after the transition depends on the current state, the input letter and whether or not the input data value is already stored in any of the registers. If the data value is not stored in any of the registers, the automaton can choose to write it in a register. The transition may also depend on which register contains the encountered data value.

Below, fix an alphabet $\Sigma \times D$, with Σ finite and D countable. Let $D_\perp = D \cup \{\perp\}$, where $\perp \notin D$ is a special symbol.

Definition 17.1. A **k-register finite memory automaton** is given by $R_A = (Q, \Sigma, \Delta, \tau_0, U, q_0, F)$, where Q is a finite set of states, $q_0 \in Q$ is the initial state and $F \subseteq Q$ is the set of final states. τ_0 is the initial register configuration given by $\tau_0 : [k] \to D_\perp$, and U is a partial *update* function: $(Q \times \Sigma) \to [k]$. The transition relation is $\Delta \subseteq (Q \times \Sigma \times [k] \times Q)$. We assume that for all $i, j \in [k]$ if $i \neq j$ and $\tau_0(i), \tau_0(j) \in D$ then $\tau_0(i) \neq \tau_0(j)$ (that is, registers initially contain distinct data values).

Note that the description of R_A is finite, D appears only in the specification of the initial register configuration. It is reasonable to suppose that bounded precision suffices for this purpose. Thus the infinite alphabet plays only an implicit (albeit critical) role in the behaviour of R_A.

$\perp$ is used above to denote an uninitialized register. The working of the automaton is as follows. Suppose that R_A is in state p, with each of the registers r_i holding data value d_i, $i \in [k]$, and its input is of the form (a, d) where $a \in \Sigma$ and $d \in D$. Now there are two cases:

- If $d \neq d_i$ for all i, then a register update is enabled. If $U(p, a)$ is undefined, this is deemed to be an error, and the machine halts. Otherwise the value d is put into r_j where $U(p, a) = j$; other registers are left untouched, and the state does not change.
- Suppose that $d = d_i$, for some $i \in [k]$, and $(p, a, i, q) \in \Delta$. Then this transition is enabled, and when applying the transition, the registers are left untouched.

This is formalized as follows. A *configuration* of R_A is of the form (q, τ) where $q \in Q$ and $\tau : [k] \to D_\perp$. Let C_A denote the set of all configurations of R_A. Δ and U define a transition $\Delta(C_A) \subseteq (C_A \times \Sigma \times D \times C_A)$ as follows: there is a transition from (q, τ) to (q', τ') in $\Delta(C_A)$ on (a, d) iff (a) $\tau = \tau'$, $d = \tau(i)$, $(i \in [k])$, $(q, a, i, q') \in \Delta$, or (b) for all $i \in [k]$, $d \neq \tau(i)$, $\tau' = \tau \oplus (U(q, a), d)$, $(q, a, U(q, a), q') \in \Delta$.

A *run* of R_A on a data word $w = (a_1, d_1)(a_2, d_2) \dots (a_n, d_n)$ is a sequence $\gamma = (q_0, \tau_0)(q_1, \tau_1) \dots (q_n, \tau_n)$, where (q_0, τ_0) is the initial configuration of R_A, and for every $i \in [n]$, there is a transition from (q_{i-1}, τ_{i-1}) to (q_i, τ_i) on (a_i, d_i) in $\Delta(C_A)$. γ is *accepting* if $q_n \in F$. The language accepted by R_A, denoted $L(R_A) = \{w \in (\Sigma \times D)^* \mid R_A \text{ has an accepting run on } w\}$.

Example 17.1. Recall the language $L_{fd(a)}$ mentioned earlier: it is the set of all data words in which all the data values in context a are distinct. Formally:

$$L_{fd(a)} = \left\{ w = (a_1, d_1)(a_2, d_2) \dots (a_n, d_n) \;\middle|\; \begin{array}{c} w \in (\Sigma \times D)^*, \\ \forall ij. a_i = a_j = a \implies d_i \neq d_j \end{array} \right\}$$

The language $\overline{L_{fd(a)}}$, can be accepted by a 2-register finite memory automaton $A = (Q, q_0, \Delta, \tau_0, U, F)$, where

- $Q = \{q_0, q_1, q_f\}$
- $\tau_0 = (\perp, \perp)$
- $U(q_0, \Sigma) = 1, U(q_1, \Sigma) = U(q_f, \Sigma) = 2^*$
- $F = \{q_f\}$
- Δ consists of,
    - $(q_0, \Sigma, 1, q_0)$
    - $(q_0, a, 1, q_1)$
    - $(q_1, \Sigma, 2, q_1)$
    - $(q_1, a, 1, q_f)$
    - $(q_f, \Sigma, \{1, 2\}, q_f)$

The automaton A is shown in Fig. 17.1 and works as follows. Initially A is in state q_0 and stores new input data in the first register. When reading the data value with label a, which appears twice, A changes the state to q_1 and waits there storing the new data in the second register. When the data value stored in the first register appears the second time with label a, A changes state to q_f and continues to be there.

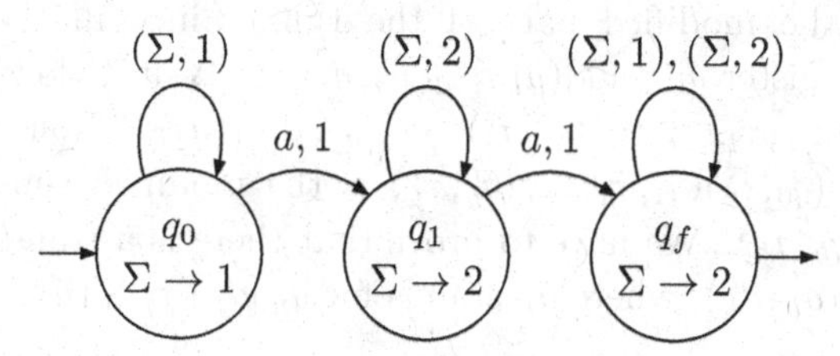

Fig. 17.1. Automaton in the Example 17.1.

Note that a finite memory automaton uses only finitely many registers to deal with infinitely many symbols, and hence we get something analogous to the pumping lemma for regular languages which asserts that a finite state automaton which accepts sufficiently long words also accepts infinitely many words. Suppose there are k registers and the automaton sees $k+1$ data values; since the only places where it can store these data values are in the registers, it is bound to forget one of the data values. This is made precise by the following lemma.

Lemma 17.1. *If a k-register automaton A accepts any word at all, then it accepts a word containing at most k distinct data values.*

Proof. Let $w = (a_1, d_1)(a_2, d_2) \dots (a_n, d_n)$ be a data word accepted by A and $(q_0, \tau_0)(q_1, \tau_1) \dots (q_n, \tau_n)$ be an accepting run of A on w. Let $U_{min}(w) = j$ be the least integer such that $d_j \notin range(\tau_{j-1})$ and $\tau_{j-1}(U(q_{j-1}, a_j)) \neq \perp$ (∞, if it is not defined). The previous condition says that j is the first position in the run, where

*We abuse the notation here, $U(q_0, \Sigma) = 1$, denotes that for all $a \in \Sigma$, $U(q_0, a) = 1$. We use similar shorthands later on whenever they are handy and intuitive, for instance in writing the transition relation of an automaton.

A is going to replace a data value in the register which is not $\perp$. If such a j does not exist, then the register values are never replaced (other than $\perp$'s) and therefore w contains at most k distinct data values. If $j \leq n$, then we will construct a data word w' of length n accepted by A, with $j < U_{min}(w')$. This implies that by finite iteration (at most n times) of the process described below, we can construct a word with at most k distinct symbols. Let $u = U(q_{j-1}, a_j)$ and let $d_u = \tau_{j-1}(u)$.

Consider

$$w' = \begin{matrix} a_1\, a_2 \ldots a_{j-1} \\ d_1\, d_2 \ldots d_{j-1} \end{matrix} \left\{ \begin{matrix} a_j\, a_{j+1} \ldots a_n \\ d_j\, d_{j+1} \ldots d_n \end{matrix} \right\} [d_j \mid d_u, d_u \mid d_j]$$

which is got by replacing all the occurrences of d_j with d_u and replacing d_u by d_j in the string, after the position j. It is clear that, when the automaton reaches the position j, the data value is present in the register $U(u)$, and therefore no replacement will be done. Hence $j < U_{min}(w')$.

However, in order to show that w' is accepted by A, it remains to be proved that $(q_0, \tau_0)(q_1, \tau_1) \ldots (q'_j, \tau'_j)(q'_{j+1}, \tau'_{j+1}) \ldots (q'_n, \tau'_n)$ is an accepting run for w', where, $q'_i = q_i$ and $\tau'_i = \tau_i [d_j \mid \tau_{j-1}(u), \tau_{j-1}(u) \mid d_j]$. We prove this using induction on the length of the modified part of the run. Since in the original run d_j is stored in the register u and $(u, q_{j-1}, a_j, q_j) \in \Delta$ it follows that there is a transition from (q_{j-1}, τ_{j-1}) to (q'_j, τ'_j) on $(a_j, \tau_{j-1}(u))$. Now for the inductive step, assume that $(q_0, \tau_0)(q_1, \tau_1) \ldots (q'_h, \tau'_h)$ is the modified run corresponding to $(q_0, \tau_0)(q_1, \tau_1) \ldots (q_h, \tau_h)$. We have to prove that there is a transition from (q'_h, τ'_h) to (q'_{h+1}, τ'_{h+1}) on (a_h, d'_h), where d'_h stands for $d_h [d_j \mid \tau_{j-1}(u), \tau_{j-1}(u) \mid d_j]$.

- If for some $t \in [k]$, $d_h = \tau_h(t)$ then $(t, q_h, a_h, q_{h+1}) \in \Delta$ and $\tau_{h+1} = \tau_h$. Therefore d'_h appears in the register t. Hence there is a transition from (q'_h, τ'_h) to (q'_{h+1}, τ'_{h+1}) on (a_h, d'_h) also.
- If $d_h \notin range(\tau_h)$, then τ_{h+1} is obtained from τ_h by replacing the contents of the register $U(q_h, a_h)$ with d_h. Since $d'_h \notin range(\tau'_h)$, τ'_{h+1} is obtained from τ'_h by replacing the contents of register $U(q_h, a_h)$. Again there is a transition from (q'_h, τ'_h) to (q'_{h+1}, τ'_{h+1}) on (a_h, d'_h).

This completes the proof. □

Note that the language $L_{fd(a)}$ requires unboundedly many data values to occur with a, and hence by the above lemma, it cannot be recognized by any register automaton. On the other hand, since $\overline{L_{fd(a)}}$ can accepted by a register automaton, we see that languages recognized by register automata are not closed under complementation. As this suggests, non-deterministic register automata are more powerful than deterministic ones.

While the lemma demonstrates a limitation of register machines in terms of computational power, it also shows the way for algorithms on these machines.

Theorem 17.1. *Emptiness checking of register automata is decidable.*

Proof. Let A be a register automaton with k registers, which we want to check for emptiness. Let $range(\tau_0) \subseteq D' \subseteq D, |D'| = k$ be a subset of D containing k different symbols including those in the register's initialization. We claim that $L(A) \neq \emptyset$ if and only if $L(A) \cap (\Sigma \times D')^* \neq \emptyset$. The if direction is trivial. The other direction follows from the preceding lemma. Thus a classical finite state automaton working on a finite alphabet can be employed for checking emptiness of A. □

The emptiness problem for register automata is in NP, since we can guess a word of length polynomial in the size of the automaton and verify that it is accepted. It has also been shown that the problem is complete for NP in [9]. The problem is no less hard for the deterministic subclass of these automata. Though, as we mentioned earlier, register automata are not closed under complementation, they are closed under intersection, union, Kleene iteration and homomorphisms.

There are many extensions of the register automaton model. An obvious one is to consider *two-way* machines: interestingly, this adds considerable computational power and the emptiness problem becomes undecidable [4, 5, 10].

17.5. Automata with counters

Automata with counters have a long history. It is well known that a finite state automaton equipped with two counters is Turing-complete, whereas with only one counter, it has the computational power of a pushdown automaton. Can the counter mechanism be employed in the context of unbounded data? While a register mechanism is used to note down data values, counters are used to record the number of occurrences of some (pre-determined) events.

When automata with counters are considered on words over finite alphabets, one typical use of counters is to note the multiplicity of a letter in the input word. On infinite alphabets, we could similarly count the number of occurrences of data values, or letter - value pairs, or these subject to constraints. But then, each such 'event type' needs a counter for itself, which implies the need for unboundedly many counters. On the other hand, as we observed above, two counter machines are already Turing-complete. This suggests restraint on the allowed counter operations. While there are many possible restrictions, a natural one is to consider *monotone* counters, which can be incremented, reset and compared against constants, but never decremented.

A model with such characteristics is presented below. The automaton includes a bag of infinitely many monotone counters, one for each possible data value. When it encounters a letter - data pair, say (a, d), the multiplicity of d is checked against a given constraint, and accordingly updated, the transition causing a change of state, as well as possible updates for other data as well. We can think of the bag as a hash table, with elements of D as keys, and counters as hash values. Transitions depend only on hash values (subject to constraints) and not keys.

Such counting is used in many infinite state systems. For instance, in systems of unbounded processes, a task enters the system with low priority which increases in the course of a computation until the task is scheduled, resetting the priority. In the context of web services, a server not only offers data dependent services, but also needs information on how many clients are at any time using a particular service. For instance, the loan requests granted by a server depend on how many clients are requesting low credit and how many need high credit at that time.

A **constraint** is a pair $c = (op, e)$, where $op \in \{<, =, \neq, >\}$ and $e \in \mathbb{N}$. When $v \in \mathbb{N}$, we say $v \models c$ if $v \,\mathsf{op}\, e$ holds. Let $\mathbb{C}$ denote the set of all constraints. Define a *bag* to be a map $h : \mathbb{D} \to \mathbb{N}$. Let $\mathbb{B}$ denote the set of bags.

Below, let $Inst = \{\uparrow^+, \downarrow\}$ stand for the set of *instructions*: $\uparrow^+$ tells the automaton to increment the counter, whereas $\downarrow$ asks for a reset.

Definition 17.2. A **class counting automaton**, abbreviated as CC_A, is a tuple $CC_A = (Q, \Delta, I, F)$, where Q is a finite set of states, $I \subseteq Q$ is the set of initial states, $F \subseteq Q$ is the set of final states. The transition relation is given by: $\Delta \subseteq (Q \times \Sigma \times C \times Inst \times U \times Q)$, where C is a finite subset of $\mathbb{C}$ and U is a finite subset of $\mathbb{N}$.

Let A be a CC_A. A configuration of A is a pair (q, h), where $q \in Q$ and $h \in \mathbb{B}$. The initial configuration of A is given by (q_0, h_0), where $\forall d \in \mathbb{D}, h_0(d) = 0$ and $q_0 \in I$.

Given a data word $w = (a_1, d_1), \ldots (a_n, d_n)$, a run of A on w is a sequence $\gamma = (q_0, h_0)(q_1, h_1) \ldots (q_n, h_n)$ such that $q_0 \in I$ and for all $i, 0 \leq i < n$, there exists a transition $t_i = (q, a, c, \pi, n, q') \in \Delta$ such that $q = q_i$, $q' = q_{i+1}$, $a = a_{i+1}$ and:

- $h_i(d_{i+1}) \models c$.
- h_{i+1} is given by:

$$h_{i+1} = \begin{cases} h_i \oplus (d, n') \text{ if } \pi = \uparrow^+, n' = h_i(d) + n \\ h_i \oplus (d, n) \;\; \text{if } \pi = \downarrow \end{cases}$$

γ is an accepting run above if $q_n \in F$. The language accepted by A is given by $\mathcal{L}(A) = \{w \in \Sigma \times \mathbb{D}^* \mid A \text{ has an accepting run on } w\}$. $\mathcal{L} \subseteq (\Sigma \times \mathbb{D})^*$ is said to be recognizable if there exists a CC_A A such that $\mathcal{L} = \mathcal{L}(A)$. Note that the counters are either incremented or reset to fixed values.

Note that the instruction $(\uparrow^+, 0)$ says that we do not wish to make any update, and $(\uparrow^+, 1)$ causes a unit increment; we use the notation [0] and [+1] for these instructions below.

Example 17.2. The language $L_{fd(a)}$ = *"Data values under a are all distinct"* is accepted by a CC_A. The CC_A accepting this language is the automaton $A = (Q, \Delta, q_0, F)$ where

$Q = \{q_0, q_1\}$, q_0 is the only initial state and $F = \{q_0\}$. Δ consists of:

- $(q_0, a, (=, 0), q_0, [+1])$

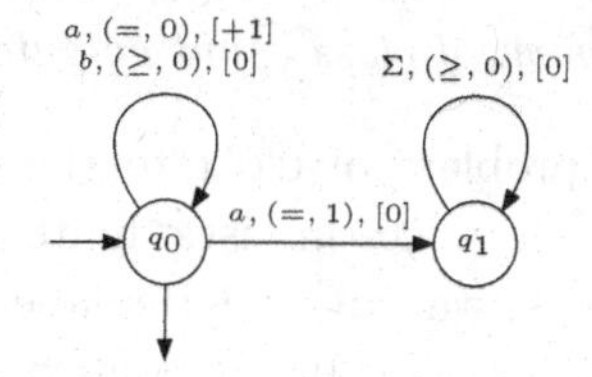

Fig. 17.2. Automaton in the Example 17.2.

- $(q_0, a, (=, 1), q_1, [0])$
- $(q_0, b, (\geq, 0), q_0, [0])$
- $(q_1, \Sigma, (\geq, 0), q_1, [0])$

The automaton is shown in Fig. 17.2. Since the automaton above is deterministic, by complementing it, that is, setting $F = \{q_1\}$, we can accept the language $\overline{L_{fd(a)}}$ = *"There exists a data value appearing at least twice under a"*.

It is easily seen that by a similar argument, the language $L_{\forall a, \leq n}$ = *"All data values under a occur at most n times"* can be accepted by a CC_A. So also is the language $L_{\exists a, \geq n}$ = *" There exists a data value appearing under a occurring more than n times"*.

Given a CC_A $A = (Q, \Delta, q_0, F)$ let m be the maximum constant used in Δ. We define the following equivalence relation on $\mathbb{N}$, $c \simeq_{m+1} c'$ iff $c < (m+1) \vee c' < (m+1) \Rightarrow c = c'$. Note that if $c \simeq_{m+1} c'$ then a transition is enabled at c if and only if it is enabled at c'. We can extend this equivalence to configurations of the CC_A as follows. Let $(q_1, h_1) \simeq_{m+1} (q_2, h_2)$ iff $q_1 = q_2$ and $\forall d \in \mathbb{D}, h_1(d) \simeq_{m+1} h_2(d)$.

Lemma 17.2. *If C_1, C_2 are two configurations of the CC_A such that $C_1 \simeq_{m+1} C_2$, then $\forall w \in (\Sigma \times \mathbb{D})^*$, $C_1 \vdash_w^* C_1' \implies \exists C_2', C_2 \vdash_w^* C_2'$ and $C_1' \simeq_{m+1} C_2'$.*

Proof. Proof by induction on the length of w. For the base case observe that any transition enabled at C_1 is enabled at C_2 and the counter updates respects the equivalence. For the inductive case consider the word $w.a$. By induction hypothesis $C_1 \vdash_w^* C_1' \implies \exists C_2', C_2 \vdash_w^* C_2'$ and $C_1' \simeq_{m+1} C_2'$. If $C_1' \vdash_a C_1''$ then using the above argument there exists C_2'' such that $C_2' \vdash_a C_2''$ and $C_1'' \simeq_{m+1} C_2''$. □

In fact the lemma holds for any $N \geq m+1$, where m is the maximum constant used in Δ. This observation paves the way for the route to decidability of the emptiness problem.

Before we proceed to discuss decidability, we observe that the model admits many extensions. For instance, instead of working with one bag of counters, the automaton can use several bags of counters, much as multiple registers are used in the register automaton. Another strengthening involves checking for the presence of *any* counter satisfying a given constraint and updating it. Moreover, the language of constraints can be strengthened: any syntax that can specify a finite or co-finite subset of $\mathbb{N}$ will do. We do not pursue these generalizations here, but merely remark that the theory extends straightforwardly.

Theorem 17.2. *The emptiness problem of class counting automata is decidable.*

Proof. We reduce the emptiness problem of CC_A to the covering problem on Petri nets. Recall that in the case of nondeterministic finite state automata over finite alphabets, for checking emptiness, we can omit the labels on transitions and reduce the problem to reachability in graphs. In the context of CC_A, the similar simplification is to omit $\Sigma \times D$ labels from the configuration graph; we are then left with counter behaviour. However reachability is no longer trivial, since we have unboundedly many counters, leading us to the realm of vector addition systems.

Definition 17.3. An ω-counter machine B is a tuple (Q, Δ, q_0) where Q is a finite set of states, $q_0 \in Q$ is the initial state and $\Delta \subseteq (Q \times C \times Inst \times U \times Q)$, where C is a finite subset of $\mathbb{C}$ and U is a finite subset of $\mathbb{N}$.

A configuration of B is a pair (q, h), where $q \in Q$ and $h : \mathbb{N} \to \mathbb{N}$. The initial configuration of B is (q_0, h_0) where $h_0(i) = 0$ for all i in $\mathbb{N}$. A run of B is a sequence $\gamma = (q_0, h_0)(q_1, h_1) \ldots (q_n, h_n)$ such that for all i such that $0 \leq i < n$, there exists a transition $t_i = (p, c, \pi, n, q) \in \Delta$ such that $p = q_i$, $q = q_{i+1}$ and there exists j such that $h(j) \models c$, and the counters are updated in a similar fashion to that of CC_A.

The reachability problem for B asks, given $q \in Q$, whether there exists a run of B from (q_0, h_0) ending in (q, h) for some h ("Can B reach q?").

Lemma 17.3. *Checking emptiness for CC_A can be reduced to checking reachability for ω-counter machines.*

Proof. It suffices to show, given a CC_A, $A = (Q, \Delta, q_0, F)$, where $F = \{q\}$, that there exists a counter machine $B_A = (Q, \Delta', q_0)$ such that A has an accepting run on some data word exactly when B_A can reach q. (When F is not a singleton, we simply repeat the construction.) Δ' is obtained from Δ by converting every transition (p, a, c, π, n, q) to (p, c, π, n, q). Now, let $L(A) \neq \emptyset$. Then there exists a data word w and an accepting run $\gamma = (q_0, h_0)(q_1, h_1) \ldots (q_n, h_n)$ of A on w, with $q_n = q$. Let $g : \mathbb{N} \to D$ be an enumeration of data values. It is easy to see that $\gamma' = (q_1, h_0 \circ g)(q_1, h_1 \circ g) \ldots (q_n, h_n \circ g)$ is a run of B_A reaching q.

($\Leftarrow$) Suppose that B_A has a run $\eta = (q_0, h_0)(q_1, h_1) \ldots (q_n, h_n)$, $q_n = q$. It can be seen that $\eta' = (q_0, h_0 \circ g^{-1})(q_1, h_1 \circ g^{-1}) \ldots (q_n, h_n \circ g^{-1})$ is an accepting run of A on $w = (a_1, d_1) \ldots (a_n, d_n)$ where w satisfies the following. Let (p, c, π, n, q) be the transition of B_A taken in the configuration (q_i, h_i). Then by the definition of B_A there exists a transition (p, a, c, π, n, q) in Δ. Then it should be the case that $a_{i+1} = a$ and $d_{i+1} = g(m)$. □

Proposition 17.1. *Checking non-emptiness of ω-counter machines is decidable.*

Let $s \subseteq \mathbb{N}$, and c a constraint. We say $s \models c$, if for all $n \in s$, $n \models c$.

We define the following partial function Bnd on all finite and cofinite subsets of $\mathbb{N}$. Given $s \subseteq_{fin} \mathbb{N}$, $Bnd(s)$ is defined to be the least number greater than all the elements in s. Given $s \subseteq_{cofinite} \mathbb{N}$, $Bnd(s)$ is defined to be $Bnd(\mathbb{N}\backslash s)$. Given an ω-counter machine $B = (Q, \Delta, q_0)$ let $m_B = max\{Bnd(s) \mid s \models c, c \text{ is used in } \Delta\}$.

We construct a Petri net $N_B = (S, T, F, M_0)$ where,

- $S = Q \cup \{i \mid i \in \mathbb{N}, 1 \leq i \leq m_B\}$.
- T is defined according to Δ as follows. Let $(p, c, \pi, n, q) \in \Delta$ and let i be such that $0 \leq i \leq m_B$ and $i \models c$. Then we add a transition t such that ${}^\bullet t = \{p, i\}$ and $t^\bullet = \{q, i'\}$, where (i) if π is $\uparrow^+$ then $i' = min\{m_B, i + n\}$, and (ii) if π is $\downarrow$ then $i' = n$.
- The flow relation F is defined according to ${}^\bullet t$ and $t^\bullet$ for each $t \in T$.
- The initial marking is defined as follows. $M_0(q_0) = 1$ and for all p in S, if $p \neq q_0$ then $M_0(p) = 0$.

The construction above glosses over some detail: Note that elements of these sets can be zero, in which case we add edges only for the places in $[m_B]$ and ignore the elements which are zero.

Let M be any marking of N_B. We say that M is a *state marking* if there exists $q \in Q$ such that $M(q) = 1$ and $\forall p \in Q$ such that $p \neq q$, $M(p) = 0$. When M is a state marking, and $M(q) = 1$, we speak of q as the state marked by M. For $q \in Q$, define $M_f(q)$ to be set of state markings that mark q. It can be shown, from the construction of N_B, that in any reachable marking M of N_B, if there exists $q \in Q$ such that $M(q) > 0$, then M is a state marking, and q is the state marked by M.

We now show that the counter machine B can reach a state q iff N_B has a reachable marking which covers a marking in $M_f(q)$. We define the following equivalence relation on $\mathbb{N}$, $m \simeq_{m_B} n$ iff $(m < m_B) \vee (n < m_B) \Rightarrow m = n$. We can lift this to the hash functions (in ω-counters) in the natural way: $h \simeq_{m_B} h'$ iff $\forall i\, (h(i) < m_B) \vee (h'(i) < m_B) \Rightarrow h(i) = h'(i)$. It can be easily shown that if $h \simeq_{m_B} h'$ then a transition is enabled at h if and only if it is enabled at h'.

Let μ be a mapping B-configurations to N_B-configurations as follows: given $\chi = (q, h)$, define $\mu(\chi) = M_\chi$, where

$$M_\chi(p) = \left\{ \begin{array}{ll} 1 & \text{iff } p = q \\ 0 & \text{iff } p \in Q\backslash\{q\} \\ |[p]| & \text{iff } p \in P\backslash Q, p \neq 0 \end{array} \right\}$$

Above $[p]$ denotes the equivalence class of p under $\simeq_{m_B}$ on $\mathbb{N}$ in h. Now suppose that B reaches q. Let the resulting configuration be $\chi = (q, h)$. We claim that the marking $\mu(\chi)$ of N_B is reachable (from M_0) and covers $M_f(q)$. Conversely if a reachable marking M of N_B covers $M_f(q)$, for some $q \in Q$, then there exists a reachable configuration $\chi = (q, h)$ of B such that $\mu(\chi) = M$. This is proved by a simple induction on the length of the run.

Since the covering problem for Petri nets is decidable, so is reachability for ω-counter machines and hence emptiness checking for CC_A is decidable. $\square$

The decision procedure above runs in EXPSPACE, and thus we have elementary decidability, though CC_A configurations form an infinite state system. The problem is complete for EXPSPACE by an easy reduction from covering problem of Petri nets. CC_A are not closed under complementation, but they are closed under union and intersection. The details can be found in [11].

17.6. Automata with hash tables

A hash table is a data structure containing 'keys' and 'values'. It provides random access to the stored 'value' corresponding to 'key'. In the case of infinite data, we can employ a hash table with the elements of D as the keys. The values have to be from a finite set, since a finite state automaton can only distinguish only finitely many symbols in the transition relation. Thus, the hash values impose an equivalence relation of finite index on data.

The main idea is as follows. On reading a (a, d), the automaton reads the table entry corresponding to d and makes a transition dependent on the table entry, the input letter a and the current state. The transition causes a change of state as well as updating of the table entry. Such a model has been termed a class memory automaton [6].

Definition 17.4. A **class memory automaton** is a tuple $CM_A = (Q, \Sigma, \Delta, q_0, F_\ell, F_g)$ where Q is a finite set of states, q_0 is the initial state and $F_g \subseteq F_\ell \subseteq Q$ are the sets of **global** and **local** accepting states respectively. The transition relation is $\Delta \subseteq (Q \times \Sigma \times (Q \cup \{\bot\}) \times Q)$.

The working of the automaton is as follows. The finite set of hash values is simply the set of automaton states. A transition of the form (p, a, s, q) on input (a, d) stands for the state transition of the automaton from p to q as well as the updating of the hash value for d from s to q. The acceptance condition has two parts. The global acceptance set F_g is as usual: after reading the input the automaton state should be in F_g. The local acceptance condition refers to the state of the hash table: the image of the hash function should be contained in F_ℓ. Thus acceptance depends on the memory of the data encountered.

Formally, a hash function is a map $h : D \to (Q \cup \{\bot\})$ such that $h(d) = \bot$ for all but finitely many data values. h holds the hash value (the state) which is assigned to the data value d when it was read the last time. A configuration of the automaton is of the form (q, h) where h is a hash function. The initial configuration of the automaton is (q_0, h_0) where $h_0(d) = \bot$ for all $d \in D$.

Transition on configurations is defined as follows: a transition from a configuration (p, h) on input (a, d) to (q, h') is enabled if $(p, a, h(d), q) \in \Delta$ $h' = h \oplus (d, q)$.

A *run* of CM_A on a data word $w = (a_1, d_1)(a_2, d_2) \ldots (a_n, d_n)$ is, as usual, a sequence $\gamma = (q_0, h_0)(q_1, h_1) \ldots (q_n, h_n)$, where h_0 is the initial configuration of CM_A, and for every $i \in [n]$, there is a transition from (q_{i-1}, h_{i-1}) to (q_i, h_i) on

(a_i, d_i) in $\Delta(C_A)$. γ is *accepting* if $q_n \in F_g$ and for all $d \in D$, $f(d) \in F_l \cup \{\perp\}$. The language accepted by CM_A, denoted $L(CM_A) = \{w \in (\Sigma \times D)^* \mid CM_A$ has an accepting run on $w\}$.

Example 17.3. The language $L_{fd(a)}$ can be accepted by the following class memory automaton $A = (Q, \Sigma, \Delta, q_0, F_l, F_g)$ where $Q = \{q_0, q_a, q_b\}$ and Δ contains the tuples $\{(p, a, \perp, q_a), (p, b, \perp, q_b), (p, b, q_a, q_a), (p, b, q_b, q_b), (p, a, q_b, q_a) \mid p \in \{q_0, q_a, q_b\}\}$. F_l is the set $\{q_b\}$ and F_g is the set Q.

Theorem 17.3. *The emptiness problem for CM_A is decidable.*

Proof. Let $A = (Q, \Sigma, \Delta, q_0, F_l, F_g)$ be a given CM_A. We construct a Petri net N_A and a set of configurations M_A such that A accepts a string if and only if N_A can reach any of M_A.

Define $N_A = (S, T, F)$ where $S = Q \cup \{q^c \mid q \in Q\}$, and the transition relation T is as follows. For each $\delta = (p, a, s, q)$ where $s \neq \perp$ we add a new transition t_δ such that ${}^\bullet t_\delta = \{p, s^c\}$ and $t_\delta^\bullet = \{q, q^c\}$. For each $\delta = (p, a, \perp, q)$ where we add a new transition t_δ such that ${}^\bullet t_\delta = \{p\}$ and $t_\delta^\bullet = \{q, q^c\}$. We add additional transitions $t_{(p,q)}$ for each $p \in F_g, q \in F_l$ such that ${}^\bullet t_{(p,q)} = \{p, q^c\}$ and $t_{(p,q)}^\bullet = \{p\}$. The flow relation is defined accordingly.

The initial marking of the net is M_0 where q_0 has a single token and all other places are empty. M_A is the set of configurations in which exactly one of $q \in F_g$ has a single token and all other places are empty.

The details are routine. The place q^c keeps track of the number of data values with state q. Using induction it can be easily shown that a run of the automata gives a firing sequence in the net and vice versa. Finally when we reach a global state we can use the additional transitions to pump out all the tokens in the local final states. The only subtlety is that the additional transitions in the net can be used even before reaching an accepting configuration in the net, in which case it amounts to abandoning certain data classes in the run of the automaton (these are data values which are not going to be used again). □

Thus emptiness for CM_A is reduced to reachability in Petri nets. As it happens, it is also as hard as Petri net reachability [7]. Since the latter problem is not even known to be elementary, we need to look for subclasses with better complexity. CM_A are not closed under complementation, but they are closed under union, intersection, homomorphisms. It also happens that they admit a natural logical characterization to which we will return later.

17.7. Automata with stacks

Another memory structure that has played a significant role in theory of computation is the stack, or pushdown mechanism. In automata theory, use of the pushdown mechanism is related to acceptance of context free languages. The main idea is that we can remember unbounded information but can access the memory only in a limited fashion. This gives the same power as a finite state machine with a single counter which can be incremented, decremented and checked for zero. In the context of data words, such a mechanism can be employed either to remember data values, or positions in the word, (denoting data classes), or both.

An elegant way of modelling such memory of positions is the concept of a pebble: whenever we wish to remember a position (perhaps one where we say a particular data value), we place a pebble on it. How many pebbles we can use, and when several pebbles have been placed, which one can be accessed first, determines the memory structure. Below, we consider a model where a stack discipline is used to access the pebbles.

17.7.1. *Pebble automata*

Below, let $Ins = \{\uparrow, \downarrow, \leftarrow, \rightarrow\}$. These are instructions to the automaton: while $\leftarrow$ and $\rightarrow$ tell the machine to move right or left along the data word, $\uparrow$ and $\downarrow$ are for pushing on to or popping up the stack.

Definition 17.5. A pebble automaton [5] P_A is a system $(Q, \Sigma, k, \Delta, q_0, F)$ where Q is a finite set of states, $q_0 \in Q$ is the initial state and $F \subseteq Q$ is the set of final states; $k > 0$ is called the *stack height*, and $\Delta : (Q \times \Sigma \times [k] \times 2^{[k]} \times 2^{[k]}) \rightarrow (Q \times Ins)$.

When a transition is of the form $\alpha \rightarrow \beta$ where $\alpha = (p, a, i, P, V)$ and $\beta = (q, \pi)$ where $\pi \in Ins$, the automaton is in state p, reading letter a, and depending on the "pointer stack" (i, P, V) transits to state q, moving the control head or operating on the stack according to Ins. The stack information, which will be clearer as we study configurations, is as follows: i is the current height of the stack, P is used to collect the pointers at a position, whereas V collects pointers having the same data value.

Let $m \geq 0$. The set of *m-configurations* of P_A is given by $C_m = (Q \times [k] \times ([k] \rightarrow [m]_0))$. Given a string $w \in (\Sigma \times D)^*$, we call $\chi \in C_{|w|}$ a w-configuration. $\chi_0 = (q, i, \theta) \in C_0$ is said to be *initial* if $q = q_0$, $i = 1$ and $\theta(1) = 1$. A configuration (q, i, θ) is said to be *final* if $q \in F$ and $\theta(i) = |w|$.

Consider a data word $(a_1, d_1) \ldots (a_m, d_m)$. A transition $(p, i, a, P, V) \rightarrow (q, \pi)$ *is enabled* at a w-configuration (q, j, θ) if $p = q$, $i = j$, $a_{\theta(i)} = a$, $P = \{\ell < i \mid \theta(\ell) = \theta(i)\}$, and $V = \{\ell < i \mid d_{\theta(\ell)} = d_{\theta(i)}\}$. Thus P is the set of all pointers pointing to the position pointed by $\theta(i)$, and V is the set of all positions which contains the same data value pointed by $\theta(i)$.

P_A can go from a w-configuration (p, i, θ) to a w-configuration (q, i', θ') if there is a transition $\alpha \rightarrow (q, \pi)$ enabled at (p, i, θ) such that $\theta'(j) = \theta(i)$ for all $j < i$ and:

(1) if $\pi = \rightarrow$ then $i = i'$ and $\theta'(i) = \theta(i) + 1$.
(2) if $\pi = \leftarrow$ then $\theta(i) > 1$, $i = i'$ and $\theta'(i) = \theta(i) - 1$.
(3) if $\pi = \downarrow$, then $i < k$, $i' = i + 1$ and $\theta'(i') = \theta(i)$.

(4) if $\pi = \uparrow$, then $i > 1$, and $i' = i - 1$.

Above, whenever the side conditions are not met (as for instance, when it tries to pop the empty stack), the machine halts. As one may expect, the pebble mechanism adds considerable computational power.

Example 17.4. The language $\overline{L_{fd(a)}}$, can be accepted by a pebble automaton $A = (Q, \Sigma, 2, q_0, F, \Delta)$, where $Q = \{q_0, q_1, q_{\rightarrow}, q_f\}$ and $F = \{q_f\}$. Δ consists of the following transitions: $\Delta = \{(q_0, \Sigma, 1, \emptyset, \emptyset) \rightarrow (q_0, \rightarrow), (q_0, a, 1, \emptyset, \emptyset) \rightarrow (q_{\rightarrow}, \downarrow)$, $(q_{\rightarrow}, a, 2, \{1\}, \{1\}) \rightarrow (q_1, \rightarrow), (q_1, \Sigma, 2, \emptyset, \emptyset) \rightarrow (q_1, \rightarrow), (q_1, a, 2, \emptyset, \{1\}) \rightarrow (q_f, \rightarrow)$, $(q_f, \Sigma, 2, *, *) \rightarrow (q_f, \rightarrow)\}$

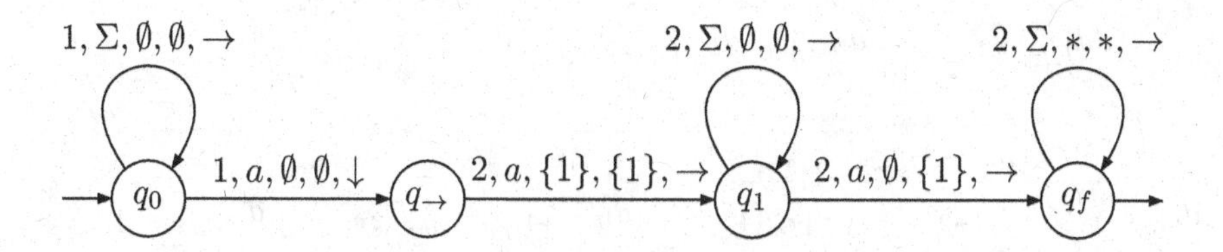

Fig. 17.3. Pebble automaton in the Example 17.4.

Automaton A is shown in Fig. 17.3 and works in the following way. It stays in state q_0 while moving to the right and non-deterministically places a new pebble at a position with label a. After placing the new pebble the automaton moves one position to the right. A continues moving to the right and after reaching a position with label a and having the same data value (as that under the first pebble) it enters the final state.

Example 17.5. The language $L_{\forall a \exists b}$, can be accepted by a pebble automaton. The automaton works the following way. A starts in state q_0, it continues moving to the right, whenever it sees an a, it drops a new pebble at the first position and enters the state $q_{\rightarrow}$. In the state $q_{\rightarrow}$ the automaton goes all the way to the left and from there starts searching for a position with a label b and having the same data value as the data value under the first pebble. Once it sees that, it lifts the pebble, continues moving right in state q_0 until it reaches the end of the input.

Indeed, the pebble automaton is too powerful, as the following theorem demonstrates.

Theorem 17.4. *Emptiness checking is undecidable for deterministic pebble automata.*

Proof. We reduce the Post's Correspondence Problem to emptiness checking deterministic pebble automata. The pebble automaton first checks whether the input is of desired form and then accepts the input if it is a solution of the PCP instance.

An instance of PCP consists of a finite number of pairs (u_i, v_i), $u_i, v_i \in \Sigma^*$. The question is whether there exists a non-empty, finite sequence $i_0, i_i, \ldots i_n$ such that:

$$w = u_{i_0} u_{i_1} \ldots u_{i_n} = v_{i_0} v_{i_1} \ldots v_{i_n}$$

This is done as follows. We take $\Sigma' = \Sigma \cup \overline{\Sigma}$, two disjoint copies of Σ, as our letter alphabet and D as the data alphabet. The words over Σ are denoted by u, v etc. and the words over $\overline{\Sigma}$ are represented as $\overline{u}, \overline{v}$ etc. The solution for the PCP can be represented in $(\Sigma' \times D)^*$ as:

$$\hat{w}_d = \begin{array}{cccccccc} a_1^{i_0} & a_2^{i_0} & .. & a_{|u_{i_0}|}^{i_0} & \overline{a_1^{i_0}} & \overline{a_2^{i_0}} & .. & \overline{a_{|v_{i_0}|}^{i_0}} \\ d_1 & d_2 & .. & d_{|u_{i_0}|} & d_1 & d_2 & .. & d_{|v_{i_0}|} \end{array} ..$$

$$\begin{array}{cccccccc} a_1^{i_1} & a_2^{i_1} & .. & a_{|u_{i_1}|}^{i_1} & \overline{a_1^{i_1}} & \overline{a_2^{i_1}} & .. & \overline{a_{|v_{i_1}|}^{i_1}} \\ d_{|u_{i_0}|+1} & d_{|u_{i_0}|+2} & .. & d_{|u_{i_0}|+|u_{i_1}|} & d_{|v_{i_0}|+1} & d_{|v_{i_0}|+2} & .. & d_{|v_{i_0}|+|v_{i_1}|} \end{array} ..$$

The string projection of $\hat{w}_d$ is $str(\hat{w}_d) = u_{i_0}\overline{v_{i_0}}u_{i_1}\overline{v_{i_1}} \ldots u_{i_n}\overline{v_{i_n}}$. Every data value appearing in the word appears exactly twice, once associated with a label in Σ and once with the corresponding label in $\overline{\Sigma}$.

We can verify these two properties by a pebble automaton A which has subroutines to check the following,

- the string projection belongs to $\{u_i\overline{v_i} \mid 1 \leq i \leq k\}^+$. In fact, this can be done by a finite automaton over a finite alphabet.
- each data value occurs exactly twice, once labelled with a letter from Σ and once with the corresponding letter in Σ'. This can be done by checking, for each a, w belongs to the languages $L_{\forall a \exists \overline{a}}$, $L_{\forall \overline{a} \exists a}$.
- the sequence of data values are the same in w and $\overline{w}$. This can be done by checking every consecutive pair occurring in the Σ-labelled part occurs in the Σ'-labelled part, and vice versa.

Thus the automaton A, by checking each property above, can verify the solution of the PCP instance. Now it is clear that if there is a data word w such that w is accepted by the automaton A then there is a solution for the PCP instance and vice versa.

Therefore emptiness of deterministic pebble automata is undecidable. □

Undecidability suggests that the machine model under consideration has strong computational power, which is reinforced by the fact that pebble automata are robust. The class is closed under logical operations (complementation, union, intersection) and its expressive power is invariant under nondeterminism and alternation.

17.8. Logics for data languages

The celebrated theorem of Büchi [12] asserts the equivalence of languages recognized by finite state automata and those defined by sentences of Monadic Second Order Logic (MSO). We consider the language of First order logic with a single binary relation $<$ and a unary relation Q_a for each $a \in \Sigma$, extended by set quantifiers. The idea is that such formulas are interpreted over words in Σ^*, with first order variables denoting positions in words, and second order (monadic) variables denoting sets of positions. MSO logics have been extensively studied not only over finite and infinite words (over finite alphabets) but also over trees and graphs.

Considering the variety of automata models we have been discussing for languages of words over infinite alphabets, a natural criterion for assessing such models is in terms of how they relate to such logical presentations. This is an extensive area of research, and we merely sketch the main ideas here.

In the same spirit as Σ-words are defined as Σ-labels over positions, we can consider data words over $(\Sigma \times D)$ as well. Guided by our decisions to consider data values only implicitly and restricting operations on data to only checking for equality, we are led to an equivalence relation on word positions: two positions are equivalent if they carry the same data value. Thus, we can consider data words to be structures with positions labelled by Σ and an equivalence relation on word positions.

This discussion suggests that we consider enriching the standard syntax of MSO with equivalence on first order terms. The syntax of $MSO(\Sigma, <, +1, \sim)$ is given by:

$$\varphi ::= a(x) \mid x = y \mid x = y + 1 \mid x < y \mid x \sim y \mid x \in X \mid \neg\varphi \mid \varphi \vee \varphi \mid \exists x.\varphi \mid \exists X.\varphi$$

Note that $<$ and $+1$ are inter-definable in the logic. The first order fragment of this logic is denoted $\mathrm{FO}(\Sigma, <, +1, \sim)$.

Let FV denote the collection of first order variables, and SV that of set variables. Given a word $w = (a_1, d_1)(a_2, d_2) \dots (a_n, d_n)$ and an interpretation $I = (I_f, I_s)$ of the variables $I_f : FV \to [n]$ and $I_s : SV \to 2^{[n]}$, we can define the semantics of formulas as follows.

$$\begin{array}{ll}
w, I \models a(x) & \text{if } a_{I_f(x)} = a \\
w, I \models x = y & \text{if } I_f(x) = I_f(y) \\
w, I \models x = y + 1 & \text{if } I_f(x) = I_f(y) + 1 \\
w, I \models x < y & \text{if } I_f(x) < I_f(y) \\
w, I \models x \in X & \text{if } I_f(x) \in I_s(X) \\
w, I \models x \sim y & \text{if } d_{I_f(x)} = d_{I_f(y)} \\
w, I \models \neg\varphi & \text{if } w, I \not\models \varphi \\
w, I \models \varphi_1 \vee \varphi_1 & \text{if } w, I \models \varphi_1 \text{ or } w, I \models \varphi_2 \\
w, I \models \exists x.\varphi & \text{if there is an } i \text{ in } [n] \text{ s.t. } w, (I_f[x \to i], I_s) \models \varphi \\
w, I \models \exists X.\varphi & \text{if there exists } J \subseteq [n] \text{ s.t. } w, (I_f, I_s[X \to J]) \models \varphi
\end{array}$$

A sentence σ is a formula with no free variables, and it is easily seen that for any w, either $w \models \sigma$ or $w \models \neg\sigma$ independent of any interpretation I. Let $L(\sigma) = \{w \mid w \models \sigma\}$.

Let $L \subseteq (\Sigma \times D)^*$. We say L is definable when there exists a sentence σ such that $L = L(\sigma)$. We are interesting in relating definable languages of data words with those recognized by automata models.

The sentence $\forall x \forall y. x \neq y \rightarrow x \not\sim y$ defines the language of data words in which no data value repeats. Similarly $\forall x \forall y. a(x) \wedge a(y) \wedge x \neq y \rightarrow x \not\sim y$ defines the language $L_{fd(a)}$. The sentence $\forall x \exists y. a(x) \rightarrow b(y) \wedge x \not\sim y$ defines the language $L_{\forall a \exists b}$.

The above examples show that definability is computationally powerful, as asserted by the following proposition.

Proposition 17.2. *The satisfiability problem for* $\mathrm{FO}(\Sigma, <, +1, \sim)$ *is undecidable [7].*

For a proof of the above claim, note that the PCP coding which we used in the undecidability of pebble automata can in fact be carried out in $\mathrm{FO}(\Sigma, <, +1, \sim)$.

The above proposition suggests that for restricting definability to implementability by devices, we should look for some fragment of the logic. There are many ways to obtain decidable fragments of quantificational logic: for example, by specifying the form of quantifier prefixes allowed [13], restricting the number of variables which can be used in the formula [14], or restricting the syntactic structure of formulas, such as in Guarded fragments ([15, 16]). The study of decidable fragments of first order logics over special classes of structures is an interesting area of study in its own right ([13]).

The restriction by number of variables used turns out to be especially interesting for infinite alphabets. A careful look at the undecidability proof above shows that *three variables* suffice. On the other hand, the example languages we have been working with are all defined above using only *two variables.* This provides sufficient motivation to focus attention on the two variable fragment of the logic above.

$\mathrm{FO}^2(\Sigma, <, +1, \sim)$, is the set of all formulas in the language of $\mathrm{FO}(\Sigma, <, +1, \sim)$ using at most two variables. As the following theorem shows, such a restriction pays off.

Theorem 17.5. *Satisfiability of* $\mathrm{FO}^2(\Sigma, <, +1, \sim)$ *is decidable [7].*

Proof. Here we only sketch the proof, the details are intricate and require a more elaborate presentation. We refer to the logic simply as FO^2.

We denote by α, β etc unary quantifier free formulas (which are called *types*). It is easy to see that FO^2 can express the following properties.

- data-blind properties, i.e. properties not using the predicate $\sim$. These are the word languages over Σ expressible by FO^2.

- Each class contains at most one occurrence of α: This can be expressed by the formula $\forall xy.\alpha(x) \wedge \alpha(y) \wedge x \sim y \rightarrow x = y$.
- In each class every α occurs before every β: This can be expressed by the formula $\forall xy.\alpha(x) \wedge \beta(y) \wedge x \sim y \rightarrow x < y$.
- Each class with an α has also a β: This can be expressed by the formula $\forall x \exists y.\alpha(x) \rightarrow \beta(y) \wedge x \sim y$.
- If a position is in a different class than its successor then it has type α: This can be expressed by $\forall xy.x \not\sim y \wedge y = x + 1 \rightarrow \alpha(x)$.

The idea behind the proof is to convert any given formula to a disjunction of conjunction of formulas of the above kinds (which is called Data Normal Form). In order to do this we first convert the given formula to the so-called Scott Normal Form:

$$\left(\forall xy.\chi \wedge \bigwedge_i \forall x \exists y.\chi_i \right)$$

A formula in this form is then converted to the form

$$\bigwedge_i \theta_i$$

where each θ_i belongs to one of the five kinds.

The crucial next step is the fact that each of these formulas can be recognized by a Class Memory automaton. The existential predicates correspond to nondeterminism in the automaton. Since CM_A are closed under union, intersection and projection we can compose automata corresponding to the subformulas to get an automaton corresponding to a formula. Hence the satisfiability of the formula reduces to non-emptiness problem of the automaton, which is decidable. □

This connection can be strengthened as the following corollary shows.

Corollary 17.1. *Languages recognized by CM_A are the languages expressed by* $\mathrm{EMSO}^2(\Sigma, <, +1, \sim, \oplus)$.

Here EMSO^2 stands for existential monadic second order formulas with their first order fragment containing only two variables and $\oplus$ stands for the *class successor* relation.

17.8.1. *Temporal logics for data languages*

If Büchi's theorem relates MSO definability and recognizability by automata, another celebrated theorem due to Kamp [17] relates FO definability to temporal logics. These are modal logics with modalities such as *eventually, until, since* etc. The success of formal methods in system verification has to do with the ease of temporal logics both in terms of algorithmic tools and reasoning.

Hence, when we consider description of data languages, temporal logic is a natural candidate. We need modalities to mark positions, and recall marks when needed. A temporal logic developed by [8] adds **freeze operators** to propositional linear time temporal logic. These come in two forms: unary modalities $\downarrow_i$, $i > 0$, and atomic formulas $\uparrow_i$. Informally, the semantics is as follows. $\downarrow_i$ stores the current data value in register i and $\uparrow_i$ checks whether the current data value equals to the data stored in the register i.

The syntax of the logic is given by:

$$\varphi ::= \top \mid a \mid \uparrow_i \mid \varphi \wedge \varphi \mid \neg\varphi \mid X\varphi \mid \varphi U \varphi \mid \downarrow_i \varphi$$

The semantics (over finite words) is defined in the obvious manner.

The formula $F \downarrow_1 XF \uparrow_1$ expresses the set of data words in which at least one data value repeats. Hence its negation expresses the language of all data words in which no data value repeats. $G(a \rightarrow \downarrow_1 XF(\uparrow_1 \wedge b)$ expresses the set of data words in which in every class, if an a occurs there is a b occurring later.

Again, the logic is too powerful: the satisfiability problem for LTL with freeze operators is undecidable. However satisfiability of LTL-freeze with one register is decidable [8]. Also, there is a natural definition of register automata corresponding to LTL-freeze, though this definition is not equivalent to the register machines we discussed earlier.

LTL-freeze cannot express the language $L_{\forall a \exists b}$, and this is precisely because of the absence of past operators. Similarly FO^2 cannot express the following language, which LTL with freeze can: between two a of the same class there is an a which belongs to some other class. This is due to the shortage of variables in FO^2. Hence in general FO^2 and LTL-freeze are incomparable.

17.9. Related models and logics

Though register automata are relatively weak, they possess an interesting theory, and the algebraic and rational aspects of register automata have been studied. As an extension of recognizability by register automata, in [18] a Myhill-Nerode theorem for data strings is shown. In [19] the authors propose a notion of regular expressions for data languages.

In [10], an extension of register automata is offered in which the automaton can guess the data values, called look-ahead register automata. An equivalent characterization in terms of regular expressions and grammar is presented for this class of automata. One interesting property of look-ahead register automata is that they are closed under reversal.

The work in [20] seeks a notion of context-free data languages. They extend register automata with a stack which can hold data values, called pushdown register automata. The automaton has the ability to rewrite the registers with arbitrary data values. The stack operations transfer symbols from the registers to the stack. The transition is based on the register which matches the top of the stack. The automaton is able to accept some analogous languages in this way, for example $w\#w^r$, where w is a data string. A definition of context-free grammars over infinite alphabets is proposed and is shown equivalent to pushdown register automata. A lemma similar to that of register automata holds in the case of pushdown automata which says that given a pushdown register automaton there exists an n such that if the automaton accepts a word it accepts a word with at most n data values. While it is not apparent how to determine n from the definition of the automata, the observation shows that the emptiness problem for this class of automata is decidable.

In [21], a notion of monoid recognizability for data languages is introduced. A corresponding extension of register automata is shown to be equivalent to recognition by monoids. This model is an extension of register automata with an equivalence relation of finite index. During a transition the automaton, in addition to reading the register, checks to which equivalence class the register configuration belongs to, selects the next state and updates the register accordingly. Though this certainly enhances computational‘ power, the automaton can still remember only finitely many data values. Moreover, decidability of the emptiness problem is obtained only in the case where register updates respect the equivalence. In [22] an EMSO characterization of monoid recognizablity is shown.

Another simple computational model, based on transducers is the Data Automaton model introduced in [7]. A Data Automaton $A = (B, C)$ consists of:

- a non-deterministic letter-to-letter string transducer B, the base automaton, with input alphabet Σ and some output alphabet Γ, and
- a non-deterministic automaton C, the class automaton, with input alphabet Γ

A data word w is accepted by A if there is an accepting run of B on the string projection of w, giving an output string $\gamma_1\gamma_2\dots\gamma_n \in \Gamma^*$, such that for each class $\{x_1, x_2, \dots, x_k\} \subseteq \{1, 2, \dots, n\}$, $x_1 < x_2 < \dots < x_k$, the class automaton accepts $\gamma_{x_1}\gamma_{x_2}\cdots\gamma_{x_k}$.

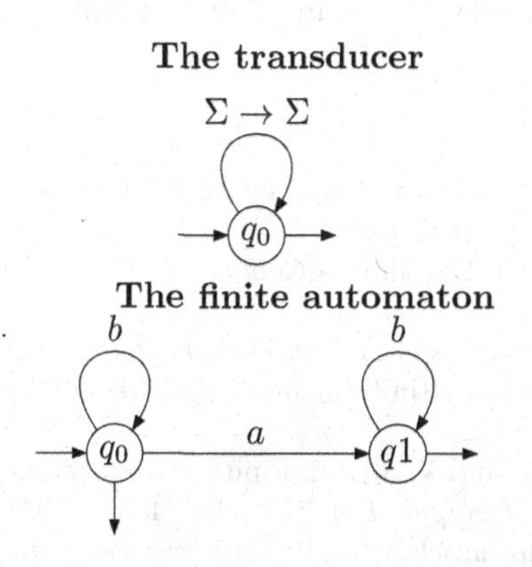

Fig. 17.4. Automaton in the Example 17.6.

Example 17.6. The language $L_{fd(a)}$, can be accepted by a data automaton $A = (B, C)$ as follows.

- The base automaton $B \colon \Sigma \to \Sigma$ is a copy automaton which copies the input (string part) to the output.
- The class automaton C accepts $b^* + b^*ab^*$, words containing at most one a. The automaton is shown in Fig. 17.4.

Theorem 17.6 ([6]). *Class Memory automata and Data automata are expressively equivalent.*

17.10. Conclusion

We have taken the reader on a touristic journey of automata models over infinite alphabets, pointing out interesting sights very briefly, without studying any of the models as they really should be. We have looked at the models only from the perspective of decidability of the emptiness problem, and if there is any moral to the story, it is only this: systems with unbounded data give rise to infinite state systems, and identifying machine classes with manageable complexity is a challenge. Our bag of tools for addressing such problems needs considerable enrichment, and it is hoped that automata theory and logic will contribute substantially towards this. Developing an underlying algebraic theory may be an important step in this direction.

On a positive note, automata over infinite alphabets provide an abstract theoretical model in which we seem to be able to represent situations that arise in a wide variety of contexts: systems with unboundedly many processes such as those studied in infinite state verification, communicating systems such as Petri nets, the verification of software programs with recursion and pointers, navigation over trees representing semistructured data, web services handling unboundedly many clients, to list a few. Such diversity suggests that use of techniques across areas may lead to new techniques in the theory of automata over infinite alphabets.

References

[1] J. E. Hopcroft and J. D. Ullman, *Introduction to Automata Theory, Languages and Computation.* (Addison-Wesley, 1979).
[2] A. V. Aho, R. Sethi, and J. D. Ullman, *Compilers: Principles, Techniques, and Tools.* (Addison-Wesley, 1986).
[3] E. M. Clarke, O. Grumberg, and D. A. Peled, *Model Checking.* (MIT Press, 2000).
[4] M. Kaminski and N. Francez, Finite-memory automata, *Theor. Comput. Sci.* **134**(2), 329–363, (1994).
[5] F. Neven, T. Schwentick, and V. Vianu, Finite state machines for strings over infinite alphabets, *ACM Trans. Comput. Log.* **5**(3), 403–435, (2004).
[6] H. Björklund and T. Schwentick. On notions of regularity for data languages. In *Proc. 16th FCT*, vol. 4639, *LNCS*, pp. 88–99. Springer, (2007).

[7] M. Bojanczyk, A. Muscholl, T. Schwentick, L. Segoufin, and C. David. Two-variable logic on words with data. In *Proc. 21st LICS*, pp. 7–16. IEEE Computer Society, (2006).

[8] S. Demri and R. Lazic. LTL with the freeze quantifier and register automata. In *Proc. 21st LICS*, pp. 17–26. IEEE Computer Society, (2006).

[9] H. Sakamoto and D. Ikeda, Intractability of decision problems for finite-memory automata, *Theor. Comput. Sci.* **231**(2), 297–308, (2000).

[10] D. Zeitlin. Look-ahead finite-memory automata. Master's thesis, Technion - Israel Institute of Technology, Tamuz, 5766, Haifa, Israel (July, 2006).

[11] M. Amaldev and R. Ramanujam. Counting multiplicity over infinite alphabets. In *Proc. 3rd Workshop on Reachability Problems*, vol. 5797, *LNCS*, pp. 141–153. Springer, (2009).

[12] J.R. Büchi, Weak second order arithmetic and finite automata, *Z. Math. Logik Grundlagen Math.* **6**, 66–92, (1960).

[13] E. Börger, E. Grädel, and Y. Gurevich, *The Classical Decision Problem.* Perspectives in Mathematical Logic, (Springer, 1997).

[14] E. Grädel, P. G. Kolaitis, and M. Y. Vardi, On the decision problem for two-variable first-order logic, *Bulletin of Symbolic Logic.* **3**(1), 53–69, (1997).

[15] H. Andréka, I. Németi, and J. van Benthem, Modal logic and bounded fragments of predicate logic, *Journal of Philosophical Logic.* **27**(3), 217–274, (1998).

[16] E. Grädel, On the restraining power of guards, *Journal of Symbolic Logic.* **64**(4), 1719–1742, (1999).

[17] H. Kamp. *On tense logic and the theory of order.* PhD thesis, UCLA, (1968).

[18] N. Francez and M. Kaminski, An algebraic characterization of deterministic regular languages over infinite alphabets, *Theor. Comput. Sci.* **306**(1-3), 155–175, (2003).

[19] M. Kaminski and T. Tan. Regular expressions for languages over infinite alphabets. In *Proc. 10th COCOON*, vol. 3106, *LNCS*, pp. 171–178. Springer, (2004).

[20] E. Y. C. Cheng and M. Kaminski, Context-free languages over infinite alphabets, *Acta Inf.* **35**(3), 245–267, (1998).

[21] P. Bouyer, A. Petit, and D. Thérien, An algebraic approach to data languages and timed languages, *Information and Computation.* **182**(2), 137–162 (May, 2003).

[22] P. Bouyer, A logical characterization of data languages, *Information Processing Letters.* **84**(2), 75–85, (2002).

Chapter 18

Automata and Logics over Signals

Fabrice Chevalier

Laboratory for Specification and Verification, ENS de Cachan, France
fabrice.chevalier@sfr.fr

Deepak D'Souza

Department of Computer Science and Automation,
Indian Institute of Science, Bangalore, India
deepakd@csa.iise.ernet.in

Raj Mohan M.

Department of Computer Science and Automation,
Indian Institute of Science, Bangalore, India
raj@csa.iisc.ernet.in

Pavithra Prabhakar

Department of Computer Science,
University of Illinois at Urbana-Champaign, USA
pprabha2@uiuc.edu

We extend some of the classical connections between automata and logic due to Büchi [1] and McNaughton and Papert [2], to languages of finitely varying functions or "signals". In particular we introduce a natural class of automata for generating finitely varying functions called ST-NFA's, and show that it coincides in terms of language-definability with a natural monadic second-order logic interpreted over finitely varying functions [3]. We also identify a "counter-free" subclass of ST-NFA's which characterize the first-order definable languages of finitely varying functions. Our proofs mainly factor through the classical results for word languages. These results have applications in automata characterisations for continuously interpreted real-time logics like Metric Temporal Logic (MTL) [4, 5].

18.1. Introduction

The classical literature contains a rich theory connecting automata and logic over words. Büchi showed that languages definable in monadic second logic (MSO) over words are precisely the class of languages accepted by finite state automata [1].

Kamp [6] showed that languages definable in Linear-Time Temporal Logic (LTL) were precisely the languages definable in the first-order (FO) fragment of Büchi's MSO. And McNaughton and Papert [2] showed that the class of counter-free finite state automata (where a "counter" in an automaton is a loop with at least two hops, each hop being on a common word u) define exactly the FO-definable languages. This last result uses a characterisation due to Schutzenberger [7] of the class of counter-free languages in terms of star-free regular expressions.

Our aim in this article is to lift these connections to languages of finitely varying functions from the non-negative reals to a finite alphabet. These functions are finitely varying in that they have only a finite number of discontinuities in any bounded interval of time. Such functions are often called "signals" in the literature, are of interest to the computer science community as they model the behaviour of timed and hybrid systems [8–10]. For example, non-zeno timed words [8] are special kinds of signals, as are the piece-wise constant behaviours of [10].

We first introduce a class of automata called ST-NFA's that run over signals and hence accept languages of signals. We should point out here that unlike timed automata we are interested in formalisms without a "metric" or operators that measure time distance. As a consequence these languages are essentially "untimed" in that they can be characterised as the set of all possible "timings" of a (regular) language of classical words. We then consider a natural monadic second-order logic introduced earlier by Rabinovich [3], and called here MSO^s, which is interpreted over signals, and in which the second-order quantification is restricted to subsets of non-negative reals whose characteristic functions are finitely-varying. We show that the class of signal languages defined by sentences in this logic is precisely the class of signal languages defined by ST-NFA's. This gives an automata-theoretic proof of a similar result obtained in [3] using logical techniques. We note that this proof also gives us a simple automata-theoretic proof of the fact that the monadic second-order logic of the non-negative reals (where second-order quantification ranges over finitely-varying subsets) is decidable.

Next, along the lines of the Schutzenberger and McNaughton-Papert results, we identify a counter-free subclass of ST-NFA's and show that they precisely characterise the class of signal languages definable by the first-order fragment FO^s of MSO^s. The notion of a counter in an ST-NFA is similar to the classical one, except that we require the ST-NFA to be "proper" in a certain sense. Our proof of this result factors transparently through the afore-mentioned results of Schutzenberger, McNaughton-Papert, and Kamp for word languages. The main difficulty, in a series of steps we perform, is to translate an LTL formula θ interpreted over word models, into one interpreted over signals which accepts precisely the timings of the word models of θ. As in the classical case for words, we show that this characterisation allows us to decide whether a given MSO^s sentence is FO^s-definable.

As a minor by-product we re-prove the expressive completeness of LTL interpreted over signals (i.e. its expressiveness coincides with FO^s over signals). This result also follows from Kamp's result showing the expressive completeness of LTL over arbitrary functions over the reals [6]. Nonetheless, our proof gives a more accessible proof of this result, since it uses only Kamp's result for classical words, for which there are simpler proofs in the literature (see [11, 12]).

Turning now to more details on related work, as already mentioned this paper builds on the classical results due to Büchi [1], Schutzenberger [7], McNaughton and Papert [2], and Kamp [6] for word languages. The work of Rabinovich contains many relevant results on signal languages. Rabinovich and Trakhtenbrot [13] introduce automata similar to ST-NFA's called signal acceptors. In [3] Rabinovich shows how to translate an MSO sentence φ to a MSO^s sentence that accepts precisely the timings of φ, and vice versa. This leads to a proof of the claim in [13] that signal languages definable by signal acceptors and MSO^s coincide. In contrast our equivalence of ST-NFA's and MSO^s uses an automata-theoretic argument similar to the proof of Büchi's result (see [14] and Chapters 1 and 2 in this volume), and helps us identify the counter-free fragment. In [15] Rabinovich also shows a star-free regular expression characterisation of FO^s-definable signal languages, along the lines of McNaughton and Papert [2].

Though we are mainly concerned with expressiveness in this work, there are a number of related decidability results in the literature. Rabin [16] shows that MSO over reals, with second-order quantification over subsets which are essentially countable unions of closed sets, is decidable. The decidability of MSO^s follows from this result. Shelah [17] showed that MSO over reals with second-order quantification over arbitrary subsets of reals is undecidable.

Preliminary versions of the basic results in this paper appeared in [4, 5] where they were used to obtain logical characterisations of versions of timed automata with a continuous interpretation, as well as a counter-free timed automata characterisation of several real-time temporal logics, including MTL [18, 19], in their continuous semantics.

18.2. Preliminaries

For an alphabet A, we use A^* to denote the set of finite words over A. For a word w in A^*, we use $|w|$ to denote its length. The set of non-negative reals and rationals will be denoted by $\mathbb{R}_{\geq 0}$ and $\mathbb{Q}_{\geq 0}$ respectively. We will deal with intervals of non-negative reals, i.e. convex subsets of $\mathbb{R}_{\geq 0}$, and denote by $\mathcal{I}_{\mathbb{R}_{\geq 0}}$ the set of such intervals with end-points in $\mathbb{R}_{\geq 0} \cup \{\infty\}$. Two intervals I and J will be called *adjacent* if $I \cap J = \emptyset$ and $I \cup J$ is an interval.

Let A be a finite alphabet and let $f : [0, r] \rightarrow A$ be a function, where $r \in \mathbb{R}_{\geq 0}$. We use $dur(f)$ to denote the duration of f, which in this case is r. A point $t \in (0, r)$ is a point of *continuity* of f if there exists $\epsilon > 0$ such that f is constant in the interval $(t - \epsilon, t + \epsilon)$. All other points in $[0, r]$ are points of *discontinuity* of f. We say f is *finitely varying* if it has only a finite number of discontinuities in its domain. We will refer to such finitely varying functions as *signals* over A, and denote the set of signals over A by $Sig(A)$. Fig. 18.1 shows a signal σ over the alphabet $\{a, b, c\}$ defined on $[0, 4]$ as follows:

$$\sigma(t) = \begin{cases} a & \text{if } t \in [0, 0.5) \cup (2, 4] \\ b & \text{if } t = 0.5 \\ c & \text{if } t \in (0.5, 2]. \end{cases}$$

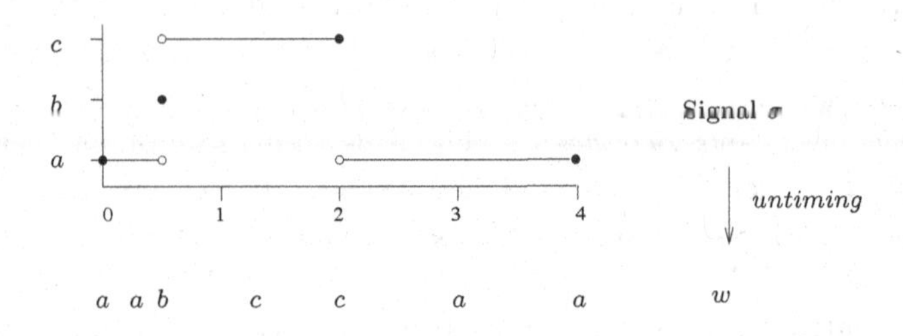

Fig. 18.1. A signal and its untiming.

An *interval representation* for a signal $\sigma : [0, r] \to A$ is a sequence of the form $(a_0, I_0) \cdots (a_n, I_n)$, with $a_i \in A$ and $I_i \in \mathcal{I}_{\mathbb{R}_{\geq 0}}$, satisfying the conditions that: $0 \in I_0$, the union of the intervals is $[0, r]$, each I_i and I_{i+1} are adjacent, and for each i, σ is constant and equal to a_i in the interval I_i. We can obtain a *canonical* interval representation for σ by putting each point of discontinuity in a singular interval by itself. Thus the above interval representation for σ is canonical if n is even, for each even i the interval I_i is singular (i.e. of the form $[t, t]$), and for no even i such that $0 < i < n$ is $a_{i-1} = a_i = a_{i+1}$. The canonical interval representation for the signal of Fig. 18.1 is $([0, 0], a)((0, 0.5), a)([0.5, 0.5], b)((0.5, 2), c)([2, 2], c)((2, 4), a)([4, 4], a)$.

A canonical interval representation for a function gives us a canonical way of "untiming" the signal: thus if $(a_0, I_0) \cdots (a_{2n}, I_{2n})$ is the canonical interval representation for a signal σ, then we define *untiming*(σ) to be the string $a_0 \cdots a_{2n}$ in A^*. The untiming thus captures explicitly the value of the function at its points of discontinuity and the open intervals between them. The untiming of the signal in Fig. 18.1 is thus *aabccaa*. Note that strings which represent the untiming of a signal will always be of odd length, and for no even position i will the letters at positions $i - 1$, i, and $i + 1$ be the same. We refer to words in A^* which satisfy these two conditions as *proper* words over A. We denote the set of proper words over A by *Prop*(A).

A canonical word w can be "timed" to get a signal in a natural way: thus a signal σ is in *timing*(w) if *untiming*$(\sigma) = w$. We extend the definition of *timing* and *untiming* to languages of signals and words in the expected way.

Finally, we say a subset X of $\mathbb{R}_{\geq 0}$ is *finitely varying* if its characteristic function $f_X : \mathbb{R}_{\geq 0} \to \{0, 1\}$ given by $f_X(t) = 1$ if $t \in X$ and 0 otherwise, is finitely varying (in the sense defined above) in every interval of the form $[0, r]$ with $r \in \mathbb{R}_{\geq 0}$.

18.3. Automata over signals – ST-NFA's

In this section we introduce a variant of classical word automata called ST-NFA's which are a convenient formalism for generating signals.

We recall that a non-deterministic finite state automaton (NFA) over an alphabet A is a structure $\mathcal{A} = (Q, S, \delta, F)$, where Q is a finite set of states, S is the set of initial states, $\delta \subseteq Q \times A \times Q$ is the transition relation, and $F \subseteq Q$ is the set of final states. A run of $\mathcal{A}$ on a word $w = a_0 \cdots a_n \in A^*$ is a sequence of states $q_0, \ldots, q_{n+1}$ such that $q_0 \in S$, and $(q_i, a_i, q_{i+1}) \in \delta$ for each $i \leq n$. The run is accepting if $q_{n+1} \in F$. The word language accepted by $\mathcal{A}$, denoted $L(\mathcal{A})$, is the set of words in A^* over which $\mathcal{A}$ has an accepting run. Languages accepted by NFA's are called *regular* languages. We say the NFA $\mathcal{A}$ is *deterministic* (and call it a DFA) if the set of start states is a singleton, and for each $p \in Q$ and $a \in A$, there is at most one out-going transition of the form (p, a, q) in δ.

A *state-transition-labeled* NFA (ST-NFA for short) over A is a structure $\mathcal{A} = (Q, S, \delta, F, l)$ similar to an NFA over A, except that $l : Q \to A$ labels states with letters from A. As a recogniser of words, the ST-NFA $\mathcal{A}$ accepts strings of the form $A(AA)^*$. A run of $\mathcal{A}$ on a string $w = a_0 a_1 \cdots a_{2n}$ in $A(AA)^*$, is a sequence of states $q_0, \ldots, q_{n+1}$ satisfying $q_0 \in S$, $(q_i, a_{2i}, q_{i+1}) \in \delta$ for each $i \in \{0, \cdots, n\}$, and $l(q_i) = a_{2i-1}$ for each $i \in \{1, \ldots, n\}$. The run is *accepting* if $q_{n+1} \in F$. We define the word language accepted by $\mathcal{A}$, denoted $L(\mathcal{A})$, to be the set of strings $w \in A^*$ on which $\mathcal{A}$ has an accepting run. Figure 18.2 shows an ST-NFA over the alphabet $\{a, b, c\}$ which accepts the word language $a(abccaa)^*$. We will use the convention that start states are indicated by sourceless incoming arrows, and final states are indicated by double circles.

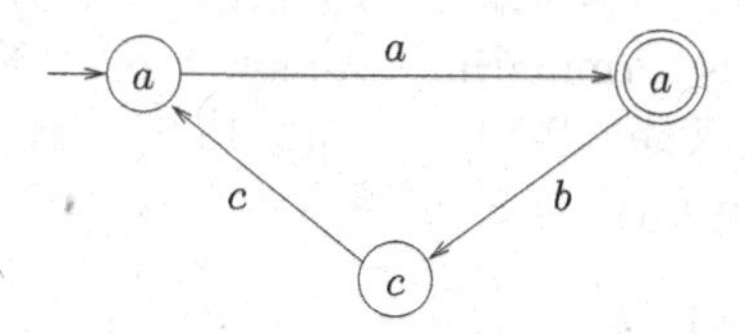

Fig. 18.2. Example ST-NFA.

An ST-NFA $\mathcal{A}$ also generates signals in a natural way: we begin by taking a transition emanating from the start state, emitting its label, and then spend time at the resulting state emitting its label all the while, before taking a transition again; and so on till we choose to stop at a final state. The language of signals generated by an ST-NFA $\mathcal{A}$ is defined to be *timing*$(L(\mathcal{A}))$, and will be denoted by $S(\mathcal{A})$. The signal shown in Fig. 18.1 is accepted by the ST-NFA shown in Fig. 18.2.

We say that an ST-NFA $\mathcal{A} = (Q, S, \delta, F, l)$ is *deterministic* if S is singleton, and there do not exist states p, q and r, with $q \neq r$, such that $(p, a, q) \in \delta$ and $(p, a, r) \in \delta$ with $l(q) = l(r)$. We denote the class of deterministic ST-NFA's by ST-DFA.

Here are some properties of proper words which will be useful in the sequel.

Proposition 18.1. *Let A be an alphabet.*

(1) If w and w' are proper words over A then $timing(w) \cap timing(w') \neq \emptyset$ iff $w = w'$.
(2) $timing(Prop(A)) = Sig(A)$.
(3) The word language $Prop(A)$ is ST-DFA-*definable.*

Proof. Parts 1 and 2 follow easily from the definitions. For part 3, the required ST-DFA $\mathcal{A}_{Prop}$ for the alphabet $\{a, b\}$ is shown in Figure 18.3 below.

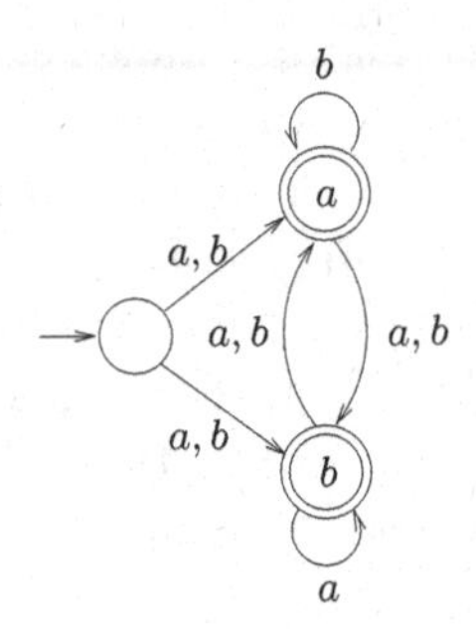

Fig. 18.3. ST-DFA $\mathcal{A}_{Prop}$ accepting the word language $Prop(\{a, b\})$.

□

Let us call a state in an ST-NFA "originating" if it has no incoming transitions, and "terminating" if it has no outgoing transitions. We say a transition $p \xrightarrow{a} q$ in an ST-NFA $\mathcal{A} = (Q, S, \rightarrow, F, l)$ is *non-proper* if $l(p) = l(q) = a$, *except* for the case when p is originating or q is terminating. We say the ST-NFA $\mathcal{A}$ is *proper* if it has no non-proper transitions. The ST-NFA of Fig. 18.2 is proper. Clearly for a proper ST-NFA $\mathcal{A}$ over A we have $L(\mathcal{A}) \subseteq Prop(A)$.

Lemma 18.1. *For every* ST-NFA $\mathcal{A}$ *over an alphabet A there is a signal language equivalent* proper ST-NFA $\mathcal{A}'$ *over A (i.e. $S(\mathcal{A}') = S(\mathcal{A})$).*

Proof. Let $\mathcal{A} = (Q, S, \rightarrow, F, l)$. We modify $\mathcal{A}$ as follows:

(1) First we transform $\mathcal{A}$ to $\mathcal{A}'$ by making the start states originating and final states terminating. This can be done by adding a new start state s' and a new final state f', and adding transitions $s' \xrightarrow{a} p$ for each transition $s \xrightarrow{a} p$ in $\mathcal{A}$ with $s \in S$, and transitions $p \xrightarrow{a} f'$ for each transition $p \xrightarrow{a} f$ in $\mathcal{A}$ with $f \in F$.
(2) Next, we transform $\mathcal{A}'$ to $\mathcal{A}''$ as follows: Pick a non-proper transition $p \xrightarrow{a} q$ and a transition $r \xrightarrow{b} p$, and add the transition $r \xrightarrow{b} q$. Repeat this till no more edges can be added.
(3) Now we drop all non-proper edges in $\mathcal{A}''$ to obtain the required proper ST-NFA $\mathcal{B}$.

Step 1 clearly preserves the word (and hence signal) language of $\mathcal{A}$. Step 2 clearly preserves the signal language of $\mathcal{A}'$. Step 3 also preserves the signal language of $\mathcal{A}''$, since any signal σ generated by $\mathcal{A}''$ using a run of non-proper edges can be simulated by using a single proper edge in $\mathcal{A}''$. □

We now want to show some closure properties of the word and signal languages accepted by ST-NFA's. For this it will be useful to go over to a class of classical NFA's which we call "bipartite" NFA's.

A *bipartite* NFA, or B-NFA for short, over an alphabet A is an NFA $\mathcal{B} = (Q, S, \delta, F)$ over A such that there exists a partition of the set of states Q into Q_1 and Q_2 satisfying

- $S \subseteq Q_1$ and $F \subseteq Q_2$, and
- $\delta \subseteq (Q_1 \times A \times Q_2) \cup (Q_2 \times A \times Q_1)$.

ST-NFA's and B-NFA's accept the same class of word languages. To see this we show how we can go from an ST-NFA to a language-equivalent B-NFA and vice-versa. Let $\mathcal{A} = (Q, S, \delta, F, l)$ be an ST-NFA over A. We define the B-NFA $\mathcal{B}$ corresponding to $\mathcal{A}$, denoted *stnfa-bnfa*($\mathcal{A}$), as follows. The states of $\mathcal{B}$ are $(\{s'\} \cup Q) \cup \{q' \mid q \in Q\}$, where s' is a new start state, the finals states are F, and we have transitions $s' \xrightarrow{a} q$ whenever $(p, a, q) \in \delta$ with $p \in S$; plus transitions of the form $p' \xrightarrow{a} q$ for each transition $(p, a, q) \in \delta$; plus transitions of the form $q \xrightarrow{l(q)} q'$ for each $q \in Q$. Figure 18.4 shows the translation applied to the example ST-NFA of Figure 18.2.

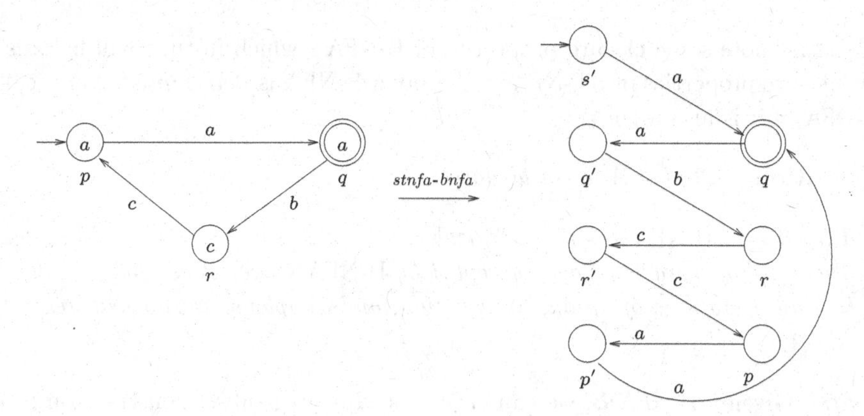

Fig. 18.4. The translation *stnfa-bnfa*.

Conversely, given a B-NFA $\mathcal{B}$, we can give an ST-NFA $\mathcal{A}$, denoted *bnfa-stnfa*($\mathcal{B}$), whose word language is the same as that of $\mathcal{B}$. Let Q_1 and Q_2 be the assumed partition of states of $\mathcal{B}$. The states of $\mathcal{A}$ comprise the *transitions* of $\mathcal{B}$ which go from Q_2 to Q_1, an initial state s, and a final state f if there is a final state of $\mathcal{B}$

which is terminating. Each state is labelled by the label of the transition to which it corresponds. The start state s and terminal final state f can be labelled by any letter as their labels will not contribute to the language of $\mathcal{A}$. The set of final states comprise the final state f above, along with the states corresponding to the transitions going out of final states of $\mathcal{B}$. There is a transition from $e_1 = (p, b, q)$ and $e_2 = (r, c, t)$ on a in $\mathcal{A}$ if there is a transition (q, a, r) in $\mathcal{B}$. There is also a transition (s, a, e_2) in $\mathcal{A}$ if there is a transition (p, a, q) out of an initial state of $\mathcal{B}$ and e_2 is a transition out of q in $\mathcal{B}$. Finally we also have a transition from (p, a, q) to f labelled b, if (q, b, r) is a transition in $\mathcal{B}$. It is not difficult to see that the word languages of $\mathcal{B}$ and $\mathcal{A}$ are the same. Figure 18.5 shows the translation *bnfa-stnfa* applied to the B-NFA on the right.

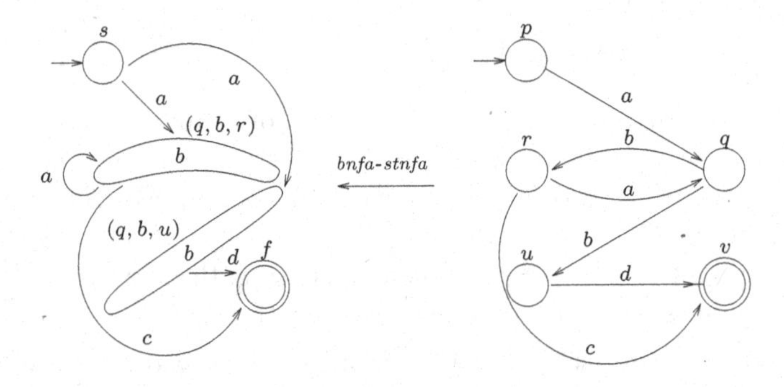

Fig. 18.5. The translation *bnfa-stnfa*.

We first note some closure properties of B-NFA's which in turn will help us to show closure properties of ST-NFA's. We say a B-NFA is *deterministic*, and call it a B-DFA, if it is also a DFA.

Proposition 18.2. *Let A be an alphabet.*

(1) The class of B-NFA*'s is determinizable.*
(2) The class of word languages accepted by B-NFA*'s over A is closed under the boolean operations of union, intersection, and complementation with respect to $A(AA)^*$.*

Proof. Given a B-NFA $\mathcal{B}$, we can apply the standard subset construction to determinize it to get a language-equivalent DFA $\mathcal{B}'$. The subset construction preserves the bipartite structure of $\mathcal{B}$, and hence $\mathcal{B}'$ is also a B-DFA.

For closure under intersection, we note that the standard product construction also preserves the bipartite structure of the two B-NFA's. For closure under complementation with respect to $A(AA)^*$, given a B-NFA $\mathcal{B}$ we can first determinize it to get a language-equivalent B-DFA $\mathcal{B}'$. Let the partition on states of $\mathcal{B}'$ be Q_1 and Q_2. We can now "complete" $\mathcal{B}'$ with respect to $A(AA)^*$ by adding two new states d_1 and d_2, with d_1 added to Q_1 and d_2 added to Q_2, and adding transitions (p_1, a, d_2) for each state p_1 in Q_1 with no outgoing transition on a, and similarly (p_2, a, d_1) for each state p_2 in Q_2 with no outgoing transition on a. We also add transitions (d_1, a, d_2) and (d_2, a, d_1) for each a in A. Finally, we simply "flip" the final states in Q_2: i.e. the new set of final states F' is $Q_2 - F$, where F is the set of final states of $\mathcal{B}'$. The resulting automaton is a B-DFA which accepts the complement (with respect to $A(AA)^*$) of the language accepted by $\mathcal{B}$. □

It follows that the class of word languages accepted by ST-NFA's over an alphabet A coincides with the class of word languages definable by B-NFA's over A, which in turn is precisely the class of regular subsets of $A(AA)^*$, and further that the class of word languages accepted by ST-NFA's over A are closed under union, intersection, and complement wrt $A(AA)^*$.

Using these observations we can now prove some closure properties of the class of ST-NFA-definable signal languages.

Lemma 18.2. *Let A be an alphabet. Then*

(1) *The class of* ST-NFA*'s over A is determinizable: that is, for any* ST-NFA *over A we can give an* ST-DFA *which accepts the same signal language.*
(2) *The class of signal languages definable by* ST-NFA*'s over an alphabet A is closed under union, intersection, and complement.*

Proof. For the first part, let $\mathcal{A}$ be an ST-NFA over A. We can determinize $\mathcal{A}$ as follows: We first go over to the word language equivalent B-NFA $\mathcal{B}$ by applying the translation *stnfa-bnfa* to $\mathcal{A}$. We now determinize $\mathcal{B}$ to get a B-DFA $\mathcal{B}'$. Finally we apply the translation *bnfa-stnfa* to $\mathcal{B}'$ to obtain an ST-NFA $\mathcal{A}'$. In fact $\mathcal{A}'$ is an ST-DFA. To see this, suppose there were states $e = (p, c, q)$, $e_1 = (p_1, b, q_1)$, and $e_2 = (p_2, b, q_2)$ in $\mathcal{A}'$, with e_1 and e_2 distinct, and transitions (e, a, e_1) and (e, a, e_2) on some $a \in A$. Then, since e_1 and e_2 are distinct, it must be the case that p_1 and p_2 are distinct, otherwise it would contradict the fact that $\mathcal{B}'$ was deterministic. But then we have transitions (q, a, p_1) (q, a, p_2) in $\mathcal{B}'$ with $p_1 \neq p_2$, which contradicts the fact that $\mathcal{B}'$ is a B-DFA.

For the second part, closure under union is immediate. For closure under complementation, let $\mathcal{A}$ be an ST-NFA over A. We first make $\mathcal{A}$ proper, to get a signal-equivalent proper ST-NFA $\mathcal{A}'$. Then $Sig(A) - S(\mathcal{A}) = Sig(A) - S(\mathcal{A}') = timing(Prop(A) – L(\mathcal{A}'))$ (using Proposition 18.1)$= timing(L(\mathcal{A}''))$ for some ST-NFA $\mathcal{A}''$ (using closure properties of ST-NFA-definable word languages) $= S(\mathcal{A}'')$. The closure under intersection follows from that of union and complementation. □

18.4. Equivalence of ST-NFA's and MSO^s

In this section we introduce a natural monadic second-order logic interpreted over signals and show that the class of signal languages it defines coincides with the class of signal languages definable by ST-NFA's.

In the logics to follow we assume a countable supply of first-order variables and second-order variables. For an alphabet A, the syntax of monadic second order logic over A, denoted $\mathrm{MSO}^s(A)$, is given by:

$$\varphi ::= Q_a(x) \mid x < y \mid x \in X \mid \neg\varphi \mid (\varphi \vee \varphi) \mid \exists x\varphi \mid \exists X\varphi,$$

where $a \in A$, x and y are first-order variables and X is a second-order variable.

We interpret a formula φ of the logic over a signal σ in $Sig(A)$, along with an interpretation $\mathbb{I}$ with respect to σ, which assigns to each first-order variable a value in $[0, dur(\sigma)]$, and to each set variable, a *finitely-varying* subset of $[0, dur(\sigma)]$. We use $X \subseteq_{fv} Y$ to denote that X is a finitely-varying subset of Y. For an interpretation $\mathbb{I}$, we use the notation $\mathbb{I}[t/x]$ to denote the interpretation which sends x to t and agrees with $\mathbb{I}$ on all other variables. Similarly, $\mathbb{I}[B/X]$ denotes the modification of $\mathbb{I}$ which maps the set variable X to a subset B of $\mathbb{R}_{\geq 0}$, and the rest to the same as that mapped by $\mathbb{I}$. We also use the notation $[t/x]$ to denote an interpretation which sends x to t when the rest of the interpretation is irrelevant.

Given a formula $\varphi \in \mathrm{MSO}^s(A)$, $\sigma \in Sig(A)$, and an interpretation $\mathbb{I}$ with respect to σ to the variables in φ, the satisfaction relation $\sigma, \mathbb{I} \models \varphi$, is defined inductively as:

$$\begin{array}{ll}
\sigma, \mathbb{I} \models Q_a(x) & \text{iff } \sigma(\mathbb{I}(x)) = a, \text{ where } a \in A. \\
\sigma, \mathbb{I} \models x < y & \text{iff } \mathbb{I}(x) < \mathbb{I}(y). \\
\sigma, \mathbb{I} \models x \in X & \text{iff } \mathbb{I}(x) \in \mathbb{I}(X). \\
\sigma, \mathbb{I} \models \neg\varphi & \text{iff } \sigma, \mathbb{I} \not\models \varphi. \\
\sigma, \mathbb{I} \models \varphi_1 \vee \varphi_2 & \text{iff } \sigma, \mathbb{I} \models \varphi_1 \text{ or } \sigma, \mathbb{I} \models \varphi_2. \\
\sigma, \mathbb{I} \models \exists x\varphi & \text{iff } \exists t \in [0, dur(\sigma)] : \sigma, \mathbb{I}[t/x] \models \varphi. \\
\sigma, \mathbb{I} \models \exists X\varphi & \text{iff } \exists B \subseteq_{fv} [0, dur(\sigma)] : \sigma, \mathbb{I}[B/X] \models \varphi.
\end{array}$$

For a sentence φ (a formula without free variables) in $\mathrm{MSO}^s(A)$, the interpretation does not play any role, and we write the satisfaction relation $\sigma, \mathbb{I} \models \varphi$ as simply $\sigma \models \varphi$. We define the language of signals defined by φ to be $S(\varphi) = \{\sigma \in Sig(A) \mid \sigma \models \varphi\}$.

As an example, the formula

$$\varphi_{cont}(x) = \exists y \exists z (y < x \wedge x < z \wedge \bigvee_{a \in A} \forall u (y < u \wedge u < z \Rightarrow Q_a(u)))$$

asserts that the point x is a point of continuity. The formula $\varphi_{disc}(x) = \neg\varphi_{cont}(x)$ asserts that x is a point of discontinuity.

We denote by $\mathrm{FO}^s(A)$ the first-order fragment of $\mathrm{MSO}^s(A)$ obtained by disallowing second-order quantification and atomic formulas of the form $x \in X$.

Theorem 18.1. *A signal language over an alphabet A is definable by a* $\mathrm{MSO}^s(A)$ *sentence iff it is definable by an* ST-NFA *over A.*

We prove this theorem in the rest of this section. The proof proceeds in a similar manner to the proof of Büchi's MSO characterization of classical automata [1] (see [14]).

We first show how to go from a formula in $\text{MSO}^s(A)$ to an equivalent ST-NFA over A. We will represent models of formulas with free variables in them, as signals with the interpretations built into them. We assume an ordering on the countable set of first-order variables given by $x_1, x_2, \ldots$, and similarly $X_1, X_2, \ldots$ for the set variables. For a formula φ with first-order free variables among $U = \{x_{i_1}, \ldots, x_{i_m}\}$ and second-order free variables among $V = \{X_{j_1}, \ldots, X_{j_n}\}$ (in order), we represent a signal σ and an interpretation $\mathbb{I}$ as a signal $\sigma_{\mathbb{I}}^{U,V} : [0, dur(\sigma)] \to A \times \{0,1\}^{m+n}$ given by $\sigma_{\mathbb{I}}^{U,V}(t) = (\sigma(t), b_1, \cdots, b_{m+n})$, where for $k \in \{1, \ldots, m\}$, $b_k = 1$ iff $\mathbb{I}(x_{i_k}) = t$, and for $k \in \{m+1, \ldots, n\}$, $b_k = 1$ iff $t \in \mathbb{I}(X_{j_{k-m}})$. Thus for a formula φ with free variables in (U, V) we have the notion of a (U, V)-model of φ.

We note that for the U, V above, not every signal over $A \times \{0,1\}^{m+n}$ is a "valid" (U, V)-model, in the sense that the components of the signal corresponding to a first-order variable in U may not have *exactly* one 1 entry. Nonetheless, the proposition below says that the valid (U, V)-models are ST-NFA-recognisable.

Proposition 18.3. *Let A be a finite alphabet and let U and V be a sets of first and second-order variables respectively. Then we can construct an* ST-NFA $\mathcal{A}_{valid}^{U,V}$ *which accepts precisely the set of signals over $A \times \{0,1\}^{|U|+|V|}$ which represent valid (U, V)-models over A.*

Proof. For each $x_{i_k} \in U$ we can construct an ST-NFA over the alphabet $A \times \{0,1\}^{|U|+|V|}$ which accepts signals which are valid encodings as far as the component corresponding to x_{i_k} is concerned. Figure 18.6 shows an ST-NFA which accepts the valid models with respect to x, when the alphabet is $\{a, b\}$, and $U = \{x, y\}$ and $V = \emptyset$. We can then take the intersection of the resulting ST-NFA's to obtain the ST-NFA $\mathcal{A}_{valid}^{U,V}$.

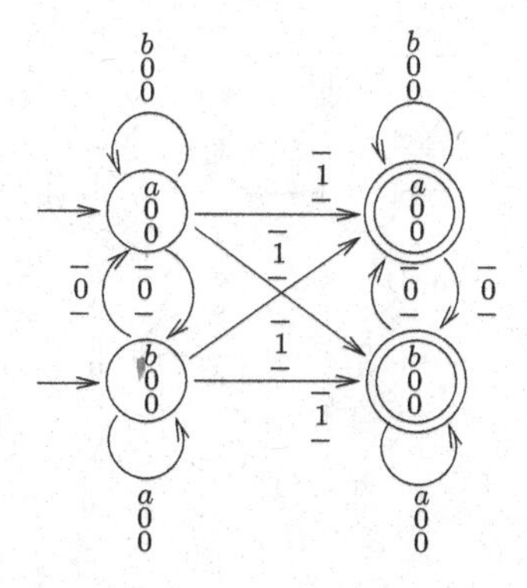

Fig. 18.6. ST-NFA accepting x-valid $(\{x, y\}, \emptyset)$-models over the alphabet $\{a, b\}$.

□

Proposition 18.4. *Let φ be an $\mathrm{MSO}^s(A)$ formula with first and second-order free variables U and V respectively. Let $\mathcal{A}$ be an* ST-NFA *accepting the (U, V)-models of φ. Then for any set of variables (U', V') such that $U \subseteq U'$ and $V \subseteq V'$, we can construct an* ST-NFA *$\mathcal{A}'$ which accepts precisely the (U', V')-models of φ.*

Proof. Let us consider the case when $U' = U \cup \{x\}$ and $V' = V$. Consider ST-NFA $\mathcal{A}''$ which has the same set of states as $\mathcal{A}$, with the same initial and final states. The labeling of states in $\mathcal{A}''$ is same as that of $\mathcal{A}$ except that they are extended with a 0 corresponding to x. For every transition in $\mathcal{A}$, there are two transitions in $\mathcal{A}''$ with the same start and target states, and the label extended with 0 in one and 1 in the other. The automaton $\mathcal{A}'$ is then obtained by taking the intersection of $\mathcal{A}''$ with $\mathcal{A}_{valid}^{U',V}$.

For the case when $U' = U$ and $V' = V \cup \{X\}$, we construct $\mathcal{A}''$ similar to the above case, except that in addition now we replace each state in $\mathcal{A}$ by two states, one labeled with an extension of the original label with a 1 corresponding to X, and the other labeled with an extension by 0 corresponding to X. The new states inherit the incoming and outgoing (extended) transitions from the corresponding original state. In this case we can avoid the intersection with $\mathcal{A}_{valid}^{U,V'}$. This construction easily extends to any U' and V'. □

Lemma 18.3. *Let φ be an $\mathrm{MSO}^s(A)$ formula and let (U, V) be the set of free variables in it. Then we can construct an* ST-NFA *$\mathcal{A}_\varphi^{U,V}$ which accepts precisely the (U, V)-models of φ.*

Proof. We construct the ST-NFA $\mathcal{A}_\varphi^{U,V}$ by induction on the structure of φ.

(1) $\varphi = Q_a(x)$: Assuming $A = \{a, b\}$, the automaton $\mathcal{A}_\varphi^{\{x\},\emptyset}$ is shown below.

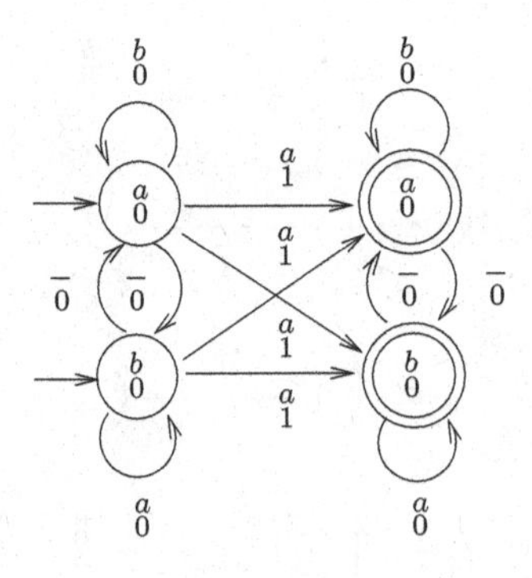

(2) $\varphi = x < y$: The automaton $\mathcal{A}_\varphi^{\{x,y\},\emptyset}$ (assuming x occurs before y in the variable ordering) is:

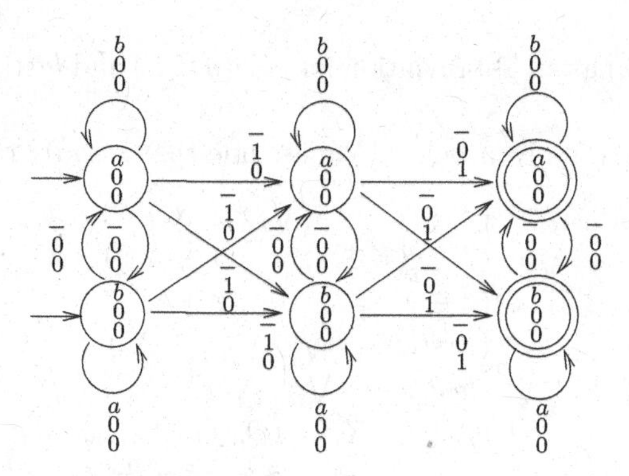

(3) For the case $\varphi = x \in X$, the automaton $\mathcal{A}_\varphi^{\{x\},\{X\}}$ is defined similarly.

(4) $\varphi = \neg\psi$: Let $\mathcal{A}_\psi^{U,V}$ be the automaton for ψ, where (U, V) is the set of free variables in ψ. Then $\mathcal{A}_\varphi^{U,V}$ is the intersection of $\mathcal{A}_{valid}^{U,V}$ with the ST-NFA that recognizes the complement of the signal language of $\mathcal{A}_\psi^{U,V}$ (cf. Lemma 18.2).

(5) $\varphi = \psi \vee \nu$: Let $\mathcal{A}_\psi^{U_1,V_1}$ be the ST-NFA for ψ, where (U_1, V_1) is the set of free variables in ψ, and let $\mathcal{A}_\nu^{U_2,V_2}$ be the ST-NFA for ν, where (U_2, V_2) is the set of free variables in ν. Let $U = U_1 \cup U_2$ and $V = (V_1 \cup V_2)$. By Prop. 18.4 we obtain ST-NFA's $\mathcal{A}_\psi^{U,V}$ and $\mathcal{A}_\nu^{U,V}$. Then $\mathcal{A}_\varphi^{U,V}$ is the ST-NFA that accepts the union of the signal languages accepted by $\mathcal{A}_\psi^{U,V}$ and $\mathcal{A}_\nu^{U,V}$.

(6) $\varphi = \exists x\psi$: Let (U', V') be the set of free variables in ψ, so that $U = U' - \{x\}$ and $V = V'$. Let $\mathcal{A}_\psi^{U',V'}$ be an ST-NFA for ψ. Now we simply project away the component corresponding to x in the symbols labelling the transitions and states of $\mathcal{A}_\psi^{U',V'}$ to obtain the required ST-NFA $\mathcal{A}_\varphi^{U,V}$.

(7) $\varphi = \exists X\psi$: Let (U', V') be the set of free variables in ψ, so that $U = U'$ and $V = V' - \{X\}$. Let $\mathcal{A}_\psi^{U',V'}$ be an ST-NFA for ψ. Again we simply project away the component corresponding to X in the symbol labelling the transitions and states of $\mathcal{A}_\psi^{U',V'}$ to obtain the required counter-free ST-NFA $\mathcal{A}_\varphi^{U,V}$.

□

From the above lemma it now follows that for a sentence $\varphi \in \mathrm{MSO}^s(A)$ we have an ST-NFA $\mathcal{A}_\varphi$ over A such that $S(\varphi) = S(\mathcal{A}_\varphi)$.

We now prove the converse direction of Theorem 18.1. Let $\mathcal{A} = (Q, S, \rightarrow, F, l)$ be an ST-NFA over A. Without loss of generality we assume that $\mathcal{A}$ is proper. We give an $\mathrm{MSO}^s(A)$ sentence $\varphi_\mathcal{A}$ such that $S(\mathcal{A}) = S(\varphi_\mathcal{A})$. The sentence $\varphi_\mathcal{A}$ describes the existence of an accepting run of the automaton on a given signal. Let $\{e_i = p_i \xrightarrow{a_i} q_i \mid i = 1, \ldots, m\}$ be the set of transitions in $\mathcal{A}$. The second order variables $X_1, \ldots, X_m$ will be used to capture the points in the signal at which the transitions $e_1, \ldots, e_m$ are taken respectively. Note that since we are assuming $\mathcal{A}$ is proper, the union of the X_i's must correspond exactly to the points of discontinuities in the given signal. We will use the abbreviation $consec(x, y, X)$ to mean that x and y are "consecutive" points in the set X, and define it to be:

$$consec(x, y, X) = x \in X \wedge y \in X \wedge \neg\exists z(x < z \wedge z < y \wedge z \in X).$$

We also use $first(x)$ as an abbreviation for $\neg\exists y(y < x)$ and $last(x)$ as an abbreviation for $\neg\exists y(x < y)$.

The formula $\varphi_{\mathcal{A}}$ is given below. We assume that i and j range over $0, \ldots, m$.

$$
\begin{aligned}
\exists X_1 \cdots \exists X_m \exists X (\forall x(& (x \in X \iff \textstyle\bigvee_i x \in X_i) \wedge \\
& (\textstyle\bigwedge_{i \neq j} (x \in X_i \Rightarrow \neg x \in X_j)) \wedge \\
& (x \in X \iff disc(x)) \wedge \\
& (first(x) \Rightarrow \textstyle\bigvee_{i:\, p_i \in S} x \in X_i) \wedge \\
& (last(x) \Rightarrow \textstyle\bigvee_{i:\, q_i \in F} x \in X_i) \wedge \\
& (\textstyle\bigwedge_i (x \in X_i \Rightarrow (Q_{a_i}(x) \wedge ((\exists y(consec(x, y, X))) \Rightarrow \\
& \qquad \forall z((x < z \wedge z < y) \Rightarrow Q_{l(q_i)}(z)))))))).
\end{aligned}
$$

This completes the proof of Theorem 18.1.

Before we close this section we observe that the version of MSO^s, called *weak* MSO^s, in which we restrict the second-order quantification to *finite* subsets of the domain of the signal (rather than finitely-varying subsets) is as expressive as the version we have defined. The justification is as follows. We note that the clause $x \in X \iff disc(x)$ forces the second-order variables X and X_i's to be interpreted as *finite* subsets of the domain since the signal model has only finitely many discontinuities. Hence quantification over finite subsets suffices to capture the ST-NFA-definable signal languages. Further, signal languages definable by second-order quantification restricted to finite subsets are clearly ST-NFA-definable (by an argument similar to the one above, where we allow components corresponding to second-order variables to have 1's only on transition (and not state) labels). Hence the expressiveness of the two variants coincide with ST-NFA-definable signal languages.

Corollary 18.1. *The class of signal languages definable in* $\mathrm{MSO}^s(A)$ *and weak* $\mathrm{MSO}^s(A)$ *coincide.*

18.5. Counter-free signal languages

In this section we introduce a counter-free version of signal languages which will be shown in the next section to characterize FO^s-definable signal languages.

We recall that a *counter* in an NFA $\mathcal{A}$ is a sequence of distinct states $q_0, \ldots, q_n$ with $n \geq 1$, along with a word $u \in A^*$, such that there is a path labeled u in $\mathcal{A}$ from q_i to q_{i+1} (for each $i \in \{0, \ldots, n-1\}$) and from q_n to q_0. We call the counter an *even counter* if u has even length (i.e $u \in (AA)^+$). An NFA is said to be *counter-free* (respectively *even-counter-free*) if it does not contain a counter (respectively even counter). A regular language is said to be *counter-free* if there exists a counter-free NFA for it.

We now define the counter-free version of ST-NFA's. A counter in an ST-NFA is similar to one in an NFA, except that by the "label" of a path in the automaton we mean the sequence of alternating transition and state labels along the path.

The label of the path $q_0 \xrightarrow{a_0} q_1 \xrightarrow{a_1} \cdots q_n \xrightarrow{a_n} q_{n+1}$ is $a_0 l(q_1) a_1 \cdots l(q_n) a_n l(q_{n+1})$ (we ignore the label of the first state in the path, but count the label of the last state in the path). Thus a *counter* in an ST-NFA $\mathcal{A}$ is a sequence of distinct states $q_0, \ldots, q_n$ with $n \geq 1$, along with a word $u \in A^*$, such that there is a path labeled u in $\mathcal{A}$ from q_i to q_{i+1} (for each $i \in \{0, \ldots, n-1\}$) and from q_n to q_0. We say an ST-NFA is *counter-free* if it does not contain a counter. We say a signal language is *counter-free* if it is definable by a counter-free *proper* ST-NFA. The proper-ness requirement is important as without it we can give counter-free ST-NFA's for signal languages we would otherwise like to consider as not being counter-free. Figure 18.7 below shows a non-proper counter-free ST-NFA and its equivalent proper version which contains a counter.

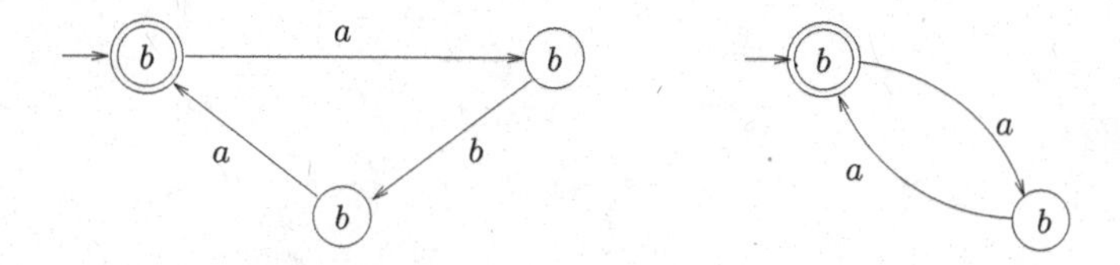

Fig. 18.7. Example showing how proper conversion may introduce counters.

We will show in the next section that the class of first-order definable signal languages coincide with the class of counter-free signal languages. Our aim in the rest of this section is to show some closure properties of counter-free signal languages that will be useful there.

We first observe that the classical subset construction for determinizing NFA's preserves counter-freeness.

Lemma 18.4. *Let $\mathcal{B}$ be an* NFA *and let $\mathcal{C}$ be the* DFA *obtained by the standard subset construction on $\mathcal{B}$. Then if $\mathcal{B}$ was counter-free, so is $\mathcal{C}$. Also, if $\mathcal{B}$ was* even-*counter-free, so is $\mathcal{C}$.*

Proof. Suppose $\mathcal{C}$ has a counter. We show that $\mathcal{B}$ has a counter.

Let S_0, S_1, ..., S_{n-1} be a counter in $\mathcal{C}$, and let w be the word associated with it, i.e, S_0 on w goes to S_1, S_1 on w goes to S_2, and so on. Choose the counter such that $|S_0| + |S_1| + \cdots + |S_{n-1}|$ is minimum. For a set X of states of $\mathcal{B}$, define *Pred*(X) to be the set of all states y of $\mathcal{B}$ such that y on w goes to some state in X. We will write *Pred*(x) for *Pred*$(\{x\})$. Similarly *Succ*(X) is the set of all states reachable from states in X on reading w.

Form a sequence of states $q_0, q_1, \ldots$ of $\mathcal{B}$ as follows. Begin with some state q_0 in S_0. Let $q_i \in S_j$. Then q_{i+1} is some state in *Pred*$(q_i) \cap S_{j-1}$ (where $j-1$ is taken modulo n). q_{i+1} is chosen to be different from q_i whenever possible. Consider the sequence of states $q_0, q_n, q_{2n} \ldots$ and let $l < k$ be such that $q_{nl} = q_{nk}$ (such an l, and k exist as each subset is finite). If for some $nk \leq i < j < nk$, $q_i \neq q_j$ then

$q_{nl}, \ldots, q_{nk}$ contains a counter on w and we are done.

Suppose this is not the case, i.e. for all $i \in \{nl, \ldots, nk\}$ q_i is equal to say p. Then from the construction of the sequence one can easily see that for all $i \in \{0, \ldots, n-1\}$ p is in every S_i. We also note that $Pred(p) \cap S_i = p$ since otherwise we would have chosen a predecessor which is different from p. Define *Reach* to be the smallest set containing p and closed with respect to *Succ*, i.e. if $q \in Reach$ and $q' \in Succ(q)$, then $q' \in Reach$. The fact that p is in every S_i implies that *Reach* is a subset of every S_i. This is depicted in Fig. 18.8.

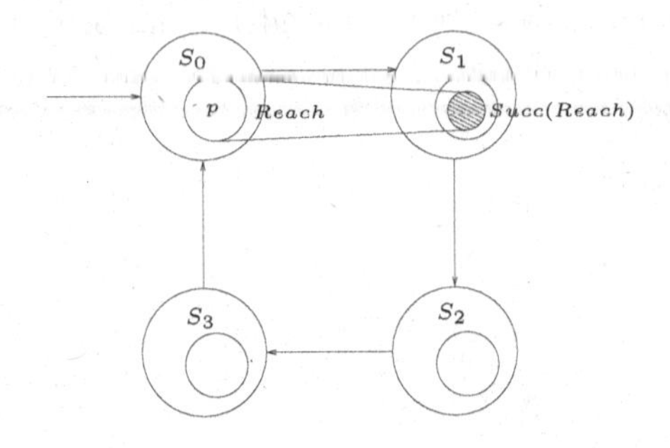

Fig. 18.8. Example depicting the set *Reach* and *Succ*(*Reach*).

Let $Y_0 = S_0 - \{p\}$ and for every $j \geq 0$ let $Y_{j+1} = Succ(Y_j)$. One can inductively argue that for every $j \geq 0$, $S_i - Reach \subseteq Y_{i+nj}$ and therefore $Y_{i+nj} \neq Y_{i+nj+1}$ because otherwise $S_i = S_{i+1}$ ($i+1$ taken modulo n) as $S_i = Reach \cup Y_{i+nj}$. Since all the Y_{i+nj}'s cannot be distinct there exist $k < l$ such that $Y_{i+nk} = Y_{i+nl}$. Let $X_0 = Y_{i+nk}$ and let $X_{j+1} = Succ(X_j)$. Note that we always maintain the invariant $p \notin Y_j$ and therefore none of the X_j's contain p. It can be seen that $X_0, \ldots, X_{n-1}$ forms a counter in $\mathcal{C}$ such that $|X_0| + \cdots + |X_{n-1}| < |S_0| + \cdots + |S_{n-1}|$, which contradicts the choice of the counter.

We note that it also follows from the above argument that if $\mathcal{B}$ had no *even* counters, $\mathcal{C}$ will also not have any even counters. □

Next, as we did for ST-NFA's, it will be convenient to characterise CF-ST-NFA's in terms of bipartite NFA's. We define the class of *even-counter-free* B-NFA's over an alphabet A, denoted ECF-B-NFA, to be the class of B-NFA's which have no even counters.

Proposition 18.5. *The class of word languages accepted by* CF-ST-NFA*'s and* ECF-B-NFA*'s over an alphabet A coincide.*

Proof. It is easy to verify that the translations *stnfa-bnfa* and *bnfa-stnfa* given in Section 18.3 already give us the required translations: that is if $\mathcal{A}$ is a CF-ST-NFA over A then *stnfa-bnfa*($\mathcal{A}$) is an ECF-B-NFA, and if $\mathcal{B}$ is an ECF-B-NFA then *bnfa-stnfa*($\mathcal{B}$) is indeed a CF-ST-NFA. □

Lemma 18.5. *The class of even-counter-free bipartite* NFA*'s over A are determinizable and closed under union, intersection, and complementation wrt $A(AA)^*$.*

Proof. Given an ECF-B-NFA $\mathcal{B}$ over A, we determinize it using the subset construction to get $\mathcal{A}'$. We have already argued that $\mathcal{A}'$ is a B-DFA over A. By Lemma 18.4 $\mathcal{A}'$ is also even-counter-free, and we are done.

To show closure under intersection we argue that the above properties are preserved by the standard product construction for intersection. Let $\mathcal{B}$ and $\mathcal{C}$ be two ECF-B-NFA's, with B_1, B_2 and C_1, C_2 being the respective partitions. In the product automaton $\mathcal{D}$ accepting the intersection of their word languages, the only reachable states are in $D_1 = B_1 \times C_1$ and $D_2 = B_2 \times C_2$, with D_1 and D_2 being the partitions. It remains to be shown that $\mathcal{D}$ does not have an even-counter. Suppose $\mathcal{D}$ has such a counter with the sequence $(s_1, t_1) \cdots (s_n, t_n)$. Then it cannot be the case that all the s_i's are the same and all the t_j's are the same, since otherwise the sequence is not a counter. Assume without loss of generality that all the s_i's are not same. Then there is a subsequence of distinct states $s_{i_1}, s_{i_2}, \ldots, s_{i_k}$ which forms an even-counter in $\mathcal{B}$. This contradicts our assumption that $\mathcal{B}$ was even-counter-free.

To show closure under complementation wrt $A(AA)^*$, we first determinize the automaton. As observed in the proof of proposition 18.2, the resulting automaton is a B-DFA. Furthermore, by lemma 18.4, it does not contain an even-counter. Once again, we "complete" the automaton as in the proof of Lemma 18.2. It is easy to see that this completion does not introduce any even counters. We can now "flip" the final states in the second partition, to obtain a ECF-B-DFA which accepts the complement of the word language of the given B-NFA wrt $A(AA)^*$. □

Using the preceding lemmas we can now argue that:

Lemma 18.6. *The class of counter-free signal languages over an alphabet A is determinizable, and closed under union, intersection, and complementation.*

Proof. To see that the class of CF-ST-NFA's is determinizable, let $\mathcal{A}$ be a counter-free ST-NFA. We go over to a word language equivalent ECF-B-NFA $\mathcal{B}$ from $\mathcal{A}$ using the translation *stnfa-bnfa*. We now determinize $\mathcal{B}$ to get $\mathcal{B}'$, and applying *bnfa-stnfa* on $\mathcal{B}'$, we get a deterministic counter-free ST-NFA.

For the closure under boolean operations, let F_1 and F_2 be two counter-free signal languages over the alphabet A. Let L_1 and L_2 be the word languages of their defining proper counter-free ST-NFA's. Once again, using proposition 18.1, we can convince ourselves that $F_1 \cup F_2 = \mathit{timing}(L_1 \cup L_2)$, $F_1 \cap F_2 = \mathit{timing}(L_1 \cap L_2)$, and $\mathit{Sig}(A) - F_1 = \mathit{timing}(\mathit{Prop}(A) - L_1) = \mathit{timing}((A(AA)^* - L_1) \cap \mathit{Prop}(A))$.

Using the closure properties of the word languages accepted by CF-ST-NFA's (or equivalently ECF-B-NFA's), and the fact that *Prop*(A) is ECF-B-NFA-definable, we can conclude that the signal languages definable by CF-ST-NFA's are closed under boolean operations. □

18.6. Counter-free characterisation of FO signal languages

In this section our aim is to show that FO^s-definable signal languages, counter-free signal languages, and temporal logic definable signal languages, all coincide.

We recall briefly the temporal logic LTL and its two interpretations, one over discrete words and the other over signals. For an alphabet A, the syntax of $LTL(A)$ is given by:

$$\theta ::= a \mid (\theta U \theta) \mid (\theta S \theta) \mid \neg\theta \mid (\theta \vee \theta),$$

where $a \in A$. The logic is interpreted over words in A^*, with the following semantics. Given a word $w = a_0 \cdots a_n$ in A^* and a position $i \in \{0, \ldots, n\}$, we say $w, i \models a$ iff $a_i = a$; and $w, i \models \theta U \eta$ iff there exists j such that $i < j \leq n$, $w, j \models \eta$ and for all k such that $i < k < j$, $w, k \models \theta$. The "since" operator S is defined in a symmetric way to U in the past, and the boolean operators in the usual way. We denote by $L(\theta)$ the set $\{w \in A^* \mid w, 0 \models \theta\}$.

The logic LTL can also be interpreted over functions as done in [6]. Here we restrict the models to finitely-varying functions in $Sig(A)$, and we denote this logic by $LTL^s(A)$. Given a signal $\sigma \in Sig(A)$, $t \in [0, dur(\sigma)]$ and $\theta \in LTL^s(A)$, the satisfaction relation $\sigma, t \models \theta$ is defined as follows:

$\sigma, t \models a$ iff $\sigma(t) = a$.
$\sigma, t \models \theta U \eta$ iff $\exists t' : t < t' \leq dur(\sigma)$, $\sigma, t' \models \eta$, and $\forall t'' : t < t'' < t', \sigma, t'' \models \theta$.
$\sigma, t \models \theta S \eta$ iff $\exists t' : 0 \leq t' < t$, $\sigma, t' \models \eta, \forall t'' : t' < t'' < t$, $\sigma, t'' \models \theta$.

The boolean operators are interpreted in the expected way. We set $S(\theta) = \{\sigma \in Sig(A) \mid \sigma, 0 \models \theta\}$.

As an example, the $LTL^s(A)$ formulas $\theta_{cont} = \bigvee_{a \in A}(a \wedge (aSa) \wedge (aUa))$ and $\theta_{disc} = \neg\theta_{cont}$ characterize the points of continuity and discontinuity respectively in a signal over A.

Theorem 18.2. *Let A be a alphabet, and let F be a signal language over A. Then the following statements are equivalent:*

(1) F is definable by an $FO^s(A)$-sentence.
(2) F is definable by a counter-free proper ST-NFA over A.
(3) F is definable by an $LTL^s(A)$-formula.

The rest of this section is devoted to a proof of this theorem. Our proof will factor through some classical results connecting counter-free languages and temporal logics. The route we follow is given schematically in the figure below, via steps labelled (a), and (b) to (e).

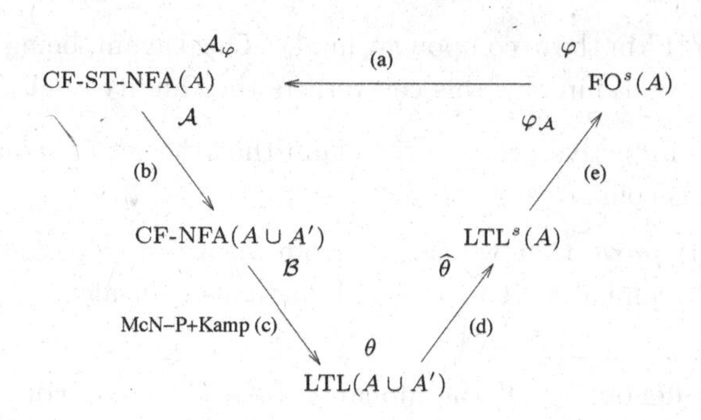

Step (a): We show how to go from a formula in $\mathrm{FO}^s(A)$ to a counter-free proper ST-NFA accepting exactly its models. It can be checked that the inductive construction carried out in Section 18.4 for Theorem 18.1 produces a counter-free proper ST-NFA at each step. This is true for the base cases $Q_a(x)$ and $x < y$. For the boolean operators, it follows by the closure properties of counter-free signal languages (cf. Lemma 18.6).

For the case of first-order quantification, let $\varphi = \exists x \psi$. Let $\mathcal{A}_\psi$ be a counter-free proper ST-NFA which accepts the valid models for ψ. Without loss of generality we assume that $\mathcal{A}_\psi$ has no unreachable or dead states, and that its start and final states are respectively originating and terminating.

We now project away the x-component in transition and state labels of $\mathcal{A}_\psi$ to get a ST-NFA $\mathcal{A}'$ accepting the valid models of $\exists x \psi$. Now we can argue that $\mathcal{A}'$ cannot have a counter. If it did, then there are two cases: either the counter is such that no symbol in it was obtained by projecting away a "1" in the x-component, or there is a symbol in it which was obtained by projecting away a "1" in the x-component. In the first case, this would mean a counter in $\mathcal{A}_\psi$ itself, contradicting the inductive assumption that $\mathcal{A}_\psi$ was counter-free. In the second it would mean $\mathcal{A}_\psi$ has a cycle containing a transition on a symbol with a "1" in the x-component, which would contradict the validity of the models generated by $\mathcal{A}_\psi$.

However, we are not yet done, as $\mathcal{A}'$ might have a non-proper edge. Now let us make $\mathcal{A}'$ proper to get $\mathcal{A}''$, using the algorithm described in Section 18.3. Recall that the algorithm adds edges and finally deletes all the non-proper edges. But it satisfies the property that the state space is the same, and every added edge from p to q has a corresponding path from p to q in $\mathcal{A}'$ which uses a non-proper transition.

Now we claim that $\mathcal{A}''$ is counter-free (and, by construction, proper). Suppose $\mathcal{A}''$ had a counter on states $q_0, \ldots, q_n$, on a string u. Now two possibilities exist:

- No u path in the counter uses an "added" edge. In this case this would be a counter in $\mathcal{A}'$ also, which is a contradiction.
- Some u path in the counter uses an "added" edge. So in $\mathcal{A}'$ the u-path has a corresponding u'-path which uses a non-proper edge in $\mathcal{A}'$. But non-proper edges in $\mathcal{A}'$ could only have come from a projection of a '1' in the x-component of a transition in $\mathcal{A}_\psi$. So the corresponding "unprojected" u'-path contains a

symbol with a "1" in the x-component in $\mathcal{A}_\psi$. Once again, being part of a loop in $\mathcal{A}''$, and hence also in $\mathcal{A}_\psi$, this contradicts the validity of $\mathcal{A}_\psi$.

This completes the inductive proof of the claim that the set of signal models of a first-order formula is counter-free.

Steps (b) to (d) prove that we can go from an arbitrary counter-free proper ST-NFA $\mathcal{A}$ over the alphabet A to a signal-language-equivalent $\mathrm{FO}^s(A)$-sentence $\varphi_{\mathcal{A}}$.

Step (b): Let us denote by A' the alphabet $\{a' \mid a \in A\}$. For a proper word $w = a_0 \ldots a_{2n} \in Prop(A)$, we define $ann(w)$ to be the word $a'_0 a_1 a'_2 \cdots a_{2n-1} a'_{2n}$ in $(A \cup A')^*$, and apply it to work on subsets of $Prop(A)$ as well. Now let $\mathcal{A}$ be a counter-free proper ST-NFA over A. By the characterisation of counter-free languages in the proof of Lemma 18.6, there is a word-language equivalent NFA $\mathcal{B}_0$ that is bipartite and has no counters except possibly on odd-length u's. However, if we annotate the labels of the edges going from left to right (with the convention that the start states are in the left partition) by replacing each label a by a', then it is easy to see that the resulting NFA is counter-free, and accepts $ann(L(A))$. Let us call this NFA over $(A \cup A')$ as $\mathcal{B}$. Thus $L(\mathcal{B}) = ann(L(\mathcal{A}))$ and is a classical counter-free word language.

Step (c): By the results due to Schutzenberger [7], McNaughton-Papert [2], and Kamp [6] for classical word languages, the class of counter-free, star-free, FO-definable, and LTL-definable word languages all coincide. Thus for a counter-free NFA $\mathcal{B}$ over $(A \cup A')$ we have an $\mathrm{LTL}(A \cup A')$ formula θ which defines the same word language as $\mathcal{B}$. Thus $L(\mathcal{B}) = L(\theta)$.

Step (d): For a formula θ in $\mathrm{LTL}(A \cup A')$ such that $L(\theta) \subseteq A'(AA')^*$ and the "un-annotation" of $L(\theta)$ (i.e. $ann^{-1}(L(\theta))$ is proper, we can construct a formula $\mathit{ltl\text{-}ltls}(\theta)$ in $\mathrm{LTL}^s(A)$ which is such that $S(\mathit{ltl\text{-}ltls}(\theta)) = \mathit{timing}(ann^{-1}(L(\theta)))$.

We will use the abbreviation $\theta_1 U_d \theta_2$ to mean that at a point of discontinuity "$\theta_1 U \theta_2$" is true in an untimed sense, and define it to be $(\theta_2 U \theta_2) \vee (\theta_1 U(\theta_{disc} \wedge (\theta_2 \vee (\theta_1 \wedge (\theta_2 U \theta_2)))))$. Symmetrically we use $\theta_1 S_d \theta_2$ for $(\theta_2 S \theta_2) \vee (\theta_1 S(\theta_{disc} \wedge (\theta_2 \vee (\theta_1 \wedge (\theta_2 S \theta_2)))))$.

The translation *ltl-ltls* is defined as follows (we use $\widehat{\eta}$ for $\mathit{ltl\text{-}ltls}(\eta)$ in some places for brevity):

$$
\begin{aligned}
\mathit{ltl\text{-}ltls}(a) &= a \wedge \theta_{cont} \text{ (where } a \in A).\\
\mathit{ltl\text{-}ltls}(a') &= a \wedge \theta_{disc} \text{ (where } a' \in A').\\
\mathit{ltl\text{-}ltls}(\neg\theta_1) &= \neg\widehat{\theta}_1.\\
\mathit{ltl\text{-}ltls}(\theta_1 \vee \theta_2) &= \widehat{\theta}_1 \vee \widehat{\theta}_2.\\
\mathit{ltl\text{-}ltls}(\theta_1 U \theta_2) &= (\theta_{disc} \Rightarrow (\widehat{\theta}_1 U_d \widehat{\theta}_2)) \wedge\\
&\qquad (\theta_{cont} \Rightarrow (\theta_{cont}\ U(\theta_{disc} \wedge (\widehat{\theta}_2 \vee (\widehat{\theta}_1 \wedge (\widehat{\theta}_1 U_d \widehat{\theta}_2)))))).\\
\mathit{ltl\text{-}ltls}(\theta_1 S \theta_2) &= (\theta_{disc} \Rightarrow (\widehat{\theta}_1 S_d \widehat{\theta}_2)) \wedge\\
&\qquad (\theta_{cont} \Rightarrow (\theta_{cont}\ S(\theta_{disc} \wedge (\widehat{\theta}_2 \vee (\widehat{\theta}_1 \wedge (\widehat{\theta}_1 S_d \widehat{\theta}_2)))))).
\end{aligned}
$$

Lemma 18.7. *Let θ be an* LTL$(A \cup A')$ *formula. Let w be a proper word over A, and let $w' = ann(w)$. Let $\sigma \in timing(w)$ with the canonical interval representation $(a_0, I_0) \cdots (a_{2n}, I_{2n})$. Then for each $i \in \{0, \cdots, 2n\}$ and for all $t \in I_i$, we have $w', i \models \theta \iff \sigma, t \models ltl\text{-}ltls(\theta)$.*

Proof. We do the proof by induction on θ. For the base case suppose $\theta = a$ with $a \in A$. Then $w', i \models a$ iff i is odd and $w'(i) = a$. This is true iff t is a point of continuity in σ and $\sigma(t) = a$. In turn this is true iff $\sigma, t \models a \wedge \theta_{cont}$.

The other base case and induction step for boolean operators are similar. We now show the induction step for $\theta = \theta_1 U \theta_2$ and omit the similar case of $\theta = \theta_1 S \theta_2$.

Left to right implication. Let $t \in I_i$ and suppose $w', i \models \theta_1 U \theta_2$. Then $\exists j > i$ such that $w', j \models \theta_2$ and $\forall i < i' < j \;\; w', i' \models \theta_1$. We note that by induction hypothesis we have that $\forall t'' \in I_j \;\; \sigma, t'' \models \widehat{\theta_2}$ and $\forall t'' \in \cup_{\{i' | i < i' < j\}} I_{i'} \;\; \sigma, t'' \models \widehat{\theta_1}$.

We distinguish two cases:

- i is even: then $\sigma, t \models \theta_{disc}$ and we have to show that $\sigma, t \models \widehat{\theta_1} U_d \widehat{\theta_2}$.
If $j = i + 1$ then $\sigma, t \models \widehat{\theta_2} U \widehat{\theta_2}$. Otherwise, let k be the greatest even integer smaller or equal to j (this is the index corresponding to the last point of discontinuity before j). Let $t_k > t$ such that $I_k = \{t_k\}$, note that $\sigma, t_k \models \theta_{disc}$ and $\forall t < t'' < t_k \;\; \sigma, t'' \models \widehat{\theta_1}$. If $k = j$ then $\sigma, t_k \models \widehat{\theta_2}$. Otherwise $k = j - 1$ and $\sigma, t_k \models \widehat{\theta_1} \wedge (\widehat{\theta_2} U \widehat{\theta_2})$. We have thus shown that in both cases $\sigma, t \models \widehat{\theta_1} U_d \widehat{\theta_2}$.
- i is odd: then $\sigma, t \models \theta_{cont}$ and we have to show that $\sigma, t \models \theta_{cont}\, U\big(\theta_{disc} \wedge (\widehat{\theta_2} \vee (\widehat{\theta_1} \wedge (\widehat{\theta_1} U_d \widehat{\theta_2})))\big)$.
Let $t_{i+1} > t$ be such that $I_{i+1} = \{t_{i+1}\}$. If $j = i+1$ then $\sigma, t_{i+1} \models \widehat{\theta_2}$. Otherwise $\sigma, t_{i+1} \models \widehat{\theta_1}$ and $w', i + 1 \models \theta_1 U \theta_2$. As $i + 1$ is even, by the previous case we have that $\sigma, t_{i+1} \models \widehat{\theta_1} U_d \widehat{\theta_2}$. We have thus proved that $\sigma, t \models \theta_{cont}\, U\big(\theta_{disc} \wedge (\widehat{\theta_2} \vee (\widehat{\theta_1} \wedge (\widehat{\theta_1} U_d \widehat{\theta_2})))\big)$, t_{i+1} being a witness for the outermost until.

Right to left implication: Let $t \in I_i$ and suppose $\sigma, t \models ltl\text{-}ltls(\theta)$. We distinguish two cases:

- i is even: then $\sigma, t \models \theta_{disc}$ so $\sigma, t \models \widehat{\theta_1} U_d \widehat{\theta_2}$. If $\sigma, t \models \widehat{\theta_2} U \widehat{\theta_2}$ then $w', i+1 \models \theta_2$ so $w', i \models \theta_1 U \theta_2$.
Otherwise $\sigma, t \models \widehat{\theta_1} U\big(\theta_{disc} \wedge (\widehat{\theta_2} \vee \widehat{\theta_1} \wedge (\widehat{\theta_2} U \widehat{\theta_2}))\big)$, so there exists $t' > t$ such that $\sigma, t' \models \theta_{disc} \wedge (\widehat{\theta_2} \vee \widehat{\theta_1} \wedge (\widehat{\theta_2} U \widehat{\theta_2}))$ and $\forall t < t'' < t' \;\; \sigma, t'' \models \widehat{\theta_1}$. As $\sigma, t' \models \theta_{disc}$ there exists $j > i$ such that $I_j = \{t'\}$. We have that $\forall i < i' < j \;\; w', i' \models \theta_1$. If $\sigma, t' \models \widehat{\theta_2}$ then $w', j \models \theta_2$. If $\sigma, t' \models \widehat{\theta_1} \wedge (\widehat{\theta_2} U \widehat{\theta_2})$ then $w', j \models \theta_1$ and $w', j+1 \models \theta_2$. In both cases we have shown that $w', i \models \theta_1 U \theta_2$.
- i is odd: let t_{i+1} such that $I_{i+1} = \{t_{i+1}\}$. Necessarily $\sigma, t_{i+1} \models \widehat{\theta_2} \vee (\widehat{\theta_1} \wedge (\widehat{\theta_1} U_d \widehat{\theta_2}))$. If $\sigma, t_{i+1} \models \widehat{\theta_2}$ then $w', i + 1 \models \theta_2$ so $w', i \models \theta_1 U \theta_2$. Otherwise $\sigma, t_{i+1} \models \widehat{\theta_1} \wedge (\widehat{\theta_1} U_d \widehat{\theta_2})$ so $w', i+1 \models \theta_1$. As $i+1$ is even and $\sigma, t_{i+1} \models \widehat{\theta_1} U_d \widehat{\theta_2}$, by the previous case, we have that $w', i+1 \models \theta_1 U \theta_2$; it follows that $w', i \models \theta_1 U \theta_2$.

□

Using the above lemma we can now show that if $\mathcal{A}$ and θ are as in the previous steps, then $S(\mathcal{A}) = S(\mathit{ltl\text{-}ltls}(\theta))$. To show $S(\mathcal{A}) \subseteq S(\mathit{ltl\text{-}ltls}(\theta))$, let $\sigma = (a_0, I_0) \cdots (a_{2n}, I_{2n}) \in S(\mathcal{A})$. Then $w = a_0 \cdots a_{2n} \in L(\mathcal{A})$ and $\sigma \in \mathit{timing}(w)$. Let $w' = \mathit{ann}(w)$. Then $w' \in L(\mathcal{B})$. Hence $w', 0 \models \theta$. By Lemma 18.7 we have that $\sigma, 0 \models \mathit{ltl\text{-}ltls}(\theta)$. Hence $\sigma \in S(\mathit{ltl\text{-}ltls}(\theta))$.

Conversely, suppose $\sigma \in S(\mathit{ltl\text{-}ltls}(\theta))$ with $\sigma = (a_0, I_0) \cdots (a_{2n}, I_{2n})$ being its canonical representation. That is $\sigma, 0 \models \mathit{ltl\text{-}ltls}(\theta)$. By Lemma 18.7 we have that $w', 0 \models \theta$, where $w = a_0 \cdots a_{2n}$ and $w' = \mathit{ann}(w)$. Hence $w' \in L(\mathcal{B})$. Hence $w \in L(\mathcal{A})$, and $\sigma \in S(\mathcal{A})$.

Step (e): A LTL$^s(A)$ formula θ can be translated to a FO$^s(A)$-formula ψ with one free variable x, such that for all $\sigma \in \mathit{Sig}(A)$, $\sigma, t \models \theta$ if and only if $\sigma, [t/x] \models \psi$. For a first-order formula φ let us denote by $\varphi[z/x]$ the formula obtained by substituting all free occurrences of x in φ by z. The translation *ltl-fo* is now given as follows:

$$\begin{aligned}
\mathit{ltl\text{-}fo}(a) &= a(x) \\
\mathit{ltl\text{-}fo}(\theta_1 U \theta_2) &= \exists z(x < z \wedge \mathit{ltl\text{-}fo}(\theta_2)[z/x] \wedge \forall y(x < y < z \Rightarrow \mathit{ltl\text{-}fo}(\theta_1)[y/z])) \\
\mathit{ltl\text{-}fo}(\theta_1 S \theta_2) &= \exists z(z < x \wedge \mathit{ltl\text{-}fo}(\theta_2)[z/x] \wedge \forall y(z < y < x \Rightarrow \mathit{ltl\text{-}fo}(\theta_1)[y/z])) \\
\mathit{ltl\text{-}fo}(\neg\theta) &= \neg(\mathit{ltl\text{-}fo}(\theta)) \\
\mathit{ltl\text{-}fo}(\theta_1 \vee \theta_2) &= \mathit{ltl\text{-}fo}(\theta_1) \vee \mathit{ltl\text{-}fo}(\theta_2).
\end{aligned}$$

We can now translate θ to the FO$^s(A)$ sentence $\psi = \forall x(\mathit{first}(x) \Rightarrow \mathit{ltl\text{-}fo}(\theta))$, so that $S(\theta) = S(\psi)$.

To summarize this direction of the proof: given a counter-free ST-NFA $\mathcal{A}$ over A by steps (b) and (c) we have an LTL($A \cup A'$) formula θ such that $\mathit{ann}(L(\mathcal{A})) = L(\theta)$. By step (d) we have LTL$^s(A)$ formula $\widehat{\theta}$ such that $S(\mathcal{A}) = S(\widehat{\theta})$. By step (e) we have an FO$^s(A)$ formula $\varphi_{\mathcal{A}}$ such that $S(\varphi_{\mathcal{A}}) = S(\widehat{\theta}) = S(\mathcal{A})$. This completes the proof of Theorem 18.2. □

To conclude this section, we show how we can decide whether a given MSO^s sentence is first-order definable, using our automata characterisation. The procedure is similar to the classical case where we first construct an automaton for the given MSO formula, determinize and minimize it to obtain the canonical DFA for the language, and simply check if the canonical DFA has a counter or not.

We can also define a *canonical* ST-NFA for a given ST-NFA $\mathcal{A}$. This is done as follows: First make $\mathcal{A}$ proper to get $\mathcal{A}'$; then translate via *stnfa-bnfa* to a B-NFA $\mathcal{B}$; determinize $\mathcal{B}$ to get a B-DFA $\mathcal{B}'$; Now minimize $\mathcal{B}'$ to get the canonical DFA $\mathcal{B}''$ for $\mathcal{B}'$. We note that $\mathcal{B}''$ is a B-DFA, *except* for a single non-final sink state d which represents the set of words in $A(AA)^*$ which have no extensions in $L(\mathcal{B}')$, as well as all words in $A^* - A(AA)^*$. This state can be dropped without affecting the language of $\mathcal{B}''$, and the resulting DFA is a B-DFA. We can now apply the translation *bnfa-stnfa* to $\mathcal{B}''$ to get a proper ST-DFA $\mathcal{A}''$. We note that for any two ST-NFA's that accept the same signal language, the canonical ST-DFA associated with them will be identical (upto isomorphism).

Now given an MSO^s formula φ, we can construct the corresponding ST-NFA $\mathcal{A}_\varphi$. Next we construct the canonical proper ST-DFA $\mathcal{A}''$ associated with $\mathcal{A}_\varphi$ as described above. Check if $\mathcal{A}''$ has a counter and return "Not FO^s-definable" if it does, otherwise return "FO^s-definable".

This procedure can be justified as follows: clearly if the procedure says "FO^s-definable", then the formula is indeed FO^s-definable (using Theorem 18.2). If it says "No", then suppose to the contrary that there does exist a counter-free proper ST-NFA $\mathcal{A}$ accepting the language $S(\varphi)$. Then we can see that each step in the canonicalisation preserves the absence of even-counters (including the minimisation step), and hence the canonical ST-DFA will not have a counter, which is a contradiction.

18.7. ST-NFA's and signal regular expressions

In this section we give a regular expression formalism for describing "regular" and "counter-free" signal languages. We will essentially define the underlying word languages via regular expressions, and say that the signal languages defined are simply the timings of the underlying word languages.

A *partial signal regular expression* over an alphabet A is given by the following syntax:

$$p ::= \emptyset \mid ab \mid p+p \mid p \cdot p \mid p^*$$

where a and b belong to A. A *signal regular expression* over A is an expression of the form

$$r ::= a \cdot p \mid r + r,$$

where $a \in A$ and p is a partial signal regular expression over A.

A partial signal regular expression p over A defines a language $L(p) \subseteq (AA)^*$ which is given inductively as follows:

- $L(\emptyset) = \emptyset$
- $L(ab) = \{ab\}$
- If p and q are partial signal regular expressions then $L(p+q) = L(p) \cup L(q)$, $L(p \cdot q) = L(p) \cdot L(q)$, and $L(p^*) = L(p)^*$.

A signal regular expression r over A defines a subset $L(r)$ of $A(AA)^*$ as follows:

- If $a \in A$ and p is a partial signal regular expression then $L(a \cdot p) = a \cdot L(p)$, and
- If r and s are signal regular expressions then $L(r+s) = L(r) \cup L(s)$.

We define the signal language associated with a signal regular expression r, denoted $S(r)$, to be $timing(L(r))$.

We now show that the class of signal languages definable by signal regular expressions is precisely the class of signal languages definable by ST-NFA's. For that it suffices to show that the word languages definable by ST-NFA's and signal regular expressions coincide.

Let us fix an alphabet A. It will be useful to consider the alphabet $\Sigma_{AA} = \{s_{ab} \mid ab \in AA\}$. We note that the homomorphism $h : \Sigma_{AA}^* \to A^*$ given by $h(s_{ab}) = ab$, is injective and onto with respect to $(AA)^*$.

Proposition 18.6. *The class of word languages defined by partial signal regular expressions over A is precisely the class of word languages which are regular subsets of $(AA)^*$.*

Proof. Languages defined by partial signal regular expressions can be seen to be regular subsets of $(AA)^*$, by a simple inductive argument. Conversely, if L is a regular subset of $(AA)^*$, then $h^{-1}(L)$ is a regular subset of Σ_{AA}^* (using a standard property of homomorphisms), and hence there exists a regular expression r over Σ_{AA} with $L(r) = h^{-1}(L)$. Now it is easy to see that if we simply replace each symbol s_{ab} by ab in r (let us denote this partial signal regular expression by $h(r)$) then $L(h(r)) = h(L(r))$ which equals L. □

Theorem 18.3. *The class of signal languages definable by signal regular expressions over A coincides with the class of signal languages definable by* ST-NFA*'s over A.*

Proof. From the definition of signal regular expressions and the preceding Prop. 18.6, it follows that the word language defined by a signal regular expression r is a finite union of languages of the form $a \cdot M$ where M is a regular subset of $(AA)^*$. Thus r defines a regular subset of $A(AA)^*$, and by Proposition 18.2, we have an ST-NFA over A accepting $L(r)$.

In the converse direction, it suffices to show that every regular subset L of $A(AA)^*$ is definable by a signal regular expression. Let L/a denote the language $\{w \in A^* \mid aw \in L\}$. Then clearly $L = \bigcup_{a \in A} a \cdot (L/a)$. Further, L/a is a regular language and a subset of $(AA)^*$. So by Prop. 18.6 each L/a is definable by a partial signal regular expression p_a. Thus, the signal regular expression r given by the sum of $a \cdot p_a$'s, describes L. □

In the rest of this section we give a characterisation of counter-free ST-NFA's in terms of star-free regular expressions.

We define a *star-free* partial signal regular expression over an alphabet A as follows:

$$p' ::= \emptyset \mid ab \mid p' + p' \mid p' \circ p' \mid \tilde{p'},$$

where $a, b \in A$. The syntax of a *star-free* signal regular expression over A is given by

$$r' ::= a \cdot p' \mid r' + r'$$

where $a \in A$ and p' is a star-free partial signal regular expression over A. We define the word language generated by a star-free partial signal regular expression p' similar to a partial signal regular expression, except that

- $L(p' \circ q') = (L(p') \cdot L(q')) \cap Prop_2(A)$,
- $L(\tilde{p'}) = Prop_2(A) - L(p')$,

where $Prop_2(A)$ is set of words in $(AA)^*$ which represent proper words – i.e. $Prop_2(A) = \bigcup_{a \in A} Prop(A)/a$. We note that $Prop_2(A)$ can be defined by (the homomorphic image of) the classical star-free expression r^A_{Prop} which is the complement of the sum, over all $a, b \in A$, of $\bar{\emptyset} \cdot s_{aa}s_{ab} \cdot \bar{\emptyset}$.

As before, we define the signal language associated with a star-free signal regular expression r' to be $S(r') = timing(L(r'))$.

Lemma 18.8. *The class of regular languages definable by* CF-ST-NFA*'s and star-free signal regular expressions over an alphabet A coincide.*

Proof. Once again, we prove that the class of word languages defined by the two formalisms coincide. Let L be the word language accepted by a counter-free ST-NFA. Then by the proof of Lemma 18.6 we know that L is a regular subset of $Prop(A)$ and is even-counter free (i.e. it is accepted by an NFA with no even-counters). Once again we can write $L = \bigcup_{a \in A} a \cdot (L/a)$, and observe that each L/a is a regular subset of $(AA)^*$ and is even-counter free. Now if we consider $h^{-1}(L/a)$, we can see that it must be *counter-free*, since a counter here would imply an even-counter in L/a. By the classical result due to Schutzenberger (see Chapter 1), we must have a star-free regular expression p_a over Σ_{AA} such that $L(p_a) = h^{-1}(L/a)$. (Recall that a star-free regular expression is built from $\emptyset$, a (for $a \in A$), sum, concatenation, and complement in A^*). Once again, if we consider the star-free partial signal regular expression $h(p_a)$ obtained from p_a by replacing each s_{ab} by ab, then we can see that $L(h(p_a)) = h(L(p_a)) = L/a$. Hence L which is the union of $a \cdot (L/a)$'s can be defined by the sum of the regular expressions $a \cdot p_a$.

Conversely let r' be a star-free signal regular expression over the alphabet A. So r is sum of expressions of the form $a \cdot p'$, where p' is a star-free partial signal regular expression. Now each p' defines a subset of $Prop_2(A)$ that is even-counter free. To see this, we observe that we can get a classical star-free regular expression from p', say $h^{-1}(p')$, by replacing every ab by s_{ab}, and the concatenation ($\circ$) and complementation ($\tilde{\ }$) by usual concatenation and complementation followed by intersection with the star-free expression r^A_{Prop}. The star-free expression $h^{-1}(p')$ is such that $h(L(h^{-1}(p')) = L(p')$. Now $h^{-1}(p')$ is star-free and hence its language has *no* counters. It follows that $L(p')$ is even-counter free. Thus $L(r)$ is a finite union of languages of the form $a \cdot M$ where M is a even-counter free subset of $Prop_2(A)$. Hence, by characterisation of counter-free ST-NFA languages in the proof of Lemma 18.6, we have that $L(r)$ is accepted by a proper counter-free ST-NFA. □

18.8. Finite variability in FO

In this section we show that the finite variability of an arbitrary function on the non-negative reals is expressible in FO^s. Apart from illustrating the expressiveness of FO, our aim is to give a correct sentence describing finite variability since such a definition seems to be missing in the literature.

In particular, the first-order sentence given by Hirshfeld and Rabinovich in [20] is incorrect. Their formula requires that every point t have an open interval to its right and to its left (in the latter case only when $t \neq 0$) in which the value of the function is constant. However this formula is satisfied by the function f below which is clearly *not* finitely varying:

$$f(t) = \begin{cases} a & \text{if } t = 1/n \text{ for some } n \in \mathbb{N} \\ b & \text{otherwise.} \end{cases}$$

To say that a function $f : [0, r] \to \Sigma$ is finitely varying, we need to say that it has a finite number of discontinuities in its domain. Since we can already say that a point x is a point of discontinuity of f via the first-order formula φ_{disc} of Sec. 18.4, it is sufficient if we can say that a given (bounded) subset of reals is finite.

This can be done as follows. It is sufficient to express that a set is infinite. We can argue using standard results in real analysis, that a subset of reals W is infinite iff it has a strictly decreasing infinite subsequence or a strictly increasing infinite subsequence. This is because we can first construct an infinite sequence $b_0, b_1, \ldots$ of distinct elements in W as follows. Pick any element in W and set it as b_0. Since W is infinite $W - \{b_0\}$ is also infinite, so we can pick another element $b_1 \in W$; and so on. The sequence $\langle b_i \rangle$ constructed this way is clearly a infinite sequence of distinct elements in W. Now every infinite sequence of distinct elements must have an infinite strictly monotonic subsequence. To see this, suppose $b_0, b_1, \ldots$ was the given infinite sequence. Let b_n be called a "peak point" if all elements in the sequence after it are strictly less than it (i.e. $b_i < b_n$ for all $i > n$). If there are infinitely many peak points, then we clearly have a strictly decreasing infinite subsequence. On the other hand if there were only finitely many peak points, let b_n be the last peak point. Then consider b_{n+1}: since it is not a peak point it must have a value b_i strictly greater that it, for some $i > n + 1$. Similarly b_i must also have a point strictly greater than it which occurs later in the sequence. Continuing in this way we have a strictly increasing infinite subsequence.

Further, to express this property in first-order logic, it is useful to observe that by the Bolzano-Weierstrass theorem, every bounded monotonic sequence converges to a limit point. Thus, for a bounded subset of reals W, the formula $decseq(W)$ below asserts that W contains an infinite strictly decreasing sequence:

$$\begin{aligned} decseq(W) = \exists l \exists a_0 (& a_0 \in W \wedge l < a_0 \wedge \\ & \forall x ((x \in W \wedge l < x) \Rightarrow \exists y (y \in W \wedge l < y \wedge y < x))). \end{aligned}$$

Similarly the formula *incseq*(W) asserts that W contains an infinite strictly increasing sequence:

$$\begin{aligned} incseq(W) = \exists l \exists a_0 (& \; a_0 \in W \wedge a_0 < l \wedge \\ & \forall x((x \in W \wedge x < l) \Rightarrow \exists y(y \in W \wedge x < y \wedge y < l))). \end{aligned}$$

Using the observations above, we can see that the formula *inf*(W) below asserts, for bounded subsets of reals W, that W is infinite:

$$inf(W) = decseq(W) \vee incseq(W).$$

Finally the required sentence asserting that a given function is finitely varying is obtained by replacing each atomic formula of the form $x \in W$ by $\varphi_{disc}(x)$, in the formula $\neg inf(W)$.

18.9. Conclusion

In this chapter we have shown that the theory of automata and logics over signals bears a close analogy to the classical theory of such formalisms over words.

Among the issues that remain to be addressed is the existence of a "canonical" minimum ST-DFA for the class of ST-NFA-definable signal languages, along the lines of the Myhill-Nerode Theorem for regular word languages. While we have identified a canonical ST-DFA for any given ST-NFA, this was done for the purpose of deciding first-order definability of signal languages. It remains to investigate a suitable definition of a "minimum" ST-DFA for a given ST-NFA-definable signal language.

Acknowledgments

We thank Kamal Lodaya and Wolfgang Thomas for several comments and suggestions which have improved the content and presentation considerably. The authors also acknowledge support from the Indo-French TIMED-DISCOVERI project.

Part of the material in this chapter is reproduced with permission from the following source:

F. Chevalier, D. D'Souza, M. Raj Mohan, P. Prabhakar, Automata and logics over finitely varying functions, Annals of Pure and Applied Logic (2009), doi:10.1016/j.apal.2009.07.007.

References

[1] J. R. Büchi, On a decision method in restricted second-order arithmetic., *Z. Math. Logik Grundlag. Math.* pp. 66–92, (1960).

[2] R. McNaughton and S. Papert, *Counter-Free Automata.* (MIT Press, 1971).

[3] A. Rabinovich, Finite variability interpretation of monadic logic of order, *Theor. Comput. Sci.* **275**(1-2), 111–125, (2002).
[4] F. Chevalier, D. D'Souza, and P. Prabakhar. On continuous timed automata with input-determined guards. In *Proc. 26th FSTTCS*, vol. 4337, *LNCS*, pp. 369–380. Springer, (2006).
[5] F. Chevalier, D. D'Souza, and P. Prabhakar. Counter-free input-determined timed automata. In *Proc. 5th FORMATS*, vol. 4763, *LNCS*, pp. 82–97. Springer, (2007).
[6] J. A. W. Kamp. *Tense Logic and the Theory of Linear Order.* PhD thesis, University of California, (1968).
[7] M. P. Schtzenberger. On finite monoids having only trivial subgroups. In *Information and Control*, pp. 8:190–194, (1965).
[8] R. Alur and D. L. Dill, A theory of timed automata, *Theor. Comput. Sci.* **126**(2), 183–235, (1994).
[9] R. Alur, C. Courcoubetis, N. Halbwachs, T. A. Henzinger, P.-H. Ho, X. Nicollin, A. Olivero, J. Sifakis, and S. Yovine, The algorithmic analysis of hybrid systems, *Theor. Comput. Sci.* **138**(1), 3–34, (1995).
[10] E. Asarin, O. Bournez, T. Dang, O. Maler, and A. Pnueli, Effective synthesis of switching controllers for linear systems, *Proc. IEEE.* **88**(7), 1011–1025, (2000).
[11] T. Wilke. Classifying discrete temporal properties. In *Proc. 16th STACS*, vol. 1563, *LNCS*, pp. 32–46. Springer, (1999).
[12] I. Hodkinson, *Temporal Logics and Automata*, In eds. D. M. Gabbay, M. Finger, and M. A. Reynolds, *Temporal Logic: Mathematical Foundations and Computational Aspects*, vol. 2, pp. 30–72. Clarendon Press, (2000).
[13] A. Rabinovich and B. A. Trakhtenbrot. From finite automata toward hybrid systems (extended abstract). In *Proc. 11th FCT*, vol. 1279, *LNCS*, pp. 411–422. Springer, (1997).
[14] W. Thomas. Automata on infinite objects. In *Handbook of Theoretical Computer Science, Volume B: Formal Models and Semantics*, pp. 133–192. Elsevier, (1990).
[15] A. Rabinovich, Star free expressions over the reals., *Theor. Comput. Sci.* **233**(1-2), 233–235, (2000).
[16] M. O. Rabin, Decidability of second order theories and automata on infinite trees, *Trans. American Math. Monthly.* **141**, 1–35, (1969).
[17] M. Shelah, The monadic theory of order, *Annals of Mathematics.* **102**, 379–419, (1975).
[18] R. Koymans, Specifying Real-Time Properties with Metric Temporal Logic., *Real-Time Systems.* **2**(4), 255–299, (1990).
[19] R. Alur, T. Feder, and T. A. Henzinger, The Benefits of Relaxing Punctuality, *J. ACM.* **43**(1), 116–146, (1996).
[20] Y. Hirshfeld and A. M. Rabinovich, Timer formulas and decidable metric temporal logic, *Inf. Comput.* **198**(2), 148–178, (2005).

Part IV

Compression

Chapter 19

Syntax Directed Compression of Trees Using Pushdown Automata

Priti Shankar

Department of Computer Science and Automation
Indian Institute of Science, Bangalore
priti@csa.iisc.ernet.in

We describe two techniques for the adaptive compression of tree-structured files using a syntactic specification of the input. We first propose a technique for the generation of compressors for files whose syntax is described by a regular tree grammar, which generates ranked trees. The technique is an adaptation of the $LR(0)$ bottom-up strategy for parsing context free languages. We then describe a modification of a top-down parsing technique for generating compressors for unranked trees, such as XML files specified by a Document Type Definition (DTD). Both techniques build devices that are effectively pushdown automata for the generation of probabilities that are used to encode the input. The compressors are automatically generated from a syntactic description of the input to be compressed.

19.1. Introduction

Specialized techniques for the efficient compression of source files whose structure is governed by a set of syntactic rules are of practical interest for two main reasons. The first is that such files (examples being files consisting of mobile code or XML data) are frequently transmitted over computer networks, and there is a considerable gain in bandwidth by transmitting compressed versions. The second has to do with the explosive growth of massive XML databases and the need to archive such data. General purpose compressors are unable to completely exploit the structure inherent in the source files, and therefore cannot deliver the same performance as syntax directed compressors that can take full advantage of such structure. At a higher level of abstraction, it would be useful to have such compressors generated directly from a description of the syntax, just as parsers are now routinely generated from grammar based specifications.

Compression techniques are typically either dictionary based, an important example being the Ziv-Lempel scheme LZ77 [1], or based on statistical modeling such as the Huffman coding [2, 3] and the arithmetic coding schemes [4, 5]. Another distinction can be made between static and adaptive schemes. In the former, the dictionary or the model is constructed ahead of compression time, by making a preliminary pass over the file to be compressed. In the latter, compression is performed on-line, as the file is read in, and the dictionary or the model is modified adaptively.

In this chapter we restrict our attention to input files consisting of a single tree or a sequence of trees. We look at both ranked trees, an example of which are files representing intermediate code for computer programs, and unranked trees such as trees representing the structure of XML documents [6]. The scheme we use is an adaptive statistical compression scheme, namely, adaptive arithmetic coding. The success of statistical schemes is based on an accurate prediction of symbol probabilities in the input stream of symbols. These probability values are used to determine the code for the symbol to be used by the encoder or compressor. The model that is used to generate these probabilities may be simple, using minimal information, or fairly complex, such as an order-k Markov model. In adaptive schemes the model is built up as the input is scanned, and often takes the form of frequency tables that are updated as symbols are read in. These frequency tables are then used to compute the probabilities used by the encoder. Sometimes it is profitable to have separate models for different parts of the input file, and to shift from one model to another depending on which part of the file is being processed, thereby achieving a *multiplexing* of models. In the schemes proposed in this chapter, whenever such multiplexing is performed, it is under the direction of an automaton which is automatically generated from the predefined syntax. The automaton is effectively a pushdown automaton, and as we shall see, it is able to track arbitrarily distant past context that can be used to generate probabilities.

Section 19.2 contains a short introduction to the arithmetic coding technique which is central to the compression strategies we describe later. Section 19.3, describes a technique for the compression of a sequence of ranked trees. Towards this end we describe a modification of the well known LR-parsing technique [7] to parse ranked trees and show how the underlying pushdown automaton can be used to generate contexts. Section 19.4 describes a technique for the compression of XML files using an adaptation of a simple top-down parsing technique for context-free grammars. Both the techniques presented in this chapter have been used to generate compressors from a syntactic description of source files [8, 9].

19.2. Arithmetic Coding and Finite Context Modeling

In this section, we first briefly present the arithmetic coding scheme with the help of an example. We then describe how a finite context is maintained adaptively with the help of frequency tables.

19.2.1. *Arithmetic Coding*

Arithmetic coding does not replace every input symbol with a specific code. Instead it processes a stream of input symbols and encodes it with a single fractional output

Table 19.1. Symbol Probabilities.

Symbol	Probability	Range
a	0.2	[0, 0.2)
b	0.3	[0.2, 0.5)
c	0.1	[0.5, 0.6)
d	0.2	[0.6, 0.8)
e	0.1	[0.8, 0.9)
!	0.1	[0.9, 1)

number. This number can be uniquely decoded to recreate the exact stream of symbols that resulted in its construction. In order to construct the output number, the symbols being encoded need to have a set of probabilities assigned to them. For the present let us assume that these probabilities are known. Initially the range of the message is the interval [0, 1). As each symbol is processed, the current range is narrowed to that portion of it allocated to the symbol. In the example that follows (adapted from [5]), suppose the alphabet is $\{a, b, c, d, e, !\}$, and a static model is used with probabilities shown in Table 19.1.

Suppose we wish to send the message *bacc!*. Initially, both the encoder and the decoder know that the range is [0, 1). After seeing the first symbol b the encoder narrows it down to [0.2, 0.5) (this is the range that the model allocates to symbol b). After reading the second symbol a, the interval is further narrowed to one-fifth of itself, since a has been allocated [0, 0.2). Thus the new interval is [0.2, 0.26). After seeing the first c the narrowed interval is [0.23, 0.236), and after seeing the second c the new interval is [0.233, 0.2336). Finally on seeing !, the interval is [0.23354, 0.2336). Knowing this to be the final range the decoder can immediately deduce that the first character was b. Now the decoder simulates the action of the encoder, and since the decoder knows that the interval [0.2, 0.5) belonged to b, the range is expanded to [0.2, 0.26). Continuing this way, the decoder can completely decode the transmitted message. It is not really necessary for the decoder to know both ends of the range produced by the encoder. Instead, a single number in the range (say 0.23355 in our example) will suffice. We assume that each message ends with a special end of file symbol eof known to both the encoder and the decoder. From the discussion above, it should be clear, that for this scheme to be effective, the model should produce probabilities that deviate from a uniform distribution. The better the model is at making such predictions, the smaller will be the number of digits required to represent the fraction.

19.2.1.1. *Finite Context Modeling*

In a finite context scheme, the probabilities of each symbol are calculated based on the *context* the symbol appears in. In its traditional setting, the context is just the symbols that have been previously encountered. The *order* of the model refers to the number of previous symbols that make up the context. In an adaptive order k

model, both the compressor and the decompresser start with the same model. The compressor encodes a symbol using the existing model and then updates the model to account for the new symbol. The decompresser similarly decodes a symbol using the existing model and then updates the model. As mentioned earlier, frequency counts are used to approximate probabilities, and the scheme is adaptive because this is being done as the symbols are being scanned. Since there are potentially q^k possibilities for level k contexts where q is the size of the symbol space, update can be a costly process, and the tables consume a large amount of space. This causes arithmetic coding to be somewhat slower than dictionary based schemes like the Ziv-Lempel scheme [1].

We will abstract the procedures of encoding and decoding by the functions $ArithEncode(context, symbol)$ and $ArithDecode(context, symbol)$. The first of these reads a symbol, consults a table specified by the context, and appends a sequence of zero or more digits to the encoded representation generated so far. The second reads the encoded representation, consults the table specified by the context, outputs a symbol, and appropriately modifies the encoded representation. For purposes of this discussion, the encoded representation is stored in a global data structure which is possibly modified by each call to it.

19.3. Compression of Ranked Trees

Intermediate code generated by a compiler has rich syntactic and semantic information which can be exploited for the purpose of compression. Compression techniques for intermediate code are of interest in the transmission of mobile code over networks. The compression of tree intermediate code for the LCC compiler [10], and of Java byte-code has been the subject of some experimentation in the past [11, 12, 13, 14]. Intermediate code structure often takes the form of a sequence of attributed trees with the tree structure being specified by a ranked regular tree grammar. Our aim in this section, is to describe a compression scheme which is *syntax aware*, and which is *automatic*, in the sense that the compressor can be generated from a syntactic description of the source. Thus it is similar in spirit to the technique used in the well known parser generator tool YACC [15]. The strategy that we describe here, uses linearized versions of the trees and works hand in hand with the parser. During tree parsing, the parsing stack effectively stores all relevant syntactical context. We will use parsing stack snapshots instead of character sequences to generate probability distributions for an arithmetic coder. As we will observe, the modified notion effectively encompasses unbounded context. This is useful when we want to correlate parts of the tree which are far apart in the linearized coding but which play an important role in constraining the set of symbols expected at a point. The context we will track can be viewed as a *condensed history* of the parse. The first compression technique we describe is therefore termed *P*rediction by *C*ondensed *H*istory (PCH).

We use the definition of a regular tree grammar from Chapter 3. However in the first part of the chapter we restrict our attention to ranked trees, and there is therefore an additional constraint on the regular tree grammar – the right hand sides of productions are not arbitrary regular expressions, but expressions encoding trees that respect the ranks of symbols.

We introduce a few new symbols used in this chapter. Let Σ be a finite alphabet. We further divide the alphabet into symbols of rank greater than 0 denoted by the set Σ_g and the set of symbols of rank 0 denoted by the set Σ_0. The rank of each symbol σ in Σ_g is denoted by $|\sigma|$. The set T_Σ consists of all trees with internal nodes labeled with elements of Σ_g, and leaves with labels from Σ_0. Such trees are called *subject trees* in this section. Special symbols called *variables* are assumed to have rank 0. If N is a set of variables, the set $T_{(\Sigma \cup N)}$ is the set of all trees with variables also allowed as labels of leaves. We refer to the elements of Σ as *terminals.*

Definition 19.1 (Ranked regular tree grammar). *A* ranked regular tree grammar *G is a four tuple* (N, Σ, S, P) *where:*

(1) N is a finite set of variable *symbols.*
(2) $\Sigma = \Sigma_0 \cup \Sigma_g$ *is a* ranked alphabet, *with the ranking function denoted by arity.*
(3) S is the start symbol *of the grammar.*
(4) P is a finite set of production rules *of the form* $X \rightarrow t$ *where* $X \in N$ *and t is* an encoding *of a tree in* $T_{(\Sigma \cup N)}$.

A *tree pattern* is thus represented by the right hand side of a production of P in the grammar above. We assume that the grammar does not have *chain rules*, that is, productions of the form $A \rightarrow B$, where both A and B are variables. (Chain rules can be handled fairly easily, but we omit them for the sake of simplicity.)

Definition 19.2 (Normal form). *A production is said to be in normal form if it is in one of the following forms below.*

(1) $A \rightarrow \sigma(B_1, B_2 \ldots, B_k)$ *where* $A, B_i, i = 1, 2, \ldots k$ *are variables,* $\sigma \in \Sigma_g$ *and* $|\sigma| = k$.
(2) $B \rightarrow b$ *, where b is an element of* Σ_0.

A grammar is in normal form if all its productions are in normal form. Any regular tree grammar can be put into normal form by the introduction of extra nonterminals. Below is an example of a regular tree grammar in normal form. Arities of symbols in the alphabet are shown in parentheses next to the symbol.

Example 19.1.
$G = (\{V, B, G, C\}, \{a(2), e(2), b(0), c(0)\}, V, P)$
Where P consists of:

$V \to a(V, B)$
$V \to a(B, V)$
$V \to e(V, C)$
$V \to b$
$C \to c$
$B \to b$

Definition 19.3 (Derivation). *For $t, t' \in T_{(\Sigma \cup N)}$, t directly derives t', written as $t \Longrightarrow t'$ if t' can be obtained from t by replacement of a leaf of t labeled X by a tree p where $X \to p \in P$. We write $\Longrightarrow_r$ if we wish to specify that rule r is used in a derivation step. The relations $\Longrightarrow^+$ and $\Longrightarrow^*$ are the transitive closure and reflexive-transitive closure respectively of $\Longrightarrow$.*

The tree language defined by the grammar $G = (N, \Sigma, S, P)$ is the set

$$L(G) = \{t \mid t \in T_\Sigma, \text{ and } S \Longrightarrow^* t\}$$

Definition 19.4 (X-derivation tree). *An X-derivation tree, D_X, for G has the following properties:*

- *The root of the tree has label X.*
- *If Z is an internal node, then the subtree rooted at Z is one of the following three types; (For describing trees we use the usual list notation)*
  - *(1) $Z(a)$ if $Z \to a, a \in \Sigma_0$ is a production of P.*
  - *(2) $Z(\sigma(D_{X_1}, D_{X_2}, \ldots D_{X_k}))$ if $Z \to \sigma(X_1, X_2 \ldots X_k)$ is an element of P.*

Regular tree grammars (for example, those used to describe machine instructions [16]) are often ambiguous by design. Thus several derivation trees may exist for a single subject tree. The grammar of Example 19.1 has two derivation trees for the subject tree $a(b, b)$.

The *tree-parsing problem* for a ranked regular tree grammar is defined as follows:

Given a ranked regular tree grammar $G = (N, \Sigma, S, P)$, and a subject tree t in T_Σ, find (a representation of) all S-derivation trees for t.

The reason we introduce the derivation tree here, is because the device we will use to generate context for the compressor is a pushdown automaton encoding the parser, which implicitly constructs all derivation trees for an input string. The derivation tree for a given subject tree can be constructed using several strategies, based on tree pattern matching techniques, either top down or bottom up, originally described in [17] and [18]. The one we employ here proposes a modification of the bottom up strategy employed in LR parsing [7] for context-free languages [19]. Such a strategy can be employed using either a postfix or a prefix linearization of the subject tree. Because arities of symbols are fixed, there is a unique correspondence between any tree and such a linearization. For the sake of completeness, we give a brief introduction to LR parsing. We assume that the reader is familiar with the notions of derivation sequences, sentential forms and pushdown automata.

19.3.1. *LR Parsing*

Given a context free grammar and an input string in the language generated by the grammar, an LR parsing sequence constructs a rightmost derivation sequence in reverse for the string while scanning the input from left to right. An LR parsing sequence is a sequence of *shift* and *reduce* moves on the parsing stack. A *shift* move shifts the next input symbol on the stack. A *reduce* move, replaces the right-hand side of a production appearing as a string on top of the parsing stack by the left-hand side non-terminal of the production. Parsing is complete when the string is reduced to the start symbol of the grammar. The contents on stack at any point in the parse are called *viable prefixes* as they are prefixes of right sentential forms of the grammar. A string of terminals can always be appended to a viable prefix to yield a right sentential form of the grammar. The set of all viable prefixes for a given context free grammar is known to be a regular set [7]. A finite state automaton that recognizes all viable prefixes can be constructed and used to guide the parse. The set of parsing stack symbols is the set of states of this automaton. If the grammar satisfies the $LR(0)$ property, the state on top of stack determines at each step, whether a *shift* move or a *reduce* move is to be performed, and if a reduction is called for, which production is to be used. Thus for an $LR(0)$ grammar, parsing is deterministic and can be performed in time linear in the size of the input. At each *shift* state, of the finite state machine, there is a set of symbols that is legal at that point in the parse. This is usually a subset of the alphabet set of the grammar. *Reduce* states do not consume any input and merely modify the contents of the parsing stack.

We include a few definitions from the theory of $LR(0)$ parsers for a better understanding of the material to follow. The reader is referred to [7] for a complete treatment. We use the following conventions for notation. Lower case English letters at the beginning of the alphabet represent terminals, those at the end represent strings of terminals. Upper case English letters at the beginning of the alphabet represent variables; those at the end of the alphabet represent either variables or terminals. Lower case Greek letters represent strings (possibly empty,) over terminals and variables.

Definition 19.5 (Item). *Given a context-free grammar $G = (N, \Sigma, S, P)$ an* item *of G is is an entity of the form $[A \longrightarrow \alpha.\beta]$ where $A \longrightarrow \alpha\beta \in P$. The dot can occur anywhere on the right hand side of the production. An item with the dot at the end is called a* complete *item.*

Definition 19.6 (Valid item). *An item $[A \longrightarrow \alpha.\beta]$ is said to be valid for viable prefix γ if there is a rightmost derivation sequence $S \Longrightarrow^* \delta Aw \Longrightarrow \delta\alpha\beta w$ and $\gamma = \delta\alpha$.*

Given a viable prefix and an item that is valid for it, the production associated with the item is one of the possible ones that can be used to continue a rightmost derivation sequence in reverse, and the dot indicates how much of the production has been matched in a left to right scan of the input string.

Definition 19.7 (Handle of a right sentential form). *A handle of a right sentential form is a string α representing the right hand side of a production $A \rightarrow \alpha$ and a position in the right sentential form where it may be replaced by A to obtain a previous right sentential form in a rightmost derivation sequence.*

The problem in LR parsing may thus be viewed as one of finding and reducing handles. We observe that a viable prefix of a right sentential form γ ends no further right than the right end of a handle for γ.

The $LR(0)$ parsing strategy precomputes a DFA called the *DFA for canonical sets of $LR(0)$ items* from an input context free grammar. The input alphabet set of this DFA is the union of the set of terminals and variables of the grammar for which the parser is generated. We refer to this as the $LR(0)$ DFA. The $LR(0)$ DFA recognizes viable prefixes of the grammar. It is created by collecting *sets* of items valid for a viable prefix, each distinct set of items being identified with a unique state of the DFA. The start state of the $LR(0)$ DFA contains all items of the form $[S \rightarrow .\alpha]$ where S is the start symbol of the grammar, augmented with all rules got by applying the function *closure* defined below on items already in the set. Let *itemset* be a set of items. The *closure* is computed by repeating the updation of *itemset* as given below, until there is no further change to *itemset*.

$$itemset = itemset \cup \{[A \rightarrow .\alpha] \mid [B \rightarrow \delta.A\beta] \in itemset\}$$

Next the function *goto* is applied to all items in a state to get new sets of items that define possibly new states. The function *goto* is defined as follows.

$$goto(itemset, X) = \{[A \rightarrow \alpha X.\beta] \mid [A \rightarrow \alpha.X\beta] \in itemset\}$$

The $LR(0)$ DFA is constructed by beginning with a set initially consisting of just the start state, and then applying the *goto* operation on each state (equivalently, item set) in the current set of states, to get a possibly new item set, and then the *closure* operation to complete the set of items, until no new item sets can be created.

A state with a complete item is a *reduce* state, and if reduction is performed by the production associated with a complete item, one is guaranteed to get a previous right sentential form by the property stated in Definition 19.6. A state with an item of the form $[A \to \alpha.a\beta]$, is a *shift* state, as the viable prefix can be extended by a. The state on top of the parsing stack is the state of the $LR(0)$ automaton reached on a viable prefix induced by an input string. The content on stack is the sequence of states induced by the viable prefix.

Definition 19.8 ($LR(0)$ Grammar). *A grammar G is $LR(0)$ iff*

(1) its start symbol does not occur on the right side of any production, and
(2) a state does not allow both shift *and* reduce *moves (in other words, there are no* shift/reduce *conflicts), and if a state contains a* reduce *item, then that is the only* reduce *item associated with the state (i.e. there are no* reduce/reduce *conflicts).*

The $LR(0)$ grammars define the class of deterministic context free languages having the prefix property, i.e languages satisfying the property that no proper prefix of a string in the language is also in the language. The addition of an endmarker eof converts a language into a language with the prefix property. Though linear representations of ranked trees are prefix free, some of the files we deal with, might consist of a sequence of such trees. So we add an end marker eof to indicate the end of file to the arithmetic compressor. Figure 19.1 is the DFA for the canonical sets of $LR(0)$ items for the grammar of Example 19.1, augmented by the production $Z \to V\ eof$. All *reduce* states are enclosed in double circles and the associated complete items are displayed alongside. To illustrate the construction, we simulate a few steps of the well known algorithm. The start state of the automation (state 1) is associated with the item set $\{[S \to .Veof], [V \to .aBV], [V \to .aVB], [V \to .eVC], [V \to .b]\}$. Note that the last four items are obtained by applying the function *closure* on the set containing the first item. Applying the function *goto* on the item set in state 1 on symbol a followed by a *closure* operation on the resulting items yields state 2 with item set $\{[V \to a.VB], [V \to a.BV], [V \to .aVB], [V \to .aBV], [V \to .eVC], [V \to .b], [B \to .b]\}$ where the last five items are obtained by using the function *closure* on the set consisting of the first two items. Applying the function *goto* on symbol b to state 2 yields the *reduce* state 6 which has a *reduce/reduce* conflict. If one chooses to reduce by the production corresponding to the first complete item in state 6, the parser would transit to state 8. On the other hand, if the other option is chosen, the transition would be to state 7, this choice illustrating the ambiguity the parser is confronted with. The next subsection describes how to deal with this ambiguity. The item sets corresponding to each state of the $LR(0)$ automaton are given below.

(1) $\{[S \to .Veof], [V \to .aBV], [V \to .aVB], [V \to .eVC], [V \to .b]\}$.
(2) $\{[V \to a.VB], [V \to a.BV], [V \to .aVB], [V \to .aBV], [V \to .eVC], [V \to .b], [B \to .b]\}$
(3) $\{[V \to b.]\}$
(4) $\{[V \to e.VC], [V \to .aBV], [V \to .aVB], [V \to .eVC], [V \to .b]\}$

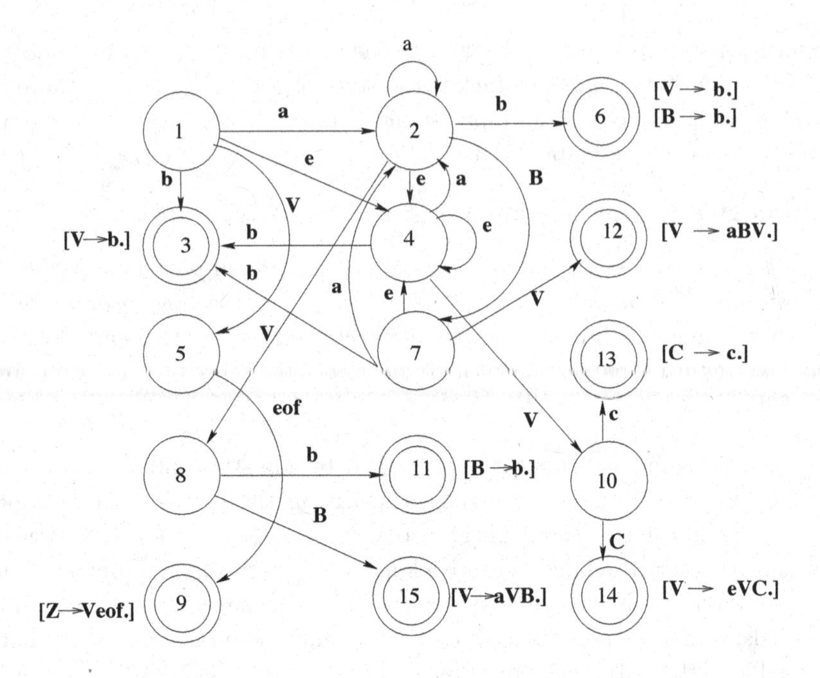

Fig. 19.1. $LR(0)$ automaton for the grammar of Example 19.1.

(5) $\{[S \to V.eof]\}$
(6) $\{[V \to b.], [B \to b.]\}$
(7) $\{[V \to aB.V], [V \to .aBV], [V \to .aVB], [V \to .eVC], [V \to .b]\}$
(8) $\{[V \to aV.B], [B \to .b]\}$
(9) $\{[S \to Veof.]\}$
(10) $\{[V \to eV.C], [C \to .c]\}$
(11) $\{[B \to b.]\}$
(12) $\{[V \to aBV.]\}$
(13) $\{[C \to c.]\}$
(14) $\{[V \to eVC.]\}$
(15) $\{[V \to aVB.]\}$

19.3.2. *Parsing Ambiguous Ranked Regular Tree Gramms*

The device we propose here is a generalization of the $LR(0)$ automaton to work for trees specified by a possibly ambiguous tree grammar. As mentioned earlier, the trees will be linearized and therefore our starting point is the context free grammar obtained from a linearized regular ranked tree grammar in normal form.

Definition 19.9 (PLG). *A preorder linearized grammar (PLG) is a context free grammar derived from a ranked regular tree grammar where the encoding of the tree* *on the right hand side of each production is a preorder listing of the nodes of the tree pattern on the right hand side.*

Note that the PLG right hand sides have no meta symbols like parentheses and commas, and the ranks of symbols are implicit in the representation. Since the PLG may be ambiguous there may be more than one viable prefix induced by an input string. Each induced viable prefix corresponds to a distinct derivation sequence for the input string. It would be useful to construct, if possible, an automaton that keeps track of *sets* of viable prefixes. We show that this is indeed possible because of the restricted form of the grammar. Our immediate aim is therefore, the design of a technique to construct such an automaton which we will henceforth refer to as the *controlling automaton.*

In order to construct the controlling automaton, we make use of two properties of the PLG which we prove below. We say that a PLG is in normal form if it is derived from a ranked regular tree grammar in normal form.

Lemma 19.1. *The $LR(0)$ automaton constructed for a PLG G in normal form, has no* shift/reduce *conflicts.*

Proof. Given the restricted form of the grammar a shift reduce conflict would manifest itself as a pair of items $I_1 = [A \rightarrow aX_1X_2 \ldots X_k.]$ with $|a| = k$ and $I_2 = [B \rightarrow .bY_1Y_2 \ldots Y_l]$ where $|b| = l$. The item I_2 arises from a closure operation on an item of the form $I_3 = [C \rightarrow cZ_1Z_2 \ldots Z_i.B \ldots Z_m]$ where $|c| = m$. Since both I_1 and I_3 are valid for the same viable prefix, say γ we have a pair of derivation sequences $S \Longrightarrow \delta_1 A w_1 \Longrightarrow \delta_1 \alpha_1 w_1$ where $\alpha_1 = aX_1X_2 \ldots X_k$, with $\gamma = \delta_1\alpha_1$, and $S \Longrightarrow \delta_2 B w_2 \Longrightarrow \delta_2\alpha_2\beta_2 w_2$ where $\alpha_2 = cZ_1Z_2 \ldots Z_i$ and $\gamma = \delta_2\alpha_2$. Since the grammar is in normal form and recalling that a handle is a suffix of γ, neither α_1 nor α_2 can be a suffix of the other and hence they are identical. Hence $a = c$ and $\beta_2 = \epsilon$ which implies that I_3 is a complete item and cannot generate I_2 via a closure operation. Hence the $LR(0)$ automaton for a grammar in this class can have no shift/reduce conflicts. □

Lemma 19.2. *If the $LR(0)$ automaton constructed for a PLG G in normal form has* reduce/reduce *conflicts, then all handles associated with the reductions are identical.*

Proof. A *reduce/reduce* conflict manifests itself as a pair of items $[A \rightarrow aX_1X_2 \ldots X_n.]$ and $[B \rightarrow bY_1Y_2 \ldots Y_k.]$. Since both of them are valid for a viable prefix, say γ we have $S \Longrightarrow \delta_1 A w_1 \Longrightarrow \delta_1\alpha_1 w_1$ where $\alpha_1 = aX_1X_2 \ldots X_n$, with $\gamma = \delta_1\alpha_1$, and $S \Longrightarrow \delta_2 B w_2 \Longrightarrow \delta_2\alpha_2 w_2$ where $\alpha_2 = bY_1Y_2 \ldots Y_k$ and $\gamma = \delta_2\alpha_2$. As the grammar is in normal form neither α_1 nor α_2 can be a suffix of the other, and are hence identical. Hence the lemma. □

Given these two results, it is possible to construct an automaton, called the controlling automaton, which will guide the parse, using a technique reminiscent of a subset construction on an NFA. All shift moves are performed as usual. If there are reduce/reduce conflicts in a *reduce* state *all reductions are performed in synchrony*, popping the same number of symbols off the stack, and a *set* of left hand side non-terminals is pushed on stack. Thus a path from the start state of the controlling automaton to any state encodes *sets* of viable prefixes all of the same length, and each item in the state is valid for some viable prefix in the set. One could either construct the controlling automaton from the $LR(0)$ automaton, using a two pass approach (where the $LR(0)$ automaton is constructed first and then a kind of subset construction is performed on it), or directly from the grammar. We employ the latter approach.

19.3.3. *Construction of the Controlling Automaton*

We assume that the PLG $G = (N, \Sigma, S, P)$ is augmented by a production $Z \rightarrow S\ eof$ (where $eof \notin \Sigma$). The controlling automaton is $M = (Q, \Sigma_c, \delta_c, q_0, \{f\})$ where each state of Q contains a set of items of the grammar.

- $\Sigma_c = \Sigma \cup \{eof\} \cup 2^N$.
- q_0 is the start state.
- The final state f is the state containing the item $[Z \rightarrow S\ eof.]$
- δ_c is a map: $Q \times (\Sigma \cup \{eof\} \cup 2^N) \mapsto Q$.

The precomputation of M is similar to the precomputation of the states of the DFA for canonical sets of $LR(0)$ items for a context free grammar [7]. However there is one important difference. In the DFA for $LR(0)$ items, transitions on variables are determined by the *goto* operation which just inspects the sets of items in any state. Here we have transitions on *sets* of variables. These sets can not be determined in advance, as we do not know a priori, which reductions are simulaneously applicable while parsing. Therefore, transitions on sets of variables are added as and when these sets are determined.
The construction of the controlling automaton proceeds as follows. The start state of the automaton contains the same set of items as would the start state of the $LR(0)$ DFA. From each state, say q, identified to be a *shift* state of the controlling automaton, we find the state say m entered on a symbol of Σ, say a. (This depends only on the set of items in the first state). We set $\delta_c(q, a)$ to m. If the state m contains complete items, we compute $m.rules$ and $m.S_l$, where $m.rules$ is the set of rules of the PLG that correspond to complete items in m, and $m.S_l$ is the set of left-hand side variables of $m.rules$. Next, given the states and transitions computed so far, we determine all states that can be exposed on stack while performing a reduction, after the set of handles is popped off the stack via the function $validlc(p, m)$. Such states are termed *valid left context* states. For each such state p, we compute the item set got by a *goto* operation on items associated with p on variables in $m.S_l$. This is obtained by applying the function *reduction* defined below. The *closure* operation on resulting items completes the new item set associated with the state say r.

The *reduction* operation on a set of complete items I_1 (representing a *reduce* state) with respect to another set of items I_2 (representing a valid left context state), is defined as follows:

$reduction(I_2, I_1) = \bigcup\{[A \to \alpha B.\beta] \mid \exists[B \to \gamma.] \in I_1, \text{ and } [A \to \alpha.B\beta] \in I_2\}$

The composition of the *closure* and *reduction* operations is defined as

$$ClosureReduction(I_2, I_1) = closure(reduction(I_2, I_1))$$

We need to specify a condition to determine whether a state is a valid left context state for a *reduce state*. To simplify the notation we assume that we identify states with the item sets associated with them. Let m be a *reduce* state, $m.rhs$ be the set of right hand sides of complete items in m, p be any *shift* state.

Define $VSet(p, m.rhs) = \{B \mid B \longrightarrow .\alpha \in p, \alpha \in m.rhs\}$. Then a necessary condition for p to be a valid left context state for a *reduce* state m is $VSet(p, m.rhs) = m.S_l$. (The condition is only necessary, but not sufficient, because there may be another production that always matches in this left context when the others do, but which is not in the matchset.) Using this condition instead of an exact one, may yield some spurious states which are unreachable at parse time. An necessary and sufficient condition is somewhat expensive to compute in comparison [19]. Note that the necessary condition defined above is a more refined condition than one that checks that for each complete item $A \longrightarrow \beta.$ in q there is a corresponding item $A \longrightarrow .\beta$ in p. The function $validlc(p, m)$ is defined as follows.

$validlc(p, m) = \textbf{if } (VSet(p, m.rhs)) == m.S_l \textbf{ then} \text{ true } \textbf{else} \text{ false}$

The controlling automaton is constructed using Algorithm 19.1.

The function $AddR(r)$ adds r to the set of *reduce* states if it is not already a member, sets $r.arity$ to the arity of operator in r, $r.rules$ to the set of rules corresponding to complete items in r, $r.S_l$ to the set consisting of the left hand side variables of rules in $r.rules$ and $r.rhs$ to the set of right hand sides of rules in $r.rules$.

Theorem 19.1. *The function δ_c satisfies the following property:*
Let $S_1S_2 \ldots S_k = \{X_1X_2 \ldots X_k \mid X_i \in S_i\}$ where $S_i \subseteq 2^N$ or $S_i = \{a\}, a \in \Sigma \cup \{eof\}$.
$LR(0)$ item $[A \to \alpha.\beta]$ is valid for viable prefix $\gamma = X_1X_2 \ldots X_k \in S_1S_2 \ldots S_k$ if and only if $\delta(q_0, S_1S_2 \ldots S_k)$ contains $[A \to \alpha.\beta]$.

Proof. We give a sketch of the proof, based on the well known proof that a similar property (differing from the one above only in that the function δ is applied on a single viable prefix instead of a set of viable prefixes) holds for the DFA for the canonical set of $LR(0)$ items for a context-free grammar [7]. The construction using algorithm *Preprocess_Grammar* may be viewed as a subset construction beginning with the $LR(0)$ DFA. We begin with the start state. Suppose m is a reduce state reached during some point in the algorithm, with $m.S_l = \{A_1, A_2, \ldots A_m\}$. In the $LR(0)$ DFA we would have individual transitions on $A_1, A_2, \ldots A_n$ to different

Algorithm 19.1 Algorithm for building the controlling automaton

1: **function** *Preprocess_Grammar*
2: //The PLG G which is the input and M the controlling automaton to be constructed, are assumed to be in global data structures
3: $shift\_states = \emptyset$
4: $reduce\_states = \emptyset$
5: $q_0 = closure(\{[Z \to .S\ eof]\})$
6: initialize *list* to contain q_0
7: **while** *list* is not empty **do**
8: delete next element q from *list* and add it to $shift\_states$
9: **for** each $a \in \Sigma_c$ such that $goto(q, a)$ is not empty **do**
10: $m = closure(goto(q, a))$
11: $\delta_c(q, a) = m$
12: **if** m contains complete items **then**
13: $AddR(m)$
14: **for** each state s in $shift\_states$ **do**
15: $AddT(s, m)$
16: **end for**
17: **else** if m is not in *list* or $shift\_states$
18: append m to *list*
19: **end if**
20: **end for**
21: **for** each state r in $reduce\_states$ **do**
22: $AddT(q, r)$
23: **end for**
24: **end while**
25: **end function**

states. Here, we collect the items in all those states into a single state. (This is exactly what *ClosureReduction* does given a *reduce* state and a *shift* state that is a valid left context state). Thus the state we reach in the controlling automaton on a label sequence $S_1S_2 \ldots S_k$ is the union of states reached in the LR(0) DFA on each viable prefix in the set, and as the property holds for each of the viable prefixes in the $LR(0)$ automaton, it holds for the set consisting of the union of all the viable prefixes. □

We also have the following lemma whose proof in omitted as it follows directly from the construction.

Lemma 19.3. *Given a path in M corresponding to a set S of viable prefixes leading to some state of M, every viable prefix in the set has the same subsequence of terminals.*

Theorem 19.2. *The controlling automaton has no* shift/reduce *conflicts.*

Algorithm 19.2 The function *AddT*

```
1: function AddT(s, r)
2:     //the function checks if the shift state s is a valid left context state for
   reduce state r and if so creates the appropriate new states and adds the corre-
   sponding transitions
3:     if validlc(s, r) == true then
4:         p = ClosureReduction(s, r)
5:         δc(s, r.Sl) = p
6:         if p contains complete items then
7:             if p ∉ reduce_states then
8:                 AddR(p)
9:                 AddT(s, p)
10:            end if
11:        else
12:            if p is not in list or shift_states then
13:                append p to list
14:            end if
15:        end if
16:    end if
17: end function
```

Proof. Since the controlling automaton is constructed using a subset construction on the $LR(0)$ automaton in the manner described, a *shift/reduce* conflict can occur only if a state of the controlling automation is the union of *shift* and *reduce* states of the $LR(0)$ automaton. Thus it must contain a *shift* item of the form $I_1 = [A \rightarrow .aX_1X_2 \ldots X_n]$ where $|a| = n$ and a *reduce* item of the form $I_2 = [B \rightarrow bY_1Y_2 \ldots Y_m.]$, $|b| = m$. The item I_1 arises from a closure operation on an item of the form $I_3 = [C \rightarrow cZ_1Z_1 \ldots Z_i.A \ldots Z_l], |c| = l$. From Lemma 19.3 $b = c$ and the length of the string of variables after b and before the dot in both items must be the same. Since the arity of b is fixed I_3 is also a complete item, and therefore I_1 cannot be generated by a closure operation. □

Theorem 19.3. *If the controlling automaton has* reduce/reduce *conflicts, all handles are of the same length.*

Proof. From Lemma 19.3 a *reduce/reduce* conflict would manifest itself as two or more complete items of the form $[A \longrightarrow aX_1X_2 \ldots X_k.]$, $[B \longrightarrow aY_1Y_2 \ldots Y_k.]$ where $|a| = k$, in the same *reduce* state, as every symbol has a fixed arity and the grammar is in normal form. Therefore all handles have the same length. □

Given the fact that there are no *shift/reduce* conflicts and that all handles are of the same length, whenever a set of reductions is applicable, deterministic parsing is possible by ensuring that all reductions are carried out in synchronization and transitions are made on a *set* of left hand side non-terminals instead of a single non-terminal.

The controlling automaton for the grammar of Example 19.1 is shown in Figure 19.2. Labels on DFA states indicate which $LR(0)$ automaton states they are composed of. For a better understanding of the algorithm, and of how ambiguity is

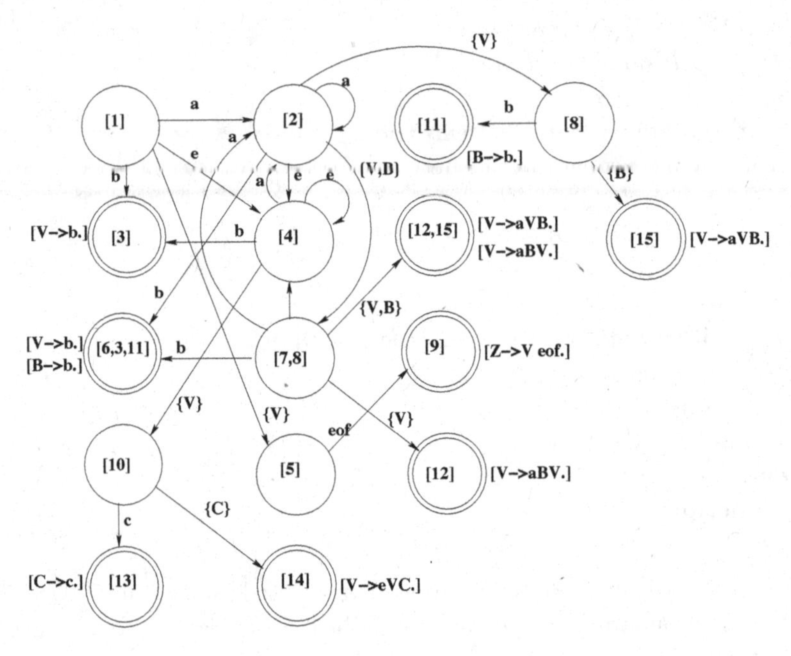

Fig. 19.2. Controlling automaton for the grammar of Example 19.1.

dealt with, we illustrate some steps of the construction. The item sets associated with states [1] and [2] of the controlling automaton are identical to those associated with states 1, and 2 of the $LR(0)$ automaton. Consider a transition out of state [2] on symbol b. The *goto* function on the set of items in state [2] yields the item set $\{[V \to b.], [B \to b.]\}$ corresponding to a *reduce* state. Let us call this state [6]. Here $[6].S_l = \{V, B\}$. Also $validlc([2],[6])$ is true as $VSet([2],[6].rhs) = \{V, B\}$. Next, $reduction([2],[6]) = \{[V \to aV.B], [V \to aB.V]\}$ and $ClosureReduction([2],[6]) = \{[V \to aV.B], [V \to aB.V], [V \to .aBV], [V \to .aVB], [V \to .eVC], [V \to .b], [B \to .b]\}$. This set of items corresponds to state [7,8] and is the union of the sets of items in states 7 and 8 of the $LR(0)$ automaton. Now $goto([7,8], b) = \{[V \to b.], [B \to b.]\}$. This is actually the union of states 3 and 11 of the $LR(0)$ automaton and the set of items is the same as that associated with state [6]. Thus we label this state [6,3,11]. Now this is a *reduce* state, and it can be checked that $validlc([2],[3,6,11])$, and $validlc([7,8],[3,6,11])$ are true. The transition on $\{V, B\}$ from state [7,8] yields state [12,15] which contains the union of the sets of items in states 12 and 15 of the $LR(0)$ DFA. Note that if S is a set of variables associated with items on which a *goto* operation could be defined from a state, not all elements of the powerset of S label edges out of the state. For example in Figure 19.2 there is no edge labeled $\{B\}$ out of state [2]. This is because state [2] is a valid left context state for the following three reduce states: [6,3,11], with $VSet = \{V, B\}$, [12,15] with $VSet = \{V\}$ and [12] with $VSet = \{V\}$. There is no reduce state r for which state [2] is a valid left context state with $VSet([2], r.rhs) = \{B\}$.

A question that naturally arises here is with regard to the size of the controlling automaton and the complexity of the preprocessing algorithm. The controlling automaton state space size is bounded by a quantity that is exponential in the grammar size, though in practice the size is usually tractable [19]. Let $|G|$ be the size of the grammar, i.e. the sum of the lengths of all the right hand sides of productions. It can be shown that the number of states $|Q|$ of the controlling automaton is loosely bounded by $2^{|G|}$, and the preprocessing time is $O(|Q|^2|\Sigma||N||G|$ [19].

19.3.4. *The PCH Compression and Decompression Procedures*

We assume that the input has gone through a lexical analyzer so that the symbols that are read by the compressor are tokens returned by the lexical analyzer. The algorithms PCH_Compressor and PCH_Decompressor use a function *GetContext(stack)* which converts a sequence of states on the parsing stack *stack* to a context. We assume that the parsing stack and the controlling automaton M are stored in global data structures and that the order of the arithmetic compressor is fixed and not part of the input. The functions *push*, *pop* and *top* perform the usual stack operations. The function *reduce* used during compression(not to be confused with the function *reduction* which is used during the preprocessing phase) is defined in Algorithm 19.3

Algorithm 19.3 The function *reduce*

```
1: function reduce(state)
2:     //this function performs simultaneously, a set of reductions corresponding
   to a set of complete items in state state and returns the state reached after
   reduction
3:     for i == 1 to state.arity do
4:         pop(stack)
5:     end for
6:     lcstate = top(stack)
7:     //The state lcstate is the left context state exposed on stack after popping
   off a set of handles during reduction
8:     state := δ_c(lcstate, state.S_l)
9:     push(state, stack)
10: end function
```

Algorithm 19.4 The PCH Compressor

```
1: function PCH_Compressor(w)
2:     //the input string w is assumed to be in a file accessed via FH
3:     CurrentState = q0
4:     push(CurrentState, stack)
5:     // f is the final state of the controlling automaton
6:     while (CurrentState ≠ f) do
7:         Symbol = GetNextSymbol(FH)
8:         CurrentContext = GetContext(stack)
9:         ArithEncode(CurrentContext, Symbol)
10:        CurrentState = δc(CurrentState, Symbol)
11:        while (CurrentState is a reduce state) do
12:            CurrentState = reduce(CurrentState)
13:        end while
14:    end while
15:    //Write the final fraction F which encodes the input
16:    Write(F)
17: end function
```

Algorithm 19.5 The PCH Decompressor

```
1: function PCH_Decompressor
2:     //the output w is written into a file accessed by FH
3:     CurrentState = q0
4:     push(CurrentState, stack)
5:     while (CurrentState ≠ f) do
6:         CurrentContext = GetContext(stack)
7:         ArithDecode(CurrentContext, Symbol)
8:         write(Symbol, FH)
9:         CurrentState = δc(CurrentState, Symbol)
10:        while (CurrentState is a reduce state) do
11:            CurrentState = reduce(CurrentState)
12:        end while
13:    end while
14: end function
```

We now show how the parser can be interfaced with the arithmetic coder using the functions *ArithEncode* and *ArithDecode*, assuming that the fraction representing the encoded representation that is modified by both the functions is stored in a global variable F. We assume that $M = (Q, \Sigma_c, \delta_c, q_0, \{f\})$ is the controlling automaton. Algorithms 19.4 and 19.5 describe the compressor and the decompressor.

We now illustrate with a small example. To keep things simple, we identify state sequences used as contexts, with viable prefixes, and therefore will use the latter to illustrate our example.

Consider the toy grammar of Example 19.1 augmented with the production $Z \to V\ eof$ where Z is the new start symbol. We refer to the controlling automaton given in Figure 19.2. Consider the two subject trees with preorder listing $ea\underline{abbb}c$

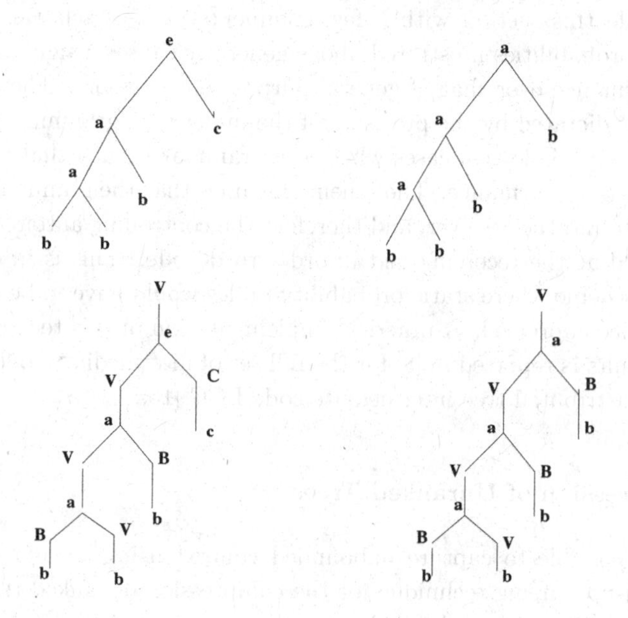

Fig. 19.3. Two subject trees along with their derivation trees.

and $aa\underline{abbb}b$ shown in Figure 19.3, along with an S-derivation tree in the tree grammar for each. The context for a conventional order-4 arithmetic compressor just prior to encoding the last symbol is shown underlined, from which it follows that the compressor will use the same table for encoding the c in the first tree and the last b in the second. However the scheme proposed here using the PLG for the tree grammar, will generate a stack context eV for the symbol c and aV for the symbol b. These take the controlling automaton to states [10] and [8] respectively of the controlling automaton in Figure 19.2. At each of these states the symbols c and b are the only ones expected and hence do not need to be encoded at all, as they will be generated by the decoder. Further, the modified context which permits this refinement in probabilities has length just one symbol, namely the state on top of stack.

An observation worth mentioning here is that the viable prefix that is used as the context for encoding and decoding the symbol at node n encodes both, the symbols on the path from the root to n in the subject tree, (these constituting precisely the subsequence of terminals in the viable prefix), as well as the maximal subtrees to the "left" of n (these being condensed into the subsequence of variables in the viable prefix). We could as well term the context "the top left context" of a node in the derivation tree. It summarizes all the information in nodes scanned before the node at which the symbol is to be encoded. Since arbitrarily large subtrees to the left are condensed into variables, the compressor is, in fact, tracking unbounded context.

We conclude this section with a few comments on this scheme. Firstly, the refinement of probabilities illustrated above generally causes a significant improvement in performance over that of general purpose compressors. The extent of the improvement is dictated by the precision of the underlying grammar. One could of course construct pathological cases where the grammar is such that it could cause a deterioration in performance. The scheme assumes that the grammar is known to both the sender and the receiver, and therefore the controlling automaton will have to be generated at the receiving end in order to decode. This is in contrast to a non-adaptive scheme where static probability tables would have to be prepended to the file to be decompressed. A practical implementation of this technique with experimental results is reported in [8] for the full set of intermediate code instructions of the ranked attributed tree intermediate code LCC [10].

19.4. Compression of Unranked Trees

While it was possible to capture unbounded context using an adaptation of the $LR(0)$ bottom-up parsing technique for the compression of ranked trees, the technique cannot be directly used for the compression of unranked trees, unless one uses a different encoding. In this section we adapt a top-down parsing technique for the compression of unranked trees represented as a textual stream. In particular, we concentrate on streaming XML documents. In general, unranked trees are generated by a regular tree grammar $G = (N, \Sigma, S, P)$ where the elements of Σ are unranked and the right hand side of a production is of the form $a(r)$ where a is an element of Σ and r is a regular expression over N. As mentioned in Chapter 3, restrictions of regular tree grammars to local grammars correspond to Document Type Definitions (DTD's) used to define XML documents [6]. Here the nonterminals are identified with the symbols of Σ and the rules are of the form $a \longrightarrow a(r)$ where r is a regular expression over Σ. The hierarchical structure of XML files is effectively defined by such grammars.

XML documents contain *element tags* which include start tags like `<name>` and end tags like `</name>`. Elements can nest other elements and therefore a tree structure can be associated with an XML document. Since XML documents are governed by a rather restrictive set of rules, it is possible to use the rules to predict what tags to expect. Further if the rules are already known a-priori, then the compressor which is tuned to take advantage of the rules can be generated directly from the rules themselves. We describe such a scheme in this section. We assume that the DTD describing the data is known to both the sender and the receiver. In addition to an element of a DTD consisting of distinct beginning and ending tags enclosing regular expressions over other elements, it can also contain plain text, comments and special instructions for XML processors ("processing instructions"). Opening element tags can additionally have attributes with values. These together constitute the *content* of the document. An example of an XML DTD and a document conforming to the DTD is given below:

Example 19.2. Consider a DTD defined as follows:

```
<!DOCTYPE addressBook[
<!ELEMENT addressBook(card*)>
<!ELEMENT card((name|(givenName,familyName)),email, note?)>
<!ELEMENT name(#PCDATA)>
<!ELEMENT givenName(#PCDATA)>
<!ELEMENT familyName(#PCDATA)>
<!ELEMENT email(#PCDATA)>
<!ELEMENT note(#PCDATA)>
]>
```

The *structural* content of the DTD is given below.

```
<!DOCTYPE addressBook[
<!ELEMENT addressBook(card*)>
<!ELEMENT card((name|(givenName,familyName)),email, note?)>
<!ELEMENT name>
<!ELEMENT givenName>
<!ELEMENT familyName>
<!ELEMENT email>
<!ELEMENT note>
]>
```

It is easy to put this structural part in the form of a local grammar where the first element in each rule can be associated with the left hand side of a production and the rest of it is the regular expression on the right hand side. There is also the textual part which contains information like attributes and PCDATA. For our purpose these components are streams of characters.

Below is an instance of an XML document conforming to this DTD.

```
<addressBook>
<card>
<givenName>John</givenName>
<familyName>Smith</familyName>
<email>jsmith@rediffmail.com</email>
</card>
<card>
```

```
<name>Mary King</name>
<email>mking@gmail.com</email>
<note>John's advisor</note>
</card>
</addressBook>
```

The structural content of the document is given below:

```
<addressBook>
<card>
<givenName></givenName>
<familyName></familyName>
<email></email>
</card>
<card>
<name></name>
<email></email>
<note></note>
</card>
</addressBook>
```

The part that does not make up structure is the content of the document.

Every well formed XML document corresponds to a tree structure. Figure 19.4 is the tree for the XML document in Example 19.2.

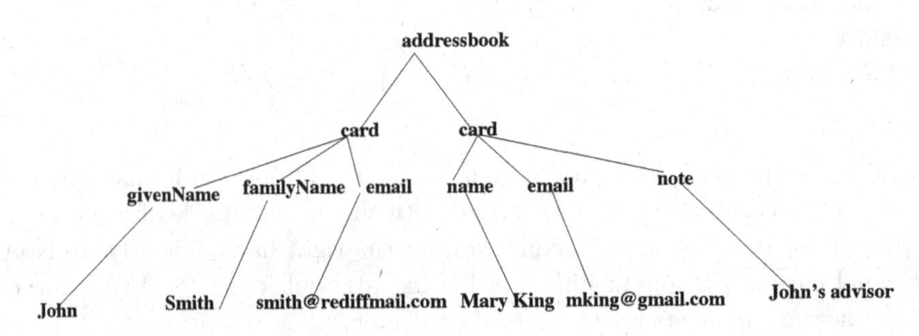

Fig. 19.4. Tree structure of the XML document in Example 19.2.

It is useful to have the compressor satisfy the following properties:

(1) It should work in a streaming environment,that is, the document should be compressed as its textual form is read in.
(2) Structure should be separated from content and each should be compressed separately.

(3) There should be a facility for multiplexing models for content, that is, it should be possible to shift from one statistical model to another, depending on which part of the input file is being compressed. The models could be *path sensitive*, i.e. be distinct for each root to element path, or be *path agnostic*, i.e. be unique for content associated with an element, but not distinct for every *instance* of the element in the DTD. Note that this classification of models has nothing to do with the *order* of the model, which we recall is the number of preceding symbols that are kept track of to compute probabilities.

It turns out that there is a rather simple top-down parsing model that can handle all these requirements in a most natural manner. The parser is very similar to *recursive descent parsers* which are among the oldest known class of parsers [20]. The notation for a DTD is highly reminiscent of that for *extended context free grammars* [20], where the left hand sides of rules are variables and right hand sides regular expressions over terminals and variables. If an extended context-free grammar satisfies a condition known as the extended LL(1) condition, then a recursive descent parser can be used to deterministically parse the input. This technique maps each variable of the grammar into a procedure, and the body of the procedure mimics the regular expression structure on the right hand side of the production associated with the variable. Variables are mapped to procedure calls, terminals are mapped into statements that check for that terminal in the input, and other meta symbols like $*, +$ and ? map to interative and control flow constructs. Procedures recursively call one another, the call stack effecting proper transfer of control whenever a return is executed. The extended LL(1) condition just stipulates that if there are alternates on the right hand side of a rule, then it should be possible to disambiguate a choice by just inspecting the the next input symbol.

The compression strategy we propose here is specifically tuned to parsing XML file structure, keeping in mind the desired properties of the compressor listed earlier. The technique converts the regular expression associated with each element into a DFA. The DFA takes the place of the body of the procedure in a recursive descent parser, and transitions on element opening tags take the place of procedure calls. Since each element has a unique open tag, disambiguation of choices of alternates is guaranteed. A call stack is maintained to effect proper returns which are executed when element closing tags are seen.

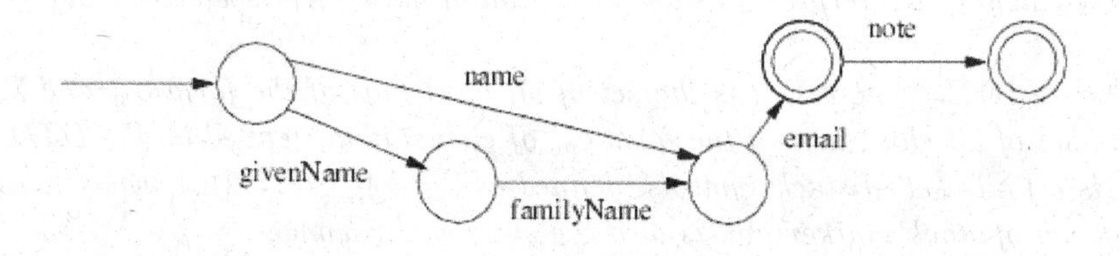

Fig. 19.5. DFA for the element card in Example 19.2.

The DFA for the right hand side of the rule for element card is shown in Figure 19.5. There are two kinds of states in this automaton, those having a single output transition and those with multiple output transitions. If one were to traverse such a DFA to generate probabilities while parsing, opening tags of symbols that label single output transitions need not be encoded as their probability is 1. Thus encoding of symbols by the arithmetic compressor needs to be performed only at states with more than one outgoing transition. A statistical model can be maintained at each such state for each element.

19.4.1. *Recursive Finite Automata*

A streaming XML document structure can be parsed by a by a set of mutually recursive finite automata constructed from the DTD which we now define. (The model we propose appears to be closely related to the modular k-SEVPA's defined in [21, 22].) The automata are defined as mutually recursive as they can recursively invoke one another.

For an arbitrary XML document represented as a string w of text, let $h(w)$ represent the homomorphism that maps the string into the subsequence that defines its structure. Thus $h(w)$ consists of a well formed sequence of open/close tags. If there are n distinct elements in the DTD definition, associate with each element of the DTD a unique index i, $1 \leq i \leq n$ with index 1 associated with the root element. For the rest of this section, we identify elements with their indices. The model we define below is specific to alphabet sets associated with the structure of XML documents.

Definition 19.10 (Recursive Finite Automata). *A set of mutually recursive finite automata R is defined as $R = \{M, \Sigma, \Gamma, \Delta, < 1, 0 >, \bot, \{a\}\}$ where:*

(1) M is a tuple $\langle M_1, M_2, \ldots M_n \rangle$, where each component of the n-tuple is a deterministic finite state automaton $M_i = (Q_i, \Sigma_i, \delta_i, \langle i, 0 \rangle, F_i)$. DFA M_i is associated with element i. The set Q_i is a finite set of states of the form $\langle i, l \rangle, 1 \leq i \leq n$, $0 \leq l < n_i$ where $n_i = |Q_i|$, Σ_i is the input alphabet and is the subset of elements occurring in the regular expression for element i. The state $\langle i, 0 \rangle$ is the start state of M_i, F_i is a subset of Q_i and is the set of final states of M_i; the transition function δ_i is a mapping from $Q_i \times \Sigma_i$ to Q_i. A transition on element e is interpreted as an invocation of the DFA associated with element e.

(2) $\Sigma = \Sigma_{op} \cup \Sigma_{cl}$ where Σ_{op} is the set of all open tags of the form o_m and Σ_{cl} is the set of all close tags of the form c_m of elements m defined by the DTD.

(3) Γ is a finite set of stack symbols, defined as $\Gamma = \bigcup_{i=1}^{n} Q_i \cup \{\bot\}$ where $\bot$ is the bottom of stack marker and is also the start stack symbol.

(4) Δ is a transition function which is a map: $(Q \times \Sigma \times (\Gamma \cup \{\epsilon\})) \mapsto (Q \times (\Gamma \cup \{\epsilon\}))$ where $Q = \bigcup_{i=1}^{n} Q_i$. The transition function is the union of three types of

transitions, i.e. $\Delta = \Delta_{op} \cup \Delta_{cl} \cup \Delta_f$ *where*

(a) *The set* Δ_{op} *is the set of* push *transitions on open tags.* $\Delta_{op}(\langle i, j\rangle, o_m, \epsilon) = (\langle m, 0\rangle, \langle i, j\rangle)$ *whenever* $\delta_i(\langle i, j\rangle, m)$ *is defined. Such transitions on encountering the open tag for element* m*, while in state* j *of DFA* i*, are independent of the stack top, push the current state on the call stack and transit to the start state of the DFA for element* m*.*

(b) *The set* Δ_{cl} *is the set of pop transitions on close tags and is defined as follows:* $\Delta_{cl}(\langle i, f\rangle, c_i, \langle l, r\rangle) = (\delta_l(\langle l, r\rangle, i), \epsilon)$ *whenever* $\delta_l(\langle l, r\rangle, i)$ *is defined. Such transitions are taken when the closing tag of element* i *is seen in the final state* f *of element with index* i*. In such a case the DFA state on top of stack,* $\langle l, r\rangle$ *is popped and a transition is taken in the DFA for element* l*, from state* r *on element* i*.*

(c) *The function* $\Delta_f(\langle 1, f\rangle, c_1, \perp) = (a, \perp)$ *where* f *is a final state of the DFA corresponding to the root element of the XML document (which by assumption has index 1)*

(5) *The state* a *is the accepting state of* R*.*

(6) *State* $\langle 1, 0\rangle$ *is the start state of the DFA representing the root element and is also the start state of* R*.*

A *configuration* of R is a triple (*state, stack, string*) where *state* is the state where R is currently stationed, *stack* represents the content of the call stack, which initially contains only the bottom of stack marker, and *string* represents $h(w)$ where w is the unconsumed suffix of the input string (which is the streaming XML document to be compressed.)

A *move* $\vdash$ is a binary relation on configurations. We say that $(q, aw, \alpha Z) \vdash (p, w, \alpha\ Y)$ if $\Delta(q, a, Z) = (p, Y), a \in \Sigma,\ Y, Z \in \Gamma \cup \{\epsilon\}$ where the top of stack is assumed to be at the right end of the string representing the stack contents. Note that the transition function defined ensures that the device is deterministic. Let $\vdash^*$ denote the reflexive and transitive closure of $\vdash$.

Assume that the current configuration of R is $(\langle i, l\rangle, o_m s, \alpha)$, where o_m is an open tag for element m and s is $h(v)$, where v is the suffix of the input XML string after o_m. When the open tag for element m is encountered in the stream, the pair $\langle i, l\rangle$ is pushed on the call stack and the start state 0 of the DFA for the element m is entered. The current configuration of R now becomes $(\langle m, 0\rangle, s, \alpha\langle i, l\rangle)$. When the matching closing tag c_m is encountered for element m, the stack is popped and the new configuration of R becomes $(\langle i, k\rangle, s', \alpha)$ where $\delta_i(\langle i, l\rangle, m) = \langle i, k\rangle$, and s' is the suffix of $h(v)$ following c_m. Each transition of a DFA on an element m thus encapsulates a *call* to the program simulating the DFA for m and a subsequent *return* after parsing the portion of the input stream corresponding to the instance of element m in the structure. We note that on a closing tag c_m, there are two implicit transitions: the first from a final state of the DFA for element m to the state say, $\langle i, l\rangle$ on top of the call stack, and then one from the state $\langle i, l\rangle$ in the DFA for element i to the state transited to on m. The first transition depends on the calling context and cannot be determined statically.

The compression for an XML document proceeds hand in hand with the parsing. A statistical model is maintained at each state of each DFA of R. These models are used to compress the *structure* of the document. In addition a model for *content* is associated with each element of the DTD. Symbols are directed to the appropriate model under the direction of R, that is, the current model is always the one associated with the current state of R. One can, in addition, use the content of the parsing stack to define a state which encodes the root to node path in the tree representing the structure of the document.

Definition 19.11 ($OP(\gamma)$). *Let* $\gamma = \langle i_1, j_1\rangle\langle i_2, j_2\rangle \ldots \langle i_l, j_l\rangle$ *be a sequence of states of* R. *Define* $OP(\gamma)$ *as the sequence of opening tags for elements* $i_1, i_2, \ldots i_l$.

Definition 19.12 ($TR(\langle i, j\rangle)$). *Define* $TR(\langle i, j\rangle) = \{\beta | \delta_i(\langle i, 0\rangle, \beta) = \langle i, j\rangle\}$, *that is the set of all strings of elements that take DFA* i *from state 0 to state* j.

The following theorem shows that the recursive finite automata defined above for a given DTD recognize all and only the well formed structural subsequences of XML documents specified by the DTD.

Theorem 19.4. *Let* D *be a DTD for an XML document with structure defined by a local grammar* G, *and let* R *be the set of recursive finite automata defined for* G. *Then an input XML stream* w *has* $h(w) \in L(G)$ *iff* $\Delta(\langle 1, 0\rangle, h(w), \perp) \vdash^* (a, \epsilon, \perp))$.

Proof. Let x be a prefix of $h(w)$, with $xy = h(w)$ where w is a syntactically well formed XML document. We show that $(\langle 1, 0\rangle, xy, \perp) \vdash^* (\langle i, j\rangle, y, \perp\gamma)$ where $\gamma = \langle i_1, j_1\rangle\langle i_2, j_2\rangle \ldots \langle i_l, j_l\rangle$ iff

(1) $OP(\gamma)o_i$ is the sequence of unmatched open tags in x
(2) the sequence of elements u corresponding to matched open and close tags after unmatched open tag o_k and before the next unmatched open tag o_{k+1} if any, $1 \leq k < l$ is in $TR(\langle i_k, j_k\rangle)$, i.e. $\delta_{i_k}(\langle i_k, 0\rangle, u) = \langle i_k, j_k\rangle$.

The only if part can be shown by an easy induction on the number of moves l of R. The if part can be shown by induction on the length l of $h(w)$.

We therefore have $x = h(w)$ where $h(w)$ is a well formed structural subsequence of an XML document iff $(\langle 1, 0\rangle, x, \perp) \vdash^* (\langle 1, f\rangle, c_1, \perp)$. Finally via a move in Δ_f $(\langle 1, f\rangle, c_1, \perp) \vdash (a, \epsilon, \perp)$ proving the theorem. □

19.4.2. *Compression and Decompression of XML Documents*

The compression and decompression schemes are specified in Algorithms 19.4.2 and 19.4.2 respectively.

A typical sequence of actions for a path-agnostic compression scheme could then be as follows: The start state of a DFA representing the right side of a rule is

Algorithm 19.6 The XML Compression Algorithm

```
1: function XML_COMPRESSOR(w)
2:     //The input w is stored in a file accessed by FH
3:     ExitLoop = false
4:     //CurrentState is a record with fields ElemIndex and StateIndex
5:     //ElementIndex is the index for each DFA and is 1 for the root element
6:     //StateIndex is the index of a state in each DFA and is 0 for the start state
7:     //begin in the start state of the root element
8:     CurrentState.ElemIndex = 1
9:     CurrentState.StateIndex = 0
10:    while ExitLoop == false do
11:        Symbol = GetNextSymbol(FH)
12:        //Symbol is a record containing a field Type indicating whether the
           symbol represents an open or close tag or PCDATA and a field Value containing
           the element name or the actual character content.
13:        CurrentContext = GetContext(CurrentState)
14:        ArithEncode(CurrentContext, Symbol)
15:        if Symbol.Type == OPENTAG then
16:            // make transition of R on an open tag
17:            push(CurrentState, stack)
18:            CurrentState.ElementIndex = GetElementIndex(Symbol)
19:            CurrentState.StateIndex = 0
20:        else
21:            if Symbol.Type == CLOSETAG then
22:                e = CurrentState.ElementIndex
23:                //e is the element instance that has just been parsed
24:                //if the bottom of stack element is reached parsing is over
25:                if (top(stack) == ⊥) then
26:                    ExitLoop = true
27:                else
28:                    CurrentState = pop(stack)
29:                    i = CurrentState.ElementIndex
30:                    //R is now stationed at a state of DFA i
31:                    CurrentState.StateIndex = δ_i(CurrentState.StateIndex, e)
32:                end if
33:            else
34:                if Symbol.Type == PCDATA then
35:                    Text = Symbol.Value
36:                    CurrentContext = GetPcdataContext(CurrentState)
37:                    ArithEncode(CurrentContext, Text)
38:                end if
39:            end if
40:        end if
41:    end while
42: end function
```

Algorithm 19.7 The XML Decompression Algorithm

```
1: function XML_DECOMPRESSOR(R)
2:     //F, the fraction representing the compressed file is assumed to be stored
   in a global variable
3:     ExitLoop = false
4:     CurrentState.ElementIndex = 1
5:     CurrentState.StateIndex = 0
6:     CurrentContext = GetContext(CurrentState)
7:     while ExitLoop == false do
8:         //Decode and modify the fraction F using CurrentContext and store
   the decoded symbol in Symbol
9:         ArithDecode(CurrentContext, Symbol)
10:        if Symbol.Type == OPENTAG then
11:            push(CurrentState, stack)
12:            CurrentState.ElementIndex = GetElementIndex(Symbol)
13:            WriteOpenTag(ElementIndex)
14:            CurrentState.StateIndex = 0
15:            CurrentContext = GetContext(CurrentState)
16:        else
17:            if Symbol.Type == CLOSETAG then
18:                e = CurrentState.ElementIndex
19:                //e is an element that has just been parsed
20:                WriteCloseTag(e)
21:                if Stack.top == ⊥ then
22:                    ExitLoop = true
23:                else
24:                    CurrentState = pop(stack)
25:                    i = CurrentState.ElementIndex
26:                    // R is now stationed at a state of DFA i
27:                    CurrentState.StateIndex = δ_i(CurrentState.StateIndex, e)
28:                end if
29:            else
30:                if Symbol.Type = PCDATA then
31:                    CurrentContext = GetPcdataContext(CurrentState)
32:                    ArithDecode(CurrentContext, Text)
33:                    WriteText(Text)
34:                end if
35:            end if
36:        end if
37:    end while
38: end function
```

entered; if there is only one edge out of the state there is no action; if that element has a `#PCDATA` attribute then the string of symbols associated with PCDATA for that element is encoded using the frequency tables for PCDATA for that element; if there is more than one edge out of the state, the open tag of the element labeling the edge taken is encoded using the statistical model for that state, and a transition is made to the the start state of the DFA for the element whose open tag is just seen. When a close tag of an element is seen, it is always at the final state of the DFA for that element, and labels the transition out of it to the state on top of the call stack. The DFA state, say $\langle i, j \rangle$ on top of stack is popped and a transition is made from $\langle i, j \rangle$ on the element whose closing tag has just been seen. The decoder mimics the action of the encoder generating symbols that are certain and using the appropriate model for decoding symbols that are not. If a path sensitive scheme is used, then each distinct context on the parsing stack leads to a different statistical model for data associated with an element. (As observed in [9] this tends to blow up the dynamic memory needed to store the context tables for large files.)

The properties desired of the compressor are satisfied in a natural manner so to speak. A symbol stream is the form in which parsers expect to see the input. Separation of structure from content and multiplexing of models is achieved by just having separate models, the direction of symbols to the correct model being effected by R. The compressor can be generated from the DTD and is *generic* in that it can be used to compress any XML document that conforms to the DTD. In this respect the scheme is similar in spirit to the parser generator schemes like YACC [15]. Experimental results from an implementation of this technique are reported in [9].

19.5. Conclusion

In this chapter we propose syntax directed compression schemes for the compression of files consisting of trees whose structure is governed by a well defined syntax. We consider both ranked and unranked trees, and show that bottom-up $LR(0)$-like parsers simulating pushdown automata are appropriate devices for compressing ranked trees, and top-down recursive-descent-like parsers simulating recursive finite automata are suitable for the compression of unranked trees. Multiplexing of statistical models can be achieved in a natural manner guided by the automata. There are several schemes that underlie special purpose compressors for tree sequences representing intermediate code [12, 23, 24, 11]. Some schema-aware techniques for compressing XML documents are described in [25, 26]. The scheme described here works at a level of abstraction, where the compressor is generated directly from a syntactical specification of the file structure, with no extra user inputs. The tools DTDPPM [27] and `rngzip` [25] use the DTD and schema respectively to generate an XML conscious compressor. An important issue is that of querying a compressed XML file without having to decompress it completely, while using an adaptive compression scheme. How to modify the approach described here to achieve this in a streaming environment is an interesting problem for future work.

Acknowledgements

The author thanks V. Vinay, Helmut Seidl and M. Raj Mohan for their helpful comments and suggestions on the chapter.

References

[1] J. Ziv and A. Lempel, A universal algorithm for sequential data compression, *Trans. Information Theory.* **23**(3), 337–343, (1977).
[2] D. Huffman, A Method for the Construction of Minimum Redundancy Codes, *Proc. Institute of Radio Engineers.* **40**, 1098–1101, (1952).
[3] D. Knuth, Dynamic Huffman Coding, *J. Algorithms.* **6**, 163–180, (1985).
[4] J. Rissanen and G. G. Langdon, Arithmetic coding, *IBM J. Research And Development.* **23**(2), 149–162, (1979).
[5] I. H. Witten, R. M. Neal, and J. G. Cleary, Arithmetic coding for data compression, *Commun. ACM.* **30**(6), 520–540, (1987).
[6] XML. W3C recommendation. http://www.w3.org/TR/REC-xml, (2004).
[7] J. E. Hopcroft and J. D. Ullman, *Introduction to Automata Theory, Languages and Computation.* (Addison-Wesley, 1979).
[8] S. Rai and P. Shankar, Efficient Statistical Modeling for the Compression of Tree Structured Intermediate Code, *The Computer Journal.* **45**(6), 476–486, (2003).
[9] H. Subramanian and P. Shankar. Compressing XML documents using recursive finite automata. In *Proc. 11th CIAA*, vol. 3845, *LNCS*. Springer, (2006).
[10] C. Fraser and D. Hanson, *A Retargetable C Compiler: Design and Implementation.* (Addison-Wesley Longman, 1995).
[11] W. Evans. Compression via guided parsing. In *Proc. 8th DCC*, p. 544. IEEE, (1998).
[12] C. W. Fraser. Automatic inference of models for statistical code compression. In *Proc. PLDI 2009*, pp. 242–246. ACM, (1999).
[13] W. Pugh. Compressing Java class files. In *Proc. PLDI 1999*, pp. 247–258. ACM, (1999).
[14] R. N. Horspool and J. Corless, Tailored compression of Java class files, *Software–Practice and Experience.* **28**(12), 53–68, (1998).
[15] S. Johnson. Yacc – yet another compiler compiler. Technical Report TR 32, AT & T, Bell Laboratories, New Jersey, USA, (1975).
[16] A. Balachandran, D. Dhamdhere, and S. Biswas, Efficient retargetable code generation using bottom up tree pattern matching, *Computer Languages.* **3**(15), 127–140, (1990).
[17] C. Hoffman and M. J. O'Donnell, Pattern matching in trees, *J. ACM.* **29**(1), 68–95, (1982).
[18] D. Chase. An improvement to bottom up tree pattern matching. In *Proc. 14th POPL*, pp. 168–177. ACM, (1987).
[19] P. Shankar, A. Gantait, A.Yuvaraj, and M. Madhavan, A new algorithm for linear regular tree pattern matching, *Theoretical Computer Science A.* **242**, 125–142, (2000).

[20] R. C. Backhouse, *Syntax of Programming Languages - Theory and Practice.* (Prentice-Hall, 1979).

[21] R. Alur, V. Kumar, P. Madhusudan, and M. Viswanathan. Congruences for Visibly Pushdown Languages. In *Proc. 32nd ICALP*, vol. 3580, *LNCS*, pp. 1102–1114. Springer, (2005).

[22] V. Kumar, P. Madhusudan, and M. Viswanathan. Visibly Pushdown Automata for Streaming XML. In *Proc. 16th WWW*, pp. 1053–1062. ACM, (2007).

[23] M. Franz. Adaptive compression of syntax trees and iterative dynamic code optimization: Two basic technologies for mobile object systems. In *Mobile Object Systems: Towards the Programmable Internet*, pp. 263–276. Springer, (1997).

[24] J. Ernst, W. Evans, C. Fraser, S. Lucco, and T. Proebsting. Code Compression. In *Proc. PLDI*, pp. 358–365. ACM, (1997).

[25] C. League and K. Eng, Schema-based compression of XML data with relax NG, *J. Computers.* **2**(10), 9–17, (2007).

[26] M.Girardot and N. Sundaresan, Millau: An encoding format for efficient representation and exchange of data over the Web, *Computer Networks.* **33**(1-6), 747–765, (2000).

[27] DTDPPM. http://xmlppm.sourceforge.net/dtdppm/index.html.

Chapter 20

Weighted Finite Automata and Digital Images

Kamala Krithivasan
Department of Computer Science and Engineering
Indian Institute of Technology Madras, Chennai 600 036.
kamala@iitm.ac.in

Y. Sivasubramanyam
Infosys Technologies Limited
Chennai 600 119.

A finite resolution digital image (gray-scale) is considered to be a function with domain the set of strings of a finite alphabet, and range the grayness value (real) of a pixel. A finite automaton with weights can be used to represent such a function. The use of weighted finite automata (WFA) for the representation of real valued functions was proposed by Culik and Karmuhaki in 1991. Subsequently, the compression of digital images using these devices was studied by Culik and Kari. In this paper we first consider the representation of black-white digital images using WFAs. We show how finite state transducers can be used to effect transformations such as scaling, translation, rotation, etc, on images represented by WFAs. We then show how WFAs can be used to represent digital gray-scale images. We discuss the inference and de-inference algorithms and factors affecting the compression obtained by this representation. We finally demonstrate how WFA can be used to represent 3D objects and effect transformations on them.

20.1. Introduction

Weighted finite automata (WFA) and their mathematical properties were studied in the work of Culik and Karhumaki in the early nineties [1]. In a subsequent paper Culik and Kari [2] showed how these devices could be used to define the grayness functions of graytone images and their application to digital image compression. Culik and Fris [3] proposed the use of multiple tape WFA which they called weighted finite transducers (WFT) for performing image transformations on digital images.

We begin this chapter with some basic definitions and notations for weighted finite automata and the representation of digital images using WFA. We then show how these devices can be used for transformations and compression of digital images.

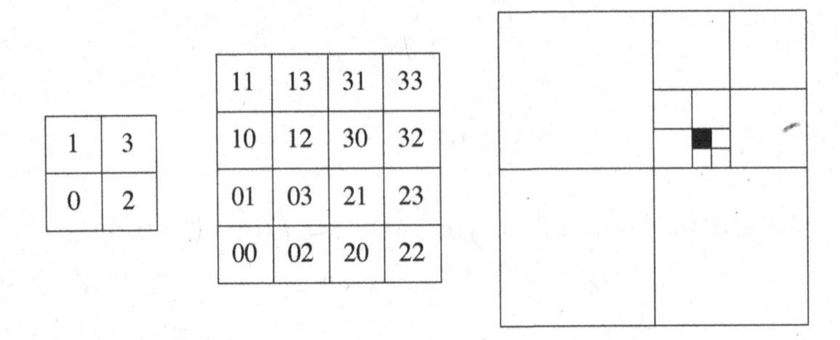

Fig. 20.1. Addressing the subsquares.

A digital image of finite resolution $m \times n$ consists of $m \times n$ pixels each of which is assigned a value corresponding to its colour or grayness. In this section we consider only square images of resolution $2^n \times 2^n$.

The $2^n \times 2^n$ pixels can be considered to form a bound square on a two dimensional space with x and y orthogonal axes. Thus, the location of each of the $2^n \times 2^n$ pixels can be specified by a tuple (x, y) representing its x and y coordinates. Hereafter we will call this tuple, the address of the pixel. The address tuple (x, y) is such that $x \in [0, 2^n - 1]$ and $y \in [0, 2^n - 1]$. Hence we can specify the x (y) coordinate as an n-bit binary number.

In our representation, the address of any pixel at (x, y) is specified as a string $w \in \Sigma^n$, where $\Sigma = \{0, 1, 2, 3\}$. If the n-bits representation of x coordinate is $x_{n-1}x_{n-2}\ldots x_1x_0$ and the n-bits representation of y coordinate is $y_{n-1}y_{n-2}\ldots y_1y_0$, then the address string $w = a_{n-1}a_{n-2}\ldots a_1a_0$ such that $a_i \in \Sigma$ and $a_i = 2x_i + y_i$, $\forall i \in [0, 2^n - 1]$. The addresses of the pixels of a 4×4 image are as shown in Figure 20.1.

Another way of getting the same address is to consider a unit square whose subsquares are addressed using 0, 1, 2, 3 as shown in Figure 20.1 and the address of the subsubsquare as a concatenation of the address of the subsquare and the subsubsquare. For example, the address of the darkened square in Figure 20.1 would be 3021. Its coordinates are (10, 9) equal to (1010, 1001) in binary. Putting $a_i = 2x_i + y_i$ we get 3021.

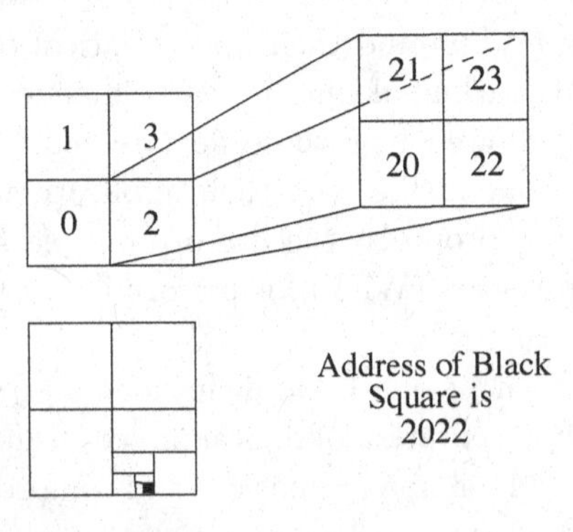

20.2. Finite Automata and Black-White Images

We give here the representation of black-white images using finite automata [4].

In order to represent a black-white square image using a finite state automaton, we specify a Boolean function $f : \Sigma \to \{0,1\}$ such that $f(w) = 1$ if the value of the pixel addressed by w is black and $f(w) = 0$ if the value of the pixel addressed by w is white for $w \in \Sigma = \{0,1,2,3\}$.

Definition 20.1. The the FSA representing the $2^n \times 2^n$ resolution black-white image is a nondeterministic FSA, $A = (K, \Sigma, \delta, I, F)$ where

Σ *is a finite set of alphabet* $\{0,1,2,3\}$.
K *is a finite set of states.*
δ *is a transition function as defined for a non-deterministic FSA.*
$I \subseteq K$ *is a set of initial states. This can be equivalently represented as single initial state* q_0 *with* ϵ *transitions to all states in* I.
$F \subseteq K$ *is a set of final states.*

The language recognized by A, $L(A) = \{w | w \in \Sigma^n, f(w) = 1\}$ i.e. the language recognized by the FSA consists of the addresses of the black pixels.

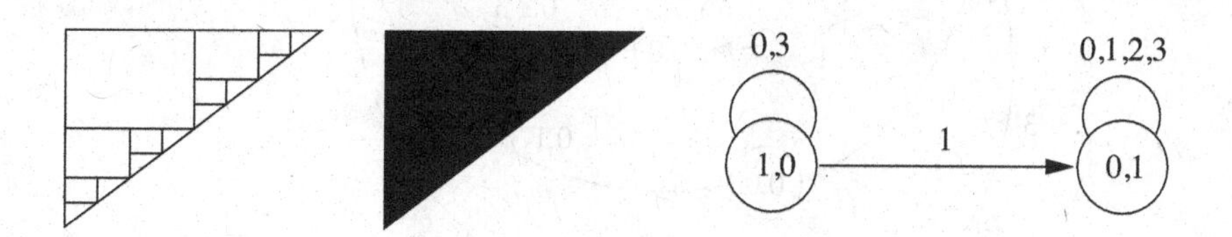

Fig. 20.2. Finite State Automaton for a Triangle.

For example, consider the image shown in Figure 20.2. The addresses of the black squares form a language $L = \{0,3\}^*1\{0,1,2,3\}^*$. Thus the FSA representing this image is an FSA which accepts the language L as shown in Figure 20.2.

Definition 20.2. Another way to represent the same non-deterministic FSA of m states is as follows:

(1) a row vector $I^A \in \{0,1\}^{1\times m}$ *called the initial distribution* ($I^A_q = 1$ *if* q *is an initial state, 0 otherwise).*
(2) a column vector $F^A \in \{0,1\}^{m\times 1}$ *called the final distribution* ($F^A_q = 1$ *if* q *is a final state, 0 otherwise).*
(3) a matrix $W^A_a \in \{0,1\}^{m\times m}, \forall a \in \Sigma$ *called the transition matrix* ($W^A_{a_{p,q}} = 1$ *if* $q \in \delta(p,a)$, *0 otherwise).*

This FSA, A, defines the function $f : \Sigma^n \to \{0,1\}$ by

$$f(a_{n-1}a_{n-2}\dots a_1 a_0) = I^A \cdot W^A_{a_{n-1}} \cdot W^A_{a_{n-2}} \cdot \dots \cdot W^A_{a_1} \cdot W^A_{a_0} \cdot F^A$$

where the operation '$\cdot$' indicates binary multiplication.

We will describe an FSA using diagrams where states are circles and the transitions are arcs labeled with the alphabet. The initial and final distributions are written inside the circles for each state (see Figure 20.2).

Example 20.1.

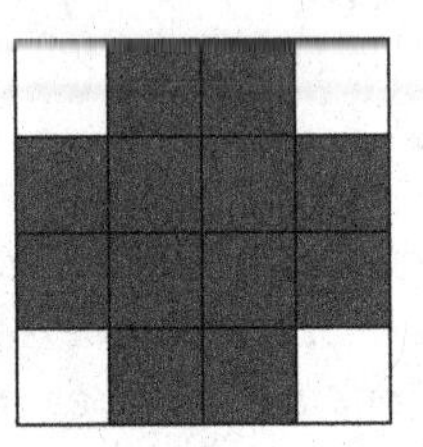

The above picture is represented by

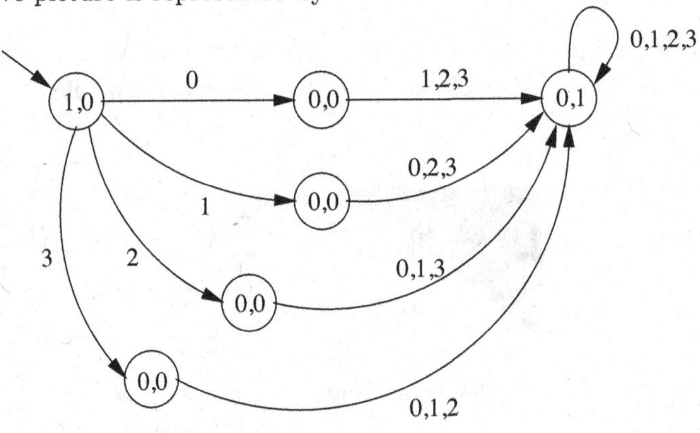

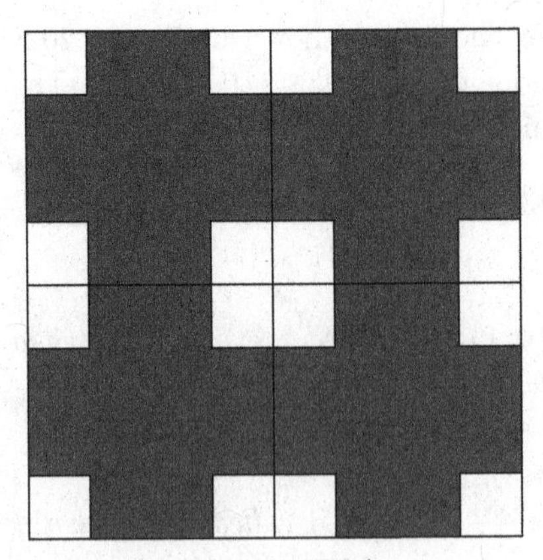

The above picture is represented by the FSA

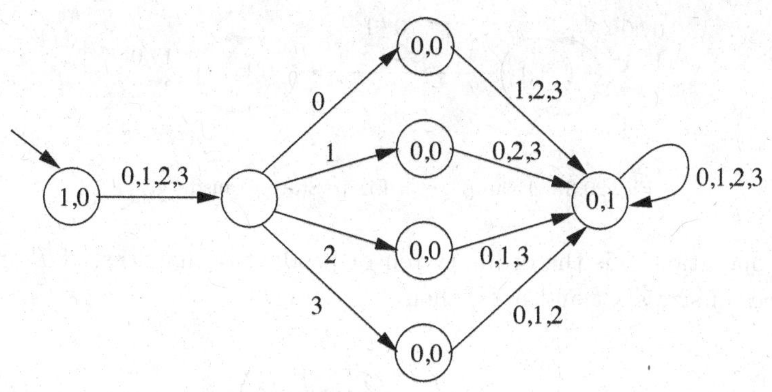

The FSA for the first figure is appended with

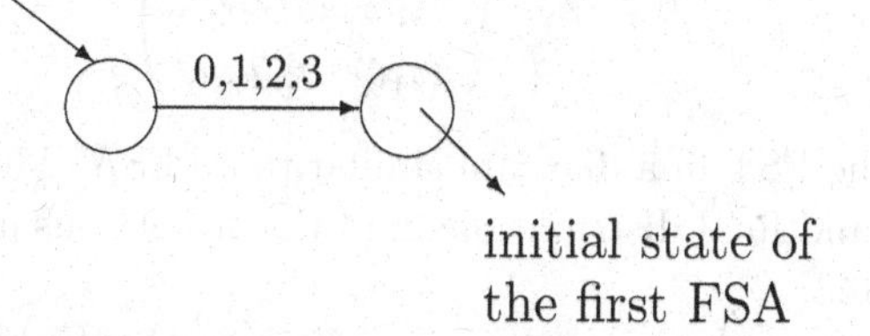

to get the FSA for the second figure.
The first figure is superimposed on the four subsquares.

20.2.1. *Transformations on Black and White Images*

Transformations like scaling up, scaling down, translation, rotation can be achieved using finite state transducers (FST). A finite state transducer represents a transformation from an alphabet Σ_1 to another alphabet Σ_2.

Definition 20.3. An m state finite state transducer FST from an alphabet Σ_1 to an alphabet Σ_2 is specified by

(1) a row vector $I \in \{0,1\}^{1\times m}$ called the initial distribution,
(2) a column vector $F \in \{0,1\}^{m\times 1}$ called the final distribution and
(3) binary matrices $W_{a,b} \in \{0,1\}^{m\times m}$ $\forall a \in \Sigma_1$ and $b \in \Sigma_2$ called transition matrices.

Here we consider only FSTs from Σ to Σ, $\Sigma = \{0,1,2,3\}$. In order to obtain a transformation of an image, we apply the corresponding FST to the FSA representing the image to get a new FSA. The application of an n state FST to an m state FSA $A = (I^A, F^A, W_a^A, a \in \Sigma)$ produces an $n \times m$ state FSA $B = (I^B, F^B, W_b^B, b \in \Sigma)$ as follows.

$$I^B = I \otimes I^A$$
$$F^B = F \otimes F^A$$
$$W_b^B = \sum_{a\in\Sigma} W_{a,b} \otimes W_a^A, \ \forall\, b \in \Sigma$$

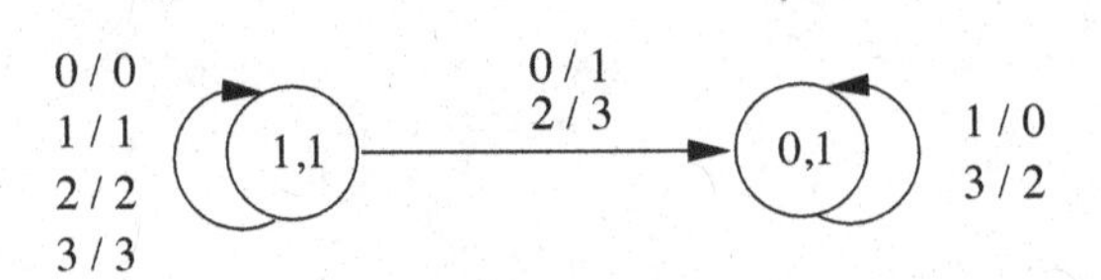

Fig. 20.3. Example of a Finite State Transducer.

where the operation $\otimes$ is the ordinary tensor product of matrices. If T and Q are two matrices of size $s \times t$ and $p \times q$ then

$$T \otimes Q = \begin{pmatrix} T_{11}Q \dots T_{1t}Q \\ \vdots \quad\quad \vdots \\ T_{s1}Q \dots T_{st}Q \end{pmatrix}$$

We represent the FST in a diagram similar to an FSA. The values in the circles denote the initial and final distributions and the transitions are labeled as 'a/b' as shown in Figure 20.3.

Translation

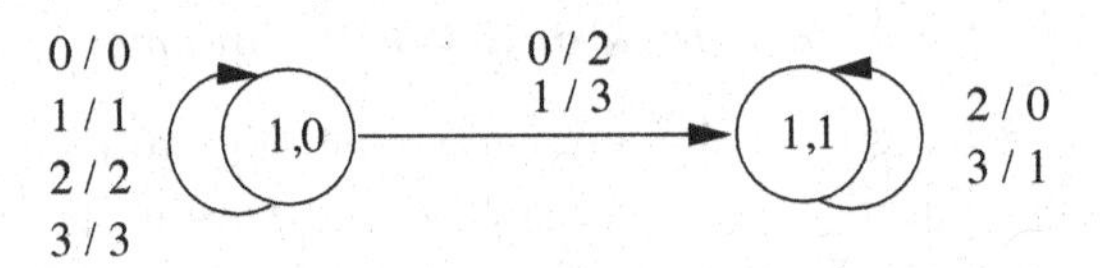

Fig. 20.4. Translation by 1 unit.

Let the finite state automaton, M, represent a black-white image of resolution $2^n \times 2^n$. Consider the translation along the x-axis from left to right by one pixel. It is assumed that the picture is wrapped around on translation i.e., in this case, the pixels in the $(2^n - 1)^{th}$ column are moved to the 0th column. This translation is equivalent to adding 1 to the x coordinate of each pixel. If the n-bit x coordinate of a pixel is $w01^r$ then the address of this pixel after translation would be $w10^r$, $0 \leq r \leq n-1$. There is no change in the value of y coordinate. The finite state transducer for this translation is shown in Figure 20.4. The FSA M' representing the translated image can be obtained by applying this FST to M. Note that in the FST, transformation in a bit of x coordinate from 0 to 1 (1 to 0) would be represented as 0/2 and 1/3 (2/0 and 3/1).

In order to translate the image by 2 units, we have to add 2 to the x coordinate of each pixel. This is same as adding 1 to the binary representation of x coordinate after ignoring the 0th bit. Similarly a translation by 4 units can be achieved by adding 1 to the x coordinate after ignoring both 0th and 1st bits. The FSTs for these operations are shown in Figure 20.5. Now we will consider translation in much

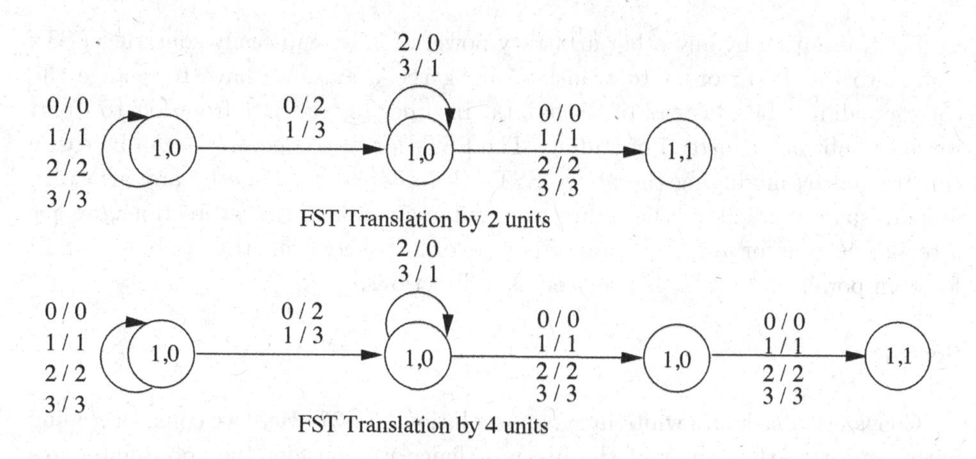

Fig. 20.5. Translation by 2, 4 unit.

larger units. For example if we want to translate the image by half the size of the image i.e., by 2^{n-1} units, we simply have to change the most significant bit from 0 to 1 or from 1 to 0. Similarly in order to translate by one fourth of the entire square (by 2^{n-2} units) we have to replace the two most significant bits of the x coordinate from 00 to 01, 01 to 10, 10 to 11, 11 to 00. Figure 20.5 gives the finite state transducers for these operations. If we want a translation without wraparound (when we want to translate by 1 unit) then we have to change the initial distribution of the second state in the FST in Figure 20.5 from 1 to 0. Figure 20.6 shows the triangle image first described in Figure 20.2. The image has been translated through $\frac{1}{2}$ and $\frac{1}{4}$ the size of entire image. It is seen that the translated right end gets wrapped on to the left beginning.

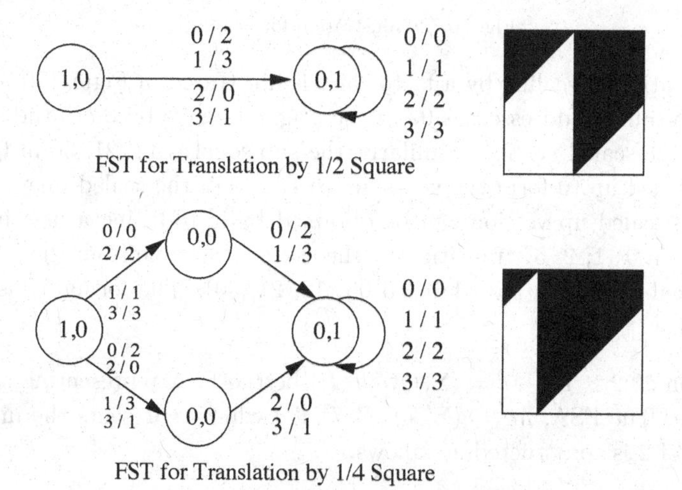

Fig. 20.6. Translation by 1/2, 1/4 square.

For translation by any other arbitrary power of 2, we can easily construct FSTs similar to these. In order to translate along the y axis, we have to change the corresponding y bits instead of the x bits. In order to translate from left to right, we must subtract 1 instead of adding. The FSTs for these operations can be easily constructed by modifying the above FSTs. In order to translate by any arbitrary amount say a units along the x direction and b units along the y directions, we get a representation for $a(b)$ as a sum of powers of 2 and continuously apply the FSTs for each power of 2 for both x as well as y directions.

Scaling

Consider a black and white image of resolution $2^n \times 2^n$. Here we consider scaling with respect to the center of the image. Hence we consider the coordinate axes shifted to the center of the image. Thus the new coordinate axes are $x' = x - 2^{n-1}$ and $y' = y - 2^{n-1}$. The operation of scaling by a factor k is defined as follows

$$x' \leftarrow \frac{x'}{k} \qquad y' \leftarrow \frac{y'}{k}$$

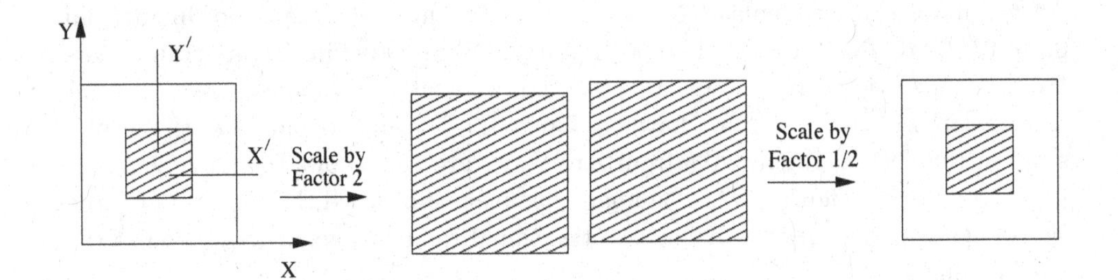

Fig. 20.7. Illustration for Scaling.

The operation of scaling by a factor of 2 is illustrated in Figure 20.7. It is seen that the sub-square addressed as 03 in the original image becomes the bigger sub-square 0 in the scaled version. Similarly, the sub-squares 12, 21, 30 in the original image are scaled up to form the sub-squares 1, 2, 3 of the scaled image. Thus the FSA for the scaled up version can be obtained by introducing a new initial state which makes a 0 (1, 2, 3) transition to the states reachable from the initial states of the original FSA by a path labeled 03 (12, 21, 30). The formal construction is as given below.

Construction 20.2.1. Let $M = (\Sigma, Q, \delta, I, F)$ be the FSA representing a black and white image. The FSA, $M' = (\Sigma', Q', \delta', I', F')$ which represents the image scaled by a factor of 2 is constructed as follows.

(1) $\Sigma' = \Sigma$
(2) $Q' = Q \cup \{q_0\}, I' = \{q_0\}.F' = F$

(3) $\delta'(q_0, 0) = \delta(\delta(q_0, 0), 3)$
(4) $\delta'(q_0, 1) = \delta(\delta(q_0, 1), 2)$
(5) $\delta'(q_0, 2) = \delta(\delta(q_0, 2), 1)$
(6) $\delta'(q_0, 3) = \delta(\delta(q_0, 3), 0)$
(7) $\delta'(q, a) = \delta(q, a),\ \forall\ a \in \Sigma,\ \forall\ q \in Q$

Next we illustrate this construction.

Consider the black and white picture and the FSA given in Figure 20.2. Scaling down by a factor of $\frac{1}{4}$ (a factor of $\frac{1}{2}$ in each direction), we get

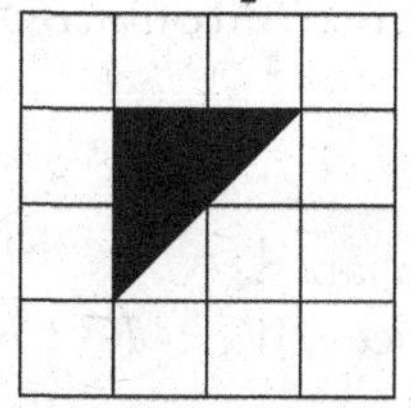

By our construction this is represented by

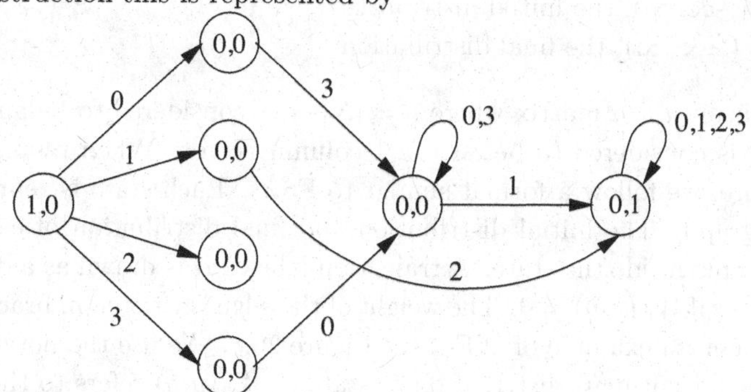

Scaling up this by a factor of 4 (a factor of 2 in each direction), we get Figure 20.2. The FSA obtained by our construction is

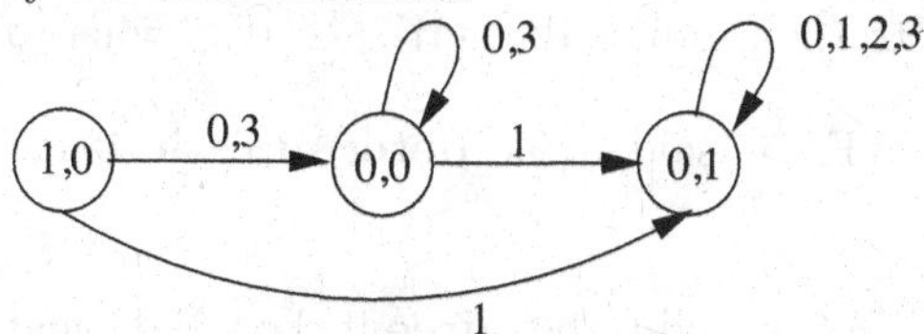

Construction 20.2.2. Let $M = (\Sigma, Q, \delta, I, F)$ be the FSA representing a black and white image. The FSA, $M' = (\Sigma', Q', \delta', I', F')$ which represents the image scaled by a factor of 2 is constructed as follows.

(1) $\Sigma' = \Sigma$
(2) $Q' = Q \cup \{q'\} \cup \{q_0, q_1, q_2, q_3\}, I' = \{q'\}.F' = F$
(3) $\delta'(q', 0) = \{q_0\}, \delta'(q_0, 3) = \{q | \exists i, i \in I, \delta(i, 0) = q\}$
(4) $\delta'(q', 1) = \{q_1\}, \delta'(q_1, 2) = \{q | \exists i, i \in I, \delta(i, 1) = q\}$
(5) $\delta'(q', 2) = \{q_2\}, \delta'(q_2, 1) = \{q | \exists i, i \in I, \delta(i, 2) = q\}$

(6) $\delta'(q', 3) = \{q_3\}, \delta'(q_3, 0) = \{q | \exists i, i \in I, \delta(i, 3) = q\}$
(7) $\delta'(q, a) = \delta(q, a), \; \forall \, a \in \Sigma, \; \forall \, q \in Q$

20.3. Weighted Finite Automata and Gray-Scale Images

In this subsection we present the basic definitions related to the weighted finite automata and give the representation of gray-scale images using the weighted finite automata. A study of weighted FSA is done in [4, 5].

Definition 20.4. A **weighted finite automaton** $M = (K, \Sigma, W, I, F)$ is specified by

(1) K is a finite set of states.
(2) Σ is a finite set called the alphabet.
(3) W is the set of weight matrices, $W_\alpha : K \times K \to \Re$ for all $\alpha \in \Sigma \bigcup \{\epsilon\}$, the weights of edges labeled α.
(4) $I : K \to (-\infty, \infty)$, the initial distribution.
(5) $F : K \to (-\infty, \infty)$, the final distribution.

Here W_α is an $n \times n$ matrix where $n = |K| \cdot I$ is considered to be an $1 \times n$ row vector and F is considered to be an $n \times 1$ column vector. When representing the WFAs as figure, we follow a format similar to FSAs. Each state is represented by a node in a graph. The initial distribution and final distribution of each state is written as a tuple inside the state. A transition labeled α is drawn as a directed arc from state p to q if $W_\alpha(p, q) \neq 0$. The weight of the edge is written in brackets on the directed arc. For an example of WFA, see Figure 20.8. We use the notation $I_q(F_q)$ to refer to the initial(final) distribution of state q. $W_\alpha(p, q)$ refers to the weight of the transition from p to q. $W_\alpha(p)$ refers to the pth row vector of the weight matrix W_α. It gives the weights of all the transitions from state p labeled α in a vector form. Also W_x refers to the product $W_{\alpha_1} \cdot W_{\alpha_2} \cdots W_{\alpha_k}$ where $x = \alpha_1 \alpha_2 \cdots \alpha_k$.

Definition 20.5. A WFA is said to be **deterministic** if its underlying FSA is deterministic.

Definition 20.6. A WFA is said to be **ϵ-free** if the weight matrix $W_\epsilon \equiv 0$ where 0 is the zero matrix of order $n \times n$.

Hereafter, whenever we use the term WFA, we refer to an ϵ-free WFA only unless otherwise specified.

A WFA M as in Definition 20.4 defines a function $f : \Sigma^* \to \Re$, where for all $x \in \Sigma^*$ and $x = \alpha_1 \alpha_2 \ldots \alpha_k$,

$$f(x) = I \cdot W_{\alpha_1} \cdot W_{\alpha_2} \cdot \cdots \cdot W_{\alpha_k} \cdot F$$

where the operation '$\cdot$' is matrix multiplication.

Definition 20.7. A **path** P of length k is defined as a tuple $(q_0q_1\ldots q_k, \alpha_1\alpha_2\ldots\alpha_k)$ where $q_i \in K, 0 \le i \le k$ and $\alpha_i \in \Sigma, 1 \le i \le k$ such that α_i denotes the label of the edge traversed while moving from q_{i-1} to q_i.

Definition 20.8. The **weight** of a path P is defined as

$$W(P) = I_{q_0} \cdot W_{\alpha_1}(q_0, q_1) \cdot W_{\alpha_2}(q_1, q_2) \cdot \cdots \cdot W_{\alpha_k}(q_{k-1}, q_k) \cdot F_{q_k}.$$

The function $f : \Sigma^* \to \Re$ represented by a WFA M can be equivalently defined as follows

$$f(x) = \sum_{P \text{ is a path of } M \text{ labeled } x} W(P), x \in \Sigma^*.$$

Definition 20.9. A function $f : \Sigma^* \to \Re$ is said to be **average preserving** if

$$f(w) = \frac{1}{m} \sum_{\alpha \in \Sigma} f(w\alpha)$$

for all $w \in \Sigma^*$ where $m = |\Sigma|$.

Definition 20.10. A WFA M is said to be **average preserving** if the function that it represents is average preserving.

The general condition to check whether a WFA is average preserving is as follows. A WFA M is average preserving if and only if

$$\sum_{\alpha \in \Sigma} W_\alpha \cdot F = mF,$$

where $m = |\Sigma|$.
We also consider the following definitions.

Definition 20.11. A WFA is said to be **i-normal** if the initial distribution of every state is 0 or 1 i.e. $I_{q_i} = 0$ or $I_{q_i} = 1$ for all $q_i \in K$.

Definition 20.12. A WFA is said to be **f-normal** if the final distribution of every state is 0 or 1 i.e. $F_{q_i} = 0$ or $F_{q_i} = 1$ for all $q_i \in K$.

Definition 20.13. A WFA is said to be **I-normal** if there is only one state with non-zero initial distribution.

Definition 20.14. A WFA is said to be **F-normal** if there is only one state with non-zero final distribution.

20.3.1. *Representation of Gray-Scale Images*

A gray-scale digital image of finite resolution consists of $2^m \times 2^m$ pixels, where each pixel takes a real grayness value (in reality the value ranges as 0, 1, ..., 256). By a multi-resolution image, we mean a collection of compatible $2^n \times 2^n$ resolution images for $n \geq 0$. Similar to black and white images, we will assign a word $x \in \Sigma^k$ where $\Sigma = \{0, 1, 2, 3\}$ to address each pixel. A word x of length less than k will address a subsquare as in black and white images.

Then we can define our finite resolution image as a function $f_I : \Sigma^k \to \Re$, where $f_I(x)$ gives the value of the pixel at address x. A multi-resolution image is a function $f_I : \Sigma^* \to \Re$. It is shown in [4] that for compatibility, the function f_I should be average preserving i.e.

$$f_I(x) = \frac{1}{4}[f_I(x0) + f_I(x1) + f_I(x2) + f_I(x3)].$$

A WFA M is said to represent a multi-resolution image if the function f_M represented by M is the same as the function f_I of the image.

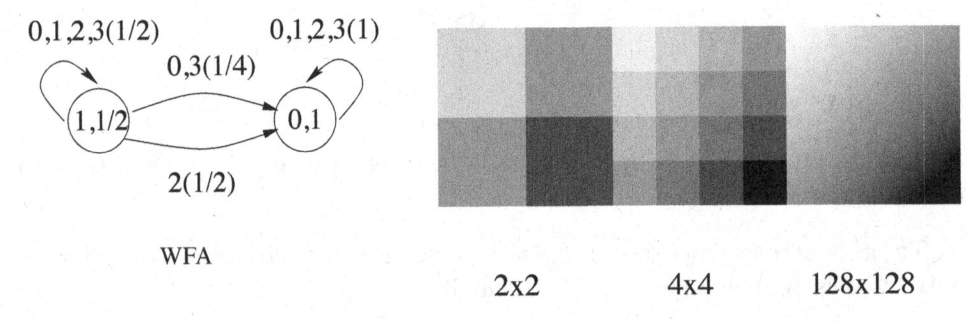

Fig. 20.8. Example: WFA computing linear grayness function.

Example 20.2. Consider the 2 state WFA shown in Figure 20.8.
Let $I = (1, 0)$ and $F = (\frac{1}{2}, 1)$ and the weight matrices are $W_0 = \begin{pmatrix} \frac{1}{2} & \frac{1}{4} \\ 0 & 1 \end{pmatrix}$, $W_1 = \begin{pmatrix} \frac{1}{2} & 0 \\ 0 & 1 \end{pmatrix}$ $W_2 = \begin{pmatrix} \frac{1}{2} & \frac{1}{2} \\ 0 & 1 \end{pmatrix}$ $W_3 = \begin{pmatrix} \frac{1}{2} & \frac{1}{4} \\ 0 & 1 \end{pmatrix}$.

$$f(22) = (1, 0) \begin{pmatrix} \frac{1}{2} & \frac{1}{2} \\ 0 & 1 \end{pmatrix} \begin{pmatrix} \frac{1}{2} & \frac{1}{2} \\ 0 & 1 \end{pmatrix} \begin{pmatrix} \frac{1}{2} \\ 1 \end{pmatrix}$$

$$= (1, 0) \begin{pmatrix} \frac{1}{4} & \frac{3}{4} \\ 0 & 1 \end{pmatrix} \begin{pmatrix} \frac{1}{2} \\ 1 \end{pmatrix}$$

$$= (1, 0) \begin{pmatrix} \frac{1}{8} + \frac{3}{4} \\ 1 \end{pmatrix} = \frac{1}{8} + \frac{6}{8} = \frac{7}{8}$$

$$f(11) = (1, 0) \begin{pmatrix} \frac{1}{2} & 0 \\ 0 & 1 \end{pmatrix} \begin{pmatrix} \frac{1}{2} & 0 \\ 0 & 1 \end{pmatrix} \begin{pmatrix} \frac{1}{2} \\ 1 \end{pmatrix}$$

$$= (1, 0) \begin{pmatrix} \frac{1}{4} & 0 \\ 0 & 1 \end{pmatrix} \begin{pmatrix} \frac{1}{2} \\ 1 \end{pmatrix}$$

$$= (1, 0) \begin{pmatrix} \frac{1}{8} \\ 1 \end{pmatrix} = \frac{1}{8}$$

Then we can calculate the values of pixels as follows. $f(03)$ = sum of weights all paths labeled 03.

$$f(03) = \left(1 \times \frac{1}{2} \times \frac{1}{2} \times \frac{1}{2}\right) + \left(1 \times \frac{1}{2} \times \frac{1}{4} \times 1\right) + \left(1 \times \frac{1}{4} \times 1 \times 1\right)$$
$$= \frac{1}{8} + \frac{1}{8} + \frac{1}{4} = \frac{1}{2}.$$

Similarly for $f(123)$ we have $f(123) = \frac{1}{16} + \frac{1}{4} + \frac{1}{16} = \frac{3}{8}$. The images obtained by this WFA are shown for resolutions 2×2, 4×4 and 128×128 in Figure 20.8.

The RGB format of colour images are such that the image in pixel form contains the red, green and blue values of the pixels. Analogous to gray-scale images, we can use the WFA to represent a colour image by extracting the red, green, blue pixel values and storing the information in a WFA.

20.3.2. *Inferencing and De-Inferencing*

We described how every digital gray-scale multi-resolution image can be represented by an average-preserving WFA . Algorithms are given for both converting a WFA into a digital image and for inferencing the WFA representing a digital image. The WFA consists of an $I_{1 \times n}$ row vector, $F_{n \times 1}$ column vector and $W_{0_{n \times n}}, W_{1_{n \times n}}, W_{2_{n \times n}}$ and $W_{3_{n \times n}}$ weight matrices.

De-Inferencing

Assume, we are given a WFA M, $(I, F, W_0, W_1, W_2, W_3)$, and we want to construct a finite resolution approximation of the multi-resolution image represented by M. Let the image to be constructed be $\mathcal{I}$ of resolution $2^k \times 2^k$. Then for all $x \in \Sigma^k$, we have to compute $f(x) = I \cdot W_x \cdot F$. The algorithm is as follows. The algorithm computes $\phi_p(x)$ for $p \in Q$ for all $x \in \Sigma^i$, $0 \leq i \leq k$. Here ϕ_p is the image of state p.

Algorithm : De_Infer_WFA

Input : WFA $M = (I, F, W_0, W_1, W_2, W_3)$.
Output : $f(x)$, for all $x \in \Sigma^k$.
begin
Step 1 : Set $\phi_p(\epsilon) \leftarrow F_p$ for all $p \in Q$
Step 2 : for $i = 1, 2, \ldots, k$, do the following
begin
Step 3 : for all $p \in Q$, $x \in \Sigma^{i-1}$ and $\alpha \in \Sigma$ compute
$$\phi_p(\alpha x) \leftarrow \sum_{q \in Q} W_\alpha(p, q) \cdot \phi_q(x)$$
end for

Step 4 : for each $x \in \Sigma^k$, compute

$$f(x) = \sum_{q \in Q} I_q \cdot \phi_q(x).$$

Step 5 : stop.

end

The time complexity of the above Algorithm is $O(n^2 4^k)$, where n is the number of states in the WFA and $4^k = 2^k \cdot 2^k$ is the number of pixels in the image. We know that $f(x)$ can be computed either by summing the weights of all the paths labeled x or by computing $I \cdot W_x \cdot F$. Finding all paths labeled of length k takes $k \cdot (4k)^n$ time. Since $n \gg k$ we prefer the matrix multiplication over this.

Inferencing

Let $\mathcal{I}$ be the digital gray-scale image of finite resolution $2^k \times 2^k$. In [5], an iterative algorithm is proposed to obtain the WFA M representing the image $\mathcal{I}$. It is also shown in [5], that the WFA so obtained is a minimum state WFA. The inference algorithm is given in Algorithm. In the algorithm Infer_WFA, N is the index of the last state created, i is the index of the first unprocessed state, ϕ_p is the image represented by state p, f_x represents the sub-image at the sub square labeled x, while $f_{avg}(x)$ represents the average pixel value of the sub-image at the sub-square labeled x, and $\gamma : Q \leftarrow \Sigma^*$ is a mapping of states to sub-squares.

Algorithm Infer_WFA

Input : Image $\mathcal{I}$ of size $2^k \times 2^k$.

Output : WFA M representing image $\mathcal{I}$.

begin

Step 1 : Set $N \leftarrow 0$, $i \leftarrow 0$, $F_{q0} \leftarrow f_{avg}(\epsilon)$, $\gamma(q_0) \leftarrow \epsilon$.

Step 2 : Process q_i, i.e., for $x = \gamma(q_i)$ and each $\alpha \in \{0, 1, 2, 3\}$ do

begin

Step 3 : If there are $c_0, c_1, \ldots, c_N$ such that $f_{k\alpha} = c_0\phi_0 + c_1\phi_1 + \cdots + c_N\phi_N$ where $\phi_j = f_{\gamma(q_j)}$ for $0 \leq j \leq N$ then set $W_\alpha(q_i, q_j) \leftarrow c_j$ for $0 \leq j \leq N$.

Step 4 : else set $\gamma(q_{N+1}) \leftarrow x\alpha$, $F_{qN+1} \leftarrow f_{avg}(x\alpha)$, $W_\alpha(q_i, q_{N+1}) \leftarrow q$ and $N \leftarrow N + 1$.

end for

Step 5 : Set $i \leftarrow i + 1$ and goto Step 2.

Step 6 : Set $I_{q0} \leftarrow 1$ and $I_{qj} \leftarrow 0$ for all $1 \leq j \leq N$.

end

Example 20.3. Consider the linearly sloping ap-function f introduced in Example 20.2. Let us apply the inference algorithm to find a minimal ap-WFA generating f.

First, the state q_0 is assigned to the square ϵ and we define $F_{q_0} = \frac{1}{2}$. Consider then the four sub-squares 0, 1, 2, 3. The image in the sub-square 1 can be expressed as $\frac{1}{2}f$ (it is obtained from the original image by decreasing the gray-scale by $\frac{1}{2}$) so that we define $W_1(q_0, q_0) = 0.5$.

The image in sub-square 0 cannot be expressed as a linear combination of f so that we have to use a second state q_1. Define $W_0(q_0, q_1) = 1$ and $F_{q_1} = \frac{1}{2}$ (the average grayness of sub-square 0 is $\frac{1}{2}$). Let f_1 denote the image in sub-square 0 of f.

The image in sub-square 3 is the same as in sub-square 0, so that $W_3(q_0, q_1) = 1$. In the quadrant 2 we have an image which can be expressed as $2f_1 - \frac{1}{2}f$. We define $W_2(q_0, q_0) = -\frac{1}{2}$ and $W_2(q_0, q_1) = 2$. The outgoing transitions from state q_0 are now ready.

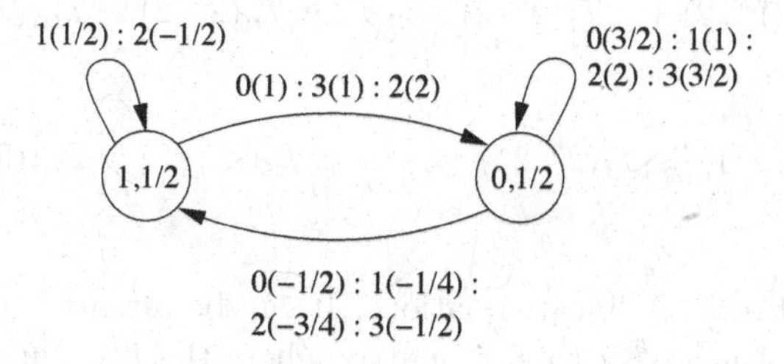

Fig. 20.9. Example: Inference of WFA.

Consider then the images in the squares 00, 01, 02 and 03. They can be expressed as $\frac{3}{2}f_1 - \frac{1}{2}f, f_1 - \frac{1}{4}f, 2f_1 - \frac{3}{4}f$ and $\frac{3}{2}f_1 - \frac{1}{2} \cdot f$. This gives us the ap-WFA shown in Figure 20.9. The initial distribution is $(1, 0)$.

It is shown in [5], that in Algorithm Infer_WFA, each state is independent of the other state, i.e., ϕ_i, where $1 \leq i \leq n$ cannot be expressed as a linear combination of other states.

$$c_1\phi_1 + c_2\phi_2 + \cdots + c_n\phi_n = 0$$

implies that $c_i = 0$ for all $1 \leq i \leq n$. Hence the WFA obtained by Algorithm Infer_WFA is a minimum state WFA.

Now consider Step 3 of Algorithm Infer_WFA. This step asks for finding $c_0, c_1, \ldots, c_n$ such that $f_\alpha = c_1\phi_1 + c_2\phi_2 + \cdots + c_n\phi_n$, i.e., to express the sub-image of the sub-square $x\alpha$ as a linear combinations of the images represented by the states so far created. Let the size of the sub image be $2^k \times 2^k$. Then the above equation can be restated as follows

$$\mathcal{I}_{x\alpha, 2^k \times 2^k} = c_0 \cdot \mathcal{I}_{0, 2^k \times 2^k} + c_1 \cdot \mathcal{I}_{1, 2^k \times 2^k} + \cdots + c_n \cdot \mathcal{I}_{n, 2^k \times 2^k}$$

where $\mathcal{I}_{q_i,2^k\times2^k}$ is the $2^k \times 2^k$ image represented by the state q_i and $\mathcal{I}_{x\alpha,2^k\times2^k}$ is the $2^k \times 2^k$ sub-image at the sub-square addressed by $x\alpha$. The equations can be rewritten as follows.

$$\begin{aligned}
c_0\mathcal{I}_0(1,1) + c_1\mathcal{I}_1(1,1) + \cdots + c_n\mathcal{I}_n(1,1) &= \mathcal{I}_{x\alpha}(1,1)\\
c_0\mathcal{I}_0(1,2) + c_1\mathcal{I}_1(1,2) + \cdots + c_n\mathcal{I}_n(1,2) &= \mathcal{I}_{x\alpha}(1,2)\\
\vdots &= \vdots\\
c_0\mathcal{I}_0(1,2^k) + c_1\mathcal{I}_1(1,2^k) + \cdots + c_n\mathcal{I}_n(1,2^k) &= \mathcal{I}_{x\alpha}(1,2^k)\\
\vdots &= \vdots\\
c_0\mathcal{I}_0(2^k,1) + c_1\mathcal{I}_1(2^k,1) + \cdots + c_n\mathcal{I}_n(2^k,1) &= \mathcal{I}_{x\alpha}(2^k,1)\\
c_0\mathcal{I}_0(2^k,2) + c_1\mathcal{I}_1(2^k,2) + \cdots + c_n\mathcal{I}_n(2^k,2) &= \mathcal{I}_{x\alpha}(2^k,2)\\
\vdots &= \vdots\\
c_0\mathcal{I}_0(2^k,2^k) + c_1\mathcal{I}_1(2^k,2^k) + \cdots + c_n\mathcal{I}_n(2^k,2^k) &= \mathcal{I}_{x\alpha}(2^k,2^k)
\end{aligned}$$

This is nothing but a set of linear equations. It can be rewritten in matrix form as $A \cdot C = B$, where A is a $4^k \times (n+1)$ matrix where the ith column represents the $2^k \times 2^k = 4^k$ pixels of the image $\mathcal{I}_i$ represented by state q_i. C is an $(n+1) \times 1$ column vector of the coefficients. B is a $4^k \times 1$ vector containing the pixels of the image $I_{x\alpha}$. Thus the Step 3 reduces to solving a set of linear equations.

One well known method of attacking this problem is using the Gaussian Elimination technique. But for using this technique, the rank of the matrix A should be $min(4^k, n)$. But in the general case, $rank(A)$ is found to be $\leq min(4^k, n)$. Hence this method cannot be used in our case.

Another standard method for solving a set of linear equations is Singular Value Decomposition. This method not only gives the solution if it exists, but also in the case of nonexisting solution, gives us the least mean square approximate solution. The computed coefficients are such that $||B - AC||$ is minimum, where $||M|| = \sqrt{M^T M}$.

Approximate Representation

In image processing applications, it is not always required that the images be exactly represented. We can see that by introducing an error parameter in the above algorithm, the number of states required for representation can be reduced. While solving the linear equations in Step 3 of Algorithm Infer_WFA, we get a solution with least mean square error. We can accept the solution if this error is greater than a positive quantity δ. This way we can represent the image approximately with a smaller automaton.

Compression

In a gray scale image eight bits are required per pixel. If the resolution is 512×512, 512×512 bytes are required to store the image. If the number of states is less in a WFA, it can be stored with lesser file space than the image, and we get a compressed representation of the image. It is not difficult to see that any WFA can be made f-normal; further it can be made I-normal since we need not bother about representing images of size 0. Further it was observed that in most cases, the weight matrices obtained are sparse. Hence, it is enough to store only the weights on the edges in a file. This may help to keep an image in compressed form.

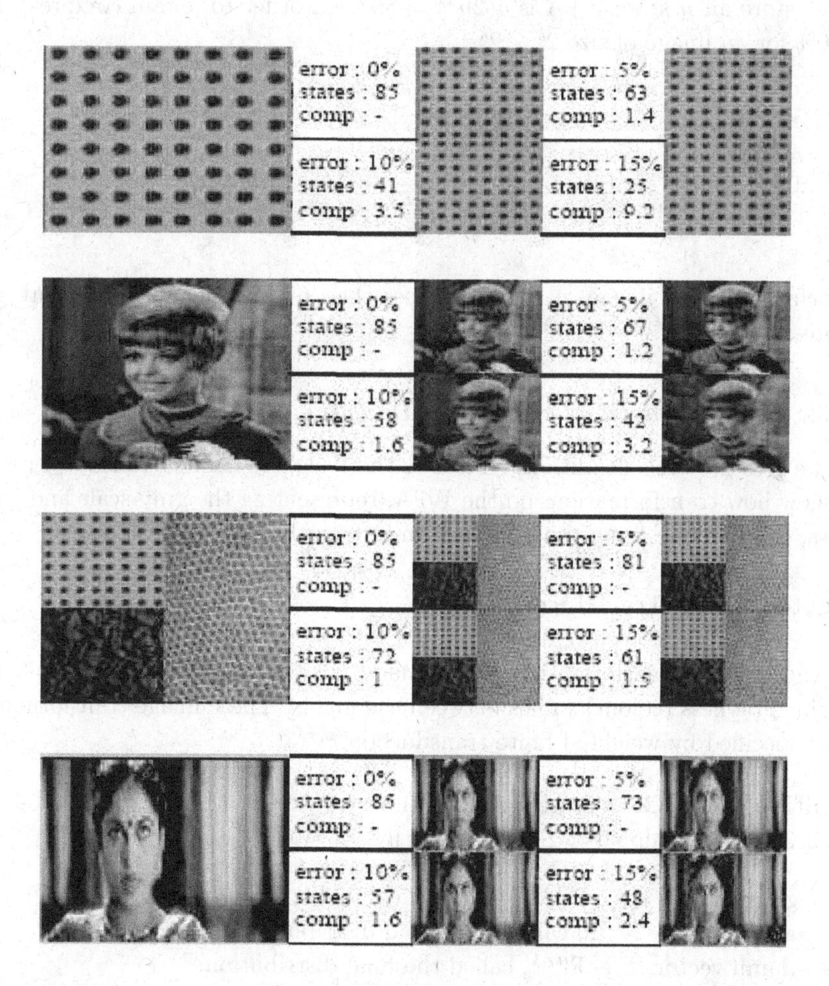

Fig. 20.10.

Observation

The inference algorithm applied to four images of size 256×256 is given in Figure 20.10. It shows the compression obtained for the images in cases where the error factor was equal to $0\%, 5\%, 10\%$ and 15%. The reconstructed images are also shown in Figure 20.10. It can be observed that while error up to 10% does not disturb the figure much, using error of 15% distorts the figure badly. Also it was observed that the number of states obtained depends on the regularity of the image itself.

The image can perhaps be further compressed by a smarter way of storing the WFA in a file. Currently for each edge, 4 bytes are needed to store the weight (type float). On an average, an n state WFA has $4\frac{n^2}{2} = 2n^2$ edges. The number of bytes used to store an n state WFA is $4(2n^2) = 8n^2$. In order to obtain compression of say 50% for an image of size $2^k \times 2^k$.

$$8n^2 \leq \frac{4^k}{2}$$
$$\Rightarrow\ n^2 \leq \frac{4^k}{16} = 4^{k-2}$$
$$\Rightarrow\ n \leq 2^{k-2}$$

For a 256×256 image, n should be less than 64 in order to obtain any good compression.

20.3.3. *Transformations on Digital Images*

Next we give the basic definitions related to the Weighted Finite Transducer (WFT) and show how transformations on the WFA, representing the gray-scale and colour images, can be done using the weighted finite transducers.

Weighted Finite Transducers

Almost every transformation of an image involves moving (scaling) pixels or changing grayness (colour) values between the pixels. These image transformations can be specified by weighted finite transducers.

Definition 20.15. An n-state weighted finite transducer M from an alphabet $\Sigma = \{0, 1, 2, 3\}$ into the alphabet Σ is specified by

(1) weight matrices $W_{a,b} \in \Re^{n \times n}$ for all $a \in \Sigma \cup \{\epsilon\}$ and $b \in \Sigma \cup \{\epsilon\}$,
(2) a row vector $I \in \Re^{1 \times n}$, called the initial distribution, and
(3) a column vector, $F \in \Re^{n \times 1}$, called the final distribution.

The WFT M is called ϵ-free if the weight matrices $W_{\epsilon,\epsilon}, W_{a,\epsilon}$, and $W_{\epsilon,b}$ are zero matrices for all $a \in \Sigma$ and $b \in \Sigma$.

Definition 20.16. The WFT M defines a function $f_M : \Sigma^* \times \Sigma^* \longrightarrow \Re$, called weighted relation between Σ^* and Σ^*, by

$$f_M(u, v) \;=\; I \cdot W_{u,v} \cdot F, \text{ for all } u \in \Sigma^*, v \in \Sigma^*$$

where

$$W_{u,v} = \sum_{\substack{a_1 \cdots a_k \\ b_1 \cdots b_k}} W_{a_1,b_1} \cdot W_{a_2,b_2} \cdot \cdots \cdot W_{a_k,b_k}, \tag{20.1}$$

if the sum converges. (If the sum does not converge, $f_M(u, v)$ remains undefined).

In (20.1) the sum is taken over all decompositions of u and v into symbols $a_i \in \Sigma \cup \{\epsilon\}$ and $b_i \in \Sigma \cup \{\epsilon\}$, respectively.

In the special case of ϵ-free transducers,

$$f_M(a_1a_2 \ldots a_k, b_1b_2 \ldots b_k) \;=\; I \cdot W_{a_1,b_1} \cdot W_{a_2,b_2} \cdot \cdots \cdot W_{a_k,b_k} \cdot F,$$

for $a_1a_2 \ldots a_k \in \Sigma^k, b_1b_2 \ldots b_k \in \Sigma^k$, and $f_M(u, v) \;=\; 0$, if $|u| \;\neq\; |v|$.

We recall that a WFA defines a multi-resolution function $f : \Sigma^* \longrightarrow \Re$ where for all $x \in \Sigma^*$ and $x = \alpha_1\alpha_2 \ldots \alpha_k$,

$$f(x) \;=\; \sum (I \cdot W_{\alpha_1} \cdot W_{\alpha_2} \cdot \cdots \cdot W_{\alpha_k} \cdot F)$$

where the summation is over all possible paths for x in the various components, the operation '$\cdot$' is the matrix multiplication.

Definition 20.17. Let $\rho : \Sigma^* \times \Sigma^* \longrightarrow \Re$ be a weighted relation and $f : \Sigma^* \longrightarrow \Re$ a multi-resolution function represented by a WFA. The application of ρ to f is the multi-resolution function $g \;=\; \rho(f) : \Sigma^* \longrightarrow \Re$ over Σ defined by

$$g(v) = \sum_{u \in \Sigma^*} f(u)\rho(u, v), \text{ for all } v \in \Sigma^*, \tag{20.2}$$

provided the sum converges. The application $M(f)$ of WFT M to f is defined as the application of the weighted relation f_M to f, i.e. $M(f) \;=\; f_M(f)$.

Equation (20.2) defines an application of a WFT to an image in the pixel form. When the image is available in the WFA-compressed form we can apply a WFT directly to it and compute the regenerated image again from the transformed WFA.

Here we define the application of the ϵ-free n-state WFT to a k-WFA. The application of an ϵ-free n-state WFT M to an m-state k-WFA Γ over the alphabet Σ specified by initial distribution I^Γ, final distribution F^Γ and weight matrices $W^\Gamma_\alpha, \alpha \in \Sigma$, is the mn-state k-WFA $\Gamma' \;=\; M(\Gamma)$ over the alphabet Σ with initial distribution $I^{\Gamma'} \;=\; I \otimes I^\Gamma$, final distribution $F^{\Gamma'} \;=\; F \otimes F^\Gamma$ and weight matrices

$$W^{\Gamma'}_b \;=\; \sum_{a \in \Sigma} W_{a,b} \otimes W^\Gamma_a, \text{ for all } b \in \Sigma.$$

Clearly $f_{\Gamma'} = M(f_\Gamma)$, i.e. the multi-resolution function defined by Γ' is the same as the application of the WFT M to the multi-resolution function computed by the WFA Γ.

We note that every WFT M is a linear operator from $\Re^{\Sigma^*} \longrightarrow \Re^{\Sigma^*}$. In other words,

$$M(r_1 f_1 + r_2 f_2) = r_1 M(f_1) + r_2 M(f_2),$$

for all $r_1, r_2 \in \Re$ and $f_1, f_2 : \Sigma^* \longrightarrow \Re$. More generally, any weighted relation acts as a linear operator.

Original Image

Scale factor of 4

Fig. 20.11. Scaling of the colour image.

We give procedures to do transformations such as scaling, translation and rotation on gray-scale and colour images. We give constructions for scaling up the gray-scale image by a factor of 4, scaling down the gray-scale image by a factor of 4, translation of the gray-scale image by units of 2, units of 4, units of $\frac{1}{2}$ and units of $\frac{1}{4}$ of the gray-scale square image. We also illustrate with examples the above mentioned transformations on gray-scale and colour images.

Scaling and Translation

Scaling and translation can be achieved by FSTs defined in the case of black and white images. Figure 20.13 given below illustrates translation of gray level images.

The colour images in RGB format are stored in 24 bits per pixel with one byte each for the value of the three primary colours, namely, red, green and blue. In representing the colour image using the WFA we use three functions one each corresponding to the three primary colours. So the transformation of scaling for colour images is same as that mentioned for the gray-scale images except that the WFA corresponding to the colour image, defines three functions one each for the three primary colours. We illustrate the scaling up of the colour image by a factor of 4 and the scaling of the colour image by a factor of $\frac{1}{4}$ using an example in Figure 20.11. Figure 20.14 illustrates translation of coloured images.

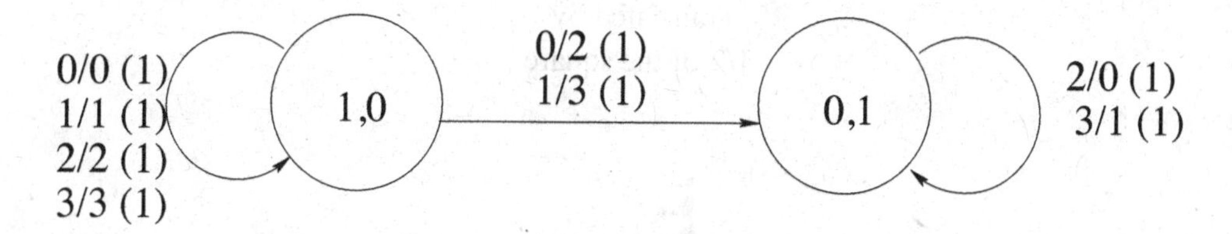

Fig. 20.12. Translation by 1 unit

Rotation

In this section we show how rotation of gray-scale images through an angle of θ can be performed using the weighted finite transducers. We consider the coordinate axes to be shifted to the center of the gray-scale image so that the rotation would be in the anti clockwise direction about the center of the image. The rotation matrix for the clockwise rotation of angle θ about the coordinate axes is given by

$$\begin{pmatrix} \cos\theta - \sin\theta \\ \sin\theta \ \cos\theta \end{pmatrix}$$

Here we discuss the case of rotation through an angle of 45°. Let x, y be the address of the pixel in the original coordinate system of the gray-scale image x' and y' represent the address of the pixel in the new coordinate system. Then the rotation by an angle of 45° can be specified as follows.

$$x'' \leftarrow \frac{x' - y'}{\sqrt{2}} \qquad\qquad y'' \leftarrow \frac{x' + y'}{\sqrt{2}}$$

The division by $\sqrt{2}$ is not easily achievable by the FST. So we scale the rotated

Original Image

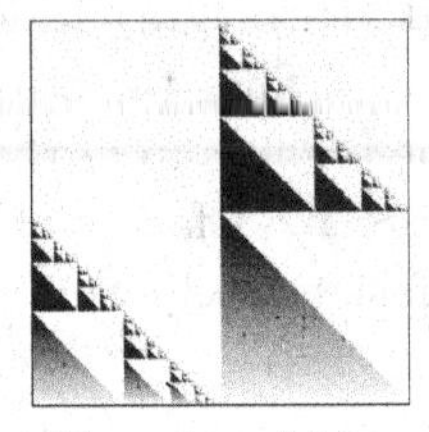

Translated by
1/2 of the square

Translated by
1/4 of the square

Fig. 20.13. Translation by 1/2 and 1/4 of square.

image by a factor of $\frac{1}{\sqrt{2}}$ so as to give the following transformation

$$x'' \leftarrow \frac{x' - y'}{2} \qquad\qquad y'' \leftarrow \frac{x' + y'}{2}$$

This not only rotates the given gray-scale image by an angle of 45° but also scales the image by a factor of $\frac{1}{\sqrt{2}}$. Let $\Gamma = (Q, \Sigma, W_\alpha, I, F)$ be the WFA representing the original gray-scale image. The operation of rotation of the gray-scale image by 45° and scaling by a factor of $\frac{1}{\sqrt{2}}$ is given by the following steps.

(1) The WFA Γ is first scaled by a factor of $\frac{1}{2}$ to get the WFA $\Gamma' = (Q', \Sigma, W'_\alpha, I', F')$.
(2) The transformation equations given in terms of the old axes would be

$$x_1 \leftarrow x - (y - 2^{n-1}) \qquad\qquad y_1 \leftarrow x + (y - 2^{n-1})$$

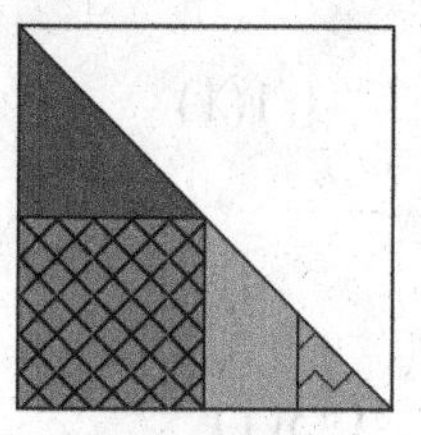

Original Image

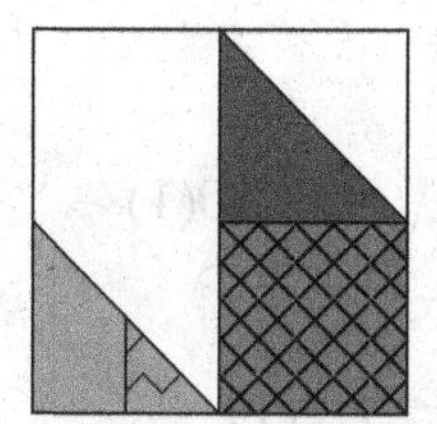

Translated by
1/2 of the square

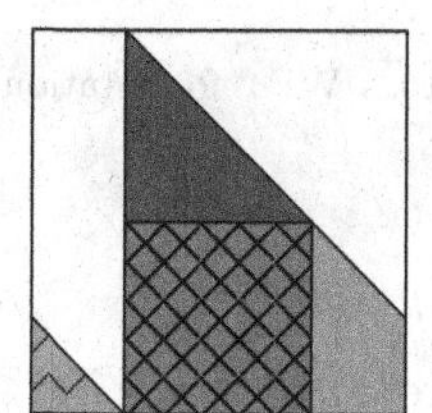

Translated by
1/4 of the square

Fig. 20.14. Translation of colour image by 1/2 and 1/4 of square.

where (x_1, y_1) is obtained from (x, y) after performing the following operations: scaling the given image, shifting the axes to the centre of the image, rotating the image by an angle of 45°, and then shifting back to the old axes. So we first need to subtract 2^{n-1} from the y coordinate. This can be easily achieved by changing the most significant bit of the y coordinate from 0 to 1 and 1 to 0. Hence we construct the WFA $\Gamma'' = (Q'', \Sigma, W''_\alpha, I'', F'')$ from Γ' as follows.

(a) $Q'' = Q', I'' = I', F'' = F'$
(b) $W''_0(q', p) = W'_1(q', p)$
(c) $W''_1(q', p) = W'_0(q', p)$
(d) $W''_2(q', p) = W'_3(q', p)$

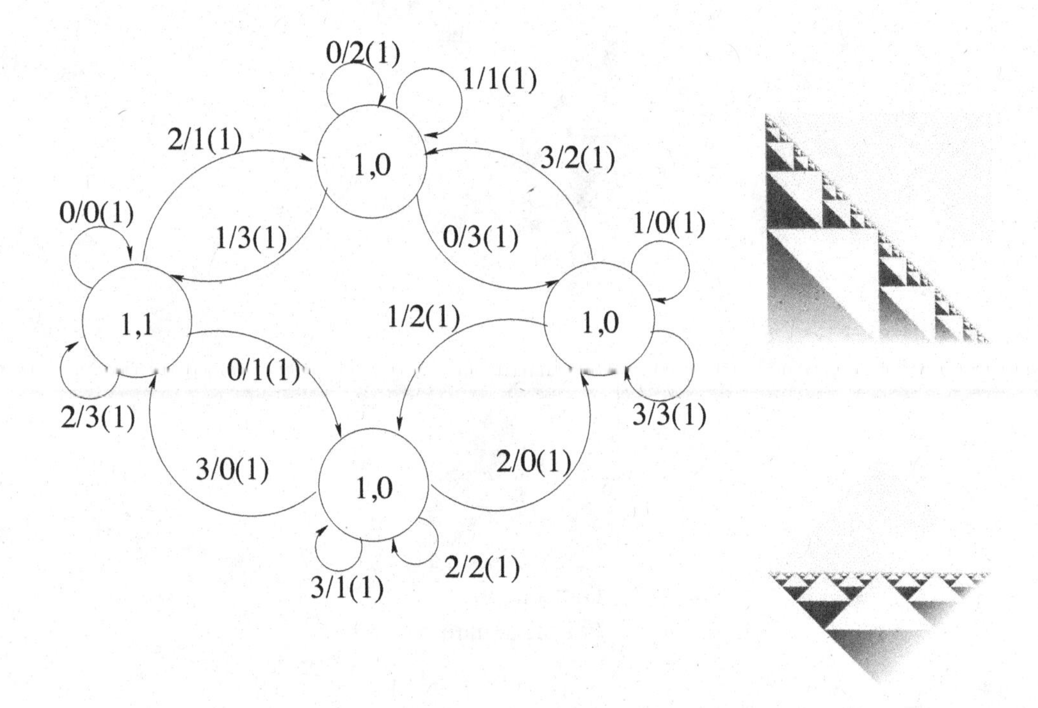

Fig. 20.15. WFT for rotation by 45°.

(e) $W_3''(q',p) = W_2'(q',p)$
(f) $W_\alpha''(q,p) = W_\alpha'(q,p) \forall q,p \in Q', \alpha \in \Sigma$

We apply the WFT in Figure 20.15 to the WFA Γ'' to obtain the rotated WFA Γ'''. The WFT in the Figure 20.15 transforms the x coordinate to $x - y$ and the y coordinate to $x + y$. The addition and subtraction can be done bit by bit using a four state WFT wherein each state represents a carry over from the previous calculation of 0 or 1 for each coordinate x and y. Since the ith alphabet $a_i = 2x_i + y_i$, we know both the x and y bit at each stage. Hence the addition and subtraction can be done on them directly.

(3) After scaling the image by a factor of $\frac{1}{\sqrt{2}}$ we have applied the rotation transformation. The rotation transformation has a scale factor of $\frac{1}{\sqrt{2}}$ but the WFT can only perform a one-one transformation. Hence there are pixels in the transformed image whose values have not been set. It is found that the transformation sets only pixels whose addresses end in 0 or 3. Hence in this final step we add transitions to the WFA Γ''' so that the pixels with address ending with 0 or 3 will set their value to pixels whose addresses end with 1 or 2. Hence

$$W_1'''(q,P) = W_0'''(q,p), \text{ for all } p \text{ with non-zero final distribution, } q \in Q'''$$

$$W_2'''(q,P) = W_3'''(q,p), \text{ for all } p \text{ with non-zero final distribution,} q \in Q'''$$

The WFA Γ''' constructed at the end of the fourth step represents the image rotated by an angle $45°$ and scaled by a factor of $\frac{1}{\sqrt{2}}$.

The rotation of colour images can be done in a similar manner as that of the gray-scale images.

20.4. Representation of Three Dimensional Objects

A three dimensional object can be represented as an FSA using an alphabet, $\Sigma = \{0, 1, 2, 3, 4, 5, 6, 7\}$. The construction for obtaining the images of the projections of the three dimensional object onto the three coordinate planes and the construction to obtain the three dimensional object from its projections are given in [6]. We show how transformations such as scaling, translation and rotation can be performed on three dimensional objects using the finite state transducers and the finite state automata representing the object.

A solid object is considered to be a three dimensional array. Hence any point in the solid can be addressed as a 3-tuple (x, y, z). In order to represent the three dimensional solid object in the form of a FSA we have to extend the alphabet set Σ, of the FSA to $\Sigma = \{0, 1, 2, 3, 4, 5, 6, 7\}$. Now any string $w \in \Sigma^n$ gives the address of a point in the three dimensional space of size $2^n \times 2^n \times 2^n$ enclosing the solid object. If the bit representation of the x coordinate is $x_{n-1}x_{n-2}\ldots x_1x_0$, y coordinate is $y_{n-1}y_{n-2}\ldots y_1y_0$,
z coordinate is $z_{n-1}z_{n-2}\ldots z_1z_0$, then the address of the point is the string $w = a_{n-1}a_{n-2}\ldots a_1a_0$ such that

$$a_i = 4x_i + 2y_i + z_i, \forall i \in [0, n-1]$$

The Figure 20.16 shows how the 8 sub-cubes of a cube are addressed. The figure also gives the FSA which generates the right angled prism.

Note that whenever we say size of the object we refer to the three dimensional space enclosed by the three dimensional object.

20.4.1. *Transformations on Three Dimensional Objects*

Scaling of Three Dimensional Objects

Consider a three dimensional object of resolution $2^n \times 2^n \times 2^n$. We consider the scaling of this three dimensional object with respect to the center of the object i.e. we shift the coordinate axes to the center of the object. The new coordinate axes are defined by $x' = x - 2^{n-1}$, $y' = y - 2^{n-1}$, $z' = z - 2^{n-1}$. The operation of scaling the object by a factor is defined as follows,

$$x' \leftarrow \frac{x'}{k},\ y' \leftarrow \frac{y'}{k},\ z' \leftarrow \frac{z'}{k}$$

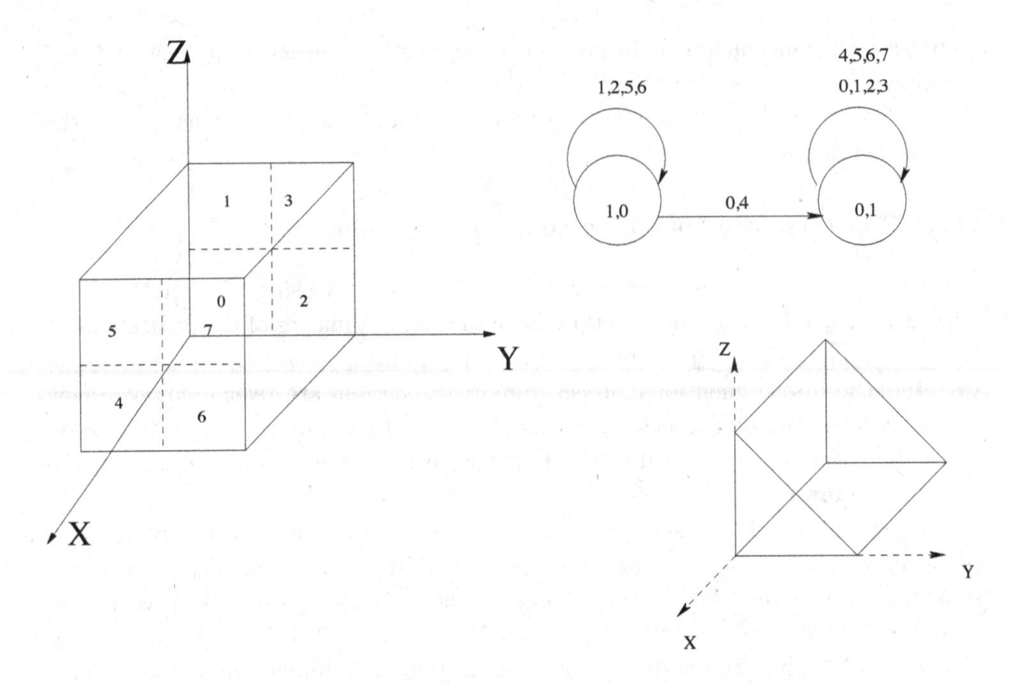

Fig. 20.16. Addressing scheme and an example automaton.

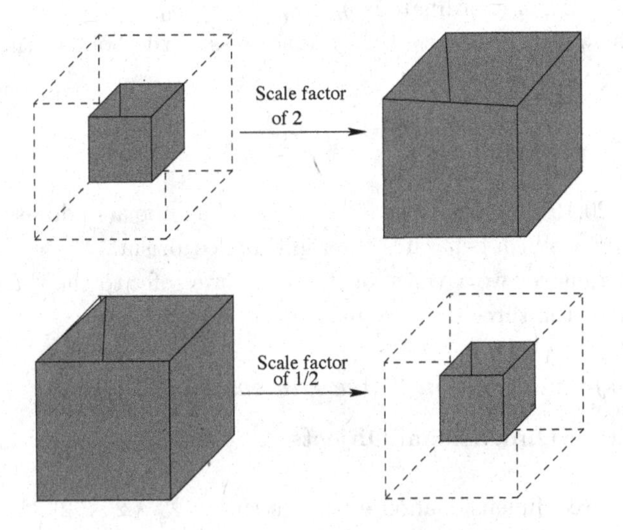

Fig. 20.17. Illustration of 3D scaling.

The operation of scaling a three dimensional object by a factor of 2 in each direction is illustrated in Figure 20.17. It can be seen that the subsquares addressed as 03, 47 in the original three dimensional object become the bigger subsquares 0 and 4 in the scaled up version of factor 2. Similarly, the subsquares 12, 56, 21, 65, 30, 74 in the original three dimensional object are scaled up to form the subsquares 1,

5, 2, 6, 3, 7. Thus the FSA for the scaled up version of the object can be obtained by introducing a new initial state q'_0 which makes a 0(1,2,3,4,5,6,7) transition to the states reachable from the initial states of the original FSA by a path labeled 03(12,21,30,47,56,65,74). The formal construction of the required FSA representing the scaled up object with a scale factor of 2 in each direction is as follows.

Construction 20.4.1. Let $M = (\Sigma = \{0,1,2,3,4,5,6,7\}, Q, \delta, I = \{q_0\}, F)$ be an FSA representing the three dimensional object. The FSA, $M' = (\Sigma', Q', \delta', I', F')$ which represents the scaled up three dimensional object by a factor of 2 in each direction is constructed as follows,

(1) $\Sigma' = \Sigma = \{0,1,2,3,4,5,6,7\}$
(2) $Q' = Q \cup \{q'_0\}$, $I' = \{q'_0\}$, $F' = F$
(3) $\delta'(q'_0, 0) = \delta(\delta(q_0, 0), 3)$
(4) $\delta'(q'_0, 1) = \delta(\delta(q_0, 1), 2)$
(5) $\delta'(q'_0, 2) = \delta(\delta(q_0, 2), 1)$
(6) $\delta'(q'_0, 3) = \delta(\delta(q_0, 3), 0)$
(7) $\delta'(q'_0, 4) = \delta(\delta(q_0, 4), 7)$
(8) $\delta'(q'_0, 5) = \delta(\delta(q_0, 5), 6)$
(9) $\delta'(q'_0, 6) = \delta(\delta(q_0, 6), 5)$
(10) $\delta'(q'_0, 7) = \delta(\delta(q_0, 7), 4)$

Figure 20.17 illustrates this.

When the three dimensional object is scaled down a factor of $\frac{1}{2}$ in each direction, the subsquare addressed as 0 in the three dimensional object becomes the subsquare 03 in the scaled down version of the three dimensional object. Similarly the subsquares addressed as 1,2,3,4,5,6,7 become the subsquares 12,21,30,47,56,65,74 in the scaled version of the three dimensional object. The FSA for the scaled down version of the three dimensional object can be obtained from the original FSA by introducing new states. Any transition from an initial state labeled 0 in the original FSA is replaced by a path labeled 03. Similarly transitions from initial state labeled as 1,2,3,4,5,6,7 are replaced by paths labeled 12,21,30,47,56,65,74 using the new states. The formal construction of the required FSA representing the scaled down version of the three dimensional object by a factor of $\frac{1}{2}$ is given below.

Construction 20.4.2. Let $M = (\Sigma, Q, \delta, I, F)$ be the FSA representing the three dimensional object. The FSA, $M' = (\Sigma', Q', \delta', I', F')$ representing the scaled down version of the three dimensional object by a factor of $\frac{1}{2}$ in each direction is constructed as follows.

(1) $\Sigma' = \Sigma = \{0,1,2,3,4,5,6,7\}$
(2) $Q' = Q \cup \{q'\} \cup \{q'_0, q'_1, q'_2, q'_3, q'_4, q'_5, q'_6, q'_7\}$, $I' = \{q'\}$, $F' = F$
(3) $\delta'(q', 0) = \{q'_0\}, \delta'(q'_0, 3) = \{q \mid \exists q_i, q_i \in I, \delta(q_i, 0) = q\}$
(4) $\delta'(q', 1) = \{q'_1\}, \delta'(q'_1, 2) = \{q \mid \exists q_i, q_i \in I, \delta(q_i, 1) = q\}$
(5) $\delta'(q', 2) = \{q'_2\}, \delta'(q'_2, 1) = \{q \mid \exists q_i, q_i \in I, \delta(q_i, 2) = q\}$

$$
\begin{aligned}
(6)\ \delta'(q',3) &= \{q_3'\}, \delta'(q_0',0) = \{q \mid \exists q_i, q_i \in I, \delta(q_i,3) = q\}\\
(7)\ \delta'(q',4) &= \{q_4'\}, \delta'(q_4',7) = \{q \mid \exists q_i, q_i \in I, \delta(q_i,4) = q\}\\
(8)\ \delta'(q',5) &= \{q_5'\}, \delta'(q_5',6) = \{q \mid \exists q_i, q_i \in I, \delta(q_i,5) = q\}\\
(9)\ \delta'(q',6) &= \{q_6'\}, \delta'(q_6',5) = \{q \mid \exists q_i, q_i \in I, \delta(q_i,6) = q\}\\
(10)\ \delta'(q',7) &= \{q_7'\}, \delta'(q_7',4) = \{q \mid \exists q_i, q_i \in I, \delta(q_i,7) = q\}\\
(11)\ \delta'(q,a) &= \delta(q,a), \forall q \in Q, \forall a \in \Sigma
\end{aligned}
$$

Translation of Three Dimensional Objects

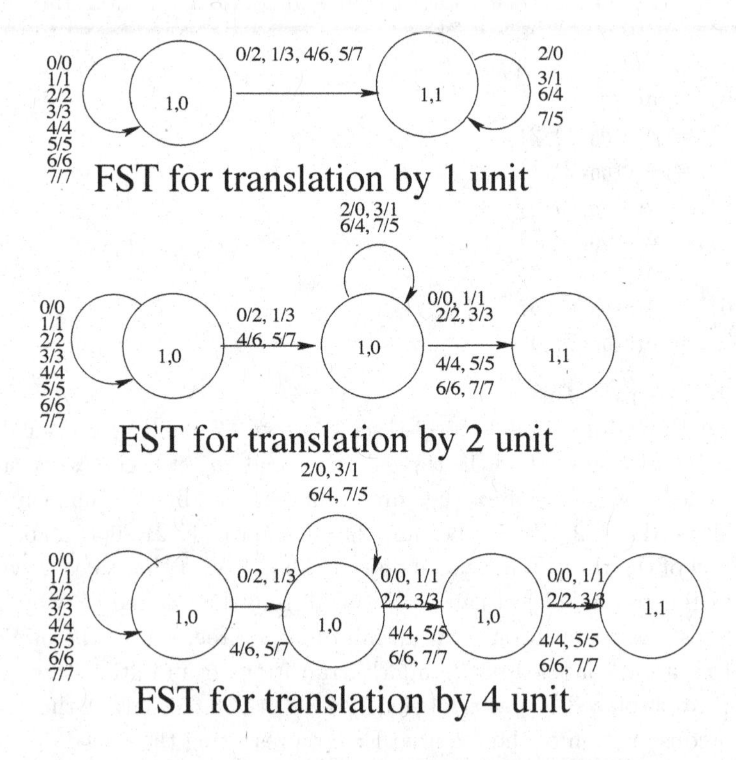

Fig. 20.18. FST's for translation of the three dimensional object.

Next we describe how translations along any of the three axes can be done on the FSA representing the three dimensional object.

Let the finite state automata, M, represent a three dimensional object enclosing a three dimensional space of size $2^n \times 2^n \times 2^n$. Consider the translation along the y-axis from left to right by one pixel. We assume that the object is wrapped around on translation i.e. the pixels in the $(2^n - 1)$ column are moved to the 0th column. This translation is equivalent to adding 1 to the y coordinate of each pixel. If the n-bit y coordinate of pixel is $w01^r$ then the address coordinate of this pixel after translation would be $w10^r, 0 \leq r \leq n-1$. There is no change in the value of the x

and z coordinates. The finite state transducer for this translation is found in Figure 20.18. The FSA M' representing the translated image can be obtained from M by applying this FST to M. In the FST, the transformation in a bit of y coordinate from 0 to 1 (1 to 0) would be represented as 0/2, 1/3, 4/6 and 5/7 (2/0, 3/1, 6/4 and 7/5).

In order to translate the image by 2 units, we have to add 2 to the y coordinate of each pixel. This is same as adding 1 to the binary representation of the y coordinate after ignoring the 0th bit. Similarly a translation by 4 units can be achieved by adding 1 to the y coordinate after ignoring both the 0th and 1st bits. The FST for these operations are shown in Figure 20.18.

Now we consider the translation of the three dimensional object in much larger units. Suppose we want to translate the object by half the size of the object i.e. by 2^{n-1} units, then we have to simply change the most significant bit from 0 to 1 or from 1 to 0. Similarly in order to translate the object by one fourth i.e. by 2^{n-2} units, we have to replace the two most significant bits of the y coordinate from 00 to 01, 01 to 10, 10 to 11 and 11 to 00. The Figure 20.19 gives the finite state transducers for these operations.

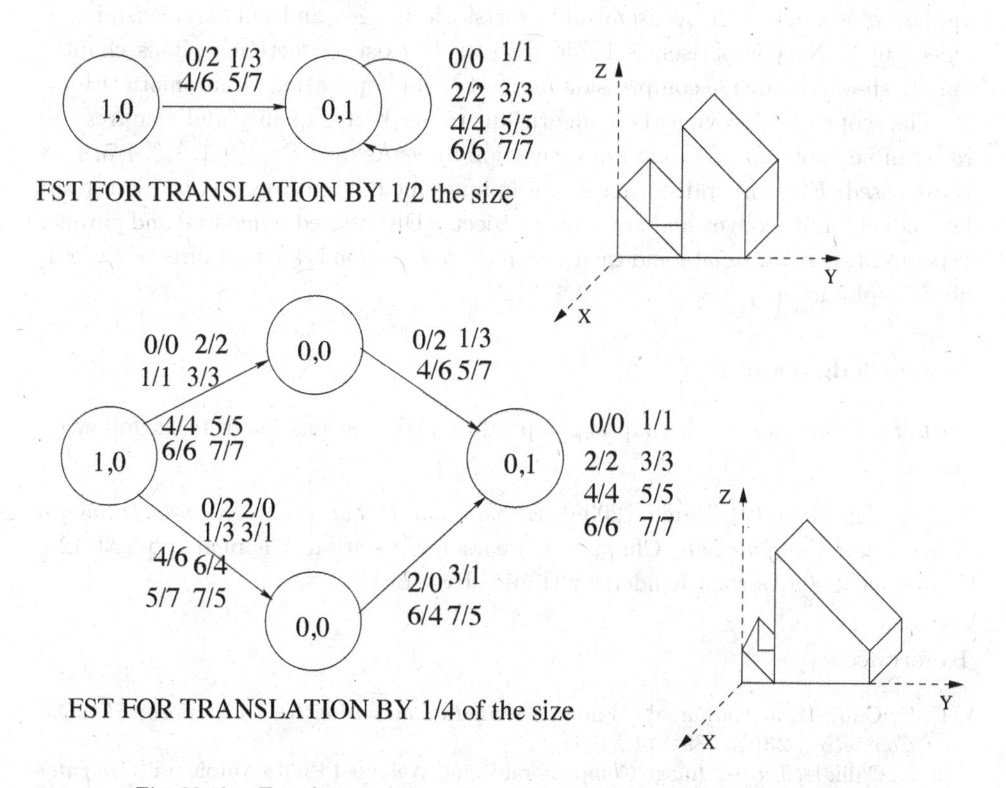

Fig. 20.19. Translation of three dimensional object by 1/2 , 1/4 of its size.

For translation by an arbitrary power of 2, we can construct FSTs similar to the above one. To translate the object along the $x(z)$-axis, we have to change the corresponding $x(z)$ bits. Also to translate in the other direction (right to left) we have to subtract 1 instead of adding. The FSTs for these operations can be easily constructed from the above FSTs. In order to translate the object by any arbitrary value, say a units along the x axis, b units along the y axis and c units along the z axis, we write a(b,c) as a sum of powers of 2 and continuously apply the FSTs for each power of 2 along the $x(y, z)$ axis.

Rotation of Three Dimensional Objects

Rotation of a three dimensional object about any one of the axes can be done in a similar manner as in the two dimensional case.

Conclusion

Representation of black and white images by FSAs over $\Sigma = \{0, 1, 2, 3\}$ and representation of gray level images by weighted FSA have been considered. Inference and de-inference algorithms for gray level images is considered. The image compression method given here will be useful only for static images and not for transmitting video clips. No comparison is made with other existing methods. This chapter mainly shows the image compression method as an application of automata theory.

The scope of improving these algorithms to get better quality and compression rate can be explored. 3D solid representation by FSAs over $\Sigma = \{0, 1, 2, 3, 4, 5, 6, 7\}$ is discussed. Efficient transformation techniques including projection on planes can be studied for 3D convex and nonconvex objects. Distributed sequential and parallel versions of these automata and their use in representation is another direction which needs exploration.

Acknowledgments

Part of the material in this chapter is reproduced with permission from the following source:

K. Krithivasan and R. Rama, (2009), *Introduction to Formal Languages, Automata Theory and Computation*, Chapter 6, Pearson Education, Chennai, pp 123–128.

References

[1] K. Culik II, J. Karhumaki, Finite Automata Computing Real Functions, *SIAM J. Computing.* **23**(4), 789–814, (1994).

[2] K. Culik II, J. Kari, Image Compression Using Weighted Finite Automata, *Computer Graphics.* **17**(3), 305–313, (1993).

[3] K. Culik II and I. Fris, Weighted finite transducers in image processing, *Discrete Applied Mathematics.* **58**(3), 223–237, (1995).

[4] K. Culik II, J. Kari, Digital Images and Formal Languages. In *Handbook of Formal Languages, Vol. 3*, G. Rozenberg and A. Salomaa (Eds.), pp. 599–616, Springer, (1997).

[5] K. Culik II, J. Kari, Inference algorithms for WFA and image compression. In *Fractal image compression*, Y. Fisher (Ed.), Chapter 13, pp. 243–258, Springer, (1994).

[6] S. V. Ramasubramaniam and K. Krithivasan, Finite automata and digital images, *Intl. J. Pattern Recognition and Artificial Intelligence*, **14**(4), 501–524, (2000).

[7] S. V. Ramasubramaniam, Finite automata and digital images, B.Tech. project report, Dept. of CSE, IIT Madras, (1999).

[8] Y. Sivasubramanyam, Studies in weighted automata with application to image processing, M.S. Thesis, Dept. of CSE, IIT Madras, 2002.

Index